MATERIALS SCIENCE AND ENGINEERING

An Introduction

Values of Selected Physical Constants

Quantity	Symbol	SI Units	cgs Units
Avogadro's number	N_A	6.023×10^{23} molecules/mol	6.023×10^{23} molecules/mol
Boltzmann's constant	k	1.38×10^{-23} J/atom-K	1.38×10^{-16} erg/atom-K 8.62×10^{-5} eV/atom-K
Bohr magneton	μ_B	9.27×10^{-24} A-m^2	9.27×10^{-21} erg/gauss[a]
Electron charge	e	1.602×10^{-19} C	4.8×10^{-10} statcoul[b]
Electron mass	—	9.11×10^{-31} kg	9.11×10^{-28} g
Gas constant	R	8.31 J/mol-K	1.987 cal/mol-K
Permeability of a vacuum	μ_0	1.257×10^{-6} henry/m	unity[a]
Permittivity of a vacuum	ϵ_0	8.85×10^{-12} farad/m	unity[b]
Planck's constant	h	6.63×10^{-34} J-s	6.63×10^{-27} erg-s 4.13×10^{-15} eV-s
Velocity of light in a vacuum	c	3×10^8 m/s	3×10^{10} cm/s

[a] In cgs-emu units.
[b] In cgs-esu units.

UNIT ABBREVIATIONS

A = ampere
Å = angstrom
Btu = British thermal unit
C = Coulomb
°C = degrees Celsius
cal = calorie (gram)
cm = centimeter
eV = electron volt
°F = degrees Fahrenheit
ft = foot
g = gram

in. = inch
J = joule
K = degrees Kelvin
kg = kilogram
lb$_f$ = pound force
lb$_m$ = pound mass
m = meter
Mg = megagram
mm = millimeter
mol = mole
MPa = megapascal

N = newton
nm = nanometer
P = poise
Pa = pascal
s = second
T = temperature
μm = micrometer (micron)
W = watt
psi = pounds per square inch

SI multiple and submultiple prefixes

Factor by Which Multiplied	Prefix	Symbol
10^9	giga	G
10^6	mega	M
10^3	kilo	k
10^{-2}	centi[a]	c
10^{-3}	milli	m
10^{-6}	micro	μ
10^{-9}	nano	n
10^{-12}	pico	p

[a] Avoided when possible

MATERIALS SCIENCE AND ENGINEERING

An Introduction

WILLIAM D. CALLISTER, JR.

Department of Materials Science and Engineering
The University of Utah

JOHN WILEY & SONS, INC.

New York   Chichester   Brisbane   Toronto   Singapore

Acquisitions Editor: Cliff Robichaud
Production Supervisor: Elizabeth Austin
Copyediting Supervisor: Deborah Herbert
Designer: Laura Nicholls

Library of Congress Cataloging in Publication Data:

Callister, William, 1940–
Materials science and engineering: an introduction/William D.
Callister, Jr.—2nd ed.
   p. cm.
  Includes bibliographical references.
  ISBN 0-471-50488-2
  1. Materials. I. Title.
TA403.C23 1990   90-33829
620. 1′1—dc20   CIP

Printed in the United States of America

10 9 8 7 6 5 4 3 2

To my parents,
William D. and Genevieve J. Callister

PREFACE

This edition has a number of changes in both organization and content that have been made in response to the feedback received from users of the first edition. It is also up to date with respect to the most recent developments in the materials discipline. In making these alterations and additions, I have adhered throughout to the objectives and philosophy of the first edition as outlined in its preface, which is reprinted in this edition.

Minor modifications have been made in Chapter 2 (Atomic Structure and Interatomic Bonding), to include a more detailed discussion of electronegativity and a more quantitative treatment of interatomic bonding. The topics of X-ray diffraction and linear and planar atomic densities have been added to Chapter 3 (The Structure of Crystalline Solids). The discussion of dislocations as crystalline defects has been moved from Chapter 7 to Chapter 4 (Imperfections in Solids); also added to this chapter is a more extensive treatment of interfacial defects. Chapter 6 (Mechanical Properties of Metals) has been expanded to include nonlinear elastic behavior, anelasticity, resilience, and a more complete discussion of hardness testing, as well as new sections on the variability of material properties and safety factors.

To improve continuity, the recovery, recrystallization, and grain growth sections found in Chapter 9 of the first edition have been shifted to Chapter 7 (Dislocations and Strengthening Mechanisms)—recovery and recrystallization phenomena must be preceded by strain hardening, which is also discussed in this chapter. In addition, a short section on deformation by twinning has also been included.

Possibly the most significant addition is a complete chapter on failure (Chapter 8), in which, on an introductory level and in moderate detail, the following topics are discussed: fracture modes, principles of fracture mechanics, impact fracture testing, fatigue, and creep.

Several new sections have been added to the phase diagram chapter (Chapter 9), which treat peritectic reactions, congruent phase transformations, and the Gibbs phase rule. A more thorough discussion on the development of microstructure on cooling is also given. Chapter 10 on phase transformations has been reorganized; the topics now proceed in a more logical sequence. Furthermore, a section on temper embrittlement has been added.

The ceramic phase diagrams section has been moved to Chapter 13 (Structures and Properties of Ceramics); discussions of the ZrO_2–CaO and SiO_2–Al_2O_3 systems have also been added. Some of the new "advanced ceramics" that are now being utilized are treated in the subsequent chapter.

Minor changes have been made in the two polymer chapters—Chapters 15 and 16. The discussion on viscoelastic behavior has been revised; in addition, a brief treatment describing the manner by which fracture occurs in these materials has been added.

Chapter 17, on composite materials, has been updated, upgraded, and reorganized. Coverage of the characteristics of discontinuous fiber-reinforced composites has been expanded, and a section has been added in which some of the processing techniques used for fiber-reinforced composites are discussed.

A rather extensive treatment of corrosion rates and the roles of activation and concentration polarizations in the prediction of these rates has been included in Chapter 18. The principles of oxidation and its kinetics are discussed in another new section that has been added to this chapter.

No significant changes have been made in the latter chapters of the book, which deal with electrical, thermal, magnetic, and optical properties of materials. The section on superconductivity was moved from electrical properties to magnetic properties, since the critical temperature is dependent on the applied magnetic field. Brief discussions relating to the new high-temperature superconductors and conducting polymers have also been included.

This edition also includes a new appendix that contains the electron configurations of all of the elements, the expansion of another appendix to include a more complete tabulation of the properties of some common engineering materials, and more than 400 new homework questions and problems.

I express my appreciation both to the individuals who critically evaluated the first edition and to those who reviewed the manuscript and offered suggestions for this edition. Photographs were provided by various organizations and individuals to whom I am also indebted; credit to each is given in the relevant figure captions. Special thanks are extended to the following for their contributions: James R. Callister, Hewlett-Packard; James A. Clum, State University of New York at Binghamton; J. Gerald Byrne, Willard D. Bascom, Richard M. Cohen, Jiri Janata, John A. Nairn, Anil V. Virkar, and Donald J. Lyman, The University of Utah; and W. Roger Cannon, Rutgers—The State University. Heartfelt thanks to my good parents for their tireless and unselfish assistance with the seemingly unending proofreading. My thanks and apologies to others whom I may have inadvertently forgotten to mention.

The continual encouragement and support of my parents, family, and friends has been a source of strength, which is sincerely appreciated.

William D. Callister, Jr.
Salt Lake City, Utah

January 1990

PREFACE TO THE FIRST EDITION

I suppose that nearly every university professor at one time or another has had the inclination to write a textbook that expresses his or her unique pedagogical approach to a particular discipline. I am no different, and this book represents the actualization of my aspiration.

I undertook this writing project with several objectives in mind. The first was to present the basic fundamentals of materials science and engineering on a level that can be understood by university students who have successfully completed the normal engineering prerequisites (engineering physics, calculus, differential equations, and at least one semester of college inorganic chemistry). In this same vein, I have strived to use terminology that is familiar to the student encountering the discipline of materials science for the first time.

Another of my objectives was to present the material in a logical order, such that each chapter builds on the previous ones. This objective is reflected in the sequencing of the chapters. Chapters 1 and 2 review the discipline of materials science and engineering and atomic bonding in solids. Chapters 3 through 10 discuss the following topics in this order: crystal structures, imperfections, diffusion, mechanical properties, dislocations and strengthening mechanisms, phase diagrams, phase transformations, and the development of microstructure and its relationship to mechanical behavior. In most instances, principles and concepts are presented in the context of metallic systems, which are, in general, the most simple. Chapter 11 treats fabrication methods for metals and the various metal alloy types. The structure, mechanical behavior, types, fabrication methods, and uses of ceramic materials are the discussion topics of the subsequent two chapters, whereas Chapters 14 and 15 treat these same subjects for polymers. Composite materials are discussed in Chapter 16. The final five chapters deal with the degradation, electrical, thermal, magnetic, and optical properties of materials. Within these discussions I have drawn attention to the relationships that exist between the structural elements of materials and their properties.

I believe that if a topic or concept merits inclusion, it is worth discussing in sufficient detail that a student can discern its practical utility. As a third objective, I have adhered to this philosophy, as evidenced in Chapters 7 through 10, which deal with the techniques used to strengthen metals.

This book could be used as a text for virtually any introductory materials science and engineering course. It is suitable not only for materials majors, but also for students studying the disciplines of chemical, civil, electrical, and mechanical engineering. Because the book covers the gamut of materials topics, it is lengthy. It would be naive to presume that all of the material could be presented in one semester. To do justice to the content, a minimum of two semesters (three quarters) is required.

My fourth objective was to expedite the learning process. To do so I incorporated into the text several student aids. These include a series of questions and problems at the end of each chapter, answers to selected problems, example problems, references, a glossary, and a list of symbols.

Regarding the questions and problems, virtually all of the problems require computations leading to numerical solutions. Many of the concepts within the discipline of materials science and engineering are descriptive in nature. By including questions that require written, descriptive answers, students better comprehend and understand these concepts. The questions are of two types: for some, the student needs only to restate in his or her own words an explanation provided in the text material. Other questions require the student to reason and/or synthesize before coming to a conclusion or solution.

Units are always a problem in engineering textbooks. I have elected to use both SI and English units concurrently; that is, values and units for both systems appear together in the text material (one set in parentheses) and, when convenient, in tables and with figures. A review of the SI unit system is contained in Appendix A; an extensive unit conversion table is located inside the back cover.

Even though I am the sole author of this work, it nevertheless includes the contributions of many others. I am indebted to a number of organizations, publications, and individuals from whom I obtained data, illustrations, and photographs; credit is given in the figure captions. I also thank associates and students who have provided suggestions and comments. Specific acknowledgment is given to the following individuals who reviewed and criticized portions of the manuscript: Carl Wood, Brigham Young University; Professors J. Gerald Byrne, Gerald B. Stringfellow, and Richard H. Boyd, the University of Utah; Professor W. Roger Cannon, Rutgers University; Professor J. Tom Tielking, Texas A & M University; Deepak Parikh, the University of Utah; Grant E. Head, Hewlett-Packard; and N. Lawrence Head, IBM Corporation. Notwithstanding this assistance, I assume the responsibility for any errors that remain. Also, the continued support and encouragement of my parents, family, and friends is sincerely appreciated.

William D. Callister, Jr.
Salt Lake City, Utah
September 1984

CONTENTS

LIST OF SYMBOLS

The number of the section in which a symbol is introduced or explained is given in parentheses.

A = area

$Å$ = angstrom unit

A_i = atomic weight of element i (2.2)

APF = atomic packing factor (3.4)

%AR = ductility, in percent area reduction (6.6)

a = lattice parameter: unit cell x-axial length (3.4)

a = crack length of a surface crack (8.5)

at% = atom percent (4.3)

B = magnetic flux density (induction) (21.2)

B_r = magnetic remanence (21.7)

BCC = body-centered cubic crystal structure (3.4)

b = lattice parameter: unit cell y-axial length (3.7)

$\mathbf{b}$ = Burgers vector (4.4)

C = capacitance (19.16)

C_i = concentration of component i

C_v, C_p = heat capacity at constant volume, pressure (20.2)

CPR = corrosion penetration rate (18.3)

CVN = Charpy V-notch (8.6)

%CW = percent cold work (7.10)

c = lattice parameter: unit cell z-axial length (3.7)

c = velocity of electromagnetic radiation in a vacuum (22.2)

D = diffusion coefficient (5.3)

D = dielectric displacement (19.17)

d = diameter

d = average grain diameter (7.8)

d_{hkl} = interplanar spacing for planes of Miller indices h, k, and l (3.15)

E = energy (2.5)

E = modulus of elasticity or Young's modulus (6.3)

$\mathscr{E}$ = electric field intensity (19.3)

E_f = Fermi energy (19.5)

E_g = band gap energy (19.6)

$E_r(t)$ = relaxation modulus (16.6)

%EL = ductility, in percent elongation (6.6)

e = electric charge per electron (19.7)

e^- = electron (18.2)

erf = Gaussian error function (5.4)

exp = e, the base for natural logarithms

F = force, interatomic or mechanical (2.5, 6.2)

$\mathscr{F}$ = Faraday constant (18.2)

FCC = face-centered cubic crystal structure (3.4)

G = shear modulus (6.3)

H = magnetic field strength (21.2)

H_c = magnetic coercivity (21.7)

HB = Brinell hardness (6.10)

HCP = hexagonal close-packed crystal structure (3.4)

HK = Knoop hardness (6.10)

HRB, HRC, HRF = Rockwell hardness: B, C, and F scales (6.10)

HR15N, HR30T, HR45W = superficial Rockwell hardness: 15N, 30T, and 45W scales (6.10)

HV = Vickers hardness (6.10)

h = Planck's constant (22.2)

(hkl) = Miller indices for a crystallographic plane (3.9)

I = electric current (19.2)

I = intensity of electromagnetic radiation (22.3)

i = current density (18.3)

i_C = corrosion current density (18.4)

J = diffusion flux (5.3)

J = electric current density (19.3)

K = stress intensity factor (8.5)

K_c = fracture toughness (8.5)

K_{Ic} = plane strain fracture toughness for mode I crack surface displacement (8.5)

k = Boltzmann's constant (4.2)

k = thermal conductivity (20.4)

l = length

l_c = critical fiber length (17.4)

ln = natural logarithm

log = logarithm taken to base 10

M = magnetization (21.2)

$\overline{M}_n$ = polymer number-average molecular weight (15.5)

$\overline{M}_w$ = polymer weight-average molecular weight (15.5)

mol% = mole percent

N = number of fatigue cycles (8.8)

N_A = Avogadro's number (3.5)

N_f = fatigue life (8.8)

n = principal quantum number (2.3)

n = number of atoms per unit cell (3.5)

n = strain-hardening exponent (6.7)

n = number of electrons in an electrochemical reaction (18.2)

n = number of conducting electrons per cubic meter (19.7)

n = index of refraction (22.5)

n' = for ceramics, the number of formula units per unit cell (13.2)

n_n = number-average degree of polymerization (15.5)

n_w = weight-average degree of polymerization (15.5)

P = dielectric polarization (19.17)

P–B ratio = Pilling–Bedworth ratio (18.10)

p = number of holes per cubic meter (19.10)

Q = activation energy

Q = magnitude of charge stored (19.16)

R = atomic radius (3.4)

R = gas constant

r = interatomic distance (2.5)

r = reaction rate (10.3, 18.3)

r_A, r_C = anion and cation ionic radii (13.2)

S = fatigue stress amplitude (8.8)

SEM = scanning electron microscopy or microscope

T = temperature

T_c = Curie temperature (21.6)

T_C = superconducting critical temperature (21.11)

T_g = glass transition temperature (14.2)

T_m = melting temperature

TEM = transmission electron microscopy or microscope

TS = tensile strength (6.6)

t = time

t_r = rupture lifetime (8.13)

U_r = modulus of resilience (6.6)

$[uvw]$ = indices for a crystallographic direction (3.8)

V = electrical potential difference (voltage) (18.2)

V_C = unit cell volume (3.4)

V_C = corrosion potential (18.4)

V_i = volume fraction of phase i (9.7)

v = velocity

vol% = volume percent

W_i = mass fraction of phase i (9.6)

wt% = weight percent (4.3)

x = length

x = space coordinate

Y = dimensionless parameter in fracture toughness expression (8.5)

y = space coordinate

z = space coordinate

α = lattice parameter: unit cell y–z interaxial angle (3.7)

α, β, γ = phase designations

α_l = linear coefficient of thermal expansion (20.3)

β = lattice parameter: unit cell x–z interaxial angle (3.7)

γ = lattice parameter: unit cell x–y interaxial angle (3.7)

γ = shear strain (6.2)

Δ = finite change in a parameter the symbol of which it precedes

ϵ = engineering strain (6.2)

ϵ = dielectric permittivity (19.16)

ϵ_r = dielectric constant or relative permittivity (19.16)

$\dot{\epsilon}_s$ = steady-state creep rate (8.13)

ϵ_T = true strain (6.7)

η = viscosity (13.8)

η = overvoltage (18.4)

θ = Bragg diffraction angle (3.15)

θ_D = Debye temperature (20.2)

λ = wavelength of electromagnetic radiation (3.15)

μ = magnetic permeability (21.2)

μ_B = Bohr magneton (21.2)

μ_r = relative magnetic permeability (21.2)

μ_e = electron mobility (19.7)

μ_h = hole mobility (19.10)

v = Poisson's ratio (6.5)

v = frequency of electromagnetic radiation (22.2)

ρ = density (3.5)

ρ = electrical resistivity (19.2)

ρ_t = radius of curvature at the tip of a crack (8.5)

σ = engineering stress, tensile or compressive (6.2)

σ = electrical conductivity (19.3)

σ_c = critical stress for crack propagation (8.5)

σ_m = maximum stress (8.5)

σ_m = mean stress (8.7)

σ_{mr} = modulus of rupture (13.7)

σ_T = true stress (6.7)

σ_w = safe or working stress (6.12)

σ_y = yield strength (6.6)

τ = shear stress (6.2)

τ_c = fiber–matrix bond strength (17.4)

τ_{crss} = critical resolved shear stress (7.5)

χ_m = magnetic susceptibility (21.2)

SUBSCRIPTS

c = composite

f = final

f = at fracture

f = fiber

i = instantaneous

m = matrix

m, max = maximum

min = minimum

0 = original

0 = at equilibrium

0 = in a vacuum

CHAPTER 1

INTRODUCTION

A familiar item that is fabricated from all three material types is the soda pop container. Soda pop is marketed in aluminum (metal) cans (top), glass (ceramic) bottles (center), and plastic (polymer) bottles (bottom). (Permission to use these photographs was granted by the Coca-Cola Company.)

1.1 HISTORICAL PERSPECTIVE

Materials are probably more deep-seated in our culture than most of us realize. Transportation, housing, clothing, communication, recreation, and food production— virtually every segment of our everyday lives is influenced to one degree or another by materials. Historically, the development and advancement of societies have been intimately tied to the members' ability to produce and manipulate materials to fill their needs. In fact, early civilizations have been designated by the level of their materials development (i.e., Stone Age, Bronze Age).

The earliest humans had access to only a very limited number of materials, those that occur naturally: stone, wood, clay, skins, and so on. With time they discovered techniques for producing materials that had properties superior to those of the natural ones; these new materials included pottery and various metals. Furthermore, it was discovered that the properties of a material could be altered by heat treatments and by the addition of other substances. At this point, materials utilization was totally a selection process, that is, deciding from a given, rather limited set of materials the one that was best suited for an application by virtue of its characteristics. It was not until relatively recent times that scientists came to understand the relationships between the structural elements of materials and their properties. This knowledge, acquired in the past 50 years or so, has empowered them to fashion, to a large degree, the characteristics of materials. Thus, tens of thousands of different materials have evolved with rather specialized characteristics which meet the needs of our modern and complex society; these include metals, plastics, glasses, and fibers.

The development of many technologies that make our existence so comfortable has been intimately associated with the accessibility of suitable materials. An advancement in the understanding of a material type is often the forerunner to the stepwise progression of a technology. For example, automobiles would not have been possible without the availability of inexpensive steel or some other comparable substitute. In our contemporary era, sophisticated electronic devices rely on components that are made from what are called semiconducting materials.

1.2 MATERIALS SCIENCE AND ENGINEERING

The discipline of *materials science* involves investigating the relationships that exist between the structures and properties of materials. In contrast, *materials engineering* is, on the basis of these structure–property correlations, designing or engineering the structure of a material to produce a predetermined set of properties. Throughout this text we draw attention to the relationships between material properties and structural elements.

"Structure" is at this point a nebulous term that deserves some explanation. In brief, the structure of a material usually relates to the arrangement of its internal components. Subatomic structure involves electrons within the individual atoms and interactions with their nuclei. On an atomic level, structure encompasses the organization of atoms or molecules relative to one another. The next larger structural realm, which contains large groups of atoms that are normally agglomerated together, is termed "microscopic," meaning that which is subject to direct observation using some type of microscope. Finally, structural elements that may be viewed with the naked eye are termed "macroscopic."

The notion of "property" deserves elaboration. While in service use, all materials are exposed to external stimuli that evoke some type of response. For example, a specimen subjected to forces will experience deformation; or a polished metal surface will reflect light. Property is a material trait in terms of the kind and magnitude of response to a specific imposed stimulus. Generally, definitions of properties are made independent of material shape and size.

properties

Virtually all important properties of solid materials may be grouped into six different categories: mechanical, electrical, thermal, magnetic, optical, and deteriorative. For each there is a characteristic type of stimulus capable of provoking different responses. Mechanical properties relate deformation to an applied load or force; examples include elastic modulus and strength. For electrical properties, such as electrical conductivity and dielectric constant, the stimulus is an electric field. The thermal behavior of solids can be represented in terms of heat capacity and thermal conductivity. Magnetic properties demonstrate the response of a material to the application of a magnetic field. For optical properties, the stimulus is electromagnetic or light radiation; index of refraction and reflectivity are representative optical properties. Finally, deteriorative characteristics indicate the chemical reactivity of materials. The chapters that follow discuss properties that fall within each of these six classifications.

Why do we study materials? Many an applied scientist or engineer, whether mechanical, civil, chemical, or electrical, will at one time or another be exposed to a design problem involving materials. Examples might include a transmission gear, the superstructure for a building, an oil refinery component, or a microprocessor "chip." Of course, materials scientists and engineers are specialists who are totally involved in the investigation and design of materials.

Many times, a materials problem is one of selecting the right material from the many thousands that are available. There are several criteria on which the final decision is normally based. First of all, the in-service conditions must be characterized, for these will dictate the properties required of the material. On only rare occasions does a material possess the maximum or ideal combination of properties. Thus it may be necessary to trade off one characteristic for another. The classic example involves strength and ductility; normally, a material having a high strength will have only a limited ductility. In such cases a reasonable compromise between two or more properties may be necessary.

A second selection consideration is any deterioration of material properties that may occur during service operation. For example, significant reductions in mechanical strength may result from exposure to elevated temperatures or corrosive environments.

Finally, probably the overriding consideration is that of economics: What will the finished product cost? A material may be found that has the ideal set of properties but is prohibitively expensive. Here again, some compromise is inevitable. The cost of a finished piece also includes any expense incurred during fabrication to produce the desired shape.

The more familiar an engineer or scientist is with the various characteristics and structure–property relationships, as well as processing techniques of materials, the more proficient and confident he or she will be to make judicious materials choices based on these criteria.

1.3 CLASSIFICATION OF MATERIALS

Solid materials have been conveniently grouped into three basic classifications: metals, ceramics, and polymers. This scheme is based primarily on chemical makeup and atomic structure, and most materials fall into one distinct grouping or another, although there are some intermediates. In addition, there are two other groups of important engineering materials—composites and semiconductors. Composites consist of combinations of two or more different materials, whereas semiconductors are utilized because of their unusual electrical characteristics. A brief explanation of the material types and representative characteristics is offered next. Subsequent chapters explore in some detail the various structural elements and properties for each.

Metals

Metallic materials are normally combinations of metallic elements. They have large numbers of nonlocalized electrons; that is, these electrons are not bound to particular atoms. Many properties of metals are directly attributable to these electrons. Metals are extremely good conductors of electricity and heat and are not transparent to visible light; a polished metal surface has a lustrous appearance. Furthermore, metals are quite strong, yet deformable, which accounts for their extensive use in structural applications.

Ceramics

Ceramics are compounds between metallic and nonmetallic elements; they are most frequently oxides, nitrides, and carbides. The wide range of materials that falls within this classification includes ceramics that are composed of clay minerals, cement, and glass. These materials are typically insulative to the passage of electricity and heat, and are more resistant to high temperatures and harsh environments than metals and polymers. With regards to mechanical behavior, ceramics are hard but very brittle.

Polymers

Polymers include the familiar plastic and rubber materials. Many of them are organic compounds that are chemically based on carbon, hydrogen, and other nonmetallic elements; furthermore, they have very large molecular structures. These materials typically have low densities and may be extremely flexible.

Composites

A number of composite materials have been engineered that consist of more than one material type. Fiberglass is a familiar example, in which glass fibers are embedded within a polymeric material. A composite is designed to display a combination of the best characteristics of each of the component materials. Fiberglass acquires strength from the glass and flexibility from the polymer. Many of the recent material developments have involved composite materials.

Semiconductors

Semiconductors have electrical properties that are intermediate between the electrical conductors and insulators. Furthermore, the electrical characteristics of these materials are extremely sensitive to the presence of minute concentrations of impurity atoms, which concentrations may be controlled over very small spatial regions. The semiconductors have made possible the advent of integrated circuitry that has totally revolutionized the electronics and computer industries (not to mention our lives) over the past two decades.

1.4 MODERN MATERIALS NEEDS

In spite of the tremendous progress that has been made in the understanding and development of materials within the past few years, there remain technological challenges requiring even more sophisticated and specialized materials. Some comment is appropriate in this regard to round out the materials perspective.

Energy is a current concern. There is a recognized need to find new, economical sources of energy and, in addition, to use the present resources more efficiently. Materials will undoubtedly play a significant role in these developments. For example, the direct conversion of solar into electrical energy has been demonstrated. Solar cells employ some rather complex and expensive materials. To ensure a viable technology, materials that are highly efficient in this conversion process yet less costly must be developed.

Nuclear energy holds some promise, but the solutions to the many problems that remain will necessarily involve materials, from fuels to containment structures to facilities for the disposal of radioactive waste.

Furthermore, environmental quality depends on our ability to control air and water pollution. Pollution control techniques employ various materials. In addition, materials processing and refinement methods need to be improved so that they produce less environmental degradation, that is, less pollution and less despoilage of the landscape from the mining of raw materials.

Significant quantities of energy are involved in transportation. Reducing the weight of transportation vehicles (automobiles, aircraft, trains, etc.), as well as increasing engine operating temperatures, will enhance fuel efficiency. New high-strength, low-density structural materials remain to be developed, as well as materials that have higher temperature capabilities, for use in engine components.

Many materials that we use are derived from resources that are nonrenewable, that is, not capable of being regenerated. These include polymers, for which the prime raw material is oil, and some metals. These nonrenewable resources are gradually becoming depleted, which necessitates either the discovery of additional reserves or the development of new materials having comparable properties and less adverse environmental impact. The latter alternative is a major challenge for the materials scientist and engineer.

REFERENCES

The references for Chapter 1 are textbooks that cover the basic fundamentals of the field of materials science and engineering.

ASKELAND, D. R., *The Science and Engineering of Materials,* 2nd edition, PWS-Kent Publishing Co., Boston, 1989.

FLINN, R. A. and P. K. TROJAN, *Engineering Materials and Their Applications,* 3rd edition, Houghton Mifflin Co., Boston, 1986.

SHACKELFORD, J. F., *Introduction to Materials Science for Engineers,* 2nd edition, Macmillan Publishing Company, New York, 1988.

SMITH, C. O., *The Science of Engineering Materials,* 3rd edition, Prentice-Hall, Inc., Englewood Cliffs, NJ, 1986.

THORNTON, P. A. and V. J. COLANGELO, *Fundamentals of Engineering Materials,* Prentice-Hall, Inc., Englewood Cliffs, NJ, 1985.

VAN VLACK, L. H., *Elements of Materials Science and Engineering,* 6th edition, Addison-Wesley Publishing Co., Reading, MA, 1989.

CHAPTER 2

ATOMIC STRUCTURE AND INTERATOMIC BONDING

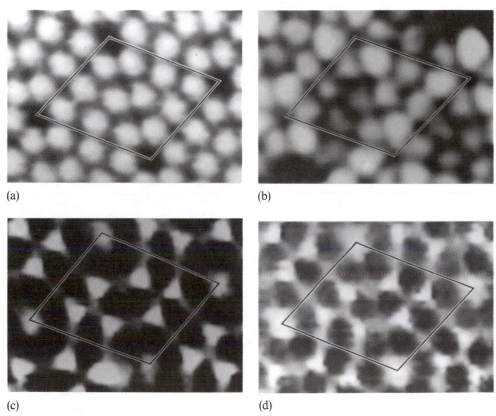

(a) (b)

(c) (d)

These four photographs were taken of the same region of the surface of a silicon specimen using a very sophisticated scanning tunneling microscope (STM). The bright regions correspond to (*a*) the positions of the surface atoms; (*b*) "dangling" bonds associated with this top layer of atoms; (*c*) "dangling" bonds that project out from the second layer of atoms below the surface; and (*d*) bonds that project sideways from those atoms in this second layer. The diamond shape that has been drawn in each of the photographs corresponds to a silicon unit cell. [Photograph supplied by R. J. Hamers, IBM Corporation. From R. J. Hamers, R. M. Tromp, and J. E. Demuth, "Surface Electronic Structure of Si (111)-(7 × 7) Resolved in Real Space," *Phys. Rev. Lett.*, **56,** 18 (1986). Copyright 1986 by The American Physical Society.]

2.1 INTRODUCTION

Some of the important properties of solid materials depend on geometrical atomic arrangements, and also the interactions that exist among the constituent atoms or molecules. This chapter, by way of preparation for subsequent discussions, considers several fundamental and important concepts, namely: atomic structure, electron configurations in atoms and the periodic table, and the various types of primary and secondary interatomic bonds that hold together the atoms comprising a solid. These topics are reviewed briefly, under the assumption that some of the material is familiar to the reader.

ATOMIC STRUCTURE

2.2 FUNDAMENTAL CONCEPTS

Each atom consists of a very small nucleus composed of protons and neutrons, which is encircled by moving electrons. Both electrons and protons are electrically charged, the charge magnitude being 1.60×10^{-19} C, which is negative in sign for electrons and positive for protons; neutrons are electrically neutral. Masses for these subatomic particles are infinitesimally small; protons and neutrons have approximately the same mass, 1.67×10^{-27} kg, which is significantly larger than that for an electron, 9.11×10^{-31} kg.

Each chemical element is characterized by the number of protons in the nucleus, or the **atomic number** (Z).[1] For an electrically neutral or complete atom, the atomic number also equals the number of electrons. This atomic number ranges in integral units from 1 for hydrogen to 94 for plutonium, the highest of the naturally occurring elements.

The *atomic mass* (A) of a specific atom may be expressed as the sum of the masses of protons and neutrons within the nucleus. Although the number of protons is the same for all atoms of a given element, the number of neutrons (N) may be variable. Thus atoms of some elements have two or more different atomic masses, which are called **isotopes.** The **atomic weight** corresponds to the weighted average of the atomic masses of the atom's naturally occurring isotopes. The **atomic mass unit (amu)** may be used for computations of atomic weight. A scale has been established whereby 1 amu is defined as $\frac{1}{12}$ of the atomic mass of the most common isotope of carbon, carbon 12 (^{12}C) ($A = 12.00000$). Within this scheme, the masses of protons and neutrons are slightly greater than unity, and

$$A \cong Z + N \tag{2.1}$$

The atomic weight of an element or the molecular weight of a compound may be specified on the basis of amu's per atom (molecule) or mass per mole of material. In one **mole** of a substance there are 6.023×10^{23} (Avogadro's number) atoms or molecules. These two atomic weight schemes are related through the following equation:

$$1 \text{ amu/atom (or molecule)} = 1 \text{ g/mol}$$

---

[1] Terms appearing in boldface type are defined in the Glossary, which follows Appendix C.

For example, the atomic weight of iron is 55.85 amu/atom, or 55.85 g/mol. Sometimes use of amu per atom or molecule is convenient; on other occasions g (or kg)/mol is preferred; the latter is used in this book.

2.3 ELECTRONS IN ATOMS

Bohr Atomic Model

During the latter part of the nineteenth century it was realized that many phenomena involving electrons in solids could not be explained in terms of classical mechanics. What followed was the establishment of a set of principles and laws that govern systems of atomic and subatomic entities, which came to be known as **quantum mechanics.** An understanding of the behavior of electrons in atoms and crystalline solids necessarily involves the discussion of quantum-mechanical concepts. However, a detailed exploration of these principles is beyond the scope of this book, and only a very superficial and simplified treatment is given.

One early outgrowth of quantum mechanics was the simplified **Bohr atomic model, in which electrons are assumed to revolve around the atomic nucleus in discrete orbitals,** and the position of any particular electron is more or less well defined in terms of its orbital. This model of the atom is represented in Figure 2.1.

Another important quantum-mechanical principle stipulates that the energies of electrons are quantized; that is, electrons are permitted to have only specific values of energy. An electron may change energy, but in doing so it must make a quantum jump either to an allowed higher energy (with absorption of energy) or to a lower energy (with the emission of energy). Often, it is convenient to think of these allowed electron energies as being associated with *energy levels* or *states*. These states do not vary continuously with energy, that is, adjacent states are separated by finite energies. For example, allowed states for the Bohr hydrogen atom are represented in Figure 2.2a. These energies are taken to be negative, whereas the zero reference is the unbound or free electron. Of course, the single electron associated with the hydrogen atom will fill only one of these states.

Thus the Bohr model represents an early attempt to describe electrons in atoms, in terms of both position (electron orbitals) and energy (quantized energy levels).

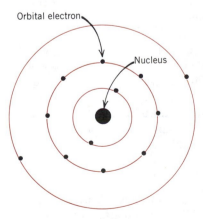

Figure 2.1 Schematic representation of the Bohr atom.

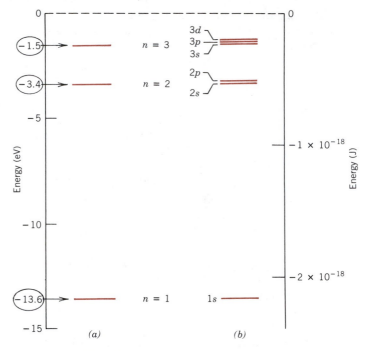

Figure 2.2 (*a*) The first three electron energy states for the Bohr hydrogen atom. (*b*) Electron energy states for the first three shells of the wave-mechanical hydrogen atom. (Adapted from W. G. Moffatt, G. W. Pearsall, and J. Wulff, *The Structure and Properties of Materials*, Vol. I, *Structure*, p. 10. Copyright © 1964 by John Wiley & Sons, New York. Reprinted by permission of John Wiley & Sons, Inc.)

Wave-Mechanical Atomic Model

The Bohr atomic model was eventually found to have some significant limitations because of its inability to explain several phenomena involving electrons. Resolution of these deficiencies was achieved with the development of what has become to be known as wave mechanics (a subdivision of quantum mechanics), and a more accurate model of the atom. In the **wave-mechanical model,** an electron is considered to exhibit both wavelike and particlelike characteristics, and the motion of an electron is described by mathematics that govern wave motion.

One important consequence of wave mechanics is that electrons are no longer treated as particles moving in discrete orbitals; but rather, position is considered to be the probability of an electron's being at various locations around the nucleus. In other words, position is described by a probability distribution or electron cloud. Figure 2.3 compares Bohr and wave-mechanical models for the hydrogen atom. Both these models are used throughout the course of this book; the choice depends on which model allows the more simple explanation.

Quantum Numbers

Using wave mechanics, every electron in an atom is characterized by four parameters called **quantum numbers.** The size, shape, and spatial orientation of an electron's

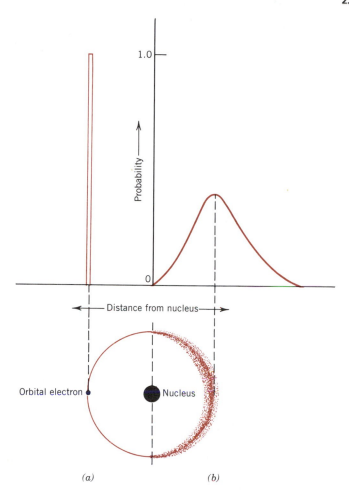

Figure 2.3 Comparison of the (*a*) Bohr and (*b*) wave-mechanical atom models in terms of electron distribution. (Adapted from Z. D. Jastrzebski, *The Nature and Properties of Engineering Materials*, 3rd edition, p. 4. Copyright © 1987 by John Wiley & Sons, New York. Reprinted by permission of John Wiley & Sons, Inc.)

probability density are specified by three of these quantum numbers. Furthermore, Bohr energy levels separate into electron subshells, and quantum numbers dictate the number of states within each subshell. Shells are specified by a *principal quantum number n,* which may take on integral values beginning with unity; sometimes these shells are designated by the letters *K, L, M, N, O,* and so on, which correspond, respectively, to $n = 1, 2, 3, 4, 5, \ldots$, as indicated in Table 2.1. It should also be noted that this quantum number, and it only, is also associated with the Bohr model.

The second quantum number, *l,* signifies the subshell, which is denoted by a lowercase letter—an *s, p, d,* or *f.* In addition, the number of these subshells is restricted by the magnitude of *n.* Allowable subshells for the several *n* values are also presented in Table 2.1. The number of energy states for each subshell is determined by the third

TABLE 2.1 The Number of Available Electron States in Some of the Electron Shells and Subshells

Principal Quantum Number n	Shell Designation	Subshells	Number of States	Number of Electrons	
				Per Subshell	Per Shell
1	K	s	1	2	2
2	L	s	1	2	8
		p	3	6	
3	M	s	1	2	18
		p	3	6	
		d	5	10	
4	N	s	1	2	32
		p	3	6	
		d	5	10	
		f	7	14	

quantum number, m_l. For an s subshell there is a single energy state, whereas for p, d, and f subshells, three, five, and seven states exist, respectively (Table 2.1). In the absence of an external magnetic field, the states within each subshell are identical. However, when a magnetic field is applied these subshell states split, each state assuming a slightly different energy.

Associated with each electron is a *spin moment,* which must be oriented either up or down. Related to this spin moment is the fourth quantum number, m_s, for which two values are possible ($+\frac{1}{2}$ and $-\frac{1}{2}$), one for each of the spin orientations.

Thus the Bohr model was further refined by wave mechanics, in which the introduction of three new quantum numbers gives rise to electron subshells within each shell. A comparison of these two models on this basis is illustrated, for the hydrogen atom, in Figures 2.2a and 2.2b.

A complete energy level diagram for the various shells and subshells using the wave-mechanical model is shown in Figure 2.4. Several features of the diagram are worth noting. First, the smaller the principal quantum number, the lower the energy level; for example, the energy of a 1s state is less than that of a 2s state, which in turn is lower than the 3s. Second, within each shell, the energy of a subshell level increases with the value of the l quantum number. For example, the energy of a 3d state is greater than a 3p, which is larger than 3s. Finally, there may be overlap in energy of a state in one shell with states in an adjacent shell, which is especially true of d and f states; for example, the energy of a 3d state is greater than that for a 4s.

Electron Configurations

The preceding discussion has dealt primarily with **electron states**—values of energy that are permitted for electrons. To determine the manner in which these states are filled with electrons, we use the **Pauli exclusion principle,** another quantum-mechanical concept. This principle stipulates that each electron state can hold no more than two electrons, which must have opposite spins. Thus, s, p, d, and f subshells may each accommodate, respectively, a total of 2, 6, 10, and 14 electrons; Table 2.1 summarizes the maximum number of electrons that may occupy each of the first four shells.

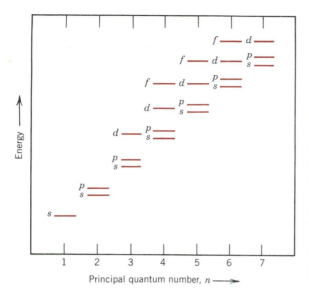

Figure 2.4 Schematic representation of the relative energies of the electrons for the various shells and subshells. (From K. M. Ralls, T. H. Courtney, and J. Wulff, *Introduction to Materials Science and Engineering*, p. 22. Copyright © 1976 by John Wiley & Sons, New York. Reprinted by permission of John Wiley & Sons, Inc.)

Of course, not all possible states in an atom are filled with electrons. For most atoms, the electrons fill up the lowest possible energy states in the electron shells and subshells, two electrons (having opposite spins) per state. The energy structure for a sodium atom is represented schematically in Figure 2.5. When all the electrons occupy the lowest possible energies in accord with the foregoing restrictions, an atom is said to be in its **ground state.** However, electron transitions to higher energy states are possible, as discussed in Chapters 19 and 22. The **electron configuration** or structure of an atom represents the manner in which these states are occupied. In the conventional notation the number of electrons in each subshell is indicated by a superscript after the shell–subshell designation. For example, the electron configurations for hydrogen, helium, and sodium are, respectively, $1s^1$, $1s^2$, and $1s^2 2s^2 2p^6 3s^1$. Electron configurations for some of the more common elements are listed in Table 2.2; a tabulation for all the elements is contained in Appendix B.

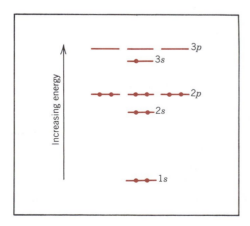

Figure 2.5 Schematic representation of the filled energy states for a sodium atom.

TABLE 2.2 A Listing of the Electron Configurations for Some of the Common Elements

Element	Symbol	Atomic Number	Electron Configuration
Hydrogen	H	1	$1s^1$
Helium	He	2	$1s^2$
Lithium	Li	3	$1s^2 2s^1$
Beryllium	Be	4	$1s^2 2s^2$
Boron	B	5	$1s^2 2s^2 2p^1$
Carbon	C	6	$1s^2 2s^2 2p^2$
Nitrogen	N	7	$1s^2 2s^2 2p^3$
Oxygen	O	8	$1s^2 2s^2 2p^4$
Fluorine	F	9	$1s^2 2s^2 2p^5$
Neon	Ne	10	$1s^2 2s^2 2p^6$
Sodium	Na	11	$1s^2 2s^2 2p^6 3s^1$
Magnesium	Mg	12	$1s^2 2s^2 2p^6 3s^2$
Aluminum	Al	13	$1s^2 2s^2 2p^6 3s^2 3p^1$
Silicon	Si	14	$1s^2 2s^2 2p^6 3s^2 3p^2$
Phosphorus	P	15	$1s^2 2s^2 2p^6 3s^2 3p^3$
Sulfur	S	16	$1s^2 2s^2 2p^6 3s^2 3p^4$
Chlorine	Cl	17	$1s^2 2s^2 2p^6 3s^2 3p^5$
Argon	Ar	18	$1s^2 2s^2 2p^6 3s^2 3p^6$
Potassium	K	19	$1s^2 2s^2 2p^6 3s^2 3p^6 4s^1$
Calcium	Ca	20	$1s^2 2s^2 2p^6 3s^2 3p^6 4s^2$
Scandium	Sc	21	$1s^2 2s^2 2p^6 3s^2 3p^6 3d^1 4s^2$
Titanium	Ti	22	$1s^2 2s^2 2p^6 3s^2 3p^6 3d^2 4s^2$
Vanadium	V	23	$1s^2 2s^2 2p^6 3s^2 3p^6 3d^3 4s^2$
Chromium	Cr	24	$1s^2 2s^2 2p^6 3s^2 3p^6 3d^5 4s^1$
Manganese	Mn	25	$1s^2 2s^2 2p^6 3s^2 3p^6 3d^5 4s^2$
Iron	Fe	26	$1s^2 2s^2 2p^6 3s^2 3p^6 3d^6 4s^2$
Cobalt	Co	27	$1s^2 2s^2 2p^6 3s^2 3p^6 3d^7 4s^2$
Nickel	Ni	28	$1s^2 2s^2 2p^6 3s^2 3p^6 3d^8 4s^2$
Copper	Cu	29	$1s^2 2s^2 2p^6 3s^2 3p^6 3d^{10} 4s^1$
Zinc	Zn	30	$1s^2 2s^2 2p^6 3s^2 3p^6 3d^{10} 4s^2$
Gallium	Ga	31	$1s^2 2s^2 2p^6 3s^2 3p^6 3d^{10} 4s^2 4p^1$
Germanium	Ge	32	$1s^2 2s^2 2p^6 3s^2 3p^6 3d^{10} 4s^2 4p^2$
Arsenic	As	33	$1s^2 2s^2 2p^6 3s^2 3p^6 3d^{10} 4s^2 4p^3$
Selenium	Se	34	$1s^2 2s^2 2p^6 3s^2 3p^6 3d^{10} 4s^2 4p^4$
Bromine	Br	35	$1s^2 2s^2 2p^6 3s^2 3p^6 3d^{10} 4s^2 4p^5$
Krypton	Kr	36	$1s^2 2s^2 2p^6 3s^2 3p^6 3d^{10} 4s^2 4p^6$

At this point, comments regarding these electron configurations are necessary. First, the **valence electrons** are those that occupy the outermost filled shell. These electrons are extremely important; as will be seen, they participate in the bonding between atoms to form atomic and molecular aggregates. Furthermore, many of the physical and chemical properties of solids are based on these valence electrons.

In addition, some atoms have what are termed "stable electron configurations"; that is, the states within the outermost or valence electron shell are completely filled.

Normally this corresponds to the occupation of just the *s* and *p* states for the outermost shell by a total of eight electrons, as in neon, argon, and krypton; one exception is helium, which contains only two 1*s* electrons. These elements (Ne, Ar, Kr, and He) are the inert, or noble, gases, which are virtually unreactive chemically. Some atoms of the elements that have unfilled valence shells assume stable electron configurations by gaining or losing electrons to form charged ions, or by sharing electrons with other atoms. This is the basis for some chemical reactions, and also for atomic bonding in solids, as explained in Section 2.6.

2.4 THE PERIODIC TABLE

All the elements have been classified according to electron configuration in the **periodic table** (Figure 2.6). Here, the elements are situated, with increasing atomic number, in seven horizontal rows called periods. The arrangement is such that all elements that are arrayed in a given column or group have similar valence electron structures, as well as chemical and physical properties. These properties change gradually and systematically, moving horizontally across each period.

The elements positioned in Group 0, the right-most group, are the inert gases, which have filled electron shells and stable electron configurations. Group VIIA and VIA elements are one and two electrons deficient, respectively, from having stable structures. The Group VIIA elements (F, Cl, Br, I, and At) are sometimes termed the halogens. The alkali and the alkaline earth metals (Li, Na, K, Be, Mg, Ca, etc.) are

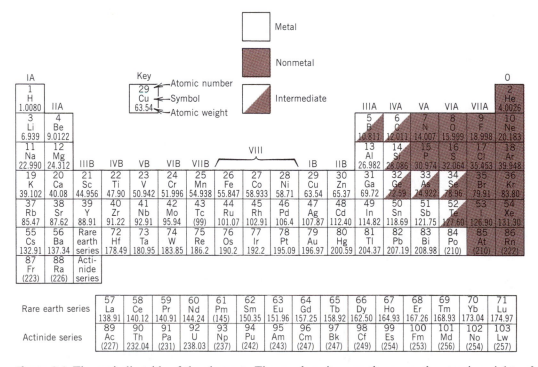

Figure 2.6 The periodic table of the elements. The numbers in parentheses are the atomic weights of the most stable or common isotopes.

IA	IIA	IIIB	IVB	VB	VIB	VIIB	VIII			IB	IIB	IIIA	IVA	VA	VIA	VIIA	0
1 H 2.1																	2 He –
3 Li 1.0	4 Be 1.5											5 B 2.0	6 C 2.5	7 N 3.0	8 O 3.5	9 F 4.0	10 Ne –
11 Na 0.9	12 Mg 1.2											13 Al 1.5	14 Si 1.8	15 P 2.1	16 S 2.5	17 Cl 3.0	18 Ar –
19 K 0.8	20 Ca 1.0	21 Sc 1.3	22 Ti 1.5	23 V 1.6	24 Cr 1.6	25 Mn 1.5	26 Fe 1.8	27 Co 1.8	28 Ni 1.8	29 Cu 1.9	30 Zn 1.6	31 Ga 1.6	32 Ge 1.8	33 As 2.0	34 Se 2.4	35 Br 2.8	36 Kr –
37 Rb 0.8	38 Sr 1.0	39 Y 1.2	40 Zr 1.4	41 Nb 1.6	42 Mo 1.8	43 Tc 1.9	44 Ru 2.2	45 Rh 2.2	46 Pd 2.2	47 Ag 1.9	48 Cd 1.7	49 In 1.7	50 Sn 1.8	51 Sb 1.9	52 Te 2.1	53 I 2.5	54 Xe –
55 Cs 0.7	56 Ba 0.9	57–71 La–Lu 1.1–1.2	72 Hf 1.3	73 Ta 1.5	74 W 1.7	75 Re 1.9	76 Os 2.2	77 Ir 2.2	78 Pt 2.2	79 Au 2.4	80 Hg 1.9	81 Tl 1.8	82 Pb 1.8	83 Bi 1.9	84 Po 2.0	85 At 2.2	86 Rn –
87 Fr 0.7	88 Ra 0.9	89–102 Ac–No 1.1–1.7															

Figure 2.7 The electronegativity values for the elements. (Reprinted from Linus Pauling, *The Nature of the Chemical Bond*, 3rd edition. Copyright 1939 and 1940, 3rd edition copyright © 1960, by Cornell University. Used by permission of the publisher, Cornell University Press.)

labeled as Groups IA and IIA, having, respectively, one and two electrons in excess of stable structures. The elements in the three long periods, Groups IIIB through IIB, are termed the transition metals, which have partially filled *d* electron states and in some cases one or two electrons in the next higher energy shell. Groups IIIA, IVA, and VA (B, Si, Ge, As, etc.) display characteristics that are intermediate between the metals and nonmetals by virtue of their valence electron structures.

As may be noted from the periodic table, most of the elements really come under the metal classification. These are sometimes termed **electropositive** elements, indicating that they are capable of giving up their few valence electrons to become positively charged ions. Furthermore, the elements situated on the right-hand side of the table are **electronegative;** that is, they readily accept electrons to form negatively charged ions, or sometimes they share electrons with other atoms. Figure 2.7 displays electronegativity values that have been assigned to the various elements arranged in the periodic table. As a general rule, electronegativity increases in moving from left to right and from bottom to top.

ATOMIC BONDING IN SOLIDS

2.5 BONDING FORCES AND ENERGIES

An understanding of many of the physical properties of materials is predicated on a knowledge of the interatomic forces that bind the atoms together. Perhaps the principles of atomic bonding are best illustrated by considering the interaction between two isolated atoms as they are brought into close proximity from an infinite separation. At large distances the interactions are negligible; but as the atoms approach, each exerts forces on the other. These forces are of two types, attractive and repulsive, and the magnitude of each is a function of the separation or interatomic distance. The origin of an attractive force F_A depends on the particular type of bonding that exists between the two atoms. Its magnitude varies with the distance, as represented schematically in Figure 2.8a. Ultimately, the outer electron shells of the two atoms begin to overlap, and a strong repulsive force F_R comes into play. The net force F_N between the two atoms is just the sum of both attractive and repulsive components; that is,

$$F_N = F_A + F_R \tag{2.2}$$

which is also a function of the interatomic separation as also plotted in Figure 2.8a.

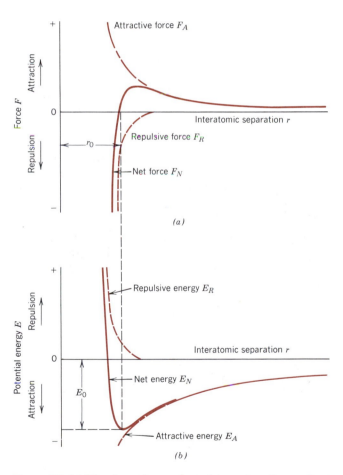

Figure 2.8 (a) The dependence of repulsive, attractive, and net forces as a function of interatomic separation for two isolated atoms. (b) The dependence of repulsive, attractive, and net potential energies as a function of interatomic separation for two isolated atoms.

When F_A and F_R balance, or become equal, there is no net force; that is,

$$F_A + F_R = 0 \tag{2.3}$$

Then a state of equilibrium exists. The centers of the two atoms will remain separated by the equilibrium spacing r_0, as indicated in Figure 2.8a. For many atoms r_0 is approximately 0.3 nm (3 Å). Once in this position, the two atoms will counteract any attempt to separate them by an attractive force, or to push them together by a repulsive action.

Sometimes it is more convenient to work with the potential energies between two atoms instead of forces. Mathematically, energy (E) and force (F) are related as

$$E = \int F \, dr \tag{2.4}$$

Or, for atomic systems,

$$E_N = \int_{\infty}^{r} F_N \, dr \tag{2.5}$$

$$= \int_{\infty}^{r} F_A \, dr + \int_{\infty}^{r} F_R \, dr \tag{2.6}$$

$$= E_A + E_R \tag{2.7}$$

in which E_N, E_A, and E_R are respectively the net, attractive, and repulsive energies for two isolated and adjacent atoms.

Figure 2.8b plots attractive, repulsive, and net potential energies as a function of interatomic separation for two atoms. The net curve, which is again the sum of the other two, has a potential energy trough or well around its minimum. Here, the same equilibrium spacing, r_0, corresponds to the separation distance at the minimum of the potential energy curve. The **bonding energy** for these two atoms, E_0, corresponds to the energy at this minimum point (also shown in Figure 2.8b); it represents the energy that would be required to separate these two atoms to an infinite separation.

Although the preceding treatment has dealt with an ideal situation involving only two atoms, a similar yet more complex condition exists for solid materials because force and energy interactions among many atoms must be considered. Nevertheless, a bonding energy, analogous to E_0 above, may be associated with each atom. The magnitude of this bonding energy and the shape of the energy versus interatomic separation curve vary from material to material, both variables depending on the type of atomic bonding. Solid substances are formed for large bonding energies, whereas for small energies the gaseous state is favored; liquids prevail when the energies are of intermediate magnitude. In general, for solid materials, melting temperature as well as cohesive properties reflect the magnitude of the bonding energy.

Three different types of primary or chemical bond are found in solids—ionic, covalent, and metallic. For each type, the bonding necessarily involves the valence electrons; furthermore, the nature of the bond depends on the electron structures of the constituent atoms. In general, each of these three types of bonding arises from the tendency of the atoms to assume stable electron structures, like those of the inert gases, by completely filling the outermost electron shell.

Secondary or physical forces and energies are also found in many solid materials; they are weaker than the primary ones, but nonetheless influence the physical properties of some materials. The sections that follow explain the several kinds of primary and secondary interatomic bonds.

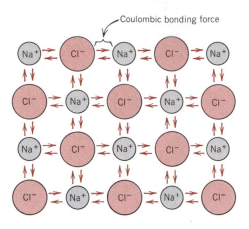

Coulombic bonding force

Figure 2.9 Schematic representation of ionic bonding in sodium chloride (NaCl).

2.6 PRIMARY INTERATOMIC BONDS

Ionic Bonding

Perhaps **ionic bonding** is the easiest to describe and visualize. It is always found in compounds that are composed of both metallic and nonmetallic elements, elements that are situated at the horizontal extremities of the periodic table. Atoms of a metallic element easily give up their valence electrons to the nonmetallic atoms. In the process all the atoms acquire stable or inert gas configurations and, in addition, an electrical charge; that is, they become ions. Sodium chloride (NaCl) is the classical ionic material. A sodium atom can assume the electron structure of neon (and a net single positive charge) by a transfer of its one valence $3s$ electron to a chlorine atom. After such a transfer, the chlorine ion has a net negative charge and an electron configuration identical to that of argon. In sodium chloride, all the sodium and chlorine exist as ions. This type of bonding is illustrated schematically in Figure 2.9.

The attractive bonding forces are coulombic; that is, positive and negative ions, by virtue of their net electrical charge, attract one another. For two isolated ions, the attractive energy E_A is a function of the interatomic distance according to[2]

$$E_A = -\frac{A}{r} \tag{2.8}$$

An analogous equation for the repulsive energy is

$$E_R = \frac{B}{r^n} \tag{2.9}$$

---

[2] The constant A in Equation 2.8 is equal to

$$\frac{1}{4\pi\varepsilon_0}(Z_1 e)(Z_2 e)$$

where ε_0 is the permittivity of a vacuum (8.85×10^{-12} F/m), Z_1 and Z_2 are the valences of the two ion types, and e is the electronic charge (1.6×10^{-19} C).

TABLE 2.3 Bonding Energies and Melting Temperatures for Various Substances

Bonding Type	Substance	Bonding Energy		Melting Temperature (°C)
		kJ/mol (kcal/mol)	eV/atom, ion, molecule	
Ionic	NaCl	640 (153)	3.3	801
	MgO	1000 (239)	5.2	2800
Covalent	Si	450 (108)	4.7	1410
	C (diamond)	713 (170)	7.4	>3550
Metallic	Hg	68 (16)	0.7	−39
	Al	324 (77)	3.4	660
	Fe	406 (97)	4.2	1538
	W	849 (203)	8.8	3410
van der Waals	Ar	7.7 (1.8)	0.08	−189
	Cl_2	31 (7.4)	0.32	−101
Hydrogen	NH_3	35 (8.4)	0.36	−78
	H_2O	51 (12.2)	0.52	0

In these expressions, A, B, and n are constants whose values depend on the particular ionic system. The value of n is approximately 8.

Ionic bonding is termed nondirectional, that is, the magnitude of the bond is equal in all directions around an ion. It follows that for ionic materials to be stable, all positive ions must have as nearest neighbors negatively charged ions in a three-dimensional scheme, and vice versa. The predominant bonding in ceramic materials is ionic. Some of the ion arrangements for these materials are discussed in Chapter 13.

Bonding energies, which generally range between 600 and 1500 kJ/mol (3 and 8 eV/atom), are relatively large, as reflected in high melting temperatures.[3] Table 2.3 contains bonding energies and melting temperatures for several ionic materials. Ionic materials are characteristically hard and brittle and, furthermore, electrically and thermally insulative. As discussed in subsequent chapters, these properties are a direct consequence of electron configurations and/or the nature of the ionic bond.

Covalent Bonding

In **covalent bonding** stable electron configurations are assumed by the sharing of electrons between adjacent atoms. Two atoms that are covalently bonded will each contribute at least one electron to the bond, and the shared electrons may be considered to belong to both atoms. Covalent bonding is schematically illustrated in Figure 2.10

---

[3] Sometimes bonding energies are expressed per atom or per ion. Under these circumstances the electron volt (eV) is a conveniently small unit of energy. It is, by definition, the energy imparted to an electron as it falls through an electric potential of one volt. The joule equivalent of the electron volt is as follows: 1.602×10^{-19} J $= 1$ eV.

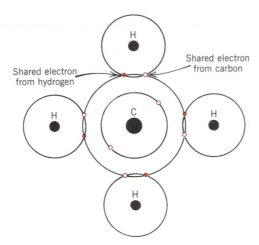

Figure 2.10 Schematic representation of covalent bonding in a molecule of methane (CH_4).

for a molecule of methane (CH_4). The carbon atom has four valence electrons, whereas each of the four hydrogen atoms has a single valence electron. Each hydrogen atom can acquire a helium electron configuration (two $1s$ valence electrons) when the carbon atom shares with it one electron. The carbon now has four additional shared electrons, one from each hydrogen, for a total of eight valence electrons, and the electron structure of neon. The covalent bond is directional; that is, it is between specific atoms and may exist only in the direction between one atom and another that participates in the electron sharing.

Many nonmetallic elemental molecules (H_2, Cl_2, F_2, etc.) as well as molecules containing dissimilar atoms, such as CH_4, H_2O, HNO_3, and HF, are covalently bonded. Furthermore, this type of bonding is found in elemental solids such as diamond (carbon), silicon, and germanium and other solid compounds composed of elements that are located on the right-hand side of the periodic table, such as gallium arsenide (GaAs), indium antimonide (InSb), and silicon carbide (SiC).

The number of covalent bonds that are possible for a particular atom is determined by the number of valence electrons. For N' valence electrons, an atom can covalently bond with at most $8 - N'$ other atoms. For example, $N' = 7$ for chlorine, and $8 - N' = 1$, which means that one Cl atom can bond to only one other atom, as in Cl_2. Similarly, for carbon, $N' = 4$, and each carbon atom has $8 - 4$, or four, electrons to share. Diamond is simply the three-dimensional interconnecting structure wherein each carbon atom covalently bonds with four other carbon atoms. This arrangement is represented in Figure 13.5.

Covalent bonds may be very strong, as in diamond, which is very hard and has a very high melting temperature, $> 3550°C$ ($6400°F$), or they may be very weak, as with bismuth, which melts at about $270°C$ ($518°F$). Bonding energies and melting temperatures for a few covalently bonded materials are presented in Table 2.3. Polymeric materials typify this bond, the basic molecular structure being a long chain of carbon atoms that are covalently bonded together with two of their available four bonds per atom. The remaining two bonds normally are shared with other atoms, which also covalently bond. Polymeric molecular structures are discussed in detail in Chapter 15.

It is possible to have interatomic bonds that are partially ionic and partially covalent, and, in fact, very few compounds exhibit pure ionic or covalent bonding. For a compound, the degree of either bond type depends on the relative positions of the constituent atoms in the periodic table (Figure 2.6). The wider the separation, (both horizontally—relative to Group IVA—and vertically) from the lower left to the upper right-hand corner, the more ionic the bond is; or, the closer the atoms are together, the greater the degree of covalency.

Metallic Bonding

Metallic bonding, the final primary bonding type, is found in metals and their alloys. A relatively simple model has been proposed that very nearly approximates the bonding scheme. Metallic materials have one, two, or at most, three valence electrons. With this model, these valence electrons are not bound to any particular atom in the solid and are more or less free to drift throughout the entire metal. They may be thought of as belonging to the metal as a whole, or forming a "sea of electrons" or an "electron cloud." The remaining nonvalence electrons and atomic nuclei form what are called *ion cores,* which possess a net positive charge, equal in magnitude to the total valence electron charge per atom. Figure 2.11 is a schematic illustration of metallic bonding. The free electrons shield the positively charged ion cores from mutually repulsive electrostatic forces, which they would otherwise exert upon one another; consequently the metallic bond is nondirectional in character. In addition, these free electrons act as a "glue" to hold the ion cores together. Bonding energies and melting temperatures for several metals are listed in Table 2.3. Bonding may be weak or strong; energies range from 68 kJ/mol (0.7 eV/atom) for mercury to 850 kJ/mol (8.8 eV/atom) for tungsten. Their respective melting temperatures are −39 and 3410°C (−38 and 6170°F).

This type of bonding is found for Group IA and IIA elements in the periodic table and, in fact, for all elemental metals. These materials are good conductors of both electricity and heat, as a consequence of the free valence electrons.

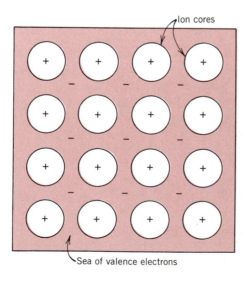

Ion cores

Sea of valence electrons

Figure 2.11 Schematic illustration of metallic bonding.

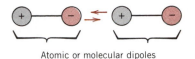

Figure 2.12 Schematic illustration of van der Waals bonding between two dipoles.

Atomic or molecular dipoles

2.7 SECONDARY BONDING OR VAN DER WAALS BONDING

Secondary, van der Waals, or physical bonds are weak in comparison to the primary or chemical ones; bonding energies are typically on the order of only 10 kJ/mol (0.1 eV/atom). Secondary bonding exists between virtually all atoms or molecules, but its presence may be obscured if any of the three primary bonding types is present. Secondary bonding is evidenced for the inert gases, which have stable electron structures, and, in addition, between molecules in molecular structures that are covalently bonded.

Secondary bonding forces arise from atomic or molecular **dipoles.** In essence, an electric dipole exists whenever there is some separation of positive and negative portions of an atom or molecule. The bonding results from the coulombic attraction between the positive end of one dipole and the negative region of an adjacent one, as indicated in Figure 2.12. Dipole interactions occur between induced dipoles, between induced dipoles and polar molecules (which have permanent dipoles), and between polar molecules. **Hydrogen bonding,** a special type of secondary bonding, is found to exist between some molecules that have hydrogen as one of the constituents. These bonding mechanisms are now discussed briefly.

Fluctuating Induced Dipole Bonds

A dipole may be created or induced in an atom or molecule that is normally electrically symmetric; that is, the overall spatial distribution of the electrons is symmetric with respect to the positively charged nucleus, as shown in Figure 2.13a. All atoms are experiencing constant vibrational motion, which can cause instantaneous and short-lived distortions of this electrical symmetry for some of the atoms or molecules, and the creation of small electric dipoles, as represented in Figure 2.13b. One of these

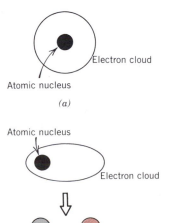

Figure 2.13 Schematic representations of (a) an electrically symmetric atom and (b) an induced atomic dipole.

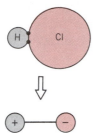

Figure 2.14 Schematic representation of a polar hydrogen chloride (HCl) molecule.

dipoles can in turn produce a displacement of the electron distribution of an adjacent molecule or atom, which induces the second one also to become a dipole that is then weakly attracted or bonded to the first; this is one type of van der Waals bonding. These attractive forces may exist between large numbers of atoms or molecules, which forces are temporary and fluctuate with time.

The liquefaction and, in some cases, the solidification of the inert gases and other electrically neutral and symmetric molecules such as H_2 and Cl_2 are realized because of this type of bonding. Melting and boiling temperatures are extremely low in materials for which induced dipole bonding predominates; of all possible intermolecular bonds, these are the weakest. Bonding energies and melting temperatures for argon and chlorine are also tabulated in Table 2.3.

Polar Molecule-Induced Dipole Bonds

Permanent dipole moments exist in some molecules by virtue of an asymmetrical arrangement of positively and negatively charged regions; such molecules are termed **polar molecules.** Figure 2.14 is a schematic representation of a hydrogen chloride molecule; a permanent dipole moment arises from net positive and negative charges that are respectively associated with the hydrogen and chlorine ends of the HCl molecule.

Polar molecules can also induce dipoles in adjacent nonpolar molecules, and a bond will form as a result of attractive forces between the two molecules. Furthermore, the magnitude of this bond will be greater than for fluctuating induced dipoles.

Permanent Dipole Bonds

Van der Waals forces will also exist between adjacent polar molecules. The associated bonding energies are significantly greater than for bonds involving induced dipoles.

The strongest secondary bonding type, the hydrogen bond, is a special case of polar molecule bonding. It occurs between molecules in which hydrogen is covalently bonded to fluorine (as in HF), oxygen (as in H_2O), and nitrogen (as in NH_3). For each H—F, H—O, or H—N bond, the single hydrogen electron is shared with the other atom. Thus the hydrogen end of the bond is essentially a positively charged bare proton, which is unscreened by any electrons. This highly positively charged end of the molecule is capable of a strong attractive force with the negative end of an adjacent molecule, as demonstrated in Figure 2.15 for HF. In essence, this single proton forms a bridge between two negatively charged atoms. The magnitude of the hydrogen bond is generally greater than that for the other types of secondary bonds,

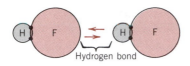

Figure 2.15 Schematic representation of hydrogen bonding in hydrogen fluoride (HF).

and may be as high as 51 kJ/mol (0.52 eV/molecule) as shown in Table 2.3. Melting and boiling temperatures for hydrogen fluoride and water are abnormally high in light of their low molecular weights, as a consequence of hydrogen bonding.

2.8 MOLECULES

At the conclusion of this chapter, let us take a moment to discuss the concept of a molecule in terms of solid materials. A molecule may be defined as a group of atoms that are bonded together by strong primary bonds. Within this context, the entirety of ionic and metallically bonded solid specimens may be considered as a single molecule. However, this is not the case for many substances in which covalent bonding predominates; these include elemental diatomic molecules (F_2, O_2, H_2, etc.), as well as a host of compounds (H_2O, CO_2, HNO_3, C_6H_6, CH_4, etc.). In the condensed liquid and solid states, bonds between molecules are weak secondary ones. Consequently, molecular materials have relatively low melting and boiling temperatures. Most of those that have small molecules composed of a few atoms are gases at ordinary, or ambient, temperatures and pressures. On the other hand, many of the modern polymers, being molecular materials composed of extremely large molecules, exist as solids; some of their properties are strongly dependent on the presence of van der Waals and hydrogen secondary bonds.

SUMMARY

This chapter began with a survey of the fundamentals of atomic structure, presenting the Bohr and wave-mechanical models of electrons in atoms. Whereas the Bohr model assumes electrons to be particles orbiting the nucleus in discrete paths, in wave mechanics we consider them to be wavelike and treat electron position in terms of a probability distribution.

Electron energy states are specified in terms of quantum numbers that give rise to electron shells and subshells. The electron configuration of an atom corresponds to the manner in which these shells and subshells are filled with electrons in compliance with the Pauli exclusion principle. The periodic table of the elements is generated by arrangement of the various elements according to valence electron configuration.

Atomic bonding in solids may be considered in terms of attractive and repulsive forces and energies. The three types of primary bond in solids are ionic, covalent, and metallic. For ionic bonds, electrically charged ions are formed by the transference of valence electrons from one atom type to another; forces are coulombic. There is a sharing of valence electrons between adjacent atoms when bonding is covalent. With metallic bonding, the valence electrons form a "sea of electrons" that is uniformly dispersed around the metal ion cores and acts as a form of glue for them.

Both van der Waals and hydrogen bonds are termed secondary, being weak in comparison to the primary ones. They result from attractive forces between electric dipoles, of which there are two types—induced and permanent. For the hydrogen bond, highly polar molecules form when hydrogen covalently bonds to a nonmetallic element such as fluorine.

IMPORTANT TERMS AND CONCEPTS

Atomic mass unit (amu)	Electronegative	Periodic table
Atomic number	Electropositive	Polar molecule
Atomic weight	Ground state	Primary bonding
Bohr atomic model	Hydrogen bond	Quantum mechanics
Bonding energy	Ionic bond	Quantum number
Coulombic force	Isotope	Secondary bonding
Covalent bond	Metallic bond	Valence electron
Dipole (electric)	Mole	van der Waals bond
Electron configuration	Molecule	Wave-mechanical model
Electron state	Pauli exclusion principle	

REFERENCES

Most of the material in this chapter is covered in college-level chemistry textbooks. Below, two are listed as references.

KOTZ, J. C. and K. F. PURCELL, *Chemistry and Chemical Reactivity,* Saunders College Publishing, Philadelphia, 1987.

MASTERTON, W. L. and C. N. HURLEY, *Chemistry, Principles and Reactions,* Saunders College Publishing, Philadelphia, 1989.

QUESTIONS AND PROBLEMS

2.1   **(a)** What is an isotope? **(b)** Why are the atomic weights of the elements not integers? Cite two reasons.

2.2   Cite the difference between atomic mass and atomic weight.

2.3   **(a)** How many grams are there in 1 amu of a material? **(b)** Mole, in the context of this book, is taken in units of gram-mole. On this basis, how many atoms are there in a pound-mole of a substance?

Note: In each chapter, most of the terms listed in the "Important Terms and Concepts" section are defined in the Glossary, which follows Appendix C. The others are important enough to warrant treatment in a full section of the text and can be referenced from the table of contents or the index.

2.4 **(a)** Cite two important quantum-mechanical concepts associated with the Bohr model of the atom. **(b)** Cite two important additional refinements that resulted from the wave-mechanical atomic model.

2.5 Relative to electrons and electron states, what does each of the four quantum numbers specify?

2.6 Allowed values for the quantum numbers of electrons are as follows:

$$n = 1, 2, 3, \ldots$$
$$l = 0, 1, 2, 3, \ldots, n - 1$$
$$m_l = 0, \pm 1, \pm 2, \pm 3, \ldots, \pm l$$
$$m_s = \pm \tfrac{1}{2}$$

The relationships between n and the shell designations are noted in Table 2.1. Relative to the subshells,

$l = 0$ corresponds to an s subshell

$l = 1$ corresponds to a p subshell

$l = 2$ corresponds to a d subshell

$l = 3$ corresponds to an f subshell

For the K shell, the four quantum numbers for each of the two electrons in the $1s$ state, in the order of nlm_lm_s, are $100(\tfrac{1}{2})$ and $100(-\tfrac{1}{2})$.

Write the four quantum numbers for all of the electrons in the L and M shells, and note which correspond to the s, p, and d subshells.

2.7 Give the electron configurations for the following ions: Fe^{2+}, Fe^{3+}, Cu^+, Ba^{2+}, Br^-, and S^{2-}.

2.8 Calcium oxide (CaO) exhibits predominantly ionic bonding. The Ca^{2+} and O^{2-} ions have electron structures that are identical to which two inert gases?

2.9 With regard to electron configuration, what do all the elements in Group IA of the periodic table have in common?

2.10 Without consulting Figure 2.6 or Table 2.2, determine whether each of the electron configurations given below is an inert gas, a halogen, an alkali metal, an alkaline earth metal, or a transition metal. Justify your choices.
(a) $1s^2 2s^2 2p^6 3s^2 3p^6 3d^7 4s^2$.
(b) $1s^2 2s^2 2p^6 3s^2 3p^6$.
(c) $1s^2 2s^2 2p^5$.
(d) $1s^2 2s^2 2p^6 3s^2$.
(e) $1s^2 2s^2 2p^6 3s^2 3p^6 3d^2 4s^2$.
(f) $1s^2 2s^2 2p^6 3s^2 3p^6 4s^1$.

2.11 **(a)** What electron subshell is being filled for the rare earth series of elements on the periodic table? **(b)** What electron subshell is being filled for the actinide series?

2.12 Calculate the force of attraction between a Ca^{2+} and an O^{2-} ion the centers of which are separated by a distance of 1.0 nm.

2.13 The net potential energy between two adjacent ions, E_N, may be represented by the sum of Equations 2.8 and 2.9, that is,

$$E_N = -\frac{A}{r} + \frac{B}{r^n} \tag{2.10}$$

Calculate the bonding energy E_0 in terms of the parameters A, B, and n using the following procedure:

1. Differentiate E_N with respect to r, and then set the resulting expression equal to zero, since the curve of E_N versus r is a minimum at E_0.
2. Solve for r in terms of A, B, and n, which yields r_0, the equilibrium interionic spacing.
3. Determine the expression for E_0 by substitution of r_0 into Equation 2.10.

2.14 For a Na^+–Cl^- ion pair, attractive and repulsive energies E_A and E_R, respectively, depend on the distance between the ions r, according to

$$E_A = -\frac{1.436}{r}$$

$$E_R = \frac{7.32 \times 10^{-6}}{r^8}$$

For these expressions, energies are expressed in electron volts per Na^+–Cl^- pair, and r is the distance in nanometers. The net energy E_N is just the sum of the two expressions above.

(a) Superimpose on a single plot E_N, E_R, and E_A versus r up to 1.0 nm.

(b) On the basis of this plot, determine (i) the equilibrium spacing r_0 between the Na^+ and Cl^- ions, and (ii) the magnitude of the bonding energy E_0 between the two ions.

(c) Mathematically determine the r_0 and E_0 values using the solutions to Problem 2.13 and compare these with the graphical results from part b.

2.15 The net potential energy E_N between two adjacent ions is sometimes represented by the expression

$$E_N = -\frac{C}{r} + D \exp\left(-\frac{r}{\rho}\right) \tag{2.11}$$

in which r is the interionic separation and C, D, and ρ are constants whose values depend on the specific material.

(a) Derive an expression for the bonding energy E_0 in terms of the equilibrium interionic separation r_0 and the constants D and ρ using the following procedure:

1. Differentiate E_N with respect to r and set the resulting expression equal to zero.
2. Solve for C in terms of D, ρ, and r_0.
3. Determine the expression for E_0 by substitution for C in Equation 2.11.

(b) Derive another expression for E_0 in terms of r_0, C, and ρ using a procedure analogous to the one outlined in part a.

2.16 (a) Briefly cite the main differences between ionic, covalent, and metallic bonding.
(b) State the Pauli exclusion principle.

2.17 Offer an explanation as to why covalently bonded materials are generally less dense than ionically or metallically bonded ones.

2.18 The percent ionic character of a bond between elements A and B (A being the most electronegative) may be approximated by the expression

$$\% \text{ ionic character} = (1 - e^{-(0.25)(X_A - X_B)^2}) \times 100 \qquad (2.12)$$

where X_A and X_B are the electronegativities for the respective elements. Compute the percents ionic character of the interatomic bonds for the following compounds: MgO, GaP, CsF, CdS, and FeO.

2.19 Make a plot of bonding energy versus melting temperature for the metals listed in Table 2.3. Using this plot, approximate the bonding energy for copper, which has a melting temperature of 1085°C.

2.20 Using Table 2.2 determine the number of covalent bonds that are possible for atoms of the following elements: silicon, bromine, nitrogen, and sulfur.

2.21 What type(s) of bonding would be expected for each of the following materials: brass (a copper–zinc alloy), rubber, barium sulfide (BaS), solid xenon, bronze, nylon, and aluminum phosphide (AlP)?

2.22 Explain why hydrogen fluoride (HF) has a higher boiling temperature than hydrogen chloride (HCl) (19.4 vs. −85°C), even though HF has a lower molecular weight.

2.23 On the basis of the hydrogen bond, explain the anomalous behavior of water when it freezes. That is, why is there volume expansion upon solidification?

CHAPTER 3

THE STRUCTURE OF CRYSTALLINE SOLIDS

An x-ray diffraction photograph (Laue) for a single crystal of magnesium. The large light spot in the center is from the incident beam which is parallel to a [0001] crystallographic direction. The peripheral light spots result from x-ray beams that are diffracted by specific crystallographic planes. The hexagonal symmetry of magnesium's hexagonal close-packed crystal structure is indicated by the pattern generated by these spots. (Photograph courtesy of J. G. Byrne, University of Utah.)

3.1 INTRODUCTION

Chapter 2 was concerned primarily with the various types of atomic bonding, which are determined by the electron structure of the individual atoms. The present discussion is devoted to the next level of the structure of materials, specifically, to some of the arrangements that may be assumed by atoms in the solid state. Within this framework, concepts of crystallinity and noncrystallinity are introduced. For crystalline solids the notion of crystal structure is presented, specified in terms of a unit cell. The three common crystal structures found in metals are then detailed, along with the scheme by which crystallographic directions and planes are expressed. Single crystals, poly-crystalline, and noncrystalline materials are considered.

CRYSTAL STRUCTURES

3.2 FUNDAMENTAL CONCEPTS

Solid materials may be classified according to the regularity with which atoms or ions are arranged with respect to one another. A **crystalline** material is one in which the atoms are situated in a repeating or periodic array over large atomic distances; that is, long-range order exists, such that upon solidification, the atoms will position themselves in a repetitive three-dimensional pattern, in which each atom is bonded to its nearest-neighbor atoms. All metals, many ceramic materials, and certain poly-mers form crystalline structures under normal solidification conditions. For those that do not crystallize, this long-range atomic order is absent; these *noncrystalline* or *amorphous* materials are discussed briefly at the end of this chapter.

Some of the properties of crystalline solids depend on the **crystal structure** of the material, the manner in which atoms, ions, or molecules are spatially arranged. There is an extremely large number of different crystal structures all having long-range atomic order; these vary from relatively simple structures for metals, to exceedingly complex ones, as displayed by some of the ceramic and polymeric materials. The present discussion deals with several common metallic crystal structures. Chapters 13 and 15 are devoted to crystal structures for ceramics and polymers, respectively.

When describing crystalline structures, atoms (or ions) are thought of as being solid spheres having well-defined diameters. This is termed the *atomic hard sphere model* in which spheres representing nearest-neighbor atoms touch one another. An example of the hard sphere model for the atomic arrangement found in some of the common elemental metals is displayed in Figure 3.1c. In this particular case all the atoms are identical. Sometimes the term **lattice** is used in the context of crystal structures; in this sense "lattice" means a three-dimensional array of points coinciding with atom positions (or sphere centers).

3.3 UNIT CELLS

The atomic order in crystalline solids indicates that small groups of atoms form a repetitive pattern. Thus, in describing crystal structures, it is often convenient to subdivide the structure into small repeat entities called **unit cells.** Unit cells for most crystal structures are parallelepipeds or prisms having three sets of parallel faces; one

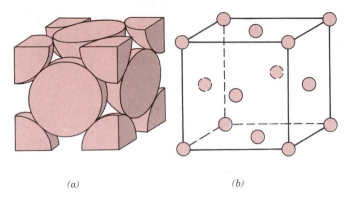

(a)   (b)

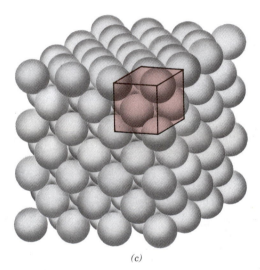

(c)

Figure 3.1 For the face-centered cubic crystal structure: (a) a hard sphere unit cell representation, (b) a reduced-sphere unit cell, and (c) an aggregate of many atoms. (Figure c adapted from W. G. Moffatt, G. W. Pearsall, and J. Wulff, *The Structure and Properties of Materials,* Vol. I, *Structure,* p. 51. Copyright © 1964 by John Wiley & Sons, New York. Reprinted by permission of John Wiley & Sons, Inc.)

is drawn within the aggregate of spheres (Figure 3.1c), which in this case happens to be a cube. A unit cell is chosen to represent the symmetry of the crystal structure, wherein all the atom positions in the crystal may be generated by translations of the unit cell integral distances along each of its edges. Thus the unit cell is the basic structural unit or building block of the crystal structure and defines the crystal structure by virtue of its geometry and the atom positions within. Convenience usually dictates that parallelepiped corners coincide with centers of the hard sphere atoms. Furthermore, more than a single unit cell may be chosen for a particular crystal structure; however, we generally use the unit cell having the highest level of geometrical symmetry.

3.4 METALLIC CRYSTAL STRUCTURES

The atomic bonding in this group of materials is metallic, and thus nondirectional in nature. Consequently, there are no restrictions as to the number and position of nearest-neighbor atoms; this leads to relatively large numbers of nearest neighbors and dense atomic packings for most metallic crystal structures. Also, for metals, using

TABLE 3.1 Atomic Radii and Crystal Structures for 16 Metals

Metal	Crystal Structure[a]	Atomic Radius[b] (nm)	Metal	Crystal Structure	Atomic Radius (nm)
Aluminum	FCC	0.1431	Molybdenum	BCC	0.1363
Cadmium	HCP	0.1490	Nickel	FCC	0.1246
Chromium	BCC	0.1249	Platinum	FCC	0.1387
Cobalt	HCP	0.1253	Silver	FCC	0.1445
Copper	FCC	0.1278	Tantalum	BCC	0.1430
Gold	FCC	0.1442	Titanium (α)	HCP	0.1445
Iron (α)	BCC	0.1241	Tungsten	BCC	0.1371
Lead	FCC	0.1750	Zinc	HCP	0.1332

[a] FCC = face-centered cubic; HCP = hexagonal close-packed; BCC = body-centered cubic.
[b] A nanometer (nm) equals 10^{-9} m; to convert from nanometers to angstrom units (Å), multiply the nanometer value by 10.

the hard sphere model for the crystal structure, each sphere represents an ion core. Table 3.1 presents the atomic radii for a number of metals. Three relatively simple crystal structures are found for most of the common metals: face-centered cubic, body-centered cubic, and hexagonal close-packed.

The Face-Centered Cubic Crystal Structure

The crystal structure found for many metals has a unit cell of cubic geometry, with atoms located at each of the corners and the centers of all the cube faces. It is aptly called the **face-centered cubic (FCC)** crystal structure. Some of the familiar metals having this crystal structure are copper, aluminum, silver, and gold (see also Table 3.1). Figure 3.1a shows a hard sphere model for the FCC unit cell, whereas in Figure 3.1b the atom centers are represented by small circles to provide a better perspective of atom positions. The aggregate of atoms in Figure 3.1c represents a section of crystal consisting of many FCC unit cells. These spheres or ion cores touch one another across a face diagonal; the cube edge length a and the atomic radius R are related through

$$a = 2R\sqrt{2} \tag{3.1}$$

This result is obtained as an example problem.

For the FCC crystal structure, each corner atom is shared among eight unit cells, whereas a face-centered atom belongs to only two. Therefore, one eighth of each of the eight corner atoms and one half of each of the six face atoms, or a total of four whole atoms, may be assigned to a given unit cell. This is depicted in Figure 3.1a, where only sphere portions are represented within the confines of the cube. The cell comprises the volume of the cube, which is generated from the centers of the corner atoms as shown in the figure.

Corner and face positions are really equivalent; that is, translation of the cube corner from an original corner atom to the center of a face atom will not alter the cell structure.

Two other important characteristics of a crystal structure are the **coordination number** and the **atomic packing factor (APF).** For metals, each atom has the same number of nearest-neighbor or touching atoms, which is the coordination number. For face-centered cubics, the coordination number is 12. This may be confirmed by examination of Figure 3.1a; the front face atom has four corner nearest-neighbor atoms surrounding it, four face atoms that are in contact from behind, and four other equivalent face atoms residing in the next unit cell to the front, which is not shown.

The APF is the fraction of solid sphere volume in a unit cell, assuming the atomic hard sphere model, or

$$APF = \frac{\text{volume of atoms in a unit cell}}{\text{total unit cell volume}} \tag{3.2}$$

For the FCC structure, the atomic packing factor is 0.74, which is the maximum packing possible for spheres all having the same diameter. Computation of this APF is also included as an example problem. Metals typically have relatively large atomic packing factors to maximize the shielding provided by the free electron cloud.

The Body-Centered Cubic Crystal Structure

Another common metallic crystal structure also has a cubic unit cell with atoms located at all eight corners and a single atom at the cube center. This is called a **body-centered cubic (BCC) crystal** structure. A collection of spheres depicting this crystal structure is shown in Figure 3.2c, whereas Figures 3.2a and 3.2b are diagrams of BCC unit cells with the atoms represented by hard sphere and reduced-sphere models, respectively. Center and corner atoms touch one another along cube diagonals, and unit cell length a and atomic radius R are related through

$$a = \frac{4R}{\sqrt{3}} \tag{3.3}$$

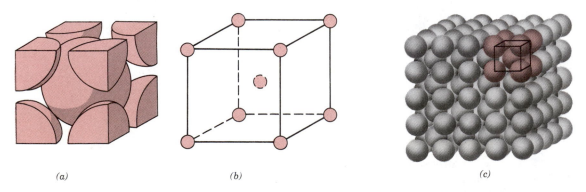

(a)   (b)   (c)

Figure 3.2 For the body-centered cubic crystal structure, (a) a hard sphere unit cell representation, (b) a reduced-sphere unit cell, and (c) an aggregate of many atoms. (Figure (c) from W. G. Moffatt, G. W. Pearsall, and J. Wulff, *The Structure and Properties of Materials,* Vol. I, *Structure,* p. 51. Copyright © 1964 by John Wiley & Sons, New York. Reprinted by permission of John Wiley & Sons, Inc.)

Chromium, iron, tungsten, as well as several other metals listed in Table 3.1 exhibit a BCC structure.

Two atoms are associated with each BCC unit cell: the equivalence of one atom from the eight corners, each of which is shared among eight unit cells, and the single center atom, which is wholly contained within its cell. In addition, corner and center atom positions are equivalent. The coordination number for the BCC crystal structure is 8; each center atom has as nearest neighbors its eight corner atoms. Since the coordination number is less for BCC than FCC, so also is the atomic packing factor for BCC lower—0.68 versus 0.74.

The Hexagonal Close-Packed Crystal Structure

Not all metals have unit cells with cubic symmetry; the final common metallic crystal structure to be discussed has a unit cell that is hexagonal. Figure 3.3a shows a reduced-sphere unit cell for this structure, which is termed **hexagonal close-packed (HCP);** an assemblage of several HCP unit cells is presented in Figure 3.3b. The top and bottom faces of the unit cell consist of six atoms that form regular hexagons and surround a single atom in the center. Another plane that provides three additional atoms to the unit cell is situated between the top and bottom planes. The atoms in this midplane have as nearest neighbors atoms in both of the adjacent two planes. The equivalence of six atoms is contained in each unit cell; one-sixth of each of the 12 top and bottom face corner atoms, one-half of each of the 2 center face atoms, and all the 3 midplane interior atoms. If a and c represent, respectively, the short and long unit cell dimensions

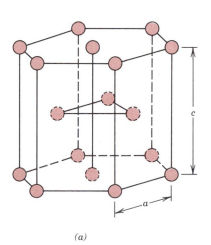

(a)

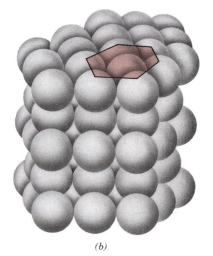

(b)

Figure 3.3 For the hexagonal close-packed crystal structure, (*a*) a reduced-sphere unit cell (*a* and *c* represent the short and long edge lengths, respectively), and (*b*) an aggregate of many atoms. (Figure (*b*) from W. G. Moffatt, G. W. Pearsall, and J. Wulff, *The Structure and Properties of Materials,* Vol. I, *Structure,* p. 51. Copyright © 1964 by John Wiley & Sons, New York. Reprinted by permission of John Wiley & Sons, Inc.)

of Figure 3.3*a*, the *c/a* ratio should be 1.633; however, for some HCP metals this ratio deviates from the ideal value.

The coordination number and the atomic packing factor for the HCP crystal structure are the same as for FCC: 12 and 0.74, respectively. The HCP metals include cadmium, magnesium, titanium, and zinc; some of these are listed in Table 3.1.

EXAMPLE PROBLEM 3.1

Calculate the volume of an FCC unit cell in terms of the atomic radius *R*.

SOLUTION

In the FCC unit cell illustrated,

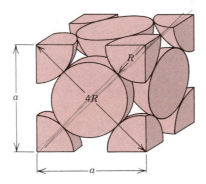

the atoms touch one another across a face-diagonal the length of which is 4*R*. Since the unit cell is a cube, its volume is a^3, where *a* is the cell edge length. From the right triangle on the face,

$$a^2 + a^2 = (4R)^2$$

or, solving for *a*,

$$a = 2R\sqrt{2} \tag{3.1}$$

The FCC unit cell volume V_C may be computed from

$$V_C = a^3 = (2R\sqrt{2})^3 = 16R^3\sqrt{2} \tag{3.4}$$

EXAMPLE PROBLEM 3.2

Show that the atomic packing factor for the FCC crystal structure is 0.74.

SOLUTION

The APF is defined as the fraction of solid sphere volume in a unit cell, or

$$APF = \frac{\text{total sphere volume}}{\text{total unit cell volume}} = \frac{V_S}{V_C}$$

Both the total sphere and unit cell volumes may be calculated in terms of the atomic radius R. The volume for a sphere is $\frac{4}{3}\pi R^3$ and since there are four atoms per FCC unit cell, the total FCC sphere volume is

$$V_S = (4)\frac{4}{3}\pi R^3 = \frac{16}{3}\pi R^3$$

From Example Problem 3.1, the total unit cell volume is

$$V_C = 16R^3\sqrt{2}$$

Therefore, the atomic packing factor is

$$\text{APF} = \frac{V_S}{V_C} = \frac{(\frac{16}{3})\pi R^3}{16R^3\sqrt{2}} = 0.74$$

3.5 DENSITY COMPUTATIONS

A knowledge of the crystal structure of a metallic solid permits computation of its true density ρ through the relationship

$$\rho = \frac{nA}{V_C N_A} \tag{3.5}$$

where

n = number of atoms associated with each unit cell
A = atomic weight
V_C = volume of the unit cell
N_A = Avogadro's number (6.023×10^{23} atoms/mol)

EXAMPLE PROBLEM 3.3

Copper has an atomic radius of 0.128 nm (1.28 Å), an FCC crystal structure, and an atomic weight of 63.5 g/mol. Compute its density and compare the answer with its measured density.

SOLUTION
Equation 3.5 is employed in the solution of this problem. Since the crystal structure is FCC, n, the number of atoms per unit cell, is 4. Furthermore, the atomic weight A_{Cu} is given as 63.5 g/mol. The unit cell volume V_C for FCC was determined in Example Problem 3.1 as $16R^3\sqrt{2}$, where R, the atomic radius, is 0.128 nm.
   Substitution for the various parameters into Equation 3.5 yields

$$\rho = \frac{nA_{\text{Cu}}}{V_C N_A} = \frac{nA_{\text{Cu}}}{(16R^3\sqrt{2})N_A}$$

$$= \frac{(4 \text{ atoms/unit cell})(63.5 \text{ g/mol})}{[16\sqrt{2}(1.28 \times 10^{-8} \text{ cm})^3/\text{unit cell}](6.023 \times 10^{23} \text{ atoms/mol})}$$

$$= 8.89 \text{ g/cm}^3$$

The literature value for the density of copper is 8.94 g/cm^3, which is in very close agreement with the foregoing result.

3.6 POLYMORPHISM AND ALLOTROPY

Some metals, as well as nonmetals, may have more than one crystal structure, a phenomenon known as **polymorphism.** When found in elemental solids, the condition is often termed **allotropy.** The prevailing crystal structure depends on both the temperature and the external pressure. One familiar example is found in carbon: graphite is the stable polymorph at ambient conditions, whereas diamond is formed at extremely high pressures. Also, pure iron has a BCC crystal structure at room temperature, which changes to FCC iron at 912°C (1674°F). Most often a modification of the density and other physical properties accompanies a polymorphic transformation.

3.7 CRYSTAL SYSTEMS

Since there are many different possible crystal structures, it is sometimes convenient to divide them into groups according to unit cell configurations and/or atomic arrangements. One such scheme is based on the unit cell geometry, that is, the shape of the appropriate unit cell parallelepiped without regard to the atomic positions in the cell. Within this framework, an x, y, z coordinate system is established with its origin at one of the unit cell corners; each of the x, y, and z axes coincides with one of the three parallelepiped edges that extend from this corner, as illustrated in Figure 3.4. The unit cell geometry is completely defined in terms of six parameters: the three edge lengths a, b, and c, and the three interaxial angles α, β, and γ. These are indicated in Figure 3.4, and are sometimes termed the **lattice parameters** of a crystal structure.

On this basis, there are found crystals having seven different possible combinations of a, b, and c, and α, β, and γ, each of which represents a distinct **crystal system.** These seven crystal systems are cubic, tetragonal, hexagonal, orthorhombic, rhombohedral, monoclinic, and triclinic. The lattice parameter relationships and unit cell sketches for each are represented in Table 3.2. The cubic system, for which $a = b = c$ and

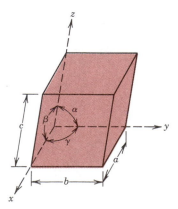

Figure 3.4 A unit cell with x, y, and z coordinate axes, showing axial lengths (a, b, and c) and interaxial angles (α, β, and γ).

TABLE 3.2 Lattice Parameter Relationships and Figures Showing Unit Cell Geometries for the Seven Crystal Systems

Crystal System	Axial Relationships	Interaxial Angles	Unit Cell Geometry
Cubic	$a = b = c$	$\alpha = \beta = \gamma = 90°$	
Hexagonal	$a = b \neq c$	$\alpha = \beta = 90°, \gamma = 120°$	
Tetragonal	$a = b \neq c$	$\alpha = \beta = \gamma = 90°$	
Rhombohedral	$a = b = c$	$\alpha = \beta = \gamma \neq 90°$	
Orthorhombic	$a \neq b \neq c$	$\alpha = \beta = \gamma = 90°$	
Monoclinic	$a \neq b \neq c$	$\alpha = \gamma = 90° \neq \beta$	
Triclinic	$a \neq b \neq c$	$\alpha \neq \beta \neq \gamma \neq 90°$	

$\alpha = \beta = \gamma = 90°$, has the greatest degree of symmetry. Least symmetry is displayed by the triclinic system, since $a \neq b \neq c$ and $\alpha \neq \beta \neq \gamma$.

From the discussion of metallic crystal structures, it should be apparent that both FCC and BCC structures belong to the cubic crystal system, whereas HCP falls within hexagonal. The conventional hexagonal unit cell really consists of three parallelepipeds situated as shown in Table 3.2.

CRYSTALLOGRAPHIC DIRECTIONS AND PLANES

When dealing with crystalline materials, it often becomes necessary to specify some particular crystallographic plane of atoms or a crystallographic direction. Labeling conventions have been established in which three integers or indices are used to designate directions and planes. The basis for determining index values is the unit cell, with a coordinate system consisting of three (x, y, and z) axes situated at one of the corners and coinciding with the unit cell edges, as shown in Figure 3.4. For some crystal systems—namely, hexagonal, rhombohedral, monoclinic, and triclinic—the three axes are *not* mutually perpendicular, as in the familiar Cartesian coordinate scheme.

3.8 CRYSTALLOGRAPHIC DIRECTIONS

A crystallographic direction is defined as a line between two points, or a vector. The following steps are utilized in the determination of the three directional indices:

1. A vector of convenient length is positioned such that it passes through the origin of the coordinate system. Any vector may be translated throughout the crystal lattice without alteration, if parallelism is maintained.

2. The length of the vector projection on each of the three axes is determined; *these are measured in terms of the unit cell dimensions a, b*, and *c*.

3. These three numbers are multiplied or divided by a common factor to reduce them to the smallest integer values.

4. The three indices, not separated by commas, are enclosed in square brackets, thus: [uvw]. The u, v, and w integers correspond to the reduced projections along the x, y, and z axes, respectively.

For each of the three axes, there will exist both positive and negative coordinates. Thus negative indices are also possible, which are represented by a bar over the appropriate index. For example, the [$1\bar{1}1$] direction would have a component in the $-y$ direction. Also, changing the signs of all indices produces an antiparallel direction; that is, [$\bar{1}1\bar{1}$] is directly opposite to [$1\bar{1}1$]. If more than one direction or plane is to be specified for a particular crystal structure, it is imperative for the maintaining of consistency that a positive–negative convention, once established, not be changed.

The [100], [110], and [111] directions are common ones; they are drawn in the unit cell shown in Figure 3.5.

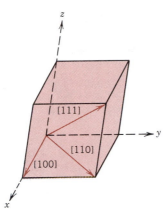

Figure 3.5 The [100], [110], and [111] directions within a unit cell.

EXAMPLE PROBLEM 3.4

Determine the indices for the direction shown in the accompanying figure.

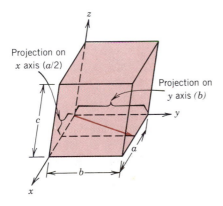

SOLUTION

The vector, as drawn, passes through the origin of the coordinate system, and therefore no translation is necessary. Projections of this vector onto the x, y, and z axes are, respectively, $a/2$, b, and $0c$, which become $\frac{1}{2}$, 1, and 0 in terms of the unit cell parameters (i.e., when the a, b, and c are dropped). Reduction of these numbers to the lowest set of integers is accompanied by multiplication of each by the factor 2. This yields the integers 1, 2, and 0, which are then enclosed in brackets as [120].

This procedure may be summarized as follows:

	x	y	z
Projections	$a/2$	b	$0c$
Projections (in terms of a, b, and c)	$\frac{1}{2}$	1	0
Reduction	1	2	0
Enclosure		[120]	

EXAMPLE PROBLEM 3.5

Draw a [1$\bar{1}$0] direction within a cubic unit cell.

SOLUTION
First construct an appropriate unit cell and coordinate axes system. In the accompanying figure the unit cell is cubic, and the origin of the coordinate system, point O, is located at one of the cube corners.

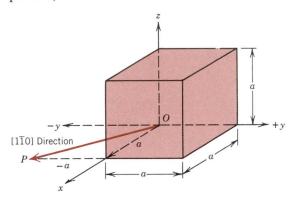

This problem is solved by reversing the procedure of the preceding example. For this [1$\bar{1}$0] direction, the projections along x, y, and z axes are a, $-a$, and $0a$, respectively. This direction is defined by a vector passing from the origin to point P, which is located by first moving along the x axis a units, and from this position, parallel to the y axis $-a$ units, as indicated in the figure. There is no z component to the vector, since the z projection is zero.

For some crystal structures, several nonparallel directions with different indices are actually equivalent; this means that the spacing of atoms along each direction is the same. For example, in cubic crystals, all the directions represented by the following indices are equivalent: [100], [$\bar{1}$00], [010], [0$\bar{1}$0], [001], and [00$\bar{1}$]. Furthermore, directions in cubic crystals having the same indices without regard to order or sign, for example, [123] and [$\bar{2}$1$\bar{3}$], are equivalent. This is, in general, not true for other crystal systems. For example, for crystals of tetragonal symmetry, [100] and [010] directions are equivalent, whereas [100] and [001] are not. As a convenience, equivalent directions are grouped together into a *family*, which are enclosed in angle brackets, thus: ⟨100⟩.

Hexagonal Crystals

A problem arises for crystals having hexagonal symmetry in that some crystallographic equivalent directions will not have the same set of indices. This is circumvented by utilizing a four-axis, or *Miller–Bravais*, coordinate system as shown in Figure 3.6. The three a_1, a_2, and a_3 axes are all contained within a single plane (called the basal plane), and at 120° angles to one another. The z axis is perpendicular to this basal plane. Directional indices, which are obtained as described above, will be denoted by four indices, as [$uvtw$]; by convention, the first three indices pertain to projections along the respective a_1, a_2, and a_3 axes in the basal plane.

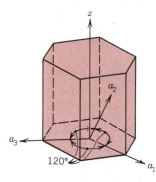

Figure 3.6 Coordinate axis system for a hexagonal unit cell (Miller–Bravais scheme).

Conversion from the three-index system to the four index system,

$$[u'v'w'] \longrightarrow [uvtw]$$

is accomplished by the following formulas:

$$u = \frac{n}{3}(2u' - v') \tag{3.6a}$$

$$v = \frac{n}{3}(2v' - u') \tag{3.6b}$$

$$t = -(u + v) \tag{3.6c}$$

$$w = nw' \tag{3.6d}$$

where primed indices are associated with the three-index scheme and unprimed, with the new Miller–Bravais four-index system; n is a factor that may be required to reduce u, v, t, and w to the smallest integers. For example, using this conversion the $[010]$ direction becomes $[\bar{1}2\bar{1}0]$. Several different directions are indicated in the hexagonal unit cell (Figure 3.7a).

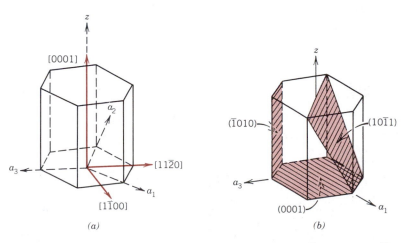

Figure 3.7 For the hexagonal crystal system, (a) $[0001]$, $[1\bar{1}00]$, and $[11\bar{2}0]$ directions, and (b) the (0001), $(10\bar{1}1)$, and $(\bar{1}010)$ planes.

3.9 CRYSTALLOGRAPHIC PLANES

The orientations of planes for a crystal structure are represented in a similar manner. Again, the unit cell is the basis, with the three-axis coordinate system as represented in Figure 3.4. In all but the hexagonal crystal system, crystallographic planes are specified by three Miller indices as (*hkl*). Any two planes parallel to each other are equivalent and have identical indices. The procedure employed in determination of the *h*, *k*, and *l* index numbers is as follows:

1. If the plane passes through the selected origin, either another parallel plane must be constructed within the unit cell by an appropriate translation, or a new origin must be established at the corner of another unit cell.

2. At this point the crystallographic plane either intersects or parallels each of the three axes; the length of the planar intercept for each axis is determined in terms of the lattice parameters *a*, *b*, and *c*.

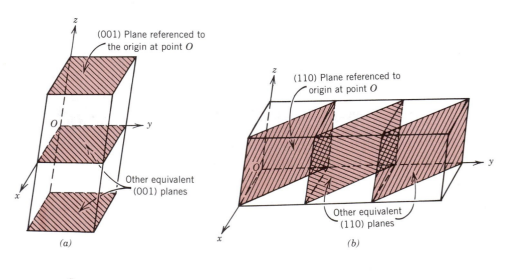

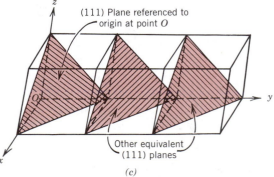

Figure 3.8 Representations of a series each of (*a*) (001), (*b*) (110), and (*c*) (111) crystallographic planes.

3. The reciprocals of these numbers are taken. A plane that parallels an axis may be considered to have an infinite intercept, and, therefore, a zero index.

4. If necessary, these three numbers are changed to the set of smallest integers by multiplication or division using a common factor.

5. Finally, the integer indices, not separated by commas, are enclosed within parentheses, thus: (hkl).

An intercept on the negative side of the origin is indicated by a bar or minus sign positioned over the appropriate index. Furthermore, reversing the directions of all indices specifies another plane parallel to, on the opposite side of and equidistant from, the origin. Several low-index planes are represented in Figure 3.8.

One interesting and unique characteristic of cubic crystals is that planes and directions having the same indices are perpendicular to one another; however, for other crystal systems there are no simple geometrical relationships between planes and directions having the same indices.

EXAMPLE PROBLEM 3.6

Determine the Miller indices for the plane shown in the accompanying sketch (a).

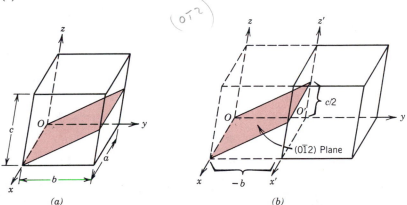

SOLUTION

Since the plane passes through the selected origin O, a new origin must be chosen at the corner of an adjacent unit cell, taken as O', and shown in sketch (b). This plane is parallel to the x axis, and the intercept may be taken as ∞a. The y and z axes intersections, referenced to the new origin O', are $-b$, and $c/2$, respectively. Thus, in terms of the lattice parameters a, b, and c, these intersections are ∞, -1, and $\frac{1}{2}$. The reciprocals of these numbers are 0, -1, and 2; and since all are integers, no further reduction is necessary. Finally, enclosure in parentheses yields $(0\bar{1}2)$.

These steps are briefly summarized below:

	x	y	z
Intercepts	∞a	$-b$	$c/2$
Intercepts (in terms of lattice parameters)	∞	-1	$\frac{1}{2}$
Reciprocals	0	-1	2
Reductions (unnecessary)			
Enclosure		$(0\bar{1}2)$	

EXAMPLE PROBLEM 3.7

Construct a $(0\bar{1}1)$ plane within a cubic unit cell.

SOLUTION

To solve this problem, carry out the procedure used in the preceding example in reverse order. To begin, the indices are removed from the parentheses, and reciprocals are taken, which yields ∞, -1, and 1. This means that the particular plane parallels the x axis while intersecting the y and z axes at $-b$ and c, respectively, as indicated in the accompanying sketch (a). This plane has been drawn in sketch (b). A plane is indicated by lines representing its intersections with the planes that constitute the faces of the unit cell or their extensions. For example, in this figure, line ef is the intersection between the $(0\bar{1}1)$ plane and the top face of the unit cell; also, line gh represents the intersection between this same $(0\bar{1}1)$ plane and the plane of the bottom unit cell face extended. Similarly, lines eg and fh are the intersections between $(0\bar{1}1)$ and back and front cell faces, respectively.

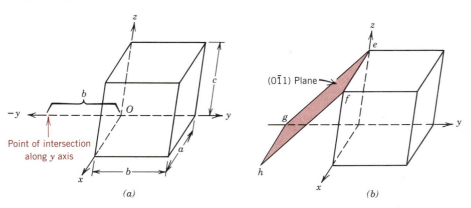

(a)   (b)

Atomic Arrangements

The atomic arrangement for a crystallographic plane, which is often of interest, depends on the crystal structure. The (110) atomic planes for FCC and BCC crystal structures are represented in Figures 3.9 and 3.10; reduced-sphere unit cells are also included. Note that the atomic packing is different for each case. The circles represent atoms lying in the crystallographic planes as would be obtained from a slice taken through the centers of the full-sized hard spheres.

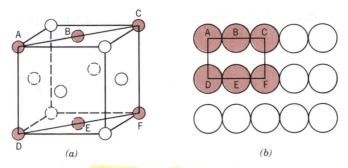

Figure 3.9 (*a*) Reduced-sphere FCC unit cell with (110) plane. (*b*) Atomic packing of an FCC (110) plane. Corresponding atom positions from (*a*) are indicated.

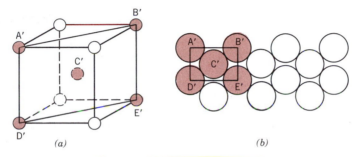

Figure 3.10 (*a*) Reduced-sphere BCC unit cell with (110) plane. (*b*) Atomic packing of a BCC (110) plane. Corresponding atom positions from (*a*) are indicated.

Atomic packing may be the same for various crystallographic planes having different indices, which will depend on the symmetry of the particular crystal structure; such planes belong to a family of equivalent planes. A family of planes is designated by enclosure in braces. For example, in cubic crystals the (111), ($\bar{1}$11), (1$\bar{1}$1), (11$\bar{1}$), (11$\bar{1}$), ($\bar{1}$1$\bar{1}$), ($\bar{1}$$\bar{1}$1), and (1$\bar{1}$$\bar{1}$) planes all belong to the {111} family. Also, in the cubic system only, planes having the same indices, irrespective of order and sign, are equivalent. For example, both (1$\bar{2}$3) and (3$\bar{1}$2) belong to the {123} family.

Hexagonal Crystals

For crystals having hexagonal symmetry, it is desirable that equivalent planes have the same indices; as with directions, this is accomplished by the Miller–Bravais system shown in Figure 3.6. This convention leads to the four-index (*hkil*) scheme, which is favored in most instances, since it more clearly identifies the orientation of a plane in a hexagonal crystal. There is some redundancy in that *i* is determined by the sum of *h* and *k* through

$$i = -(h + k) \tag{3.7}$$

Otherwise the three h, k, and l indices are identical for both indexing systems. Figure 3.7b presents several of the common planes that are found for crystals having hexagonal symmetry.

3.10 LINEAR AND PLANAR ATOMIC DENSITIES

The two previous sections discussed the equivalency of nonparallel directions and planes, where equivalency is related to the degree of atomic spacing or atomic packing. It is felt appropriate at this time to introduce the concepts of atomic linear and planar densities. *Linear density* corresponds to the fraction of line length in a particular crystallographic direction that passes through atom centers. Similarly, *planar density* is simply the fraction of total crystallographic plane area that is occupied by atoms (represented as circles); the plane must pass through an atom's center for the particular atom to be included. These concepts, the one- and two-dimensional analogs of the atomic packing factor, are illustrated in the following example problems.

EXAMPLE PROBLEM 3.8

Calculate the linear density of the [100] direction for BCC.

SOLUTION

A BCC unit cell (reduced sphere) and the [100] direction therein are shown in Figure 3.11a; represented in Figure 3.11b is the linear packing in this direction. As a basis for our computation let us use the line length within the unit cell, L_l, which in this case is the lattice parameter a—the distance between the centers of atoms M and N. In terms of the atomic radius R,

$$L_l = a = \frac{4R}{\sqrt{3}} \qquad \text{(see Equation 3.3)}$$

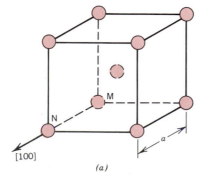

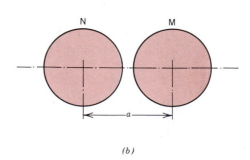

Figure 3.11 (a) Reduced sphere BCC unit cell with the [100] direction indicated. (b) Atomic spacing in the [100] direction for the BCC crystal structure—between atoms M and N in (a).

Now, the total line length intersecting circles (atoms M and N), L_c, is equal to 2R. And, the linear density LD is just the following ratio:

$$LD = \frac{L_c}{L_l} = \frac{2R}{4R\sqrt{3}} = 0.866$$

EXAMPLE PROBLEM 3.9

Calculate the planar density of the (110) plane for FCC.

SOLUTION
The atomic packing of this plane is represented in Figure 3.9b. Consider that portion of the plane that intersects a unit cell (Figure 3.9b), and then compute both this planar area and total circle area in terms of the atomic radius R. Planar density, then, is just the ratio of these two areas.

The unit cell plane area, A_p, is simply that of the rectangle circumscribed by the centers of the atoms A, C, D, and F (Figure 3.9b). The rectangle length $(\overline{AC})$ and width $(\overline{AD})$ are, respectively,

$$\overline{AC} = 4R$$

$$\overline{AD} = 2R\sqrt{2} \qquad \text{(see Equation 3.1)}$$

Therefore,

$$A_p = (\overline{AC})(\overline{AD})$$
$$= (4R)(2R\sqrt{2}) = 8R^2\sqrt{2}$$

Now, for the total circle area, one fourth of each of atoms A, C, D, and F and one half of atoms B and E reside within this rectangle, which gives a total of 2 equivalent circles. Thus the total circle area A_c is just

$$A_c = (2)\pi R^2$$

Finally, the planar density PD is just

$$PD = \frac{A_c}{A_p} = \frac{2\pi R^2}{8R^2\sqrt{2}} = 0.555$$

Linear and planar densities are important considerations relative to the process of slip—that is, the mechanism by which metals plastically deform (Section 7.4). Slip occurs on the most densely packed crystallographic planes and, in those planes, along directions having the greatest atomic packing.

3.11 CLOSE-PACKED CRYSTAL STRUCTURES

It may be remembered from the discussion on metallic crystal structures that both face-centered cubic and hexagonal close-packed crystal structures have atomic packing factors of 0.74, which is the most efficient packing of equal-sized spheres or atoms. In addition to unit cell representations, these two crystal structures may be described in terms of close-packed planes of atoms (i.e., planes having a maximum atom or sphere-packing density); a portion of one such plane is illustrated in Figure 3.12*a*. Both crystal structures may be generated by the stacking of these close-packed planes on top of one another; the difference between the two structures lies in the stacking sequence.

Let the centers of all the atoms in one close-packed plane be labeled *A*. Associated with this plane are two sets of equivalent triangular depressions formed by three adjacent atoms, into which the next close-packed plane of atoms may rest. Those having the triangle vertex pointing up are arbitrarily designated as *B* positions, while the remaining depressions are those with the down vertices, which are marked *C* in Figure 3.12*a*.

A second close-packed plane may be positioned with the centers of its atoms over either *B* or *C* sites; at this point both are equivalent. Suppose that the *B* positions are arbitrarily chosen; the stacking sequence is termed *AB*, which is illustrated in Figure 3.12*b*. The real distinction between FCC and HCP lies in where the third close-packed layer is positioned. For HCP, the centers of this layer are aligned directly above the original *A* positions. This stacking sequence, *ABABAB*..., is repeated over and over. Of course, the *ACACAC*... arrangement would be equivalent. These

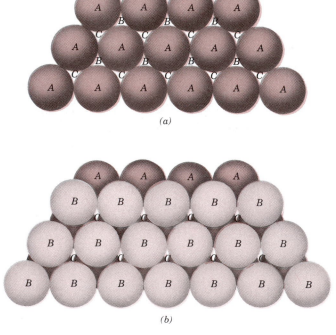

(a)

(b)

Figure 3.12 (*a*) A portion of a close-packed plane of atoms; *A*, *B*, and *C* positions are indicated. (*b*) The *AB* stacking sequence for close-packed atomic planes. (Adapted from W. G. Moffatt, G. W. Pearsall, and J. Wulff, *The Structure and Properties of Materials,* Vol. I, *Structure,* p. 50. Copyright © 1964 by John Wiley & Sons, New York. Reprinted by permission of John Wiley & Sons, Inc.)

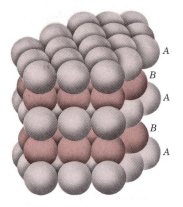

Figure 3.13 Close-packed plane stacking sequence for hexagonal close-packed. (Adapted from W. G. Moffatt, G. W. Pearsall, and J. Wulff, *The Structure and Properties of Materials,* Vol. I, *Structure,* p. 51. Copyright © 1964 by John Wiley & Sons, New York. Reprinted by permission of John Wiley & Sons, Inc.)

close-packed planes for HCP are (0001)-type planes, and the correspondence between this and the unit cell representation is shown in Figure 3.13.

For the face-centered crystal structure, the centers of the third plane are situated over the *C* sites of the first plane (Figure 3.14*a*). This yields an *ABCABCABC* . . . stacking sequence; that is, the atomic alignment repeats every third plane. It is more

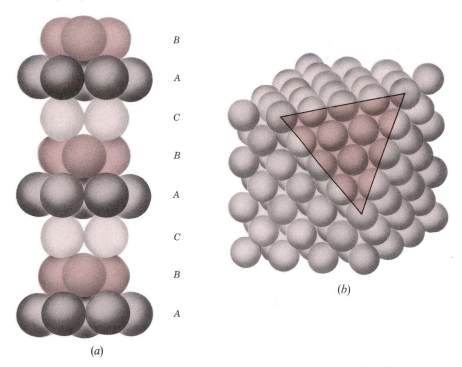

Figure 3.14 (*a*) Close-packed stacking sequence for face-centered cubic. (*b*) A corner has been removed to show the relation between the stacking of close-packed planes of atoms and the FCC crystal structure; the heavy triangle outlines a (111) plane. (Figure (*b*) from W. G. Moffatt, G. W. Pearsall, and J. Wulff, *The Structure and Properties of Materials,* Vol. I, *Structure,* p. 51. Copyright © 1964 by John Wiley & Sons, New York. Reprinted by permission of John Wiley & Sons, Inc.)

difficult to correlate the stacking of close-packed planes to the FCC unit cell. However, this relationship is demonstrated in Figure 3.14b; these planes are of the (111) type. The significance of these FCC and HCP close-packed planes will become apparent in Chapter 7.

CRYSTALLINE AND NONCRYSTALLINE MATERIALS

3.12 SINGLE CRYSTALS

For a crystalline solid, when the periodic and repeated arrangement of atoms is perfect or extends throughout the entirety of the specimen without interruption, the result is a **single crystal.** All unit cells interlock in the same way and have the same orientation. Single crystals exist in nature, but they may also be produced artificially. They are ordinarily difficult to grow, because the environment must be carefully controlled.

If the extremities of a single crystal are permitted to grow without any external constraint, the crystal will assume a regular geometric shape having flat faces, as with some of the gem stones; the shape is indicative of the crystal structure. A photograph of several single crystals is shown in Figure 3.15. Within the past few years, ceramic single crystals have become extremely important in many of our modern technologies, in particular electronic microcircuits.

Figure 3.15 Photograph showing several single crystals of fluorite, CaF_2. (Smithsonian Institution photograph number 38181P.)

3.13 POLYCRYSTALLINE MATERIALS

Virtually all the familiar crystalline solids are composed of a collection of many small crystals or **grains;** such materials are termed **polycrystalline.** Various stages in the solidification of a polycrystalline specimen are represented schematically in Figure 3.16. Initially, small crystals or nuclei form at various positions. These have random crystallographic orientations, as indicated by the square grids. The small grains grow by the successive addition from the surrounding liquid of atoms to the structure of each. The extremities of adjacent grains impinge on one another as the solidification process approaches completion. As indicated in Figure 3.16, the crystallographic orientation varies from grain to grain. Also, there exists some atomic mismatch within the region where two grains meet; this area, called a **grain boundary,** is discussed in more detail in Section 4.5.

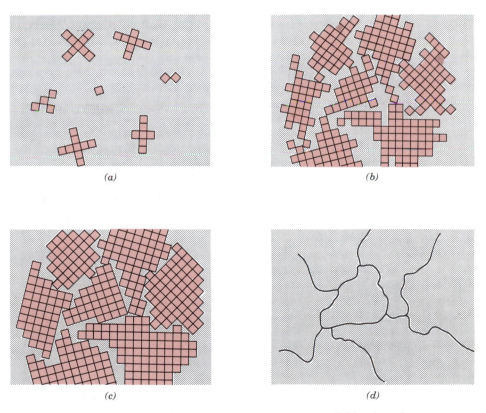

(a)   (b)

(c)   (d)

Figure 3.16 Schematic diagrams of the various stages in the solidification of a polycrystalline material; the square grids depict unit cells. (a) Small crystallite nuclei. (b) Growth of the crystallites; the obstruction of some grains that are adjacent to one another is also shown. (c) Upon completion of solidification, grains having irregular shapes have formed. (d) The grain structure as it would appear under the microscope; dark lines are the grain boundaries. (Adapted from W. Rosenhain, *An Introduction to the Study of Physical Metallurgy,* 2nd edition, Constable & Company Ltd., London, 1915.)

3.14 ANISOTROPY

The physical properties of single crystals of some substances depend on the crystallographic direction in which measurements are taken. For example, the elastic modulus, the electrical conductivity, and the index of refraction may have different values in the [100] and [111] directions. This directionality of properties is termed **anisotropy**, and it is associated with the variance of atomic or ionic spacing with crystallographic direction. Substances in which measured properties are independent of the direction of measurement are **isotropic.** The extent and magnitude of anisotropic effects in crystalline materials are functions of the symmetry of the crystal structure; the degree of anisotropy increases with decreasing structural symmetry—triclinic structures normally are highly anisotropic. The modulus of elasticity values at [100], [110], and [111] orientations for several materials are presented in Table 3.3.

For many polycrystalline materials, the crystallographic orientations of the individual grains are totally random. Under these circumstances, even though each grain may be anisotropic, a specimen composed of the grain aggregate behaves isotropically. Also, the magnitude of a measured property represents some average of the directional values. Techniques are available for producing polycrystalline materials for which the grains have a preferential crystallographic orientation. These are utilized when anisotropic characteristics are desirable.

3.15 X-RAY DIFFRACTION: DETERMINATION OF CRYSTAL STRUCTURES

The reader may wonder how crystal structures are studied, since atoms are of a size that virtually precludes their direct observation. Much of our understanding regarding the atomic and molecular arrangements in solids has resulted from x-ray diffraction investigations. A brief overview of the diffraction phenomenon and how, using x-rays, atomic interplanar distances and crystal structures are deduced will now be given.

TABLE 3.3 Modulus of Elasticity Values for Several Metals at Various Crystallographic Orientations

Metal	Modulus of Elasticity [psi $\times 10^6$ (MPa $\times 10^3$)]		
	[100]	[110]	[111]
Aluminum	9.2 (63.7)	10.5 (72.6)	11.0 (76.1)
Copper	9.7 (66.7)	18.9 (130.3)	27.7 (191.1)
Iron	18.1 (125.0)	30.5 (210.5)	39.6 (272.7)
Tungsten	55.8 (384.6)	55.8 (384.6)	55.8 (384.6)

Source: R. W. Hertzberg, *Deformation and Fracture Mechanics of Engineering Materials*, 3rd edition. Copyright © 1989 by John Wiley & Sons, New York. Reprinted by permission of John Wiley & Sons, Inc.

The Diffraction Phenomenon

Diffraction occurs when a wave encounters a series of regularly spaced obstacles, that (1) are capable of scattering the wave, and (2) have spacings that are comparable in magnitude to the wavelength. Furthermore, diffraction is a consequence of specific phase relationships that are established between two or more waves that have been scattered by the obstacles.

Consider waves 1 and 2 in Figure 3.17a, which have the same wavelength (λ) and are in phase at point O–O'. Now let us suppose that both waves are scattered in such a way that they traverse different paths. The phase relationship between the scattered waves, which will depend upon the difference in path length, is important. One possibility results when this path length difference is an integral number of wavelengths. As noted in Figure 3.17a, these scattered waves (now labeled 1' and 2') are still in phase. They are said to mutually reinforce (or constructively interfere with) one

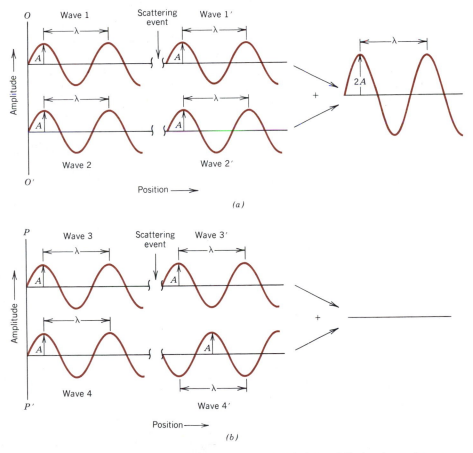

Figure 3.17 (a) Demonstration of how two waves (labeled 1 and 2) that have the same wavelength λ and remain in phase after a scattering event (waves 1' and 2') constructively interfere with one another. The amplitudes of the scattered waves add together in the resultant wave. (b) Demonstration of how two waves (labeled 3 and 4) that have the same wavelength and become out of phase after a scattering event (waves 3' and 4') destructively interfere with one another. The amplitudes of the two scattered waves cancel one another.

another; and, when amplitudes are added, the wave shown on the right side of the figure results. This is a manifestation of **diffraction,** and we refer to a diffracted beam as one composed of a large number of scattered waves that mutually reinforce one another.

Other phase relationships are possible between scattered waves that will not lead to this mutual reinforcement. The other extreme is that demonstrated in Figure 3.17b, wherein the path length difference after scattering is some integral number of *half* wavelengths. The scattered waves are out of phase—that is, corresponding amplitudes cancel or annul one another, or destructively interfere (i.e., the resultant wave has zero amplitude), as indicated on the extreme right side of the figure. Of course, phase relationships intermediate between these two extremes exist, resulting in only partial reinforcement.

X-Ray Diffraction and Bragg's Law

X-Rays are a form of electromagnetic radiation which have high energies and short wavelengths—wavelengths on the order of the atomic spacings for solids. When a beam of x-rays impinges on a solid material, a portion of this beam will be scattered in all directions by the electrons associated with each atom or ion that lies within the beam's path. Let us now examine the necessary conditions for diffraction of x-rays by a periodic arrangement of atoms.

Consider the two parallel planes of atoms A–A′ and B–B′ in Figure 3.18, which have the same h, k, and l Miller indices and are separated by the interplanar spacing d_{hkl}. Now assume that a parallel, monochromatic, and coherent (in-phase) beam of x-rays of wavelength λ is incident on these two planes at an angle θ. Two rays in this beam, labeled 1 and 2, are scattered by atoms P and Q. Constructive interference of

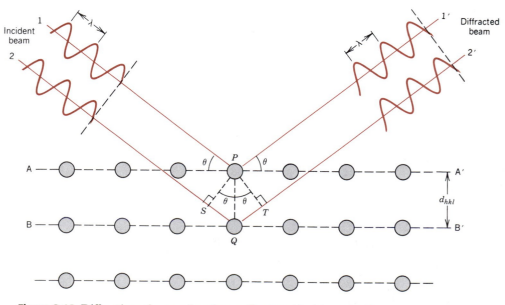

Figure 3.18 Diffraction of x-rays by planes of atoms (A–A′ and B–B′).

the scattered rays 1′ and 2′ occurs also at an angle θ to the planes, if the path length difference between 1–P–1′ and 2–Q–2′ (i.e., $\overline{SQ} + \overline{QT}$) is equal to a whole number, n, of wavelengths. That is, the condition for diffraction is

$$n\lambda = \overline{SQ} + \overline{QT} \tag{3.8}$$

or

$$n\lambda = d_{hkl} \sin \theta + d_{hkl} \sin \theta = 2d_{hkl} \sin \theta \tag{3.9}$$

Equation 3.9 is known as **Bragg's law**; also, n is the order of reflection, which may be any integer $(1, 2, 3, \ldots)$ consistent with $\sin \theta$ not exceeding unity. Thus we have a simple expression relating the x-ray wavelength and interatomic spacing to the angle of the diffracted beam. If Bragg's law is not satisfied, then the interference will be nonconstructive in nature so as to yield a very low-intensity diffracted beam.

The magnitude of the distance between two adjacent and parallel planes of atoms (i.e., the interplanar spacing d_{hkl}) is a function of the Miller indices (h, k, and l) as well as the lattice parameter(s). For example, for crystal structures having cubic symmetry,

$$d_{hkl} = \frac{a}{\sqrt{h^2 + k^2 + l^2}} \tag{3.10}$$

in which a is the lattice parameter (unit cell edge length). Relationships similar to Equation 3.10, but which are more complex, exist for the other six crystal systems noted in Table 3.2.

Bragg's law, Equation 3.9, is a necessary but not sufficient condition for diffraction by real crystals. It specifies when diffraction will occur for unit cells having atoms positioned only at cell corners. However, atoms situated at other sites (e.g., face and interior unit cell positions as with FCC and BCC) act as extra scattering centers, which can produce out-of-phase scattering at certain Bragg angles. The net result is the absence of some diffracted beams which, according to Equation 3.9, should be present. For example, for the BCC crystal structure, $h + k + l$ must be even if diffraction is to occur, whereas for FCC, h, k, and l must all be either odd or even.

Diffraction Techniques

One common diffraction technique employs a powdered or polycrystalline specimen consisting of many fine and randomly oriented particles that are exposed to monochromatic x-radiation. Each powder particle is a crystal, and having a large number of them with random orientations ensures that some particles are properly oriented such that every possible set of crystallographic planes will be available for diffraction.

The *diffractometer* is an apparatus used to determine the angles at which diffraction occurs for powdered specimens; its features are represented schematically in Figure 3.19. A specimen S in the form of a flat plate is supported so that rotations about the axis labeled O are possible; this axis is perpendicular to the plane of the page. The monochromatic x-ray beam is generated at point T, and the intensities of diffracted beams are detected with a counter labeled C in the figure. The specimen, x-ray source, and counter are all coplanar.

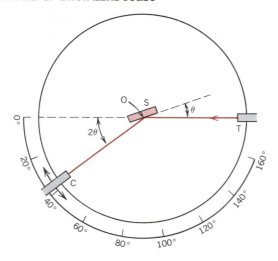

Figure 3.19 Schematic diagram of an x-ray diffractometer; T = x-ray source, S = specimen, C = detector, and O = the axis around which the specimen and detector rotate.

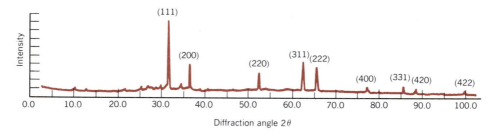

Figure 3.20 Diffraction pattern for powdered lead. (Courtesy of Wesley L. Holman.)

The counter is mounted on a movable carriage which may also be rotated about the O axis; its angular position in terms of 2θ is marked on a graduated scale.[1] Carriage and specimen are mechanically coupled such that a rotation of the specimen through θ is accompanied by a 2θ rotation of the counter; this assures that the incident and reflection angles are maintained equal to one another (Figure 3.19). Collimators are incorporated within the beam path to produce a well-defined and focused beam. Utilization of a filter provides a near-monochromatic beam.

As the counter moves at constant angular velocity, a recorder automatically plots the diffracted beam intensity (monitored by the counter) as a function of 2θ; 2θ is termed the *diffraction angle,* which is measured experimentally. Figure 3.20 shows a diffraction pattern for a powdered specimen of lead. The high-intensity peaks result when the Bragg diffraction condition is satisfied by some set of crystallographic planes. These peaks are plane-indexed in the figure.

---

[1] It should be pointed out that the symbol θ has been used in two different contexts for this discussion. Here, θ represents the angular locations of both x-ray source and counter relative to the specimen surface. Previously (e.g., Equation 3.9), it denoted the angle at which the Bragg criterion for diffraction is satisfied.

Other powder techniques have been devised wherein diffracted beam intensity and position are recorded on a photographic film instead of being measured by a counter.

One of the primary uses of x-ray diffractometry is for the determination of crystal structure. The unit cell size and geometry may be resolved from the angular positions of the diffraction peaks, whereas arrangement of atoms within the unit cell is associated with the relative intensities of these peaks.

X-Rays, as well as electron and neutron beams, are also used in other types of material investigations. For example, crystallographic orientations of single crystals are possible using x-ray diffraction (or Laue) photographs. On page 30 one such photograph is shown for a magnesium crystal; each bright spot (with the exception of the bright center one) results from an x-ray beam that was diffracted by a specific set of crystallographic planes. Other uses of x-rays include qualitative and quantitative chemical identifications, and the determination of residual stresses and crystal size.

EXAMPLE PROBLEM 3.10

For BCC iron, compute (a) the interplanar spacing, and (b) the diffraction angle for the (211) set of planes. The lattice parameter for Fe is 0.2866 nm (2.866 Å). Also, assume that monochromatic radiation having a wavelength of 0.1542 nm (1.542 Å) is used, and the order of reflection is 1.

SOLUTION

(a) The value of the interplanar spacing d_{hkl} is determined using Equation 3.10, with $a = 0.2866$ nm, and $h = 2$, $k = 1$, and $l = 1$, since we are considering the (211) planes. Therefore,

$$d_{hkl} = \frac{a}{\sqrt{h^2 + k^2 + l^2}}$$

$$= \frac{0.2866 \text{ nm}}{\sqrt{(2)^2 + (1)^2 + (1)^2}} = 0.1170 \text{ nm } (1.170 \text{ Å})$$

(b) The value of θ may now be computed using Equation 3.9, with $n = 1$, since this is a first-order reflection:

$$\sin \theta = \frac{n\lambda}{2d_{hkl}} = \frac{(1)(0.1542 \text{ nm})}{(2)(0.1170 \text{ nm})} = 0.659$$

$$\theta = \sin^{-1}(0.659) = 41.22°$$

The diffraction angle is 2θ, or

$$2\theta = (2)(41.22°) = 82.44°$$

3.16 NONCRYSTALLINE SOLIDS

It has been mentioned that **noncrystalline** solids lack a systematic and regular arrangement of atoms over relatively large atomic distances. Sometimes such materials

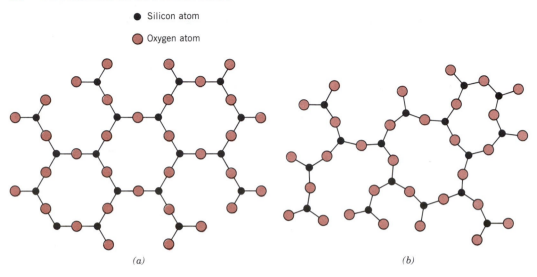

Figure 3.21 Two-dimensional schemes of the structure of (*a*) crystalline silicon dioxide and (*b*) noncrystalline silicon dioxide.

are also called **amorphous** (meaning literally without form), or supercooled liquids inasmuch as their atomic structure resembles that of a liquid.

An amorphous condition may be illustrated by comparison of the crystalline and noncrystalline structures of the ceramic compound silicon dioxide (SiO_2), which may exist in both states. Figures 3.21*a* and 3.21*b* present two-dimensional schematic diagrams for both structures of SiO_2. Even though each silicon ion bonds to three oxygen ions for both states, beyond this, the structure is much more disordered and irregular for the noncrystalline structure.

Whether a crystalline or amorphous solid forms depends on the ease with which a random atomic structure in the liquid can transform to an ordered state during solidification. Amorphous materials, therefore, are characterized by atomic or molecular structures that are relatively complex and become ordered only with some difficulty. Furthermore, rapidly cooling through the freezing temperature favors the formation of a noncrystalline solid, since little time is allowed for the ordering process.

All metals form crystalline solids; but some ceramic materials are crystalline, whereas others, the inorganic glasses, are amorphous. Polymers may be completely crystalline, entirely noncrystalline, or a mixture of the two. More about the structure and properties of amorphous ceramics and polymers is contained in Chapters 13 and 15.

SUMMARY

Atoms in crystalline solids are positioned in an orderly and repeated pattern, which is in contrast to the random and disordered atomic distribution found in noncrystalline or amorphous materials. Atoms may be represented as solid spheres, and, for crystalline solids, crystal structure is just the spatial arrangement of these spheres. The various crystal structures are specified in terms of parallelepiped unit cells, which are characterized by geometry and atom positions within.

Most common metals exist in at least one of three relatively simple crystal structures: face-centered cubic (FCC), body-centered cubic (BCC), and hexagonal close-packed (HCP). Two features of a crystal structure are coordination number (or number of nearest-neighbor atoms) and atomic packing factor (the fraction of solid sphere volume in the unit cell). Coordination number and atomic packing factor are the same for both FCC and HCP crystal structures, each of which may be generated by the stacking of close-packed planes of atoms.

Crystallographic planes and directions are specified in terms of an indexing scheme. The basis for the determination of each index is a coordinate axis system defined by the unit cell for the particular crystal structure. Directional indices are computed in terms of vector projections on each of the coordinate axes, whereas planar indices are determined from the reciprocals of axial intercepts. For hexagonal unit cells, a four-index scheme for both directions and planes is found to be more convenient.

Crystallographic directional and planar equivalencies are related to atomic linear and planar densities, respectively. The atomic packing (i.e., planar density) of spheres in a crystallographic plane depends on both the indices of the plane as well as the crystal structure. For a given crystal structure, planes having identical atomic packing yet different Miller indices belong to the same family.

Single crystals are materials in which the atomic order extends uninterrupted over the entirety of the specimen; under some circumstances, they may have flat faces and regular geometric shapes. The vast majority of crystalline solids, however, are polycrystalline, being composed of many small crystals or grains having different crystallographic orientations.

X-Ray diffractometry is used for crystal structure and interplanar spacing determinations. A beam of x-rays directed on a crystalline material may experience diffraction (constructive interference) as a result of its interaction with a series of parallel atomic planes according to Bragg's law. Interplanar spacing is a function of the Miller indices and lattice parameter(s) as well as the crystal structure.

IMPORTANT TERMS AND CONCEPTS

Allotropy

Amorphous

Anisotropy

Atomic packing factor (APF)

Body-centered cubic (BCC)

Bragg's law

Coordination number

Crystal structure

Crystal system

Crystalline

Diffraction

Face-centered cubic (FCC)

Grain

Grain boundary

Hexagonal close-packed (HCP)

Isotropic

Lattice

Lattice parameters

Miller indices

Noncrystalline

Polycrystalline

Polymorphism

Single crystal

Unit cell

REFERENCES

BARRETT, C. S. and T. B. MASSALSKI, *Structure of Metals,* 3rd edition, Pergamon Press, Oxford, 1980.

BRAGG, W. L., *The Crystalline State, Vol. I: A General Survey,* The Macmillan Co., New York, 1934.

BUERGER, M. J., *Elementary Crystallography,* John Wiley & Sons, New York, 1956.

COHEN, J. B., *Diffraction Methods in Materials Science,* The Macmillan Co., New York, 1966.

CULLITY, B. D., *Elements of X-Ray Diffraction,* 2nd edition, Addison-Wesley Publishing Co., Reading, MA, 1978.

WYCKOFF, R. W. G., *Crystal Structures,* 2nd edition, Interscience Publishers, 1963.

QUESTIONS AND PROBLEMS

3.1 What is the difference between atomic structure and crystal structure?

3.2 What is the difference between a crystal structure and a crystal system?

3.3 If the atomic radius of aluminum is 0.143 nm, calculate the volume of its unit cell in cubic meters.

3.4 Show for the body-centered cubic crystal structure that the unit cell edge length a and the atomic radius R are related through $a = 4R/\sqrt{3}$.

3.5 For the HCP crystal structure, show that the ideal c/a ratio is 1.633.

3.6 Show that the atomic packing factor for BCC is 0.68.

3.7 Show that the atomic packing factor for HCP is 0.74.

3.8 Assume that a metal exists having a simple cubic crystal structure, that is, atoms located at the corners of a cube and touching one another along the cube edges (Figure 3.22). **(a)** What is the coordination number for this crystal structure? **(b)** Calculate the atomic packing factor.

3.9 Iron has a BCC crystal structure, an atomic radius of 0.124 nm, and an atomic weight of 55.9 g/mol. Compute and compare its density with the experimental value found in Table C.1, Appendix C.

3.10 Calculate the radius of an iridium atom, given that Ir has an FCC crystal structure, a density of 22.4 g/cm^3, and an atomic weight of 192.2 g/mol.

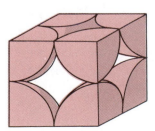

Figure 3.22 Hard sphere unit cell representation of the simple cubic crystal structure.

3.11 Calculate the radius of a vanadium atom, given that V has a BCC crystal structure, a density of 5.96 g/cm^3, and an atomic weight of 50.9 g/mol.

3.12 Some hypothetical metal has the simple cubic crystal structure shown in Figure 3.22. If its atomic weight is 74.5 g/mol and the atomic radius is 0.145 nm, compute its density.

3.13 Titanium has an HCP crystal structure and a density of 4.51 g/cm^3. **(a)** What is the volume of its unit cell in cubic meters? **(b)** If the c/a ratio is 1.58, compute the values of c and a.

3.14 Using atomic weight, crystal structure, and atomic radius data tabulated inside the front cover, compute the theoretical densities of aluminum, nickel, titanium, and tungsten, and then compare these values with the measured densities listed in this same table. The c/a ratio for titanium is 1.58.

3.15 Niobium has an atomic radius of 0.1430 nm (1.430 Å) and a density of 8.57 g/cm^3. Determine whether it has an FCC or BCC crystal structure.

3.16 Below are listed the atomic weight, density, and atomic radius for three hypothetical alloys. For each determine whether its crystal structure is FCC, BCC, or simple cubic and then justify your determination. A simple cubic unit cell is shown in Figure 3.22.

Alloy	Atomic Weight (g/mol)	Density (g/cm^3)	Atomic Radius (nm)
A	43.1	6.4	0.122
B	184.4	12.3	0.146
C	91.6	9.6	0.137

3.17 The unit cell for uranium has orthorhombic symmetry, with a, b, and c lattice parameters of 0.286, 0.587, and 0.495 nm, respectively. If its density, atomic weight, and atomic radius are 19.05 g/cm^3, 238.03 g/mol, and 0.1385 nm, respectively, compute the atomic packing factor.

3.18 Show that the ideal c/a ratio for the HCP crystal structure is 1.633.

3.19 Demonstrate that the atomic packing factor for HCP is 0.74.

3.20 Indium has a tetragonal unit cell for which the a and c lattice parameters are 0.459 and 0.495 nm, respectively.
   (a) If the atomic packing factor and atomic radius are 0.69 and 0.1625 nm, respectively, determine the number of atoms in each unit cell.
   (b) The atomic weight for indium is 114.82 g/mol; compute its density.

3.21 Zinc has an HCP unit cell for which the ratio of the lattice parameters c/a is 1.85. If the radius of the Zn atom is 0.1332 nm, **(a)** determine the unit cell volume, and **(b)** calculate the density of Zn and compare it with the literature value.

3.22 Magnesium has an HCP crystal structure, a c/a ratio of 1.624, and a density of 1.74 g/cm^3. Compute the atomic radius for Mg.

3.23 Cobalt has an HCP crystal structure, an atomic radius of 0.1253 nm, and a c/a ratio of 1.623. Compute the volume of the unit cell for Co.

3.24 This is a unit cell for a hypothetical metal:

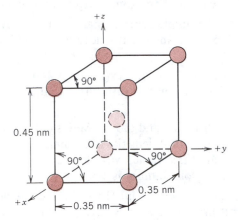

(a) To which crystal system does this unit cell belong?

(b) What would this crystal structure be called?

(c) Calculate the density of the material, given that its atomic weight is 141 g/mol.

3.25 Sketch a unit cell for the face-centered orthorhombic crystal structure.

3.26 Draw an orthorhombic unit cell, and within that cell a $[12\bar{1}]$ direction and a (210) plane.

3.27 Sketch a monoclinic unit cell, and within that cell a $[0\bar{1}1]$ direction and an (002) plane.

3.28 Here are unit cells for two hypothetical metals:

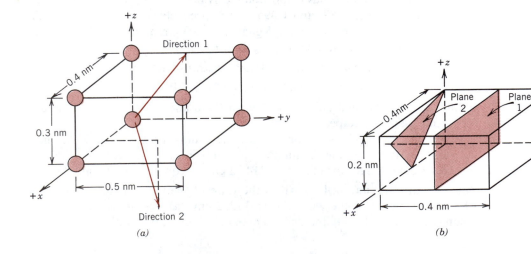

(a)   (b)

(a) What are the indices for the directions indicated by the two vectors in sketch (a)?

(b) What are the indices for the two planes drawn in sketch (b)?

3.29 Within a cubic unit cell, sketch the following directions: **(a)** $[10\bar{1}]$; **(b)** $[211]$; **(c)** $[10\bar{2}]$; **(d)** $[3\bar{1}3]$; **(e)** $[\bar{1}1\bar{1}]$; **(f)** $[\bar{2}12]$; **(g)** $[3\bar{1}2]$; **(h)** $[301]$.

3.30 Determine the indices for the directions shown in the following cubic unit cell:

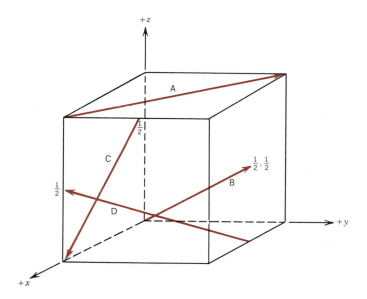

3.31 Determine the indices for the directions shown in the following cubic unit cell:

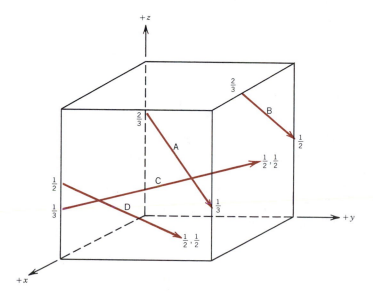

3.32 For tetragonal crystals, cite the indices of directions that are equivalent to each of the following directions: **(a)** $[011]$; **(b)** $[100]$.

3.33 (a) Convert the [100] and [111] directions into the four-index Miller–Bravais scheme for hexagonal unit cells. **(b)** Make the same conversion for the (010) and (101) planes.

3.34 Determine the Miller indices for the planes shown in the following unit cell:

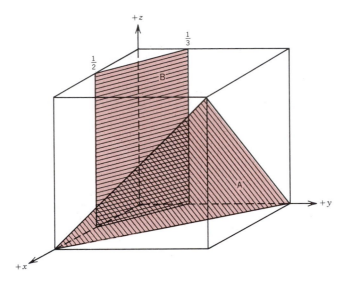

3.35 Determine the Miller indices for the planes shown in the following unit cell:

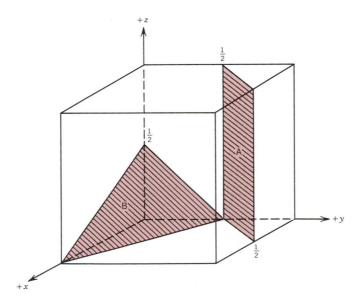

3.36 Determine the Miller indices for the planes shown in the following unit cell:

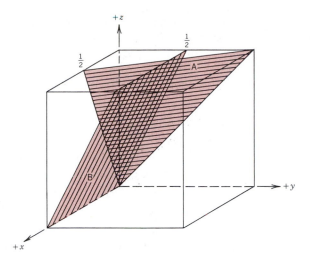

3.37 Sketch the (01$\bar{1}$1) and (2$\bar{1}$$\bar{1}$0) planes in a hexagonal unit cell.

3.38 Determine the indices for the planes shown in the hexagonal unit cells shown below.

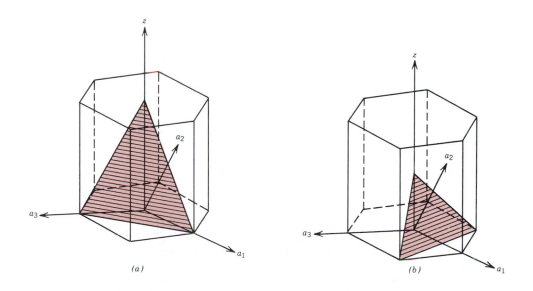

(a)   *(b)*

3.39 Sketch within a cubic unit cell the following planes: **(a)** (10$\bar{1}$); **(b)** (2$\bar{1}$1); **(c)** (012); **(d)** (3$\bar{1}$3); **(e)** ($\bar{1}$1$\bar{1}$); **(f)** ($\bar{2}$12); **(g)** (3$\bar{1}$2); **(h)** (301).

3.40 For the simple cubic crystal structure shown in Figure 3.22, sketch the atomic packing of the following planes (represent atoms with circles drawn to full size): **(a)** 100; **(b)** (110); **(c)** (111).

3.41 Sketch the atomic packing of the (100) plane for both FCC and BCC crystal structures (similar to Figures 3.9*b* and 3.10*b*).

3.42 Consider the reduced-sphere unit cell shown in Problem 3.24, having an origin of the coordinate system positioned at the atom labeled with an O. For the following sets of planes, determine which are equivalent:
(a) (100), (0$\bar{1}$0), and (001).
(b) (110), (101), (011), and ($\bar{1}$01).
(c) (111), (1$\bar{1}$1), (11$\bar{1}$), and ($\bar{1}$1$\bar{1}$).

3.43 Cite the indices of the direction that results from the intersection of each of the following pair of planes within a cubic crystal: **(a)** (110) and (111) planes; **(b)** (110) and (1$\bar{1}$0) planes; and **(c)** (11$\bar{1}$) and (001) planes.

3.44 Compute and compare the linear densities of the [100], [110], and [111] directions for FCC.

3.45 Compute and compare the linear densities of the [110] and [111] directions for BCC.

3.46 Calculate and compare the planar densities of the (100) and (111) planes for FCC.

3.47 Calculate and compare the planar densities of the (100) and (110) planes for BCC.

3.48 Calculate the planar density of the (0001) plane for HCP.

3.49 Here are three different crystallographic planes for a unit cell of a hypothetical metal; the circles represent atoms:

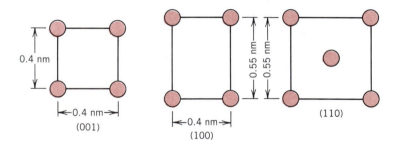

(a) To what crystal system does the unit cell belong?
(b) What would this crystal structure be called?

3.50 Below are shown three different crystallographic planes for a unit cell of some hypothetical metal; the circles represent atoms:

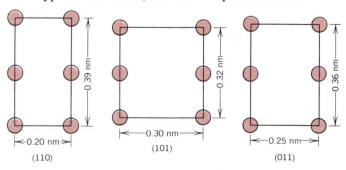

(a) To what crystal system does the unit cell belong?

(b) What would this crystal structure be called?

(c) If the density of this metal is 18.91 g/cm³, determine its atomic weight.

3.51 Explain why the properties of polycrystalline materials are most often isotropic.

3.52 Using the data for aluminum in Table 3.1, compute the interplanar spacing for the (110) set of planes.

3.53 Determine the expected diffraction angle for the first-order reflection from the (310) set of planes for BCC chromium when monochromatic radiation of wavelength 0.0711 nm is used.

3.54 Using the data for α-iron in Table 3.1, compute the interplanar spacings for the (111) and (211) sets of planes.

3.55 The metal rhodium has a FCC crystal structure. If the angle of diffraction for the (311) set of planes occurs at 36.12° (first-order reflection) when monochromatic x-radiation having a wavelength of 0.0711 nm is used, compute **(a)** the interplanar spacing for this set of planes, and **(b)** the atomic radius for a rhodium atom.

3.56 The metal niobium has a BCC crystal structure. If the angle of diffraction for the (211) set of planes occurs at 75.99° (first-order reflection) when monochromatic x-radiation having a wavelength of 0.1659 nm is used, compute **(a)** the interplanar spacing for this set of planes, and **(b)** the atomic radius for the niobium atom.

3.57 For which set of crystallographic planes will a first-order diffraction peak occur at a diffraction angle of 44.53° for FCC nickel when monochromatic radiation having a wavelength of 0.1542 nm is used?

3.58 Figure 3.20 shows an x-ray diffraction pattern for lead taken using a diffractometer and monochromatic x-radiation having a wavelength of 0.1542 nm; each

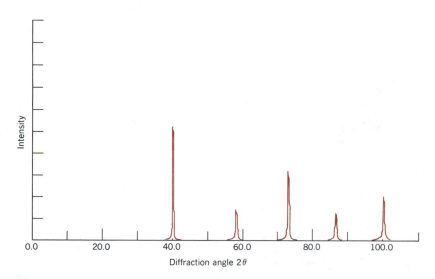

Figure 3.23 Diffraction pattern for powdered tungsten. (Courtesy of Wesley L. Holman.)

diffraction peak on the pattern has been indexed. Compute the interplanar spacing for each set of planes indexed; also determine the lattice parameter of Pb for each of the peaks.

3.59 The diffraction peaks shown in Figure 3.20 are indexed according to the reflection rules for FCC (i.e., h, k, and l must all be either odd or even). Cite the h, k, and l indices of the first four diffraction peaks for BCC crystals consistent with $h + k + l$ being even.

3.60 Figure 3.23 shows the first five peaks of the x-ray diffraction pattern for tungsten which has a BCC crystal structure; monochromatic x-radiation having a wavelength of 0.1542 nm was used.
(a) Index (i.e., give h, k, and l indices) for each of these peaks.
(b) Determine the interplanar spacing for each of the peaks.
(c) For each peak, determine the atomic radius for W and compare these with the value presented in Table 3.1.

3.61 Would you expect a material in which the atomic bonding is predominantly ionic in nature to be more or less likely to form a noncrystalline solid upon solidification than a covalent material? Why? (See Section 2.6.)

CHAPTER 4

IMPERFECTIONS IN SOLIDS

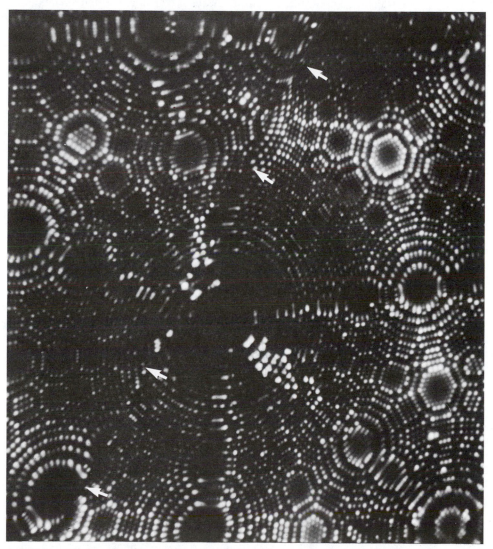

A field ion micrograph taken at the tip of a pointed tungsten specimen. Field ion microscopy is a sophisticated and fascinating technique that permits observation of individual atoms in a solid, which are represented by white spots. The symmetry and regularity of the atom arrangements is evident from the positions of the spots in this micrograph. A disruption of this symmetry occurs along a grain boundary which is traced by the arrows. Approximately 3,460,000 ×. (Photomicrograph courtesy of J. J. Hren and R. W. Newman.)

4.1 INTRODUCTION

Thus far it has been tacitly assumed that perfect order exists throughout crystalline materials on an atomic scale. However, such an idealized solid does not exist: all contain large numbers of various defects or imperfections. As a matter of fact, many of the properties of materials are profoundly sensitive to deviations from crystalline perfection; the influence is not always adverse, and often specific characteristics are deliberately fashioned by the introduction of controlled amounts or numbers of particular defects, as detailed in succeeding chapters.

By "crystalline defect" is meant a lattice irregularity having one or more of its dimensions on the order of an atomic diameter. Classification of crystalline imperfections is frequently made according to geometry or dimensionality of the defect. Several different imperfections are discussed in this chapter, including point defects (those associated with one or two atomic positions), linear (or one-dimensional) defects, as well as interfacial defects, or boundaries, which are two-dimensional. Impurities in solids are also discussed, since impurity atoms may exist as point defects. Finally, techniques for the microscopic examination of defects and the structure of materials are briefly described.

POINT DEFECTS

4.2 VACANCIES AND SELF-INTERSTITIALS

The simplest of the point defects is a **vacancy** or vacant lattice site, one normally occupied from which an atom is missing (Figure 4.1). Vacancies are formed during solidification, and also as a result of atomic vibrations, which can cause the displacement of atoms from their normal lattice sites.

The equilibrium number of vacancies N_v for a given quantity of material depends on and increases with temperature according to

$$N_v = N \exp\left(-\frac{Q_v}{kT}\right) \tag{4.1}$$

In this expression, N is the total number of atomic sites, Q_v is the energy of activation (vibrational energy required for the formation of a vacancy), T is the absolute temperature,[1] in kelvins, and k is the gas or **Boltzmann's constant.** The value of k is 1.38×10^{-23} J/atom-K, or 8.62×10^{-5} eV/atom-K, depending on the units of Q_v.[2] Thus the number of vacancies increases exponentially with temperature; that is, as T in Equation 4.1 increases, so does also the expression $\exp -(Q_v/kT)$. For most metals, the fraction of vacancies N_v/N just below the melting temperature is on the order of 10^{-4}; that is, one lattice site out of 10,000 will be empty. As ensuing discussions indicate, a number of other material parameters have an exponential dependence on temperature similar to that of Equation 4.1.

A **self-interstitial** is an atom from the crystal that is crowded into an interstitial site, a small void space that under ordinary circumstances is not occupied. This kind

---

[1] Absolute temperature in kelvins (K) is equal to °C + 273.
[2] Boltzmann's constant per mole of atoms becomes the gas constant R; in such a case $R = 8.31$ J/mol-K or 1.987 cal/mol-K.

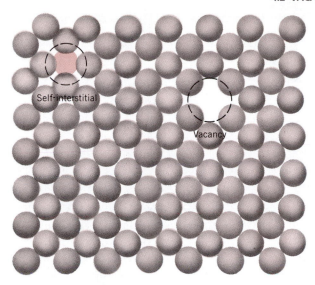

Figure 4.1 Two-dimensional representations of a vacancy and a self-interstitial. (Adapted from W. G. Moffatt, G. W. Pearsall, and J. Wulff, *The Structure and Properties of Materials,* Vol. I, *Structure,* p. 77. Copyright © 1964 by John Wiley & Sons, New York. Reprinted by permission of John Wiley & Sons, Inc.)

of defect is also represented in Figure 4.1. In metals, a self-interstitial introduces relatively large distortions in the surrounding lattice because the atom is substantially larger than the interstitial position in which it is situated. Consequently, the formation of this defect is not highly probable, and it exists in very small concentrations, which are significantly lower than for vacancies.

EXAMPLE PROBLEM 4.1

Calculate the equilibrium number of vacancies per cubic meter for copper at 1000°C. The activation energy for vacancy formation is 0.9 eV/atom; the atomic weight and density (at 1000°C) for copper are 63.5 g/mol and 8.4 g/cm³, respectively.

SOLUTION

This problem may be solved by using Equation 4.1; it is first necessary, however, to determine the value of N, the number of atomic sites per cubic meter for copper from its atomic weight A_{Cu}, its density ρ, and Avogadro's number N_A, according to

$$N = \frac{N_A \rho}{A_{Cu}} \tag{4.2}$$

$$= \frac{(6.023 \times 10^{23} \text{ atoms/mol})(8.4 \text{ g/cm}^3)(10^6 \text{ cm}^3/\text{m}^3)}{63.5 \text{ g/mol}}$$

$$= 8.0 \times 10^{28} \text{ atoms/m}^3$$

Thus the number of vacancies at 1000°C (1273 K) is equal to

$$N_v = N \exp\left(-\frac{Q_v}{kT}\right)$$

$$= (8.0 \times 10^{28} \text{ atoms/m}^3) \exp\left[-\frac{(0.9 \text{ eV})}{(8.62 \times 10^{-5} \text{ eV/K})(1273 \text{ K})}\right]$$

$$= 2.2 \times 10^{25} \text{ vacancies/m}^3$$

4.3 IMPURITIES IN SOLIDS

A pure metal consisting of only one type of atom just isn't possible; impurity or foreign atoms will always be present, and some will exist as crystalline point defects. In fact, even with relatively sophisticated techniques, it is difficult to refine metals to a purity in excess of 99.9999%. At this level, on the order of 10^{22} to 10^{23} impurity atoms will be present in one cubic meter of material. Most familiar metals are not highly pure; rather, they are **alloys,** in which impurity atoms have been added intentionally to impart specific characteristics to the material. Ordinarily alloying is used in metals to improve mechanical strength and corrosion resistance. For example, sterling silver is a 92.5% silver–7.5% copper alloy. In normal ambient environments, pure silver is highly corrosion resistant, but also very soft. Alloying with copper enhances the mechanical strength significantly, without depreciating the corrosion resistance appreciably.

The addition of impurity atoms to a metal will result in the formation of a **solid solution** and/or a new *second phase,* depending on the kinds of impurity, their concentrations, and the temperature of the alloy. The present discussion is concerned with the notion of a solid solution; treatment of the formation of a new phase is deferred to Chapter 9.

Several terms relating to impurities and solid solutions deserve mention. With regard to alloys, **solute** and **solvent** are terms that are commonly employed. "Solvent" represents the element or compound that is present in the greatest amount; on occasion, solvent atoms are also called *host atoms.* "Solute" is used to denote an element or compound present in a minor concentration.

Solid Solutions

A solid solution forms when, as the solute atoms are added to the host material, the crystal structure is maintained, and no new structures are formed. Perhaps it is useful to draw an analogy with a liquid solution. If two liquids, soluble in each other (such as water and alcohol) are combined, a liquid solution is produced as the molecules intermix, and its composition is homogeneous throughout. A solid solution is also compositionally homogeneous; the impurity atoms are randomly and uniformly dispersed within the solid.

Impurity point defects are found in solid solutions, of which there are two types: **substitutional** and **interstitial.** For substitutional, solute or impurity atoms replace or substitute for the host atoms (Figure 4.2). There are several features of the solute and solvent atoms that determine the degree to which the former dissolves in the latter. One is the atomic size factor; appreciable quantities of a solute may be accommodated

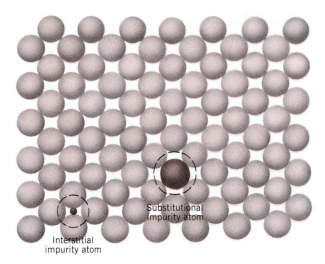

Figure 4.2 Two-dimensional schematic representations of substitutional and interstitial impurity atoms. (Adapted from W. G. Moffatt, G. W. Pearsall, and J. Wulff, *The Structure and Properties of Materials*, Vol. I, *Structure*, p. 77. Copyright © 1964 by John Wiley & Sons, New York. Reprinted by permission of John Wiley & Sons, Inc.)

in this type of solid solution only when the difference in atomic radii between the two atom types is less than about ±15%. Otherwise the solute atoms will create substantial lattice distortions and a new phase will form. Another feature is termed the electrochemical factor; the more electropositive one element and the more electronegative the other, the greater is the likelihood that they will form an intermetallic compound instead of a substitutional solid solution. In addition, the relative valencies of the two atom types will also have an influence. Other factors being equal, a metal will have more of a tendency to dissolve another metal of higher valency than one of a lower valency. A final requirement for complete solid solubility is that the crystal structures for metals of both atom types be the same.

An example of a substitutional solid solution is found for copper and nickel. These two elements are completely soluble in one another at all proportions. With regard to the aforementioned rules that govern degree of solubility, the atomic radii for copper and nickel are 0.128 and 0.125 nm (1.28 and 1.25 Å), respectively; their electronegativities are 1.9 and 1.8 (Figure 2.7); and the most common valencies are +1 for copper (although it sometimes can be +2) and +2 for nickel. Finally, both have the FCC crystal structure.

For interstitial solid solutions, impurity atoms fill the voids or interstices among the host atoms (see Figure 4.2). For metallic materials that have relatively high atomic packing factors, these interstitial positions are relatively small. Consequently, the atomic diameter of an interstitial impurity must be substantially smaller than that of the host atoms. Normally, the maximum allowable concentration of interstitial impurity atoms is low (less than 10%). Even very small impurity atoms are ordinarily larger than the interstitial sites, and as a consequence they introduce some lattice strains on the adjacent atoms.

Carbon forms an interstitial solid solution when added to iron; the maximum concentration of carbon is about 2%. The atomic radius of the carbon atom is much less than that for iron: 0.071 nm (0.71 Å) versus 0.124 nm (1.24 Å). Solid solutions are also possible for ceramic materials, as discussed in Section 13.4.

Specification of Composition

It is often necessary to express the overall **composition** of an alloy in terms of **concentrations** of its constituent elements. The two most common ways to specify concentration are weight (or mass) percent and atom percent. The basis for **weight percent** (wt%) is the weight of a particular element relative to the total alloy weight. For an alloy that contains only hypothetical A and B atoms, the concentration of A in wt%, C_A, is defined as

$$C_A = \frac{m_A}{m_A + m_B} \times 100 \tag{4.3}$$

where m_A and m_B represent the weight (or mass) of elements A and B, respectively. The concentration of B would be computed in an analogous manner.

The basis for **atom percent** (at%) calculations is the number of moles of an element in relation to the total moles of all the elements in the alloy. The number of moles in some specified mass of a hypothetical element D, $N_m(D)$, may be computed as follows:

$$N_m(D) = \frac{m'_D}{A_D} \tag{4.4}$$

Here, m'_D and A_D denote the mass (in grams) and the atomic weight, respectively, for element D.

Concentration in terms of atom percent of element D in an alloy containing D and E atoms, C'_D, is defined by

$$C'_D = \frac{N_m(D)}{N_m(D) + N_m(E)} \times 100 \tag{4.5}$$

In like manner, the atom percent of E may be determined.

Atom percent computations also can be carried out on the basis of the number of atoms instead of moles, since one mole of all substances contains the same number of atoms.

MISCELLANEOUS IMPERFECTIONS

4.4 DISLOCATIONS–LINEAR DEFECTS

A *dislocation* is a linear or one-dimensional defect around which some of the atoms are misaligned. One type of dislocation is represented in Figure 4.3: an extra portion of a plane of atoms, or half-plane, the edge of which terminates within the crystal. This is termed an **edge dislocation;** it is a linear defect that centers around the line that is defined along the end of the extra half-plane of atoms. This is sometimes termed the **dislocation line,** which, for the edge dislocation in Figure 4.3, is perpendicular to the plane of the page. Within the region around the dislocation line there

Burgers vector

b

Edge
dislocation
line

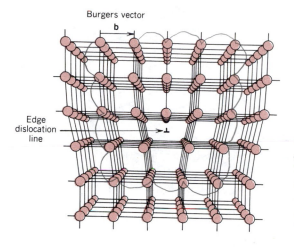

Figure 4.3 The atom positions around an edge dislocation; extra half-plane of atoms shown in perspective. (Adapted from A. G. Guy, *Essentials of Materials Science*, McGraw-Hill Book Company, New York, 1976, p. 153.)

is some localized lattice distortion. The atoms above the dislocation line in Figure 4.3 are squeezed together, and those below are pulled apart; this is reflected in the slight curvature for the vertical planes of atoms as they bend around this extra half-plane. The magnitude of this distortion decreases with distance away from the dislocation line; at positions far removed, the crystal lattice is virtually perfect. Sometimes the edge dislocation in Figure 4.3 is represented by the symbol ⊥, which also indicates the position of the dislocation line. An edge dislocation may also be formed by an extra half-plane of atoms that is included in the bottom portion of the crystal; its designation is a ⊤.

Another type of dislocation, called a **screw dislocation,** exists, which may be thought of as being formed by a shear stress that is applied to produce the distortion shown in Figure 4.4*a*: the upper front region of the crystal is shifted one atomic distance to the right relative to the bottom portion. The atomic distortion associated with a screw dislocation is also linear and along a dislocation line, line *AB* in Figure 4.4*b*. The screw dislocation derives its name from the spiral or helical path or ramp that is traced around the dislocation line by the atomic planes of atoms. Sometimes the symbol ↻ is used to designate a screw dislocation.

Most dislocations found in crystalline materials are probably neither pure edge nor pure screw, but exhibit components of both types; these are termed **mixed dislocations.** All three dislocation types are represented schematically in Figure 4.5; the lattice distortion that is produced away from the two faces is mixed, having varying degrees of screw and edge character.

The magnitude and direction of the lattice distortion associated with a dislocation is expressed in terms of a **Burgers vector,** denoted by a **b**. Burgers vectors are indicated in Figures 4.3 and 4.4 for edge and screw dislocations, respectively. Furthermore, the nature of a dislocation (i.e., edge, screw, or mixed) is defined by the relative orientations of dislocation line and Burgers vector. For an edge, they are perpendicular (Figure 4.3), whereas for a screw, both are parallel (Figure 4.4); they are neither perpendicular nor parallel for a mixed dislocation. Also, even though a dislocation changes direction and nature within a crystal (e.g., from edge to mixed to screw), the Burgers vector will be the same at all points along its line. For example, all positions of the curved dislocation in Figure 4.5 will have the Burgers vector shown. For metallic materials,

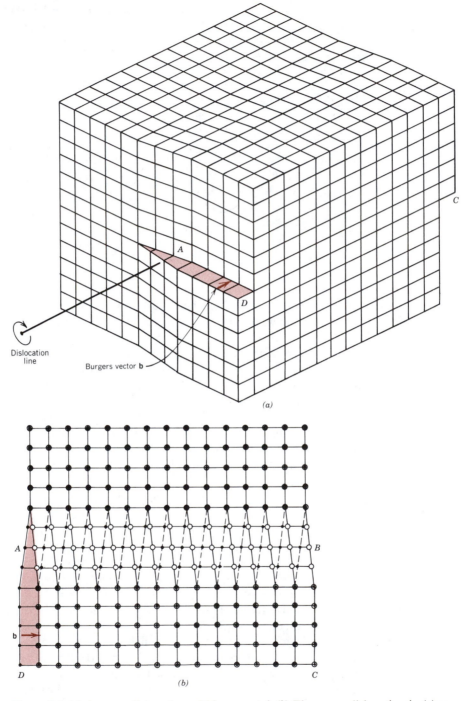

Figure 4.4 (*a*) A screw dislocation within a crystal. (*b*) The screw dislocation in (*a*) as viewed from above. The dislocation line extends along line *AB*. Atom positions above the slip plane are designated by open circles, those below by solid circles. (Figure (*b*) from W. T. Read, Jr., *Dislocations in Crystals,* McGraw-Hill Book Company, New York, 1953.)

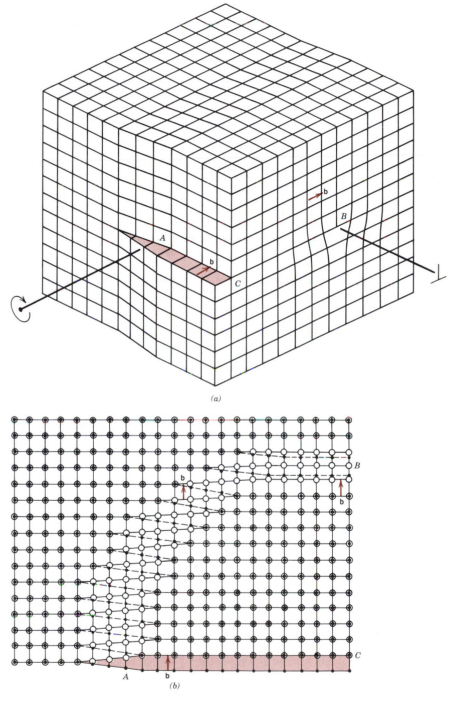

Figure 4.5 (*a*) Schematic representation of a dislocation that has edge, screw, and mixed character. (*b*) Top view, where open circles denote atom positions above the slip plane. Solid circles, atom positions below. At point *A*, the dislocation is pure screw, while at point *B*, it is pure edge. For regions in between where there is curvature in the dislocation line, the character is mixed edge and screw. (Figure (*b*) from W. T. Read, Jr., *Dislocations in Crystals,* McGraw-Hill Book Company, New York, 1953.)

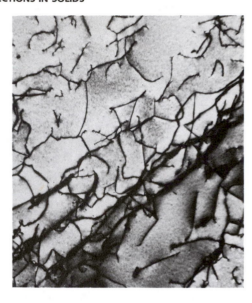

Figure 4.6 A transmission electron micrograph of a titanium alloy in which the dark lines are dislocations. $51,450\times$. (Courtesy of M. R. Plichta, Michigan Technological University.)

the Burgers vector for a dislocation will point in a close-packed crystallographic direction and will be of magnitude equal to the interatomic spacing.

Dislocations can be observed in crystalline materials using electron-microscopic techniques. In Figure 4.6, a high magnification transmission electron micrograph, the dark lines are the dislocations.

Virtually all crystalline materials contain some dislocations that were introduced during solidification, during plastic deformation, and as a consequence of thermal stresses that result from rapid cooling. Dislocations are involved in the plastic deformation of crystalline materials, the discussion of which is deferred to Chapter 7.

4.5 INTERFACIAL DEFECTS

Interfacial defects are boundaries that have two dimensions and normally separate regions of the materials that have different crystal structures and/or crystallographic orientations. These imperfections include external surfaces, grain boundaries, twin boundaries, stacking faults, and phase boundaries.

External Surfaces

One of the most obvious boundaries is the external surface, which is considered to be an imperfection inasmuch as it represents the boundary along which the crystal structure terminates. Surface atoms are not bonded to the maximum number of nearest neighbors, and are therefore in a higher energy state than the atoms at interior positions. The bonds of these surface atoms that are not satisfied give rise to a surface energy, expressed in units of energy per unit area (J/m^2 or erg/cm^2). To reduce this energy, materials tend to minimize, if at all possible, the total surface area. For example, liquids assume a shape having a minimum area—the droplets become spherical. Of course, this is not possible with solids, which are mechanically rigid.

Grain Boundaries

Another interfacial defect, the grain boundary, was introduced in Section 3.13 as the boundary separating two small grains or crystals having different crystallographic orientations in polycrystalline materials. A grain boundary is represented schematically from an atomic perspective in Figure 4.7. Within the boundary region, which is probably just several atom distances wide, there is some atomic mismatch along which there is a transition from the crystalline orientation of one grain to that of an adjacent one.

Various degrees of crystallographic misalignment between adjacent grains are possible (Figure 4.7). When this orientation mismatch is slight, on the order of a few degrees, then the term *small-angle grain boundary* is used. These boundaries can be described in terms of dislocation arrays. One simple small-angle grain boundary is formed when edge dislocations are aligned in the manner of Figure 4.8. This type is called a *tilt boundary;* the angle of misorientation, θ, is also indicated in the figure. *Twist* small-angle grain boundaries result from arrays of screw dislocations. Frequently, regions of material separated by small-angle grain boundaries are termed *subgrains*.

Not all atoms are bonded to other atoms along a grain boundary, and, consequently, there is an interfacial or grain boundary energy similar to the surface energy described above. The magnitude of this energy is a function of the degree of misorientation, being larger for high-angle boundaries. Grain boundaries are more chemically reactive than the grains themselves as a consequence of this boundary energy. Furthermore, impurity atoms often preferentially segregate along these boundaries because of their higher energy state. The total interfacial energy is lower in large or

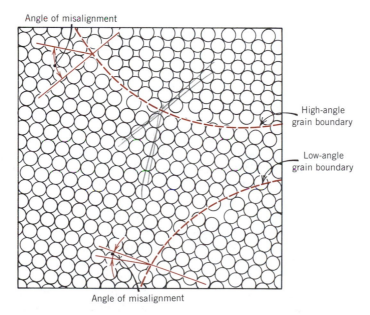

Figure 4.7 Schematic diagram showing low- and high-angle grain boundaries and the adjacent atom positions.

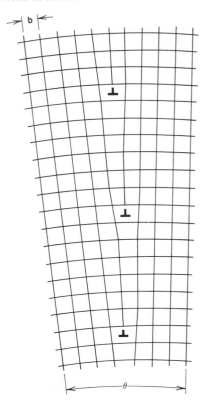

Figure 4.8 Demonstration of how a tilt boundary having an angle of misorientation θ results from an alignment of edge dislocations.

coarse-grained materials than in fine-grained ones, since there is less total boundary area in the former. Grains grow at elevated temperatures to reduce the total boundary energy, a phenomenon explained in Section 7.13.

In spite of this disordered arrangement of atoms and lack of complete bonding along grain boundaries, a polycrystalline material is still very strong; cohesive forces within and across the boundary are present. Furthermore, the density of a polycrystalline specimen is virtually identical to that of a single crystal of the same material.

Twin Boundaries

A *twin boundary* is a special type of grain boundary across which there is a specific mirror lattice symmetry; that is, atoms on one side of the boundary are located in mirror image positions of the atoms on the other side (Figure 4.9). The region of material between these boundaries is appropriately termed a *twin*. Twins result from atomic displacements that are produced from applied mechanical shear forces (mechanical twins), and also during annealing heat treatments following deformation (annealing twins). Twinning occurs on a definite crystallographic plane and in a specific direction, both of which depend on the crystal structure. Annealing twins are typically found in metals that have the FCC crystal structure, while mechanical twins are observed in BCC and HCP metals. The role of mechanical twins in the deformation process is discussed in Section 7.7. Annealing twins may be observed in the photomicrograph of the polycrystalline brass specimen shown in Figure 4.11*c*. The twins

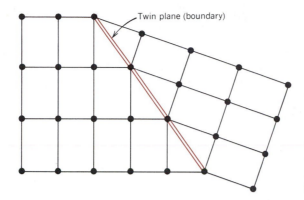

Figure 4.9 Schematic diagram showing a twin plane or boundary and the adjacent atom positions (dark circles).

correspond to those regions having relatively straight and parallel sides and a different visual contrast than the untwinned regions of the grains within which they reside. An explanation for the variety of textural contrasts in this photomicrograph is provided in Section 4.9.

Miscellaneous Interfacial Defects

Other interfacial defects are possible to include stacking faults, phase boundaries, and ferromagnetic domain walls. Stacking faults are found in FCC metals when there is an interruption in the *ABCABCABC*... stacking sequence of close-packed planes (Section 3.11). Phase boundaries exist in multiphase materials (Section 9.3) across which there is a sudden change in physical and/or chemical characteristics. For ferromagnetic and ferrimagnetic materials, the boundary that separates regions having different directions of magnetization is termed a domain wall, which is discussed in Section 21.7.

Associated with each of the defects discussed in this section is an interfacial energy, the magnitude of which depends on boundary type, and which will vary from material to material. Normally, the interfacial energy will be greatest for external surfaces and least for domain walls.

4.6 BULK OR VOLUME DEFECTS

Other defects exist in all solid materials that are much larger than those heretofore discussed. These include pores, cracks, foreign inclusions, and other phases. They are normally introduced during processing and fabrication steps. Some of these defects and their effect on the properties of materials are discussed in subsequent chapters.

4.7 ATOMIC VIBRATIONS

Every atom in a solid material is vibrating very rapidly about its lattice position within the crystal. In a sense, these vibrations may be thought of as imperfections or defects. At any instant of time not all atoms vibrate at the same frequency and amplitude, nor with the same energy. At a given temperature there will exist a distribution of energies for the constituent atoms about an average energy. Over time the vibrational energy of any specific atom will also vary in a random manner. With rising temperature, this average energy increases, and, in fact, the temperature of a solid is really just a measure of the average vibrational activity of atoms and molecules. At room temperature, a typical vibrational frequency is on the order of 10^{13} vibrations per second, whereas the amplitude is a few thousandths of a nanometer.

Many properties and processes in solids are manifestations of this vibrational atomic motion. For example, melting occurs when the vibrations are vigorous enough to rupture large numbers of atomic bonds. A more detailed discussion of atomic vibrations and their influence on the properties of materials is presented in Chapter 20.

MICROSCOPIC EXAMINATION

4.8 GENERAL

On occasion it is necessary or desirable to examine the structural elements and defects that influence the properties of materials. The capacity to perform such examinations is important, first to ensure that the associations between the properties and structure (and defects) are properly understood, and second to predict the properties of materials once these relationships have been established. Several of the techniques that are commonly used in such investigations are discussed next.

Some structural elements are of *macroscopic* dimensions, that is, are large enough to be observed with the unaided eye. For example, the shape and average size or diameter of the grains for a polycrystalline specimen are important structural elements. Macroscopic grains are often evident on aluminum street light posts and also on garbage cans. Relatively large grains having different textures are clearly visible on the surface of the sectioned lead ingot shown in Figure 4.10. However, in most materials the constituent grains are of *microscopic* dimensions, having diameters that may be

Figure 4.10 High-purity polycrystalline lead ingot in which the individual grains may be discerned. $0.7\times$. (Reproduced with permission from *Metals Handbook,* Vol. 9, 9th Edition, *Metallography and Microstructures,* American Society for Metals, Metals Park, OH, 1985.)

on the order of microns,[3] and their details must be investigated using some type of microscope. Grain size and shape are only two features of what is termed the **microstructure;** these and other microstructural characteristics are discussed in subsequent chapters.

4.9 MICROSCOPY

Both optical and electron microscopes are commonly used in **microscopy.** These instruments aid in investigations of the microstructural features of all three material types (metals, ceramics, and polymers). Most of these techniques employ photographic equipment in conjunction with the microscope; the photograph on which the image is recorded is called a **photomicrograph.**

Optical Microscopy

With optical microscopy, the light microscope is used to study the microstructure; optical and illumination systems are its basic elements. For materials that are opaque to visible light (all metals and many ceramics and polymers), only the surface is subject to observation, and the light microscope must be used in a reflecting mode. Contrasts in the image produced result from differences in reflectivity of the various regions of the microstructure. Investigations of this type are often termed *metallographic,* since metals were first examined using this technique.

Normally, careful and meticulous surface preparations are necessary to reveal the important details of the microstructure. The specimen surface must first be ground and polished to a smooth and mirrorlike finish. This is accomplished by using successively finer abrasive papers and powders. The microstructure is revealed by a surface treatment using an appropriate chemical reagent in a procedure termed *etching.* The chemical reactivity of the grains of some single-phase materials depends on crystallographic orientation. Consequently, in a polycrystalline specimen, etching characteristics vary from grain to grain. Figure 4.11b shows how normally incident light is reflected by three etched surface grains, each having a different orientation. Figure 4.11a depicts the surface structure as it might appear when viewed with the microscope; the luster or texture of each grain depends on its reflectance properties. A photomicrograph of a polycrystalline specimen exhibiting these characteristics is shown in Figure 4.11c.

Also, small grooves form along grain boundaries as a consequence of etching. Since atoms along grain boundary regions are more chemically active, they dissolve at a greater rate than those within the grains. These grooves become discernible when viewed under a microscope because they reflect light at an angle different from that of the grains themselves; this effect is displayed in Figure 4.12a. Figure 4.12b is a photomicrograph of a polycrystalline specimen in which the grain boundary grooves are clearly visible as dark lines.

When the microstructure of a two-phase alloy is to be examined, an etchant is chosen that produces a different texture for each phase so that the different phases may be distinguished from each other.

---

[3] A micron (μm), sometimes called a micrometer, is 10^{-6} m.

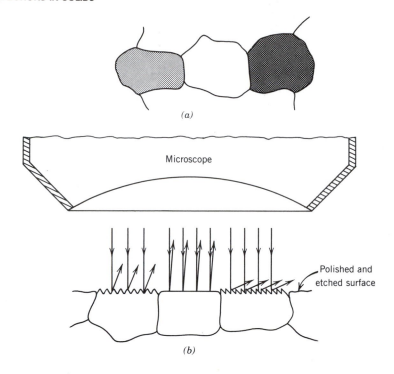

Figure 4.11 (*a*) Polished and etched grains as they might appear when viewed with an optical microscope. (*b*) Section taken through these grains showing how the etching characteristics and resulting surface texture vary from grain to grain because of differences in crystallographic orientation. (*c*) Photomicrograph of a polycrystalline brass specimen. 60×. (Photomicrograph courtesy of J. E. Burke, General Electric Co.)

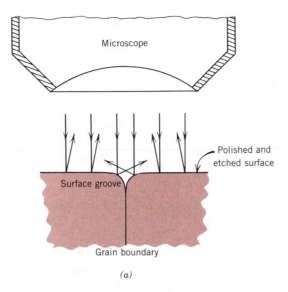

Microscope

Polished and etched surface

Surface groove

Grain boundary

(a)

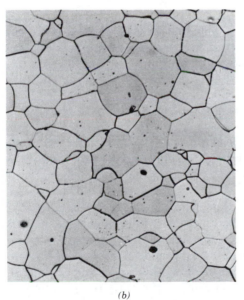

(b)

Figure 4.12 *(a)* Section of a grain boundary and its surface groove produced by etching; the light reflection characteristics in the vicinity of the groove are also shown. *(b)* Photomicrograph of the surface of a polished and etched polycrystalline specimen of an iron-chromium alloy in which the grain boundaries appear dark. 100×. (Photomicrograph courtesy of L. C. Smith and C. Brady, the National Bureau of Standards, Washington, DC.)

Electron Microscopy

The upper limit to the magnification possible with an optical microscope is approximately 2000 diameters. Consequently, some structural elements are too fine or small to permit observation using optical microscopy. Under such circumstances the electron microscope, which is capable of much higher magnifications, may be employed.

An image of the structure under investigation is formed using beams of electrons instead of light radiation. According to quantum mechanics, a high velocity electron will become wavelike, having a wavelength that is inversely proportional to its velocity. When accelerated across large voltages, electrons can be made to have wavelengths on the order of 0.003 nm (3 pm). High magnifications and resolving powers of these microscopes are consequences of the short wavelengths of electron beams; in fact, atomic features may be resolved. The electron beam is focused and the image formed with magnetic lenses; otherwise the geometry of the miscroscope components is essentially the same as with optical systems. Both transmission and reflection beam modes of operation are possible for electron microscopes.

Transmission Electron Microscopy.   The image seen with a **transmission electron microscope (TEM)** is formed by an electron beam that passes through the specimen. Details of internal microstructural features are accessible to observation; contrasts in the image are produced by differences in beam scattering or diffraction produced between various elements of the microstructure or defect. Since solid materials are highly absorptive to electron beams, a specimen to be examined must be prepared in the form of a very thin foil; this ensures transmission through the specimen of an appreciable fraction of the incident beam. The transmitted beam is projected onto a fluorescent screen or a photographic film so that the image may be viewed. Magnifications approaching 1,000,000 × are possible with transmission electron microscopy, which is frequently utilized in the study of dislocations.

Scanning Electron Microscopy.   A more recent innovation, having proved to be an extremely useful investigative tool, is the **scanning electron microscope (SEM).** The surface of a specimen to be examined is scanned with an electron beam, and the reflected (or back-scattered) beam of electrons is collected, then displayed at the same scanning rate on a cathode ray tube (similar to a TV screen). The image that appears on the screen, which may be photographed, represents the surface features of the specimen. The surface may or may not be polished and etched, but it must be electrically conductive; a very thin metallic surface coating must be applied to nonconductive materials. Magnifications ranging from 10 to in excess of 50,000 diameters are possible, as are also very great depths-of-field. Accessory equipment permits qualitative and semiquantitative analysis of the elemental composition of very localized surface areas.

Microscopic examination is an extremely useful tool in the study and characterization of materials. This will become apparent in subsequent chapters that correlate the microstructure with various characteristics and properties. Examination of microstructure is also used to determine the mode of mechanical fracture, to predict the mechanical properties of alloys, to show whether an alloy has been correctly heat treated, and also to design alloys with new property combinations.

4.10 GRAIN SIZE DETERMINATION

The grain size is often determined when the properties of a polycrystalline material are under consideration. In this regard, there exist a number of techniques by which size is specified in terms of average grain volume, diameter, or area. Grain size may be estimated by using an intercept method, described as follows. Straight lines all the same length are drawn through several photomicrographs that show the grain structure. The grains intersected by each line segment are counted; the line length is then divided by an average of the number of grains intersected, taken over all the line segments. The average grain diameter is found by dividing this result by the linear magnification of the photomicrographs.

Probably the most common method utilized, however, is that devised by the American Society for Testing and Materials (ASTM). The ASTM has prepared 10 standard comparison charts, all having different average grain sizes. To each is assigned a number ranging from 1 to 10, which is termed the *grain size number;* the larger this number, the smaller the grains. A specimen must be properly prepared to reveal the grain structure, which is photographed at a magnification of $100\times$. Grain size is expressed as the grain size number of the chart that most nearly matches the grains in the micrograph. Thus a relatively simple and convenient visual determination of grain size number is possible. Grain size number is used extensively in the specification of steels.

The rationale behind the assignment of the grain size number to these various charts is as follows. Let n represent the grain size number, and N the average number of grains per square inch at a magnification of $100\times$. These two parameters are related to each other through the expression

$$N = 2^{n-1} \tag{4.6}$$

SUMMARY

All solid materials contain large numbers of imperfections or deviations from crystalline perfection. The several types of imperfection are categorized on the basis of their geometry and size. Point defects are those associated with one or two atomic positions, including vacancies (or vacant lattice sites), self-interstitials (host atoms that occupy interstitial sites), and impurity atoms.

A solid solution may form when impurity atoms are added to a solid, in which case the original crystal structure is retained and no new phases are formed. For substitutional solid solutions, impurity atoms substitute for host atoms, and appreciable solubility is possible only when atomic diameters and electronegativities for both atom types are similar, and when both elements have the same crystal structure and same valence. Interstitial solid solutions form for relatively small impurity atoms that occupy interstitial sites among the host atoms.

Composition of an alloy may be specified in weight percent or atom percent. The basis for weight percent computations is the weight (or mass) of each alloy constitutent relative to the total alloy weight. Atom percents are calculated in terms of the number of moles for each constituent relative to the total moles of all the elements in the alloy.

Dislocations are one-dimensional crystalline defects of which there are two pure types: edge and screw. An edge may be thought of in terms of the lattice distortion along the end of an extra half-plane of atoms; a screw, by a helical planar ramp. For mixed dislocations, components of both pure edge and screw are found. The magnitude and direction of lattice distortion associated with a dislocation is specified by its Burgers vector. The relative orientations of Burgers vector and dislocation line are (1) perpendicular for edge, (2) parallel for screw, and (3) neither perpendicular nor parallel for mixed.

Other imperfections include interfacial defects [external surfaces, grain boundaries (both low- and high-angle), twin boundaries, etc.], volume defects (cracks, pores, etc.), and atomic vibrations. Each type of imperfection has some influence on the properties of a material.

Many of the important defects and structural elements of materials are of microscopic dimensions, and observation is possible only with the aid of a microscope. Both optical and electron microscopes are employed, usually in conjunction with photographic equipment. Transmissive and reflective modes are possible for each microscope type; preference is dictated by the nature of the specimen as well as the structural element or defect to be examined.

Grain size of polycrystalline materials is frequently determined using photomicrographic techniques. Two methods are commonly employed: intercept and standard comparison charts.

IMPORTANT TERMS AND CONCEPTS

Activation energy	Imperfection	Self-Interstitial
Alloy	Interstitial solid solution	Solid solution
Atom percent	Microscopy	Solute
Atomic vibration	Microstructure	Solvent
Boltzmann's constant	Mixed dislocation	Substitutional solid solution
Burgers vector	Photomicrograph	
Composition	Point defect	Transmission electron microscopy (TEM)
Concentration	Scanning electron microscopy (SEM)	Vacancy
Dislocation line	Screw dislocation	Weight percent
Edge dislocation		
Grain size		

REFERENCES

KEHL, G. L., *Principles of Metallographic Laboratory Practice,* McGraw-Hill Book Co., New York, 1949.

MOFFATT, W. G., G. W. PEARSALL, and J. WULFF, *The Structure and Properties of Materials,* Vol. 1, *Structure,* John Wiley & Sons, New York, 1964.

PHILLIPS, V. A., *Modern Metallographic Techniques and Their Applications,* Wiley-Interscience, New York, 1971.

VAN BUEREN, H. G., *Imperfections in Crystals,* North-Holland Publishing Co., Amsterdam (Wiley-Interscience, New York), 1960.

VANDER VOORT, G. F., *Metallography, Principles and Practice,* McGraw-Hill Book Co., New York, 1984.

QUESTIONS AND PROBLEMS

4.1   Calculate the fraction of atom sites that are vacant for lead at its melting temperature of 327°C (600 K). Assume an activation energy of 0.55 eV/atom.

4.2 Calculate the number of vacancies per cubic meter in gold at 900°C. The activation energy for vacancy formation is 0.98 eV/atom. Furthermore, the density and atomic weight for Au are 19.32 g/cm^3 and 196.9 g/mol, respectively.

4.3 Calculate the activation energy for vacancy formation in silver, given that the equilibrium number of vacancies at 800°C (1073 K) is 3.6 × 10^{23} m^{-3}. The atomic weight and density (at 800°C) for silver are, respectively, 107.9 g/mol and 9.5 g/cm^3.

4.4 Below, atomic radius, crystal structure, electronegativity, and the most common valence are tabulated, for several elements; for those that are nonmetals, only atomic radii are indicated.

Element	Atomic Radius (nm)	Crystal Structure	Electro-negativity	Valence
Ni	0.1246	FCC	1.8	2+
C	0.071			
H	0.046			
O	0.060			
Ag	0.1445	FCC	1.9	1+
Al	0.1431	FCC	1.5	3+
Co	0.1253	HCP	1.8	2+
Cr	0.1249	BCC	1.6	3+
Fe	0.1241	BCC	1.8	2+
Pt	0.1387	FCC	2.2	2+
Zn	0.1332	HCP	1.6	2+

Which of these elements would you expect to form the following with nickel:
(a) A substitutional solid solution having complete solubility?
(b) A substitutional solid solution of incomplete solubility?
(c) An interstitial solid solution?

4.5 For the FCC crystal structure, interstitial sites that may be occupied by impurity atoms are located at the center of each of the unit cell edges. Compute the radius r of an impurity atom that will just fit into one of these sites, in terms of the atomic radius R of the host atom. Consult Section 3.4.

4.6 Calculate the composition, in weight percent, of an alloy that contains 105 kg iron, 0.2 kg carbon, and 1.0 kg chromium.

4.7 What is the composition, in atom percent, of an alloy that contains 33 g copper and 47 g zinc?

4.8 What is the composition, in atom percent, of an alloy that contains 95 lb iron, 10 lb chromium, and 8 lb nickel?

4.9 What is the composition, in atom percent, of an alloy that consists of 92.5 wt% Ag and 7.5 wt% Cu?

4.10 Convert the atom percent composition in Problem 4.8 to weight percent.

4.11 Calculate the number of atoms per cubic meter in aluminum.

4.12 Silicon carbide, SiC, has a density of 3.22 g/cm^3. How many atoms of Si and C are there in 1.0 cm^3 of SiC?

4.13 Nickel forms a substitutional solid solution with copper. Compute the number of nickel atoms per cubic centimeter for a copper–nickel alloy that contains 1.0 wt% Ni and 99.0 wt% Cu. The densities of pure nickel and copper are 8.90 and 8.93 g/cm^3, respectively.

4.14 Zinc forms a substitutional solid solution with copper. Compute the weight percent zinc that must be added to copper to yield an alloy that contains 1.75×10^{21} Zn atoms per cubic centimeter. The densities of pure Zn and Cu are 7.13 and 8.93 g/cm^3, respectively.

4.15 An iron–carbon alloy, with a BCC crystal structure, contains 0.1 at% C. How many carbon atoms are there in 1 m^3 of this alloy?

4.16 Cite the relative Burgers vector–dislocation line orientations for edge, screw, and mixed dislocations.

4.17 For both FCC and BCC crystal structures, the Burgers vector **b** may be expressed as

$$\mathbf{b} = \frac{a}{2}[hkl]$$

where a is the unit cell edge length and $[hkl]$ is the crystallographic direction having the greatest linear atomic density.
  (a) What are the Burgers vector representations for FCC, BCC, and simple cubic crystal structures? See Homework Problems 3.44 and 3.45.
  (b) If the magnitude of the Burgers vector $|\mathbf{b}|$ is

$$|\mathbf{b}| = \frac{a}{2}(h^2 + k^2 + l^2)^{1/2}$$

determine the values of $|\mathbf{b}|$ for copper and iron. You may want to consult Table 3.1.

4.18 **(a)** The surface energy of a single crystal depends on the crystallographic orientation with respect to the surface. Explain why this is so. **(b)** For an FCC crystal, such as aluminum, would you expect the surface energy for a (100) plane to be greater or less than that for a (111) plane? Why?

4.19 **(a)** For a given material, would you expect the surface energy to be greater than, the same as, or less than the grain boundary energy? Why? **(b)** The grain boundary energy of a low-angle grain boundary is less than for a high-angle one. Why is this so?

4.20 **(a)** Briefly describe a twin and a twin boundary. **(b)** Cite the difference between mechanical and annealing twins.

4.21 For each of the following stacking sequences found in FCC metals, cite the type of planar defect that exists:
  (a) . . . A B C A B C B A C B A . . .
  (b) . . . A B C A B C B C A B C . . .
  Now, copy the stacking sequences and indicate the position(s) of planar defect(s) with a vertical dashed line.

4.22 Using the intercept method, determine the average grain size, in millimeters, of the specimen whose microstructure is shown in Figure 4.12b; assume that the magnification is 100×, and use at least seven straight line segments.

4.23 Employing the intercept technique, determine the average grain size for the steel specimen whose microstructure is shown in Figure 9.21*a*; use at least seven straight line segments.

4.24 **(a)** For an ASTM grain size of 4, approximately how many grains would there be per square inch in a micrograph taken at a magnification of $100 \times$? **(b)** Estimate the grain size number for the photomicrograph in Figure 4.12*b*, assuming a magnification of $100 \times$.

4.25 A photomicrograph was taken of some metal at a magnification of $100 \times$ and it was determined that the average number of grains per square inch is 24. Compute the ASTM grain size number for this alloy.

CHAPTER 5

DIFFUSION

Photograph of a steel gear that has been "case hardened." The outer surface layer was selectively hardened by a high-temperature heat treatment during which carbon from the surrounding atmosphere diffused into the surface. The "case" appears as the dark outer rim of that segment of the gear that has been sectioned. Actual size. (Photograph courtesy of Surface Division Midland-Ross.)

5.1 INTRODUCTION

Many reactions and processes that are important in the treatment of materials rely on the transfer of mass either within a specific solid (ordinarily on a microscopic level) or from a liquid, a gas, or another solid phase. This is necessarily accomplished by **diffusion,** the phenomenon of material transport by atomic motion. This chapter discusses the atomic mechanisms by which diffusion occurs, the mathematics of diffusion, and the influence of temperature and diffusing species on the rate of diffusion.

The phenomenon of diffusion may be demonstrated with the use of a *diffusion couple,* which is formed by joining bars of two different metals together so that there is intimate contact between the two faces, as illustrated for copper and nickel in Figure 5.1, which includes schematic representations of atom positions and composition across the interface. This couple is heated for an extended period at an elevated temperature (but below the melting temperature of both metals), and cooled to room temperature. Chemical analysis will reveal a condition similar to that represented in Figure 5.2, namely, pure copper and nickel at the two extremities of the couple, separated by an alloyed region. Concentrations of both metals vary with position as shown in Figure 5.2c. This result indicates that copper atoms have migrated or diffused into the nickel, and that nickel has diffused into copper. This process, whereby atoms of one metal diffuse into another, is termed **interdiffusion** or **impurity diffusion.**

Interdiffusion may be discerned from a macroscopic perspective by changes in concentration which occur over time, as in the example for the Cu–Ni diffusion couple. There is a net drift or transport of atoms from high to low concentration regions.

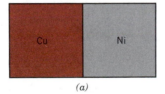

(a)

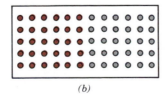

(b)

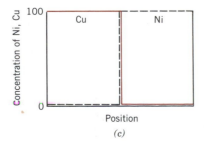

(c)

Figure 5.1 (*a*) A copper-nickel diffusion couple before a high-temperature heat treatment. (*b*) Schematic representations of Cu (colored circles) and Ni (black circles) atom locations within the diffusion couple. (*c*) Concentrations of copper and nickel as a function of position across the couple.

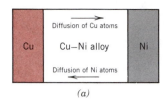

(a)

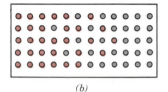

(b)

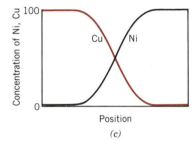

Position

(c)

Figure 5.2 (a) A copper-nickel diffusion couple after a high-temperature heat treatment, showing the alloyed diffusion zone. (b) Schematic representations of Cu (colored circles) and Ni (black circles) atom locations within the couple. (c) Concentrations of copper and nickel as a function of position across the couple.

Diffusion also occurs for pure metals, but all atoms exchanging positions are of the same type; this is termed **self-diffusion.** Of course, self-diffusion is not normally subject to observation by noting compositional changes.

5.2 DIFFUSION MECHANISMS

From an atomic perspective, diffusion is just the stepwise migration of atoms from lattice site to lattice site. In fact, the atoms in solid materials are in constant motion, rapidly changing positions. For an atom to make such a move, two conditions must be met: (1) there must be an empty adjacent site, and (2) the atom must have sufficient energy to break bonds with its neighbor atoms and then cause some lattice distortion during the displacement. This energy is vibrational in nature (Section 4.7). At a specific temperature some small fraction of the total number of atoms are capable of diffusive motion, by virtue of the magnitudes of their vibrational energies. This fraction increases with rising temperature.

Several different models for this atomic motion have been proposed; of these possibilities, two dominate for metallic diffusion.

Vacancy Diffusion

One mechanism involves the interchange of an atom from a normal lattice position to an adjacent vacant lattice site or vacancy, as represented schematically in Figure

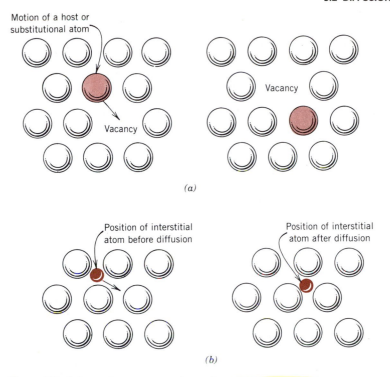

Figure 5.3 Schematic representations of (*a*) vacancy diffusion and (*b*) interstitial diffusion.

5.3*a*. This mechanism is aptly termed **vacancy diffusion.** Of course, this process necessitates the presence of vacancies, and the extent to which vacancy diffusion can occur is a function of the number of these defects that are present; significant concentrations of vacancies may exist in metals at elevated temperatures (Section 4.2). Since diffusing atoms and vacancies exchange positions, the diffusion of atoms in one direction corresponds to the motion of vacancies in the opposite direction. Both self-diffusion and interdiffusion occur by this mechanism; for the latter, the impurity atoms must substitute for host atoms.

Interstitial Diffusion

The second type of diffusion involves atoms that migrate from an interstitial position to a neighboring one that is empty. This mechanism is found for interdiffusion of impurities such as hydrogen, carbon, nitrogen, and oxygen, which have atoms that are small enough to fit into the interstitial positions. Host or substitutional impurity atoms rarely form interstitials and do not normally diffuse via this mechanism. This phenomenon is appropriately termed **interstitial diffusion** (Figure 5.3*b*).

In most metal alloys, interstitial diffusion occurs much more rapidly than diffusion by the vacancy mode, since the interstitial atoms are smaller, and thus more mobile. Furthermore, there are more empty interstitial positions than vacancies; hence, the probability of interstitial atomic movement is greater than for vacancy diffusion.

5.3 STEADY-STATE DIFFUSION

Diffusion is a time-dependent process—that is, in a macroscopic sense, the quantity of an element that is transported within another is a function of time. Often, it is necessary to know how fast diffusion occurs, or the rate of mass transfer. This rate is frequently expressed as a **diffusion flux** (J), defined as the mass (or, equivalently, the number of atoms) M diffusing through and perpendicular to a unit cross-sectional area of solid per unit of time. In mathematical form, this may be represented as

$$J = \frac{M}{At} \qquad (5.1a)$$

where A denotes the area across which diffusion is occurring and t is the elapsed diffusion time. In differential form, this expression becomes

$$J = \frac{1}{A}\frac{dM}{dt} \qquad (5.1b)$$

The units for J are kilograms or atoms per meter squared per second (kg/m²-s or atoms/m²-s).

If the diffusion flux does not change with time, a steady-state condition exists. One common example of **steady-state diffusion** is the diffusion of atoms of a gas through a plate of metal for which the concentrations (or pressures) of the diffusing species on both surfaces of the plate are held constant. This is represented schematically in Figure 5.4a.

When concentration C is plotted versus position (or distance) within the solid x, the resulting curve is termed the **concentration profile;** the slope at a particular point

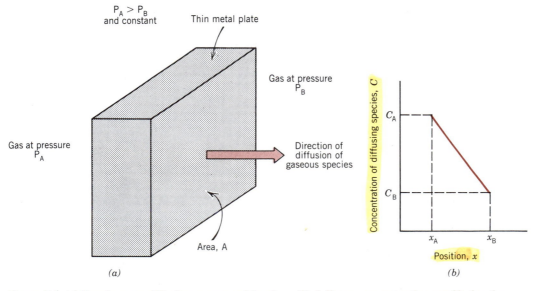

Figure 5.4 (a) Steady-state diffusion across a thin plate. (b) A linear concentration profile for the diffusion situation in (a).

on this curve is the **concentration gradient**:

$$\text{concentration gradient} = \frac{dC}{dx} \tag{5.2a}$$

In the present treatment, the concentration profile is assumed to be linear, as depicted in Figure 5.4b, and

$$\text{concentration gradient} = \frac{\Delta C}{\Delta x} = \frac{C_A - C_B}{x_A - x_B} \tag{5.2b}$$

For diffusion problems, it is usually most convenient to express concentration in terms of mass of diffusing species per unit volume of solid (kg/m^3 or g/cm^3).

The mathematics of steady-state diffusion in a single (x) direction are relatively simple, in that the flux is proportional to the concentration gradient through the expression

$$J = -D\frac{dC}{dx} \tag{5.3}$$

The constant of proportionality D is called the **diffusion coefficient,** which is expressed in square meters per second. The negative sign in this expression indicates that the direction of diffusion is down the concentration gradient, from a high to a low concentration. Equation 5.3 is sometimes called **Fick's first law.**

Sometimes the term **driving force** is used in the context of what compels a reaction to occur. For diffusion reactions, several such forces are possible; but when diffusion is according to Equation 5.3, the concentration gradient is the driving force.

One practical example of steady-state diffusion is found in the purification of hydrogen gas. One side of a thin sheet of palladium metal is exposed to the impure gas composed of hydrogen and other gaseous species such as nitrogen, oxygen, and water vapor. The hydrogen selectively diffuses through the sheet to the opposite side, which is maintained at a constant and lower hydrogen pressure.

EXAMPLE PROBLEM 5.1

A plate of iron is exposed to a carburizing (carbon-rich) atmosphere on one side and a decarburizing (carbon-deficient) atmosphere on the other side at 700°C (1300°F). If a condition of steady state is achieved, calculate the diffusion flux of carbon through the plate if the concentrations of carbon at positions of 5 and 10 mm (5×10^{-3} and 10^{-2} m) beneath the carburizing surface are 1.2 and 0.8 kg/m^3, respectively. Assume a diffusion coefficient of 3×10^{-11} m^2/s at this temperature.

SOLUTION

Fick's first law, Equation 5.3, is utilized to determine the diffusion flux. Substitution of the values above into this expression yields

$$J = -D\frac{C_A - C_B}{x_A - x_B} = -(3 \times 10^{-11} \text{ m}^2/\text{s})\frac{(1.2 - 0.8) \text{ kg/m}^3}{(5 \times 10^{-3} - 10^{-2}) \text{ m}}$$

$$= 2.4 \times 10^{-9} \text{ kg/m}^2\text{-s}$$

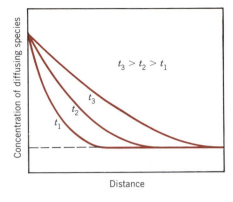

Figure 5.5 Concentration profiles for nonsteady-state diffusion taken at three different times, t_1, t_2, and t_3.

5.4 NONSTEADY-STATE DIFFUSION

Most practical diffusion situations are nonsteady-state ones. That is, the diffusion flux and the concentration gradient at some particular point in a solid vary with time, with a net accumulation or depletion of the diffusing species resulting. This is illustrated in Figure 5.5, which shows concentration profiles at three different diffusion times. Under conditions of nonsteady state, use of Equation 5.3 is no longer convenient; instead, the partial differential equation

$$\frac{\partial C}{\partial t} = \frac{\partial}{\partial x}\left(D\,\frac{\partial C}{\partial x}\right) \tag{5.4a}$$

known as **Fick's second law,** is used. If the diffusion coefficient is independent of composition (which should be verified for each particular diffusion situation), Equation 5.4a simplifies to

$$\frac{\partial C}{\partial t} = D\,\frac{\partial^2 C}{\partial x^2} \tag{5.4b}$$

Solutions to this expression (concentration in terms of both position and time) are possible when physically meaningful boundary conditions are specified. Comprehensive collections of these are given by Crank and Carslaw and Jaegar (see References).

One practically important solution is for a semi-infinite solid[1] in which the surface concentration is held constant. Frequently, the source of the diffusing species is a gas phase, the partial pressure of which is maintained at a constant value. Furthermore, the following assumptions are made:

1. Before diffusion, any of the diffusing solute atoms in the solid are uniformly distributed with concentration of C_0.
2. The value of x at the surface is zero and increases with distance into the solid.
3. The time is taken to be zero the instant before the diffusion process begins.

---

[1] A bar of solid is considered to be semi-infinite if none of the diffusing atoms reaches the bar end during the time over which diffusion takes place. A bar of length ℓ is considered to be semi-infinite when $\ell > 10\sqrt{Dt}$.

TABLE 5.1 Tabulation of Error Function Values

z	erf(z)	z	erf(z)	z	erf(z)
0	0	0.55	0.5633	1.3	0.9340
0.025	0.0282	0.60	0.6039	1.4	0.9523
0.05	0.0564	0.65	0.6420	1.5	0.9661
0.10	0.1125	0.70	0.6778	1.6	0.9763
0.15	0.1680	0.75	0.7112	1.7	0.9838
0.20	0.2227	0.80	0.7421	1.8	0.9891
0.25	0.2763	0.85	0.7707	1.9	0.9928
0.30	0.3286	0.90	0.7970	2.0	0.9953
0.35	0.3794	0.95	0.8209	2.2	0.9981
0.40	0.4284	1.0	0.8427	2.4	0.9993
0.45	0.4755	1.1	0.8802	2.6	0.9998
0.50	0.5205	1.2	0.9103	2.8	0.9999

These boundary conditions are simply stated as

For $t = 0$, $C = C_0$ at $0 \leq x \leq \infty$

For $t > 0$, $C = C_s$ (the constant surface concentration) at $x = 0$

$$C = C_0 \text{ at } x = \infty$$

Application of these boundary conditions to Equation 5.4b yields the solution

$$\frac{C_x - C_0}{C_s - C_0} = 1 - \text{erf}\left(\frac{x}{2\sqrt{Dt}}\right) \tag{5.5}$$

where C_x represents the concentration at depth x after time t. The expression erf($x/2\sqrt{Dt}$) is the Gaussian error function,[2] values of which are given in mathematical tables for various ($x/2\sqrt{Dt}$) values; a partial listing is given in Table 5.1. The concentration parameters that appear in Equation 5.5 are noted in Figure 5.6, a concentration profile taken at a specific time. Equation 5.5 thus demonstrates the relationship between concentration, position, and time, namely, that C_x, being a function of the dimensionless parameter $x/\sqrt{Dt}$, may be determined at any time and position if the parameters C_0, C_s, and D are known.

Suppose that it is desired to achieve some specific concentration of solute, C_1, in an alloy; the left-hand side of Equation 5.5 now becomes

$$\frac{C_1 - C_0}{C_s - C_0} = \text{constant}$$

This being the case, the right-hand side of this same expression is also a constant, and subsequently

$$\frac{x}{2\sqrt{Dt}} = \text{constant} \tag{5.6a}$$

---

[2] This Gaussian error function is defined by

$$\text{erf}(z) = \frac{2}{\sqrt{\pi}} \int_0^z e^{-y^2} \, dy$$

where $x/2\sqrt{Dt}$ has been replaced by the variable z.

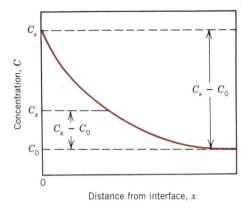

Figure 5.6 Concentration profile for non-steady-state diffusion; concentration parameters relate to Equation 5.5.

or

$$\frac{x^2}{Dt} = \text{constant} \tag{5.6b}$$

Some diffusion computations are thus facilitated on the basis of this relationship, as demonstrated in Example Problem 5.3.

EXAMPLE PROBLEM 5.2

For some applications, it is necessary to harden the surface of a steel (or iron–carbon alloy) above that of its interior. One way this may be accomplished is by increasing the surface concentration of carbon in a process termed **carburizing;** the steel piece is exposed, at an elevated temperature, to an atmosphere rich in a hydrocarbon gas, such as methane (CH_4).

Consider one such alloy that initially has a uniform carbon concentration of 0.25 wt% and is to be treated at 950°C (1750°F). If the concentration of carbon at the surface is suddenly brought to and maintained at 1.20 wt%, how long will it take to achieve a carbon content of 0.80 wt% at a position 0.5 mm below the surface? The diffusion coefficient for carbon in iron at this temperature is 1.6×10^{-11} m^2/s; assume that the steel piece is semi-infinite.

SOLUTION
Since this is a nonsteady-state diffusion problem in which the surface composition is held constant, Equation 5.5 is used. Values for all the parameters in this expression except time t are specified in the problem as follows:

$$C_0 = 0.25 \text{ wt\% C}$$

$$C_s = 1.20 \text{ wt\% C}$$

$$C_x = 0.80 \text{ wt\% C}$$

$$x = 0.50 \text{ mm} = 5 \times 10^{-4} \text{ m}$$

$$D = 1.6 \times 10^{-11} \text{ m}^2/\text{s}$$

Thus

$$\frac{C_x - C_0}{C_s - C_0} = \frac{0.80 - 0.25}{1.20 - 0.25} = 1 - \text{erf}\left[\frac{(5 \times 10^{-4}\ \text{m})}{2\sqrt{(1.6 \times 10^{-11}\ \text{m}^2/\text{s})(t)}}\right]$$

$$0.4210 = \text{erf}\left(\frac{62.5\ \text{s}^{1/2}}{\sqrt{t}}\right)$$

We must now determine from Table 5.1 the value of z for which the error function is 0.4210. An interpolation is necessary, as

z	erf(z)
0.35	0.3794
z	0.4210
0.40	0.4284

$$\frac{z - 0.35}{0.40 - 0.35} = \frac{0.4210 - 0.3794}{0.4284 - 0.3794}$$

or

$$z = 0.392$$

Therefore,

$$\frac{62.5\ \text{s}^{1/2}}{\sqrt{t}} = 0.392$$

and solving for t,

$$t = \left(\frac{62.5\ \text{s}^{1/2}}{0.392}\right)^2 = 25{,}400\ \text{s} = 7.1\ \text{h}$$

EXAMPLE PROBLEM 5.3

The diffusion coefficients for copper in aluminum at 500 and 600°C are 4.8×10^{-14} and $5.3 \times 10^{-13}\ \text{m}^2/\text{s}$, respectively. Determine the approximate time at 500°C that will produce the same diffusion result (in terms of concentration of Cu at some specific point in Al) as a 10-h heat treatment at 600°C.

SOLUTION
This is a diffusion problem in which Equation 5.6b may be employed. The composition in both diffusion situations will be equal at the same position (i.e., x is also a constant), thus

$$Dt = \text{constant} \tag{5.7}$$

at both temperatures. That is,

$$(Dt)_{500} = (Dt)_{600}$$

or

$$t_{500} = \frac{(Dt)_{600}}{D_{500}} = \frac{(5.3 \times 10^{-13}\ \text{m}^2/\text{s})(10\ \text{h})}{4.8 \times 10^{-14}\ \text{m}^2/\text{s}} = 110.4\ \text{h}$$

5.5 FACTORS THAT INFLUENCE DIFFUSION

Diffusing Species

The magnitude of the diffusion coefficient D is indicative of the rate at which atoms diffuse. Coefficients, both self- and interdiffusion, for several metallic systems are listed in Table 5.2. The diffusing species as well as the host material influence the diffusion coefficient. For example, there is a significant difference in magnitude between self- and carbon interdiffusion in α iron at 500°C, the D value being greater for the carbon interdiffusion (1.1×10^{-20} vs. 2.3×10^{-12} m^2/s). This comparison also provides a contrast between rates of diffusion via vacancy and interstitial modes as discussed above. Self-diffusion occurs by a vacancy mechanism, whereas carbon diffusion in iron is interstitial.

Temperature

Temperature has a most profound influence on the coefficients and diffusion rates. For example, for the self-diffusion of Fe in α-Fe, the diffusion coefficient increases approximately five orders of magnitude (from 1.1×10^{-20} to 3.9×10^{-15} m^2/s) in rising temperature from 500 to 900°C (Table 5.2). The temperature dependence of diffusion coefficients is related to temperature according to

$$D = D_0 \exp\left(-\frac{Q_d}{RT}\right) \tag{5.8}$$

where

D_0 = a temperature-independent preexponential (m^2/s)
Q_d = the **activation energy** for diffusion (J/mol, cal/mol, or eV/atom)
R = the gas constant, 8.31 J/mol-K, 1.987 cal/mol-K, or 8.62×10^{-5} eV/atom
T = absolute temperature (K)

The activation energy may be thought of as that energy required to produce the diffusive motion of one mole of atoms. A large activation energy results in a relatively small diffusion coefficient. Table 5.2 also contains a listing of D_0 and Q_d values for several diffusion systems.

Taking natural logarithms of Equation 5.8 yields

$$\ln D = \ln D_0 - \frac{Q_d}{R}\left(\frac{1}{T}\right) \tag{5.9}$$

Since D_0, Q_d, and R are all constants, this expression takes on the form of an equation of a straight line:

$$y = b + mx$$

where y and x are analogous, respectively, to the variables $\ln D$ and $1/T$. Thus, if $\ln D$ is plotted versus the reciprocal of the absolute temperature, a straight line should result, having slope and intercept of $-Q_d/R$ and $\ln D_0$, respectively. This is, in fact, the manner in which the values of Q_d and D_0 are determined experimentally. From such a plot for several alloy systems (Figure 5.7), it may be noted that linear relationships exist for all cases shown.

TABLE 5.2 A Tabulation of Diffusion Data

Diffusing Species	Host Metal	$D_0(m^2/s)$	Activation Energy Q_d			Calculated Values	
			kJ/mol	kcal/mol	eV/atom	$T(°C)$	$D(m^2/s)$
Fe	α-Fe (BCC)	2.0×10^{-4}	241	57.5	2.49	500	1.1×10^{-20}
						900	3.9×10^{-15}
Fe	γ-Fe (FCC)	5.0×10^{-5}	284	67.9	2.94	900	1.1×10^{-17}
						1100	7.8×10^{-16}
C	α-Fe	6.2×10^{-7}	80	19.2	0.83	500	2.3×10^{-12}
						900	1.6×10^{-10}
C	γ-Fe	1.0×10^{-5}	136	32.4	1.40	900	9.2×10^{-12}
						1100	7.0×10^{-11}
Cu	Cu	7.8×10^{-5}	211	50.4	2.18	500	4.4×10^{-19}
Zn	Cu	3.4×10^{-5}	191	45.6	1.98	500	4.3×10^{-18}
Al	Al	1.7×10^{-4}	142	34.0	1.47	500	4.1×10^{-14}
Cu	Al	6.5×10^{-5}	135	32.3	1.40	500	4.8×10^{-14}
Mg	Al	1.2×10^{-4}	131	31.2	1.35	500	1.8×10^{-13}
Cu	Ni	2.7×10^{-5}	255	61.0	2.64	500	1.5×10^{-22}

Source: C. J. Smithells and E. A. Brandes (Editors), *Metals Reference Book,* 5th edition, Butterworths, London, 1976.

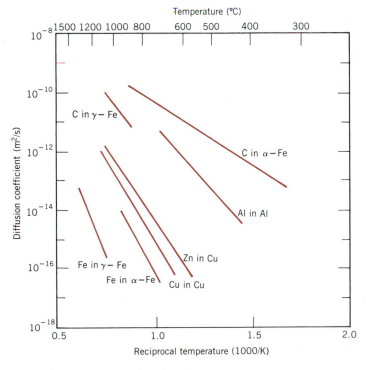

Figure 5.7 Plot of the logarithm of the diffusion coefficient versus the reciprocal of absolute temperature for several metals. [Data taken from C. J. Smithells and E. A. Brandes (Editors), *Metals Reference Book,* 5th edition, Butterworths, London, 1976.]

EXAMPLE PROBLEM 5.4

Using the data in Table 5.2, compute the diffusion coefficient for magnesium in aluminum at 400°C.

SOLUTION

This diffusion coefficient may be determined by applying Equation 5.8; the values of D_0 and Q_d from Table 5.2 are 1.2×10^{-4} m^2/s and 131 kJ/mol, respectively. Thus

$$D = (1.2 \times 10^{-4}\ \text{m}^2/\text{s}) \exp\left[-\frac{(131{,}000\ \text{J/mol})}{(8.31\ \text{J/mol-K})(400 + 273\ \text{K})}\right]$$

$$= 8.1 \times 10^{-15}\ \text{m}^2/\text{s}$$

5.6 OTHER DIFFUSION PATHS

Atomic migration may also occur along dislocations, grain boundaries, and external surfaces. These are sometimes called "*short-circuit*" *diffusion paths* inasmuch as rates are much faster than for bulk diffusion. However, in most situations short-circuit contributions to the overall diffusion flux are insignificant because the cross-sectional areas of these paths are extremely small.

5.7 MATERIALS PROCESSING AND DIFFUSION

Some properties of materials are subject to alteration and improvement as a result of processes and transformations that involve atomic diffusion. For these transformations to occur over reasonable time periods (usually on the order of hours), they are ordinarily carried out at elevated temperatures at which diffusion rates are comparatively rapid. These high-temperature procedures, often termed *heat treatments*, are utilized at least once during the production of almost all common metallic, ceramic, and polymeric materials. For example, the strength of some steels is reliant on appropriate heat treatments (Chapter 11), as is also the mechanical integrity of many ceramics (Section 14.9).

SUMMARY

Solid-state diffusion is a means of mass transport within solid materials by stepwise atomic motion. The term "self-diffusion" refers to the migration of host atoms; for impurity atoms, the term "interdiffusion" is used. Two mechanisms are possible: vacancy and interstitial. For a given host metal, interstitial atomic species generally diffuse more rapidly.

For steady-state diffusion, the concentration profile of the diffusing species is time independent, and the flux or rate is proportional to the negative of the concentration gradient according to Fick's first law. The mathematics for nonsteady state are described by Fick's second law, a partial differential equation. The solution for a constant surface composition boundary condition involves the Gaussian error function.

The magnitude of the diffusion coefficient is indicative of the rate of atomic motion, being strongly dependent on and increasing exponentially with increasing temperature.

IMPORTANT TERMS AND CONCEPTS

Activation energy	**Diffusion flux**	**Interstitial diffusion**
Carburizing	**Driving force**	**Nonsteady-state**
Concentration gradient	**Fick's first and second**	**diffusion**
Concentration profile	**laws**	**Self-diffusion**
Diffusion	**Interdiffusion**	**Steady-state diffusion**
Diffusion coefficient	**(impurity diffusion)**	**Vacancy diffusion**

REFERENCES

BORG, R. J. and G. J. DIENES (Editors), *An Introduction to Solid State Diffusion,* Academic Press, San Diego, 1988.

CARSLAW, H. S. and J. C. JAEGER, *Conduction of Heat in Solids,* Clarendon Press, Oxford, 1959.

CRANK, J., *The Mathematics of Diffusion,* 2nd edition, Clarendon Press, Oxford, 1975.

GIRIFALCO, L. A., *Atomic Migration In Crystals,* Blaisdell Publishing Company, New York, 1964.

SHEWMON, P. G., *Diffusion in Solids,* McGraw-Hill Book Company, New York, 1963.

QUESTIONS AND PROBLEMS

5.1 Briefly explain the difference between self-diffusion and interdiffusion.

5.2 Self-diffusion involves the motion of atoms that are all of the same type; therefore it is not subject to observation by compositional changes, as with interdiffusion. Suggest one way in which self-diffusion may be monitored.

5.3 **(a)** Compare interstitial and vacancy atomic mechanisms for diffusion. **(b)** Cite two reasons why interstitial diffusion is normally more rapid than vacancy diffusion.

5.4 Briefly explain the concept of steady state as it applies to diffusion.

5.5 **(a)** Briefly explain the concept of a driving force. **(b)** What is the driving force for steady-state diffusion?

5.6 The purification of hydrogen gas by diffusion through a palladium sheet was discussed in Section 5.3. Compute the number of kilograms of hydrogen that pass per hour through a 5 mm thick sheet of palladium having an area of 0.2 m^2 at 500°C. Assume a diffusion coefficient of $1.0 \times 10^{-8} \text{ m}^2/\text{s}$, that the concentrations at the high- and low-pressure sides of the plate are 2.4 and 0.6 kg of hydrogen per cubic meter of palladium, and that steady-state conditions have been attained.

5.7 A sheet of steel 1.5 mm thick has nitrogen atmospheres on both sides at 1200°C and is permitted to achieve a steady-state diffusion condition. The diffusion coefficient for nitrogen in steel at this temperature is $6 \times 10^{-11} \text{ m}^2/\text{s}$, and the diffusion flux is found to be $1.2 \times 10^{-7} \text{ kg/m}^2$-s. Also, it is known that the concentration of nitrogen in the steel at the high-pressure surface is 4 kg/m^3. How far into the sheet from this high-pressure side will the concentration be 2 kg/m^3? Assume a linear concentration profile.

5.8   A sheet of BCC iron 1 mm thick was exposed to a carburizing gas atmosphere on one side and a decarburizing atmosphere on the other side at 725°C. After having reached steady state, the iron was quickly cooled to room temperature. The carbon concentrations at the two surfaces of the sheet were determined to be 0.012 and 0.0075 wt%. Compute the diffusion coefficient if the diffusion flux is 1.40×10^{-8} kg/m^2-s. *Hint:* Convert the concentrations from weight percent to kilograms of carbon per cubic meter of iron.

5.9   Show that Fick's second law (Equation 5.4b) takes on the form of Fick's first law (Equation 5.3) for conditions of steady state, that is,

$$\frac{\partial C}{\partial t} = 0$$

5.10   Show that

$$C_x = \frac{B}{\sqrt{Dt}} \exp\left(-\frac{x^2}{4Dt}\right)$$

is also a solution to Equation 5.4b. The parameter B is a constant, being independent of both x and t.

5.11   Determine the carburizing time necessary to achieve a carbon concentration of 0.45 wt% at a position 2 mm into an iron–carbon alloy that initially contains 0.20 wt% C. The surface concentration is to be maintained at 1.30 wt% C, and the treatment is to be conducted at 1000°C. Use the diffusion data for γ-Fe in Table 5.2.

5.12   An FCC iron–carbon alloy initially containing 0.55 wt% C is exposed to an oxygen-rich and virtually carbon-free atmosphere at 1325 K (1052°C). Under these circumstances the carbon diffuses from the alloy and reacts at the surface with the oxygen in the atmosphere, that is, the carbon concentration at the surface position is maintained essentially at 0 wt% C. (This process of carbon depletion is termed *decarburization*.) At what position will the carbon concentration be 0.25 wt% after a 10-h treatment? The value of D at 1325 K is 4.3×10^{-11} m^2/s.

5.13   Nitrogen from a gaseous phase is to be diffused into pure iron at 700°C. If the surface concentration is maintained at 0.1 wt% N, what will be the concentration 1 mm from the surface after 10 h? The diffusion coefficient for nitrogen in iron at 700°C is 2.5×10^{-11} m^2/s.

5.14   Simplify Equation 5.5 for the situation when C_x is halfway between C_s and C_0.

5.15   For a steel alloy it has been determined that a carburizing heat treatment of 10 h duration will raise the carbon concentration to 0.45 wt% at a point 2.5 mm from the surface. Estimate the time necessary to achieve the same concentration at a 5.0-mm position for an identical steel and at the same carburizing temperature.

5.16   Cite the values of the diffusion coefficients for the interdiffusion of carbon in both α-iron (BCC) and γ-iron (FCC) at 900°C. Which is larger? Explain why this is the case.

5.17   Using the data in Table 5.2, compute the value of D for the diffusion of zinc in copper at 650°C.

5.18 At what temperature will the diffusion coefficient for the diffusion of copper in nickel have a value of 6.5×10^{-17} m^2/s? Use the diffusion data in Table 5.2.

5.19 The preexponential and activation energy for the diffusion of chromium in nickel are 1.1×10^{-4} m^2/s and 272,000 J/mol, respectively. At what temperature will the diffusion coefficient have a value of 1.2×10^{-14} m^2/s?

5.20 The activation energy for the diffusion of copper in silver is 193,000 J/mol. Calculate the diffusion coefficient at 1200 K (927°C), given that D at 1000 K (727°C) is 1.0×10^{-14} m^2/s.

5.21 The diffusion coefficients for iron in nickel are given at two temperatures:

T (K)	D (m^2/s)
1273	9.4×10^{-16}
1473	2.4×10^{-14}

(a) Determine the values of D_0 and the activation energy Q_d.
(b) What is the magnitude of the D at 1100°C (1373 K)?

5.22 The diffusion coefficients for carbon in nickel are given at two temperatures:

T (°C)	D (m^2/s)
600	5.5×10^{-14}
700	3.9×10^{-13}

(a) Determine the values of D_0 and Q_d.
(b) What is the magnitude of D at 850°C?

5.23 Below is shown a plot of the logarithm (to the base 10) of the diffusion coefficient versus reciprocal of the absolute temperature, for the diffusion of gold in silver. Determine values for the activation energy and preexponential.

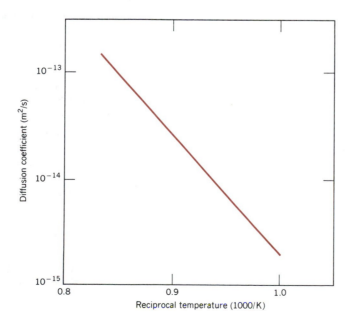

5.24 Carbon is allowed to diffuse through a steel plate 10 mm thick. The concentrations of carbon at the two faces are 0.85 and 0.40 kg C/cm^3 Fe, which are maintained constant. If the preexponential and activation energy are 6.2×10^{-7} m^2/s and 80,000 J/mol, respectively, compute the temperature at which the diffusion flux is 6.3×10^{-10} kg/m^2-s.

5.25 The steady-state diffusion flux through a metal plate is 7.8×10^{-8} kg/m^2-s at a temperature of 1200°C (1473 K) and when the concentration gradient is -500 kg/m^4. Calculate the diffusion flux at 1000°C (1273 K) for the same concentration gradient and assuming an activation energy for diffusion of 145,000 J/mol.

5.26 At approximately what temperature would a specimen of γ-iron have to be carburized for 2 h to produce the same diffusion result as at 900°C for 15 h?

5.27 **(a)** Calculate the diffusion coefficient for copper in aluminum at 500°C. **(b)** What time will be required at 600°C to produce the same diffusion result (in terms of concentration at a specific point) as for 10 h at 500°C?

5.28 A copper–nickel diffusion couple similar to that shown in Figure 5.1*a* is fashioned. After a 500-h heat treatment at 1000°C (1273 K) the concentration of Ni is 3.0 wt% at the 1.0-mm position within the copper. At what temperature must the diffusion couple need to be heated to produce this same concentration (i.e., 3.0 wt% Ni) at a 2.0-mm position after 500 h? The preexponential and activation energy for the diffusion of Ni in Cu are 2.7×10^{-4} m^2/s and 236,000 J/mol, respectively.

5.29 A diffusion couple similar to that shown in Figure 5.1*a* is prepared using two hypothetical metals A and B. After a 20-h heat treatment at 800°C (and subsequently cooling to room temperature) the concentration of B in A is 2.5 wt% at the 5.0-mm position within metal A. If another heat treatment is conducted on an identical diffusion couple, only at 1000°C for 20 h, at what position will the composition be 2.5 wt% B? Assume that the preexponential and activation energy for the diffusion coefficient are 1.5×10^{-4} m^2/s and 125,000 J/mol, respectively.

5.30 The outer surface of a steel gear is to be hardened by increasing its carbon content; the carbon is to be supplied from an external carbon-rich atmosphere which is maintained at an elevated temperature. A diffusion heat treatment at 600°C (873 K) for 100 min increases the carbon concentration to 0.75 wt% at a position 0.5 mm below the surface. Estimate the diffusion time required at 900°C (1173 K) to achieve this same concentration also at a 0.5-mm position. Assume that the surface carbon content is the same for both heat treatments, which is maintained constant. Use the diffusion data in Table 5.2 for C diffusion in α-Fe.

5.31 An FCC iron–carbon alloy initially containing 0.10 wt% C is carburized at an elevated temperature and in an atmosphere wherein the surface carbon concentration is maintained at 1.10 wt%. If after 48 h the concentration of carbon is 0.30 wt% at a position 3.5 mm below the surface, determine the temperature at which the treatment was carried out.

5.32 A diffusion couple was formed between pure copper and a copper–nickel alloy. After heating the couple to 1273 K (1000°C) for 30 days, the concentration of nickel in the copper is 10.0 wt% at a position 0.50 mm from the initial copper–alloy interface. What is the original composition of the copper–nickel alloy? The preexponential and activation energy for the diffusion of Ni in Cu are 2.7×10^{-4} m^2/s and 236,000 J/mol, respectively.

CHAPTER 6

MECHANICAL PROPERTIES OF METALS

A modern tensile testing apparatus with associated instrumentation. (Photograph courtesy the Instron Corporation.)

6.1 INTRODUCTION

Many materials, when in service, are subjected to forces or loads; examples include the aluminum alloy from which an airplane wing is constructed and the steel in an automobile axle. In such situations it is necessary to know the characteristics of the material and to design the member from which it is made such that any resulting deformation will not be excessive and fracture will not occur. The mechanical behavior of a material reflects the relationship between its response or deformation to an applied load or force. Important mechanical properties are strength, hardness, ductility, and stiffness.

The mechanical properties of materials are ascertained by performing carefully designed laboratory experiments that replicate as nearly as possible the service conditions. Factors to be considered include the nature of the applied load and its duration, as well as the environmental conditions. It is possible for the load to be tensile, compressive, or shear, and its magnitude may be constant with time, or it may fluctuate continuously. Application time may be for only a fraction of a second, or it may extend over a period of many years. Service temperature may be an important factor.

The role of structural engineers is to determine stresses and stress distributions within members that are subjected to well-defined loads. This may be accomplished by experimental testing techniques and/or by theoretical and mathematical stress analyses. These topics are treated in traditional stress analysis and strength of materials texts.

Materials and metallurgical engineers, on the other hand, are concerned with producing and fabricating materials to meet service requirements as predicted by these stress analyses. This necessarily involves an understanding of the relationships between the microstructure (i.e., internal features) of materials and their mechanical properties.

Materials are frequently chosen for structural applications because they have desirable combinations of mechanical characteristics. The present discussion is confined primarily to the mechanical behavior of metals; polymers and ceramics are treated separately because they are, to a large degree, mechanically dissimilar to metals. This chapter discusses the stress–strain behavior of metals and the principal mechanical properties related thereto, and examines other mechanical characteristics that are important. Treatments relating to the microscopic aspects of deformation mechanisms and methods to strengthen and regulate the mechanical behavior of metals are deferred to later chapters.

6.2 CONCEPTS OF STRESS AND STRAIN

If a load is static or changes relatively slowly with time and is applied uniformly over a cross section or surface of a member, the mechanical behavior may be ascertained by a simple stress–strain test; these are most commonly conducted for metals at room temperature. There are three principal ways in which a load may be applied: namely, tension, compression, and shear (Figures 6.1a,b,c). In engineering practice many loads are torsional rather than pure shear; this type of loading is illustrated in Figure 6.1d.

Tension Tests

One of the most common mechanical stress–strain tests is performed in *tension*. As will be seen, the tension test can be used to ascertain several mechanical properties

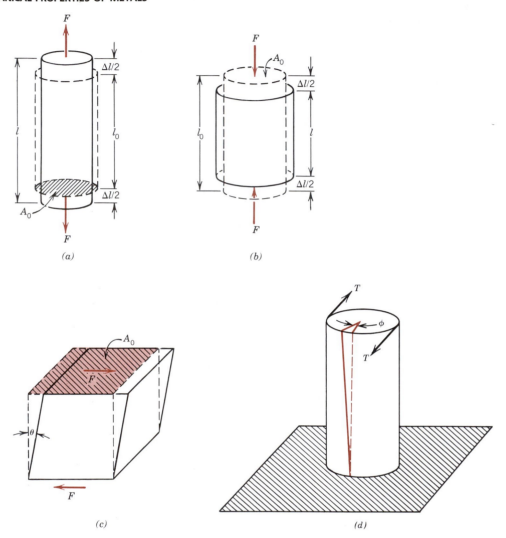

Figure 6.1 (*a*) Schematic illustration of how a tensile load produces an elongation and positive linear strain. Dashed lines represent the shape before deformation; solid lines, after deformation. (*b*) Schematic illustration of how a compressive load produces contraction and a negative linear strain. (*c*) Schematic representation of shear strain γ, where $\gamma = \tan \theta$. (*d*) Schematic representation of torsional deformation (i.e., angle of twist ϕ) produced by an applied torque T.

of materials that are important in design. A specimen is deformed, usually to fracture, with a gradually increasing tensile load that is applied uniaxially along the long axis of a specimen. A standard tensile specimen is shown in Figure 6.2. Normally, the cross section is circular, but rectangular specimens are also used. During testing, deformation is confined to the narrow center region, which has a uniform cross section along its length. The standard diameter is approximately 0.5 in. (12.8 mm), whereas the reduced section length should be at least four times this diameter; $2\frac{1}{4}$ in. (60 mm)

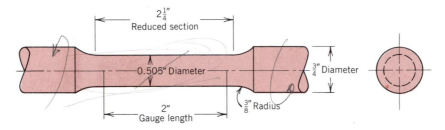

Figure 6.2 A standard tensile specimen with circular cross section.

is common. Gauge length is used in ductility computations, as discussed in Section 6.6; the standard value is 2.0 in. (50 mm). The specimen is mounted by its ends into the holding grips of the testing apparatus (Figure 6.3). The tensile testing machine is designed to elongate the specimen at a constant rate, and to continuously and simultaneously measure the instantaneous applied load (with a load cell) and the resulting elongations (using an extensometer). A stress–strain test typically takes several minutes to perform and is destructive; that is, the test specimen is permanently deformed and usually fractured.

The output of such a tensile test is recorded on a strip chart as load or force versus elongation. These load–deformation characteristics are dependent on the specimen size. For example, it will require twice the load to produce the same elongation if the cross-sectional area of the specimen is doubled. To minimize these geometrical factors, load and elongation are normalized to the respective parameters of **engineering stress** and **engineering strain**. Engineering stress σ is defined by the relationship

$$\sigma = \frac{F}{A_0} \tag{6.1}$$

in which F is the instantaneous load applied perpendicular to the specimen cross section, in units of pounds force (lb$_f$) or newtons (N), and A_0 is the original cross-sectional area before any load is applied (in.^{2} or m^2). The units of engineering stress

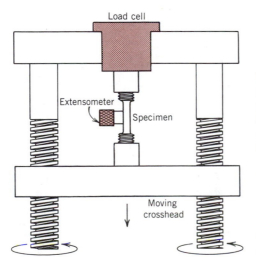

Figure 6.3 Schematic representation of the apparatus used to conduct tensile stress–strain tests. The specimen is elongated by the moving crosshead; load cell and extensometer measure, respectively, the magnitude of the applied load and the elongation. (Adapted from H. W. Hayden, W. G. Moffatt, and J. Wulff, *The Structure and Properties of Materials,* Vol. III, *Mechanical Behavior,* p. 2. Copyright © 1965 by John Wiley & Sons, New York. Reprinted by permission of John Wiley & Sons, Inc.)

(referred to subsequently as just stress) are pounds force per square inch, psi (Customary U.S.), or megapascals, MPa (SI); 1 MPa = 10^6 N/m^2.[1]

Engineering strain ϵ is defined according to

$$\epsilon = \frac{l_i - l_0}{l_0} = \frac{\Delta l}{l_0} \qquad (6.2)$$

in which l_0 is the original length before any load is applied, and l_i is the instantaneous length. Sometimes the quantity $l_i - l_0$ is denoted as Δl, and is the deformation elongation or change in length at some instant, as referenced to the original length. Engineering strain (subsequently called just strain) is unitless, but inches per inch or meters per meter are often used; the value of strain is obviously independent of the unit system. Sometimes strain is also expressed as a percentage, in which the strain value is multiplied by 100.

Compression Tests

Compression stress–strain tests may be conducted if in-service forces are of this type. A compression test is conducted in a manner similar to the tensile test, except that the force is compressive and the specimen contracts along the direction of the stress. Equations 6.1 and 6.2 are utilized to compute compressive stress and strain, respectively. By convention, a compressive force is taken to be negative, which yields a negative stress. Furthermore, since l_0 is greater than l_i, compressive strains computed from Equation 6.2 are necessarily also negative. Tensile tests are more common because they are easier to perform; also, for most materials used in structural applications, very little additional information is obtained from compressive tests since a material behaves the same way in each test. Compressive tests are used when a material's behavior under large and permanent (i.e., plastic) strains is desired, as in manufacturing applications.

Shear and Torsional Tests

For tests performed using a pure shear force as shown in Figure 6.1c, the shear stress τ is computed according to

$$\tau = \frac{F}{A_0} \qquad (6.3)$$

where F is the load or force imposed parallel to the upper and lower faces, each of which has an area of A_0. The shear strain γ is defined as the tangent of the strain angle θ, as indicated in the figure. The units for shear stress and strain are the same as for their tensile counterparts.

Torsion is a variation of pure shear, wherein a structural member is twisted in the manner of Figure 6.1d; torsional forces produce a rotational motion about the longitudinal axis of one end of the member relative to the other end. Examples of torsion are found for machine axles and drive shafts, and also for twist drills. Torsional

---

[1] Conversion from one system of stress units to the other is accomplished by the relationship

145 psi = 1 MPa

tests are normally performed on cylindrical solid shafts or tubes. A shear stress τ is a function of the applied torque T, whereas shear strain γ is related to the angle of twist, ϕ in Figure 6.1*d*.

ELASTIC DEFORMATION

6.3 STRESS–STRAIN BEHAVIOR

The degree to which a structure deforms or strains depends on the magnitude of an imposed stress. For most metals that are stressed in tension and at relatively low levels, stress and strain are proportional to each other through the relationship

$$\sigma = E\epsilon \qquad E = \text{YOUNG'S MODULUS} \quad (6.4)$$

This is known as Hooke's law, and the constant of proportionality E (psi or MPa) is the **modulus of elasticity** or *Young's modulus*. For most typical metals the magnitude of this modulus ranges between 6.5×10^6 psi (4.5×10^4 MPa), for magnesium, and 59×10^6 psi (40.7×10^4 MPa), for tungsten. Modulus of elasticity values for several metals at room temperature are presented in Table 6.1.

Deformation in which stress and strain are proportional is called **elastic deformation;** a plot of stress (ordinate) versus strain (abscissa) results in a linear relationship, as shown in Figure 6.4. The slope of this linear segment corresponds to the modulus of elasticity E. This modulus may be thought of as stiffness, or a material's resistance to elastic deformation. The greater the modulus, the stiffer the material, or the smaller the elastic strain that results from the application of a given stress. The modulus is an important design parameter used for computing elastic deflections.

Elastic deformation is nonpermanent, which means that when the applied load is released, the piece returns to its original shape. As shown in the stress–strain plot (Figure 6.4), application of the load corresponds to moving from the origin up and along the straight line. Upon release of the load, the line is traversed in the opposite direction, back to the origin.

There are some materials (e.g., gray cast iron and concrete) for which this initial elastic portion of the stress–strain curve is not linear (Figure 6.5); hence, it is not

TABLE 6.1 Room Temperature Elastic and Shear Moduli, and Poisson's Ratio for Various Metal Alloys

Metal Alloy	Modulus of Elasticity E		Shear Modulus G MODULUS OF RIGIDITY		Poisson's Ratio
	psi $\times 10^6$	MPa $\times 10^4$	psi $\times 10^6$	MPa $\times 10^4$	
Aluminum	10.0	6.9	3.8	2.6	0.33
Brass	14.6	10.1	5.4	3.7	0.35
Copper	16.0	11.0	6.7	4.6	0.35
Magnesium	6.5	4.5	2.5	1.7	0.29
Nickel	30.0	20.7	11.0	7.6	0.31
Steel	30.0	20.7	12.0	8.3	0.27
Titanium	15.5	10.7	6.5	4.5	0.36
Tungsten	59.0	40.7	23.2	16.0	0.28

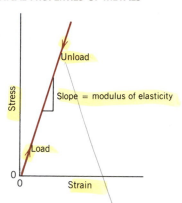

Figure 6.4 Schematic stress–strain diagram showing linear elastic deformation for loading and unloading cycles.

possible to determine a modulus of elasticity as described above. For this nonlinear behavior, either *tangent* or *secant modulus* is normally used. Tangent modulus is taken as the slope of the stress–strain curve at some specified level of stress, while secant modulus represents the slope of a secant drawn from the origin to some given point of the σ–ϵ curve. The determination of these moduli is illustrated in Figure 6.5.

On an atomic scale, macroscopic elastic strain is manifested as small changes in the interatomic spacing and the stretching of interatomic bonds. As a consequence, the magnitude of the modulus of elasticity is a measure of the resistance to separation of adjacent atoms, that is, the interatomic bonding forces. Furthermore, this modulus

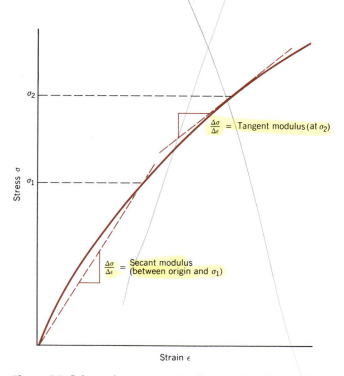

Figure 6.5 Schematic stress–strain diagram showing nonlinear elastic behavior, and how secant and tangent moduli are determined.

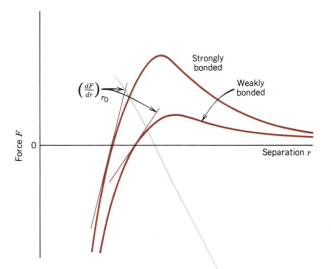

Figure 6.6 Force versus interatomic separation for weakly and strongly bonded atoms. The magnitude of the modulus of elasticity is proportional to the slope of each curve at the equilibrium interatomic separation r_0.

is proportional to the slope of the interatomic force–separation curve (Figure 2.8a) at the equilibrium spacing:

$$E \propto \left(\frac{dF}{dr}\right)_{r_0} \tag{6.5}$$

Figure 6.6 shows the force–separation curves for materials having both strong and weak interatomic bonds; the slope at r_0 is indicated for each.

Values of the modulus of elasticity for ceramic materials are characteristically higher than for metals; for polymers, they are lower. These differences are a direct consequence of the different types of atomic bonding in the three materials types. Furthermore, with increasing temperature, the modulus of elasticity diminishes, as is shown for several metals in Figure 6.7.

As would be expected, the imposition of compressive, shear, or torsional stresses also evokes elastic behavior. The stress–strain characteristics at low stress levels are virtually the same for both tensile and compressive situations, to include the magnitude of the modulus of elasticity. Shear stress and strain are proportional to each other through the expression

$$\tau = G\gamma \tag{6.6}$$

where G is the *shear modulus*, the slope of the linear elastic region of the shear stress–strain curve. Table 6.1 gives the shear moduli for a number of the common metals.

6.4 ANELASTICITY

Up to this point, it has been assumed that elastic deformation is time independent, that is, that an applied stress produces an instantaneous elastic strain which remains constant over the period of time the stress is maintained. It has also been assumed

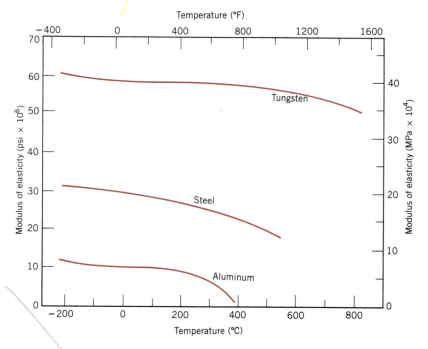

Figure 6.7 Plot of modulus of elasticity versus temperature for tungsten, steel, and aluminum. (Adapted from K. M. Ralls, T. H. Courtney, and J. Wulff, *Introduction to Materials Science and Engineering.* Copyright © 1976 by John Wiley & Sons, New York. Reprinted by permission of John Wiley & Sons, Inc.)

that upon release of the load the strain is totally recovered, that is, that the strain immediately returns to zero. In most engineering materials, however, there will also exist a time-dependent elastic strain component. That is, elastic deformation will continue after the stress application, and upon load release some finite time is required for complete recovery. This time-dependent elastic behavior is known as **anelasticity,** and it is due to time-dependent microscopic and atomistic processes that are attendant to the deformation. For metals the anelastic component is normally small and is often neglected. However, for some polymeric materials its magnitude is significant; in this case it is termed *viscoelastic behavior,* which is the discussion topic of Section 16.6.

EXAMPLE PROBLEM 6.1

A piece of copper originally 12 in. (305 mm) long is pulled in tension with a stress of 40,000 psi (276 MPa). If the deformation is entirely elastic, what will be the resultant elongation?

SOLUTION
Since the deformation is elastic, strain is dependent on stress according to Equation 6.4. Furthermore, the elongation Δl is related to the original length l_0 through

Equation 6.2. Combining these two expressions and solving for Δl yields

$$\sigma = \epsilon E = \left(\frac{\Delta l}{l_0}\right)E$$

$$\Delta l = \frac{\sigma l_0}{E}$$

The values of σ and l_0 are given as 40,000 psi and 12 in., respectively, and the magnitude of E for copper from Table 6.1 is 16×10^6 psi (11.0×10^4 MPa). Elongation is obtained by substitution into the expression above as

$$\Delta l = \frac{(40,000 \text{ psi})(12 \text{ in.})}{(16 \times 10^6 \text{ psi})} = 0.030 \text{ in. } (0.76 \text{ mm})$$

6.5 ELASTIC PROPERTIES OF MATERIALS

When a tensile stress is imposed on a metal specimen, an elastic elongation and accompanying strain ϵ_z result in the direction of the applied stress (arbitrarily taken to be the z direction), as indicated in Figure 6.8. As a result of this elongation, there will be constrictions in the lateral (x and y) directions perpendicular to the applied stress; from these contractions, the compressive strains ϵ_x and ϵ_y may be determined. If the applied stress is uniaxial (only in the z direction), then $\epsilon_x = \epsilon_y$. A parameter termed **Poisson's ratio** v is defined as the ratio of the lateral and axial strains, or

$$v = -\frac{\epsilon_x}{\epsilon_z} = -\frac{\epsilon_y}{\epsilon_z} \qquad (6.7)$$

The negative sign is included in the expression so that v will always be positive, since ϵ_x and ϵ_z will always be of opposite sign. Theoretically, Poisson's ratio for isotropic materials should be $\frac{1}{4}$; furthermore, the maximum value for v (or that value for which there is no net volume change) is 0.50. For many metals and other alloys, values of

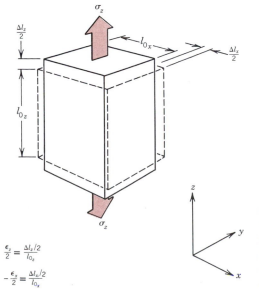

$$\frac{\epsilon_z}{2} = \frac{\Delta l_z/2}{l_{0_z}}$$

$$-\frac{\epsilon_x}{2} = \frac{\Delta l_x/2}{l_{0_x}}$$

Figure 6.8 Axial (z) elongation (positive strain) and lateral (x and y) contractions (negative strains) in response to an imposed tensile stress. Solid lines represent dimensions after stress application; dashed lines, before.

Poisson's ratio range between 0.25 and 0.35. Table 6.1 shows v values for several common metallic materials.

Shear and elastic moduli are related to each other and to Poisson's ratio according to

$$E = 2G(1 + v) \tag{6.8}$$

In most metals G is about $0.4E$; thus, if the value of one modulus is known, the other may be approximated.

Many materials are elastically anisotropic; that is, the elastic behavior (e.g., the magnitude of E) varies with crystallographic direction (see Table 3.3). For these materials the elastic properties are completely characterized only by the specification of several elastic constants, their number depending on characteristics of the crystal structure. Even for isotropic materials, for complete characterization of the elastic properties, at least two constants must be given. Since the grain orientation is random in most polycrystalline materials, these may be considered to be isotropic; inorganic ceramic glasses are also isotropic. The remaining discussion of mechanical behavior assumes isotropy and polycrystallinity because such is the character of most engineering materials.

EXAMPLE PROBLEM 6.2

A tensile stress is to be applied along the long axis of a cylindrical brass rod that has a diameter of 0.4 in. (10 mm). Determine the magnitude of the load required to produce a 10^{-4} in. (2.5×10^{-3} mm) change in diameter if the deformation is entirely elastic.

SOLUTION
This deformation situation is represented in the accompanying drawing.

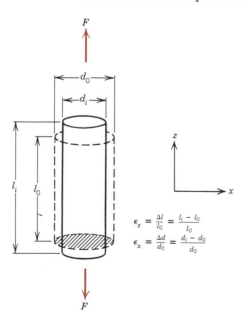

$$\epsilon_z = \frac{\Delta l}{l_0} = \frac{l_i - l_0}{l_0}$$

$$\epsilon_x = \frac{\Delta d}{d_0} = \frac{d_i - d_0}{d_0}$$

When the force F is applied, the specimen will elongate in the z direction and at the same time experience a reduction in diameter, Δd, of 10^{-4} in. in the x direction. For the strain in the x direction,

$$\epsilon_x = \frac{\Delta d}{d_0} = \frac{-10^{-4} \text{ in.}}{0.4 \text{ in.}} = -2.5 \times 10^{-4}$$

which is negative, since the diameter is reduced.

It next becomes necessary to calculate the strain in the z direction using Equation 6.7. The value for Poisson's ratio for brass is 0.35 (Table 6.1), and thus

$$\epsilon_z = -\frac{\epsilon_x}{v} = -\frac{(-2.5 \times 10^{-4})}{0.35} = 7.14 \times 10^{-4}$$

The applied stress may now be computed using Equation 6.4 and the modulus of elasticity, given in Table 6.1 as 14.6×10^6 psi (10.1×10^4 MPa), as

$$\sigma = \epsilon_z E = (7.14 \times 10^{-4})(14.6 \times 10^6 \text{ psi}) = 10,400 \text{ psi}$$

Finally, from Equation 6.1, the applied force may be determined as

$$F = \sigma A_0 = \sigma \left(\frac{d_0}{2}\right)^2 \pi$$

$$= (10,400 \text{ psi})\left(\frac{0.4 \text{ in.}}{2}\right)^2 \pi = 1310 \text{ lb}_f \text{ (5820 N)}$$

PLASTIC DEFORMATION

For most metallic materials, elastic deformation persists only to strains of about 0.005. As the material is deformed beyond this point, the stress is no longer proportional to strain (Hooke's law, Equation 6.4, ceases to be valid), and permanent, nonrecoverable, or **plastic deformation** occurs. Figure 6.9a plots schematically the tensile stress–strain behavior into the plastic region for a typical metal. The transition from elastic to plastic is a gradual one for most metals; some curvature results at the onset of plastic deformation, which increases more rapidly with rising stress.

From an atomic perspective, plastic deformation corresponds to the breaking of bonds with original atom neighbors and then reforming bonds with new neighbors as large numbers of atoms or molecules move relative to one another; upon removal of the stress they do not return to their original positions. The mechanism of this deformation is different for crystalline and amorphous materials. For crystalline solids, deformation is accomplished by means of a process called slip, which involves the motion of dislocations as discussed in Section 7.2. Plastic deformation in noncrystalline solids (as well as liquids) occurs by a viscous flow mechanism, which is outlined in Section 13.8.

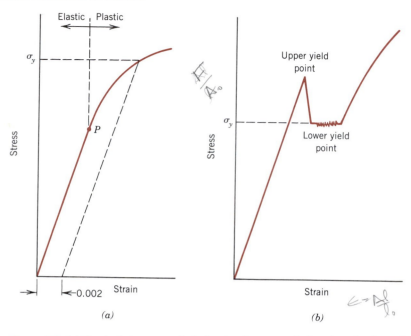

Figure 6.9 (*a*) Typical stress–strain behavior for a metal showing elastic and plastic deformations, the proportional limit *P*, and the yield strength σ_y, as determined using the 0.002 strain offset method. (*b*) Representative stress–strain behavior found for some steels demonstrating the yield point phenomenon.

6.6 TENSILE PROPERTIES

Yielding and Yield Strength

Most structures are designed to ensure that only elastic deformation will result when a stress is applied. It is therefore desirable to know the stress level at which plastic deformation begins, or where the phenomenon of **yielding** occurs. For metals that experience this gradual elastic–plastic transition, the point of yielding may be determined as the initial departure from linearity of the stress–strain curve; this is sometimes called the **proportional limit,** as indicated by point *P* in Figure 6.9*a*. In such cases the position of this point may not be determined precisely. As a consequence, a convention has been established wherein a straight line is constructed parallel to the elastic portion of the stress–strain curve at some specified strain offset, usually 0.002. The stress corresponding to the intersection of this line and the stress–strain curve as it bends over in the plastic region is defined as the **yield strength** σ_y.[2] This is demonstrated in Figure 6.9*a*.

For those materials having a nonlinear elastic region (Figure 6.5), use of the strain offset method is not possible, and the usual practice is to define the yield strength as the stress required to produce some amount of strain (e.g., $\epsilon = 0.005$).

---

[2] "Strength" is used in lieu of "stress" because strength is a property of the metal, whereas stress is related to the magnitude of the applied load.

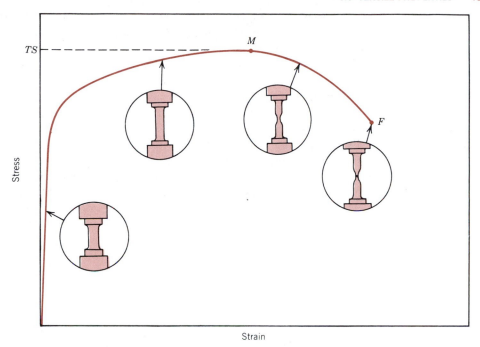

Figure 6.10 Typical engineering stress–strain behavior to fracture, point F. The tensile strength TS is indicated at point M. The circular insets represent the geometry of the deformed specimen at various points along the curve.

Some steels and other materials exhibit the tensile stress–strain behavior as shown in Figure 6.9b. The elastic–plastic transition is very well defined and occurs abruptly in what is termed a *yield point phenomenon*. At the upper yield point, plastic deformation is initiated with an actual decrease in stress. Continued deformation fluctuates slightly about some constant stress value, termed the lower yield point; stress subsequently rises with increasing strain. For metals that display this effect, the yield strength is taken as the average stress that is associated with the lower yield point, since it is well defined and relatively insensitive to the testing procedure.[3] Thus it is not necessary to employ the strain offset method for these materials.

The magnitude of the yield strength for a metal is just a measure of its resistance to plastic deformation. Yield strengths may range from 5000 psi (35 MPa) for a low-strength aluminum to over 200,000 psi (1400 MPa) for high-strength steels.

Tensile Strength

After yielding, the stress necessary to continue plastic deformation increases to a maximum, point M in Figure 6.10, and then decreases to the eventual fracture, point F. The **tensile strength** TS (psi or MPa) is the stress at the maximum on the engineering

---

[3] It should be pointed out that to observe the yield point phenomenon, a "stiff" tensile-testing apparatus must be used; by stiff is meant that there is very little elastic deformation of the machine during loading.

stress–strain curve (Figure 6.10). This corresponds to the maximum stress that can be sustained by a structure in tension; if this stress is applied and maintained, fracture will result. All deformation up to this point is uniform throughout the narrow region of the tensile specimen. However, at this maximum stress, a small constriction or neck begins to form at some point, and all subsequent deformation is confined at this neck, as indicated by the schematic specimen insets in Figure 6.10. This phenomenon is termed "necking," and fracture ultimately occurs at the neck. The fracture or rupture strength corresponds to the stress at fracture.

Tensile strengths may vary anywhere from 7000 psi (50 MPa) for an aluminum to as high as 450,000 psi (3000 MPa) for the high-strength steels. Ordinarily, when the strength of a metal is cited for design purposes, the yield strength is used. This is because by the time a stress corresponding to the tensile strength has been applied, often a structure has experienced so much plastic deformation that it is useless. Furthermore, fracture strengths are not normally specified for engineering design purposes.

EXAMPLE PROBLEM 6.3

From the tensile stress–strain behavior for the brass specimen shown in Figure 6.11, determine the following:

(a) The modulus of elasticity.

(b) The yield strength at a strain offset of 0.002.

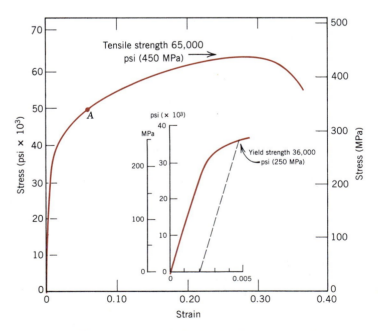

Figure 6.11 The stress–strain behavior for the brass specimen discussed in Example Problem 6.3.

(c) The maximum load that can be sustained by a cylindrical specimen having an original diameter of 0.505 in. (12.8 mm).

(d) The change in length of a specimen originally 10 in. (254 mm) long which is subjected to a tensile stress of 50,000 psi (345 MPa).

SOLUTION

(a) The modulus of elasticity is the slope of the elastic or initial linear portion of the stress–strain curve. The strain axis has been expanded in the inset, Figure 6.11, to facilitate this computation. The slope of this linear region is the rise over the run, or the change in stress divided by the corresponding change in strain; in mathematical terms,

$$E = \text{slope} = \frac{\Delta\sigma}{\Delta\epsilon} = \frac{\sigma_2 - \sigma_1}{\epsilon_2 - \epsilon_1} \tag{6.9}$$

Inasmuch as the line segment passes through the origin, it is convenient to take both σ_1 and ϵ_1 as zero. If σ_2 is arbitrarily taken as 20,000 psi, then ϵ_2 will have a value of 0.0014. Therefore,

$$E = \frac{(20,000 - 0) \text{ psi}}{0.0014 - 0} = 14.3 \times 10^6 \text{ psi } (9.85 \times 10^4 \text{ MPa})$$

which is very close to the value of 14.6×10^6 psi (10.1×10^4 MPa) given for brass in Table 6.1.

(b) The 0.002 strain offset line is constructed as shown in the inset; its intersection with the stress–strain curve is at approximately 36,000 psi (250 MPa), which is the yield strength of the brass.

(c) The maximum load that can be sustained by the specimen is calculated by using Equation 6.1, in which σ is taken to be the tensile strength, from Figure 6.11, 65,000 psi (450 MPa). Solving for F, the maximum load, yields

$$F = \sigma A_0 = \sigma \left(\frac{d_0}{2}\right)^2 \pi = (65,000 \text{ psi})\left(\frac{0.505}{2} \text{ in.}\right)^2 \pi$$

$$= 13,000 \text{ lb}_f \ (5.77 \times 10^4 \text{ N})$$

(d) To compute the change in length, Δl, in Equation 6.2, it is first necessary to determine the strain that is produced by a stress of 50,000 psi. This is accomplished by locating the stress point on the stress–strain curve, point A, and reading the corresponding strain from the strain axis, which is approximately 0.06. Inasmuch as $l_0 = 10$ in., we have

$$\Delta l = \epsilon l_0 = (0.06)(10 \text{ in.}) = 0.6 \text{ in. (15.2 mm)}$$

Ductility

Ductility is another important mechanical property. It is a measure of the degree of plastic deformation that has been sustained at fracture. A material that experiences very little or no plastic deformation upon fracture is termed *brittle*. The tensile stress–strain behaviors for both ductile and brittle materials are schematically illustrated in Figure 6.12.

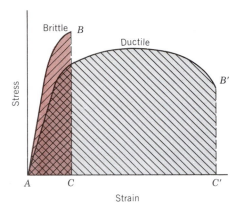

Figure 6.12 Schematic representations of tensile stress-strain behavior for brittle and ductile materials loaded to fracture.

Ductility may be expressed quantitatively as either *percent elongation* or *percent area reduction.* The percent elongation %EL is the percentage of plastic strain at fracture, or

$$\%EL = \left(\frac{l_f - l_0}{l_0}\right) \times 100 \tag{6.10}$$

where l_f is the fracture length[4] and l_0 is the original gauge length as above. Inasmuch as a significant proportion of the plastic deformation at fracture is confined to the neck region, the magnitude of %EL will depend on specimen gauge length. The shorter l_0, the greater is the fraction of total elongation from the neck and, consequently, the higher the value of %EL. Therefore, l_0 should be specified when percent elongation values are cited; it is commonly 2 in. (50 mm).

Percent area reduction %AR is defined as

$$\%AR = \left(\frac{A_0 - A_f}{A_0}\right) \times 100 \tag{6.11}$$

where A_0 is the original cross-sectional area and A_f is the cross-sectional area at the point of fracture.[4] Percent area reduction values are independent of both l_0 and A_0. Furthermore, for a given material the magnitudes of %EL and %AR will, in general, be different. Most metals possess at least a moderate degree of ductility at room temperature; however, some become brittle as the temperature is lowered (Section 8.6).

A knowledge of the ductility of materials is important for at least two reasons. First, it indicates to a designer the degree to which a structure will deform plastically before fracture. Second, it specifies the degree of allowable deformation during fabrication operations. We sometimes refer to relatively ductile materials as being "forgiving," in the sense that they may experience local deformation without fracture should there be an error in the magnitude of the design stress calculation.

Brittle materials are *approximately* considered to be those having a fracture strain of less that about 5%.

---

[4] Both l_f and A_f are measured subsequent to fracture, and after the two broken ends have been repositioned back together.

TABLE 6.2 Typical Mechanical Properties of Several Metals in an Annealed State, and of Commercial Purity

Metal	Yield Strength [psi (MPa)]	Tensile Strength [psi (MPa)]	Ductility, %EL (in 2 in.)
Gold	Nil	19,000 (130)	45
Aluminum	4,000 (28)	10,000 (69)	45
Copper	10,000 (69)	29,000 (200)	45
Iron	19,000 (130)	38,000 (262)	45
Nickel	20,000 (138)	70,000 (480)	40
Titanium	35,000 (240)	48,000 (330)	30
Molybdenum	82,000 (565)	95,000 (655)	35

Thus several important mechanical properties of metals may be determined from tensile stress–strain tests. Table 6.2 presents some typical room temperature values of yield strength, tensile strength, and ductility for several of the common metals, which have been heat treated to render them relatively low strength. These properties are sensitive to any prior deformation, the presence of impurities, and/or any heat treatment to which the metal has been subjected. The modulus of elasticity is one mechanical parameter that is insensitive to these treatments. As with modulus of elasticity, the magnitudes of both yield and tensile strengths decline with increasing temperature; just the reverse holds for ductility—it increases with temperature. Figure 6.13 shows how the stress–strain behavior of iron varies with temperature.

Resilience

Resilience is the capacity of a material to absorb energy when it is deformed elastically and then, upon unloading, to have this energy recovered. The associated property is

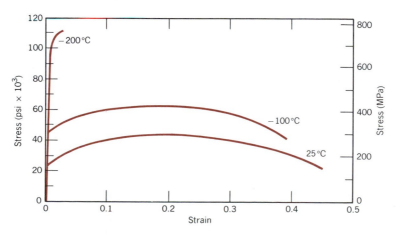

Figure 6.13 Engineering stress–strain behavior for iron at three temperatures.

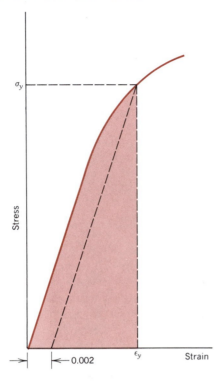

Figure 6.14 Schematic representation showing how modulus of resilience (corresponding to the shaded area) is determined from the tensile stress–strain behavior of a material.

the *modulus of resilience, U_r,* which is the strain energy per unit volume required to stress a material from an unloaded state up to the point of yielding.

Computationally, the modulus of resilience for a specimen subjected to a uniaxial tension test is just the area under the engineering stress–strain curve taken to yielding (Figure 6.14), or

$$U_r = \int_0^{\epsilon_y} \sigma d\epsilon \tag{6.12a}$$

Assuming a linear elastic region,

$$U_r = \tfrac{1}{2}\sigma_y\epsilon_y \tag{6.12b}$$

in which ϵ_y is the strain at yielding.

The units of resilience are the product of the units from each of the two axes of the stress–strain plot. For Customary U.S. units this is inch-pounds force per cubic inch (in.-lb$_f$/in.^{3}, equivalent to psi), whereas with SI units it is joules per cubic meter (J/m^3, equivalent to Pa). Both inch-pounds force and joules are units of energy, and thus this area under the stress–strain curve represents energy absorption per unit volume (in cubic inches or cubic meters) of material.

Incorporation of Equation 6.4 into Equation 6.12b yields

$$U_r = \tfrac{1}{2}\sigma_y\epsilon_y = \tfrac{1}{2}\sigma_y\left(\frac{\sigma_y}{E}\right) = \frac{\sigma_y^2}{2E} \tag{6.13}$$

Thus resilient materials are those having high yield strengths and low moduli of elasticity; such alloys would be used in spring applications.

Toughness

Toughness is a mechanical term that is used in several contexts; loosely speaking, it is a measure of the ability of a material to absorb energy up to fracture. Specimen geometry as well as the manner of load application are important in toughness determinations. For dynamic (high strain rate) loading conditions and when a notch (or point of stress concentration) is present, *notch toughness* is assessed by using an impact test, as discussed in Section 8.6. Furthermore, fracture toughness is a property indicative of a material's resistance to fracture when a crack is present (Section 8.5).

For the static (low strain rate) situation, toughness may be ascertained from the results of a tensile stress–strain test. It is the area under the σ–ϵ curve up to the point of fracture. The units for toughness are the same as for resilience (i.e., energy per unit volume of material). For a material to be tough, it must display both strength and ductility; and often, ductile materials are tougher than brittle ones. This is demonstrated in Figure 6.12, in which the stress–strain curves are plotted for both material types. Hence, even though the brittle material has higher yield and tensile strengths, by virtue of lack of ductility, it has a lower toughness than the ductile one; this is deduced by comparing the areas ABC and $AB'C'$ in Figure 6.12.

6.7 TRUE STRESS AND STRAIN

From Figure 6.10, the decline in the stress necessary to continue deformation past the maximum, point M, seems to indicate that the material is becoming weaker. This is not at all the case; as a matter of fact, it is increasing in strength. However, the cross-sectional area is decreasing rapidly within the neck region, where deformation is occurring. This results in a reduction in the load-bearing capacity of the specimen. The stress, as computed from Equation 6.1, is on the basis of the original cross-sectional area before any deformation, and does not take into account this diminution in area at the neck.

Sometimes it is more meaningful to use a true stress–true strain scheme. **True stress** σ_T is defined as the load F divided by the instantaneous cross-sectional area A_i over which deformation is occurring (i.e., the neck, past the tensile point), or

$$\sigma_T = \frac{F}{A_i} \tag{6.14}$$

Furthermore, it is occasionally more convenient to represent strain as **true strain** ϵ_T, defined by

$$\epsilon_T = \ln \frac{l_i}{l_0} \tag{6.15}$$

If no volume change occurs during deformation, that is, if

$$A_i l_i = A_0 l_0 \tag{6.16}$$

true and engineering stress and strain are related according to

$$\sigma_T = \sigma(1 + \epsilon) \tag{6.17a}$$

$$\epsilon_T = \ln(1 + \epsilon) \tag{6.17b}$$

Equations 6.17a and 6.17b are valid only to the onset of necking; beyond this point true stress and strain should be computed from actual load, cross-sectional area, and gauge length measurements.

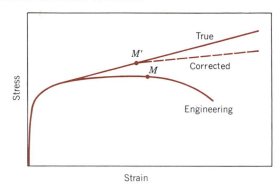

Figure 6.15 A comparison of typical tensile engineering stress–strain and true stress–strain behaviors. Necking begins at point M on the engineering curve, which corresponds to M' on the true curve. The "corrected" true stress–strain curve takes into account the complex stress state within the neck region.

A schematic comparison of engineering and true stress–strain behavior is made in Figure 6.15. It is worth noting that the true stress necessary to sustain increasing strain continues to rise past the tensile point M'.

Coincident with the formation of a neck is the introduction of a complex stress state within the neck region (i.e., the existence of other stress components in addition to the axial stress). As a consequence, the correct stress (*axial*) within the neck is slightly lower than the stress computed from the applied load and neck cross-sectional area. This leads to the "corrected" curve in Figure 6.15.

For some metals and alloys the region of the true stress–strain curve from the onset of plastic deformation to the point at which necking begins may be approximated by

$$\sigma_T = K\epsilon_T^n \qquad (6.18)$$

In this expression K and n are constants, which values will vary from alloy to alloy,

TABLE 6.3 Tabulation of n and K Values (Equation 6.18) for Several Alloys

Material	n	K psi	K MPa
Low-carbon steel (annealed)	0.26	77,000	530
Alloy steel (Type 4340, annealed)	0.15	93,000	640
Stainless steel (Type 304, annealed)	0.45	185,000	1275
Aluminum (annealed)	0.20	26,000	180
Aluminum alloy (Type 2024, heat treated)	0.16	100,000	690
Copper (annealed)	0.54	46,000	315
Brass (70 Cu–30 Zn, annealed)	0.49	130,000	895

Source: S. Kalpakuian, *Manufacturing Processes for Engineering Materials.* Copyright © 1984 Addison-Wesley Publishing Co. Reprinted by permission of Addison-Wesley Publishing Co., Inc. Reading, MA.

and will also depend on the condition of the material (i.e., whether it has been plastically deformed, heat treated, etc.). The parameter n is often termed the *strain-hardening exponent* and has a value less than unity. Values of n and K for several alloys are contained in Table 6.3.

EXAMPLE PROBLEM 6.4

A cylindrical specimen of steel having an original diameter of 0.505 in. (12.8 mm) is tensile tested to fracture and found to have an engineering fracture strength σ_f of 67,000 psi (460 MPa). If its cross-sectional diameter at fracture is 0.422 in. (10.7 mm), determine:

(a) The ductility in terms of percent area reduction.

(b) The true stress at fracture.

SOLUTION

(a) Ductility is computed using Equation 6.11, as

$$\%AR = \frac{(0.505 \text{ in.}/2)^2\pi - (0.422 \text{ in.}/2)^2\pi}{(0.505 \text{ in.}/2)^2\pi} \times 100$$

$$= \frac{0.20 \text{ in.}^2 - 0.14 \text{ in.}^2}{0.20 \text{ in.}^2} \times 100 = 30\%$$

(b) True stress is defined by Equation 6.14, where in this case the area is taken as the fracture area A_f. However, the load at fracture must first be computed from the fracture strength as

$$F = \sigma_f A_0 = (67,000 \text{ psi})(0.20 \text{ in.}^2) = 13,400 \text{ lb}_f$$

Thus the true stress is calculated as

$$\sigma_T = \frac{F}{A_f} = \frac{13,400 \text{ lb}_f}{0.14 \text{ in.}^2} = 95,700 \text{ psi (660 MPa)}$$

EXAMPLE PROBLEM 6.5

Compute the strain-hardening exponent n in Equation 6.18 for an alloy in which a true stress of 60,000 psi (415 MPa) produces a true strain of 0.10; assume a value of 150,000 psi (1035 MPa) for K.

SOLUTION
This requires some algebraic manipulation of Equation 6.18 so that n becomes the independent parameter. This is accomplished by taking logarithms and rearranging. Solving for n yields

$$n = \frac{\log \sigma_T - \log K}{\log \epsilon_T}$$

$$= \frac{\log(60,000 \text{ psi}) - \log(150,000 \text{ psi})}{\log(0.1)} = 0.40$$

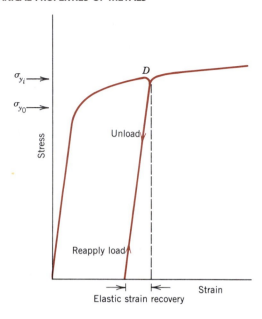

Figure 6.16 Schematic tensile stress–strain diagram showing the phenomena of elastic strain recovery and strain hardening. The initial yield strength is designated as σ_{y_0}; σ_{y_i} is the yield strength after releasing the load at point D, and then upon reloading.

6.8 ELASTIC RECOVERY DURING PLASTIC DEFORMATION

Upon release of the load during the course of a stress–strain test, some fraction of the total deformation is recovered as elastic strain. This behavior is demonstrated in Figure 6.16, a schematic engineering stress–strain plot. During the unloading cycle, the curve traces a near straight-line path from the point of unloading (point D), and its slope is virtually identical to the modulus of elasticity, or parallel to the initial elastic portion of the curve. The magnitude of this elastic strain, which is regained during unloading, corresponds to the strain recovery, as shown in Figure 6.16. If the load is reapplied, the curve will traverse essentially the same linear portion in the direction opposite to unloading; yielding will again occur at the unloading stress level where the unloading began. There will also be an elastic strain recovery associated with fracture.

6.9 COMPRESSIVE, SHEAR, AND TORSIONAL DEFORMATION

Of course, metals may experience plastic deformation under the influence of applied compressive, shear, and torsional loads. The resulting stress–strain behavior into the plastic region will be similar to the tensile counterpart (Figure 6.9a: yielding and the associated curvature). However, for compression, there will be no maximum, since necking does not occur; furthermore, the mode of fracture will be different from that for tension.

6.10 HARDNESS

Another mechanical property that may be important to consider is **hardness,** which is a measure of a material's resistance to localized plastic deformation (e.g., a small dent or a scratch). Early hardness tests were based on natural minerals with a scale constructed solely on the ability of one material to scratch another that was softer. A qualitative and somewhat arbitrary hardness indexing scheme was devised, termed the Mohs scale, which ranged from 1 on the soft end for talc to 10 for diamond. Quantitative hardness techniques have been developed over the years in which a small indenter is forced into the surface of a material to be tested, under controlled conditions of load and rate of application. The depth or size of the resulting indentation is measured, which in turn is related to a hardness number; the softer the material, the larger and deeper the indentation, and the lower the hardness index number. Measured hardnesses are only relative (rather than absolute), and care should be exercised when comparing values determined by different techniques.

Rockwell Hardness Tests

The Rockwell tests constitute the most common method used to measure hardness because they are so simple to perform and require no special skills. Several different scales may be utilized from possible combinations of various indenters and different loads, which permit the testing of virtually all metals and alloys, from the hardest to the softest. Indenters include spherical and hardened steel balls having diameters of $\frac{1}{16}$, $\frac{1}{8}$, $\frac{1}{4}$, and $\frac{1}{2}$ in. (1.588, 3.175, 6.350, and 12.70 mm), and a conical diamond (Brale) indenter, which is used for the hardest materials.

With this system, a hardness number is determined by the difference in depth of penetration resulting from the application of an initial minor load followed by a larger major load; utilization of a minor load enhances test accuracy. On the basis of the magnitude of both major and minor loads, there are two types of tests: Rockwell and superficial Rockwell. For Rockwell, the minor load is 10 kg, whereas major loads are 60, 100, and 150 kg. Each scale is represented by a letter of the alphabet; several are listed with the corresponding indenter and load in Tables 6.4 and 6.5a. For superficial tests, 3 kg is the minor load; 15, 30, and 45 kg are the possible major load values. These scales are identified by a 15, 30, or 45 (according to load), followed by N, T, W, X, or Y, depending on indenter. Superficial tests are frequently performed on thin specimens. Table 6.5b presents several superficial scales.

When specifying Rockwell and superficial hardnesses, both hardness number and scale symbol must be indicated. The scale is designated by the symbol HR followed by the appropriate scale identification.[5] For example, 80 HRB represents a Rockwell hardness of 80 on the B scale, and 60 HR30W indicates a superficial hardness of 60 on the 30W scale.

For each scale, hardnesses may range up to 130; however, as hardness values rise above 100 or drop below 20 on any scale, they become inaccurate; and because the scales have some overlap, in such a situation it is best to utilize the next harder or softer scale.

Inaccuracies also result if the test specimen is too thin, if an indentation is made too near a specimen edge, or if two indentations are made too close to one another. Specimen thickness should be at least ten times the indentation depth, whereas allowance should be made for at least three indentation diameters between the center

---

[5] Rockwell scales are also frequently designated by an R with the appropriate scale letter as a subscript, for example, R_C denotes the Rockwell C scale.

TABLE 6.4 Hardness Testing Techniques

Test	Indenter	Shape of Indentation		Load	Formula for Hardness Number[a]
		Side View	Top View		
Brinell	10-mm sphere of steel or tungsten carbide	D d	d	P	$HB = \dfrac{2P}{\pi D[D - \sqrt{D^2 - d^2}]}$
Vickers microhardness	Diamond pyramid	$136°$		P	$HV = 1.854P/d_1^2$
Knoop microhardness	Diamond pyramid	t $l/b = 7.11$ $b/t = 4.00$	b l	P	$HK = 14.2P/l^2$
Rockwell and Superficial Rockwell	Diamond cone $\tfrac{1}{16}, \tfrac{1}{8}, \tfrac{1}{4}, \tfrac{1}{2}$ in. diameter steel spheres	$120°$		$\left.\begin{array}{l} 60\text{ kg} \\ 100\text{ kg} \\ 150\text{ kg} \end{array}\right\}$ Rockwell $\left.\begin{array}{l} 15\text{ kg} \\ 30\text{ kg} \\ 45\text{ kg} \end{array}\right\}$ Superficial Rockwell	

[a] For the hardness formulas given, P (the applied load) is in kg, while D, d, d_1, and l are all in mm.

Source: Adapted from H. W. Hayden, W. G. Moffatt, and J. Wulf, *The Structure and Properties of Materials*, Vol. III, *Mechanical Behavior*. Copyright © 1965 by John Wiley & Sons, New York. Reprinted by permission of John Wiley & Sons, Inc.

TABLE 6.5a Rockwell Hardness Scales

Scale Symbol	Indenter	Major Load (kg)
A	Diamond	60
B	$\frac{1}{16}$ in. ball	100
C	Diamond	150
D	Diamond	100
E	$\frac{1}{8}$ in. ball	100
F	$\frac{1}{16}$ in. ball	60
G	$\frac{1}{16}$ in. ball	150
H	$\frac{1}{8}$ in. ball	60
K	$\frac{1}{8}$ in. ball	150

TABLE 6.5b Superficial Rockwell Hardness Scales

Scale Symbol	Indenter	Major Load (kg)
15N	Diamond	15
30N	Diamond	30
45N	Diamond	45
15T	$\frac{1}{16}$ in. ball	15
30T	$\frac{1}{16}$ in. ball	30
45T	$\frac{1}{16}$ in. ball	45
15W	$\frac{1}{8}$ in. ball	15
30W	$\frac{1}{8}$ in. ball	30
45W	$\frac{1}{8}$ in. ball	45

of one indentation and the specimen edge, or to the center of a second indentation. Furthermore, testing of specimens stacked one on top of another is not recommended. Also, accuracy is dependent on the indentation being made into a smooth flat surface.

The modern apparatus for making Rockwell hardness measurements is automated and very simple to use; hardness is read directly, and each measurement requires only a few seconds.

The modern testing apparatus also permits a variation in the time of load application. This variable must also be considered in interpreting hardness data.

Brinell Hardness Tests

In Brinell tests, as in Rockwell measurements, a hard, spherical indenter is forced into the surface of the metal to be tested. The diameter of the hardened steel (or tungsten carbide) indenter is 10.00 mm (0.394 in.). Standard loads range between 500 and 3000 kg in 500-kg increments; during a test, the load is maintained constant for a specified time (between 10 and 30 s). Harder materials require greater applied loads. The Brinell hardness number, HB, is a function of both the magnitude of the load and the diameter of the resulting indentation (see Table 6.4).[6] This diameter is measured with a special low-power microscope, utilizing a scale that is etched on the eyepiece. The measured diameter is then converted to the appropriate HB number using a chart; only one scale is employed with this technique.

Maximum specimen thickness as well as indentation position (relative to specimen edges) and minimum indentation spacing requirements are the same as for Rockwell tests. In addition, a well-defined indentation is required; this necessitates a smooth flat surface in which the indentation is made.

Knoop and Vickers Microhardness Tests

Two other hardness testing techniques are Knoop (pronounced $n\bar{u}p$) and Vickers (sometimes also called diamond pyramid). For each test a very small diamond indenter having pyramidal geometry is forced into the surface of the specimen. Applied loads are much smaller than for Rockwell and Brinell, ranging between 1 and 1000 g. The resulting impression is observed under a microscope and measured; this measurement is then converted into a hardness number (Table 6.4). Careful specimen surface preparation (grinding and polishing) may be necessary to ensure a well-defined indentation which may be accurately measured. The Knoop and Vickers hardness numbers are designated by HK and HV, respectively,[7] and hardness scales for both techniques are approximately equivalent. Knoop and Vickers are referred to as microhardness testing methods on the basis of load and indenter size. Both are well suited for measuring the hardness of small, selected specimen regions; furthermore, Knoop is used for testing brittle materials such as ceramics.

Hardness Conversion

The facility to convert the hardness measured on one scale to that of another is most desirable. However, since hardness is not a well-defined material property, and because of the experimental dissimilarities among the various techniques, a comprehensive conversion scheme has not been devised. Hardness conversion data have been determined experimentally and found to be dependent on material type and characteristics. The most reliable conversion data exist for steels, and such data are presented in Figure 6.17 for Knoop, Brinell, and two Rockwell scales; the Mohs scale is also included. In light of the above discussion, care should be exercised in extrapolation of these data to other alloy systems.

---

[6] The Brinell hardness number is also represented by BHN.

[7] Sometimes KHN and VHN are used to denote Knoop and Vickers hardness numbers, respectively.

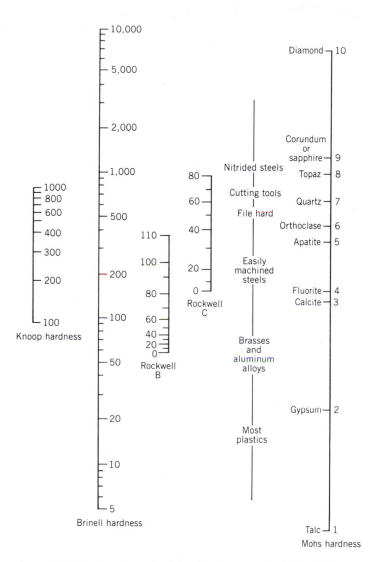

Figure 6.17 Comparison of several hardness scales. (Adapted from G. F. Kinney, *Engineering Properties and Applications of Plastics,* p. 202. Copyright © 1957 by John Wiley & Sons, New York. Reprinted by permission of John Wiley & Sons, Inc.)

Correlation Between Hardness and Tensile Strength

Both tensile strength and hardness are indicators of a metal's resistance to plastic deformation. Consequently, they are roughly proportional, as shown in Figure 6.18, for tensile strength as a function of the HB for cast iron, steel, and brass. The same proportionality relationship does not hold for all metals, as Figure 6.18 indicates. As a rule of thumb for most steels, the HB and the tensile strength are related according

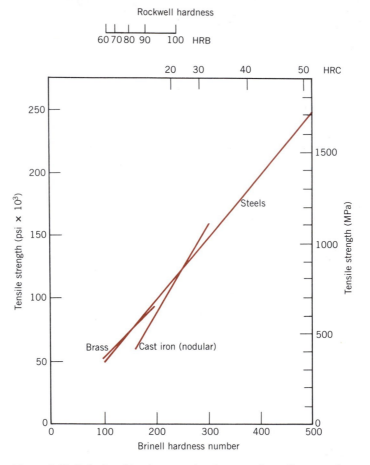

Figure 6.18 Relationships between hardness and tensile strength for steel, brass, and cast iron. (Data taken from *Metals Handbook: Properties and Selection, Irons and Steels,* Vol. 1, 9th edition, B. Bardes, Editor, American Society for Metals, 1978, pp. 36 and 461; and *Metals Handbook: Properties and Selection: Nonferrous Alloys and Pure Metals,* Vol. 2, 9th edition, H. Baker, Managing Editor, American Society for Metals, 1979, p. 327.)

to

$$TS \text{ (psi)} = 500 \times \text{HB} \tag{6.19a}$$

$$TS \text{ (MPa)} = 3.45 \times \text{HB} \tag{6.19b}$$

Hardness tests are performed more frequently than any other mechanical test for several reasons:

1. They are simple and inexpensive—ordinarily no special specimen need be prepared.

2. The test is nondestructive—the specimen is neither fractured nor excessively deformed; a small indentation is the only deformation.

3. Other mechanical properties often may be estimated from hardness data, such as tensile strength (Figure 6.18).

6.11 VARIABILITY OF MATERIAL PROPERTIES

At this point it is worthwhile to discuss an issue that sometimes proves troublesome to many engineering students, namely, that material properties are not exact quantities. That is, even if we have a most precise measuring apparatus and a highly controlled test procedure, there will always be some scatter or variability in the data that are collected from specimens of the same material. For example, consider a number of identical tensile samples that are prepared from a single bar of some metal alloy, which samples are subsequently stress–strain tested in the same apparatus. We would most likely observe that each resulting stress–strain plot is slightly different than the others. This would lead to a variety of modulus of elasticity, yield strength, and tensile strength values. A number of factors lead to uncertainties in measured data. These include the test method, variations in specimen fabrication procedures, operator bias, and apparatus calibration. Furthermore, inhomogeneities may exist within the same lot of material, and/or slight compositional and other differences from lot to lot. Of course, appropriate measures should be taken to minimize the possibility of measurement error, and also to mitigate those factors which lead to data variability.

It should also be mentioned that this scatter exists for other material properties such as density, electrical conductivity, and coefficient of thermal expansion.

It is important for the design engineer to realize that this scatter and variability of materials properties are inevitable and must be dealt with appropriately. On occasion, data must be subjected to statistical treatments and probabilities determined. For example, instead of asking the question, "What is the fracture strength of this alloy?" the engineer should become accustomed to asking the question, "What is the probability of failure of this alloy under these given circumstances?"

In spite of the variation of some measured property, specification of a "typical" value is still desirable. Most commonly, the typical value is described by taking an average of the data. This is obtained by dividing the sum of all measured values by the number of measurements taken. In mathematical terms, the average $\bar{x}$ of some parameter x is

$$\bar{x} = \frac{\sum_{i=1}^{n} x_i}{n} \tag{6.20}$$

where n is the number of observations or measurements and x_i is the value of a discrete measurement.

Sometimes it is also desirable to quantify the degree of dispersion, or scatter, of the measured data. The most common measure of this variability is the standard deviation s, which is determined using the following expression:

$$s = \left[\frac{\sum_{i=1}^{n} (x_i - \bar{x})^2}{n - 1} \right]^{1/2} \tag{6.21}$$

where x_i, $\bar{x}$, and n are defined above. A large value of the standard deviation corresponds to a high degree of scatter.

EXAMPLE PROBLEM 6.6

The following tensile strengths were measured for four specimens of the same steel alloy:

Sample Number	Tensile Strength (MPa)
1	520
2	512
3	515
4	522

(a) Compute the average tensile strength.

(b) Determine the standard deviation.

SOLUTION

(a) The average tensile strength $(\overline{TS})$ is computed using Equation 6.20 with $n = 4$:

$$\overline{TS} = \frac{\sum\limits_{i=1}^{4} (TS)_i}{4}$$

$$= \frac{520 + 512 + 515 + 522}{4}$$

$$= 517 \text{ MPa}$$

(b) And for the standard deviation, using Equation 6.21,

$$s = \left[\frac{\sum\limits_{i=1}^{4} \left((TS)_i - \overline{TS} \right)^2}{4 - 1} \right]^{1/2}$$

$$= \left(\frac{(520 - 517)^2 + (512 - 517)^2 + (515 - 517)^2 + (522 - 517)^2}{4 - 1} \right)^{1/2}$$

$$= 4.6 \text{ MPa}$$

Figure 6.19 presents the tensile strength by specimen number for this example problem, and also how the data may be represented in graphical form. The tensile strength data point (Figure 6.19b) corresponds to the average value $\overline{TS}$, whereas scatter is depicted by error bars (short horizontal lines) situated above and below the data point symbol and connected to this symbol by vertical lines. The upper error bar is positioned at a value of the average value plus the standard deviation $(\overline{TS} + s)$, whereas the lower error bar corresponds to the average minus the standard deviation $(\overline{TS} - s)$.

6.12 SAFETY FACTORS

In the previous section it was noted that virtually all engineering materials exhibit a variability in their mechanical properties. Furthermore, uncertainties will also exist

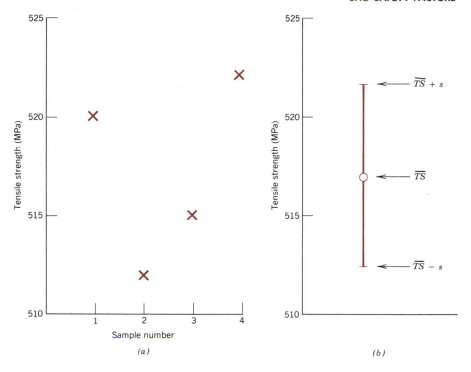

Figure 6.19 (*a*) Tensile strength data associated with Example Problem 6.6. (*b*) The manner in which these data could be plotted. The data point corresponds to the average value of the tensile strength ($\overline{TS}$); error bars which indicate the degree of scatter correspond to the average value plus and minus the standard deviation ($\overline{TS} \pm s$).

in the magnitude of applied loads for in-service applications; ordinarily, stress calculations are only approximate. Therefore, design allowances must be made to protect against unanticipated failure. This is accomplished by establishing, for the particular material used, a **safe stress** or *working stress*, usually denoted as σ_w. For static situations and ductile metals, σ_w is taken as the yield strength divided by a *factor of safety, N*, or

$$\sigma_w = \frac{\sigma_y}{N} \tag{6.22}$$

With brittle ceramic materials, there is almost always a high degree of scatter in strength values; therefore, more sophisticated statistical analyses are required.

Of course, the choice of an appropriate value of N is necessary. If N is too large, then component overdesign will result, that is, either too much material or an alloy having a higher-than-necessary strength will be used. Values normally range between 1.2 and 4.0; a good average is 2.0. Selection of N will depend on a number of factors, including economics, previous experience, the accuracy with which mechanical forces and material properties may be determined, and, most important, the consequences of failure in terms of loss of life and/or property damage.

SUMMARY

A number of the important mechanical properties of materials, predominantly metals, have been discussed in this chapter. Concepts of stress and strain were first introduced. Stress is a measure of an applied mechanical load or force, normalized to take into account cross-sectional area. Two different stress parameters were defined—engineering stress and true stress. Strain represents the amount of deformation induced by a stress; both engineering and true strains are used.

Some of the mechanical characteristics of metals can be ascertained by simple stress–strain tests. There are four test types: tension, compression, torsion, and shear. Tensile are the most common. A material that is stressed first undergoes elastic, or nonpermanent, deformation, wherein stress and strain are proportional. The constant of proportionality is the modulus of elasticity for tension and compression, and is the shear modulus when the stress is shear.

The phenomenon of yielding occurs at the onset of plastic or permanent deformation; yield strength is determined by a strain offset method from the stress–strain behavior, which is indicative of the stress at which plastic deformation begins. Tensile strength corresponds to the maximum tensile stress that may be sustained by a specimen, whereas percents elongation and area reduction are measures of ductility— the amount of plastic deformation that has occurred at fracture. Resilience is the capacity of a material to absorb energy during elastic deformation; modulus of resilience is the area beneath the engineering stress–strain curve up to the yield point. Also, static toughness represents the energy absorbed during the fracture of a material, and is taken as the area under the entire engineering stress–strain curve. Ductile materials are normally tougher than brittle ones.

Hardness is a measure of the resistance to localized plastic deformation. In several popular hardness-testing techniques (Rockwell, Brinell, Knoop, and Vickers) a small indenter is forced into the surface of the material, and an index number is determined on the basis of the size or depth of the resulting indentation. For many metals, hardness and tensile strength are approximately proportional to each other.

Mechanical properties (as well as other material properties) are not exact and precise quantities, in that there will always be some scatter for the measured data. Typical material property values are commonly specified in terms of averages, whereas magnitudes of scatter may be expressed as standard deviations.

As a result of uncertainties in both measured mechanical properties and in-service applied stresses, safe or working stresses are normally utilized for design purposes. For ductile materials, working stress is the ratio of the yield strength and the factor of safety.

IMPORTANT TERMS AND CONCEPTS

Anelasticity	Engineering strain	Plastic deformation
Ductility	Engineering stress	Poisson's ratio
Elastic deformation	Hardness	Proportional limit
Elastic recovery	Modulus of elasticity	Resilience

Safe stress	Toughness	Yielding
Shear	True strain	Yield strength
Tensile strength	True stress	

REFERENCES

DIETER, G. E., *Mechanical Metallurgy*, 3rd edition, McGraw-Hill Book Co., New York, 1986.

HAYDEN, H. W., W. G. MOFFATT, and J. WULFF, *The Structure and Properties of Materials*, Vol. III, *Mechanical Behavior*, John Wiley & Sons, New York, 1965.

McCLINTOCK, F. A. and A. S. ARGON, *Mechanical Behavior of Materials*, Addison-Wesley Publishing Co., Reading, MA, 1966.

MARIN, J., *Mechanical Behavior of Engineering Materials*, Prentice-Hall, Englewood Cliffs, NJ, 1962.

Metals Handbook, 9th edition, Vol. 8, *Mechanical Testing*, American Society for Metals, Metals Park, OH, 1985.

TEGART, W. J. M., *Elements of Mechanical Metallurgy*, The Macmillan Company, New York, 1966.

QUESTIONS AND PROBLEMS

6.1 A specimen of aluminum having a rectangular cross section 0.4 in. × 0.5 in. (10 mm × 12.7 mm) is pulled in tension with 8000 lb_f (35,500 N) force, producing only elastic deformation. Calculate the resulting strain.

6.2 A cylindrical specimen of a titanium alloy having a modulus of elasticity of 15.5×10^6 psi (10.7×10^4 MPa) and an original diameter of 0.15 in. (3.8 mm) will experience only elastic deformation when a tensile load of 450 lb_f (2000 N) is applied. Compute the maximum length of the specimen before deformation if the maximum allowable elongation is 0.0165 in. (0.42 mm).

6.3 A steel bar 4 in. (100 mm) long and having a square cross section 0.8 in. (20 mm) on an edge is pulled in tension with a load of 20,000 lb_f (8.9×10^4 N) and experiences an elongation of 4×10^{-3} in. (0.1 mm). Assuming that the deformation is entirely elastic, calculate the modulus of elasticity of the steel.

6.4 Consider a cylindrical nickel wire 2.0 mm (0.08 in.) in diameter and 3×10^4 mm (1200 in.) long. Calculate its elongation when a load of 300 N (67 lb_f) is applied. Assume that the deformation is totally elastic.

6.5 For a brass alloy, the stress at which plastic deformation begins is 50,000 psi (345 MPa), and the modulus of elasticity is 15×10^6 psi (10.3×10^4 MPa).
   (a) What is the maximum load that may be applied to a specimen with a cross-sectional area of 0.2 in.^{2} (130 mm^2) without plastic deformation?
   (b) If the original specimen length is 3 in. (76 mm), what is the maximum length to which it may be stretched without causing plastic deformation?

6.6 A cylindrical rod of copper ($E = 16 \times 10^6$ psi, 11×10^4 MPa) having a yield strength of 35,000 psi (240 MPa) is to be subjected to a load of 1500 lb_f (6660 N). If the length of the rod is 15 in. (380 mm), what must be the diameter to allow an elongation of 0.020 in. (0.50 mm)?

6.7   Consider a cylindrical specimen of steel (Figure 6.20) 8.0 mm (0.31 in.) in diameter and 50 mm (2.0 in.) long which is pulled in tension. Determine its elongation when a load of 15,000 N (3370 lb$_f$) is applied.

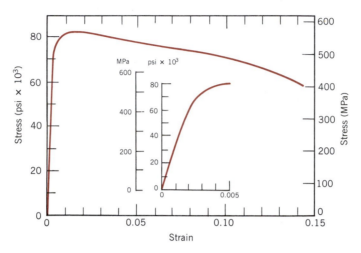

Figure 6.20 Tensile stress–strain behavior for a plain carbon steel.

6.8   Figure 6.21 shows, for a gray cast iron, the tensile engineering stress–strain curve in the elastic region. Determine **(a)** the tangent modulus at 1500 psi (10.3 MPa), and **(b)** the secant modulus taken to 1000 psi (6.9 MPa).

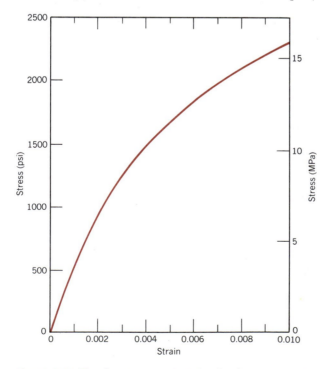

Figure 6.21 Tensile stress–strain behavior for a gray cast iron.

6.9 As was noted in Section 3.14, for single crystals of some substances, the physical properties are anisotropic, that is, they are dependent on crystallographic direction. One such property is the modulus of elasticity. For cubic single crystals, the modulus of elasticity in a general $[uvw]$ direction, E_{uvw}, is described by the relationship

$$\frac{1}{E_{uvw}} = \frac{1}{E_{\langle 100 \rangle}} - 3\left(\frac{1}{E_{\langle 100 \rangle}} - \frac{1}{E_{\langle 111 \rangle}}\right)(\alpha^2\beta^2 + \beta^2\gamma^2 + \gamma^2\alpha^2)$$

where $E_{\langle 100 \rangle}$ and $E_{\langle 111 \rangle}$ are the moduli of elasticity in $[100]$ and $[111]$ directions, respectively; α, β, and γ are the cosines of the angles between $[uvw]$ and the respective $[100]$, $[010]$, and $[001]$ directions. Verify that the $E_{\langle 110 \rangle}$ values for aluminum, copper, and iron in Table 3.3 are correct.

6.10 In Section 2.6 it was noted that the net bonding energy E_N between two isolated positive and negative ions is a function of interionic distance r as follows:

$$E_N = -\frac{A}{r} + \frac{B}{r^n} \qquad (6.23)$$

where A, B, and n are constants for the particular ion pair. The modulus of elasticity E is proportional to the slope of the interionic force–separation curve at the equilibrium interionic separation; that is,

$$E \propto \left(\frac{dF}{dr}\right)_{r_0}$$

Derive an expression for the dependence of the modulus of elasticity on these A, B, and n parameters (for the two-ion system) using the following procedure:

1. Establish a relationship for the force F as a function of r, realizing that

$$F = \frac{dE_N}{dr}$$

2. Now take the derivative dF/dr.
3. Develop an expression for r_0, the equilibrium separation. Since r_0 corresponds to the value of r at the minimum of the E_N-versus-r curve (Figure 2.8b), take the derivative dE_N/dr, set it equal to zero, and solve for r, which corresponds to r_0.
4. Finally, substitute this expression for r_0 into the relationship obtained by taking dF/dr.

6.11 Using the solution to Problem 6.10, rank the magnitudes of the moduli of elasticity for the following hypothetical X, Y, and Z materials from the greatest to the least. The appropriate A, B, and n parameters (Equation 6.23) for these three materials are tabulated below; they yield E_N in units of electron volts and r in nanometers:

Material	A	B	n
X	1.5	7.0×10^{-6}	8
Y	2.0	1.0×10^{-5}	9
Z	3.5	4.0×10^{-6}	7

6.12 A cylindrical specimen of aluminum having a diameter of 0.75 in. (19 mm) and a length of 8 in. (200 mm) is deformed elastically in tension with a force of 11,000 lb$_f$ (48,800 N). Using the data contained in Table 6.1, determine the following:

(a) The amount by which this specimen will elongate in the direction of the applied stress.

(b) The change in diameter of the specimen. Will the diameter increase or decrease?

6.13 A cylindrical bar of steel 0.4 in. (10 mm) in diameter is to be deformed elastically by application of a force along the bar axis. Using the data in Table 6.1, determine the force that will produce an elastic reduction of 1.2×10^{-4} in. (3×10^{-3} mm) in the diameter.

6.14 A cylindrical specimen of some metal alloy 0.4 in. (10.0 mm) in diameter is stressed elastically in tension. A force of 3370 lb$_f$ (15,000 N) produces a reduction in specimen diameter of 2.8×10^{-4} in. (7×10^{-3} mm). Compute Poisson's ratio for this material if its modulus of elasticity is 14.5×10^6 psi (10^5 MPa).

6.15 A cylindrical specimen of a hypothetical metal alloy is stressed in compression. If its original and final diameters are 30.00 and 30.04 mm, respectively, and its final length is 105.20 mm, compute its original length if the deformation is totally elastic. The elastic and shear moduli for this alloy are 65.5×10^3 and 25.4×10^3 MPa, respectively.

6.16 Consider a cylindrical specimen of some hypothetical metal alloy that has a diameter of 10.0 mm (0.39 in.). A tensile force of 1500 N (340 lb$_f$) produces an elastic reduction in diameter of 6.7×10^{-4} mm (2.64×10^{-5} in.). Compute the modulus of elasticity for this alloy, given that Poisson's ratio is 0.35.

6.17 A brass alloy is known to have a yield strength of 40,000 psi (275 MPa), a tensile strength of 55,000 psi (380 MPa), and a modulus of elasticity of 15×10^6 psi (10.3×10^4 MPa). A cylindrical specimen of this alloy 0.50 in. (12.7 mm) in diameter and 10 in. (250 mm) long is stressed in tension and found to elongate 0.30 in. (7.6 mm). On the basis of the information given, is it possible to compute the magnitude of the load that is necessary to produce this change in length? If so, calculate the load. If not, explain why.

6.18 A cylindrical metal specimen 0.5 in. (12.7 mm) in diameter and 10 in. (254 mm) long is to be subjected to a tensile stress of 4000 psi (28 MPa); at this stress level the resulting deformation will be totally elastic.

(a) If the elongation must be less than 3.2×10^{-3} in. (0.080 mm), which of the metals in Table 6.1 are suitable candidates? Why?

(b) If, in addition, the maximum permissible diameter decrease is 4.7×10^{-5} in. (1.2×10^{-3} mm), which of the metals in Table 6.1 may be used? Why?

6.19 Consider the brass alloy with stress–strain behavior shown in Figure 6.11. A cylindrical specimen of this material 0.39 in. (10.0 mm) in diameter and 4.0 in. (101.6 mm) long is pulled in tension with a force of 2250 lb$_f$ (10,000 N). If it is known that this alloy has a Poisson's ratio of 0.35, compute: **(a)** the specimen elongation, and **(b)** the reduction in specimen diameter.

6.20 Cite the primary differences between elastic, anelastic, and plastic deformation behaviors.

6.21 A cylindrical rod 120 mm long and having a diameter of 15.0 mm is to be deformed using a tensile load of 35,000 N. It must not experience either plastic deformation or a diameter reduction of more than 1.2×10^{-2} mm. Of the materials listed below, which are possible candidates? Justify your choice(s).

Material	Modulus of Elasticity ($MPa \times 10^3$)	Yield Strength (MPa)	Poisson's Ratio
Aluminum alloy	70	250	0.33
Titanium alloy	105	850	0.36
Steel alloy	205	550	0.27
Magnesium alloy	45	170	0.29

6.22 A cylindrical rod 15 in. (380 mm) long, having a diameter of 0.4 in. (10 mm), is to be subjected to a tensile load. If the rod is to experience neither plastic deformation nor an elongation of more than 0.035 in. (0.9 mm) when the applied load is 5500 lb$_f$ (24,500 N), which of the four metals or alloys listed below are possible candidates? Justify your choice(s).

Material	Modulus of Elasticity (psi)	Yield Strength (psi)	Tensile Strength (psi)
Aluminum alloy	10×10^6	37,000	61,000
Brass alloy	14.6×10^6	50,000	61,000
Copper	16×10^6	36,000	42,000
Steel	30×10^6	65,000	80,000

6.23 Figure 6.20 shows the tensile engineering stress–strain behavior for a steel alloy.
(a) What is the modulus of elasticity?
(b) What is the proportional limit?
(c) What is the yield strength at a strain offset of 0.002?
(d) What is the tensile strength?

6.24 A cylindrical specimen of a brass alloy having a length of 4 in. (100 mm) must elongate only 0.2 in. (5 mm) when a tensile load of 22,500 lb$_f$ (100,000 N) is applied. Under these circumstances, what must be the radius of the specimen? Consider this brass alloy to have the stress–strain behavior shown in Figure 6.11.

6.25 A load of 10,000 lb$_f$ (44,500 N) is applied to a cylindrical specimen of steel (displaying the stress–strain behavior shown in Figure 6.20) that has a cross-sectional diameter of 0.40 in. (10 mm).
(a) Will the specimen experience elastic or plastic deformation? Why?
(b) If the original specimen length is 20 in. (500 mm), how much will it increase in length when this load is applied?

6.26 A bar of a steel alloy exhibiting the stress–strain behavior shown in Figure 6.20 is subjected to a tensile load; the specimen is 12 in. (300 mm) long, and of square cross section 0.175 in. (4.5 mm) on a side.

(a) Compute the magnitude of the load necessary to produce an elongation of 0.018 in. (0.46 mm).

(b) What will be the deformation after the load has been released?

6.27 A cylindrical specimen of aluminum having a diameter of 0.505 in. (12.8 mm) and a gauge length of 2.000 in. (50.800 mm) is pulled in tension. Use the load–elongation characteristics tabulated below to complete problems a through g.

Load		Length		Load		Length	
lb$_f$	N	in.	mm	lb$_f$	N	in.	mm
1,650	7,330	2.002	50.851	10,400	46,200	2.120	53.848
3,400	15,100	2.004	50.902	10,650	47,300	2.160	54.864
5,200	23,100	2.006	50.952	10,700	47,500	2.200	55.880
6,850	30,400	2.008	51.003	10,400	46,100	2.240	56.896
7,750	34,400	2.010	51.054	10,100	44,800	2.270	57.658
8,650	38,400	2.020	51.308	9,600	42,600	2.300	58.420
9,300	41,300	2.040	51.816	8,200	36,400	2.330	59.182
10,100	44,800	2.080	52.832		Fracture		

(a) Plot the data as engineering stress versus engineering strain.
(b) Compute the modulus of elasticity.
(c) Determine the yield strength at a strain offset of 0.002.
(d) Determine the tensile strength of this alloy.
(e) What is the approximate ductility, in percent elongation?
(f) Compute the modulus of resilience.
(g) Determine an appropriate working stress for this material.

6.28 A specimen of magnesium having a rectangular cross section of dimensions $\frac{1}{8}$ in. × $\frac{3}{4}$ in. (3.2 mm × 19.1 mm) is deformed in tension. Using the load–elongation data tabulated below, complete problems a through g.

Load		Length	
lb$_f$	N	in.	mm
0	0	2.500	63.50
310	1380	2.501	63.53
625	2780	2.503	63.58
1265	5630	2.505	63.63
1670	7430	2.508	63.70
1830	8140	2.510	63.75
2220	9870	2.525	64.14
2890	12,850	2.575	65.41
3170	14,100	2.625	66.68
3225	14,340	2.675	67.95
3110	13,830	2.725	69.22
2810	12,500	2.775	70.49
	Fracture		

(a) Plot the data as engineering stress versus engineering strain.
(b) Compute the modulus of elasticity.
(c) Determine the yield strength at a strain offset of 0.002.
(d) Determine the tensile strength of the alloy.
(e) Compute the modulus of resilience.
(f) What is the ductility, in percent elongation?
(g) Determine an appropriate working stress for this material.

6.29 A cylindrical metal specimen having an original diameter of 0.505 in. (12.8 mm) and gauge length of 2.000 in. (50.80 mm) is pulled in tension until fracture occurs. The diameter at the point of fracture is 0.260 in. (6.60 mm), and the fractured gauge length is 2.840 in. (72.14 mm). Calculate the ductility in terms of percent area reduction and percent elongation.

6.30 Calculate the moduli of resilience for the materials having the stress–strain behaviors shown in Figures 6.11 and 6.20.

6.31 Determine the modulus of resilience for each of the following alloys:

Alloy	Yield Strength [psi (MPa)]
Steel	120,000 (830)
Brass	55,000 (380)
Aluminum	40,000 (275)
Titanium	100,000 (690)

Use modulus of elasticity values in Table 6.1.

6.32 A steel alloy to be used for a spring application must have a modulus of resilience of at least 300 psi (2.07 MPa). What must be its minimum yield strength?

6.33 (a) Make a schematic plot showing the tensile true stress–strain behavior for a typical metal alloy.

(b) Superimpose on this plot a schematic curve for the compressive true stress–strain behavior for the same alloy. Explain any difference between this curve and the one in part a.

(c) Now superimpose a schematic curve for the compressive engineering stress–strain behavior for this same alloy, and explain any difference between this curve and the one in part b.

6.34 Show that Equations 6.17a and 6.17b are valid when there is no volume change during deformation.

6.35 Demonstrate that Equation 6.15, the expression defining true strain, may also be represented by

$$\epsilon_T = \ln\left(\frac{A_0}{A_i}\right)$$

when specimen volume remains constant during deformation. Which of these two expressions is more valid during necking? Why?

6.36 Using the data in Problem 6.27 and Equations 6.14, 6.15, and 6.17a, generate a true stress–true strain plot for aluminum. Equation 6.17a becomes invalid past the point at which necking begins; therefore, measured diameters are given below for the last four data points, which should be used in true stress computations.

Load		Length		Diameter	
lb$_f$	N	in.	mm	in.	mm
10,400	46,100	2.240	56.896	0.461	11.71
10,100	44,800	2.270	57.658	0.431	10.95
9,600	42,600	2.300	58.420	0.418	10.62
8,200	36,400	2.330	59.182	0.370	9.40

6.37 A tensile test is performed on a metal specimen, and it is found that a true plastic strain of 0.2 is produced when a true stress of 83,500 psi (575 MPa) is applied; for the same metal, the value of K in Equation 6.18 is 125,000 psi (860 MPa). Calculate the true strain that results from the application of a true stress of 87,000 psi (600 MPa).

6.38 For some metal alloy, a true stress of 50,000 psi (345 MPa) produces a plastic true strain of 0.02. How much will a specimen of this material elongate when a true stress of 60,000 psi (415 MPa) is applied if the original length is 20 in. (500 mm)? Assume a value of 0.22 for the strain-hardening exponent n.

6.39 The following true stresses produce the corresponding true plastic strains for a brass alloy:

σ_T(psi)	ϵ_T
50,000	0.10
60,000	0.20

What true stress is necessary to produce a true plastic strain of 0.25?

6.40 For a brass alloy, the following engineering stresses produce the corresponding plastic engineering strains, prior to necking:

Engineering Stress (MPa)	Engineering Strain
315	0.105
340	0.220

On the basis of this information, compute the *engineering* stress necessary to produce an *engineering* strain of 0.28.

6.41 Find the toughness (or energy to cause fracture) for a metal that experiences both elastic and plastic deformation. Assume Equation 6.4 for elastic deformation, that the modulus of elasticity is 25×10^6 psi (17.2×10^4 MPa), and that elastic deformation terminates at a strain of 0.01. For plastic deformation, assume that the relationship between stress and strain is described by Equation 6.18, in which the values for K and n are 10^6 psi (6900 MPa) and 0.30, respectively.

Furthermore, plastic deformation occurs between strain values of 0.01 and 0.75, at which point fracture occurs.

6.42 For a tensile test, it can be demonstrated that necking begins when

$$\frac{d\sigma_T}{d\epsilon_T} = \sigma_T \tag{6.24}$$

Using Equation 6.18, determine the value of the true strain at this onset of necking.

6.43 Taking the logarithm of both sides of Equation 6.18 yields

$$\log \sigma_T = \log K + n \log \epsilon_T \tag{6.25}$$

Thus a plot of $\log \sigma_T$ versus $\log \epsilon_T$ in the plastic region to the point of necking should yield a straight line having a slope of n and an intercept (at $\log \sigma_T = 0$) of $\log K$.

Using the appropriate data tabulated in Problem 6.27, make a plot of $\log \sigma_T$ versus $\log \epsilon_T$ and determine the values of n and K. It will be necessary to convert engineering stresses and strains to true stresses and strains using Equations 6.17a and 6.17b.

6.44 A cylindrical specimen of a brass alloy 10.0 mm (0.39 in.) in diameter and 120.0 mm (4.72 in.) long is pulled in tension with a force of 11,750 N (2640 lb$_f$); the force is subsequently released.

(a) Compute the final length of the specimen at this time. The tensile stress–strain behavior for this alloy is shown in Figure 6.11.

(b) Compute the final specimen length when the load is increased to 23,500 N (5280 lb$_f$) and then released.

6.45 A brass specimen having a rectangular cross section of dimensions 0.50 in. × 0.25 in. (12.7 mm × 6.4 mm) has the stress–strain behavior shown in Figure 6.11. If this specimen is subjected to a tensile force of 4500 lb$_f$ (20,000 N), then

(a) Determine the elastic and plastic strain values once the load is completely released.

(b) If its original length is 24.0 in. (610.0 mm), what will be its final length after the load in part a is applied and then released?

6.46 (a) A 10-mm-diameter Brinell hardness indenter produced an indentation 2.50 mm in diameter in a steel alloy when a load of 1000 kg was used. Compute the HB of this material.

(b) What will be the diameter of an indentation to yield a hardness of 300 HB when a 500 kg load is used?

6.47 Estimate the Brinell and Rockwell hardnesses for the following:

(a) The naval brass for which the stress–strain behavior is shown in Figure 6.11.

(b) The steel for which the stress–strain behavior is shown in Figure 6.20.

6.48 Using the data represented in Figure 6.18, specify equations relating tensile strength and Brinell hardness for brass and nodular cast iron, similar to Equations 6.19a and 6.19b for steels.

6.49 Cite five factors that lead to scatter in measured material properties.

6.50 Below are tabulated a number of Rockwell G hardness values which were measured on a single brass specimen. Compute average and standard deviation hardness values.

47.3	48.7	47.1
52.1	50.0	50.4
45.6	46.2	45.9
49.9	48.3	46.4
47.6	51.1	48.5
50.4	46.7	49.7

6.51 Upon what three criteria are factors of safety based?

6.52 Determine the working stresses for the two alloys the stress–strain behaviors of which are shown in Figures 6.11 and 6.20.

6.53 A large tower is to be supported by a series of steel wires. It is estimated that the load on each wire will be 3000 lb$_f$ (13,300 N). Determine the minimum required wire diameter assuming a factor of safety of 2 and a yield strength of 125,000 psi (860 MPa) for the steel.

CHAPTER 7

DISLOCATIONS AND STRENGTHENING MECHANISMS

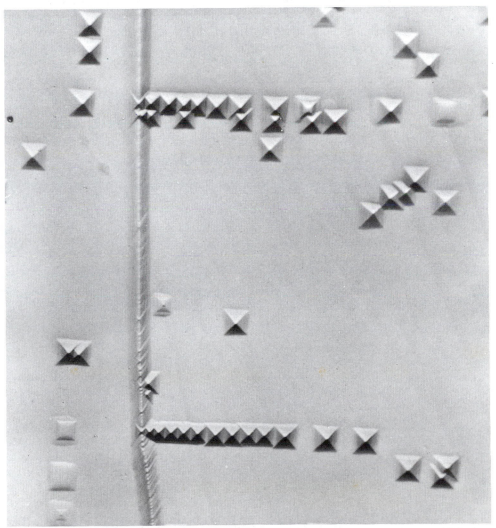

In this photomicrograph of a lithium fluoride (LiF) single crystal, the small pyramidal pits represent those positions at which dislocations intersect the surface. The surface was polished and then chemically treated; these "etch pits" result from localized chemical attack around the dislocations and indicate the distribution of the dislocations. $750 \times$. (Photomicrograph courtesy of W. G. Johnston, General Electric Co.)

7.1 INTRODUCTION

Chapter 6 explained that materials may experience two kinds of deformation: elastic and plastic. Plastic deformation is permanent, and strength and hardness are measures of a material's resistance to this deformation. On a microscopic scale, plastic deformation corresponds to the net movement of large numbers of atoms in response to an applied stress. During this process, interatomic bonds must be ruptured and then reformed. In crystalline solids, plastic deformation most often involves the motion of dislocations, linear crystalline defects which were introduced in Section 4.4. This chapter discusses the characteristics of dislocations and their involvement in plastic deformation. In addition, and probably most importantly, several techniques are presented for strengthening single-phase metals, the mechanisms of which are described in terms of dislocations. Finally, the latter sections of this chapter are concerned with recovery and recrystallization—processes that occur in plastically deformed metals, normally at elevated temperatures—and, in addition, grain growth.

DISLOCATIONS AND PLASTIC DEFORMATION

Early materials studies led to the computation of the theoretical strengths of perfect crystals, which were many times greater than those actually measured. During the 1930s it was theorized that this discrepancy in mechanical strengths could be explained by a type of linear crystalline defect that has since come to be known as a dislocation. It was not until the 1950s, however, that the existence of such dislocation defects was established by direct observation with the electron microscope. Since then a theory of dislocations has evolved that explains many of the physical and mechanical phenomena in crystalline materials, primarily metals and ceramics.

7.2 BASIC CONCEPTS

Edge and screw are the two fundamental dislocation types. In an edge dislocation, localized lattice distortion exists along the end of an extra half-plane of atoms, which also defines the dislocation line (Figure 4.3). A screw dislocation may be thought of as resulting from shear distortion; its dislocation line passes through the center of a spiral, atomic plane ramp (Figure 4.4). Many dislocations in crystalline materials have both edge and screw components; these are mixed dislocations (Figure 4.5).

Plastic deformation corresponds to the motion of large numbers of dislocations. An edge dislocation moves in response to a shear stress applied in a direction perpendicular to its line; the mechanics of dislocation motion are represented in Figure 7.1. Let the initial extra half-plane of atoms be plane A. When the shear stress is applied as indicated (Figure 7.1a), plane A is forced to the right; this in turn pushes the top halves of planes B, C, D, and so on, in the same direction. If the applied shear stress is of sufficient magnitude, the interatomic bonds of plane B are severed along the shear plane, and the upper half of plane B becomes the extra half-plane as plane A links up with the bottom half of plane B (Figure 7.1b). This process is subsequently repeated for the other planes, such that the extra half-plane, by discrete steps, moves from left to right by successive and repeated breaking of bonds and shifting by

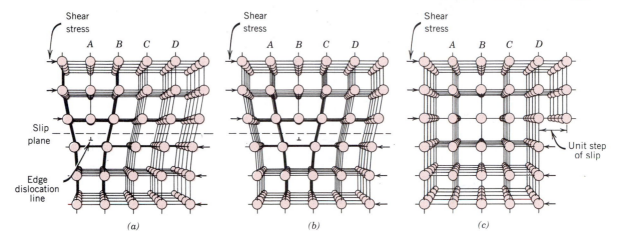

Figure 7.1 Atomic rearrangements that accompany the motion of an edge dislocation as it moves in response to an applied shear stress. (*a*) The extra half-plane of atoms is labeled *A*. (*b*) The dislocation moves one atomic distance to the right as *A* links up to the lower portion of plane *B*; in the process, the upper portion of *B* becomes the extra half-plane. (*c*) A step forms on the surface of the crystal as the extra half-plane exits. (Adapted from A. G. Guy, *Essentials of Materials Science*, McGraw-Hill Book Company, New York, 1976, p. 153.)

interatomic distances of upper half-planes. Before and after the movement of a dislocation through some particular region of the crystal, the atomic arrangement is ordered and perfect; it is only during the passage of the extra half-plane that the lattice structure is disrupted. Ultimately this extra half-plane may emerge from the right surface of the crystal, forming an edge that is one atomic distance wide; this is shown in Figure 7.1*c*.

The process by which plastic deformation is produced by dislocation motion is termed **slip;** the plane along which the dislocation line traverses is the *slip plane,* as indicated in Figure 7.1. Macroscopic plastic deformation simply corresponds to permanent deformation that results from the movement of dislocations, or slip, in response to an applied shear stress, as represented in Figure 7.2*a*.

Dislocation motion is analogous to the mode of locomotion employed by a caterpillar (Figure 7.3). The caterpillar forms a hump near its posterior end by pulling in its last pair of legs a unit leg distance. The hump is propelled forward by repeated lifting and shifting of leg pairs. When the hump reaches the anterior end, the entire caterpillar has moved forward by the leg separation distance. The caterpillar hump and its motion correspond to the extra half-plane of atoms in the dislocation model of plastic deformation.

The motion of a screw dislocation in response to the applied shear stress is shown in Figure 7.2*b*; the direction of movement is perpendicular to the stress direction. For an edge, motion is parallel to the shear stress. However, the net plastic deformation for the motion of both dislocation types is the same (see Figure 7.2). The direction of motion of the mixed dislocation line is neither perpendicular nor parallel to the applied stress, but lies somewhere in between.

Virtually all crystalline materials contain some dislocations that were introduced during solidification, during plastic deformation, and as a consequence of thermal

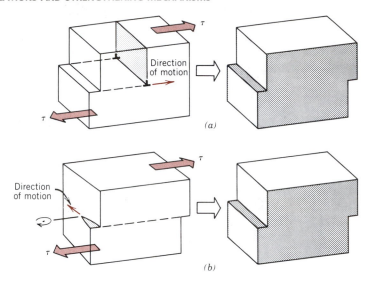

Figure 7.2 The formation of a step on the surface of a crystal by the motion of (a) an edge dislocation and (b) a screw dislocation. Note that for an edge, the dislocation line moves in the direction of the applied shear stress τ; for a screw, the dislocation line motion is perpendicular to the stress direction. (Adapted from H. W. Hayden, W. G. Moffatt, and J. Wulff, *The Structure and Properties of Materials,* Vol. III, *Mechanical Behavior,* p. 70. Copyright © 1965 by John Wiley & Sons, New York. Reprinted by permission of John Wiley & Sons.)

stresses that result from rapid cooling. The number of dislocations, or **dislocation density** in a material, is expressed as the total dislocation length per unit volume, or, equivalently, the number of dislocations that intersect a unit area of a random section. The units of dislocation density are millimeters of dislocation per cubic millimeter or just per square millimeter. Dislocation densities as low as 10^3 mm^{-2} are typically found in carefully prepared metal crystals. For heavily deformed metals, the density may run as high as 10^9 to 10^{10} mm^{-2}. Heat treating a deformed metal specimen will diminish the density to on the order of 10^5 to 10^6 mm^{-2}.

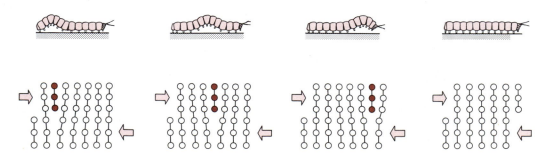

Figure 7.3 Representation of the analogy between caterpillar and dislocation motion.

7.3 CHARACTERISTICS OF DISLOCATIONS

Several characteristics of dislocations are important with regard to the mechanical properties of metals. These include strain fields that exist around dislocations, which are influential in determining the mobility of the dislocations, as well as their ability to multiply.

When metals are plastically deformed, some fraction of the deformation energy (approximately 5%) is retained internally; the remainder is dissipated as heat. The major portion of this stored energy is as strain energy associated with dislocations. Consider the edge dislocation represented in Figure 7.4. As already mentioned, some atomic lattice distortion exists around the dislocation line because of the presence of the extra half-plane of atoms. As a consequence, there are regions in which compressive, tensile, and shear **lattice strains** are imposed on the neighboring atoms. For example, atoms immediately above and adjacent to the dislocation line are squeezed together. As a result, these atoms may be thought of as experiencing a compressive strain relative to atoms positioned in the perfect crystal and far removed from the dislocation; this is illustrated in Figure 7.4. Directly below the half-plane, the effect is just the opposite; lattice atoms sustain an imposed tensile strain, which is as shown. Shear strains also exist in the vicinity of the edge dislocation. For a screw dislocation, lattice strains are pure shear only. These lattice distortions may be considered to be strain fields that radiate from the dislocation line. The strains extend into the surrounding atoms, and their magnitude decreases with radial distance from the dislocation.

The strain fields surrounding dislocations in close proximity to one another may interact such that forces are imposed on each dislocation by the combined interactions of all its neighboring dislocations. For example, consider two edge dislocations that have the same sign and the identical slip plane, as represented in Figure 7.5a. The compressive and tensile strain fields for both lie on the same side of the slip plane; the strain field interaction is such that there exists between these two isolated dislocations a mutual repulsive force that tends to move them apart. On the other hand, two dislocations of opposite sign and having the same slip plane will be attracted to one another, as indicated in Figure 7.5b, and dislocation annihilation will occur when they meet. That is, the two extra half-planes of atoms will align and become a complete plane. Dislocation interactions are possible between edge, screw, and/or mixed dislocations, and for a variety of orientations. These strain fields and associated forces are important in the strengthening mechanisms for metals.

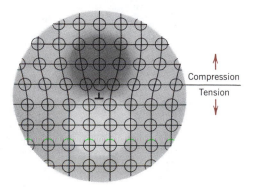

Compression
Tension

Figure 7.4 Regions of compression (dark) and tension (light) located around an edge dislocation. (Adapted from W. G. Moffatt, G. W. Pearsall, and J. Wulff, *The Structure and Properties of Materials,* Vol. I, *Structure,* p. 85. Copyright © 1964 by John Wiley & Sons, New York. Reprinted by permission of John Wiley & Sons, Inc.)

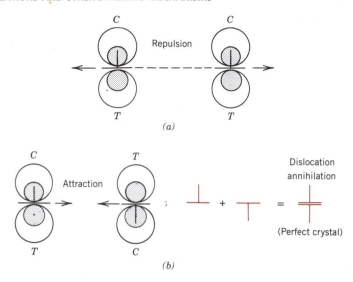

Figure 7.5 (a) Two edge dislocations of the same sign and lying
on the same slip plane exert a repulsive force on each other;
C and T denote compression and tensile regions, respectively.
(b) Edge dislocations of opposite sign and lying on the same
slip plane exert an attractive force on each other. Upon meeting,
they annihilate each other and leave a region of perfect crystal.
(Adapted from H. W. Hayden, W. G. Moffatt, and J. Wulff, *The
Structure and Properties of Materials,* Vol. III, *Mechanical
Behavior,* p. 75. Copyright © 1965 by John Wiley & Sons,
New York. Reprinted by permission of John Wiley & Sons, Inc.)

During plastic deformation, the number of dislocations increases dramatically.
We know that the dislocation density in a metal that has been highly deformed may
be as high as 10^{10} mm^{-2}. Grain boundaries, as well as internal defects and surface
irregularities such as scratches and nicks, which act as stress concentrations, may
serve as dislocation formation sites during deformation. Under some circumstances,
existing dislocations may also multiply.

7.4 SLIP SYSTEMS

Dislocations do not move with the same degree of ease on all crystallographic planes
of atoms and in all crystallographic directions. Ordinarily there is a preferred plane,
and in that plane there are specific directions along which dislocation motion occurs.
This plane is called the *slip plane;* it follows that the direction of movement is called
the *slip direction.* This combination of the slip plane and the slip direction is termed
the **slip system.** The slip system depends on the crystal structure of the metal and is
such that the atomic distortion that accompanies the motion of a dislocation is a
minimum. For a particular crystal structure, the slip plane is that plane having the
most dense atomic packing, that is, has the greatest planar density. The slip direction
corresponds to the direction, in this plane, that is most closely packed with atoms,
that is, has the highest linear density. Planar and linear atomic densities were discussed
in Section 3.10.

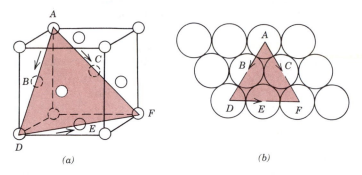

Figure 7.6 (*a*) A {111} ⟨110⟩ slip system shown within an FCC unit cell. (*b*) The (111) plane from (*a*) and three ⟨110⟩ slip directions (as indicated by arrows) within that plane comprise possible slip systems.

Consider, for example, the FCC crystal structure, a unit cell of which is shown in Figure 7.6*a*. There is a set of planes, the {111} family, all of which are closely packed. A (111)-type plane is indicated in the unit cell; in Figure 7.6*b*, this plane is positioned within the plane of the page, in which atoms are now represented as touching nearest neighbors.

Slip occurs along ⟨110⟩-type directions within the {111} planes, as also indicated in Figure 7.6. Hence, {111}⟨110⟩ represents the slip plane and direction combination, or the slip system for FCC. Figure 7.6*b* demonstrates that a given slip plane may contain more than a single slip direction. Thus several slip systems may exist for a particular crystal structure; the number of independent slip systems represents the different possible combinations of slip planes and directions. For example, for face-centered cubic, there are 12 slip systems: four unique {111} planes, and within each plane three independent ⟨110⟩ directions.

The possible slip systems for BCC and HCP crystal structures are listed in Table 7.1. For each of these structures, slip is possible on more than one family of planes

TABLE 7.1 Slip Systems for Face-Centered Cubic, Body-Centered Cubic, and Hexagonal Close-Packed Metals

Metals	Slip Plane	Slip Direction	Number of Slip Systems
Face-Centered Cubic			
Cu, Al, Ni, Ag, Au	{111}	⟨1$\bar{1}$0⟩	12
Body-Centered Cubic			
α-Fe, W, Mo	{110}	⟨$\bar{1}$11⟩	12
α-Fe, W	{211}	⟨$\bar{1}$11⟩	12
α-Fe, K	{321}	⟨$\bar{1}$11⟩	24
Hexagonal Close-Packed			
Cd, Zn, Mg, Ti, Be	{0001}	⟨11$\bar{2}$0⟩	3
Ti, Mg, Zr	{10$\bar{1}$0}	⟨11$\bar{2}$0⟩	3
Ti, Mg	{10$\bar{1}$1}	⟨11$\bar{2}$0⟩	6

(e.g., {110}, {211}, and {321} for BCC). For metals having these two crystal structures, some slip systems are often operable only at elevated temperatures.

Metals with FCC or BCC crystal structures have a relatively large number of slip systems (at least 12). These metals are quite ductile because extensive plastic deformation is normally possible along the various systems. Conversely, HCP metals, having few active slip systems, are normally quite brittle.

7.5 SLIP IN SINGLE CRYSTALS

A further explanation of slip is simplified by treating the process in single crystals, then making the appropriate extension to polycrystalline materials. As mentioned previously, edge, screw, and mixed dislocations move in response to shear stresses applied along a slip plane and in a slip direction. Even though an applied stress may be pure tensile (or compressive), shear components exist at all but parallel or perpendicular alignments to the stress direction. These are termed **resolved shear stresses,** and their magnitudes depend not only on the applied stress, but also on the orientation of both the slip plane and direction within that plane. Let ϕ represent the angle between the normal to the slip plane and the applied stress direction, and λ the angle between the slip and stress directions, as indicated in Figure 7.7; it can then be shown that for the resolved shear stress τ_R

$$\tau_R = \sigma \cos \phi \cos \lambda \tag{7.1}$$

where σ is the applied stress. In general, $\phi + \lambda \neq 90°$, since it need not be the case that the tensile axis, the slip plane normal, and the slip direction all lie in the same plane.

A metal single crystal has a number of different slip systems that are capable of operating. The resolved shear stress normally differs for each one because the orientation of each relative to the stress axis (ϕ and λ angles) also differs. However, one slip system is generally oriented most favorably, that is, has the largest resolved shear stress, $\tau_R(\text{max})$:

$$\tau_R(\text{max}) = \sigma(\cos \phi \cos \lambda)_{\text{max}} \tag{7.2}$$

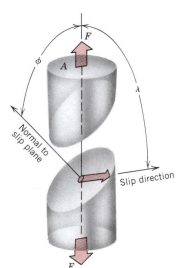

Figure 7.7 Geometrical relationships between the tensile axis, slip plane, and slip direction used in calculating the resolved shear stress for a single crystal.

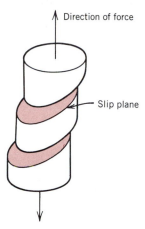

Direction of force

Slip plane

Figure 7.8 Macroscopic slip in a single crystal.

In response to an applied tensile or compressive stress, slip in a single crystal commences on the most favorably oriented slip system when the resolved shear stress reaches some critical value, termed the **critical resolved shear stress** τ_{crss}; it represents the minimum shear stress required to initiate slip, and is a property of the material that determines when yielding occurs. The single crystal plastically deforms or yields when $\tau_R(\text{max}) = \tau_{crss}$, and the magnitude of the applied stress required to initiate yielding (i.e., the yield strength σ_y) is

$$\sigma_y = \frac{\tau_{crss}}{(\cos \phi \cos \lambda)_{max}} \tag{7.3}$$

The minimum stress necessary to introduce yielding occurs when a single crystal is oriented such that $\phi = \lambda = 45°$; under these conditions,

$$\sigma_y = 2\tau_{crss} \tag{7.4}$$

For a single crystal specimen that is stressed in tension, deformation will be as in Figure 7.8, where slip occurs along a number of equivalent and most favorably oriented planes and directions at various positions along the specimen length. This slip deformation forms as small steps on the surface of the single crystal that are parallel to one another and loop around the circumference of the specimen as indicated in Figure 7.8. Each step results from the movement of a large number of dislocations along the same slip plane. On the surface of a polished single crystal specimen, these steps appear as lines, which are called slip lines. A zinc single crystal that has been plastically deformed to the degree that these slip markings are discernible is shown in Figure 7.9.

With continued extension of a single crystal, both the number of slip lines and the slip step width will increase. For FCC and BCC metals, slip may eventually begin along a second slip system, that which is next most favorably oriented with the tensile axis. Furthermore, for HCP crystals having few slip systems, if, for the most favorable slip system, the stress axis is either perpendicular to the slip direction ($\lambda = 90°$) or parallel to the slip plane ($\phi = 90°$), the critical resolved shear stress will be zero. For these extreme orientations the crystal ordinarily fractures rather than deforming plastically.

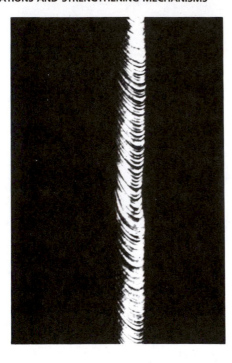

Figure 7.9 Slip in a zinc single crystal. (From C. F. Elam, *The Distortion of Metal Crystals,* Oxford University Press, London, 1935.)

EXAMPLE PROBLEM 7.1

Consider a single crystal of BCC iron oriented such that a tensile stress is applied along a [010] direction. (a) Compute the resolved shear stress along a (110) plane and in a [$\bar{1}$11] direction when a tensile stress of 7500 psi (52 MPa) is applied. (b) If slip occurs on a (110) plane and in a [$\bar{1}$11] direction, and the critical resolved shear stress is 4350 psi (30 MPa), calculate the magnitude of the applied tensile stress necessary to initiate yielding.

SOLUTION

(a) A BCC unit cell along with the slip direction and plane as well as the direction of the applied stress are all shown in the accompanying diagram (*a*).

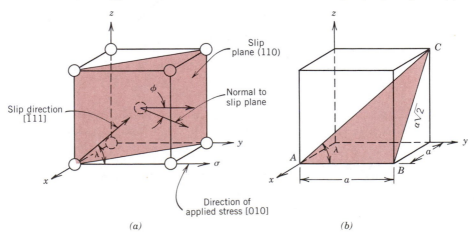

(a)   (b)

As indicated, ϕ, the angle between the (110) plane normal and the [010] direction is 45°. From the triangle ABC in diagram (b), λ, the angle between the [$\bar{1}$11] and [010] directions is $\tan^{-1}(a\sqrt{2}/a) = 54.7°$, a being the unit cell length. Thus according to Equation 7.1,

$$\tau_R = \sigma \cos \phi \cos \lambda = (7500 \text{ psi})(\cos 45°)(\cos 54.7°)$$
$$= 3060 \text{ psi } (21.1 \text{ MPa})$$

(b) The yield strength σ_y may be computed from Equation 7.3; ϕ and λ will be the same as for part a, and

$$\sigma_y = \frac{4350 \text{ psi}}{(\cos 45°)(\cos 54.7°)} = 10,600 \text{ psi } (73.1 \text{ MPa})$$

7.6 PLASTIC DEFORMATION OF POLYCRYSTALLINE MATERIALS

Deformation and slip in polycrystalline materials is somewhat more complex. Because of the random crystallographic orientations of the numerous grains, the direction of slip varies from one grain to another. For each, dislocation motion occurs along the slip system that has the most favorable orientation, as defined above. This is exemplified by a photomicrograph of a polycrystalline copper specimen that has been plastically deformed (Figure 7.10); before deformation the surface was polished. Slip lines are visible, and it appears that two slip systems operated for most of the grains, as evidenced by two sets of parallel yet intersecting sets of lines. Furthermore, variation in grain orientation is indicated by the difference in alignment of the slip lines for the several grains.

Gross plastic deformation of a polycrystalline specimen corresponds to the comparable distortion of the individual grains by means of slip. During deformation, mechanical integrity and coherency are maintained along the grain boundaries; that is, the grain boundaries do not come apart or open up. As a consequence, each individual grain is constrained, to some degree, in the shape it may assume by its neighboring grains. The manner in which grains distort as a result of gross plastic

Figure 7.10 Slip lines on the surface of a polycrystalline specimen of copper that was polished and subsequently deformed. 173×. (Photomicrograph courtesy of C. Brady, National Bureau of Standards.)

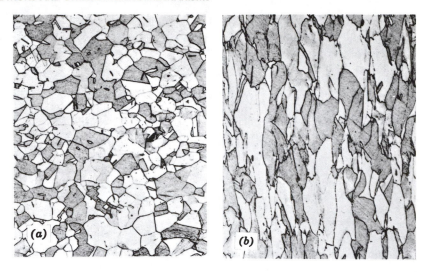

Figure 7.11 Alteration of the grain structure of a polycrystalline metal as a result of plastic deformation. (*a*) Before deformation the grains are equiaxed. (*b*) The deformation has produced elongated grains. 170×. (From W. G. Moffatt, G. W. Pearsall, and J. Wulff, *The Structure and Properties of Materials*, Vol. I, *Structure*, p. 140. Copyright © 1964 by John Wiley & Sons, New York. Reprinted by permission of John Wiley & Sons, Inc.)

deformation is indicated in Figure 7.11. Before deformation the grains are equiaxed, or have approximately the same dimension in all directions. For this particular deformation, the grains became elongated along the direction in which the specimen was extended.

Polycrystalline metals are stronger than their single-crystal equivalents, which means that greater stresses are required to initiate slip and the attendant yielding. This is, to a large degree, also a result of geometrical constraints that are imposed on the grains during deformation. Even though a single grain may be favorably oriented with the applied stress for slip, it cannot deform until the adjacent and less favorably oriented grains are capable of slip also; this requires a higher applied stress level.

7.7 DEFORMATION BY TWINNING

In addition to slip, plastic deformation in some metallic materials can occur by the formation of mechanical twins, or *twinning*. The concept of a twin was introduced in Section 4.5; that is, a shear force can produce atomic displacements such that on one side of a plane (the twin boundary), atoms are located in mirror image positions of atoms on the other side. The manner in which this is accomplished is demonstrated in Figure 7.12. Here, open circles represent atoms that did not move, and dashed and solid circles represent original and final positions, respectively, of atoms within the twinned region. As may be noted in this figure, the displacement magnitude within

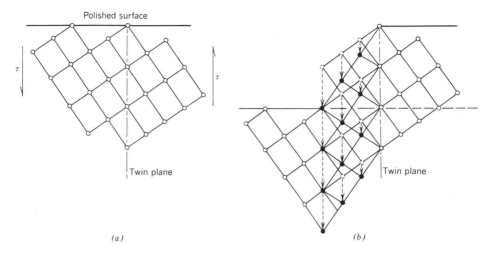

Figure 7.12 Schematic diagram showing how twinning results from an applied shear stress τ. In (*b*), open circles represent atoms that did not change position; dashed and solid circles represent original and final atom positions, respectively. (From G. E. Dieter, *Mechanical Metallurgy,* 3rd edition. Copyright © 1986 by McGraw-Hill Book Company, New York. Reproduced with permission of McGraw-Hill Book Company.)

the twin region (indicated by arrows) is proportional to the distance from the twin plane. Furthermore, twinning occurs on a definite crystallographic plane and in a specific direction that depend on crystal structure. For example, for BCC metals, the twin plane and direction are (112) and [111], respectively.

Slip and twinning deformations are compared in Figure 7.13 for a single crystal which is subjected to a shear stress τ. Slip ledges are shown in Figure 7.13*a*, the formation of which were described in Section 7.5; for twinning, the shear deformation is homogeneous (Figure 7.13*b*). These two processes differ from one another in several respects. First of all, for slip, the crystallographic orientation above and below the slip plane is the same both before and after the deformation; whereas for twinning, there will be a reorientation across the twin plane. In addition, slip occurs in distinct

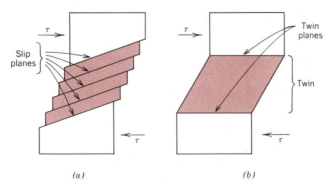

Figure 7.13 For a single crystal subjected to a shear stress τ, (*a*) deformation by slip; (*b*) deformation by twinning.

atomic spacing multiples, whereas the atomic displacement for twinning is less than the interatomic separation.

Mechanical twinning occurs in metals that have BCC and HCP crystal structures, at low temperatures, and high rates of loading (shock loading), conditions under which the slip process is restricted; that is, there are few operable slip systems. The amount of bulk plastic deformation from twinning is normally small relative to that resulting from slip. However, the real importance of twinning lies with the accompanying crystallographic reorientations; twinning may place new slip systems in orientations that are favorable relative to the stress axis such that the slip process can now take place.

MECHANISMS OF STRENGTHENING IN METALS

Metallurgical and materials engineers are often called on to design alloys having high strengths yet some ductility and toughness; ordinarily, ductility is sacrificed when an alloy is strengthened. Several hardening techniques are at the disposal of an engineer, and frequently alloy selection depends on the capacity of a material to be tailored with the mechanical characteristics required for a particular application.

Important to the understanding of strengthening mechanisms is the relation between dislocation motion and mechanical behavior of metals. Because macroscopic plastic deformation corresponds to the motion of large numbers of dislocations, *the ability of a metal to plastically deform depends on the ability of dislocations to move.* Since hardness and strength (both yield and tensile) are related to the ease with which plastic deformation can be made to occur, by reducing the mobility of dislocations, the mechanical strength may be enhanced; that is, greater mechanical forces will be required to initiate plastic deformation. In contrast, the more unconstrained the dislocation motion, the greater the facility with which a metal may deform, and the softer and weaker it becomes. Virtually all strengthening techniques rely on this simple principle: *restricting or hindering dislocation motion renders a material harder and stronger.*

The present discussion is confined to strengthening mechanisms for single-phase metals, by grain size reduction, solid-solution alloying, and strain hardening. Deformation and strengthening of multiphase alloys are more complicated, involving concepts yet to be discussed.

7.8 STRENGTHENING BY GRAIN SIZE REDUCTION

The size of the grains, or average grain diameter, in a polycrystalline metal influences the mechanical properties. Adjacent grains normally have different crystallographic orientations and, of course, a common grain boundary, as indicated in Figure 7.14. During plastic deformation, slip or dislocation motion must take place across this common boundary, say from grain A to grain B in Figure 7.14. The grain boundary acts as a barrier to dislocation motion for two reasons:

1. Since the two grains are of different orientations, a dislocation passing into grain B will have to change its direction of motion; this becomes more difficult as the crystallographic misorientation increases.

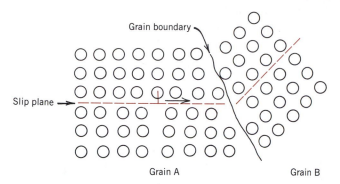

Figure 7.14 The motion of a dislocation as it encounters a grain boundary, illustrating how the boundary acts as a barrier to continued slip. Slip planes are discontinuous and change directions across the boundary. (Adapted from L. H. Van Vlack, *A Textbook of Materials Technology.* Copyright © 1973 by Addison-Wesley Publishing Co. Reprinted by permission of Addison-Wesley Publishing Co., Inc. Reading, MA.)

2. The atomic disorder within a grain boundary region will result in a discontinuity of slip planes from one grain into the other.

It should be mentioned that, for high-angle grain boundaries, it may not be the case that dislocations traverse grain boundaries during deformation; rather, a stress concentration ahead of a slip plane in one grain may activate sources of new dislocations in an adjacent grain.

A fine-grained material (one that has small grains) is harder and stronger than one that is coarse grained, since the former has a greater total grain boundary area to impede dislocation motion. For many materials, the yield strength σ_y varies with grain size according to

$$\sigma_y = \sigma_0 + k_y d^{-1/2} \tag{7.5}$$

In this expression, d is the average grain diameter, and σ_0 and k_y are constants for a particular material. Figure 7.15 demonstrates the yield strength dependence on grain size for a brass alloy. Grain size may be regulated by the rate of solidification from the liquid phase, and also by plastic deformation followed by an appropriate heat treatment, as discussed in Section 7.13.

Small-angle grain boundaries (Section 4.5) are not effective in interfering with the slip process because of the slight crystallographic misalignment across the boundary. On the other hand, twin boundaries (Section 4.5) will effectively block slip and increase the strength of the material. Boundaries between two different phases are also impediments to movements of dislocations. The sizes and shapes of the constituent phases significantly affect the mechanical properties of multiphase alloys; these are topics of discussion in Sections 10.7, 10.8, and 17.1.

7.9 SOLID-SOLUTION HARDENING

Another technique to strengthen and harden metals is alloying with impurity atoms that go into either substitutional or interstitial solid solution. Accordingly, this is

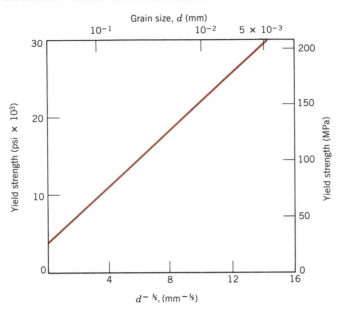

Figure 7.15 The influence of grain size on the yield strength of a 70 Cu–30 Zn brass alloy. Note that the grain diameter increases from right to left and is not linear. (Adapted from H. Suzuki, "The Relation Between the Structure and Mechanical Properties of Metals," Vol. II, *National Physical Laboratory, Symposium No. 15,* p. 524.)

called **solid-solution hardening.** High purity metals are almost always softer and weaker than alloys composed of the same base metal. Increasing the concentration of the impurity results in an attendant increase in tensile strength, and hardness, as indicated in Figures 7.16a and 7.16b for zinc in copper; the dependence of ductility on zinc concentration is presented in Figure 7.16c.

Alloys are stronger than pure metals because impurity atoms that go into solid solution ordinarily impose lattice strains on the surrounding host atoms. Lattice strain field interactions between dislocations and these impurity atoms result, and, consequently, dislocation movement is restricted. For example, an impurity atom that is smaller than a host atom for which it substitutes exerts tensile strains on the surrounding crystal lattice, as illustrated in Figure 7.17a. Conversely, a larger substitutional atom imposes compressive strains in its vicinity (Figure 7.18a). These solute atoms tend to segregate around dislocations in a way so as to reduce the overall strain energy, that is, to cancel some of the strain in the lattice surrounding a dislocation. To accomplish this, a smaller impurity atom is located where its tensile strain will partially nullify some of the dislocation's compressive strain. For the edge dislocation in Figure 7.17b, this would be adjacent to the dislocation line and above the slip plane. A larger impurity atom would be situated as in Figure 7.18b.

The resistance to slip is greater when impurity atoms are present because the overall lattice strain must increase if a dislocation is torn away from them. Furthermore, the same lattice strain interactions (Figures 7.17b and 7.18b) will exist between

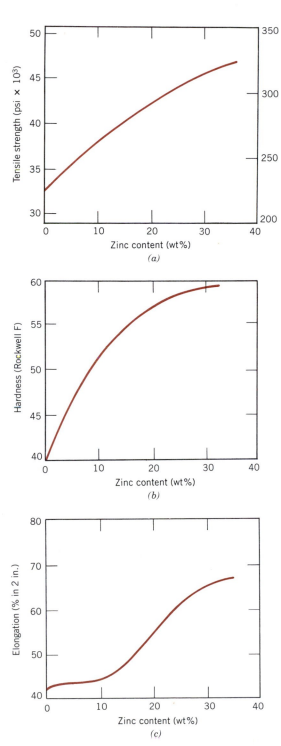

Figure 7.16 Variation with zinc content of (*a*) tensile strength, (*b*) hardness, and (*c*) ductility (%EL) for copper-zinc alloys, showing strengthening. (Adapted from *Metals Handbook: Properties and Selection: Nonferrous Alloys and Pure Metals*, Vol. 2, 9th edition, H. Baker, Managing Editor, American Society for Metals, 1979, p. 314.)

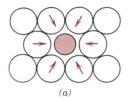

(a)

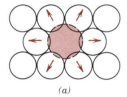

(b)

Figure 7.17 (*a*) Representation of tensile lattice strains imposed on host atoms by a smaller substitutional impurity atom. (*b*) Possible locations of smaller impurity atoms relative to an edge dislocation such that there is partial cancellation of impurity–dislocation lattice strains.

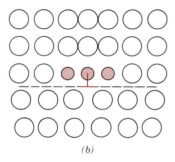

(a)

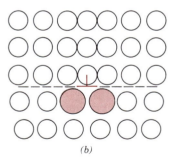

(b)

Figure 7.18 (*a*) Representation of compressive strains imposed on host atoms by a larger substitutional impurity atom. (*b*) Possible locations of larger impurity atoms relative to an edge dislocation such that there is partial cancellation of impurity–dislocation lattice strains.

impurity atoms and dislocations that are in motion during plastic deformation. Thus a greater applied stress is necessary to first initiate and then continue plastic deformation for solid-solution alloys, as opposed to pure metals; this is evidenced by the enhancement of strength and hardness.

7.10 STRAIN HARDENING

Strain hardening is the phenomenon whereby a ductile metal becomes harder and stronger as it is plastically deformed. Sometimes it is also called *work hardening*, or, because the temperature at which deformation takes place is "cold" relative to the absolute melting temperature of the metal, **cold working.** Most metals strain harden at room temperature.

It is sometimes convenient to express the degree of plastic deformation as *percent cold work* rather than as strain. Percent cold work (%CW) is defined as

$$\%CW = \left(\frac{A_0 - A_d}{A_0}\right) \times 100 \tag{7.6}$$

where A_0 is the original area of the cross section that experiences deformation, and A_d is the area after deformation.

Figures 7.19a and 7.19b demonstrate how steel, brass, and copper increase in yield and tensile strength with increasing cold work. The price for this enhancement of hardness and strength is in the ductility of the metal. This is shown in Figure 7.19c, in which the ductility, in percent elongation, experiences a reduction with increasing

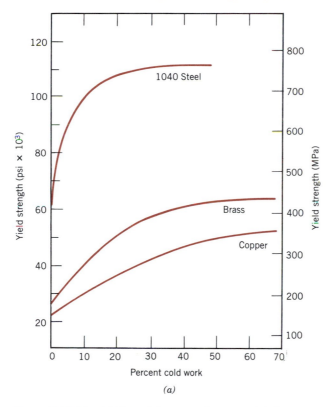

(a)

Figure 7.19 For 1040 steel, brass, and copper, (a) the increase in yield strength, (b) the increase in tensile strength, and (c) the decrease in ductility (%EL) with percent cold work. (Adapted from *Metals Handbook: Properties and Selection, Irons and Steels,* Vol. 1, 9th edition, B. Bardes, Editor, American Society for Metals, 1978, p. 226, and *Metals Handbook: Properties and Selection: Nonferrous Alloys and Pure Metals,* Vol. 2, 9th edition, H. Baker, Managing Editor, American Society for Metals, 1979, pp. 276 and 327.)

(continued)

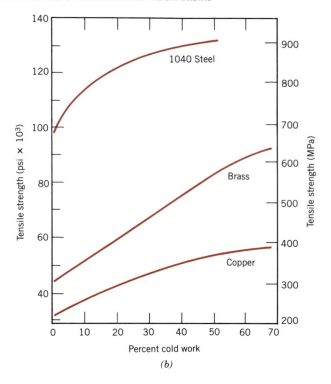

(b)

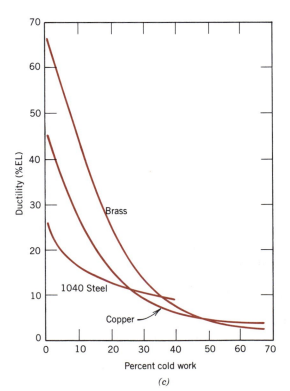

(c)

Figure 7.19 (*continued*)

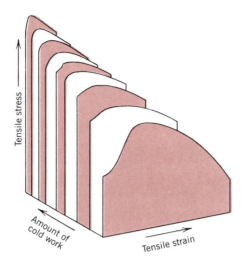

Figure 7.20 The influence of cold work on the stress–strain behavior for a low-carbon steel. (From *Metals Handbook: Properties and Selection, Irons and Steels,* Vol. 1, 9th edition, B. Bardes, Editor, American Society for Metals, 1978, p. 221.)

percent cold work for the same three alloys. The influence of cold work on the stress–strain behavior of a steel is vividly portrayed in Figure 7.20.

Strain hardening is demonstrated in a stress–strain diagram presented earlier (Figure 6.16). Initially, the metal with yield strength σ_{y_0} is plastically deformed to point D. The stress is released, then reapplied with a resultant new yield strength, σ_{y_i}. The metal has thus become stronger during the process because σ_{y_i} is greater than σ_{y_0}.

The strain hardening phenomenon is explained on the basis of dislocation–dislocation strain field interactions similar to those discussed in Section 7.3. The dislocation density in a metal increases with deformation or cold work, as already mentioned. Consequently, the average distance of separation between dislocations decreases—the dislocations are positioned closer together. On the average, dislocation–dislocation strain interactions are repulsive. The net result is that the motion of a dislocation is hindered by the presence of other dislocations. As the dislocation density increases, this resistance to dislocation motion by other dislocations becomes more pronounced. Thus the imposed stress necessary to deform a metal increases with increasing cold work.

Strain hardening is often utilized commercially to enhance the mechanical properties of metals during fabrication procedures. The effects of strain hardening may be removed by an annealing heat treatment, as discussed in Section 11.2.

In passing, for the mathematical expression relating true stress and strain, Equation 6.18, the parameter n is called the *strain hardening exponent,* which is a measure of the ability of a metal to strain harden; the larger its magnitude, the greater the strain hardening for a given amount of plastic strain.

EXAMPLE PROBLEM 7.2

Compute the tensile strength and ductility (%EL) of a cylindrical copper rod if it is cold worked such that the diameter is reduced from 0.6 in. to 0.48 in. (15.2 mm to 12.2 mm).

SOLUTION

It is first necessary to determine the percent cold work resulting from the deformation. This is possible using Equation 7.6:

$$\%CW = \frac{(0.60 \text{ in.}/2)^2\pi - (0.48 \text{ in.}/2)^2\pi}{(0.60 \text{ in.}/2)^2\pi} \times 100 = 36.0\%$$

The tensile strength is read directly from the curve for copper (Figure 7.19*b*) as 50,000 psi (345 MPa). From Figure 7.19*c*, the ductility at 36%CW is about 7%EL.

RECOVERY, RECRYSTALLIZATION, AND GRAIN GROWTH

As outlined in the preceding paragraphs of this chapter, plastically deforming a polycrystalline metal specimen at temperatures that are low relative to its absolute melting temperature produces microstructural and property changes to include (1) a change in grain shape (Section 7.6), (2) strain hardening (Section 7.10), and (3) an increase in dislocation density (Section 7.3). Some fraction of the energy expended in deformation is stored in the metal as strain energy which is associated with tensile, compressive, and shear zones around the newly created dislocations (Section 7.3). Furthermore, other properties such as electrical conductivity (Section 19.8) and corrosion resistance may be modified as a consequence of plastic deformation.

These properties and structures may revert back to the precold-worked states by appropriate heat treatment. Such restoration results from two different processes that occur at elevated temperatures: **recovery** and **recrystallization,** which may be followed by **grain growth.**

7.11 RECOVERY

During recovery, some of the stored internal strain energy is relieved by virtue of dislocation motion (in the absence of an externally applied stress), as a result of enhanced atomic diffusion at the elevated temperature. There is some reduction in the number of dislocations, and dislocation configurations (similar to that shown in Figure 4.8) are produced having low strain energies. In addition, physical properties such as electrical and thermal conductivities and the like are recovered to their precold-worked states.

7.12 RECRYSTALLIZATION

Even after recovery is complete, the grains are still in a relatively high strain energy state. Recrystallization is the formation of a new set of strain-free and equiaxed grains that have low dislocation densities and are characteristic of the precold-worked condition. The driving force to produce this new grain structure is the difference in internal energy between the strained and unstrained material. The new grains form as very small nuclei and grow until they completely replace the parent material, processes that involve short-range diffusion. Several stages in the recrystallization process are represented in Figures 7.21*a* to 7.21*d*; in these photomicrographs, the small grains are those that have recrystallized. Thus recrystallization of cold-worked metals may be used to refine the grain structure.

Also, during recrystallization, the mechanical properties that were changed as a result of cold working are restored to their precold-worked values; that is, the metal

Figure 7.21 Photomicrographs showing several stages of the recrystallization and grain growth of brass. (*a*) Cold-worked (33%CW) grain structure. (*b*) Initial stage of recrystallization after heating 3 s at 580°C (1075°F); the very small grains are those that have recrystallized. (*c*) Partial replacement of cold-worked grains by recrystallized ones (4 s at 580°C). (*d*) Complete recrystallization (8 s at 580°C). (*e*) Grain growth after 15 min at 580°C. (*f*) Grain growth after 10 min at 700°C (1290°F). All photomicrographs 75×. (Photomicrographs courtesy of J. E. Burke, General Electric Company.)

(*continued*)

Figure 7.21 (*continued*)

becomes softer, weaker, yet more ductile. Some heat treatments are designed to allow recrystallization to occur with these modifications in the mechanical characteristics (Section 11.2).

Recrystallization is a process the extent of which depends on both time and temperature. The degree (or fraction) of recrystallization increases with time, as may be noted in the photomicrographs shown in Figures 7.21a–d. The explicit time dependence of recrystallization is addressed in more detail in Section 10.3.

The influence of temperature is demonstrated in Figure 7.22, which plots tensile strength and ductility (at room temperature) of a brass alloy as a function of the temperature and for a constant heat treatment time of 1 h. The grain structures found at the various stages of the process are also presented schematically.

The recrystallization behavior of a particular metal alloy is sometimes specified in terms of a **recrystallization temperature,** the temperature at which recrystallization just reaches completion in 1 h. Thus the recrystallization temperature for the brass alloy of Figure 7.22 is about 450°C (850°F). Typically, it is between one third and one half of the absolute melting temperature of a metal or alloy and depends on several factors, including the amount of prior cold work and the purity of the alloy. Increasing the percentage of cold work enhances the rate of recrystallization, with the result that the recrystallization temperature is lowered; this effect is shown in Figure 7.23. There exists some critical degree of cold work below which recrystallization cannot be made to occur, as shown in the figure; normally, this is between 2 and 20% cold work.

Recrystallization proceeds more rapidly in pure metals than in alloys. Thus alloying raises the recrystallization temperature, sometimes quite substantially. For pure metals, the recrystallization temperature is normally $0.3T_m$, where T_m is the

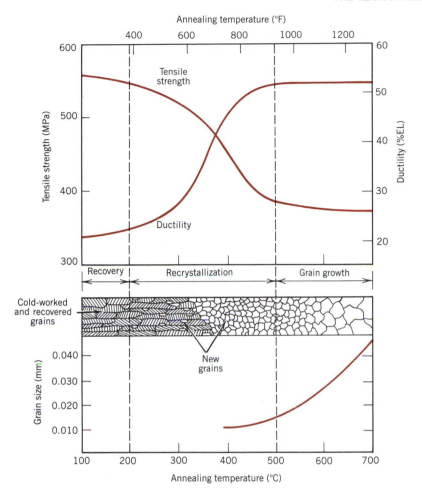

Figure 7.22 The influence of annealing temperature on the tensile strength and ductility of a brass alloy. Grain size as a function of annealing temperature is indicated. Grain structure during recovery, recrystallization, and grain growth stages is shown schematically. (Adapted from G. Sachs and K. R. Van Horn, *Practical Metallurgy, Applied Metallurgy and the Industrial Processing of Ferrous and Nonferrous Metals and Alloys,* American Society for Metals, 1940, p. 139.)

absolute melting temperature; for some commercial alloys it may run as high as $0.7T_m$. Recrystallization and melting temperatures for a number of metals and alloys are listed in Table 7.2.

Plastic deformation operations are often carried out at temperatures above the recrystallization temperature in a process termed *hot working,* described in Section 12.2. The material remains relatively soft and ductile during deformation because it does not strain harden, and thus large deformations are possible.

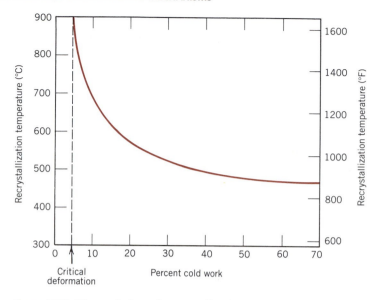

Figure 7.23 The variation of recrystallization temperature with percent cold work for iron. For deformations less than the critical (about 5%CW), recrystallization will not occur.

TABLE 7.2 Recrystallization and Melting Temperatures for Various Metals and Alloys

Metal	Recrystallization Temperature		Melting Temperature	
	°C	°F	°C	°F
Lead	−4	25	327	620
Tin	−4	25	232	450
Zinc	10	50	420	788
Aluminum (99.999 wt%)	80	176	660	1220
Copper (99.999 wt%)	120	250	1085	1985
Brass (60Cu–40Zn)	475	887	900	1652
Nickel (99.99 wt%)	370	700	1455	2651
Iron	450	840	1538	2800
Tungsten	1200	2200	3410	6170

EXAMPLE PROBLEM 7.3

A cylindrical rod of noncold-worked brass having an initial diameter of 0.25 in. (6.4 mm) is to be cold worked by drawing such that the cross-sectional area is reduced. It is required to have a cold-worked yield strength of at least 50,000 psi (345 MPa) and a ductility in excess of 20%EL; in addition, a final diameter of 0.20 in. (5 mm) is necessary. Describe the manner in which this procedure may be carried out.

SOLUTION

Let us first consider the consequences (in terms of yield strength and ductility) of cold working in which the brass specimen diameter is reduced from 0.25 in. (designated by d_0) to 0.20 in. (d_i). The %CW may be computed from Equation 7.6 as

$$\%CW = \frac{(d_0/2)^2\pi - (d_i/2)^2\pi}{(d_0/2)^2\pi} \times 100$$

$$= \frac{(0.25 \text{ in.}/2)^2\pi - (0.20 \text{ in.}/2)^2\pi}{(0.25 \text{ in.}/2)^2\pi} \times 100 = 36\%CW$$

From Figures 7.19a and 7.19c, a yield strength of 60,000 psi (414 MPa) and a ductility of 10%EL are attained from this deformation. According to the stipulated criteria, the yield strength is satisfactory; however, the ductility is too low.

Another processing alternative is a partial diameter reduction, followed by a recrystallization heat treatment in which the effects of the cold work are nullified. The required yield strength, ductility, and diameter are achieved through a second drawing step.

Again, reference to Figure 7.19a indicates that 20%CW is required to give a yield strength of 50,000 psi. On the other hand, from Figure 7.19c, ductilities greater than 20%EL are possible only for deformations of 23%CW or less. Thus during the final drawing operation, deformation must be between 20%CW and 23%CW. Let's take the average of these extremes, 21.5%CW, and then calculate the final diameter for the first drawing d_0', which becomes the original diameter for the second drawing. Again, using Equation 7.6,

$$21.5\%CW = \frac{(d_0'/2)^2\pi - (0.20 \text{ in.}/2)^2\pi}{(d_0'/2)^2\pi} \times 100$$

Now, solving for d_0' from the expression above gives

$$d_0' = 0.226 \text{ in. (5.7 mm)}$$

7.13 GRAIN GROWTH

After recrystallization is complete, the strain-free grains will continue to grow if the metal specimen is left at the elevated temperature—a phenomenon called **grain growth.** Grain growth does not need to be preceded by recovery and recrystallization; it may occur in all polycrystalline materials, metals and ceramics alike.

An energy is associated with grain boundaries, as explained in Section 4.5. As grains increase in size, the total boundary area decreases, yielding an attendant reduction in the total energy; this is the driving force for grain growth.

Grain growth occurs by the migration of grain boundaries. Obviously, not all grains can enlarge, but some grow at the expense of others that shrink. Thus the average grain size increases with time, and at any particular instant there will exist a range of grain sizes. Boundary motion is just the short-range diffusion of atoms from one side of the boundary to the other. The directions of boundary movement and atomic motion are opposite to each other, as shown in Figure 7.24.

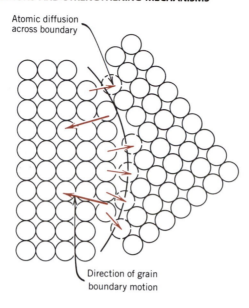

Atomic diffusion
across boundary

Direction of grain
boundary motion

Figure 7.24 Schematic representation of grain growth via atomic diffusion. (Adapted from L. H. Van Vlack, *Elements of Materials Science and Engineering*, 6th edition. Copyright © 1989 by Addison-Wesley Publishing Co. Reprinted by permission of Addison-Wesley Publishing Co., Inc. Reading, MA.)

For many polycrystalline materials, the grain diameter d varies with time t according to the relationship

$$d^n - d_0^n = Kt \qquad (7.7)$$

where d_0 is the initial grain diameter at $t = 0$, and K and n are time-independent constants; the value of n is generally equal to or greater than 2.

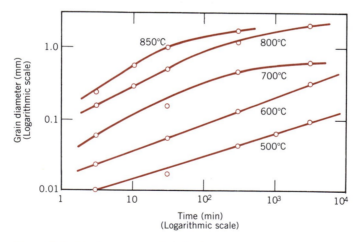

Figure 7.25 The logarithm of grain diameter versus the logarithm of time for grain growth in brass at several temperatures. (From J. E. Burke, "Some Factors Affecting the Rate of Grain Growth in Metals." Reprinted with permission from *Metallurgical Transactions*, Vol. 180, 1949, a publication of The Metallurgical Society of AIME, Warrendale, Pennsylvania.)

The dependence of grain size on time and temperature is demonstrated in Figure 7.25, a plot of the logarithm of grain size as a function of the logarithm of time for a brass alloy at several temperatures. At lower temperatures the time dependence is linear from the plot. Furthermore, grain growth proceeds more rapidly as temperature increases; that is, the curves are displaced upward to larger grain sizes. This is explained by the enhancement of diffusion rate with rising temperature.

The mechanical properties at room temperature of a fine-grained metal are usually superior to those of coarse-grained ones. If the grain structure of a single-phase alloy is coarser than that desired, refinement may be accomplished by plastically deforming the material, then subjecting it to a recrystallization heat treatment, as described above.

SUMMARY

On a microscopic level, plastic deformation corresponds to the motion of dislocations in response to an externally applied shear stress, a process termed "slip." Slip occurs on specific crystallographic planes and within these planes only in certain directions. A slip system represents a slip plane–slip direction combination, and operable slip systems depend on the crystal structure of the material.

The critical resolved shear stress is the minimum shear stress required to initiate dislocation motion; the yield strength of a single crystal depends on both the magnitude of the critical resolved shear stress and the orientation of slip components relative to the direction of the applied stress. For polycrystalline materials, slip occurs within each grain along the slip systems that are most favorably oriented with the applied stress; furthermore, during deformation, grains change shape in such a manner that coherency at the grain boundaries is maintained.

Under some circumstances limited plastic deformation may occur in BCC and HCP metals by mechanical twinning. Normally, twinning is important to the degree that accompanying crystallographic reorientations make the slip process more favorable.

Since the ease with which a material is capable of plastic deformation is a function of dislocation mobility, restricting dislocation motion increases hardness and strength. On the basis of this principle, three different strengthening mechanisms were discussed. Grain boundaries serve as barriers to dislocation motion; thus refining the grain size of a polycrystalline material renders it harder and stronger. Solid-solution hardening results from lattice strain interactions between impurity atoms and dislocations. And, finally, as a material is plastically deformed, the dislocation density increases, as does also the extent of repulsive dislocation–dislocation strain field interactions; strain hardening is just the enhancement of strength with increased plastic deformation.

The microstructural and mechanical characteristics of a plastically deformed metal specimen may be restored to their predeformed states by an appropriate heat treatment, during which recovery, recrystallization, and grain growth processes are allowed to occur. During recovery there is a reduction in dislocation density and alterations in dislocation configurations. Recrystallization is the formation of a new set of grains that are strain free; in addition, the material becomes softer and more ductile. Grain growth is the increase in average grain size of polycrystalline materials, which proceeds by grain boundary motion.

IMPORTANT TERMS AND CONCEPTS

Cold working	Lattice strain	Resolved shear stress
Critical resolved shear stress	Recovery	Slip
	Recrystallization	Slip system
Dislocation density	Recrystallization temperature	Solid-solution hardening
Grain growth		Strain hardening

REFERENCES

BYRNE, J. G., *Recovery, Recrystallization, and Grain Growth,* The Macmillan Co., New York, 1965.

HAYDEN, H. W., W. G. MOFFATT, and J. WULFF, *The Structure and Properties of Materials,* Vol. III, *Mechanical Behavior,* John Wiley & Sons, New York, 1965.

HIRTH, J. P., and J. LOTHE, *Theory of Dislocations,* 2nd edition, Wiley-Interscience, New York, 1982.

HULL, D., *Introduction to Dislocations,* 3rd edition, Pergamon Press, Inc., Elmsford, NY, 1984.

PECKNER, D. (Editor), *The Strengthening of Metals,* Reinhold Publishing Corporation, New York, 1964. Chapters 1–3.

READ, W. T., JR., *Dislocations in Crystals,* McGraw-Hill Book Company, New York, 1953.

WEERTMAN, J. and J. R. WEERTMAN, *Elementary Dislocation Theory,* The Macmillan Co., New York, 1964.

QUESTIONS AND PROBLEMS

7.1 To provide some perspective on the dimensions of atomic defects, consider a metal specimen that has a dislocation density of 10^4 mm^{-2}. Suppose that all the dislocations in 1000 mm^3 (1 cm^3) were somehow removed and linked end to end. How far (in miles) would this chain extend? Now suppose that the density is increased to 10^{10} mm^{-2} by cold working. What would be the chain length of dislocations in 1000 mm^3 of material?

7.2 Consider two edge dislocations of opposite sign and having slip planes that are separated by several atomic distances as indicated in the diagram. Briefly describe the defect that results when these two dislocations become aligned with each other.

7.3   Is it possible for two screw dislocations of opposite sign to annihilate each other? Explain your answer.

7.4   For each of edge, screw, and mixed dislocations, cite the relationship between the direction of the applied shear stress and the direction of dislocation line motion.

7.5   (a) Define a slip system. (b) Do all metals have the same slip system? Why or why not?

7.6   (a) Compare planar densities (Section 3.10) for the (100), (110), and (111) planes for FCC. (b) Compare planar densities for the (100), (110), and (111) planes for BCC.

7.7   One slip system for the BCC crystal structure is $\{110\}\langle 111\rangle$. In a manner similar to Figure 7.6b sketch a $\{110\}$-type plane for the BCC structure, representing atom positions with circles. Now, using arrows, indicate two different $\langle 111\rangle$ slip directions within this plane.

7.8   Determine the slip system for the simple cubic crystal structure, a unit cell of which is shown in Figure 3.22.

7.9   One slip system for the HCP crystal structure is $\{0001\}\langle 11\bar{2}0\rangle$. In a manner similar to Figure 7.6b, sketch a $\{0001\}$-type plane for the HCP structure, and using arrows, indicate three different $\langle 11\bar{2}0\rangle$ slip directions within this plane. You might find Figure 3.7 helpful.

7.10 Explain the difference between resolved shear stress and critical resolved shear stress.

7.11 Sometimes $\cos \phi \cos \lambda$ in Equation 7.1 is termed the *Schmid factor*. Determine the magnitude of the Schmid factor for a FCC single crystal oriented with its [100] direction parallel to the loading axis.

7.12 Consider a metal single crystal oriented such that the normal to the slip plane and the slip direction are at angles of 60° and 35°, respectively, with the tensile axis. If the critical resolved shear stress is 900 psi (6.2 MPa), will an applied stress of 1750 psi (12 MPa) cause the single crystal to yield? If not, what stress will be necessary?

7.13 A single crystal of zinc is oriented for a tensile test such that its slip plane normal makes an angle of 65° with the tensile axis. Three possible slip directions make angles of 30°, 48°, and 78° with the same tensile axis. (a) Which of these three slip directions is most favored? (b) If plastic deformation begins at a tensile stress of 355 psi (2.5 MPa), determine the critical resolved shear stress for zinc.

7.14 Consider a single crystal of aluminum oriented such that a tensile stress is applied along a [001] direction. If slip occurs on a (111) plane and in a $[\bar{1}01]$ direction, and is initiated at an applied tensile stress of 147 psi (1.0 MPa), compute the critical resolved shear stress.

7.15 The critical resolved shear stress for copper is 70 psi (0.48 MPa). Determine the maximum possible yield strength for a single crystal of Cu pulled in tension.

7.16 List four major differences between deformation by twinning and deformation by slip relative to mechanism, conditions of occurrence, and final result.

7.17 Briefly explain why small-angle grain boundaries are not as effective in interfering with the slip process as are high-angle grain boundaries.

7.18 Briefly explain why HCP metals are typically more brittle than FCC and BCC metals.

7.19 Describe in your own words the three strengthening mechanisms discussed in this chapter (i.e., grain size reduction, solid solution hardening, and strain hardening). Be sure to explain how dislocations are involved in each of the strengthening techniques.

7.20 **(a)** From the plot of yield strength versus (grain diameter)$^{-1/2}$ for a 70Cu–30Zn cartridge brass, Figure 7.15, determine values for the constants σ_0 and k_y in Equation 7.5. **(b)** Now predict the yield strength of this alloy when the average grain diameter is 1.0×10^{-3} mm.

7.21 The lower yield point for an iron that has an average grain diameter of 5×10^{-2} mm is 19,500 psi (135 MPa). At a grain diameter of 8×10^{-3} mm, the yield point increases to 37,500 psi (260 MPa). At what grain diameter will the lower yield point be 30,000 psi (205 MPa)?

7.22 If it is assumed that the plots in Figures 7.15 and 7.16 are for noncold-worked brass alloys, determine the following:
(a) The composition of the brass alloy in Figure 7.19.
(b) The grain size of the alloy in Figure 7.19; assume its composition is the same as the alloy in Figure 7.15.

7.23 When making hardness measurements, what will be the effect of making an indentation very close to a preexisting indentation? Why?

7.24 **(a)** Show, for a tensile test, that

$$\%\mathrm{CW} = \left(\frac{\epsilon}{\epsilon + 1} \right) \times 100$$

if there is no change in specimen volume during the deformation process (i.e., $A_0 l_0 = A_d l_d$). **(b)** Using the result of part a, compute the percent cold work experienced by naval brass (the stress–strain behavior of which is shown in Figure 6.11) when a stress of 60,000 psi (415 MPa) is applied.

7.25 Two previously undeformed cylindrical specimens of an alloy are to be strain hardened by reducing their cross-sectional areas (while maintaining their circular cross sections). For one specimen, the initial and deformed radii are 15 mm and 12 mm, respectively. The second specimen, with an initial radius of 11 mm, must have the same deformed hardness as the first specimen; compute the second specimen's radius after deformation.

7.26 Two previously undeformed specimens of the same metal are to be plastically deformed by reducing their cross-sectional areas. One has a circular cross section, and the other is rectangular; during deformation the circular cross section is to remain circular, and the rectangular is to remain as such. Their original and deformed dimensions are as follows:

	Circular (diameter, mm)	Rectangular (mm)
Original dimensions	15.2	125 × 175
Deformed dimensions	11.4	75 × 200

Which of these specimens will be the hardest after plastic deformation, and why?

7.27 A cylindrical specimen of cold worked copper has a ductility (%EL) of 25%. If its cold-worked radius is 0.40 in. (10 mm), what was its radius before deformation?

7.28 **(a)** What is the approximate ductility (%EL) of a brass that has a yield strength of 40,000 psi (275 MPa)? **(b)** What is the approximate HB of a 1040 steel having a yield strength of 100,000 psi (690 MPa)?

7.29 Determine whether or not it is possible to cold work steel so as to give a minimum Brinell hardness of 240 and at the same time have a ductility of at least 15%EL. Justify your decision.

7.30 Determine whether or not it is possible to cold work brass so as to give a minimum Brinell hardness of 150 and at the same time have a ductility of at least 20%EL. Justify your decision.

7.31 A cylindrical specimen of cold-worked steel has a Brinell hardness of 240.
   (a) Estimate its ductility in percent elongation.
   (b) If the specimen remained cylindrical during deformation and its uncold-worked radius was 0.40 in. (10.2 mm), determine its radius after deformation.

7.32 It is necessary to select a metal alloy for an application that requires a yield strength of at least 50,000 psi (345 MPa) while maintaining a minimum ductility (%EL) of 20%. If the metal may be cold worked, decide which of the following are candidates: copper, brass, and a 1040 steel. Why?

7.33 Experimentally, it has been observed for single crystals of a number of metals that the critical resolved shear stress τ_{crss} is a function of the dislocation density ρ_D as

$$\tau_{crss} = \tau_0 + A\sqrt{\rho_D}$$

where τ_0 and A are constants. For copper, the critical resolved shear stress is 100 psi (0.69 MPa) at a dislocation density of 10^4 mm^{-2}. If it is known that the value of τ_0 for copper is 10 psi (0.069 MPa), compute the τ_{crss} at a dislocation density of 10^6 mm^{-2}.

7.34 Briefly cite the differences between recovery and recrystallization processes.

7.35 Estimate the fraction of recrystallization from the photomicrograph in Figure 7.21c.

7.36 Explain the differences in grain structure for a metal that has been cold worked and one that has been cold worked and then recrystallized.

7.37 Briefly explain why some metals (e.g., lead and tin) do not strain harden when deformed at room temperature.

7.38 A cylindrical rod of 1040 steel originally 0.6 in. (15.2 mm) in diameter is to be cold worked by drawing; the circular cross section will be maintained during deformation. A cold-worked tensile strength in excess of 122,000 psi (840 MPa) and a ductility of at least 12%EL are desired. Furthermore, the final diameter must be 0.4 in. (10 mm). Explain how this may be accomplished.

7.39 A cylindrical rod of brass originally 0.40 in. (10.2 mm) in diameter is to be cold worked by drawing; the circular cross section will be maintained during deformation. A cold-worked yield strength in excess of 55,000 psi (380 MPa) and a ductility of at least 15%EL are desired. Furthermore, the final diameter must be 0.30 in. (7.6 mm). Explain how this may be accomplished.

7.40 A cylindrical brass rod having a minimum tensile strength of 65,000 psi (450 MPa), a ductility of at least 13%EL, and a final diameter of 0.50 in. (12.7 mm) is desired. Some 0.75 in. (19.0 mm) diameter brass stock which has been cold worked 35% is available. Describe the procedure you would follow to obtain this material. Assume that brass experiences cracking at 65%CW.

7.41 (a) What is the driving force for recrystallization? (b) For grain growth?

7.42 (a) From Figure 7.25, compute the length of time required for the average grain diameter to increase from 0.01 to 0.1 mm at 500°C for this brass material. (b) Repeat the calculation at 600°C.

7.43 The average grain diameter for a brass material was measured as a function of time at 650°C, which is tabulated below at two different times:

Time (min)	Grain Diameter (mm)
30	3.9×10^{-2}
90	6.6×10^{-2}

(a) What was the original grain diameter?
(b) What grain diameter would you predict after 150 min at 650°C?

7.44 An undeformed specimen of some alloy has an average grain diameter of 0.040 mm. You are asked to reduce its average grain diameter to 0.010 mm. Is this possible? If so, explain the procedures you would use and name the processes involved. If it is not possible, explain why.

7.45 Grain growth is strongly dependent on temperature (i.e., rate of grain growth increases with increasing temperature), yet temperature is not explicitly given as a part of Equation 7.7.
(a) Into which of the parameters in this expression would you expect temperature to be included?
(b) On the basis of your intuition, cite an explicit expression for this temperature dependence.

7.46 An uncold-worked brass specimen of average grain size 0.01 mm has a yield strength of 150 MPa (21,750 psi). Estimate the yield strength of this alloy after it has been heated to 500°C for 1000 s, if it is known that the value of σ_0 is 25 MPa (3625 psi).

FAILURE

A oil barge that fractured in a brittle manner by crack propagation around its girth. (Photography by Neal Boenzi. Reprinted with permission from *The New York Times*.)

8.1 INTRODUCTION

The failure of engineering materials is almost always an undesirable event for several reasons; these include human lives that are put in jeopardy, economic losses, and the interference with the availability of products and services. Even though the causes of failure and the behavior of materials may be known, prevention of failures is difficult to guarantee. The usual causes are improper materials selection and processing and inadequate design of the component or its misuse. It is the responsibility of the engineer to anticipate and plan for possible failure and, in the event that failure does occur, to assess its cause and then take appropriate preventative measures against future incidents.

Topics to be addressed in this chapter are the following: simple fracture (both ductile and brittle modes), fundamentals of fracture mechanics, impact fracture testing, the ductile-to-brittle transition, fatigue, and creep. These discussions include failure mechanisms, testing techniques, and methods by which failure may be prevented or controlled.

FRACTURE

8.2 FUNDAMENTALS OF FRACTURE

Simple fracture is the separation of a body into two or more pieces in response to an imposed stress that is static (i.e., constant or slowly changing with time) and at temperatures that are low relative to the melting temperature of the material. The applied stress may be tensile, compressive, shear, or torsional; the present discussion will be confined to fractures that result from uniaxial tensile loads. For engineering materials, two fracture modes are possible: **ductile** and **brittle.** Classification is based on the ability of a material to experience plastic deformation. Ductile materials typically exhibit substantial plastic deformation with high energy absorption before fracture. On the other hand, there is normally little or no plastic deformation with low energy absorption accompanying a brittle fracture. The tensile stress–strain behaviors of both fracture types may be reviewed in Figure 6.12.

"Ductile" and "brittle" are relative terms; whether a particular fracture is one mode or the other depends on the situation. Ductility may be quantified in terms of percent elongation (Equation 6.10) and percent area reduction (Equation 6.11). Furthermore, ductility is a function of temperature of the material, the strain rate, and the stress state. The disposition of normally ductile materials to fail in a brittle manner is discussed in Section 8.6.

Any fracture process involves two steps—crack formation and propagation— in response to an imposed stress. The mode of fracture is highly dependent on the mechanism of crack propagation. Ductile fracture is characterized by extensive plastic deformation in the vicinity of an advancing crack. Furthermore, the process proceeds relatively slowly as the crack length is extended. Such a crack is often said to be *stable*. That is, it resists any further extension unless there is an increase in the applied stress. In addition, there will ordinarily be evidence of appreciable gross deformation at the fracture surfaces (e.g., twisting and tearing). On the other hand, for brittle fracture, cracks may spread extremely rapidly, with very little accompanying plastic deformation. Such cracks may be said to be *unstable,* and crack propagation, once started, will continue spontaneously without an increase in magnitude of the applied stress.

Ductile fracture is almost always preferred for two reasons. First, brittle fracture occurs suddenly and catastrophically without any warning; this is a consequence of the spontaneous and rapid crack propagation. On the other hand, for ductile fracture, the presence of plastic deformation gives warning that fracture is imminent, allowing preventive measures to be taken. Second, more strain energy is required to induce ductile fracture inasmuch as ductile materials are generally tougher. Under the action of an applied tensile stress, most metal alloys are ductile, whereas ceramics are notably brittle, and polymers may exhibit both types of fracture.

8.3 DUCTILE FRACTURE

Ductile fracture surfaces will have their own distinctive features on both macroscopic and microscopic levels. Figure 8.1 shows schematic representations for two characteristic macroscopic fracture profiles. The configuration shown in Figure 8.1*a* is found for extremely soft metals, such as pure gold and lead at room temperature, and other metals, polymers, and inorganic glasses at elevated temperatures. These highly ductile materials neck down to a point fracture, showing virtually 100% area reduction.

The most common type of tensile fracture profile for ductile metals is that represented in Figure 8.1*b*, which fracture is preceded by only a moderate amount of necking. The fracture process normally occurs in several stages (Figure 8.2). First, after necking begins, small cavities, or microvoids, form in the interior of the cross section, as indicated in Figure 8.2*b*. Next, as deformation continues, these microvoids enlarge, come together, and coalesce to form an elliptical crack, which has its long axis perpendicular to the stress direction. The crack continues to grow in a direction parallel to its major axis by this microvoid coalescence process (Figure 8.2*c*). Finally, fracture ensues by the rapid propagation of a crack around the outer perimeter of the neck (Figure 8.2*d*), by shear deformation at an angle of about 45° with the tensile axis—this is the angle at which the shear stress is a maximum. Sometimes a fracture having this characteristic surface contour is termed a *cup-and-cone fracture* because one of the mating surfaces is in the form of a cup, the other like a cone. In this type of fractured specimen (Figure 8.3*a*), the central interior region of the surface has an irregular and fibrous appearance, which is indicative of plastic deformation.

Much more detailed information regarding the mechanism of fracture is available from microscopic examination, normally using electron microscopy. Studies of this

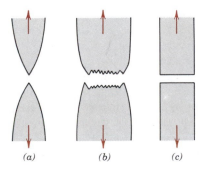

Figure 8.1 (*a*) Highly ductile fracture in which the specimen necks down to a point. (*b*) Moderately ductile fracture after some necking. (*c*) Brittle fracture without any plastic deformation.

(*a*)   (*b*)   (*c*)

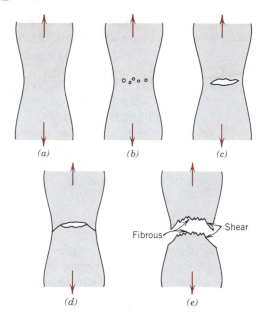

Figure 8.2 Stages in the cup-and-cone fracture. (*a*) Initial necking. (*b*) Small cavity formation. (*c*) Coalescence of cavities to form a crack. (*d*) Crack propagation. (*e*) Final shear fracture at a 45° angle relative to the tensile direction. (From K. M. Ralls, T. H. Courtney, and J. Wulff, *Introduction to Materials Science and Engineering,* p. 468. Copyright © 1976 by John Wiley & Sons, New York. Reprinted by permission of John Wiley & Sons, Inc.)

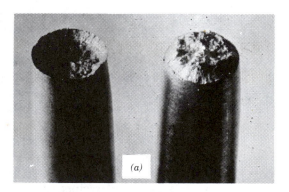

Figure 8.3 (*a*) Cup-and-cone fracture in aluminum. (*b*) Brittle fracture in a mild steel. From H. W. Hayden, W. G. Moffatt, and J. Wulff, *The Structure and Properties of Materials,* Vol. III, *Mechanical Behavior,* p. 144. Copyright © 1965 by John Wiley & Sons, New York. Reprinted by permission of John Wiley & Sons, Inc.)

type are termed *fractographic*. The electron microscope is preferred for fractographic examinations since it has a much better resolution and depth of field than does the optical microscope; these characteristics are necessary to reveal the topographical features of fracture surfaces. Ordinarily, the scanning electron microscope (SEM) is used, wherein the specimen is viewed directly.

When the fibrous central region of a cup-and-cone fracture surface is examined with the electron microscope at a high magnification, it will be found to consist of numerous spherical "dimples" (Figure 8.4*a*); this structure is characteristic of fracture resulting from uniaxial tensile failure. Each dimple is one half of a microvoid that formed and then separated during the fracture process. Dimples also form on the 45° shear lip of the cup-and-cone fracture. However, these will be elongated or C-shaped,

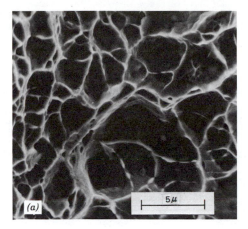 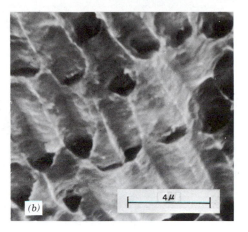

Figure 8.4 (*a*) Scanning electron fractograph showing spherical dimples characteristic of ductile fracture resulting from uniaxial tensile loads. (*b*) Scanning electron fractograph showing parabolic-shaped dimples characteristic of ductile fracture resulting from shear loading. (From R. W. Hertzberg, *Deformation and Fracture Mechanics of Engineering Materials,* 3rd edition. Copyright © 1989 by John Wiley & Sons, New York. Reprinted by permission of John Wiley & Sons, Inc.)

as shown in Figure 8.4*b*. This parabolic shape may be indicative of shear failure. Furthermore, other microscopic fracture surface features are also possible. Fractographs such as those shown in Figures 8.4*a* and 8.4*b* provide valuable information in the analyses of fracture, such as the fracture mode, the stress state, as well as the site of crack initiation.

8.4 BRITTLE FRACTURE

Brittle fracture takes place without any appreciable deformation, and by rapid crack propagation. The direction of crack motion is very nearly perpendicular to the direction of the applied tensile stress and yields a relatively flat fracture surface, as indicated in Figure 8.1*c*.

Fracture surfaces of materials that failed in a brittle manner will have their own distinctive patterns; any signs of gross plastic deformation will be absent. For example, in some steel pieces, a series of V-shaped "chevron" markings may form near the center of the fracture cross section that point back toward the crack initiation site (Figure 8.5*a*). Other brittle fracture surfaces contain lines or ridges that radiate from the origin of the crack in a fanlike pattern (Figure 8.5*b*). Often, both of these marking patterns will be sufficiently coarse to be discerned with the naked eye. For very hard and fine-grained metals, there will be no discernible fracture pattern. Brittle fracture in amorphous materials, such as ceramic glasses, yields a relatively shiny and smooth surface.

For most brittle crystalline materials, crack propagation corresponds to the successive and repeated breaking of atomic bonds along specific crystallographic planes; such a process is termed *cleavage*. This type of fracture is said to be **transgranular** (or *transcrystalline*), because the fracture cracks pass through the grains. Macroscopically, the fracture surface may have a grainy or faceted texture (Figure 8.3*b*), as a result of

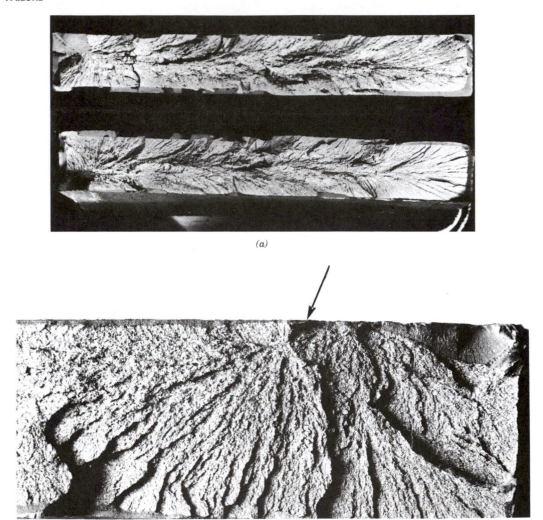

(a)

(b)

Figure 8.5 (a) Photograph showing V-shaped "chevron" markings characteristic of brittle fracture. (From R. W. Hertzberg, *Deformation and Fracture Mechanics of Engineering Materials,* 3rd edition. Copyright © 1989 by John Wiley & Sons, New York. Reprinted by permission of John Wiley & Sons, Inc. Photograph courtesy of Roger Slutter, Lehigh University.) (b) Photograph of brittle fracture surface showing radial fan-shaped ridges. Arrow indicates origin of crack. (Reproduced with permission from D. J. Wulpi, *Understanding How Components Fail,* American Society for Metals, Materials Park, OH, 1985.)

changes in orientation of the cleavage planes from grain to grain. This feature is more evident in the scanning electron micrograph shown in Figure 8.6a.

In some alloys, crack propagation is along grain boundaries; this fracture is termed **intergranular.** Figure 8.6b is a scanning electron micrograph showing a typical intergranular fracture, in which the three-dimensional nature of the grains may be seen. This type of fracture normally results subsequent to the occurrence of processes that weaken or embrittle grain boundary regions.

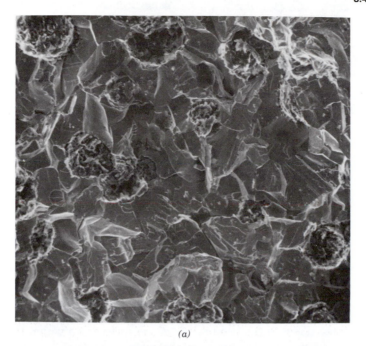

(a)

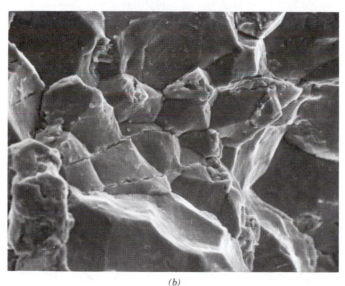

(b)

Figure 8.6 (*a*) Scanning electron fractograph of ductile cast iron showing a transgranular fracture surface. (From V. J. Colangelo and F. A. Heiser, *Analysis of Metallurgical Failures,* 2nd edition. Copyright © 1987 by John Wiley & Sons, New York. Reprinted by permission of John Wiley & Sons, Inc.) (*b*) Scanning electron fractograph showing an intergranular fracture surface. (Reproduced with permission from D. J. Wulpi, *Understanding How Components Fail,* American Society for Metals, Materials Park, OH, 1985.)

8.5 PRINCIPLES OF FRACTURE MECHANICS

Brittle fracture of normally ductile materials, such as that shown on page 189, has demonstrated the need for a better understanding of the mechanisms of fracture. Extensive research endeavors over the past several decades have led to the evolution of the field of **fracture mechanics.** Knowledge gleaned therefrom allows quantification of the relationships between material properties, stress level, the presence of crack-producing flaws, and crack propagation mechanisms. Design engineers are now better equipped to anticipate, and thus prevent, structural failures. The present discussion centers on some of the fundamental principles of the mechanics of fracture.

Stress Concentration

The fracture strength of a solid material is a function of the cohesive forces that exist between atoms. On this basis, the theoretical cohesive strength of a brittle elastic solid has been estimated to be approximately $E/10$, where E is the modulus of elasticity. The experimental fracture strengths of most engineering materials normally lie between 10 and 1000 times below this theoretical value. In the 1920s, A. A. Griffith proposed that this discrepancy between theoretical cohesive strength and observed fracture strength could be explained by the presence of very small, microscopic flaws or cracks that always exist under normal conditions at the surface and within the interior of a body of material. These flaws are a detriment to the fracture strength because an applied stress may be amplified or concentrated at the tip, the magnitude of this amplification depending on crack orientation and geometry. This phenomenon is demonstrated in Figure 8.7, a stress profile across a cross section containing an internal crack. As indicated by this profile, the magnitude of this localized stress diminishes

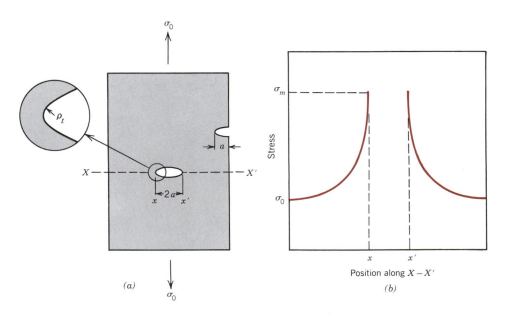

Figure 8.7 (*a*) The geometry of surface and internal cracks. (*b*) Schematic stress profile along the line $X-X'$ in (*a*), demonstrating stress amplification at crack tip positions.

with distance away from the crack tip. At positions far removed, the stress is just the nominal stress σ_0, or the load divided by the specimen cross-sectional area. Due to their ability to amplify an applied stress in their locale, these flaws are sometimes called **stress raisers.**

If it is assumed that a crack has an elliptical shape and is oriented with its long axis perpendicular to the applied stress, the maximum stress at the crack tip, σ_m, may be approximated by

$$\sigma_m = 2\sigma_0 \left(\frac{a}{\rho_t}\right)^{1/2} \tag{8.1}$$

where σ_0 is the magnitude of the nominal applied tensile stress, ρ_t is the radius of curvature of the crack tip (Figure 8.7a), and a represents the length of a surface crack, or half of the length of an internal one. Thus for a relatively long microcrack that has a small tip radius of curvature, the factor $(a/\rho_t)^{1/2}$ may be very large. This will yield a value of σ_m that is many times that of σ_0.

Sometimes the ratio σ_m/σ_0 is denoted as the *stress concentration factor K_t*:

$$K_t = \frac{\sigma_m}{\sigma_0} = 2\left(\frac{a}{\rho_t}\right)^{1/2} \tag{8.2}$$

which is simply a measure of the degree to which an external stress is amplified at the tip of a small crack.

By way of comment, it should be said that stress amplification is not restricted to these microscopic defects; it may occur at macroscopic internal discontinuities (e.g., holes), at sharp corners, and notches in large structures. Figure 8.8 shows theoretical stress concentration factor curves for several simple and common components.

Furthermore, the effect of a stress raiser is more significant in brittle than in ductile materials. For a ductile material, plastic deformation ensues when the maximum stress exceeds the yield strength. This leads to a more uniform distribution of stress in the vicinity of the stress raiser and to the development of a maximum stress concentration factor less than the theoretical value. Such yielding and stress redistribution do not occur to any appreciable extent around flaws and discontinuities in brittle materials; therefore, essentially the theoretical stress concentration will result.

Griffith then went on to propose that all brittle materials contain a population of small cracks and flaws that have a variety of sizes, geometries, and orientations. Fracture will result when, upon application of a tensile stress, the theoretical cohesive strength of the material is exceeded at the tip of one of these flaws. This leads to the formation of a crack that then rapidly propagates. If no flaws were present, the fracture strength would be equal to the cohesive strength of the material. Very small and virtually defect-free metallic and ceramic whiskers have been grown with fracture strengths that approach their theoretical values.

Griffith Theory of Brittle Fracture

During the propagation of a crack, there is a release of what is termed the *elastic strain energy,* some of the energy that is stored in the material as it is elastically deformed. Furthermore, during the crack extension process, new free surfaces are created at the faces of a crack, which give rise to an increase in surface energy of the system. Griffith developed a criterion for crack propagation of an elliptical crack

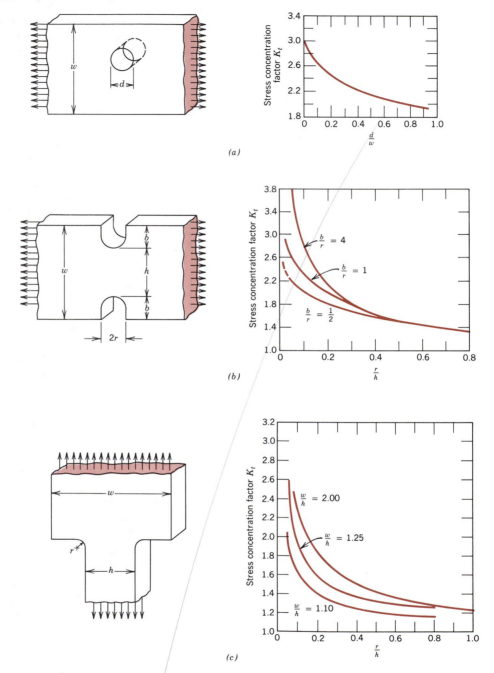

Figure 8.8 Theoretical stress concentration factor curves for three simple geometrical shapes. (From G. H. Neugebauer, *Prod. Eng.* (NY), Vol. 14, pp. 82–87, 1943.)

(Figure 8.7a) by performing an energy balance using these two energies. He demonstrated that the critical stress σ_c required for crack propagation in a brittle material is described by

$$\sigma_c = \left(\frac{2E\gamma_s}{\pi a}\right)^{1/2}$$

(8.3)

where

E = modulus of elasticity
γ_s = specific surface energy
a = one half the length of an internal crack

Worth noting is that this expression does not involve the crack tip radius ρ_t, as does the stress concentration equation (Equation 8.1); however, it is assumed that the radius is sufficiently sharp (on the order of the interatomic spacing) so as to raise the local stress at the tip above the cohesive strength of the material.

The previous development applies only to completely brittle materials for which there is no plastic deformation. Most metals and many polymers do experience some plastic deformation during fracture; this leads to a blunting of the tip of a crack, a decrease in the crack tip radius, and subsequently an increase in the fracture strength. Mathematically, this may be accommodated by replacing γ_s in Equation 8.3 by $\gamma_s + \gamma_p$, where γ_p represents a plastic deformation energy associated with crack extension. For highly ductile materials, it may be the case that $\gamma_p \gg \gamma_s$.

In the 1950s, G. R. Irwin chose to incorporate both γ_s and γ_p into a single term, $\mathscr{G}$, as

$$\mathscr{G} = 2(\gamma_s + \gamma_p)$$

(8.4)

$\mathscr{G}$ is known as the *strain energy release rate,* and crack extension occurs when it exceeds a critical value, $\mathscr{G}_c$.

Stress Analysis of Cracks

As we continue to explore the development of fracture mechanics, it is worthwhile to examine the stress distributions in the vicinity of the tip of an advancing crack. There are three fundamental ways, or modes, by which a load can operate on a crack, and each will affect a different crack surface displacement; these are illustrated in Figure 8.9. Mode I is an opening (or tensile) load, whereas modes II and III are sliding

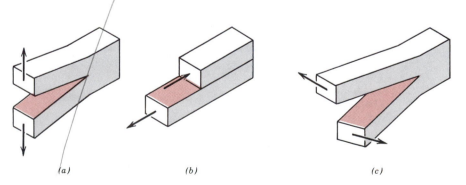

(a)   (b)   (c)

Figure 8.9 The three modes of crack surface displacement. (a) Mode I, opening or tensile mode; (b) mode II, sliding mode; and (c) mode III, tearing mode.

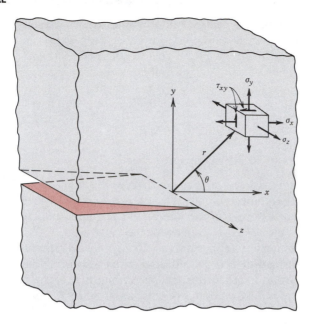

Figure 8.10 The stresses acting in front of a crack that is loaded in a tensile mode I configuration.

and tearing modes, respectively. Mode I is encountered most frequently, and only it will be treated in the ensuing discussion on fracture mechanics.

For this mode I configuration, the stresses acting on an element of material are shown in Figure 8.10. Using elastic theory principles and the notation indicated, tensile (σ_x and σ_y) and shear (τ_{xy}) stresses are functions of both radial distance r and the angle θ as follows:[1]

$$\sigma_x = \frac{K}{\sqrt{2\pi r}} f_x(\theta) \tag{8.5a}$$

$$\sigma_y = \frac{K}{\sqrt{2\pi r}} f_y(\theta) \tag{8.5b}$$

$$\tau_{xy} = \frac{K}{\sqrt{2\pi r}} f_{xy}(\theta) \tag{8.5c}$$

If the plate is thin relative to the dimensions of the crack, then $\sigma_z = 0$, or a condition of *plane stress* is said to exist. At the other extreme (a relatively thick plate),

---

[1] The $f(\theta)$ functions are as follows:

$$f_x(\theta) = \cos\frac{\theta}{2}\left(1 - \sin\frac{\theta}{2}\sin\frac{3\theta}{2}\right)$$

$$f_y(\theta) = \cos\frac{\theta}{2}\left(1 + \sin\frac{\theta}{2}\sin\frac{3\theta}{2}\right)$$

$$f_{xy}(\theta) = \sin\frac{\theta}{2}\cos\frac{\theta}{2}\cos\frac{3\theta}{2}$$

$\sigma_z = v(\sigma_x + \sigma_y)$, and the state is referred to as **plane strain** (since $\epsilon_z = 0$); v in this expression is Poisson's ratio.

In Equations 8.5, the parameter K is termed the **stress intensity factor;** its use provides for a convenient specification of the stress distribution around a flaw. It should be noted that this stress intensity factor and the stress concentration factor K_t in Equation 8.2, although similar, are not equivalent.

The value of the stress intensity factor is a function of the applied stress, the size and position of the crack, as well as the geometry of the solid piece in which the crack is located.

Fracture Toughness

In the above discussion, a criterion was developed for the crack propagation in a brittle material containing a flaw; fracture occurs when the applied stress level exceeds some critical value σ_c (Equation 8.3). Similarly, since the stresses in the vicinity of a crack tip can be defined in terms of the stress intensity factor, a critical value of this parameter exists, which may be used to specify the conditions for brittle fracture; this critical value is termed the **fracture toughness** K_c. In general, it may be expressed in the form

$$K_c = Y\sigma\sqrt{\pi a} \tag{8.6}$$

where Y is a dimensionless parameter that depends on both the specimen and crack geometries. For example, for the plate of infinite width in Figure 8.11a, $Y = 1.0$; or for a plate of semi-infinite width containing an edge crack of length a (Figure 8.11b), $Y = 1.1$.

By definition, fracture toughness is a property that is the measure of a material's resistance to brittle fracture when a crack is present. It should also be noted that fracture toughness has the unusual units of $\text{psi}\sqrt{\text{in.}}$ ($\text{MPa}\sqrt{\text{m}}$).

For relatively thin specimens, the value of K_c will depend on and decrease with increasing specimen thickness B, as indicated in Figure 8.12. Eventually, K_c becomes independent of B, at which time the condition of plane strain is said to exist.[2] The constant K_c value for thicker specimens is known as the **plane strain fracture toughness** K_{Ic}, which is also defined by

$$K_{Ic} = Y\sigma\sqrt{\pi a} \tag{8.7}$$

It is the fracture toughness normally cited since its value is always less than K_c. The I subscript for K_{Ic} denotes that this critical value of K is for mode I crack displacement, as illustrated in Figure 8.9a. Brittle materials, for which appreciable plastic deformation is not possible in front of an advancing crack, have low K_{Ic} values and are vulnerable to catastrophic failure. On the other hand, K_{Ic} values are relatively large for ductile materials. Fracture mechanics is especially useful in predicting catastrophic failure in

---

[2] Experimentally, it has been verified that for plane strain conditions

$$B \geq 2.5 \left(\frac{K_{Ic}}{\sigma_y}\right)^2 \tag{8.8}$$

where σ_y is the 0.002 strain offset yield strength of the material.

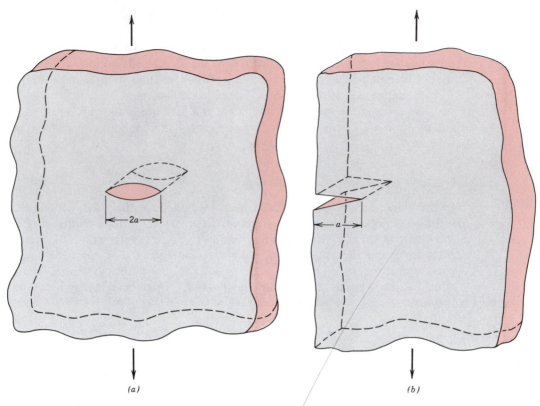

Figure 8.11 Schematic representations of (*a*) an interior crack in a plate of infinite width, and (*b*) an edge crack in a plate of semi-infinite width.

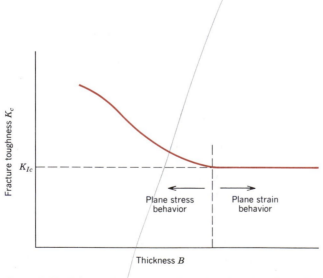

Figure 8.12 Schematic representation showing the effect of plate thickness on fracture toughness.

TABLE 8.1 Room-Temperature Yield Strength and Plane Strain Fracture Toughness Data for Selected Engineering Materials

Material	Yield Strength		K_{Ic}	
	psi $\times 10^3$	MPa	psi$\sqrt{\text{in.}} \times 10^3$	MPa$\sqrt{\text{m}}$
Metals				
Aluminum alloy[a] (2024-T351)	47	325	33	36
Aluminum alloy[a] (7075-T651)	73	505	26	29
Alloy steel[a] (4340 tempered @ 260°C)	238	1640	45.8	50.0
Alloy steel[a] (4340 tempered @ 425°C)	206	1420	80.0	87.4
Titanium alloy[a] (Ti-6Al-4V)	130	910	40–60	44–66
Ceramics				
Aluminum oxide	—	—	2.7–4.8	3.0–5.3
Soda-lime glass	—	—	0.64–0.73	0.7–0.8
Concrete	—	—	0.18–1.27	0.2–1.4
Polymers				
Polymethyl methacrylate (PMMA)	—	—	0.9	1.0
Polystyrene (PS)	—	—	0.73–1.0	0.8–1.1

[a] **Source:** Adapted with permission from *1989 Guide to Selecting Engineered Materials,* ASM INTERNATIONAL, Materials Park, OH, 1989.

materials having intermediate ductilities. Plane strain fracture toughness values for a number of different materials are presented in Table 8.1.

The stress intensity factor K in Equations 8.5 and the plane strain fracture toughness K_{Ic} are related to one another in the same sense as are stress and yield strength. A material may be subjected to many values of stress; however, there is a specific stress level at which the material plastically deforms—that is, the yield strength. Likewise, a variety of K's are possible, whereas K_{Ic} is unique for a particular material.

Several different testing techniques are used to measure K_{Ic}. Virtually any specimen size and shape consistent with mode I crack displacement may be utilized, and accurate values will be realized provided that the Y scale parameter in Equation 8.7 has been properly determined.

The plane strain fracture toughness K_{Ic} is a fundamental material property that depends on many factors, the most influential of which are temperature, strain rate, and microstructure. The magnitude of K_{Ic} diminishes with increasing strain rate and decreasing temperature. Furthermore, an enhancement in yield strength wrought by solid solution or dispersion additions or by strain hardening generally produces a corresponding decrease in K_{Ic}. Furthermore, K_{Ic} normally increases with reduction in grain size as composition and other microstructural variables are maintained constant. Yield strengths are included for some of the materials listed in Table 8.1.

Design Using Fracture Mechanics

According to Equations 8.6 and 8.7, three variables must be considered relative to the possibility for fracture of some structural component—viz the fracture toughness (K_c) or plane strain fracture toughness (K_{Ic}), the imposed stress (σ), and the flaw size (a), assuming, of course, that Y has been determined. When designing a component, it is first important to decide which of these variables are constrained by the application and which are subject to design control. For example, material selection (and hence K_c or K_{Ic}) is often dictated by factors such as density (for lightweight applications) or the corrosion characteristics of the environment. Or, the allowable flaw size is either measured or specified by the limitations of available flaw detection techniques. It is important to realize, however, that once any combination of two of the above parameters is prescribed, the third becomes fixed (Equations 8.6 and 8.7). For example, assume that K_{Ic} and the magnitude of a are specified by application constraints; therefore, the design (or critical) stress σ_c must be

$$\sigma_c \leq \frac{K_{Ic}}{Y\sqrt{\pi a}} \tag{8.9}$$

On the other hand, if stress level and plane strain fracture toughness are fixed by the design situation, then the maximum allowable flaw size a_c is

$$a_c = \frac{1}{\pi}\left(\frac{K_{Ic}}{\sigma Y}\right)^2 \tag{8.10}$$

A number of nondestructive test (NDT) techniques have been developed that permit detection and measurement of both internal and surface flaws. Such NDT methods are used to avoid the occurrence of catastrophic failure by examining structural components for defects and flaws that have dimensions approaching the critical size.

EXAMPLE PROBLEM 8.1

A structural component in the form of a very wide plate, as shown in Figure 8.11a, is to be fabricated from a 4340 steel. Two sheets of this alloy, each having a different heat treatment and thus different mechanical properties are available. One, denoted material A, has a yield strength of 860 MPa (125,000 psi) and a plane strain fracture toughness of 98.9 MPa$\sqrt{\text{m}}$ (90,000 psi$\sqrt{\text{in.}}$). For the other, material Z, σ_y and K_{Ic} values are 1515 MPa (220,000 psi) and 60.4 MPa$\sqrt{\text{m}}$ (55,000 psi$\sqrt{\text{in.}}$), respectively.

(a) For each alloy, determine whether or not plane strain conditions prevail if the plate is 10 mm (0.39 in.) thick.

(b) It is not possible to detect flaw sizes less than 3 mm, which is the resolution limit of the flaw detection apparatus. If the plate thickness is sufficient such that the K_{Ic} value may be used, determine whether or not a critical flaw is subject to detection. Assume that the design stress level is one half of the yield strength; also, for this configuration, the value of Y is 1.0.

SOLUTION

(a) Plane strain is established by Equation 8.8. For material A,

$$B = 2.5\left(\frac{K_{Ic}}{\sigma_y}\right)^2 = 2.5\left(\frac{98.9 \text{ MPa}\sqrt{m}}{860 \text{ MPa}}\right)^2$$

$$= 0.033 \text{ m} = 33 \text{ mm (1.30 in.)}$$

Thus plane strain conditions *do not* hold for material A because this value of B is greater than 10 mm, the actual plate thickness; the situation is one of plane stress and must be treated as such.

And for material Z,

$$B = 2.5\left(\frac{60.4 \text{ MPa}\sqrt{m}}{1515 \text{ MPa}}\right)^2 = 0.004 \text{ m} = 4.0 \text{ mm (0.16 in.)}$$

which is less than the actual plate thickness, and therefore the situation is one of plane strain.

(b) We need only determine the critical flaw size for material Z because the situation for material A is not plane strain, and K_{Ic} may not be used. Employing Equation 8.10 and taking σ to be $\sigma_y/2$,

$$a_c = \frac{1}{\pi}\left(\frac{60.4 \text{ MPa}\sqrt{m}}{(1)(1515/2) \text{ MPa}}\right)^2$$

$$= 0.002 \text{ m} = 2.0 \text{ mm (0.079 in.)}$$

Therefore, the critical flaw size for material Z is not subject to detection since it is less than 3 mm.

8.6 IMPACT FRACTURE TESTING

Prior to the advent of fracture mechanics as a scientific discipline, impact testing techniques were established so as to ascertain the fracture characteristics of materials. It was realized that the results of laboratory tensile tests could not be extrapolated to predict fracture behavior; for example, under some circumstances normally ductile metals fracture abruptly and with very little plastic deformation. Impact test conditions were chosen to represent those most severe relative to the potential for fracture, namely, (1) deformation at a relatively low temperature, (2) a high strain rate (i.e., rate of deformation), and (3) a triaxial stress state (which may be introduced by the presence of a notch).

Impact Testing Techniques

Two standardized tests, the **Charpy** and **Izod,** were designed and are still used to measure the **impact energy,** sometimes also termed *notch toughness*. The Charpy V-notch (CVN) technique is most commonly used in the United States. For both Charpy and Izod, the specimen is in the shape of a bar of square cross section, into which a V-notch is machined (Figure 8.13a). The apparatus for making V-notch impact tests is illustrated schematically in Figure 8.13b. The load is applied as an impact blow from a weighted pendulum hammer that is released from a cocked position at a fixed height h. The specimen is positioned at the base as shown. Upon release, a knife edge

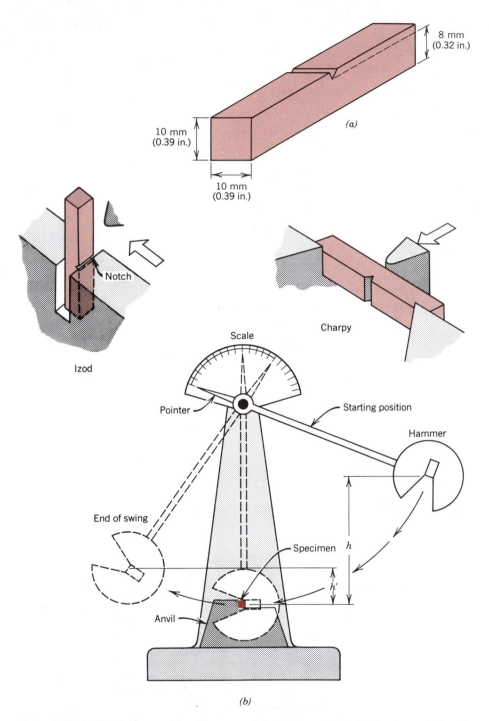

Figure 8.13 (*a*) Specimen used for Charpy and Izod impact tests. (*b*) A schematic drawing of an impact testing apparatus. The hammer is released from fixed height *h* and strikes the specimen; the energy expended in fracture is reflected in the difference between *h* and the swing height *h'*. Specimen placements for both Charpy and Izod tests are also shown. (Figure (*b*) adapted from H. W. Hayden, W. G. Moffatt, and J. Wulff, *The Structure and Properties of Materials,* Vol. III, *Mechanical Behavior,* p. 13. Copyright © 1965 by John Wiley & Sons, New York. Reprinted by permission of John Wiley & Sons, Inc.)

mounted on the pendulum strikes and fractures the specimen at the notch, which acts as a point of stress concentration for this high velocity impact blow. The pendulum continues its swing, rising to a maximum height h', which is lower than h. The energy absorption, computed from the difference between h and h', is a measure of the impact energy. The primary difference between the Charpy and Izod techniques lies in the manner of specimen support, as illustrated in Figure 8.13b. Furthermore, these are termed impact tests in light of the manner of load application. Variables including specimen size and shape as well as notch configuration and depth influence the test results.

Both plane strain fracture toughness and these impact tests determine the fracture properties of materials. The former are quantitative in nature, in that a specific property of the material is determined (i.e., K_{Ic}). The results of the impact tests, on the other hand, are more qualitative and are of little use for design purposes. Impact energies are of interest mainly in a relative sense and for making comparisons—absolute values are of little significance. Attempts have been made to correlate plane strain fracture toughnesses and CVN energies, with only limited success. Plane strain toughness tests are not as simple to perform as impact tests; furthermore, equipment and specimens are more expensive.

Ductile-to-Brittle Transition

One of the primary functions of Charpy and Izod tests is to determine whether or not a material experiences a ductile-to-brittle transition with decreasing temperature and, if so, the range of temperatures over which it occurs. The ductile-to-brittle transition is related to the temperature dependence of the measured impact energy absorption. This transition is represented for a steel by curve A in Figure 8.14. At

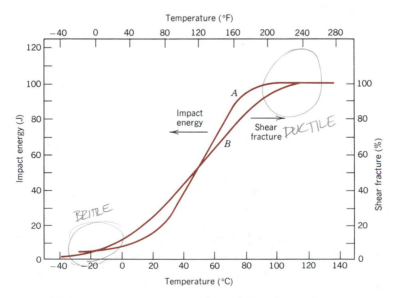

Figure 8.14 Temperature dependence of the Charpy V-notch impact energy (curve A) and percent shear fracture (curve B) for an A283 steel. (Reprinted from *Welding Journal*. Used by permission of the American Welding Society.)

Figure 8.15 Photograph of fracture surfaces of A36 steel Charpy V-notch specimens tested at indicated temperatures (in °C). (From R. W. Hertzberg, *Deformation and Fracture Mechanics of Engineering Materials,* 3rd edition, Fig. 9.6, p. 329. Copyright © 1989 by John Wiley & Sons, Inc., New York. Reprinted by permission of John Wiley & Sons, Inc.)

higher temperatures the CVN energy is relatively large, in correlation with a ductile mode of fracture. As the temperature is lowered, the impact energy drops suddenly over a relatively narrow temperature range, below which the energy has a constant but small value; that is, the mode of fracture is brittle.

Alternatively, appearance of the failure surface is indicative of the nature of fracture, and may be used in transition temperature determinations. For ductile fracture this surface appears fibrous (or of shear character); conversely, totally brittle surfaces have a granular texture (or cleavage character). Over the ductile-to-brittle transition, features of both types will exist (Figure 8.15). Frequently, the percent shear fracture is plotted as a function of temperature—curve B in Figure 8.14.

For many alloys there is a range of temperatures over which the ductile-to-brittle transition occurs (Figure 8.14); this presents some difficulty in specifying a single ductile-to-brittle transition temperature. No explicit criterion has been established, and so this temperature is often defined as that temperature at which the CVN energy assumes some value (e.g., 20 J or 15 ft-lb$_f$), or corresponding to some given fracture appearance (e.g., 50% fibrous fracture). Matters are further complicated inasmuch as a different transition temperature may be realized for each of these criteria. Perhaps the most conservative transition temperature is that at which the fracture surface becomes 100% fibrous; on this basis, the transition temperature is approximately 110°C (230°F) for the steel alloy that is the subject of Figure 8.14.

Structures constructed from alloys that exhibit this ductile-to-brittle behavior should be used only at temperatures above the transition temperature, to avoid brittle and catastrophic failure. Classic examples of this type of failure occurred, with disastrous consequences, during World War II when a number of welded transport ships, away from combat, suddenly and precipitously split in half. The vessels were constructed of a steel alloy that possessed adequate ductility from room-temperature tensile tests. The brittle fractures occurred at relatively low ambient temperatures, at about 4°C (40°F), in the vicinity of the transition temperature of the alloy. Each fracture crack originated at some point of stress concentration, probably a sharp corner or fabrication defect, which crack then propagated around the entire girth of the ships that split.

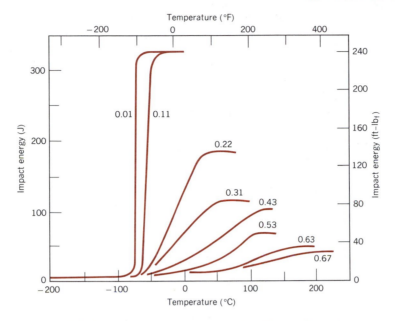

Figure 8.16 Influence of carbon content on the Charpy V-notch energy-versus-temperature behavior for steel. (Reprinted with permission from ASM International, Metals Park, OH 44073-9989, USA; Rinebolt, J. A. and Harris, W. J., Jr., "Affect of Alloying Elements on Notch Toughness of Pearlitic Steels," *Transactions of ASM*, Vol. 43, 1951.)

Not all metal alloys display a ductile-to-brittle transition. Those having FCC crystal structures (including aluminum- and copper-based alloys) remain ductile even at extremely low temperatures. However, BCC and HCP alloys experience this transition. For these materials the transition temperature is sensitive to both alloy composition and microstructure. For example, decreasing the average grain size of steels results in a lowering of the transition temperature. Also, carbon content has a decided influence on the CVN energy–temperature behavior of a steel, as indicated in Figure 8.16.

Most ceramics and polymers also experience a ductile-to-brittle transition. For ceramic materials, the transition occurs only at elevated temperatures, ordinarily in excess of 1000°C (1850°F). This behavior, as related to polymers, is discussed in Section 16.9.

FATIGUE

Fatigue is a form of failure that occurs in structures subjected to dynamic and fluctuating stresses (e.g., bridges, aircraft, and machine components). Under these circumstances it is possible for failure to occur at a stress level considerably lower than the tensile or yield strength for a static load. The term "fatigue" is used because this type of failure normally occurs after a lengthy period of repeated stress or strain cycling. Fatigue is important inasmuch as it is the single largest cause of failure in metals, estimated to comprise approximately 90% of all metallic failures; polymers and ceramics (except for glasses) are also susceptible to this type of failure. Furthermore, it is catastrophic and insidious, occurring very suddenly and without warning.

Fatigue failure is brittlelike in nature even in normally ductile metals, in that there is very little, if any, gross plastic deformation associated with failure. The process occurs by the initiation and propagation of cracks, and ordinarily the fracture surface is perpendicular to the direction of an applied tensile stress.

8.7 CYCLIC STRESSES

The applied stress may be axial (tension-compression), flexural (bending), or torsional (twisting) in nature. In general, three different fluctuating stress–time modes are possible. One is represented schematically by a regular and sinusoidal time dependence in Figure 8.17a, wherein the amplitude is symmetrical about a mean zero stress level, for example, alternating from a maximum tensile stress (σ_{max}) to a minimum compressive stress (σ_{min}) of equal magnitude; this is referred to as a *reversed stress cycle*. Another type, termed *repeated stress cycle,* is illustrated in Figure 8.17b; the maxima and minima are asymmetrical relative to the zero stress level. Finally, the stress level may vary randomly in amplitude and frequency, as exemplified in Figure 8.17c.

Also indicated in Figure 8.17b are several parameters used to characterize the fluctuating stress cycle. The stress amplitude alternates about a *mean stress σ_m*, defined as the average of the maximum and minimum stresses in the cycle, or

$$\sigma_m = \frac{\sigma_{max} + \sigma_{min}}{2} \tag{8.11}$$

Furthermore, the *range of stress σ_r* is just the difference between σ_{max} and σ_{min}, namely,

$$\sigma_r = \sigma_{max} - \sigma_{min} \tag{8.12}$$

Stress amplitude σ_a is just one half of this range of stress, or

$$\sigma_a = \frac{\sigma_r}{2} = \frac{\sigma_{max} - \sigma_{min}}{2} \tag{8.13}$$

Finally, the *stress ratio R* is just the ratio of minimum and maximum stress amplitudes:

$$R = \frac{\sigma_{min}}{\sigma_{max}} \tag{8.14}$$

By convention, tensile stresses are positive and compressive stresses are negative. For example, for the reversed stress cycle, the value of R is -1.

8.8 THE *S–N* CURVE

As with other mechanical characteristics, the fatigue properties of materials can be determined from laboratory simulation tests. A test apparatus should be designed to duplicate as nearly as possible the service stress conditions (stress level, time frequency, stress pattern, etc.). A schematic diagram of a rotating-bending test apparatus, commonly used for fatigue testing, is shown in Figure 8.18; the compression and tensile

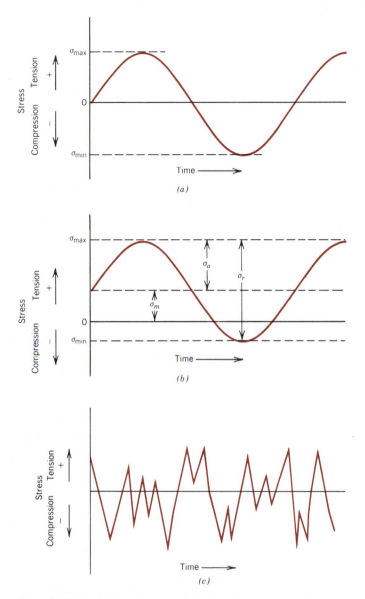

Figure 8.17 Variation of stress with time that accounts for fatigue failures. (*a*) Reversed stress cycle, in which the stress alternates from a maximum tensile stress (+) to a maximum compressive stress (−) of equal magnitude. (*b*) Repeated stress cycle, in which maximum and minimum stresses are asymmetrical relative to the zero stress level; mean stress σ_m, range of stress σ_r, and stress amplitude σ_a are indicated. (*c*) Random stress cycle.

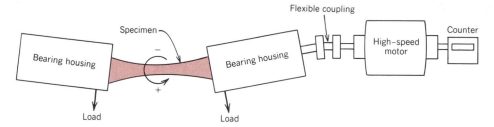

Figure 8.18 Schematic diagram of fatigue testing apparatus for making rotating-bending tests. (Adapted from C. A. Keyser, *Materials Science in Engineering,* 4th edition, Merrill Publishing Company, Columbus, OH, 1986. Reprinted by permission of the publisher.)

stresses are imposed on the specimen as it is simultaneously bent and rotated. Tests are also frequently conducted using an alternating uniaxial tension-compression stress cycle.

A series of tests are commenced by subjecting a specimen to the stress cycling at a relatively large maximum stress amplitude (σ_{max}), usually on the order of two thirds of the static tensile strength; the number of cycles to failure is counted. This procedure is repeated on other specimens at progressively decreasing maximum stress amplitudes. Data are plotted as stress S versus the logarithm of the number N of cycles to failure for each of the specimens. The values of S are normally taken as stress amplitudes (σ_a, Equation 8.13); on occasion, σ_{max} or σ_{min} values may be used.

Two distinct types of $S-N$ behavior are observed, which are represented schematically in Figures 8.19. As these plots indicate, the higher the magnitude of the stress, the smaller the number of cycles the material is capable of sustaining before failure. For some ferrous (iron base) and titanium alloys, the $S-N$ curve (Figure 8.19a) becomes horizontal at higher N values; or, there is a limiting stress level, called the **fatigue limit** (also sometimes the *endurance limit*), below which fatigue failure will not occur. This fatigue limit represents the largest value of fluctuating stress that will *not* cause failure for essentially an infinite number of cycles. For many steels, fatigue limits range between 35 and 60% of the tensile strength.

Most nonferrous alloys (e.g., aluminum, copper, magnesium) do not have a fatigue limit, in that the $S-N$ curve continues its downward trend at increasingly greater N values (Figure 8.19b). Thus fatigue will ultimately occur regardless of the magnitude of the stress. For these materials, the fatigue response is specified as **fatigue strength,** which is defined as the stress level at which failure will occur for some specified number of cycles (e.g., 10^7 cycles). The determination of fatigue strength is also demonstrated in Figure 8.19b.

Another important parameter that characterizes a material's fatigue behavior is **fatigue life** N_f. It is the number of cycles to cause failure at a specified stress level, as taken from the $S-N$ plot (Figure 8.19b).

Unfortunately, there always exists considerable scatter in fatigue data, that is, a variation in the measured N value for a number of specimens tested at the same stress level. This may lead to significant design uncertainties when fatigue life and/or fatigue limit (or strength) are being considered. The scatter in results is a consequence of the fatigue sensitivity to a number of test and material parameters that are impossible to control precisely. These parameters include specimen fabrication and surface prep-

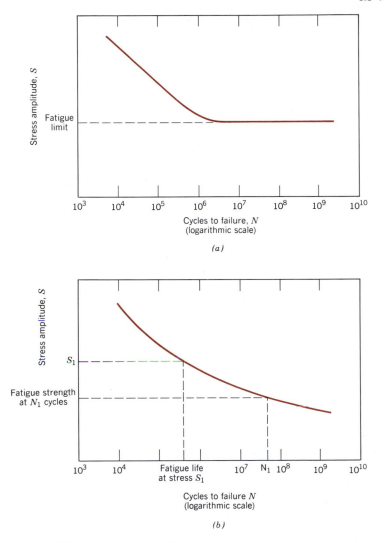

Figure 8.19 Stress amplitude (S) versus logarithm of the number of cycles to fatigue failure (N) for (a) a material that displays a fatigue limit, and (b) a material that does not display a fatigue limit.

aration, metallurgical variables, specimen alignment in the apparatus, mean stress, and test frequency.

Fatigue $S–N$ curves similar to those shown in Figure 8.19 represent "best fit" curves which have been drawn through average-value data points. It is a little unsettling to realize that approximately one half of the specimens tested actually failed at stress levels lying nearly 25% below the curve (as determined on the basis of statistical treatments).

Several statistical techniques have been developed which are used to specify fatigue life and fatigue limit in terms of probabilities. One convenient way of representing data treated in this manner is with a series of constant probability curves, several of which are plotted in Figure 8.20. The P value associated with each curve represents

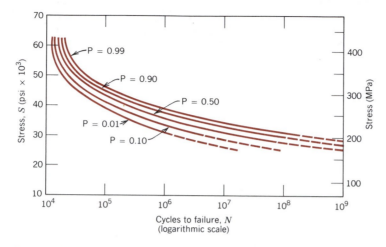

Figure 8.20 Fatigue S–N probability of failure curves for a 7075-T6 aluminum alloy; P denotes the probability of failure. (G. M. Sinclair and T. J. Dolan, *Trans., ASME,* **75,** 1953. Reprinted with permission of the American Society of Mechanical Engineers.)

the probability of failure. For example, at a stress of 30,000 psi, we would expect 1% of the specimens to fail at about 10^6 cycles and 50% to fail at about 2×10^7 cycles, and so on. It should be remembered that S–N curves represented in the literature are normally average values, unless noted otherwise.

The fatigue behaviors represented in Figures 8.19a and 8.19b may be classified into two domains. One is associated with relatively high loads that produce not only elastic strain but also some plastic strain during each cycle. Consequently, fatigue lives are relatively short; this domain is termed *low-cycle fatigue* and occurs at less than about 10^4 to 10^5 cycles. For lower stress levels wherein deformations are totally elastic, longer lives result. This is called *high-cycle fatigue* inasmuch as relatively large numbers of cycles are required to produce fatigue failure. High-cycle fatigue is associated with fatigue lives greater than about 10^4 to 10^5 cycles.

8.9 CRACK INITIATION AND PROPAGATION

The process of fatigue failure is characterized by three distinct stages: (1) crack initiation, wherein a small crack forms at some point of high stress concentration; (2) crack propagation, during which this crack advances incrementally with each stress cycle; and (3) final failure, which occurs very rapidly once the advancing crack has reached a critical size. The fatigue life N_f, the total number of cycles to failure, therefore can be taken as the sum of the number of cycles for crack initiation N_i and crack propagation N_p:

$$N_f = N_i + N_p \tag{8.15}$$

The contribution of the final failure stage to the total fatigue life is insignificant since it occurs so rapidly. Relative proportions to the total life of N_i and N_p depend on the particular material and test conditions. At low stress levels (i.e., for high-cycle fatigue), a large fraction of the fatigue life is utilized in crack initiation. With increasing

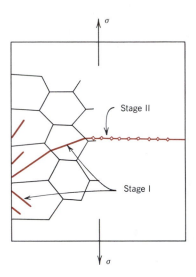

Figure 8.21 Schematic representation showing stages I and II of fatigue crack propagation in polycrystalline metals. (Copyright ASTM. Reprinted with permission.)

stress level, N_i decreases and the cracks form more rapidly. Thus for low-cycle fatigue (high stress levels), the propagation stage predominates (i.e., $N_p > N_i$).

Cracks associated with fatigue failure almost always initiate (or nucleate) on the surface of a component at some point of stress concentration. Crack nucleation sites include surface scratches, sharp fillets, keyways, threads, dents, and the like. In addition, cyclic loading can produce microscopic surface discontinuties resulting from dislocation slip steps which may also act as stress raisers, and therefore as crack initiation sites.

Once a stable crack has nucleated, it then initially propagates very slowly and, in polycrystalline metals, along crystallographic planes of high shear stress; this is sometimes termed *stage I propagation* (Figure 8.21). This stage may constitute a large or small fraction of the total fatigue life depending on stress level and the nature of the test specimen; high stresses and the presence of notches favor a short-lived stage I. In polycrystalline metals, cracks normally extend through only several grains during this propagation stage. The fatigue surface that is formed during stage I propagation has a flat and featureless appearance.

Eventually, a second propagation stage (*stage II*) takes over, wherein the crack extension rate increases dramatically. Furthermore, at this point there is also a change in propagation direction to one that is roughly perpendicular to the applied tensile stress (see Figure 8.21). During this stage of propagation, crack growth proceeds by a repetitive plastic blunting and sharpening process at the crack tip, a mechanism illustrated in Figure 8.22. At the beginning of the stress cycle (zero load), the crack tip has the shape of a sharp double-notch (Figure 8.22a). As the tensile stress is applied (Figure 8.22b), localized deformation occurs at each of these tip notches along slip planes that are oriented at 45° angles relative to the plane of the crack. With increased crack widening, the tip advances by continued shear deformation and the assumption of a blunted configuration (Figure 8.22c). During compression, the directions of shear deformation at the crack tip are reversed (Figure 8.22d) until, at the culmination of the cycle, a new sharp double-notch tip has formed (Figure 8.22e). Thus the crack tip has advanced a one-notch distance during the course of a complete cycle. This process

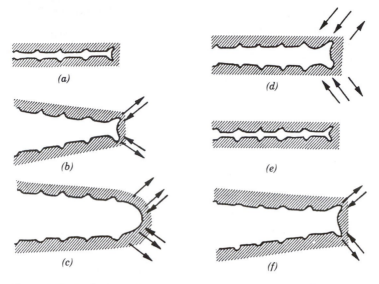

Figure 8.22 Fatigue crack propagation mechanism (stage II) by repetitive crack tip plastic blunting and sharpening: (*a*) zero load, (*b*) small tensile load, (*c*) maximum tensile load, (*d*) small compressive load, (*e*) maximum compressive load, (*f*) small tensile load. The loading axis is vertical. (Copyright ASTM. Reprinted with permission.)

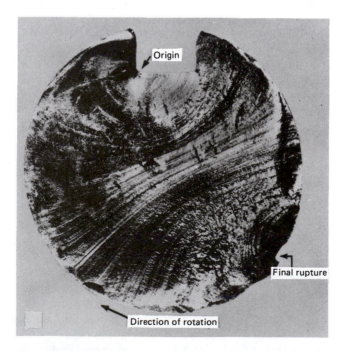

Figure 8.23 Fracture surface of a rotating steel shaft that experienced fatigue failure. Beachmark ridges are visible in the photograph. (Reproduced with permission from D. J. Wulpi, *Understanding How Components Fail,* American Society for Metals, Materials Park, OH, 1985.)

is repeated with each subsequent cycle until eventually some critical crack dimension is achieved which precipitates the final failure stage and catastrophic failure ensues.

The region of a fracture surface that formed during stage II propagation may be characterized by two types of markings termed *beachmarks* and *striations*. Both of these features indicate the position of the crack tip at some point in time and appear as concentric ridges that expand away from the crack initiation site(s), frequently in a circular or semicircular pattern. Beachmarks (sometimes also called "clamshell marks") are of macroscopic dimensions (Figure 8.23), and may be observed with the unaided eye. These markings are found for components that experienced interruptions during stage II propagation—for example, a machine that operated only during normal work-shift hours. Each beachmark band represents a period of time over which crack growth occurred.

On the other hand, fatigue striations are microscopic in size and subject to observation with the electron microscope (either TEM or SEM). Figure 8.24 is an electron fractograph which shows this feature. Each striation is thought to represent the advance distance of the crack front during a single load cycle. Striation width depends on, and increases with, increasing stress range.

At this point it should be emphasized that although both beachmarks and striations are fatigue fracture surface features having similar appearances, they are nevertheless different, both in origin and size. There may be literally thousands of striations within a single beachmark.

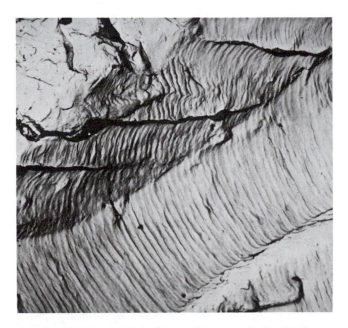

Figure 8.24 Transmission electron fractograph showing fatigue striations in aluminum. (From V. J. Colangelo and F. A. Heiser, *Analysis of Metallurgical Failures,* 2nd edition. Copyright © 1987 by John Wiley & Sons, New York. Reprinted by permission of John Wiley & Sons, Inc.)

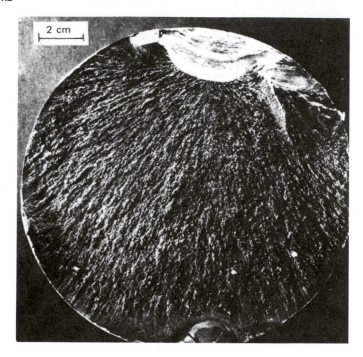

Figure 8.25 Fatigue failure surface. A crack formed at the top edge. The smooth region also near the top corresponds to the area over which the crack propagated slowly. Rapid failure occurred over the area having a dull and fibrous texture (the largest area). Approximately 0.5 ×. (Reproduced by permission from *Metals Handbook: Fractography and Atlas of Fractographs,* Vol. 9, 8th edition, H. E. Boyer (Editor), American Society for Metals, 1974.)

Often, the cause of failure may be deduced after examination of the failure surfaces. The presence of beachmarks and/or striations on a fracture surface confirms that the cause of failure was fatigue. Nevertheless, the absence of either or both does not exclude fatigue as the cause of failure.

One final comment regarding fatigue failure surfaces: Beachmarks and striations will not appear on that region over which the rapid failure occurs. Rather, the rapid failure may be either ductile or brittle; evidence of plastic deformation will be present for ductile, and absent for brittle, failure. This region of failure may be noted in Figure 8.25.

8.10 CRACK PROPAGATION RATE

Even though measures may be taken to minimize the possibility of fatigue failure, cracks and crack nucleation sites will always exist in structural components. Under the influence of cyclic stresses, cracks will inevitably form and grow; this process, if unabated, can ultimately lead to failure. The intent of the present discussion is to develop a criterion whereby fatigue life may be predicted on the basis of material and stress state parameters. Principles of fracture mechanics (Section 8.5) will be employed

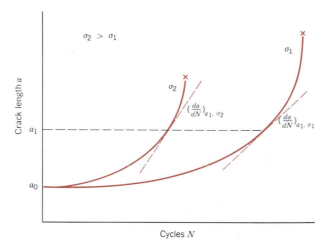

Figure 8.26 Crack length versus the number of cycles at stress levels σ_1 and σ_2 for fatigue studies. Crack growth rate da/dN is indicated at crack length a_1 for both stress levels.

inasmuch as the treatment involves determination of a maximum crack length that may be tolerated without inducing failure. It should be noted that this discussion relates to the domain of high-cycle fatigue, that is, for fatigue lives greater than about 10^4 to 10^5 cycles.

Results of fatigue studies have shown that the life of a structural component may be related to the rate of crack growth. During stage II propagation, cracks may grow from a barely perceivable size to some critical length. Experimental techniques are available which are employed to monitor crack length during the cyclic stressing. Data are recorded and then plotted as crack length a versus the number of cycles N.[3] A typical plot is shown in Figure 8.26, where curves are included from data generated at two different stress levels; the initial crack length a_0 for both sets of tests is the same. Crack growth rate da/dN is taken as the slope at some point of the curve. Two important results are worth noting: (1) initially, growth rate is small, but increases with increasing crack length; and (2) growth rate is enhanced with increasing applied stress level and for a specific crack length (a_1 in Figure 8.26).

Fatigue crack propagation rate is a function of not only stress level and crack size but also material variables. Mathematically, this rate may be expressed in terms of the stress intensity factor K (developed using fracture mechanics in Section 8.5) and takes the form

$$\frac{da}{dN} = A(\Delta K)^m \tag{8.16}$$

---

[3] The symbol N in the context of Section 8.8 represents the number of cycles to fatigue failure; in the present discussion it denotes the number of cycles associated with some crack length prior to failure.

The parameters A and m are constants for the particular material, which will also depend on environment, frequency, and the stress ratio (R in Equation 8.14). The value of m normally ranges between 1 and 6.

Furthermore, ΔK is the stress intensity factor range at the crack tip, that is,

$$\Delta K = K_{max} - K_{min} \tag{8.17a}$$

or, from Equation 8.6,

$$\Delta K = Y\Delta\sigma\sqrt{\pi a} = Y(\sigma_{max} - \sigma_{min})\sqrt{\pi a} \tag{8.17b}$$

Since crack growth stops or is negligible for a compression portion of the stress cycle, if σ_{min} is compressive, then K_{min} and σ_{min} are taken to be zero: that is, $\Delta K = K_{max}$ and $\Delta\sigma = \sigma_{max}$. Also note that K_{max} and K_{min} in Equation 8.17a represent stress

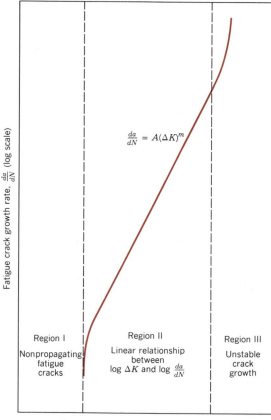

Figure 8.27 Schematic representation of logarithm fatigue crack propagation rate da/dN versus logarithm stress intensity factor range ΔK. The three regions of different crack growth response (I, II, and III) are indicated. (Reprinted with permission from ASM International, Metals Park, OH 44073-9989, Clark, W. G., Jr., "How Fatigue Crack Initiation and Growth Properties Affect Material Selection and Design Criteria," *Metals Engineering Quarterly,* Vol. 14, No. 3, 1974.)

intensity factors, not the fracture toughness K_c nor the plane strain fracture toughness K_{Ic}.

The typical fatigue crack growth rate behavior of materials is represented schematically in Figure 8.27 as the logarithm of crack growth rate da/dN versus the logarithm of the stress intensity factor range ΔK. The resulting curve has a sigmoidal shape which may be divided into three distinct regions, labeled I, II, and III. In region I (at low stress levels and/or small crack sizes), preexisting cracks will not grow with cyclic loading. Furthermore, associated with region III is accelerated crack growth, which occurs just prior to the rapid fracture.

The curve is essentially linear in region II, which is consistent with Equation 8.16. This may be confirmed by taking the logarithm of both sides of this expression, which leads to

$$\log\left(\frac{da}{dN}\right) = \log[A(\Delta K)^m] \tag{8.18a}$$

$$\log\left(\frac{da}{dN}\right) = m \log \Delta K + \log A \tag{8.18b}$$

Indeed, according to Equation 8.18b, a straight line segment will result when $\log(da/dN)$-versus-log ΔK data are plotted; the slope and intercept correspond to the values of m and $\log A$, respectively, which may be determined from test data that have been represented in the manner of Figure 8.27. Figure 8.28 is one such plot for a Ni–Mo–V steel alloy. The linearity of the data may be noted, which verifies the power law relationship of Equation 8.16. Furthermore, the slope yields a value of 3 for m; A is approximately 1.8×10^{-14}, as taken from the extrapolated intercept for da/dN in in./cycle and ΔK in psi$\sqrt{\text{in.}}$.

One of the goals of failure analysis is to be able to predict fatigue life for some component, given its service constraints and laboratory test data. We are now able to develop an analytical expression for N_f by integration of Equation 8.16. Rearrangement is first necessary as follows:

$$dN = \frac{da}{A(\Delta K)^m} \tag{8.19}$$

which may be integrated as

$$N_f = \int_0^{N_f} dN = \int_{a_0}^{a_c} \frac{da}{A(\Delta K)^m} \tag{8.20}$$

The limits on the second integral are between the initial flaw length a_0, which may be measured using nondestructive examination techniques, and the critical crack length a_c determined from fracture toughness tests.

Substitution of the expression for ΔK (Equation 8.17b) leads to

$$N_f = \int_{a_0}^{a_c} \frac{da}{A(Y \Delta\sigma \sqrt{\pi a})^m}$$

$$= \frac{1}{A\pi^{m/2}(\Delta\sigma)^m} \int_{a_0}^{a_c} \frac{da}{Y^m a^{m/2}} \tag{8.21}$$

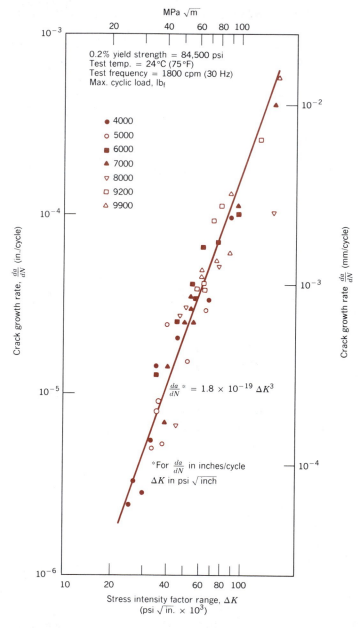

Figure 8.28 Logarithm crack growth rate versus logarithm stress intensity factor range for a Ni–Mo–V steel. (Reprinted by permission of the Society for Experimental Mechanics, Inc.)

Here it is assumed that $\Delta\sigma$ (or $\sigma_{max} - \sigma_{min}$) is constant; furthermore, in general Y will depend on crack length a and therefore cannot be removed from within the integral.

A word of caution: Equation 8.21 presumes the validity of Equation 8.16 over the entire life of the component, which may or may not hold true. Therefore, this expression should only be taken as an estimate of N_f.

EXAMPLE PROBLEM 8.2

A relatively large sheet of steel is to be exposed to cyclic tensile and compressive stresses of magnitudes 100 MPa and 50 MPa, respectively. Prior to testing, it has been determined that the length of the largest surface crack is 2.0 mm (2×10^{-3} m). Estimate the fatigue life of this sheet if its plane strain fracture toughness is 25 MPa$\sqrt{m}$ and the values of m and A in Equation 8.16 are 3.0 and 1.0×10^{-12}, respectively, for $\Delta\sigma$ in MPa and a in m. Assume that the parameter Y is independent of crack length and has a value of 1.0.

SOLUTION

It first becomes necessary to compute the critical crack length a_c, the integration upper limit in Equation 8.21. Equation 8.10 is employed for this computation, assuming a stress level of 100 MPa, since this is the maximum tensile stress. Therefore,

$$a_c = \frac{1}{\pi}\left(\frac{K_{Ic}}{\sigma Y}\right)^2$$

$$= \frac{1}{\pi}\left(\frac{25 \text{ MPa}\sqrt{m}}{(100 \text{ MPa})(1)}\right)^2 = 0.02 \text{ m}$$

We now want to solve Equation 8.21 using 0.002 m as the lower integration limit a_0, as stipulated in the problem. The value of $\Delta\sigma$ is just 100 MPa, the magnitude of the tensile stress, since σ_{min} is compressive. Therefore, integration yields

$$N_f = \frac{1}{A\pi^{m/2}(\Delta\sigma)^m}\int_{a_0}^{a_c}\frac{da}{Y^m a^{m/2}}$$

$$= \frac{1}{A\pi^{3/2}(\Delta\sigma)^3 Y^3}\int_{a_0}^{a_c} a^{-3/2}\,da$$

$$= \frac{1}{A\pi^{3/2}(\Delta\sigma)^3 Y^3}(-2)a^{-1/2}\Big|_{a_0}^{a_c}$$

$$= \frac{2}{A\pi^{3/2}(\Delta\sigma)^3 Y^3}\left(\frac{1}{\sqrt{a_0}} - \frac{1}{\sqrt{a_c}}\right)$$

$$= \frac{2}{(1.0 \times 10^{-12})(\pi)^{3/2}(100)^3(1)^3}\left(\frac{1}{\sqrt{0.002}} - \frac{1}{\sqrt{0.02}}\right)$$

$$= 5.49 \times 10^6 \text{ cycles}$$

8.11 FACTORS THAT AFFECT FATIGUE LIFE

As was mentioned in Section 8.8, the fatigue behavior of engineering materials is highly sensitive to a number of variables. Some of these factors include mean stress level, geometrical design, surface effects, metallurgical variables, as well as the environment. This section is devoted to a discussion of these factors and, in addition, to measures that may be taken to improve the fatigue resistance of structural components.

Mean Stress

The dependence of fatigue life on stress amplitude is represented on the S–N plot. Such data are taken for a constant mean stress σ_m, often for the reversed cycle situation ($\sigma_m = 0$). Mean stress, however, will also affect fatigue life, which influence may be represented by a series of S–N curves, each measured at a different σ_m; this is depicted schematically in Figure 8.29. As may be noted, increasing the mean stress level leads to a decrease in fatigue life.

Surface Effects

For many common loading situations, the maximum stress within a component or structure occurs as its surface. Consequently, most cracks leading to fatigue failure originate at surface positions, specifically at stress amplification sites. Therefore, it has been observed that fatigue life is especially sensitive to the condition and configuration of the component surface. Numerous factors influence fatigue resistance, the proper management of which will lead to an improvement in fatigue life. These include design criteria as well as various surface treatments.

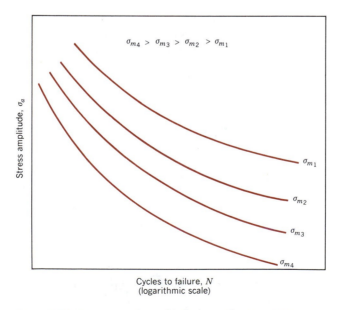

Figure 8.29 Demonstration of influence of mean stress σ_m on S–N fatigue behavior.

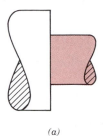

(a)

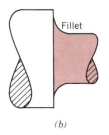

Fillet

(b)

Figure 8.30 Demonstration of how design can reduce stress amplification. (a) Poor design: sharp corner. (b) Good design: fatigue lifetime improved by incorporating rounded fillet into a rotating shaft at the point where there is a change in diameter.

Design Factors.   The design of a component can have a significant influence on its fatigue characteristics. Any notch or geometrical discontinuity can act as a stress raiser and fatigue crack initiation site; these design features include grooves, holes, keyways, threads, and so on. The sharper the discontinuity (i.e., the smaller the radius of curvature), the more severe the stress concentration. The probability of fatigue failure may be reduced by avoiding (when possible) these structural irregularities, or by making design modifications whereby sudden contour changes leading to sharp corners are eliminated—for example, calling for rounded fillets with large radii of curvature at the point where there is a change in diameter for a rotating shaft (Figure 8.30).

Surface Treatments.   During machining operations, small scratches and grooves are invariably introduced into the workpiece surface by cutting tool action. These surface markings can limit the fatigue life. It has been observed that improving the surface finish by polishing will enhance fatigue life significantly.

One of the most effective methods of increasing fatigue performance is by imposing residual compressive stresses within a thin outer surface layer. Thus a surface tensile stress of external origin will be partially nullified and reduced in magnitude by the residual compressive stress. The net effect is that the likelihood of crack formation and therefore of fatigue failure is reduced.

Residual compressive stresses are commonly introduced into ductile metals mechanically by localized plastic deformation within the outer surface region. Commercially, this is often accomplished by a process termed *shot peening*. Small, hard particles (shot) having diameters within the range of 0.1 to 1.0 mm are projected at high velocities onto the surface to be treated. The resulting deformation induces compressive stresses to a depth of between one quarter and one half of the shot diameter.

Case hardening is a technique whereby both surface hardness and fatigue life are enhanced for steel alloys. This is accomplished by a carburizing or nitriding process

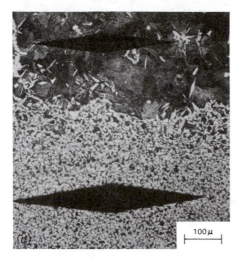

Figure 8.31 Photomicrograph showing both core (bottom) and carburized outer case (top) regions of a case-hardened steel. The case is harder as attested by the smaller microhardness indentation. (From R. W. Hertzberg, *Deformation and Fracture Mechanics of Engineering Materials,* 3rd edition. Copyright © 1989 by John Wiley & Sons, New York. Reprinted by permission of John Wiley & Sons, Inc.)

whereby a component is exposed to a carbonaceous or nitrogenous atmosphere at an elevated temperature. A carbon- or nitrogen-rich outer surface layer (or "case") is introduced by atomic diffusion from the gaseous phase. The case is normally on the order of 1 mm deep and is harder than the inner core of material. (The influence of carbon content on hardness for Fe–C alloys is demonstrated in Figure 10.21a.) The improvement of fatigue properties results from increased hardness within the case, as well as the desired residual compressive stresses the formation of which attends the carburizing or nitriding process. A carbon-rich outer case may be observed for the gear shown in the photograph on page 94; it appears as a dark outer rim within the sectioned segment. The increase in case hardness is demonstrated in the photomicrograph appearing in Figure 8.31. The dark and elongated diamond shapes are Knoop microhardness indentations. The upper indentation, lying within the carburized layer, is smaller than the core indentation.

8.12 ENVIRONMENTAL EFFECTS

Environmental factors may also affect the fatigue behavior of materials. A few brief comments will be given relative to two types of environment-assisted fatigue failure: thermal fatigue and corrosion fatigue.

Thermal fatigue is normally induced at elevated temperatures by fluctuating thermal stresses; mechanical stresses from an external source need not be present. The origin of these thermal stresses is the restraint to the dimensional expansion and/or contraction that would normally occur in a structural member with variations in temperature. The magnitude of a thermal stress developed by a temperature change ΔT is dependent on the coefficient of thermal expansion α_l and the modulus of elasticity E according to

$$\sigma = \alpha_l E \Delta T \qquad (8.22)$$

(The topics of thermal expansion and thermal stresses are discussed in Sections 20.3 and 20.5.) Of course, thermal stresses will not arise if this mechanical restraint is absent. Therefore, one obvious way to prevent this type of fatigue is to eliminate, or at least reduce, the restraint source, thus allowing unhindered dimensional changes with temperature variations, or to choose materials with appropriate physical properties.

Failure which occurs by the simultaneous action of a cyclic stress and chemical attack is termed **corrosion fatigue.** Corrosive environments have a deleterious influence and produce shorter fatigue lives. Even the normal ambient atmosphere will affect the fatigue behavior of some materials. Small pits may form as a result of chemical reactions between the environment and material, which serve as points of stress concentration, and therefore as crack nucleation sites. In addition, crack propagation rate is enhanced as a result of the corrosive environment. The nature of the stress cycles will influence the fatigue behavior; for example, lowering the load application frequency leads to longer periods during which the opened crack is in contact with the environment and to a reduction in the fatigue life.

Several approaches to corrosion fatigue prevention exist. On one hand, we can take measures to reduce the rate of corrosion by some of the techniques discussed in Chapter 18, for example, apply protective surface coatings, select a more corrosion-resistant material, and reduce the corrosiveness of the environment. And/or it might be advisable to take actions to minimize the probability of normal fatigue failure, as outlined above, for example, reduce the applied tensile stress level and impose residual compressive stresses on the surface of the member.

CREEP

Materials are often placed in service at elevated temperatures and exposed to static mechanical stresses (e.g., turbine rotors in jet engines and steam generators that experience centrifugal stresses, and high-pressure steam lines). Deformation under such circumstances is termed **creep.** Defined as the time-dependent and permanent deformation of materials when subjected to a constant load or stress, creep is normally an undesirable phenomenon and is often the limiting factor in the lifetime of a part. It is observed in all materials types; for metals it only becomes important for temperatures greater than about $0.4T_m$ (T_m = absolute melting temperature). Amorphous polymers, which include plastics and rubbers, are especially sensitive to creep deformation as discussed in Section 16.6.

8.13 GENERALIZED CREEP BEHAVIOR

A typical creep test consists of subjecting a specimen to a constant load or stress while maintaining the temperature constant; deformation or strain is measured and plotted as a function of elapsed time. Most tests are the constant load type, which yield information of an engineering nature; constant stress tests are employed to provide a better understanding of the mechanisms of creep.

Figure 8.32 is a schematic representation of the typical constant load creep behavior of metals. Upon application of the load there is an instantaneous deformation, as indicated in the figure, which is mostly elastic. The resulting creep curve consists of three regions, each of which has its own distinctive strain–time feature. *Primary* or *transient creep* occurs first, typified by a continuously decreasing creep rate; that

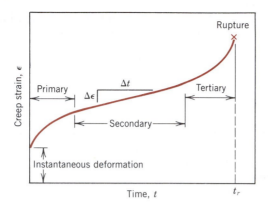

Figure 8.32 Typical creep curve of strain versus time at constant stress and elevated temperature. The minimum creep rate $\Delta\epsilon/\Delta t$ is the slope of the linear segment in the secondary region. Rupture lifetime t_r is the total time to rupture.

is, the slope of the curve diminishes with time. This suggests that the material is experiencing an increase in creep resistance or strain hardening (Section 7.10)—deformation becomes more difficult as the material is strained. For *secondary creep,* sometimes termed *steady-state creep,* the rate is constant; that is, the plot becomes linear. This is often the stage of creep that is of the longest duration. The constancy of creep rate is explained on the basis of a balance between the competing processes of strain hardening and recovery, recovery (Section 7.11) being the process whereby a material becomes softer and retains its ability to experience deformation. Finally, for *tertiary creep,* there is an acceleration of the rate and ultimate failure. This failure is frequently termed *rupture* and results from microstructural and/or metallurgical changes; for example, grain boundary separation, and the formation of internal cracks, cavities, and voids. Also, for tensile loads, a neck may form at some point within the deformation region. These all lead to a decrease in the effective cross-sectional area and an increase in strain rate.

For metallic materials most creep tests are conducted in uniaxial tension using a specimen having the same geometry as for tensile tests (Figure 6.2). On the other hand, uniaxial compression tests are more appropriate for brittle materials; these provide a better measure of the intrinsic creep properties inasmuch as there is no stress amplification and crack propagation, as with tensile loads. Compressive test specimens are usually right cylinders or parallelepipeds having length-to-diameter ratios ranging from about 2 to 4. For most materials creep properties are virtually independent of loading direction.

Possibly the most important parameter from a creep test is the slope of the secondary portion of the creep curve ($\Delta\epsilon/\Delta t$ in Figure 8.32); this is often called the minimum or *steady-state creep rate* $\dot{\epsilon}_s$. It is the engineering design parameter that is considered for long-life applications, such as a nuclear power plant component that is scheduled to operate for several decades, and when failure or too much strain is not an option. On the other hand, for many relatively short-life creep situations (e.g., turbine blades in military aircraft and rocket motor nozzles), *time to rupture,* or the *rupture lifetime* t_r, is the dominant design consideration; it is also indicated in Figure 8.32. Of course, for its determination, creep tests must be conducted to the point of failure; these are termed *creep rupture* tests. Thus a knowledge of these creep characteristics of a material allows the design engineer to ascertain its suitability for a specific application.

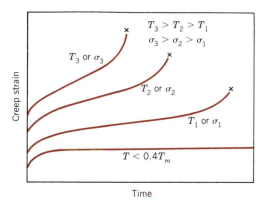

Figure 8.33 Influence of stress σ and temperature T on creep behavior.

8.14 STRESS AND TEMPERATURE EFFECTS

Both temperature and the level of the applied stress influence the creep characteristics (Figure 8.33). At a temperature substantially below $0.4T_m$, and after the initial deformation, the strain is virtually independent of time. With either increasing stress or temperature, the following will be noted: (1) the instantaneous strain at the time of stress application increases; (2) the steady-state creep rate is increased; and (3) the rupture lifetime is diminished.

The results of creep rupture tests are most commonly presented as the logarithm of stress versus the logarithm of rupture lifetime. Figure 8.34 is one such plot for a nickel alloy in which a linear relationship can be seen to exist at each temperature. For some alloys and over relatively large stress ranges, nonlinearity in these curves is observed.

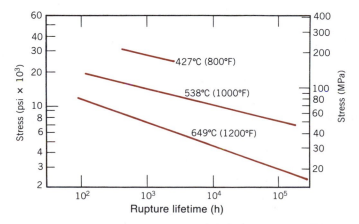

Figure 8.34 Logarithm of stress versus logarithm of rupture lifetime for a low carbon–nickel alloy at three temperatures. (From *Metals Handbook: Properties and Selection: Stainless Steels, Tool Materials and Special-Purpose Metals,* Vol. 3, 9th edition, D. Benjamin, Senior Editor, American Society for Metals, 1980, p. 130.)

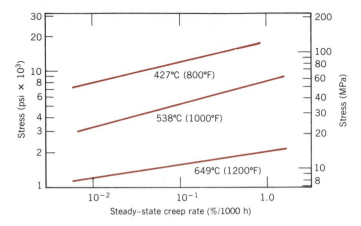

Figure 8.35 Logarithm of stress versus logarithm of steady-state creep rate for a low carbon–nickel alloy at three temperatures. (From *Metals Handbook: Properties and Selection: Stainless Steels, Tool Materials and Special-Purpose Metals,* Vol. 3, 9th edition, D. Benjamin, Senior Editor, American Society for Metals, 1980, p. 131.)

Empirical relationships have been developed in which the steady-state creep rate as a function of stress and temperature is expressed. Its dependence on stress can be written

$$\dot{\epsilon}_s = K_1 \sigma^n \qquad (8.23)$$

where K_1 and n are material constants. A plot of the logarithm of $\dot{\epsilon}_s$ versus the logarithm of σ yields a straight line with slope of n; this is shown in Figure 8.35 for a nickel alloy at three temperatures. Clearly, a straight line segment is drawn at each temperature.

Now, when the influence of temperature is included,

$$\dot{\epsilon}_s = K_2 \sigma^n \exp\left(-\frac{Q_c}{RT}\right) \qquad (8.24)$$

where K_2 and Q_c are constants; Q_c is termed the activation energy for creep.

Several theoretical mechanisms have been proposed to explain the creep behavior for various materials; these mechanisms involve stress-induced vacancy diffusion, grain boundary diffusion, dislocation motion, and grain boundary sliding. Each leads to a different value of the stress exponent n in Equation 8.23. It has been possible to elucidate the creep mechanism for a particular material by comparing its experimental n value with values predicted for the various mechanisms. In addition, correlations have been made between the activation energy for creep (Q_c) and the activation energy for diffusion (Q_d, Equation 5.8).

Creep data of this nature are represented pictorially for some well-studied systems in the form of stress–temperature diagrams, which are termed *deformation mechanism maps*. These maps indicate stress–temperature regimes (or areas) over which various mechanisms operate. Constant strain rate contours are often also included. Thus for some creep situation, given the appropriate deformation mechanism map and any two of the three parameters—temperature, stress level, and creep strain rate—the third parameter may be determined.

8.15 DATA EXTRAPOLATION METHODS

The need often arises for engineering creep data that are impractical to collect from normal laboratory tests. This is especially true for prolonged exposures (on the order of years). One solution to this problem involves performing creep and/or creep rupture tests at temperatures in excess of those required, for shorter time periods, and at a comparable stress level, and then by making a suitable extrapolation to the in-service condition. A commonly used extrapolation procedure employs the Larson–Miller parameter, defined as

$$T(C + \log t_r) \tag{8.25}$$

where C is a constant (usually on the order of 20), for T in Kelvin and the rupture lifetime t_r in hours. The rupture lifetime of a given material measured at some specific stress level will vary with temperature such that this parameter remains constant. Or, the data may be plotted as the logarithm of stress versus the Larson–Miller parameter, as shown in Figure 8.36. Utilization of this technique is demonstrated in the following example problem.

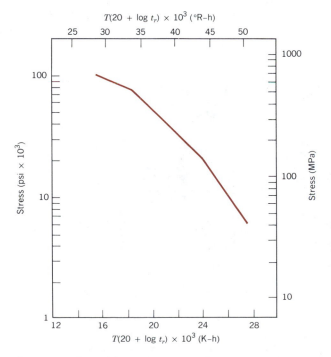

Figure 8.36 Logarithm stress versus the Larson–Miller parameter for an S-590 iron. (From F. R. Larson and J. Miller, *Trans. ASME,* **74,** 765 (1952). Reprinted by permission of ASME.)

EXAMPLE PROBLEM 8.3

Using the Larson–Miller data for S-590 iron shown in Figure 8.36, predict the time to rupture for a component that is subjected to a stress of 20,000 psi (140 MPa) at 800°C (1073 K).

SOLUTION

From Figure 8.36, at 20,000 psi (140 MPa) the value of the Larson–Miller parameter is 24.0×10^3, for T in K and t_r in h; therefore,

$$24.0 \times 10^3 = T(20 + \log t_r)$$
$$= 1073(20 + \log t_r)$$

and, solving for the time,

$$22.37 = 20 + \log t_r$$
$$t_r = 233 \text{ h (9.7 days)}$$

8.16 ALLOYS FOR HIGH-TEMPERATURE USE

There are several factors that affect the creep characteristics of metals. These include melting temperature, elastic modulus, and grain size. In general, the higher the melting temperature, the greater the elastic modulus, and the larger the grain size, the better is a material's resistance to creep. Stainless steels (Section 12.5), the refractory metals (Section 12.11), and the superalloys (Section 12.12) are especially resilient to creep

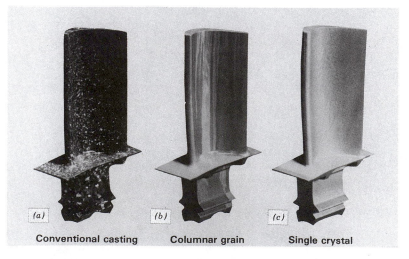

(a) Conventional casting   **(b)** Columnar grain   **(c)** Single crystal

Figure 8.37 (a) Polycrystalline turbine blade that was produced by a conventional casting technique. High-temperature creep resistance is improved as a result of an oriented columnar grain structure (b) produced by a sophisticated directional solidification technique. Creep resistance is further enhanced when single-crystal blades (c) are used. (Courtesy of Pratt & Whitney.)

and are commonly employed in high-temperature service applications. The creep resistance of the cobalt and nickel superalloys is enhanced by solid-solution alloying, and also by the addition of a dispersed phase which is virtually insoluble in the matrix. In addition, advanced processing techniques have been utilized; one such technique is directional solidification, which produces either highly elongated grains or single-crystal components (Figure 8.37). Another is the controlled unidirectional solidification of alloys having specially designed compositions wherein two-phase composites result.

SUMMARY

Fracture is one form of failure that occurs for static applied loads and at relatively low temperatures. Ductile and brittle modes are possible, both of which involve the formation and propagation of cracks. For ductile fracture, evidence will exist of gross plastic deformation at the fracture surface. In tension, highly ductile metals will neck down to essentially a point fracture; cup-and-cone mating fracture surfaces result for moderate ductility. Microscopically, dimples (spherical and parabolic) are produced. Cracks in ductile materials are said to be stable (i.e., resist extension without an increase in applied stress); and inasmuch as fracture is noncatastrophic, this fracture mode is almost always preferred.

For brittle fracture, cracks are unstable, and the fracture surface is relatively flat and perpendicular to the direction of the applied tensile load. Chevron and ridgelike patterns are possible, which indicate the direction of crack propagation. Transgranular (through-grain) and intergranular (between-grain) fractures are found in brittle polycrystalline materials.

The discipline of fracture mechanics allows for a better understanding of the fracture process and provides for structural design wherein the probability of failure is minimized. The significant discrepancy between actual and theoretical fracture strengths of brittle materials is explained by the existence of small flaws that are capable of amplifying an applied tensile stress in their vicinity, leading ultimately to crack formation. Stress amplification is greatest for long flaws that have small tip radii of curvature. Fracture ensues when the theoretical cohesive strength is exceeded at the tip of one of these flaws. Consideration of elastic strain and crack surface energies led Griffith to develop an expression for a crack propagation critical stress in brittle materials; this parameter is a function of elastic modulus, specific surface energy, and crack length.

The stress distributions in front of an advancing crack may be expressed in terms of position (as radial and angular coordinates) as well as stress intensity factor. The critical value of the stress intensity factor (i.e., that at which fracture occurs) is termed the fracture toughness, which is related to stress level, crack length, and a geometrical factor. The fracture toughness of a material is indicative of its resistance to brittle fracture when a crack is present. It depends on specimen thickness, and, for relatively thick specimens (i.e., conditions of plane strain), is termed the plane strain fracture toughness. This parameter is the one normally cited for design purposes; its value is relatively large for ductile materials (and small for brittle ones), and is a function of microstructure, strain rate, and temperature. With regard to designing against the possibility of fracture, consideration must be given to material (its fracture toughness), the stress level, and the flaw size detection limit.

Qualitatively, the fracture behavior of materials may be determined using Charpy and Izod impact testing techniques; impact energy (or notch toughness) is measured for specimens into which a V-shaped notch has been machined. On the basis of the temperature dependence of this impact energy (or appearance of the fracture surface), it is possible to ascertain whether or not a material experiences a ductile-to-brittle transition and the temperature range over which such a transition occurs. Metal alloys having BCC and HCP crystal structures experience this transition, and, for structural applications, should be used at temperatures in excess of this transition range.

Fatigue is a common type of catastrophic failure wherein the applied stress level fluctuates with time. Test data are plotted as stress versus the logarithm of the number of cycles to failure. For many materials, the number of cycles to failure increases continuously with diminishing stress. Fatigue strength represents the failure stress for a specified number of cycles. For some steels and titanium alloys, stress ceases to decrease with, and becomes independent of, the number of cycles; fatigue limit is the magnitude of this constant stress level, below which fatigue will not occur even for virtually an infinite number of cycles. Another fatigue property is fatigue life, which, for a specific stress, is the number of cycles to failure.

As a result of significant scatter in measured fatigue data, statistical analyses are performed that lead to specification of fatigue life and limit in terms of probabilities.

The processes of fatigue crack initiation and propagation were discussed. Cracks normally nucleate on the surface of a component at some point of stress concentration. Propagation proceeds in two stages, which are characterized by propagation direction and rate. The mechanism for the more rapid stage II corresponds to a repetitive plastic blunting and sharpening process at the advancing crack tip.

Two characteristic fatigue surface features are beachmarks and striations. Beachmarks form on components that experience applied stress interruptions; they normally may be observed with the naked eye. Fatigue striations are of microscopic dimensions, and each is thought to represent the crack tip advance distance over a single load cycle.

An analytical expression was proposed for fatigue crack propagation rate in terms of the stress intensity range at the crack tip. Integration of the expression yields an equation whereby fatigue life may be estimated.

Measures that may be taken to extend fatigue life include (1) reducing the mean stress level, (2) eliminating sharp surface discontinuities, (3) improving the surface finish by polishing, (4) imposing surface residual compressive stresses by shot peening, and (5) case hardening by using a carburizing or nitriding process.

The fatigue behavior of materials may also be affected by the environment. Thermal stresses may be induced in components that are exposed to elevated temperature fluctuations and when thermal expansion and/or contraction is restrained; fatigue for these conditions is termed thermal fatigue. The presence of a chemically active environment may lead to a reduction in fatigue life for corrosion fatigue; small pit crack nucleation sites form on the component surface as a result of chemical reactions.

The time-dependent plastic deformation of materials subjected to a constant load (or stress) and temperatures greater than about $0.4T_m$ is termed creep. A typical creep curve (strain versus time) will normally exhibit three distinct regions. For transient (or primary) creep, the rate (or slope) diminishes with time. The plot becomes linear (i.e., creep rate is constant) in the steady-state (or secondary) region. And finally, deformation accelerates for tertiary creep, just prior to failure (or rupture). Important design parameters available from such a plot include the steady-state creep rate (slope of the linear region) and rupture lifetime.

Both temperature and applied stress level influence creep behavior. Increasing either of these parameters produces the following effects: (1) an increase in the instantaneous initial deformation, (2) an increase in the steady-state creep rate, and (3) a diminishment of the rupture lifetime. Analytical expressions were presented which relate $\dot{\epsilon}_s$ to both temperature and stress. Creep mechanisms may be discerned on the basis of steady-state rate stress exponent and creep activation energy values.

Extrapolation of creep test data to lower temperature-longer time regimes is possible using the Larson–Miller parameter.

Metal alloys that are especially resistant to creep have high elastic moduli and melting temperatures; these include the superalloys, the stainless steels, and the refractory metals. Various processing techniques are employed to improve the creep properties of these materials.

IMPORTANT TERMS AND CONCEPTS

Brittle fracture
Case hardening
Charpy test
Corrosion fatigue
Creep
Ductile fracture
Ductile-to-brittle transition

Fatigue
Fatigue life
Fatigue limit
Fatigue strength
Fracture mechanics
Fracture toughness
Impact energy
Intergranular fracture

Izod test
Plane strain
Plane strain fracture toughness
Stress intensity factor
Stress raiser
Thermal fatigue
Transgranular fracture

REFERENCES

COLANGELO, V. J. and F. A. HEISER, *Analysis of Metallurgical Failures,* 2nd edition, John Wiley & Sons, New York, 1987.

COLLINS, J. A., *Failure of Materials in Mechanical Design,* John Wiley & Sons, New York, 1981.

DIETER, G. E., *Mechanical Metallurgy,* 3rd edition, McGraw-Hill Book Co., New York, 1986.

HERTZBERG, R. W., *Deformation and Fracture Mechanics of Engineering Materials,* 3rd edition, John Wiley & Sons, New York, 1989.

Metals Handbook, 9th edition, Vol. 11, *Failure Analysis and Prevention,* American Society for Metals, Metals Park, OH, 1986.

Metals Handbook, 9th edition, Vol. 12, *Fractography,* ASM International, Metals Park, OH, 1987.

TETELMAN, A. S. and A. J. McEVILY, *Fracture of Structural Materials,* John Wiley & Sons, New York, 1967.

WULPI, D. J., *Understanding How Components Fail,* American Society for Metals, Metals Park, OH, 1985.

QUESTIONS AND PROBLEMS

8.1 Cite the three principal causes for mechanical failure.

8.2 Estimate the theoretical cohesive strengths of the ceramic materials listed in Table 13.4.

8.3 What is the magnitude of the maximum stress that exists at the tip of an internal crack having a radius of curvature of 10^{-5} in. (2.5×10^{-4} mm) and a crack length of 10^{-3} in. (2.5×10^{-2} mm) when a tensile stress of 25,000 psi (170 MPa) is applied?

8.4 Estimate the theoretical fracture strength of a brittle material if it is known that fracture occurs by the propagation of an elliptically shaped surface crack of length 0.02 in. (0.5 mm) and having a tip radius of curvature of 2×10^{-4} in. (5×10^{-3} mm) when a stress of 150,000 psi (1035 MPa) is applied.

8.5 A specimen of a ceramic material having a modulus of elasticity of 25×10^4 MPa (36.3×10^6 psi) is pulled in tension with a stress of 750 MPa (109,000 psi). Will the specimen fail if its "most severe flaw" is an internal crack that has a length of 0.20 mm (7.87×10^{-3} in.) and a tip radius of curvature of 0.001 mm (3.94×10^{-5} in.)? Why or why not?

8.6 If the specific surface energy for aluminum oxide is $0.90 \ J/m^2$, using data contained in Table 13.4, compute the critical stress required for the propagation of an internal crack of length 0.40 mm.

8.7 An MgO component must not fail when a tensile stress of 13.5 MPa (1960 psi) is applied. Determine the maximum allowable surface crack length if the surface energy of MgO is $1.0 \ J/m^2$. Data found in Table 13.4 may prove helpful.

8.8 The parameter K in Equations 8.5a, 8.5b, and 8.5c is a function of the applied nominal stress σ and crack length a as

$$K = \sigma \sqrt{\pi a}$$

Compute the magnitudes of the normal stresses σ_x and σ_y in front of a surface crack of length 2.0 mm (0.079 in.) (as depicted in Figure 8.10) in response to a nominal tensile stress of 100 MPa (14,500 psi) at the following positions:
(a) $r = 0.1$ mm (3.9×10^{-3} in.), $\theta = 0°$
(b) $r = 0.1$ mm (3.9×10^{-3} in.), $\theta = 45°$
(c) $r = 0.5$ mm (0.02 in.), $\theta = 0°$
(d) $r = 0.5$ mm (0.02 in.), $\theta = 45°$

8.9 The parameter K in Equations 8.5a, 8.5b, and 8.5c is defined in the previous problem.
(a) For a surface crack of length 2.0 mm (7.87×10^{-2} in.), determine the radial position at an angle θ of 30° at which the normal stress σ_x is 100 MPa (14,500 psi) when the magnitude of the nominal applied stress is 150 MPa

(21,750 psi).

(b) Compute the normal stress σ_y at this same position.

8.10 Below is shown a portion of a tensile specimen.

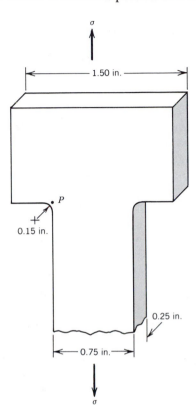

(a) Compute the magnitude of the stress at point P when the externally applied stress is 20,000 psi (140 MPa).

(b) How much will the radius of curvature at point P have to be increased to reduce this stress by 25%?

8.11 A cylindrical hole 0.75 in. (19.0 mm) in diameter passes entirely through the thickness of a steel plate 0.5 in. (12.7 mm) thick, 5 in. (127 mm) wide, and 10 in. (254 mm) long.

(a) Compute the magnitude of the stress at the edge of this hole when a tensile stress of 5000 psi (34.5 MPa) is applied in a lengthwise direction.

(b) Calculate the stress at the hole edge when this same stress is applied in a widthwise direction.

8.12 Cite the significant differences between the stress intensity factor, the plane stress fracture toughness, and the plane strain fracture toughness.

8.13 For each of the metal alloys listed in Table 8.1, compute the minimum component thickness for which the condition of plane strain is valid.

8.14 A specimen of a 4340 steel alloy having a plane strain fracture toughness of 50,000 psi$\sqrt{\text{in.}}$ (54.8 MPa$\sqrt{\text{m}}$) is exposed to a stress of 150,000 psi (1030 MPa).

Will this specimen experience fracture if it is known that the largest surface crack is 0.02 in. (0.5 mm) long? Why or why not? Assume that the parameter Y has a value of 1.0.

8.15 Some aircraft component is fabricated from an aluminum alloy that has a plane strain fracture toughness of 40 MPa$\sqrt{\text{m}}$ (3.64×10^4 psi$\sqrt{\text{in.}}$). It has been determined that fracture results at a stress of 300 MPa (43,500 psi) when the maximum (or critical) internal crack length is 4.0 mm (0.16 in.). For this same component and alloy, will fracture occur at a stress level of 260 MPa (38,000 psi) when the maximum internal crack length is 6.0 mm (0.24 in.)? Why or why not?

8.16 Suppose that a wing component on an aircraft is fabricated from an aluminum alloy that has a plane strain fracture toughness of 26 MPa$\sqrt{\text{m}}$ (23,700 psi$\sqrt{\text{in.}}$). It has been determined that fracture results at a stress of 112 MPa (16,240 psi) when the maximum internal crack length is 8.6 mm (0.34 in.). For this same component and alloy, compute the stress level at which fracture will occur for a critical internal crack length of 6.0 mm (0.24 in.).

8.17 A large plate is fabricated from a steel alloy that has a plane strain fracture toughness of 75,000 psi$\sqrt{\text{in.}}$ (82.4 MPa$\sqrt{\text{m}}$). If, during service use, the plate is exposed to a tensile stress of 50,000 psi (345 MPa), determine the minimum length of a surface crack that will lead to fracture. Assume a value of 1.0 for Y.

8.18 Calculate the maximum internal crack length allowable for a Ti–6Al–4V titanium alloy (Table 8.1) component that is loaded to a stress one half of its yield strength. Assume that the value of Y is 1.50.

8.19 A structural component in the form of a wide plate is to be fabricated from a steel alloy that has a plane strain fracture toughness of 98.9 MPa$\sqrt{\text{m}}$ (90,000 psi$\sqrt{\text{in.}}$) and a yield strength of 860 MPa (125,000 psi). The flaw size resolution limit of the flaw detection apparatus is 3.0 mm (0.12 in.). If the design stress is one half of the yield strength and the value of Y is 1.0, determine whether or not a critical flaw for this plate is subject to detection.

8.20 A structural component in the shape of a flat plate 25.4 mm (1.0 in.) thick is to be fabricated from a metal alloy for which the yield strength and plane strain fracture toughness values are 700 MPa (101,500 psi) and 49.5 MPa$\sqrt{\text{m}}$ (45,000 psi$\sqrt{\text{in.}}$), respectively; for this particular geometry, the value of Y is 1.65. Assuming a design stress of one half of the yield strength, is it possible to compute the critical length of a surface flaw? If so, determine its length; if this computation is not possible from the given data, then explain why.

8.21 For a flat plate of width w containing an internal crack of length $2a$ which is centrally positioned as illustrated in the figure below, the parameter Y for tensile loading may be determined as

$$Y = \left(\frac{w}{\pi a} \tan \frac{\pi a}{w} \right)^{1/2} \tag{8.26}$$

Assuming an internal crack length (i.e., $2a$) of 1.0 in. (25.4 mm) within a plate of width 4.0 in. (101.6 mm), determine the minimum plane strain fracture toughness necessary to ensure that fracture will not occur for a design stress of 60,000 psi (415 MPa).

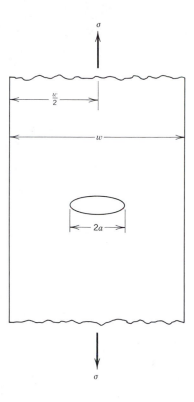

8.22 After consultation of other references, write a brief report on one or two non-destructive test techniques that are used to detect and measure internal and/or surface flaws in metal alloys.

8.23 Tabulated below are data that were gathered from a series of Charpy impact tests on a tempered 4340 steel alloy:

Temperature (°C)	Impact Energy (J)
0	105
−25	104
−50	103
−75	97
−100	63
−113	40
−125	34
−150	28
−175	25
−200	24

(a) Plot the data as impact energy versus temperature.

(b) Determine a ductile-to-brittle transition temperature as that temperature corresponding to the average of the maximum and minimum impact energies.

(c) Determine a ductile-to-brittle transition temperature as that temperature at which the impact energy is 50 J.

8.24 Tabulated below are data that were gathered from a series of Charpy impact tests on a commercial low-carbon steel alloy:

Temperature (°C)	Impact Energy (J)
50	76
40	76
30	71
20	58
10	38
0	23
−10	14
−20	9
−30	5
−40	1.5

(a) Plot the data as impact energy versus temperature.

(b) Determine a ductile-to-brittle transition temperature as that temperature corresponding to the average of the maximum and minimum impact energies.

(c) Determine a ductile-to-brittle transition temperature as that temperature at which the impact energy is 20 J.

8.25 Briefly explain why BCC and HCP metal alloys may experience a ductile-to-brittle transition with decreasing temperature, whereas FCC alloys do not experience such a transition.

8.26 A fatigue test was conducted in which the mean stress was 10,000 psi (70 MPa) and the stress amplitude was 30,000 psi (210 MPa).

(a) Compute the maximum and minimum stress levels.

(b) Compute the stress ratio.

(c) Compute the magnitude of the stress range.

8.27 A cylindrical 1045 steel bar (Figure 8.38) is subjected to repeated compression-tension stress cycling along its axis. If the load amplitude is 15,000 lb_f (66,700 N), compute the minimum allowable bar diameter to ensure that fatigue failure will not occur.

8.28 A 0.25 in. (6.4 mm) diameter cylindrical rod fabricated from a 2014-T6 aluminum alloy (Figure 8.38) is subjected to a reversed tension-compression load cycling along its axis. If the maximum tensile and compressive loads are $+1200$ lb_f ($+5340$ N) and -1200 lb_f (-5340 N), respectively, determine its fatigue life. Assume the stress plotted in Figure 8.38 is stress amplitude.

8.29 A 0.60 in. (15.2 mm) diameter cylindrical rod fabricated from a 2014-T6 alloy (Figure 8.38) is subjected to a repeated tension-compression load cycling along its axis. Compute the maximum and minimum loads that will be applied to yield a fatigue life of 1.0×10^8 cycles. Assume that the stress plotted on the vertical axis is stress amplitude, and data were taken for a mean stress of 5000 psi (35 MPa).

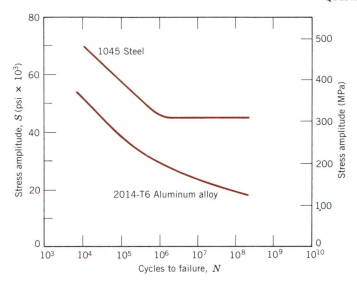

Figure 8.38 Stress magnitude S versus the logarithm of the number of cycles to fatigue failure N for an aluminum alloy and a plain carbon steel. (Adapted from H. W. Hayden, W. G. Moffatt, and J. Wulff, *The Structure and Properties of Materials*, Vol. III, *Mechanical Behavior*, p. 15. Copyright © 1965 by John Wiley & Sons, New York. Reprinted by permission of John Wiley & Sons, Inc.)

8.30 The fatigue data for a brass alloy are given below:

Stress Amplitude (MPa)	Cycles to Failure
170	3.7×10^4
148	1.0×10^5
130	3.0×10^5
114	1.0×10^6
92	1.0×10^7
80	1.0×10^8
74	1.0×10^9

    (a) Make an S–N plot (stress amplitude versus logarithm cycles to failure) using these data.
    (b) Determine the fatigue strength at 4×10^6 cycles.
    (c) Determine the fatigue life for 120 MPa.

8.31 Suppose that the fatigue data for the brass alloy in Problem 8.30 were taken from torsional tests, and that a shaft of this alloy is to be used for a coupling that is attached to an electric motor operating at 1800 rpm. Give the maximum torsional stress amplitude possible for each of the following lifetimes of the coupling: **(a)** 1 year, **(b)** 1 month, **(c)** 1 day, and **(d)** 1 hour.

8.32 The fatigue data for a steel alloy are given below:

Stress Amplitude [psi (MPa)]	Cycles to Failure
68,000 (470)	10^4
63,400 (440)	3×10^4
56,200 (390)	10^5
51,000 (350)	3×10^5
45,300 (310)	10^6
42,200 (290)	3×10^6
42,200 (290)	10^7
42,200 (290)	10^8

(a) Make an S–N plot (stress amplitude versus logarithm cycles to failure) using the data.

(b) What is the fatigue limit for this alloy?

(c) Determine fatigue lifetimes at stress amplitudes of 60,000 psi (415 MPa) and 40,000 psi (275 MPa).

(d) Estimate fatigue strengths at 2×10^4 and 6×10^5 cycles.

8.33 Suppose that the fatigue data for the steel alloy in Problem 8.32 were taken for bending-rotating tests, and that a rod of this alloy is to be used for an automobile axle that rotates at an average rotational velocity of 600 revolutions per minute. Give the maximum lifetimes of continuous driving that are allowable for the following stress levels: (a) 65,000 psi (450 MPa); (b) 55,000 psi (380 MPa); (c) 45,000 psi (310 MPa); and (d) 40,000 psi (275 MPa).

8.34 Three identical fatigue specimens (denoted A, B, and C) are fabricated from a nonferrous alloy. Each is subjected to one of the maximum-minimum stress cycles listed below; the frequency is the same for all three tests.

Specimen	σ_{max} (MPa)	σ_{min} (MPa)
A	+450	−150
B	+300	−300
C	+500	−200

(a) Rank the fatigue lifetimes of these three specimens from the longest to the shortest.

(b) Now justify this ranking using a schematic S–N plot.

8.35 Cite five factors that may lead to scatter in fatigue life data.

8.36 Make a schematic sketch of the fatigue behavior for some metal for which the stress ratio R has a value of +1.

8.37 Using Equations 8.13 and 8.14, demonstrate that increasing the value of the stress ratio R produces a decrease in stress amplitude σ_a.

8.38 Surfaces for some steel specimens that have failed by fatigue have a bright crystalline or grainy appearance. Laymen may explain the failure by saying that the metal crystallized while in service. Offer a criticism for this explanation.

8.39 Briefly explain the difference between fatigue striations and beachmarks both in terms of **(a)** size and **(b)** origin.

8.40 Consider a flat plate of some metal alloy that is to be exposed to repeated tensile-compressive cycling in which the mean stress is 25 MPa. If the initial and critical surface crack lengths are 0.25 and 5.0 mm, respectively, and the values of m and A are 4.0 and 5×10^{-15}, respectively (for $\Delta\sigma$ in MPa and a in m), estimate the maximum tensile stress to yield a fatigue life of 3.2×10^5 cycles. Assume that Y has a value of 2.0, which is independent of crack length.

8.41 Consider a large, flat plate of a metal alloy which is to be exposed to reversed tensile-compressive cycles of stress amplitude 150 MPa. If initially the length of the largest surface crack in this specimen is 0.75 mm and the plane strain fracture toughness is $35 \text{ MPa}\sqrt{m}$, whereas the values of m and A are 2.5 and 2×10^{-12}, respectively (for $\Delta\sigma$ in MPa and a in m), estimate the fatigue life of this plate. Assume that the parameter Y has a value of 1.75 which is independent of crack length.

8.42 Consider a metal component that is exposed to cyclic tensile-compressive stresses. If the fatigue lifetime must be a minimum of 5×10^6 cycles and it is known that the maximum initial surface crack length is 2.0×10^{-2} in. and the maximum tensile stress is 25,000 psi, compute the critical surface crack length. Assume that Y is independent of crack length and has a value of 2.25, and that m and A have values of 3.5 and 1.3×10^{-23}, respectively, for $\Delta\sigma$ and a in units of psi and in., respectively.

8.43 List four measures that may be taken to increase the resistance to fatigue of a metal alloy.

8.44 Give the approximate temperature at which creep deformation becomes an important consideration for each of the following metals: nickel, copper, iron, tungsten, lead, aluminum.

8.45 Superimpose on the same strain-versus-time plot schematic creep curves for both constant tensile stress and constant load, and explain the difference in behavior.

8.46 The following creep data were taken on an aluminum alloy at 400°C (750°F) and a constant stress of 3660 psi (25 MPa). Plot the data as strain versus time, then determine the steady-state or minimum creep rate. *Note:* The initial and instantaneous strain is not included.

Time (min)	Strain	Time (min)	Strain
0	0.000	16	0.135
2	0.025	18	0.153
4	0.043	20	0.172
6	0.065	22	0.193
8	0.078	24	0.218
10	0.092	26	0.255
12	0.109	28	0.307
14	0.120	30	0.368

8.47 A specimen 40 in. (1015 mm) long of a low carbon–nickel alloy (Figure 8.35) is to be exposed to a tensile stress of 10,000 psi (70 MPa) at 427°C (800°F). Determine its elongation after 10,000 h. Assume that the total of both instantaneous and primary creep elongations is 0.05 in. (1.3 mm).

8.48 For a cylindrical low carbon–nickel alloy specimen (Figure 8.35) originally 0.75 in. (19.05 mm) in diameter and 25 in. (635 mm) long, what tensile load is necessary to produce a total elongation of 0.25 in. (6.4 mm) after 5000 h at 538°C (1000°F)? Assume that the sum of instantaneous and primary creep elongations is 0.07 in. (1.8 mm).

8.49 If a component fabricated from a low carbon–nickel alloy (Figure 8.34) is to be exposed to a tensile stress of 31 MPa (4500 psi) at 649°C (1200°F), estimate its rupture lifetime.

8.50 A cylindrical component constructed from a low carbon–nickel alloy (Figure 8.34) has a diameter of 0.75 in. (19.1 mm). Determine the maximum load which may be applied for it to survive 10,000 h at 538°C (1000°F).

8.51 From Equation 8.23, if the logarithm of $\dot{\epsilon}_s$ is plotted versus the logarithm of σ, then a straight line should result, the slope of which is the stress exponent n. Using Figure 8.35, determine the value of n for the low carbon–nickel alloy at each of the three temperatures.

8.52 **(a)** Estimate the activation energy for creep (i.e., Q_c in Equation 8.24) for the low carbon–nickel alloy having the steady-state creep behavior shown in Figure 8.35. Use data taken at a stress level of 8000 psi (55 MPa) and temperatures of 427°C and 538°C. **(b)** Estimate $\dot{\epsilon}_s$ at 649°C (922 K).

8.53 Steady-state creep rate data are given below for some alloy taken at 200°C (473 K):

$\dot{\epsilon}_s$ (h^{-1})	σ [MPa (psi)]
2.5×10^{-3}	55 (8,000)
2.4×10^{-2}	69 (10,000)

If it is known that the activation energy for creep is 140,000 J/mol, compute the steady-state creep rate at a temperature of 250°C (523 K) and a stress level of 48 MPa (7000 psi).

8.54 Steady-state creep data taken for an iron alloy at a stress level of 20,000 psi (140 MPa) are given below:

$\dot{\epsilon}_s$ (h^{-1})	T (K)
6.6×10^{-4}	1090
8.8×10^{-2}	1200

If it is known that the value of the stress exponent n for this alloy is 8.5, compute the steady-state creep rate at 1300 K and a stress level of 12,000 psi (83 MPa).

8.55 An S-590 iron component (Figure 8.36) must have a creep rupture lifetime of at least 20 days at 650°C (923 K). Compute the maximum allowable stress level.

8.56 Consider an S-590 iron component (Figure 8.36) that is subjected to a stress of 8000 psi (55 MPa). At what temperature will the rupture lifetime be 200 h?

8.57 An 18-8 Mo stainless steel (Figure 8.39) must have a creep rupture lifetime of at least 5 years at 500°C (773 K). Compute the maximum allowable stress level.

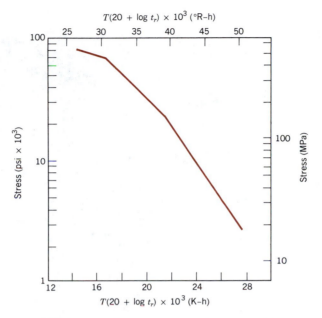

Figure 8.39 Logarithm stress versus the Larson–Miller parameter for an 18-8 Mo stainless steel. (From F. R. Larson and J. Miller, *Trans. ASME,* **74,** 765 (1952). Reprinted by permission of ASME.)

8.58 Consider an 18-8 Mo stainless steel component (Figure 8.39) that is subjected to a stress of 5000 psi (34.5 MPa). At what temperature will the rupture lifetime be 10 years? 20 years?

8.59 Cite three metallurgical/processing techniques that are employed to enhance the creep resistance of metal alloys.

CHAPTER 9

PHASE DIAGRAMS

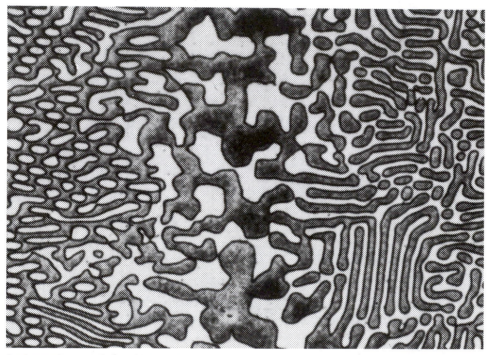

A photomicrograph in which is shown the interface between a reversible-matrix, aluminum-copper eutectic alloy. Magnification unknown. (Reproduced with permission from *Metals Handbook,* Vol. 9, 9th Edition, *Metallography and Microstructures,* American Society for Metals, Metals Park, OH, 1985.)

9.1 INTRODUCTION

The understanding of phase diagrams for alloy systems is extremely important because there is a strong correlation between microstructure and mechanical properties, and the development of microstructure of an alloy is related to the characteristics of its phase diagram. In addition, phase diagrams provide valuable information about melting, casting, crystallization, and other phenomena.

This chapter presents and discusses the following topics: (1) terminology associated with phase diagrams and phase transformations; (2) the interpretation of phase diagrams; (3) some of the common and relatively simple binary phase diagrams, including that for the iron–carbon system; and (4) the development of equilibrium microstructures, upon cooling, for several situations.

DEFINITIONS AND BASIC CONCEPTS

It is necessary to establish a foundation of definitions and basic concepts relating to alloys, phases, and equilibrium before delving into the interpretation and utilization of phase diagrams. The term **component** is frequently used in this discussion; components are pure metals and/or compounds of which an alloy is composed. For example, in a copper–zinc brass, the components are Cu and Zn. *Solute* and *solvent*, which are also common terms, were defined in Section 4.3. Another term used in this context is **system,** which has two meanings. First, "system" may refer to a specific body of material under consideration (e.g., a ladle of molten steel). Or, it may relate to the series of possible alloys consisting of the same components, but without regard to alloy composition (e.g., the iron–carbon system).

The concept of a solid solution was introduced in Section 4.3. By way of review, a solid solution consists of atoms of at least two different types; the solute atoms occupy either substitutional or interstitial positions in the solvent lattice, and the crystal structure of the solvent is maintained.

9.2 SOLUBILITY LIMIT

For many alloy systems and at some specific temperature, there is a maximum concentration of solute atoms that may dissolve in the solvent to form a solid solution; this is called a **solubility limit.** The addition of solute in excess of this solubility limit results in the formation of another solid solution or compound that has a distinctly different composition. To illustrate this concept, consider the sugar–water $(C_{12}H_{22}O_{11}–H_2O)$ system. Initially, as sugar is added to water, a sugar–water solution or syrup forms. As more sugar is introduced, the solution becomes more concentrated, until the solubility limit is reached, or the solution becomes saturated with sugar. At this time the solution is not capable of dissolving any more sugar, and further additions simply settle to the bottom of the container. Thus the system now consists of two separate substances: a sugar–water syrup liquid solution and solid crystals of undissolved sugar.

This solubility limit of sugar in water depends on the temperature of the water and may be represented in graphical form on a plot of temperature along the ordinate

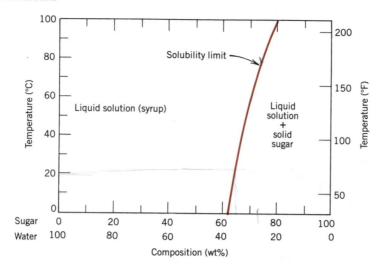

Figure 9.1 The solubility of sugar ($C_{12}H_{22}O_{11}$) in a sugar–water syrup.

and composition (in weight percent sugar) along the abscissa, as shown in Figure 9.1. Along the composition axis, increasing sugar concentration is from left to right, and percentage of water is read from right to left. Since only two components are involved (sugar and water), the sum of the concentrations at any composition will equal 100 wt%. The solubility limit is represented as the nearly vertical line in the figure. For compositions and temperatures to the left of the solubility line, only the syrup liquid solution exists; to the right of the line, syrup and solid sugar coexist. The solubility limit at some temperature is the composition that corresponds to the intersection of the given temperature coordinate and the solubility limit line. For example, at 20°C the maximum solubility of sugar in water is 65 wt%. As Figure 9.1 indicates, the solubility limit increases slightly with rising temperature.

9.3 PHASES

Also critical to the understanding of phase diagrams is the concept of a **phase.** A phase may be defined as a homogeneous portion of a system that has uniform physical and chemical characteristics. Every pure material is considered to be a phase; so also is every solid, liquid, and gaseous solution. For example, the sugar–water syrup solution just discussed is one phase, and solid sugar is another. Each has different physical properties (one is a liquid, the other is a solid); furthermore, each is different chemically (i.e., has a different chemical composition); one is virtually pure sugar, the other is a solution of H_2O and $C_{12}H_{22}O_{11}$. If more than one phase is present in a given system, each will have its own distinct properties, and a boundary separating the phases will exist across which there will be a discontinuous and abrupt change in physical and/or chemical characteristics. When two phases are present in a system, it is not necessary that there be a difference in both physical and chemical properties; a disparity in one or the other set of properties is sufficient. When water and ice are present in a container, two separate phases exist; they are physically dissimilar (one is a solid, the other is a liquid) but identical in chemical makeup. Also, when a substance can exist in two or more polymorphic forms (e.g., having both FCC and BCC structures), each of these structures is a separate phase because their respective physical characteristics differ.

Sometimes, a single-phase system is termed "homogeneous." Systems composed of two or more phases are termed "mixtures" or "heterogeneous systems." Most metallic alloys, and for that matter ceramic, polymeric, and composite systems, are heterogeneous. Ordinarily, the phases interact in such a way that the property combination of the multiphase system is different from, and more attractive than, either of the individual phases.

9.4 MICROSTRUCTURE

Many times, the physical properties and, in particular, the mechanical behavior of a material depend on the microstructure. Microstructure is subject to direct microscopic observation, using optical or electron microscopes; this topic was touched on in Section 4.9. In metal alloys, microstructure is characterized by the number of phases present, their proportions, and the manner in which they are distributed or arranged. The microstructure of an alloy depends on such variables as the alloying elements present, their concentrations, and the heat treatment of the alloy (i.e., the temperature, the heating time at temperature, and the rate of cooling to room temperature).

The procedure of specimen preparation for microscopic examination was briefly outlined in Section 4.9. After appropriate polishing and etching, the different phases may be distinguished by their appearance. For example, the photomicrograph shown on page 246 is of a two-phase aluminum–copper alloy; the one phase appears light, the other phase is dark. When only a single phase or solid solution is present, the texture will be uniform, except for grain boundaries that may be revealed (Figure 4.12b).

9.5 PHASE EQUILIBRIA

Equilibrium is another essential concept. It is best described in terms of a thermodynamic quantity called the **free energy.** In brief, free energy is a function of the internal energy of a system, and also the randomness or disorder of the atoms or molecules (or entropy). A system is at equilibrium if its free energy is at a minimum under some specified combination of temperature, pressure, and composition. In a macroscopic sense, this means that the characteristics of the system do not change with time but persist indefinitely; that is, the system is stable. A change in temperature, pressure, and/or composition for a system in equilibrium will result in an increase in the free energy and in a possible spontaneous change to another state whereby the free energy is lowered.

The term **phase equilibrium,** often used in the context of this discussion, refers to equilibrium as it applies to systems in which more than one phase may exist. Phase equilibrium is reflected by a constancy with time in the phase characteristics of a system. Perhaps an example best illustrates this concept. Suppose that a sugar–water syrup is contained in a closed vessel and the solution is in contact with solid sugar at 20°C. If the system is at equilibrium, the composition of the syrup is 65 wt% $C_{12}H_{22}O_{11}$–35 wt% H_2O (Figure 9.1), and the amounts and compositions of the syrup and solid sugar will remain constant with time. If the temperature of the system is suddenly raised—say, to 100°C—this equilibrium or balance is temporarily upset in that the solubility limit has been increased to 80 wt% $C_{12}H_{22}O_{11}$ (Figure 9.1). Thus some of the solid sugar will go into solution in the syrup. This will continue until the new equilibrium syrup concentration is established at the higher temperature.

This sugar–syrup example has illustrated the principle of phase equilibrium using a liquid–solid system. In many metallurgical and materials systems of interest, phase equilibrium involves just solid phases. In this regard the state of the system is reflected in the characteristics of the microstructure, which necessarily include not only the phases present and their compositions but, in addition, the relative phase amounts and their spatial arrangement or distribution.

Free energy considerations and diagrams similar to Figure 9.1 provide information about the equilibrium characteristics of a particular system, which is important; but they do not indicate the time period necessary for the attainment of a new equilibrium state. It is often the case, especially in solid systems, that a state of equilibrium is never completely achieved because the rate of approach to equilibrium is extremely slow; such a system is said to be in a nonequilibrium or metastable state. A metastable state or microstructure may persist indefinitely, experiencing only extremely slight and almost imperceptible changes as time progresses. Often, metastable structures are of more practical significance than equilibrium ones. For example, some steel and aluminum alloys rely for their strength on the development of metastable microstructures during carefully designed heat treatments (Sections 10.5 and 11.7).

Thus not only is an understanding of equilibrium states and structures important, but the speed or rate at which they are established and, in addition, the factors that affect the rate must be considered. This chapter is devoted almost exclusively to equilibrium structures; the treatment of reaction rates and nonequilibrium structures is deferred to Chapters 10 and 11.

EQUILIBRIUM PHASE DIAGRAMS

Much of the information about the control of microstructure or phase structure of a particular alloy system is conveniently and concisely displayed in what is called a **phase diagram,** also often termed an *equilibrium* or *constitutional diagram.* Many microstructures develop from phase transformations, the changes that occur between phases when the temperature is altered (ordinarily upon cooling). This may involve the transition from one phase to another, or the appearance or disappearance of a phase. Phase diagrams are helpful in predicting phase transformations and the resulting microstructures, which may have equilibrium or nonequilibrium character.

Equilibrium phase diagrams represent the relationships between temperature and the compositions and the quantities of phases at equilibrium. There are several different varieties; but in the present discussion, temperature and composition are the variable parameters, for binary alloys. A binary alloy is one that contains two components. If more than two components are present, phase diagrams become extremely complicated and difficult to represent. The principles of microstructural control with the aid of phase diagrams can be illustrated with binary alloys even though, in reality, most alloys contain more than two components. External pressure is also a parameter that influences the phase structure. However, in practicality, pressure remains virtually constant; thus the phase diagrams presented here are for a constant pressure of one atmosphere (1 atm).

9.6 BINARY ISOMORPHOUS SYSTEMS

Possibly the easiest type of binary phase diagram to understand and interpret is that which is characterized by the copper–nickel system (Figure 9.2a). Temperature is plotted along the ordinate, and the abscissa represents the composition of the alloy, in weight percent (bottom) and atomic percent (top) of nickel. The composition ranges from 0 wt% Ni (100 wt% Cu) on the left horizontal extremity to 100 wt% Ni (0 wt% Cu) on the right. Three different phase regions, or fields, appear on the diagram, an alpha (α) field, a liquid (L) field, and a two-phase $\alpha + L$ field. Each region is defined by the phase or phases that exist over the range of temperatures and compositions delimited by the phase boundary lines.

The liquid L is a homogeneous liquid solution composed of both copper and nickel. The α phase is a substitutional solid solution consisting of both Cu and Ni atoms, and having an FCC crystal structure. At temperatures below about 1080°C, copper and nickel are mutually soluble in each other in the solid state for all compositions. This complete solubility is explained by the fact that both Cu and Ni have the same crystal structure (FCC), nearly identical atomic radii and electronegativities, and similar valences, as discussed in Section 4.3. The copper–nickel system is termed **isomorphous** because of this complete liquid and solid solubility of the two components.

A couple of comments are in order regarding nomenclature. First, for metallic alloys, solid solutions are commonly designated by lowercase Greek letters (α, β, γ, etc.). Furthermore, with regard to phase boundaries, the line separating the L and $\alpha + L$ phase fields is termed the *liquidus line*, as indicated in Figure 9.2a; the liquid phase is present at all temperatures and compositions above this line. The *solidus line* is located between the α and $\alpha + L$ regions, below which only the solid α phase exists.

For Figure 9.2a, the solidus and liquidus lines intersect at the two composition extremities; these correspond to the melting temperatures of the pure components. For example, the melting temperatures of pure copper and nickel are 1085°C and 1455°C, respectively. Heating pure copper corresponds to moving vertically up the left-hand temperature axis. Copper remains solid until its melting temperature is reached. The solid-to-liquid transformation takes place at the melting temperature, and no further heating is possible until this transformation has been completed.

For any composition other than pure components, this melting phenomenon will occur over the range of temperatures between the solidus and liquidus lines; both solid α and liquid phases will be in equilibrium within this temperature range. For example, upon heating an alloy of composition 50 wt% Ni–50 wt% Cu (Figure 9.2a), melting begins at approximately 1280°C (2340°F); the amount of liquid phase continuously increases with temperature until about 1320°C (2410°F), at which the alloy is completely liquid.

Interpretation of Phase Diagrams

For a binary system of known composition and temperature that is at equilibrium, at least three kinds of information are available: (1) the phases that are present, (2)

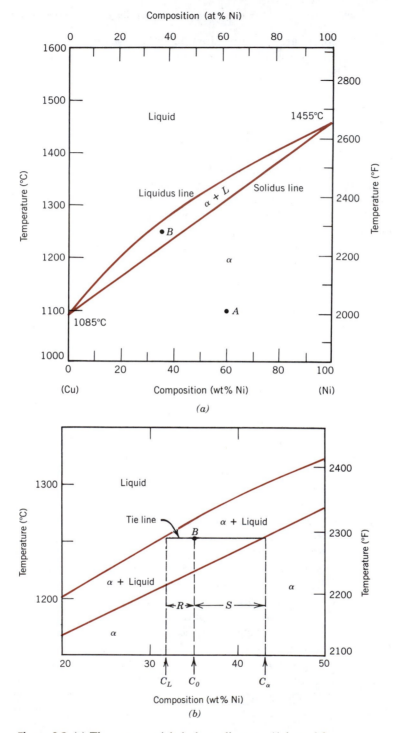

Figure 9.2 (*a*) The copper–nickel phase diagram. (Adapted from
Metals Handbook: Metallography, Structures and Phase Diagrams,
Vol. 8, 8th edition, ASM Handbook Committee, T. Lyman, Editor,
American Society for Metals, 1973, p. 294.) (*b*) A portion of the
copper–nickel phase diagram for which compositions and phase
amounts are determined at point *B*.

the compositions of these phases, and (3) the percentages or fractions of the phases. The procedures for making these determinations will be demonstrated using the copper–nickel system.

Phases Present.   The establishment of what phases are present is relatively simple. One just locates the temperature–composition point on the diagram and notes the phase(s) with which the corresponding phase field is labeled. For example, an alloy of composition 60 wt% Ni–40 wt% Cu at 1100°C would be located at point *A* in Figure 9.2*a*; since this is within the α region, only the single α phase will be present. On the other hand, a 35 wt% Ni–65 wt% Cu alloy at 1250°C (point *B*) will consist of both α and liquid phases at equilibrium.

Determination of Phase Compositions.   The first step in the determination of phase compositions (in terms of the concentrations of the components) is to locate the temperature–composition point on the phase diagram. Different methods are used for single- and two-phase regions. If only one phase is present, the procedure is trivial: the composition of this phase is simply the same as the overall composition of the alloy. For example, consider the 60 wt% Ni–40 wt% Cu alloy at 1100°C (point *A*, Figure 9.2*a*). At this composition and temperature, only the α phase is present, having a composition of 60 wt% Ni–40 wt% Cu.

For an alloy having composition and temperature located in a two-phase region, the situation is more complicated. In all two-phase regions (and in two-phase regions only), one may imagine a series of horizontal lines, one at every temperature; each of these is known as a **tie line,** or sometimes as an isotherm. These tie lines extend across the two-phase region and terminate at the phase boundary lines on either side. To compute the equilibrium concentrations of the two phases, the following procedure is used:

1. A tie line is constructed across the two-phase region at the temperature of the alloy.
2. The intersections of the tie line and the phase boundaries on either side are noted.
3. Perpendiculars are dropped from these intersections to the horizontal composition axis, from which the composition of each of the respective phases is read.

For example, consider again the 35 wt% Ni–65 wt% Cu alloy at 1250°C, located at point *B* in Figure 9.2*b* and lying within the α + *L* region. Thus the problem is to determine the composition (in wt% Ni and Cu) for both the α and liquid phases. The tie line has been constructed across the α + *L* phase region, as shown in Figure 9.2*b*. The perpendicular from the intersection of the tie line with the liquidus boundary meets the composition axis at 32 wt% Ni–68 wt% Cu, which is the composition of the liquid phase, C_L. Likewise, for the solidus–tie line intersection, we find a composition for the α solid-solution phase, $C_α$, of 43 wt% Ni–57 wt% Cu.

Determination of Phase Amounts.   The relative amounts (as fraction or as percentage) of the phases present at equilibrium may also be computed with the aid of phase diagrams. Again, the single- and two-phase situations must be treated separately. The solution is obvious in the single-phase region: Since only one phase is present, the alloy is composed entirely of that phase; that is, the phase fraction is 1.0 or, alternately, the percentage is 100%. From the previous example for the 60 wt% Ni–40 wt% Cu alloy at 1100°C (point A in Figure 9.2a), only the α phase is present; hence, the alloy is completely or 100% α.

If the composition and temperature position is located within a two-phase region, things are more complex. The tie line must be utilized in conjunction with a procedure that is often called the **lever rule** (or the *inverse lever rule*), which is applied as follows:

1. The tie line is constructed across the two-phase region at the temperature of the alloy.
2. The overall alloy composition is located on the tie line.
3. The fraction of one phase is computed by taking the length of tie line from the overall alloy composition to the phase boundary for the *other* phase, and dividing by the total tie line length.
4. The fraction of the other phase is determined in the same manner.
5. If phase percentages are desired, each phase fraction is multiplied by 100. When the composition axis is scaled in weight percent, the phase fractions computed using the lever rule are mass fractions—the mass (or weight) of a specific phase divided by the total alloy mass (or weight). The mass of each phase is computed from the product of each phase fraction and the total alloy mass. On occasion, it becomes necessary to compute volume phase fractions, which is accomplished by consideration of the phase densities, as outlined in Example Problem 9.3.

In the employment of the lever rule, tie line segment lengths may be determined either by direct measurement from the phase diagram using a linear scale, preferably graduated in millimeters, or by subtracting compositions as taken from the composition axis.

Consider again the example shown in Figure 9.2b, in which at 1250°C both α and liquid phases are present for a 35 wt% Ni–65 wt% Cu alloy. The problem is to compute the fraction of each of the α and liquid phases. The tie line has been constructed that was used for the determination of α and L phase compositions. Let the overall alloy composition be located along the tie line and denoted as C_0, and mass fractions be represented by W_L and W_α for the respective phases. From the lever rule, W_L may be computed according to

$$W_L = \frac{S}{R + S} \tag{9.1a}$$

or, by subtracting compositions,

$$W_L = \frac{C_\alpha - C_0}{C_\alpha - C_L} \tag{9.1b}$$

Composition need be specified only in terms of one of the constituents for a binary alloy; for the computation above, weight percent nickel will be used (i.e., $C_0 = 35$ wt% Ni, $C_\alpha = 43$ wt% Ni, and $C_L = 32$ wt% Ni), and

$$W_L = \frac{43 - 35}{43 - 32} = 0.73$$

Similarly, for the α phase,

$$W_\alpha = \frac{R}{R + S} \qquad (9.2a)$$

$$= \frac{C_0 - C_L}{C_\alpha - C_L} \qquad (9.2b)$$

$$= \frac{35 - 32}{43 - 32} = 0.27$$

Of course, identical answers are obtained if compositions are expressed in weight percent copper instead of nickel.

Thus the lever rule may be employed to determine the relative amounts or fractions of phases in any two-phase region for a binary alloy if the temperature and composition are known and if equilibrium has been established. Its derivation is presented as an example problem.

It is easy to confuse the foregoing procedures for the determination of phase compositions and fractional phase amounts; thus a brief summary is warranted. *Compositions* of phases are expressed in terms of concentrations of components (e.g., Cu, Ni). For any alloy consisting of a single phase, the composition of that phase is the same as the total alloy composition. If two phases are present, the tie line must be employed, the extremities of which determine the compositions of the respective phases. With regard to *relative phase amounts* (e.g., mass fraction of the α or liquid phase), when a single phase exists, the alloy is completely that phase. For a two-phase alloy, on the other hand, the lever rule is utilized, in which a ratio of tie line segment lengths is taken.

EXAMPLE PROBLEM 9.1

Derive the lever rule.

SOLUTION

Consider the phase diagram for copper and nickel (Figure 9.2b) and alloy of composition C_0 at 1250°C, and let C_α, C_L, W_α, and W_L represent the same parameters as above. This derivation is accomplished through two conservation-of-mass expressions. With the first, since only two phases are present, the sum of their mass fractions must be equal to unity, that is,

$$W_\alpha + W_L = 1 \qquad (9.3)$$

For the second, the mass of one of the components (either Cu or Ni) that is present in both of the phases must be equal to the mass of that component in the total alloy, or

$$W_\alpha C_\alpha + W_L C_L = C_0 \qquad (9.4)$$

Simultaneous solution of these two equations leads to the lever rule expressions for this particular situation, Equations 9.1b and 9.2b:

$$W_L = \frac{C_\alpha - C_0}{C_\alpha - C_L} \tag{9.1b}$$

$$W_\alpha = \frac{C_0 - C_L}{C_\alpha - C_L} \tag{9.2b}$$

Development of Microstructure in Isomorphous Alloys

At this point it is instructive to examine the development of microstructure that occurs for isomorphous alloys during solidification. We first treat the situation in which the cooling occurs very slowly, in that phase equilibrium is continuously maintained.

Let us consider the copper–nickel system (Figure 9.2a), specifically an alloy of composition 35 wt% Ni–65 wt% Cu as it is cooled from 1300°C. The region of the

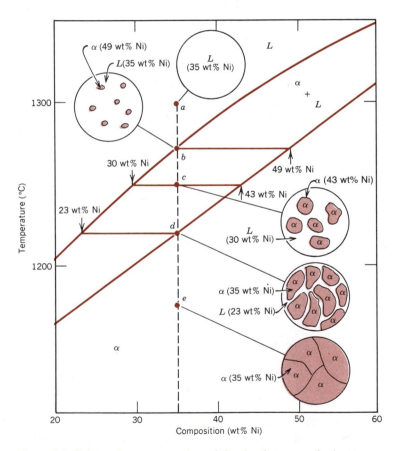

Figure 9.3 Schematic representation of the development of microstructure during the equilibrium solidification of a 35 wt% Ni–65 wt% Cu alloy.

Cu–Ni phase diagram in the vicinity of this composition is shown in Figure 9.3. Cooling of an alloy of the above composition corresponds to moving down the vertical dashed line. At 1300°C, point *a*, the alloy is completely liquid (of composition 35 wt% Ni–65 wt% Cu) and has the microstructure represented by the circle inset in the figure. As cooling begins, no microstructural or compositional changes will be realized until we reach the liquidus line (point *b*, ~1270°C). At this point, the first solid α begins to form, which has a composition dictated by the tie line drawn at this temperature (i.e., 49 wt% Ni–51 wt% Cu); the composition of the liquid is still approximately 35 wt% Ni–65 wt% Cu, which is different than that of the solid α. With continued cooling, both compositions and relative amounts of each of the phases will change. The compositions of the liquid and α phases will follow the liquidus and solidus lines, respectively. Furthermore, the fraction of the α phase will increase with continued cooling. Also, the overall alloy composition (35 wt% Ni–65 wt% Cu) remains unchanged during cooling even though there is a redistribution of copper and nickel between the phases.

At 1250°C, point *c* in Figure 9.3, the compositions of the liquid and α phases are 30 wt% Ni–70 wt% Cu and 43 wt% Ni–57 wt% Cu, respectively.

The solidification process is virtually complete at 1220°C, point *d*; the composition of the solid is approximately 35 wt% Ni–65 wt% Cu (the overall alloy composition) while that of the last remaining liquid is 23 wt% Ni–77 wt% Cu. Upon crossing the solidus line, this remaining liquid solidifies; the final product then is a polycrystalline α-phase solid solution that has a uniform 35 wt% Ni–65 wt% Cu composition (point *e*, Figure 9.3). Subsequent cooling will produce no microstructural or compositional alterations.

Conditions of equilibrium solidification are realized only for extremely slow cooling rates. The reason for this is that with changes in temperature, there must be readjustments in the compositions of the two phases in accordance with the phase diagram, as discussed in the preceding paragraphs. These readjustments are accomplished by diffusional processes, that is, diffusion in both solid and liquid phases and also across the solid–liquid interface. Inasmuch as diffusion is a time-dependent phenomenon (Section 5.3), to maintain equilibrium during cooling, sufficient time must be allowed at each temperature for the appropriate compositional readjustments. Diffusion rates (i.e., the magnitude of the diffusion coefficients) are especially low for the solid phase and, for both phases, decrease with diminishing temperature. In virtually all practical solidification situations, cooling rates are much too rapid to allow these compositional readjustments and the maintenance of equilibrium, and microstructures other than those described above result.

One important consequence of nonequilibrium solidification for isomorphous alloys is a nonuniform distribution of the two elements within the grains, which is termed *segregation*. The center region of each grain, which is the first to freeze, is rich in the high-melting element (e.g., nickel for the Cu–Ni system), whereas the concentration of the low-melting element increases with position from this region to the grain boundary; thus concentration gradients across the grains are established, which is represented schematically in Figure 9.4. This is often termed a "cored" structure, which gives rise to less than the optimal properties. As a casting having a cored structure is reheated, grain boundary regions will melt first inasmuch as they are richer in the low-melting component. This produces a sudden loss in mechanical integrity due to the thin liquid film that separates the grains. Furthermore, this melting

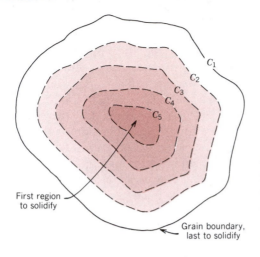

First region
to solidify

Grain boundary,
last to solidify

Figure 9.4 Schematic representation of coring in a single grain. For a cored structure, concentration gradients are established across the grains; dashed lines indicate constant concentration contours (C_1, C_2, C_3, etc.).

may begin at a temperature below the equilibrium solidus temperature of the alloy. Coring may be eliminated by a homogenization heat treatment carried out at a temperature below the solidus point for the particular alloy composition. During this process, atomic diffusion occurs, which produces compositionally homogeneous grains.

Mechanical Properties of Isomorphous Alloys

We shall now briefly explore how the mechanical properties of solid isomorphous alloys are affected by composition as other structural variables (e.g., grain size) are held constant. For all temperatures and compositions below the melting temperature of the lowest-melting component, only a single solid phase will exist. Therefore, each component will experience solid-solution hardening (Section 7.9), or an increase in strength and hardness by additions of the other component. This effect is demonstrated in Figure 9.5a as tensile strength versus composition for the copper–nickel system at room temperature; at some intermediate composition, the curve necessarily passes through a maximum. Plotted in Figure 9.5b is the ductility (%EL)–composition behavior, which is just the opposite of tensile strength; that is, ductility decreases with additions of the second component, and the curve exhibits a minimum.

9.7 BINARY EUTECTIC SYSTEMS

Another type of common and relatively simple phase diagram found for binary alloys is shown in Figure 9.6 for the copper–silver system; this is known as a binary eutectic phase diagram. A number of features of this phase diagram are important and worth noting. First of all, three single-phase regions are found on the diagram: α, β, and liquid. The α phase is a solid solution rich in copper; it has silver as the solute component and an FCC crystal structure. The β phase solid solution also has an FCC structure, but copper is the solute. Technically, pure copper and pure silver are considered to be α and β phases, respectively.

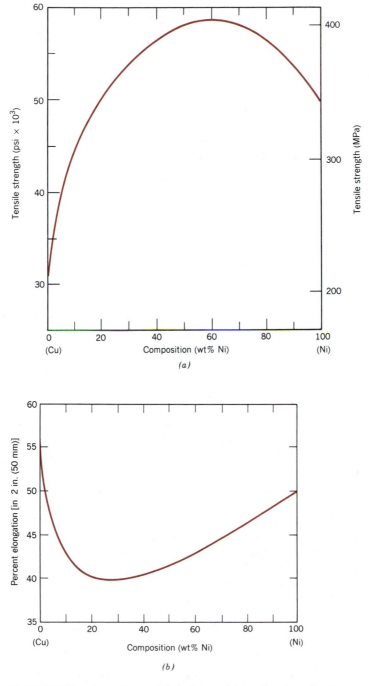

Figure 9.5 For the copper–nickel system, (*a*) tensile strength versus composition, and (*b*) ductility (%EL) versus composition. A solid solution exists over all compositions for this system.

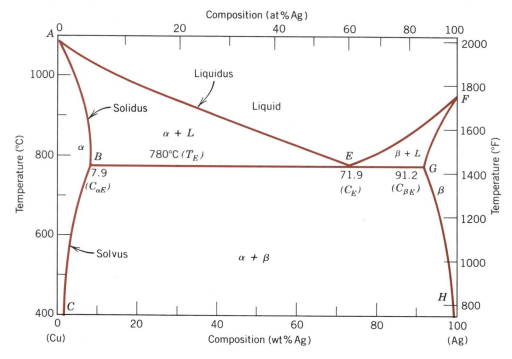

Figure 9.6 The copper–silver phase diagram. (Adapted from *Metals Handbook: Metallography, Structures and Phase Diagrams,* Vol. 8, 8th edition, ASM Handbook Committee, T. Lyman, Editor, American Society for Metals, 1973, p. 253.)

Thus the solubility in each of these solid phases is limited, in that at any temperature below line *BEG* only a limited concentration of silver will dissolve in copper (for the α phase), and similarly for copper in silver (for the β phase). The solubility limit for the α phase corresponds to the boundary line, labeled *CBA*, between the $\alpha/(\alpha + \beta)$ and $\alpha/(\alpha + L)$ phase regions; it increases with temperature to a maximum [7.9 wt% Ag at 780°C (1436°F)] at point *B*, and decreases back to zero at the melting temperature of pure copper, point *A* [1085°C (1985°F)]. At temperatures below 780°C (1436°F), the solid solubility limit line separating the α and $\alpha + \beta$ phase regions is termed a **solvus line;** the boundary *AB* between the α and $\alpha + L$ fields is the solidus line, as indicated in Figure 9.6. For the β phase, both solvus and solidus lines also exist, *HG* and *GF*, respectively, as shown. The maximum solubility of copper in the β phase, point *G* (8.8 wt% Cu), also occurs at 780°C (1436°F). This horizontal line *BEG*, which is parallel to the composition axis and extends between these maximum solubility positions, may also be considered to be a solidus line; it represents the lowest temperature at which a liquid phase may exist for any copper–silver alloy that is at equilibrium.

There are also three two-phase regions found for the copper–silver system (Figure 9.6): $\alpha + L$, $\beta + L$, and $\alpha + \beta$. The α and β phase solid solutions coexist for all compositions and temperatures within the $\alpha + \beta$ phase field; the α + liquid and β + liquid phases also coexist in their respective phase regions. Furthermore, compositions and relative amounts for the phases may be determined using tie lines and the lever rule as outlined in the preceding section.

As silver is added to copper, the temperature at which the alloys become totally liquid decreases along the liquidus line, line AE; thus the melting temperature of copper is lowered by silver additions. The same may be said for silver: the introduction of copper reduces the temperature of complete melting along the other liquidus line, FE. These liquidus lines meet at the point E on the phase diagram, through which also passes the horizontal isotherm line BEG. Point E is called an **invariant point,** which is designated by the composition C_E and temperature T_E; for the copper–silver system, the values of C_E and T_E are 71.9 wt% Ag and 780°C (1436°F), respectively.

An important reaction occurs for an alloy of composition C_E as it changes temperature in passing through T_E; this reaction may be written as follows:

$$L\,(C_E) \underset{\text{heating}}{\overset{\text{cooling}}{\rightleftharpoons}} \alpha\,(C_{\alpha E}) + \beta\,(C_{\beta E}) \tag{9.5}$$

Or, upon cooling, a liquid phase is transformed into the two solid α and β phases at the temperature T_E; the opposite reaction occurs upon heating. This is called an **eutectic reaction** (eutectic means easily melted), and C_E and T_E represent the eutectic composition and temperature, respectively; $C_{\alpha E}$ and $C_{\beta E}$ are the respective compositions of the α and β phases at T_E. Thus for the copper–silver system, Equation 9.5 may be written as follows:

$$L\,(71.9\text{ wt\% Ag}) \underset{\text{heating}}{\overset{\text{cooling}}{\rightleftharpoons}} \alpha\,(7.9\text{ wt\% Ag}) + \beta\,(91.2\text{ wt\% Ag})$$

Often, the horizontal solidus line at T_E is called the *eutectic isotherm.*

The eutectic reaction, upon cooling, is similar to solidification for pure components in that the reaction proceeds to completion at a constant temperature, or isothermally, at T_E. However, the solid product of eutectic solidification is always two solid phases, whereas for a pure component only a single phase forms. Because of this eutectic reaction, phase diagrams similar to that in Figure 9.6 are termed eutectic phase diagrams; components exhibiting this behavior comprise a eutectic system.

In the construction of binary phase diagrams, it is important to understand that one or at most two phases may be in equilibrium within a phase field. This holds true for the phase diagrams in Figures 9.2a and 9.6. For a eutectic system, three phases (α, β, and L) may be in equilibrium, but only at points along the eutectic isotherm. Another general rule is that single-phase regions are always separated from each other by a two-phase region which consists of the two single phases that it separates. For example, the $\alpha + \beta$ field is situated between the α and β single-phase regions in Figure 9.6.

Another common eutectic system is that for lead and tin; the phase diagram (Figure 9.7) has a general shape similar to that for copper–silver. For the lead–tin system the solid solution phases are also designated by α and β; in this case, α represents a solid solution of tin in lead, and for β, tin is the solvent and lead is the solute. The eutectic invariant point is located at 61.9 wt% Sn and 183°C (361°F). Of course, maximum solid solubility compositions as well as component melting temperatures will be different for the copper–silver and lead–tin systems, as may be observed by comparing their phase diagrams.

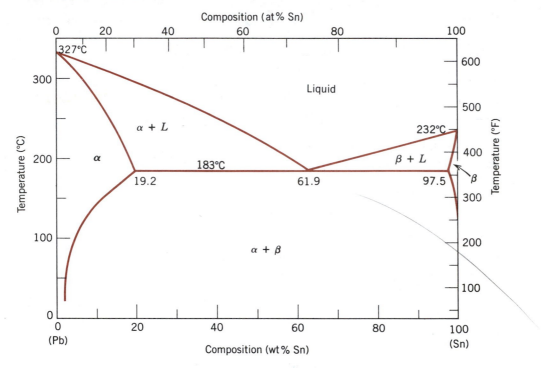

Figure 9.7 The lead–tin phase diagram. (Adapted from *Metals Handbook: Metallography, Structures and Phase Diagrams,* Vol. 8, 8th edition, ASM Handbook Committee, T. Lyman, Editor, American Society for Metals, 1973, p. 330.)

On occasion, low-melting-temperature alloys are prepared having near-eutectic compositions. A familiar example is the 60–40 solder, containing 60 wt% Sn and 40 wt% Pb. Figure 9.7 indicates that an alloy of this composition is completely molten at about 185°C (365°F), which makes this material especially attractive as a low-temperature solder, since it is easily melted.

EXAMPLE PROBLEM 9.2.

For a 40 wt% Sn–60 wt% Pb alloy at 150°C (300°F), (a) What phase(s) is (are) present? (b) What is (are) the composition(s) of the phase(s)?

SOLUTION

(a) Locate this temperature–composition point on the phase diagram (point B in Figure 9.8). Inasmuch as it is within the α + β region, both α and β phases will coexist.

(b) Since two phases are present, it becomes necessary to construct a tie line across the α + β phase field at 150°C, as indicated in Figure 9.8. The composition of the α phase corresponds to the tie line intersection with the α/(α + β) solvus phase boundary—about 11 wt% Sn–89 wt% Pb, denoted as C_α. Similarly for the β phase, which will have a composition approximately 99 wt% Sn–1 wt% Pb (C_β).

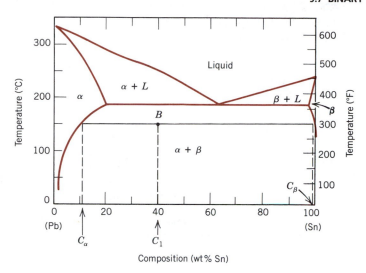

Figure 9.8 The lead–tin phase diagram. For a 40 wt% Sn–60 wt% Pb alloy at 150°C (point *B*), phase compositions and relative amounts are computed in Example Problems 9.2 and 9.3.

EXAMPLE PROBLEM 9.3

For the lead tin alloy in Example Problem 9.2, calculate the relative amount of each phase present in terms of (a) mass fraction and (b) volume fraction. Assume that the densities of the α and β phases are 11.2 and 7.3 g/cm³, respectively.

SOLUTION

(a) Since the alloy consists of two phases, it is necessary to employ the lever rule. If C_1 denotes the overall alloy composition, mass fractions may be computed by subtracting compositions, in terms of weight percent tin, as follows:

$$W_\alpha = \frac{C_\beta - C_1}{C_\beta - C_\alpha} = \frac{99 - 40}{99 - 11} = 0.67$$

$$W_\beta = \frac{C_1 - C_\alpha}{C_\beta - C_\alpha} = \frac{40 - 11}{99 - 11} = 0.33$$

(b) To compute volume fractions, first determine the volume of each phase in some arbitrary mass of alloy, using the results of part a. For example, 100 g of this alloy will be composed of 67 g of the α phase and 33 g of β. The volume of each phase—$v(\alpha)$ or $v(\beta)$—is then just this mass divided by its density:

$$v(\alpha) = \frac{67 \text{ g}}{11.2 \text{ g/cm}^3} = 5.98 \text{ cm}^3$$

and

$$v(\beta) = \frac{33 \text{ g}}{7.3 \text{ g/cm}^3} = 4.52 \text{ cm}^3$$

Finally, the volume fraction of each phase (V_α and V_β) is equal to its volume divided by the total alloy volume, or

$$V_\alpha = \frac{v(\alpha)}{v(\alpha) + v(\beta)} = \frac{5.98 \text{ cm}^3}{5.98 \text{ cm}^3 + 4.52 \text{ cm}^3} = 0.57$$

and

$$V_\beta = \frac{v(\beta)}{v(\alpha) + v(\beta)} = \frac{4.52 \text{ cm}^3}{5.98 \text{ cm}^3 + 4.52 \text{ cm}^3} = 0.43$$

Note: When the densities of the phases in a two-phase alloy differ significantly, there will be quite a disparity between mass and volume fractions, as may be substantiated by comparison of the results of parts a and b.

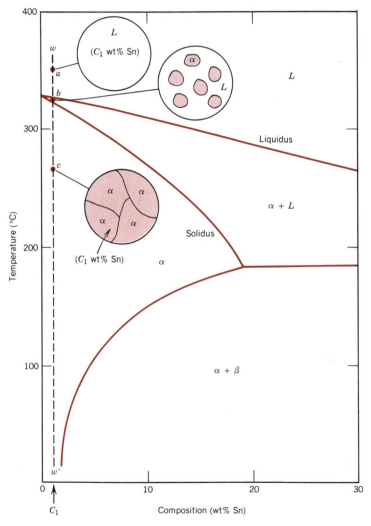

Figure 9.9 Schematic representations of the equilibrium microstructures for a lead–tin alloy of composition C_1 as it is cooled from the liquid-phase region.

Phase volume fractions are important because they (rather than phase mass fractions) may be determined from examination of the microstructure. Furthermore, the mechanical properties of an alloy may be estimated on the basis of phase volume fractions.

Development of Microstructure in Eutectic Alloys

Depending on composition, several different types of microstructures are possible for the slow cooling of alloys belonging to binary eutectic systems. These possibilities will be considered in terms of the lead–tin phase diagram, Figure 9.7.

The first case is for compositions ranging between a pure component and the maximum solid solubility for that component at room temperature [20°C (70°F)]. For the lead–tin system, this includes lead-rich alloys containing between 0 and about 2 wt% Sn (for the α phase solid solution) and also essentially pure tin, since the solubility of lead in tin (for the β phase) is negligible at room temperature. For example, consider an alloy of composition C_1 (Figure 9.9) as it is slowly cooled from a temperature within the liquid phase region, say, 350°C; this corresponds to moving down the dashed vertical line ww' in the figure. The alloy remains totally liquid and of composition C_1 until we cross the liquidus line at approximately 330°C, at which time the solid α phase begins to form. While passing through this narrow $\alpha + L$ phase region, solidification proceeds in the same manner as was described for the copper–nickel alloy in the preceding section; that is, with continued cooling more of the solid α forms. Furthermore, liquid and solid phase compositions are different, which follow along the liquidus and solidus phase boundaries, respectively. Solidification reaches completion at the point where ww' crosses the solidus line. The resulting alloy is polycrystalline with a uniform composition of C_1, and no subsequent changes will occur upon cooling to room temperature. This microstructure is represented schematically by the inset at point c in Figure 9.9.

The second case considered is for compositions that range between the room temperature solubility limit and the maximum solid solubility at the eutectic temperature. For the lead–tin system (Figure 9.7), these compositions extend from about 2 wt% Sn to 19.2 wt% Sn (for lead-rich alloys) and from 97.5 wt% Sn to virtually pure tin (for tin-rich alloys). Let us examine an alloy of composition C_2 as it is cooled along the vertical line xx' in Figure 9.10. Down to the intersection of xx' and the solvus line, changes that occur are similar to the previous case, as we pass through the corresponding phase regions (as demonstrated by the insets at points d, e, and f). Just above the solvus intersection, point f, the microstructure consists of α grains of composition C_2. Upon crossing the solvus line, the α solid solubility is exceeded, which results in the formation of small β phase particles; these are indicated in the microstructure inset at point g. With continued cooling, these particles will grow in size because the mass fraction of the β phase increases slightly with decreasing temperature.

The third case involves solidification of the eutectic composition, 61.9 wt% Sn (C_3 in Figure 9.11). Consider an alloy having this composition that is cooled from a temperature within the liquid phase region (e.g., 250°C) down the vertical line yy' in Figure 9.11. As the temperature is lowered, no changes occur until we reach the eutectic temperature, 183°C. Upon crossing the eutectic isotherm, the liquid transforms to the two α and β phases. This transformation may be represented by the reaction

$$L\,(61.9\text{ wt\% Sn}) \longrightarrow \alpha\,(19.2\text{ wt\% Sn}) + \beta\,(97.5\text{ wt\% Sn}) \tag{9.6}$$

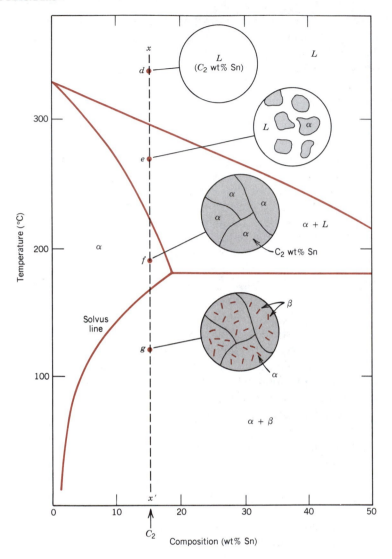

Figure 9.10 Schematic representations of the equilibrium microstructures for a lead–tin alloy of composition C_2 as it is cooled from the liquid-phase region.

in which the α and β phase compositions are dictated by the eutectic isotherm end points. During this transformation there must necessarily be a redistribution of the lead and tin components, inasmuch as the α and β phases have different compositions neither of which is the same as that of the liquid. This redistribution is accomplished by atomic diffusion.

The resulting microstructure consists of alternating layers (sometimes called la-mellae) of α and β phases that form simultaneously during the transformation. This microstructure, represented schematically in Figure 9.11, point i, is called a **eutectic structure,** and is characteristic of this reaction. A photomicrograph of this structure for the lead–tin eutectic is shown in Figure 9.12. The α and β phases form in these

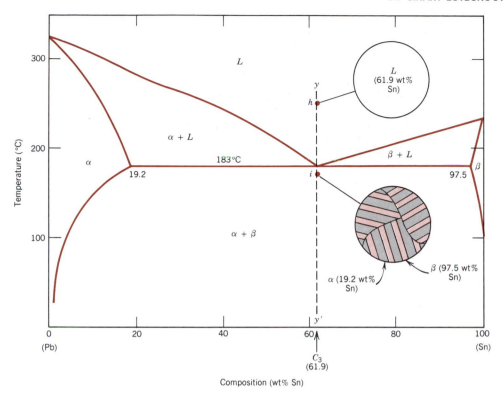

Figure 9.11 Schematic representations of the equilibrium microstructures for a lead–tin alloy of eutectic composition C_3 above and below the eutectic temperature.

alternating layers because, for this lamellar configuration, atomic diffusion need only occur over relatively short distances. Subsequent cooling of the alloy from just below the eutectic to room temperature will result in only minor microstructural changes.

The fourth and final microstructural case for this system includes all compositions other than the eutectic that, when cooled, cross the eutectic isotherm. Consider, for example, the composition C_4, Figure 9.13a, which lies to the left of the eutectic; as the temperature is lowered, we move down the line zz', beginning at point j. The microstructural development between points j and l is similar to that for the second

Figure 9.12 Photomicrograph showing the microstructure of a lead-tin alloy of eutectic composition. This microstructure consists of alternating layers of a lead-rich α-phase solid solution (dark layers), and a tin-rich β-phase solid solution (light layers). 375×. (Reproduced with permission from *Metals Handbook*, Vol. 9, 9th Edition, *Metallography and Microstructures*, American Society for Metals, Materials Park, OH, 1985.)

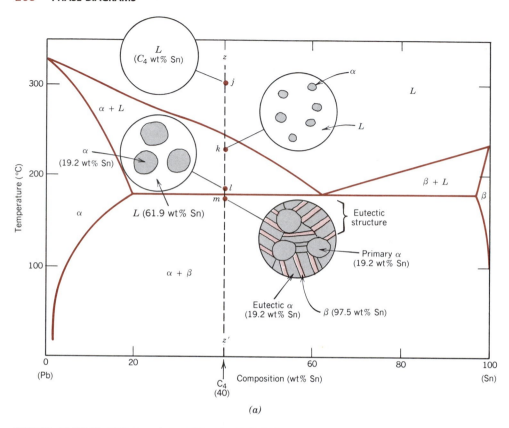

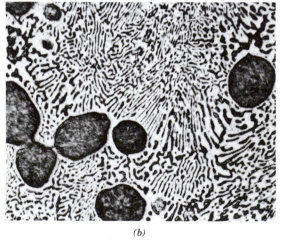

(b)

Figure 9.13 (*a*) Schematic representations of the equilibrium microstructures for a lead–tin alloy of composition C_4 as it is cooled from the liquid phase region. (*b*) Photomicrograph showing the microstructure of a lead-tin alloy of composition 50 wt% Sn–50 wt% Pb. This microstructure is composed of a primary lead-rich α phase (large dark regions) within a lamellar eutectic structure consisting of a tin-rich β phase (light layers) and a lead-rich α phase (dark layers). 400×. (Reproduced with permission from *Metals Handbook*, Vol. 9, 9th Edition, *Metallography and Microstructures*, American Society for Metals, Materials Park, OH, 1985.)

case, such that just prior to crossing the eutectic isotherm (point l), the α and liquid phases are present having compositions of approximately 19.2 and 61.9 wt% Sn, respectively, as determined from the appropriate tie line. As the temperature is lowered to just below the eutectic, the liquid phase, which is of the eutectic composition, will transform to the eutectic structure (i.e., alternating α and β lamellae); insignificant changes will occur with the α phase which formed during cooling through the $\alpha + L$ region. This microstructure is represented schematically by the inset at point m in Figure 9.13a. Thus the α phase will be present both in the eutectic structure and also as that phase that formed while cooling through the $\alpha + L$ phase field. To distinguish one α from the other, that which resides in the eutectic structure is called **eutectic** α, while the other that formed prior to crossing the eutectic isotherm is termed **primary** α; both are labeled in Figure 9.13a. The photomicrograph in Figure 9.13b is of a lead–tin alloy in which both primary α and eutectic structures are shown.

In dealing with microstructures, it is sometimes convenient to use the term **microconstituent,** that is, an element of the microstructure having an identifiable and characteristic structure. For example, in the point m inset, Figure 9.13a, there are two microconstituents, namely, primary α and the eutectic structure. Thus the eutectic structure is a microconstituent even though it is a mixture of two phases, because it has a distinct lamellar structure, with a fixed ratio of the two phases.

It is possible to compute the relative amounts of both eutectic and primary α microconstituents. Since the eutectic microconstituent always forms from the liquid having the eutectic composition, this microconstituent may be assumed to have a composition of 61.9 wt% Sn. Hence, the lever rule is applied using a tie line between the α–($\alpha + \beta$) phase boundary (19.2 wt% Sn) and the eutectic composition. For example, consider the alloy of composition C_4' in Figure 9.14. The fraction of the eutectic microconstituent W_e is just the same as the fraction of liquid W_L from which it

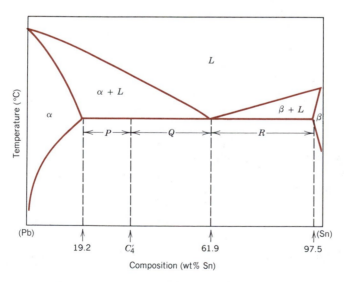

Figure 9.14 The lead–tin diagram used in computations for relative amounts of primary α and eutectic microconstituents for an alloy of composition C_4'.

transforms, or

$$W_e = W_L = \frac{P}{P + Q}$$

$$= \frac{C_4' - 19.2}{61.9 - 19.2} = \frac{C_4' - 19.2}{42.7} \tag{9.7}$$

Furthermore, the fraction of primary α, $W_{\alpha'}$, is just the fraction of the α phase that existed prior to the eutectic transformation; or, from Figure 9.14,

$$W_{\alpha'} = \frac{Q}{P + Q}$$

$$= \frac{61.9 - C_4'}{61.9 - 19.2} = \frac{61.9 - C_4'}{42.7} \tag{9.8}$$

The fractions of *total* α, W_α (both eutectic and primary), and also of total β, W_β, are determined by use of the lever rule and a tie line that extends *entirely across the $\alpha + \beta$ phase field*. Again, for an alloy having composition C_4',

$$W_\alpha = \frac{Q + R}{P + Q + R}$$

$$= \frac{97.5 - C_4'}{97.5 - 19.2} = \frac{97.5 - C_4'}{78.3} \tag{9.9}$$

and

$$W_\beta = \frac{P}{P + Q + R}$$

$$= \frac{C_4' - 19.2}{97.5 - 19.2} = \frac{C_4' - 19.2}{78.3} \tag{9.10}$$

Analogous transformations and microstructures result for alloys having compositions to the right of the eutectic (i.e., between 61.9 and 97.5 wt% Sn). However, below the eutectic temperature, the microstructure will consist of the eutectic and primary β microconstituents because, upon cooling from the liquid, we pass through the β + liquid phase field.

When, for case 4 (represented in Figure 9.13a), conditions of equilibrium are not maintained while passing through the α (or β) + liquid phase region, the following consequences will be realized for the microstructure upon crossing the eutectic isotherm: (1) grains of the primary microconstituent will be cored, that is, have a nonuniform distribution of solute across the grains; and (2) the fraction of the eutectic microconstituent formed will be greater than for the equilibrium situation.

9.8 EQUILIBRIUM DIAGRAMS HAVING INTERMEDIATE PHASES OR COMPOUNDS

The isomorphous and eutectic phase diagrams discussed thus far are relatively simple, but those for many binary alloy systems are much more complex. The eutectic copper–silver and lead–tin phase diagrams (Figures 9.6 and 9.7) have only two solid phases,

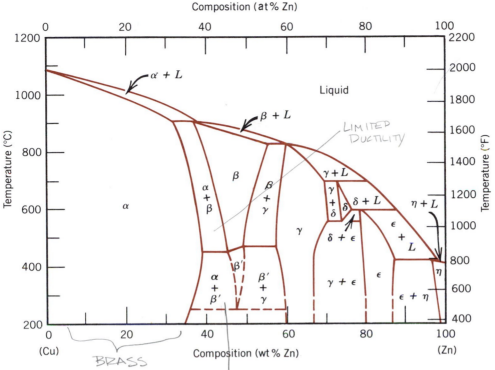

Figure 9.15 The copper–zinc phase diagram. (Adapted from *Metals Handbook: Metallography, Structures and Phase Diagrams,* Vol. 8, 8th edition, ASM Handbook Committee, T. Lyman, Editor, American Society for Metals, 1973, p. 301.)

α and β; these are sometimes termed **terminal solid solutions,** because they exist over composition ranges near the concentration extremities of the phase diagram. For other alloy systems, **intermediate solid solutions** (or *intermediate phases*) may be found at other than the two composition extremes. Such is the case for the copper–zinc system. Its phase diagram (Figure 9.15) may at first appear formidable because there are some invariant points and reactions similar to the eutectic that have not yet been discussed. In addition, there are six different solid solutions—two terminal (α and η) and four intermediate (β, γ, δ, and ϵ). (The β' phase is termed an ordered solid solution, one in which the copper and zinc atoms are situated in a specific and ordered arrangement within each unit cell.) Some phase boundary lines near the bottom of Figure 9.15 are dashed to indicate that their positions have not been exactly determined. The reason for this is that at low temperatures, diffusion rates are very slow and inordinately long times are required for the attainment of equilibrium. Again, only single- and two-phase regions are found on the diagram, and the same rules outlined in Section 9.6 are utilized for computing phase compositions and relative amounts. The commercial brasses are copper-rich copper–zinc alloys; for example, cartridge brass has a composition of 70 wt% Cu–30 wt% Zn and a microstructure consisting of a single α phase.

For some systems, discrete intermediate compounds rather than solid solutions may be found on the phase diagram, and these compounds have distinct chemical

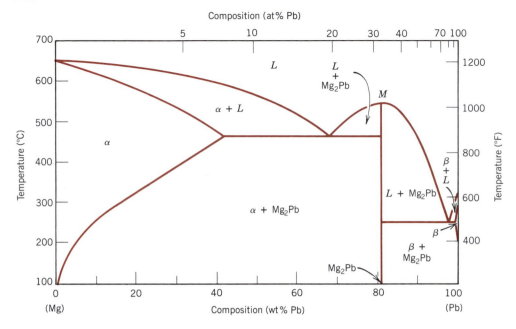

Figure 9.16 The magnesium–lead phase diagram. (Adapted from *Metals Handbook: Metallography, Structures and Phase Diagrams,* Vol. 8, 8th edition, ASM Handbook Committee, T. Lyman, Editor, American Society for Metals, 1973, p. 315.)

formulas; for metal–metal systems, they are called **intermetallic compounds.** For example, consider the magnesium–lead system (Figure 9.16). The compound Mg_2Pb has a composition of 19 wt% Mg–81 wt% Pb (33 at% Pb), and is represented as a vertical line on the diagram, rather than as a phase region of finite width; hence, Mg_2Pb can exist by itself only at this precise composition.

Several other characteristics are worth noting for this magnesium–lead system. First, the compound Mg_2Pb melts at approximately 550°C (1020°F), as indicated by point *M* in Figure 9.16. Also, the solubility of lead in magnesium is rather extensive, as indicated by the relatively large composition span for the α phase field. On the other hand, the solubility of magnesium in lead is extremely limited. This is evident from the very narrow β terminal solid-solution region on the right or lead-rich side of the diagram. Finally, this phase diagram may be thought of as two simple eutectic diagrams joined back to back, one for the Mg–Mg_2Pb system, the other for Mg_2Pb–Pb; as such, the compound Mg_2Pb is really considered to be a component. This separation of complex phase diagrams into smaller-component units may simplify them and, furthermore, expedite their interpretation.

9.9 EUTECTOID AND PERITECTIC REACTIONS

In addition to the eutectic, other invariant points involving three different phases are found for some alloy systems. One of these occurs for the copper–zinc system (Figure 9.15) at 558°C (1036°F) and 75 wt% Zn–25 wt% Cu. A portion of the phase diagram in this vicinity appears enlarged in Figure 9.17. Upon cooling, a solid δ phase trans-

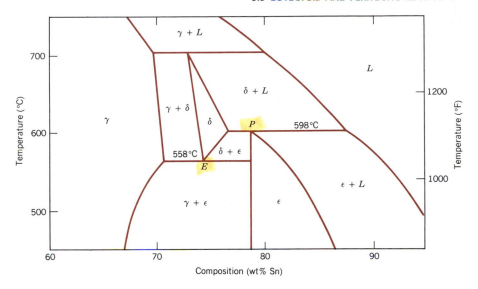

Figure 9.17 A region of the copper–zinc phase diagram which has been enlarged to show eutectoid and peritectic invariant points, labeled E (558°C, 75 wt% Zn) and P (598°C, 78.6 wt% Zn), respectively.

forms into two other solid phases (γ and ϵ) according to the reaction

$$\delta \underset{\text{heating}}{\overset{\text{cooling}}{\rightleftarrows}} \gamma + \epsilon \tag{9.11}$$

The reverse reaction occurs upon heating. It is called a **eutectoid** (or eutecticlike) **reaction,** and the invariant point (point E, Figure 9.17) and the horizontal tie line at 558°C are termed the *eutectoid* and *eutectoid isotherm,* respectively. The feature distinguishing "eutectoid" from "eutectic" is that one solid phase instead of a liquid transforms into two other solid phases at a single temperature. A eutectoid reaction is found in the iron–carbon system (Section 9.13), which is very important in the heat treating of steels.

The **peritectic reaction** is yet another invariant reaction involving three phases at equilibrium. With this reaction, upon heating, one solid phase transforms into a liquid phase and another solid phase. A peritectic exists for the copper–zinc system (Figure 9.17, point P) at 598°C (1108°F) and 78.6 wt% Zn–21.4 wt% Cu; this reaction is as follows:

$$\delta + L \underset{\text{heating}}{\overset{\text{cooling}}{\rightleftarrows}} \epsilon \tag{9.12}$$

The low-temperature solid phase may be an intermediate solid solution (e.g., ϵ in the above reaction), or it may be a terminal solid solution. One of the latter peritectics exists at about 97 wt% Zn and 425°C (797°F), wherein the η phase, when heated, transforms to ϵ and liquid phases. Three other peritectics are found for the Cu–Zn system, the reactions of which involve β, δ, and γ intermediate solid solutions as the low-temperature phases that transform upon heating.

9.10 CONGRUENT PHASE TRANSFORMATIONS

Phase transformations may be classified according to whether or not there is any change in composition for the phases involved. Those for which there are no compositional alterations are said to be **congruent transformations.** Conversely, for *incongruent transformations,* at least one of the phases will experience a change in composition. Examples of congruent transformations include allotropic transformations (Section 3.6) and melting of pure materials. Eutectic and eutectoid reactions, as well as the melting of an alloy that belongs to an isomorphous system, all represent incongruent transformations.

Intermediate solid-solution phases are sometimes classified on the basis of whether they melt congruently or incongruently. The intermetallic compound Mg_2Pb melts congruently at the point designated *M* on the magnesium–lead phase diagram, Figure 9.16. Also, for the nickel–titanium system, Figure 9.18, there is a congruent melting

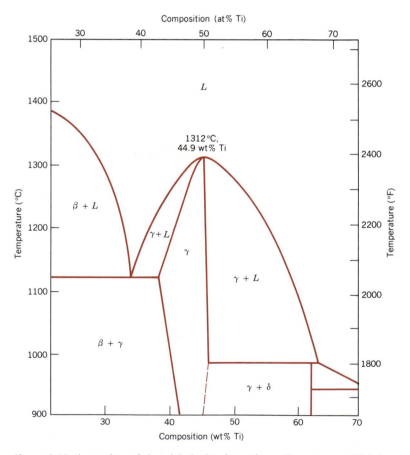

Figure 9.18 A portion of the nickel–titanium phase diagram on which is shown a congruent melting point for the γ phase solid solution at 1312°C and 44.9 wt% Ti. (Adapted with permission from *Metals Handbook,* Vol. 8, 8th ed., *Metallography, Structures and Phase Diagrams,* American Society for Metals, Metals Park, OH, 1973.)

point for the γ solid solution which corresponds to the point of tangency for the pairs of liquidus and solidus lines, at 1312°C and 44.9 wt% Ti. Furthermore, the peritectic reaction is an example of incongruent melting for an intermediate phase.

9.11 CERAMIC AND TERNARY PHASE DIAGRAMS

It need not be assumed that phase diagrams exist only for metal–metal systems; in fact, phase diagrams that are very useful in the design and processing of ceramic systems have been experimentally determined for quite a number of these materials. Ceramic phase diagrams are discussed in Section 13.5.

Phase diagrams have also been determined for metallic (as well as ceramic) systems containing more than two components; however, their representation and interpretation may be exceedingly complex. For example, a ternary, or three-component, composition–temperature phase diagram in its entirety is depicted by a three-dimensional model. Portrayal of features of the diagram or model in two dimensions is possible but somewhat difficult.

9.12 THE GIBBS PHASE RULE

The construction of phase diagrams as well as some of the principles governing the conditions for phase equilibria are dictated by laws of thermodynamics. One of these is the **Gibbs phase rule,** proposed by the nineteenth century physicist J. Willard Gibbs. This rule represents a criterion for the number of phases that will coexist within a system at equilibrium, and is expressed by the simple equation

$$P + F = C + N \tag{9.13}$$

where P is the number of phases present (the phase concept is discussed in Section 9.3). The parameter F is termed the *number of degrees of freedom* or the number of externally controlled variables (e.g., temperature, pressure, composition) which must be specified to completely define the state of the system. Or, expressed another way, F is the number of these variables that can be changed independently without altering the number of phases that coexist at equilibrium. The parameter C in Equation 9.13 represents the number of components in the system. Components are normally elements or stable compounds and, in the case of phase diagrams, are the materials at the two extremities of the horizontal compositional axis (e.g., H_2O and $C_{12}H_{22}O_{11}$, and Cu and Ni for the phase diagrams shown in Figures 9.1 and 9.2a, respectively). Finally, N in Equation 9.13 is the number of noncompositional variables (e.g., temperature and pressure).

Let us demonstrate the phase rule by applying it to binary temperature–composition phase diagrams, specifically the copper–silver system, Figure 9.6. Since pressure is constant (1 atm), the parameter N is 1—temperature is the only noncompositional variable. Equation 9.13 now takes the form

$$P + F = C + 1 \tag{9.14}$$

Furthermore, the number of components C is 2 (viz Cu and Ag), and

$$P + F = 2 + 1 = 3$$

or

$$F = 3 - P$$

Consider the case of single-phase fields on the phase diagram (e.g., α, β, and liquid regions). Since only one phase is present, $P = 1$ and

$$F = 3 - P$$
$$= 3 - 1 = 2$$

This means that to completely describe the characteristics of any alloy which exists within one of these phase fields, we must specify two parameters; these are composition and temperature, which locate, respectively, the horizontal and vertical positions of the alloy on the phase diagram.

For the situation wherein two phases coexist, for example, $\alpha + L$, $\beta + L$, and $\alpha + \beta$ phase regions, Figure 9.6, the phase rule stipulates that we have but one degree of freedom since

$$F = 3 - P$$
$$= 3 - 2 = 1$$

Thus it is necessary to specify either temperature or the composition of one of the phases to completely define the system. For example, suppose that we decide to specify temperature for the $\alpha + L$ phase region, say, T_1 in Figure 9.19. The compositions of the α and liquid phases (C_α and C_L) are thus dictated by the extremities of the tie line constructed at T_1 across the $\alpha + L$ field. It should be noted that only the nature of the phases is important in this treatment and not the relative phase amounts. This is to say that the overall alloy composition could lie anywhere along this tie line constructed at temperature T_1 and still give C_α and C_L compositions for the respective α and liquid phases.

The second alternative is to stipulate the composition of one of the phases for this two-phase situation, which thereby fixes completely the state of the system. For example, if we specified C_α as the composition of the α phase that is in equilibrium with the liquid (Figure 9.19), then both the temperature of the alloy (T_1) and the composition of the liquid phase (C_L) are established, again by the tie line drawn across the $\alpha + L$ phase field so as to give this C_α composition.

For binary systems, when three phases are present, there are no degrees of freedom, since

$$F = 3 - P$$
$$= 3 - 3 = 0$$

This means that the compositions of all three phases as well as the temperature are fixed. This condition is met for a eutectic system by the eutectic isotherm; for the Cu–Ag system (Figure 9.6), it is the horizontal line that extends between points B and G. At this temperature, 780°C, the points at which each of the α, L, and β phase fields touch the isotherm line correspond to the respective phase compositions; namely, the composition of the α phase is fixed at 7.9 wt% Ag, that of the liquid at 71.9 wt% Ag, and that of the β phase at 91.2 wt% Ag. Thus three-phase equilibrium will not be represented by a phase field, but rather by the unique horizontal isotherm line. Furthermore, all three phases will be in equilibrium for any alloy composition that

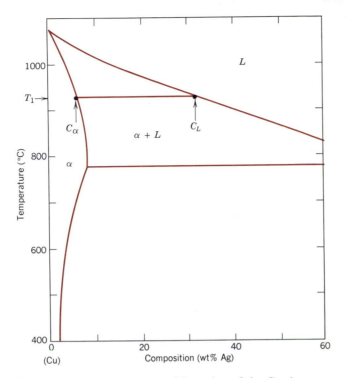

Figure 9.19 Enlarged copper-rich section of the Cu–Ag phase diagram in which the Gibbs phase rule for the coexistence of two phases (i.e., α and L) is demonstrated. Once the composition of either phase (i.e., C_α or C_L) or the temperature (i.e., T_1) is specified, values for the two remaining parameters are established by construction of the appropriate tie line.

lies along the length of the eutectic isotherm (e.g., for the Cu–Ag system at 780°C and compositions between 7.9 and 91.2 wt% Ag).

One use of the Gibbs phase rule is in analyzing for nonequilibrium conditions. For example, a microstructure for a binary alloy which developed over a range of temperatures and consisting of three phases is a nonequilibrium one; under these circumstances, three phases will exist only at a single temperature.

THE IRON–CARBON SYSTEM

Of all binary alloy systems, the one that is possibly the most important is that for iron and carbon. Both steels and cast irons, primary structural materials in every technologically advanced culture, are essentially iron–carbon alloys. This section is devoted to a study of the phase diagram for this system and the development of several of the possible microstructures. The relationships between heat treatment, microstructure, and mechanical properties are explored in Chapters 10 and 11.

9.13 THE IRON–IRON CARBIDE (Fe–Fe₃C) PHASE DIAGRAM

A portion of the iron–carbon phase diagram is presented in Figure 9.20. Pure iron, upon heating, experiences two changes in crystal structure before it melts. At room temperature the stable form, called **ferrite**, or α iron, has a BCC crystal structure. Ferrite experiences a polymorphic transformation to FCC **austenite**, or γ iron, at 912°C (1674°F). This austenite persists to 1394°C (2541°F), at which temperature the FCC austenite reverts back to a BCC phase known as δ ferrite, which finally melts at 1538°C (2800°F). All these changes are apparent along the left vertical axis of the phase diagram.

The composition axis in Figure 9.20 extends only to 6.70 wt% C; at this concentration the intermediate compound iron carbide, or **cementite** (Fe₃C), is formed, which is represented by a vertical line on the phase diagram. Thus the iron–carbon system

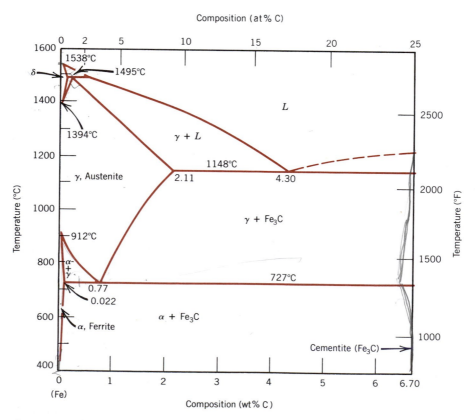

Figure 9.20 The iron–iron carbide phase diagram. (Adapted from *Metals Handbook: Metallography, Structures and Phase Diagrams,* Vol. 8, 8th edition, ASM Handbook Committee, T. Lyman, Editor, American Society for Metals, 1973, p. 275.)

may be divided into two parts: an iron-rich portion, as in Figure 9.20, and the other (not shown) for compositions between 6.70 and 100 wt% C (pure graphite). In practice, all steels and cast irons have carbon contents less than 6.70 wt% C; therefore, we consider only the iron–iron carbide system. Figure 9.20 would be more appropriately labeled the Fe–Fe₃C phase diagram, since Fe₃C is now considered to be a component. Convention and convenience dictate that composition still be expressed in "wt% C" rather than "wt% Fe₃C"; 6.70 wt% C corresponds to 100 wt% Fe₃C.

Carbon is an interstitial impurity in iron and forms a solid solution with each of α and δ ferrites, and also with austenite, as indicated by the α, δ, and γ single-phase fields in Figure 9.20. In the BCC α ferrite, only small concentrations of carbon are soluble; the maximum solubility is 0.022 wt% at 727°C (1341°F). The limited solubility is explained by the shape and size of the BCC interstitial positions, which make it difficult to accommodate the carbon atoms. Even though present in relatively low concentrations, carbon significantly influences the mechanical properties of ferrite. This particular iron–carbon phase is relatively soft, may be made magnetic at temperatures below 768°C (1414°F), and has a density of 7.88 g/cm³. Figure 9.21a is a photomicrograph of α ferrite.

The austenite, or γ phase of iron, when alloyed with just carbon, is not stable below 727°C (1341°F), as indicated in Figure 9.20. The maximum solubility of carbon in austenite, 2.11 wt%, occurs at 1148°C (2098°F). This solubility is approximately 100 times greater than the maximum for BCC ferrite, since the FCC interstitial positions are shaped such that when the carbon atoms fill them, the strains imposed on the surrounding iron atoms are much lower. As the discussions that follow demonstrate, phase transformations involving austenite are very important in the heat treating of steels. In passing, it should be mentioned that austenite is nonmagnetic. Figure 9.21b shows a photomicrograph of this austenite phase.

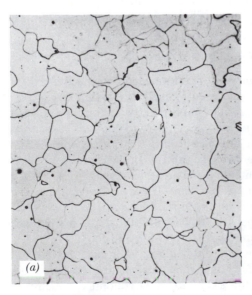 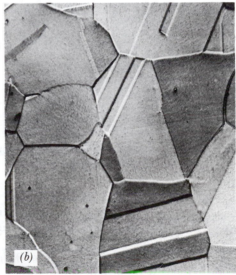

(a)   (b)

Figure 9.21 Photomicrographs of (a) α ferrite (90×) and (b) austenite (325×). (Copyright 1971 by United States Steel Corporation.)

The δ ferrite is virtually the same as α ferrite, except for the range of temperatures over which each exists. Since the δ ferrite is stable only at relatively high temperatures, it is of no technological importance and is not discussed further.

Cementite (Fe_3C) forms when the solubility limit of carbon in α ferrite is exceeded below 727°C (1341°F) (for compositions within the $\alpha + Fe_3C$ phase region). As indicated in Figure 9.20, Fe_3C will also coexist with the γ phase between 727 and 1148°C (1341 and 2098°F). Mechanically, cementite is very hard and brittle; the strength of some steels is greatly enhanced by its presence.

Strictly speaking, cementite is only metastable; that is, it will remain as a compound indefinitely at room temperature. But if heated to between 650 and 700°C (1200 and 1300°F) for several years, it will gradually change or transform into α iron and carbon, in the form of graphite, which will remain upon subsequent cooling to room temperature. Thus the phase diagram in Figure 9.20 is not a true equilibrium one because cementite is not an equilibrium compound. However, inasmuch as the decomposition rate of cementite is extremely sluggish, virtually all the carbon in steel will be as Fe_3C instead of graphite, and the iron–iron carbide phase diagram is, for all practical purposes, valid. As will be seen in Section 12.6, addition of silicon to cast irons greatly accelerates this cementite decomposition reaction to form graphite.

The two-phase regions are labeled in Figure 9.20. It may be noted that one eutectic exists for the iron–iron carbide system, at 4.30 wt% C and 1148°C (2098°F); for this eutectic reaction,

$$L \underset{\text{heating}}{\overset{\text{cooling}}{\rightleftharpoons}} \gamma + Fe_3C \tag{9.15}$$

the liquid solidifies to form austenite and cementite phases. Of course, subsequent cooling to room temperature will promote additional phase changes.

It may be noted that a eutectoid invariant point exists at a composition of 0.77 wt% C and a temperature of 727°C (1341°F). This eutectoid reaction may be represented by

$$\gamma \, (0.77 \text{ wt\% C}) \longrightarrow \alpha \, (0.022 \text{ wt\% C}) + Fe_3C \, (6.7 \text{ wt\% C}) \tag{9.16}$$

or, upon cooling, the solid γ phase is transformed into α iron and cementite. (Eutectoid phase transformations were addressed in Section 9.9.) The eutectoid phase changes described by Equation 9.16 are very important, being fundamental to the heat treatment of steels, as explained in subsequent discussions.

Ferrous alloys are those in which iron is the prime component, but carbon as well as other alloying elements may be present. In the classification scheme of ferrous alloys based on carbon content, there are three types: iron, steel, and cast iron. Commercially pure iron contains less than 0.008 wt% C and, from the phase diagram, is composed almost exclusively of the ferrite phase at room temperature. The iron–carbon alloys that contain between 0.008 and 2.11 wt% C are classified as steels. In most steels the microstructure consists of both α and Fe_3C phases. Upon cooling to room temperature, an alloy within this composition range must pass through at least a portion of the γ phase field; distinctive microstructures are subsequently produced, as discussed below. Although a steel alloy may contain as much as 2.11 wt% C, in practice, rarely do carbon concentrations exceed 1.0 wt%. The properties and various classifications of steels are treated in Section 12.5. Cast irons are classified as ferrous alloys that contain between 2.11 and 6.70 wt% C. However, commercial cast irons normally contain less than 4.5 wt% C. These alloys are discussed further in Section 12.6.

9.14 DEVELOPMENT OF MICROSTRUCTURES IN IRON–CARBON ALLOYS

Several of the various microstructures that may be produced in steel alloys and their relationships to the iron–iron carbon phase diagram are now discussed, and it is shown that the microstructure that develops depends on both the carbon content and heat treatment. The discussion is confined to very slow cooling of steel alloys, in which equilibrium is continuously maintained. A more detailed exploration of the influence of heat treatment on microstructure, and ultimately on the mechanical properties of steels, is contained in Chapter 10.

Phase changes that occur upon passing from the γ region into the $\alpha + Fe_3C$ phase field (Figure 9.20) are relatively complex and similar to those described for the eutectic systems in Section 9.7. Consider, for example, an alloy of eutectoid composition (0.77 wt% C) as it is cooled from a temperature within the γ phase region, say, 800°C, that is, beginning at point a in Figure 9.22 and moving down the vertical line

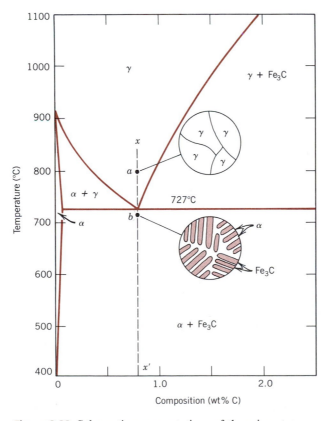

Figure 9.22 Schematic representations of the microstructures for an iron–carbon alloy of eutectoid composition (0.77 wt% C) above and below the eutectoid temperature.

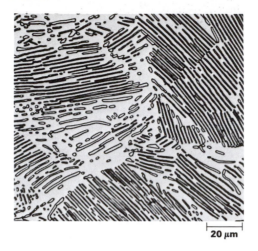

Figure 9.23 Photomicrograph of a eutectoid steel showing the pearlite microstructure consisting of alternating layers of α ferrite (the light phase) and Fe$_3$C (thin layers most of which appear dark). 500×. (Reproduced with permission from *Metals Handbook,* Vol. 9, 9th edition, *Metallography and Microstructures,* American Society for Metals, Materials Park, OH, 1985.)

20 μm

xx'. Initially, the alloy is composed entirely of the austenite phase having a composition of 0.77 wt% C and corresponding microstructure, also indicated in Figure 9.22. As the alloy is cooled, there will occur no changes until the eutectoid temperature (727°C) is reached. Upon crossing this temperature to point *b*, the austenite transforms according to Equation 9.16. Or, austenite containing 0.77 wt% C decomposes to form ferrite, which has a much lower carbon concentration (0.022 wt% C), as well as to Fe$_3$C, with a much higher carbon content (6.7 wt% C). This phase change must involve the diffusion of carbon because all three phases must have different compositions.

The microstructure for this eutectoid steel that is slowly cooled through the eutectoid temperature is similar to that for an alloy of eutectic composition (Figures 9.11 and 9.12), that is, alternating layers of lamellae of the two phases (α and Fe$_3$C) that form simultaneously during the transformation. In this case, the relative layer thickness is approximately 8 to 1. This microstructure, represented schematically in Figure 9.22, point *b*, is called **pearlite** because it has the appearance of mother of pearl when viewed under the microscope at low magnifications. Figure 9.23 is a photomicrograph of a eutectoid steel showing the pearlite. The pearlite exists as grains, often termed "colonies"; within each colony the layers are oriented in essentially the same direction, which varies from one colony to another. The thick light layers are the ferrite phase, and the cementite phase appears as thin lamellae most of which appear dark. Many cementite layers are so thin that adjacent phase boundaries are indistinguishable, which layers appear dark at this magnification. Mechanically, pearlite has properties intermediate between the soft, ductile ferrite and the hard, brittle cementite.

This alternating α-Fe$_3$C layer arrangement forms because for this structure, only relatively short carbon diffusion path lengths are required. Furthermore, subsequent cooling of the pearlite from point *b* in Figure 9.22 will produce relatively insignificant microstructural changes.

Hypoeutectoid Alloys

Microstructures for iron–iron carbide alloys having other than the eutectoid composition are now explored; these are analogous to the fourth case described in Section

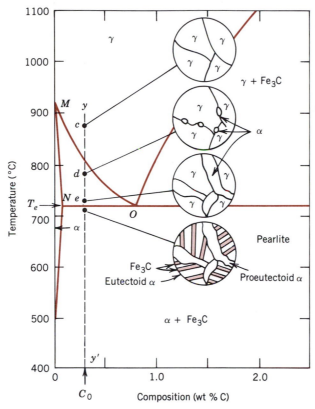

Figure 9.24 Schematic representations of the microstructures for an iron–carbon alloy of hypoeutectoid composition C_0 (containing less than 0.77 wt% C) as it is cooled from within the austenite phase region to below the eutectoid temperature.

9.7 and illustrated in Figure 9.13a for the eutectic system. Consider a composition C_0 to the left of the eutectoid, between 0.022 and 0.77 wt% C; such is termed a **hypoeutectoid** (less than eutectoid) **alloy.** Cooling an alloy of this composition is represented by moving down the vertical line yy' in Figure 9.24. At about 875°C, point c, the microstructure will consist entirely of grains of the γ phase, as shown schematically in the figure. In cooling to point d, about 775°C, which is within the $\alpha + \gamma$ phase region, both these phases will coexist as in the schematic microstructure. Most of the small α particles will form along the original γ grain boundaries. The compositions of both α and γ phases may be determined using the appropriate tie line; these correspond, respectively, to about 0.020 and 0.50 wt% C.

While cooling an alloy through the $\alpha + \gamma$ phase region, the composition of the ferrite phase changes with temperature along the α–$(\alpha + \gamma)$ phase boundary, line MN, becoming slightly richer in carbon. On the other hand, the change in composition of the austenite is more dramatic, proceeding along the $(\alpha + \gamma)$–γ boundary, line MO, as the temperature is reduced.

Cooling from point *d* to *e*, just above the eutectoid but still in the $\alpha + \gamma$ region, will produce an increased fraction of the α phase and a microstructure similar to that also shown: the α particles will have grown larger. At this point, the compositions of the α and γ phases are determined by constructing a tie line at the temperature T_e; the α phase will contain 0.022 wt% C, while the γ phase will be of the eutectoid composition, 0.77 wt% C.

As the temperature is lowered just below the eutectoid, to point *f*, all the γ phase that was present at temperature T_e (and having the eutectoid composition) will transform to pearlite, according to the reaction in Equation 9.16. There will be virtually no change in the α phase that existed at point *e* in crossing the eutectoid temperature—it will normally be present as a continuous matrix phase surrounding the isolated pearlite colonies. The microstructure at point *f* will appear as the corresponding schematic inset in Figure 9.24. Thus the ferrite phase will be present both in the pearlite and also as the phase that formed while cooling through the $\alpha + \gamma$ phase region. The ferrite that is present in the pearlite is called *eutectoid ferrite*, whereas the other, that formed above T_e, is termed **proeutectoid** (meaning pre- or before eutectoid) **ferrite,** as labeled in Figure 9.24. Figure 9.25 is a photomicrograph of a 0.38 wt% C steel; large, white regions correspond to the proeutectoid ferrite. For pearlite, the spacing between the α and Fe_3C layers varies from grain to grain; some of the pearlite appears dark because the many close-spaced layers are unresolved at the magnification of the photomicrograph. It should also be noted that two micro-constituents are present in this photomicrograph—proeutectoid ferrite and pearlite— and which will appear in all hypoeutectoid iron–carbon alloys that are slowly cooled to a temperature below the eutectoid.

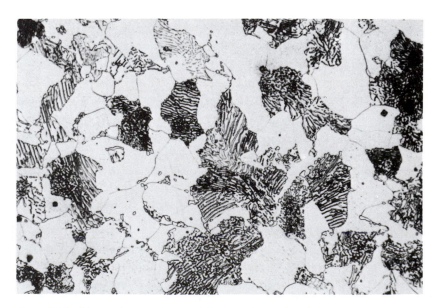

Figure 9.25 Photomicrograph of a 0.38 wt% C steel having a microstructure consisting of pearlite and proeutectoid ferrite. 635×. (Photomicrograph courtesy of Republic Steel Corporation.)

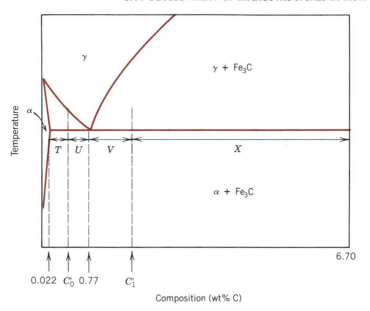

Figure 9.26 A portion of the Fe–Fe$_3$C phase diagram used in computations for relative amounts of proeutectoid and pearlite microconstituents for hypoeutectoid (C_0') and hypereutectoid (C_1') compositions.

The relative amounts of the proeutectoid α and pearlite may be determined in a manner similar to that described in Section 9.7 for primary and eutectic micro-constituents. We use the lever rule in conjunction with a tie line that extends from the α–(α + Fe$_3$C) phase boundary (0.022 wt% C) and the eutectoid composition (0.77 wt% C), inasmuch as pearlite is the transformation product of austenite having this composition. For example, let us consider an alloy of composition C_0' in Figure 9.26. Thus the fraction of pearlite, W_p, may be determined according to

$$W_p = \frac{T}{T + U}$$

$$= \frac{C_0' - 0.022}{0.77 - 0.022} = \frac{C_0' - 0.022}{0.75} \tag{9.17}$$

Furthermore, the fraction of proeutectoid α, $W_{\alpha'}$, is computed as follows:

$$W_{\alpha'} = \frac{U}{T + U}$$

$$= \frac{0.77 - C_0'}{0.77 - 0.022} = \frac{0.77 - C_0'}{0.75} \tag{9.18}$$

Of course, fractions of both total α (eutectoid and proeutectoid) and cementite are determined using the lever rule and a tie line that extends across the entirety of the α + Fe$_3$C phase region, from 0.022 to 6.7 wt% C.

Hypereutectoid Alloys

Analogous transformations and microstructures result for **hypereutectoid alloys,** those containing between 0.77 and 2.11 wt% C, which are cooled from temperatures within the γ phase field. Consider an alloy of composition C_1 in Figure 9.27 which, upon cooling, moves down the line zz'. At point g only the γ phase will be present with a composition of C_1; the microstructure will appear as shown, having only γ grains. Upon cooling into the $\gamma + Fe_3C$ phase field, say, to point h, the cementite phase will begin to form along the initial γ grain boundaries, similar to the α phase in Figure 9.24, point d. This cementite is called **proeutectoid cementite**—that which forms before the eutectoid reaction. Of course, the cementite composition remains constant (6.70 wt% C) as the temperature changes. However, the composition of the austenite phase will move along line PO toward the eutectoid. As the temperature is lowered through the eutectoid to point i, all remaining austenite of eutectoid composition is converted into pearlite; thus the resulting microstructure consists of pearlite and

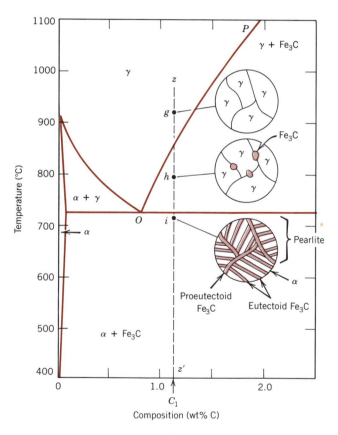

Figure 9.27 Schematic representations of the microstructures for an iron–carbon alloy of hypereutectoid composition C_1 (containing between 0.77 and 2.1 wt% C), as it is cooled from within the austenite phase region to below the eutectoid temperature.

Figure 9.28 Photomicrograph of a 1.4 wt% C steel having a microstructure consisting of a white proeutectoid cementite network surrounding the pearlite colonies. 1000×. (Copyright 1971 by United States Steel Corporation.)

proeutectoid cementite as microconstituents (Figure 9.27). In the photomicrograph of a 1.4 wt% C steel (Figure 9.28), note that the proeutectoid cementite appears light. Since it has much the same appearance as proeutectoid ferrite (Figure 9.25), there is some difficulty in distinguishing between hypoeutectoid and hypereutectoid steels on the basis of microstructure.

Relative amounts of both pearlite and proeutectoid Fe_3C microconstituents may be computed for hypereutectoid steel alloys in a manner analogous to that for hypoeutectoid materials; the appropriate tie line extends between 0.77 and 6.70 wt% C. Thus for an alloy having composition C_1' in Figure 9.26, fractions of pearlite W_p and proeutectoid cementite $W_{Fe_3C'}$ are determined from the following lever rule expressions:

$$W_p = \frac{X}{V + X} = \frac{6.70 - C_1'}{6.70 - 0.77} = \frac{6.70 - C_1'}{5.93} \tag{9.19}$$

and

$$W_{Fe_3C'} = \frac{V}{V + X} = \frac{C_1' - 0.77}{6.70 - 0.77} = \frac{C_1' - 0.77}{5.93} \tag{9.20}$$

EXAMPLE PROBLEM 9.4

For a 99.65 wt% Fe–0.35 wt% C alloy at a temperature just below the eutectoid, determine the following:

(a) The fractions of total ferrite and cementite phases.
(b) The fractions of the proeutectoid ferrite and pearlite.
(c) The fraction of eutectoid ferrite.

SOLUTION

(a) This part of the problem is solved by application of the lever rule expressions employing a tie line that extends all the way across the $\alpha + Fe_3C$ phase field. Thus C_0 is 0.35 wt% C, and

$$W_\alpha = \frac{6.70 - 0.35}{6.70 - 0.022} = 0.95$$

and

$$W_{Fe_3C} = \frac{0.35 - 0.022}{6.70 - 0.022} = 0.05$$

(b) The fractions of proeutectoid ferrite and pearlite are determined by using the lever rule, and a tie line that extends only to the eutectoid composition (i.e., Equations 9.17 and 9.18). Or

$$W_p = \frac{0.35 - 0.022}{0.77 - 0.022} = 0.44$$

and

$$W_{\alpha'} = \frac{0.77 - 0.35}{0.77 - 0.022} = 0.56$$

(c) All ferrite is either as proeutectoid or eutectoid (in the pearlite). Therefore, the sum of these two ferrite fractions will equal the fraction of total ferrite, that is,

$$W_{\alpha'} + W_{\alpha e} = W_\alpha$$

where $W_{\alpha e}$ denotes the fraction of the total alloy that is eutectoid ferrite. Values for W_α and $W_{\alpha'}$ were determined in parts a and b as 0.95 and 0.56, respectively. Therefore,

$$W_{\alpha e} = W_\alpha - W_{\alpha'} = 0.95 - 0.56 = 0.39$$

Nonequilibrium Cooling

In this discussion on the microstructural development of iron–carbon alloys it has been assumed that, upon cooling, conditions of metastable equilibrium[1] have been continuously maintained; that is, sufficient time has been allowed at each new temperature for any necessary adjustment in phase compositions and relative amounts as predicted from the Fe–Fe$_3$C phase diagram. In most situations these cooling rates are impractically slow and really unnecessary; in fact, on many occasions nonequilibrium conditions are desirable. Two nonequilibrium effects of practical importance are (1) the occurrence of phase changes or transformations at temperatures other than

---

[1] The term "metastable equilibrium" is used in this discussion inasmuch as Fe$_3$C is only a metastable compound.

those predicted by phase boundary lines on the phase diagram, and (2) the existence at room temperature of nonequilibrium phases that do not appear on the phase diagram. Both are discussed in the next chapter.

9.15 THE INFLUENCE OF OTHER ALLOYING ELEMENTS

Additions of other alloying elements (Cr, Ni, Ti, etc.) bring about rather dramatic changes in the binary iron–iron carbide phase diagram, Figure 9.20. The extent of these alterations of the positions of phase boundaries and the shapes of the phase fields depends on the particular alloying element and its concentration. One of the important changes is the shift in position of the eutectoid, with respect to temperature and to carbon concentration. These effects are illustrated in Figures 9.29 and 9.30, which plot the eutectoid temperature and composition (in wt% C) as a function of concentration for several other alloying elements. Thus other alloy additions alter not only the temperature of the eutectoid reaction but also the relative fractions of

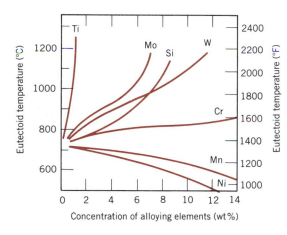

Figure 9.29 The dependence of eutectoid temperature on alloy concentration for several alloying elements in steel. (From Dr. Edgar C. Bain, *Functions of the Alloying Elements in Steel,* American Society for Metals, 1939, p. 127.)

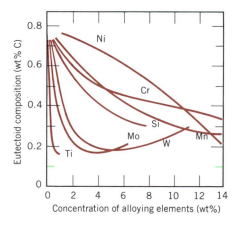

Figure 9.30 The dependence of eutectoid composition (wt% C) on alloy concentration for several alloying elements in steel. (From Dr. Edgar C. Bain, *Functions of the Alloying Elements in Steel,* American Society for Metals, 1939, p. 127.)

pearlite and the proeutectoid phase that form. Steels are normally alloyed for other reasons, however—usually to either improve their corrosion resistance or to render them amenable to heat treatment (see Chapter 11).

SUMMARY

Equilibrium phase diagrams are a convenient and concise way of representing the most stable relationships between phases in alloy systems. This discussion considered binary phase diagrams for which temperature and composition are variables. Areas, or phase regions, are defined on these temperature-versus-composition plots within which either one or two phases exist. For an alloy of specified composition and at a known temperature, the phases present, their compositions, and relative amounts under equilibrium conditions may be determined. Within two-phase regions, tie lines and the lever rule must be used for phase composition and mass fraction computations, respectively.

Several different kinds of phase diagram were discussed for metallic systems. Isomorphous diagrams are those for which there is complete solubility in the solid phase; the copper–nickel system displays this behavior. Also discussed for alloys belonging to isomorphous systems were the development of microstructure for both cases of equilibrium and nonequilibrium cooling, and also the dependence of mechanical characteristics on composition.

In a eutectic reaction, as found in some alloy systems, a liquid phase transforms isothermally to two different solid phases upon cooling. Such a reaction is noted on the copper–silver and lead–tin phase diagrams. Complete solid solubility for all compositions does not exist; instead, solid solutions are terminal—there is only a limited solubility of each component in the other. Four different kinds of microstructures that may develop for the equilibrium cooling of alloys belonging to eutectic systems were discussed.

Other equilibrium phase diagrams are more complex, having intermediate compounds and/or phases, possibly more than a single eutectic, and other reactions including eutectoid, peritectic, and congruent phase transformations. These are found for copper–zinc and magnesium–lead systems.

The Gibbs phase rule was introduced; it is a simple equation that relates the number of phases present in a system at equilibrium with the number of degrees of freedom, the number of components, and the number of noncompositional variables.

Considerable attention was given to the iron–carbon system, and specifically, the iron–iron carbide phase diagram, which technologically is one of the most important. The development of microstructure in many iron–carbon alloys and steels depends on the eutectoid reaction in which the FCC austenite phase of composition 0.77 wt% C transforms isothermally to the BCC α ferrite phase (0.022 wt% C) and the intermetallic compound, cementite (Fe_3C). The microstructural product of an iron–carbon alloy of eutectoid composition is pearlite, a microconstituent consisting of alternating layers of ferrite and cementite. The microstructures of alloys having a carbon content less than the eutectoid (hypoeutectoid) are comprised of a proeutectoid ferrite phase in addition to pearlite. On the other hand, pearlite and a proeutectoid cementite constitute the microconstituents for hypereutectoid alloys—those with carbon contents in excess of the eutectoid composition.

IMPORTANT TERMS AND CONCEPTS

Austenite

Cementite

Component

Congruent
  transformation

Equilibrium

Eutectic (reaction)

Eutectic phase

Eutectic structure

Eutectoid (reaction)

Ferrite

Free energy

Gibbs phase rule

Hypereutectoid alloy

Hypoeutectoid alloy

Intermediate solid
  solution

Intermetallic compound

Invariant point

Isomorphous

Lever rule

Liquidus line

Metastable

Microconstituent

Pearlite

Peritectic

Phase

Phase diagram

Phase equilibrium

Primary phase

Proeutectoid cementite

Proeutectoid ferrite

Solidus line

Solubility limit

Solvus line

System

Terminal solid solution

Tie line

REFERENCES

GORDON, P., *Principles of Phase Diagrams in Materials Systems,* McGraw-Hill Book Company, Inc., New York, 1968.

HANSEN, M. and K. ANDERKO, *Constitution of Binary Alloys,* 2nd edition, McGraw-Hill Book Company, New York, 1958. *First Supplement* (R. P. Elliott), 1965. *Second Supplement* (F. A. Shunk), 1969.

HUME-ROTHERY, W., *Electrons, Atoms, Metals and Alloys,* Metal Industry (Louis Cassier Company Ltd.), London, 1955. Chapters 33–39.

MASSALSKI, T. B., J. L. MURRAY, L. H. BENNETT, and H. BAKER (Editors), *Binary Alloy Phase Diagrams,* American Society for Metals, Metals Park, OH, 1986. In two volumes.

Metals Handbook, 8th edition, Vol. 8, *Metallography, Structures and Phase Diagrams,* American Society for Metals, Metals Park, OH, 1973.

Metals Handbook, 9th edition, Vol. 9, *Metallography and Microstructures,* American Society for Metals, Metals Park, OH, 1985.

RHINES, F. N., *Phase Diagrams in Metallurgy—Their Development and Application,* McGraw-Hill Book Company, Inc., New York, 1956.

QUESTIONS AND PROBLEMS

9.1 What is the difference between composition and concentration?

9.2 Cite three variables that determine the microstructure of an alloy.

9.3 What thermodynamic condition must be met for a state of equilibrium to exist?

9.4 What is the difference between the states of phase equilibrium and metastability?

9.5 Cite the phases that are present and the phase compositions for the following alloys:
(a) 15 wt% Sn–85 wt% Pb at 100°C (212°F).
(b) 25 wt% Pb–75 wt% Mg at 425°C (800°F).
(c) 85 wt% Ag–15 wt% Cu at 800°C (1470°F).
(d) 55 wt% Zn–45 wt% Cu at 600°C (1110°F).
(e) 1.25 kg Sn and 14 kg Pb at 200°C (390°F).
(f) 7.6 lb$_m$ Cu and 144.4 lb$_m$ Zn at 600°C (1110°F).
(g) 21.7 mol Mg and 35.4 mol Pb at 400°C (750°F).
(h) 4.2 mol Cu and 1.1 mol Ag at 900°C (1650°F).

9.6 What phases and mass fractions are present at 1000, 800, 500, 300, and 200°C as an alloy of composition 52 wt% Zn–48 wt% Cu is cooled?

9.7 Determine the relative amounts (mass fractions) of the phases for the alloys and temperatures given in Problem 9.5.

9.8 Below is a portion of the H_2O–NaCl phase diagram:
(a) Using this diagram, briefly explain how spreading salt on ice that is at a temperature below 0°C (32°F) can cause the ice to melt.
(b) What concentration of salt is necessary to have a 50% ice–50% liquid brine at −10°C (14°F)?

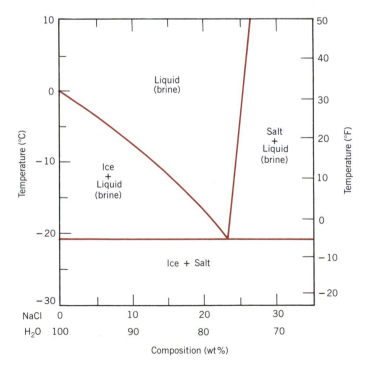

9.9 A 1.5-kg specimen of a 90 wt% Pb–10 wt% Sn alloy is heated to 250°C (480°F), at which temperature it is entirely an α phase solid solution (Figure 9.7). The alloy is to be melted to the extent that 50% of the specimen is liquid, the remainder being the α phase. This may be accomplished either by heating the alloy or changing its composition while holding the temperature constant.
(a) To what temperature must the specimen be heated?
(b) How much tin must be added to the 1.5-kg specimen at 250°C to achieve this state?

9.10 Consider the sugar–water phase diagram of Figure 9.1.
(a) How much sugar will dissolve in 1000 g water at 80°C (176°F)?
(b) If the saturated liquid solution in part a is cooled to 20°C (68°F), some of the sugar will precipitate out as a solid. What will be the composition of the saturated liquid solution (in wt% sugar) at 20°C?
(c) How much of the solid sugar will come out of solution upon cooling to 20°C?

9.11 A magnesium–lead alloy of mass 7.5 kg consists of a solid α phase which has a composition that is just slightly below the solubility limit at 300°C (570°F).
(a) What mass of lead is in the alloy?
(b) If the alloy is heated to 400°C (750°F), how much more lead may be dissolved in the α phase without exceeding the solubility limit of this phase?

9.12 **(a)** Briefly describe the phenomenon of coring and why it occurs. **(b)** Cite one undesirable consequence of coring.

9.13 It is desired to produce a copper–nickel alloy that has a minimum noncold-worked tensile strength of 55,000 psi (380 MPa) and a ductility of at least 45%EL. Is such an alloy possible? If so, what must be its composition? If this is not possible, then explain why.

9.14 Is it possible to have a copper–nickel alloy which, at equilibrium, consists of a liquid phase of composition 20 wt% Ni–80 wt% Cu and also an α phase of composition 37 wt% Ni–63 wt% Cu? If so, what will be the approximate temperature of the alloy? If this is not possible, explain why.

9.15 Is it possible to have a copper–silver alloy which, at equilibrium, consists of an α phase of composition 5 wt% Ag–95 wt% Cu, and also a β phase of composition 95 wt% Ag–5 wt% Cu? If so, what will be the approximate temperature of the alloy? If this is not possible, explain why.

9.16 A lead–tin alloy of composition 30 wt% Sn–70 wt% Pb is slowly heated from a temperature of 150°C (300°F).
(a) At what temperature does the first liquid phase form?
(b) What is the composition of this liquid phase?
(c) At what temperature does complete melting of the alloy occur?
(d) What is the composition of the last solid remaining prior to complete melting?

9.17 A 50 wt% Ni–50 wt% Cu alloy is slowly cooled from 1400°C (2550°F) to 1200°C (2190°F).
(a) At what temperature does the first solid phase form?
(b) What is the composition of this solid phase?

(c) At what temperature does the last liquid solidify?

(d) What is the composition of this last remaining liquid phase?

9.18 A 30 wt% Ni–70 wt% Cu alloy is heated to a temperature within the α + liquid phase region. If the composition of the α phase is 40 wt% Ni, determine (a) the temperature of the alloy, (b) the composition of the liquid phase, and (c) the mass fractions of both phases.

9.19 Below are given the solidus and liquidus temperatures for the copper–gold system. Construct the phase diagram for this system and label each region.

Composition (wt% Au)	Solidus Temperature (°C)	Liquidus Temperature (°C)
0	1085	1085
20	1019	1042
40	972	996
60	934	946
80	911	911
90	928	942
95	974	984
100	1064	1064

9.20 A 40 wt% Pb–60 wt% Mg alloy is heated to a temperature within the α + liquid phase region. If the mass fraction of each phase is 0.5, estimate (a) the temperature of the alloy, and (b) the compositions of the two phases.

9.21 For alloys of two hypothetical metals A and B, there exist an α, A-rich phase and a β, B-rich phase. From the mass fractions of both phases for two different alloys, which are at the same temperature, determine the composition of the phase boundary (or solubility limit) for both α and β phases at this temperature.

Alloy Composition	Fraction α Phase	Fraction β Phase
60 wt% A–40 wt% B	0.57	0.43
30 wt% A–70 wt% B	0.14	0.86

9.22 A hypothetical A–B alloy of composition 55 wt% B–45 wt% A at some temperature is found to consist of mass fractions of 0.5 for both α and β phases. If the composition of the β phase is 90 wt% B–10 wt% A, what is the composition of the α phase?

9.23 Is it possible to have a copper–silver alloy of composition 20 wt% Ag–80 wt% Cu which, at equilibrium, consists of α and liquid phases having mass fractions $W_{\alpha} = 0.80$ and $W_L = 0.20$? If so, what will be the approximate temperature of the alloy? If such an alloy is not possible, explain why.

9.24 For 5.7 kg of a magnesium–lead alloy of composition 50 wt% Pb–50 wt% Mg, is it possible, at equilibrium, to have α and Mg_2Pb phases having respective masses of 5.13 and 0.57 kg? If so, what will be the approximate temperature of the alloy? If such an alloy is not possible, explain why.

9.25 Compute the volume fractions of the α and β phases of a 75 wt% Ag–25 wt% Cu alloy at 500°C. Take the densities of the α and β phases to be 8.9 and 10.5 g/cm³, respectively.

9.26 At 700°C (1290°F), what is the maximum solubility **(a)** of Cu in Ag? **(b)** Of Ag in Cu?

9.27 A 45 wt% Pb–55 wt% Mg alloy is rapidly quenched to room temperature from an elevated temperature in such a way that the high-temperature microstructure is preserved. This microstructure is found to consist of the α phase and Mg_2Pb, having respective mass fractions of 0.65 and 0.35. Determine the approximate temperature from which the alloy was quenched.

9.28 Is it possible to have a magnesium–lead alloy in which the mass fractions of primary α and total α are 0.60 and 0.85, respectively, at 460°C (860°F)? Why or why not?

9.29 For 2.8 kg of a lead–tin alloy, is it possible to have the masses of primary β and total β of 2.21 kg and 2.53 kg, respectively, at 180°C (355°F)? Why or why not?

9.30 For a lead–tin alloy of composition 80 wt% Sn–20 wt% Pb and at 180°C (356°F) do the following:
(a) Determine the mass fractions of the α and β phases.
(b) Determine the mass fractions of primary β and eutectic microconstituents.
(c) Determine the mass fraction of eutectic β.

9.31 The microstructure of a copper–silver alloy at 775°C (1425°F) consists of primary α and eutectic structures. If the mass fractions of these two microconstituents are 0.73 and 0.27, respectively, determine the composition of the alloy.

9.32 Consider the hypothetical eutectic phase diagram for metals A and B, which is similar to that for the lead–tin system, Figure 9.7. Assume that (1) α and β phases exist at the A and B extremities of the phase diagram, respectively; (2) the eutectic composition is 36 wt% A–64 wt% B; and (3) the composition of the α phase at the eutectic temperature is 88 wt% A–12 wt% B. Determine the composition of an alloy that will yield primary β and total β mass fractions of 0.367 and 0.768, respectively.

9.33 Briefly explain why, upon solidification, an alloy of eutectic composition forms a microstructure consisting of alternating layers of the two solid phases.

9.34 For an 85 wt% Pb–15 wt% Mg alloy, make schematic sketches of the microstructure that would be observed for conditions of very slow cooling at the following temperatures: 600°C (1110°F), 500°C (930°F), 270°C (520°F), and 200°C (390°F). Label all phases and indicate their approximate compositions.

9.35 For a 68 wt% Zn–32 wt% Cu alloy, make schematic sketches of the microstructure that would be observed for conditions of very slow cooling at the following temperatures: 1000°C (1830°F), 760°C (1400°F), 600°C (1110°F), and 400°C (750°F). Label all phases and indicate their approximate compositions.

9.36 For a 30 wt% Zn–70 wt% Cu alloy, make schematic sketches of the microstructure that would be observed for conditions of very slow cooling at the following temperatures: 1100°C (2010°F), 950°C (1740°F), 900°C (1650°F), and 700°C (1290°F). Label all phases and indicate their approximate compositions.

9.37 What is the principal difference between congruent and incongruent phase transformations?

9.38 Figure 9.31 is the tin–gold phase diagram, for which only single-phase regions are labeled. Specify temperature–composition points at which all eutectics, eutectoids, peritectics, and congruent phase transformations occur. Also, for each, write the reaction upon cooling.

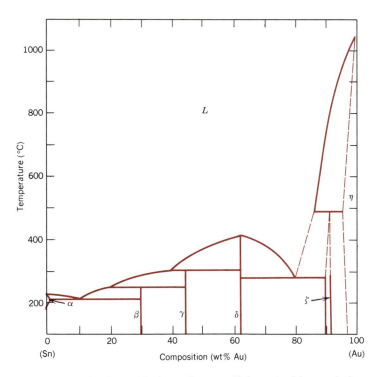

Figure 9.31 The tin–gold phase diagram. (Adapted with permission from *Metals Handbook,* Vol. 8, 8th edition, *Metallography, Structures and Phase Diagrams,* American Society for Metals, Metals Park, OH, 1973.)

9.39 Figure 9.32 is a portion of the aluminum–copper phase diagram for which only single-phase regions are labeled. Specify all temperature–composition points at which eutectics, eutectoids, peritectics, and congruent phase transformations occur. Also, for each, write the reaction upon cooling.

9.40 For a ternary system, three components are present; temperature is also a variable. Compute the maximum number of phases that may be present for a ternary system, assuming that pressure is held constant.

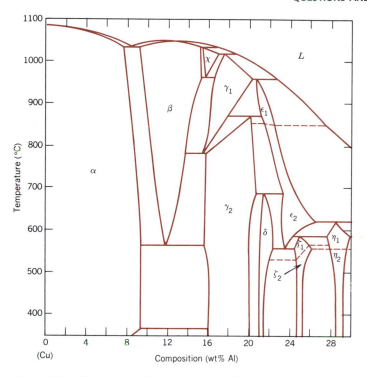

Figure 9.32 The copper–aluminum phase diagram. (Adapted with permission from *Metals Handbook,* Vol. 8, 8th edition, *Metallography, Structures and Phase Diagrams,* American Society for Metals, Metals Park, OH, 1973.)

9.41 Below is shown the pressure–temperature phase diagram for H_2O. Apply the Gibbs phase rule at points A, B, and C; that is, specify the number of degrees

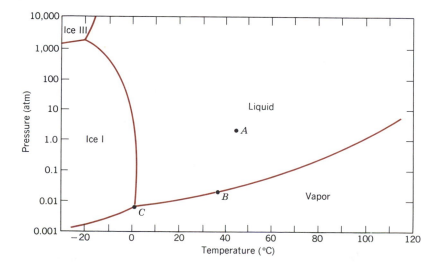

of freedom at each of the points, that is, the number of externally controllable variables that need be specified to completely define the system.

9.42 Carbon exists as an interstitial impurity in both BCC α-iron and FCC γ-iron. Briefly explain why the maximum solubility of carbon is much greater in γ-iron than in α-iron (i.e., 2.11 versus 0.022 wt% C) in spite of a higher atomic packing factor for FCC.

9.43 Compute the mass fractions of α ferrite and cementite in pearlite.

9.44 What is the difference between a phase and a microconstituent?

9.45 **(a)** What is the distinction between hypoeutectoid and hypereutectoid steels?
(b) In a hypoeutectoid steel, both eutectoid and proeutectoid ferrite exist. Explain the difference between them. What will be the carbon concentration in each?

9.46 Briefly explain why a proeutectoid phase forms along austenite grain boundaries. *Hint:* Consult Section 4.5.

9.47 What is the carbon concentration of an iron–carbon alloy for which the fraction of total ferrite is 0.94?

9.48 What is the proeutectoid phase for an iron–carbon alloy in which the mass fractions of total ferrite and total cementite are 0.86 and 0.14, respectively? Why?

9.49 Consider 1.0 kg of austenite containing 1.15 wt% C, cooled to below 727°C (1341°F).
(a) What is the proeutectoid phase?
(b) How many kilograms each of total ferrite and cementite form?
(c) How many kilograms each of pearlite and the proeutectoid phase form?
(d) Schematically sketch and label the resulting microstructure.

9.50 Consider 2.5 kg of austenite containing 0.65 wt% C, cooled to below 727°C (1341°F).
(a) What is the proeutectoid phase?
(b) How many kilograms each of total ferrite and cementite form?
(c) How many kilograms each of pearlite and the proeutectoid phase form?
(d) Schematically sketch and label the resulting microstructure.

9.51 Compute the mass fractions of proeutectoid ferrite and pearlite that form in an iron–carbon alloy containing 0.25 wt% C.

9.52 The microstructure of an iron–carbon alloy consists of proeutectoid ferrite and pearlite; the mass fractions of these two microconstituents are 0.286 and 0.714, respectively. Determine the concentration of carbon in this alloy.

9.53 The mass fractions of total ferrite and total cementite in an iron–carbon alloy are 0.91 and 0.09, respectively. Is this a hypoeutectoid or hypereutectoid alloy? Why?

9.54 The microstructure of an iron–carbon alloy consists of proeutectoid cementite and pearlite; the mass fractions of these microconstituents are 0.11 and 0.89, respectively. Determine the concentration of carbon in this alloy.

9.55 Consider 2.0 kg of a 99.6 wt% Fe–0.4 wt% C alloy that is cooled to a temperature just below the eutectoid.
(a) How many kilograms of proeutectoid ferrite form?

(b) How many kilograms of eutectoid ferrite form?

(c) How many kilograms of cementite form?

9.56 Compute the maximum mass fraction of proeutectoid cementite possible for a hypereutectoid iron–carbon alloy.

9.57 Is it possible to have an iron–carbon alloy for which the mass fractions of total cementite and proeutectoid ferrite are 0.057 and 0.36, respectively? Why or why not?

9.58 Is it possible to have an iron–carbon alloy for which the mass fractions of total ferrite and pearlite are 0.860 and 0.969, respectively? Why or why not?

9.59 Compute the mass fraction of eutectoid cementite in an iron–carbon alloy that contains 1.00 wt% C.

9.60 The mass fraction of *eutectoid* cementite in an iron–carbon alloy is 0.109. On the basis of this information, is it possible to determine the composition of the alloy? If so, what is its composition? If this is not possible, explain why.

9.61 The mass fraction of *eutectoid* ferrite in an iron–carbon alloy is 0.71. On the basis of this information, is it possible to determine the composition of the alloy? If, so, what is its composition? If this is not possible, explain why.

9.62 For an iron–carbon alloy of composition 3 wt% C–97 wt% Fe, make schematic sketches of the microstructure that would be observed for conditions of very slow cooling at the following temperatures: 1250°C (2280°F), 1145°C (2095°F), and 700°C (1290°F). Label the phases and indicate their compositions (approximate).

9.63 Often, the properties of multiphase alloys may be approximated by the relationship

$$E \text{ (alloy)} = E_\alpha V_\alpha + E_\beta V_\beta$$

where E represents a specific property (modulus of elasticity, hardness, etc.), and V is the volume fraction. The subscripts α and β denote the existing phases or microconstituents. Employ the relationship above to determine the approximate Brinell hardness of a 99.8 wt% Fe–0.2 wt% C alloy. Assume Brinell hardnesses of 80 and 280 for ferrite and pearlite, respectively, and that volume fractions may be approximated by mass fractions.

9.64 On the basis of the photomicrograph (i.e., the relative amounts of the microconstituents) for the lead-tin alloy shown in Figure 9.13b and the Pb–Sn phase diagram (Figure 9.7), estimate the composition of the alloy, and then compare this estimate with the composition given in the figure legend of Figure 9.13b. Make the following assumptions: (1) the area fraction of each phase and microconstituent in the photomicrograph is equal to its volume fraction; (2) the densities of the α and β phases as well as the eutectic structure are 11.2, 7.3, and 8.7 g/cm^3, respectively; and (3) this photomicrograph represents the equilibrium microstructure at 180°C (356°F).

9.65 A steel alloy contains 97.5 wt% Fe, 2.0 wt% Mo, and 0.5 wt% C.

(a) What is the eutectoid temperature of this alloy?

(b) What is the eutectoid composition?

(c) What is the proeutectoid phase?

Assume that there are no changes in the positions of other phase boundaries with the addition of Mo.

9.66 A steel alloy is known to contain 93.8 wt% Fe, 6.0 wt% Ni, and 0.2 wt% C.

(a) What is the approximate eutectoid temperature of this alloy?

(b) What is the proeutectoid phase when this alloy is cooled to a temperature just below the eutectoid?

(c) Compute the relative amounts of the proeutectoid phase and pearlite.

Assume that there are no alterations in the positions of other phase boundaries with the addition of Ni.

PHASE TRANSFORMATIONS IN METALS: Development of Microstructure and Alteration of Mechanical Properties

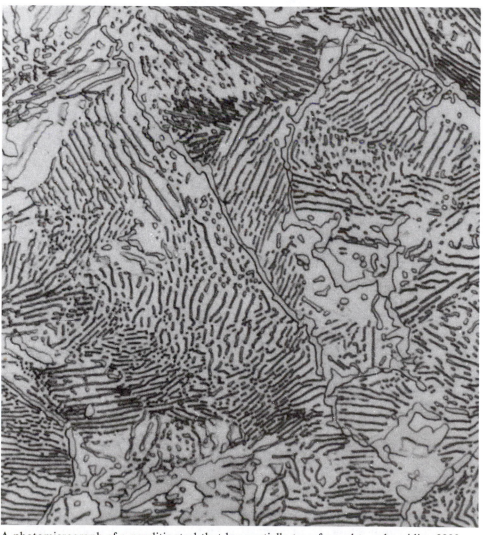

A photomicrograph of a pearlitic steel that has partially transformed to spheroidite. 2000 ×. (Courtesy of United States Steel Corporation.)

10.1 INTRODUCTION

One reason for the versatility of metallic materials lies in the wide range of mechanical properties they possess, which are accessible to management by various means. Three strengthening mechanisms were discussed in Chapter 7, namely, grain size refinement and solid-solution and strain hardening. Additional techniques are available wherein the mechanical properties are reliant on the characteristics of the microstructure.

The development of microstructure in both single- and two-phase alloys ordinarily involves some type of phase transformation—an alteration in the number and/or character of the phases. The first portion of this chapter is devoted to a brief discussion of some of the basic principles relating to transformations involving solid phases. Inasmuch as most phase transformations do not occur instantaneously, consideration is given to the dependence of reaction progress on time, or the **transformation rate.** This is followed by a discussion of the development of two-phase microstructures for iron–carbon alloys. Modified phase diagrams are introduced which permit determination of the microstructure that results from a specific heat treatment. Finally, other microconstituents in addition to pearlite are presented, and, for each, the mechanical properties are discussed.

PHASE TRANSFORMATIONS

10.2 BASIC CONCEPTS

A variety of phase transformations are important in the processing of materials, and usually they involve some alteration of the microstructure. For purposes of this discussion, these transformations are divided into three classifications. In one group are simple diffusion-dependent transformations in which there is no change in either the number or composition of the phases present. These include solidification of a pure metal, allotropic transformations, and, recrystallization and grain growth (see Sections 7.12 and 7.13).

In another type of diffusion-dependent transformation, there is some alteration in phase compositions and often in the number of phases present; the final microstructure ordinarily consists of two phases. The eutectoid reaction, described by Equation 9.16, is of this type; it receives further attention in Section 10.5.

The third kind of transformation is diffusionless, wherein a metastable phase is produced. As discussed in Section 10.5, a martensitic transformation, which may be induced in some steel alloys, falls into this category.

10.3 THE KINETICS OF SOLID-STATE REACTIONS

Most solid-state transformations do not occur instantaneously because obstacles impede the course of the reaction and make it dependent on time. For example, since most transformations involve the formation of at least one new phase that has a composition and/or crystal structure different from that of the parent one, some atomic rearrangements via diffusion are required. Diffusion is a time-dependent phenomenon, as discussed in Section 5.4. A second impediment to the formation of a new phase is the increase in energy associated with the phase boundaries that are created between parent and product phases.

From a microstructural standpoint, the first process to accompany a phase transformation is **nucleation**—the formation of very small (often submicroscopic) particles, or nuclei, of the new phase, which are capable of growing. Favorable positions for the formation of these nuclei are imperfection sites, especially grain boundaries. The second stage is *growth,* in which the nuclei increase in size; during this process, of course, some volume of the parent phase disappears. The transformation reaches completion if growth of these new phase particles is allowed to proceed until the equilibrium fraction is attained.

As would be expected, the transformation rate is an important consideration in the heat treatment of materials, the study of which is often termed **kinetics.** With most kinetic investigations, the fraction of reaction that has occurred is measured as a function of time, while the temperature is maintained constant. Transformation progress is usually ascertained by either microscopic examination or measurement of some physical property (such as electrical conductivity) the magnitude of which is distinctive of the new phase. Data are plotted as the fraction of transformed material versus the logarithm of time; an S-shaped curve similar to that in Figure 10.1 represents the typical kinetic behavior for most solid-state reactions. Nucleation and growth stages are indicated in the figure.

For solid-state transformations displaying the kinetic behavior in Figure 10.1, the fraction of transformation y is a function of time t as follows:

$$y = 1 - \exp(-kt^n) \tag{10.1}$$

where k and n are time-independent constants for the particular reaction. The above expression is often referred to as the *Avrami equation.*

By convention, the rate of a transformation r is taken as the reciprocal of time required for the transformation to proceed halfway to completion, $t_{0.5}$, or

$$r = \frac{1}{t_{0.5}} \tag{10.2}$$

This $t_{0.5}$ is also noted in Figure 10.1.

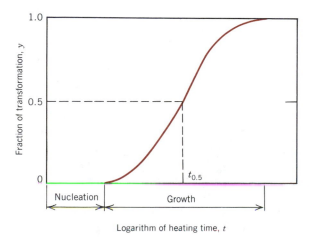

Figure 10.1 Plot of fraction reacted versus the logarithm of time typical of many solid-state transformations in which temperature is held constant.

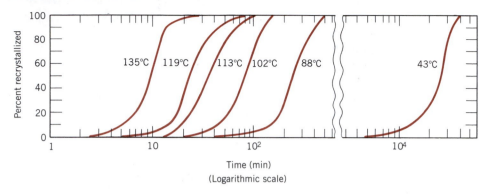

Figure 10.2 Percent recrystallization as a function of time and at constant temperature for pure copper. (Reprinted with permission from *Metallurgical Transactions,* Vol. 188, 1950, a publication of The Metallurgical Society of AIME, Warrendale, Pennsylvania. Adapted from B. F. Decker and D. Harker, "Recrystallization in Rolled Copper," *Trans. AIME,* **188,** 1950, p. 888.)

Temperature is one variable in a heat treatment process that is subject to control, and it may have a profound influence on the kinetics and thus on the rate of a transformation. This is demonstrated in Figure 10.2, where *y*-versus-log *t* S-shaped curves at several temperatures for the recrystallization of copper are shown.

For most reactions and over specific temperature ranges, rate increases with temperature according to

$$r = Ae^{-Q/RT} \tag{10.3}$$

where

R = the gas constant
T = absolute temperature
A = a temperature-independent constant
Q = an activation energy for the particular reaction

It may be recalled that the diffusion coefficient has the same temperature dependence (Equation 5.8). Processes the rates of which exhibit this relationship with temperature are sometimes termed **thermally activated.**

10.4 MULTIPHASE TRANSFORMATIONS

Phase transformations may be wrought in metal alloy systems by varying temperature, composition, and the external pressure; however, temperature changes by means of heat treatments are most conveniently utilized to induce phase transformations. This corresponds to crossing a phase boundary on the composition–temperature phase diagram as an alloy of given composition is heated or cooled.

During a phase transformation, an alloy proceeds toward an equilibrium state that is characterized by the phase diagram in terms of the product phases, their compositions, and relative amounts. Most phase transformations require some finite time to go to completion, and the speed or rate is often important in the relationship between the heat treatment and the development of microstructure. One limitation of phase diagrams is their inability to indicate the time period required for the attainment of equilibrium.

The rate of approach to equilibrium for solid systems is so slow that true equilibrium structures are rarely achieved. Equilibrium conditions are maintained only if heating or cooling is carried out at extremely slow and unpractical rates. For other than equilibrium cooling, transformations are shifted to lower temperatures than indicated by the phase diagram; for heating, the shift is to higher temperatures. These phenomena are termed **supercooling** and **superheating,** respectively. The degree of each depends on the rate of temperature change; the more rapid the cooling or heating, the greater the supercooling or superheating. For example, for normal cooling rates the iron–carbon eutectoid reaction is typically displaced 10 to 20°C (18 to 36°F) below the equilibrium transformation temperature.

For many technologically important alloys, the preferred state or microstructure is a metastable one, intermediate between the initial and equilibrium states; on occasion, a structure far removed from the equilibrium one is desired. It thus becomes imperative to investigate the influence of time on phase transformations. This kinetic information is, in many instances, of greater value than a knowledge of the final equilibrium state.

MICROSTRUCTURAL AND PROPERTY CHANGES IN IRON–CARBON ALLOYS

Some of the basic kinetic principles of solid-state transformations are now extended and applied specifically to iron–carbon alloys in terms of the relationships between heat treatment, the development of microstructure, and mechanical properties. This system has been chosen because it is familiar and because a wide variety of microstructures and mechanical properties are possible for iron–carbon (or steel) alloys.

10.5 ISOTHERMAL TRANSFORMATION DIAGRAMS

Pearlite

Consider again the iron–iron carbide eutectoid reaction

$$\gamma \; (0.77 \text{ wt\% C}) \longrightarrow \alpha \; (0.022 \text{ wt\% C}) + Fe_3C \; (6.70 \text{ wt\% C}) \qquad (9.16)$$

which is fundamental to the development of microstructure in steel alloys. Upon cooling, austenite, having an intermediate carbon concentration, transforms to a ferrite phase, having a much lower carbon content, and also to cementite, with a much higher carbon concentration. For carbon atoms to selectively segregate in the cementite phase, diffusion is necessary. Figure 10.3 illustrates schematically microstructural changes that accompany this eutectoid reaction as pearlite is formed; the direction of carbon diffusion is indicated by the arrows. Carbon atoms diffuse away from the ferrite regions and to the cementite layers, to give a concentration of 6.70 wt% C, as the pearlite extends from the grain boundary into the unreacted austenite grains. The layered pearlite forms because the carbon atoms need diffuse only minimal distances with this structure.

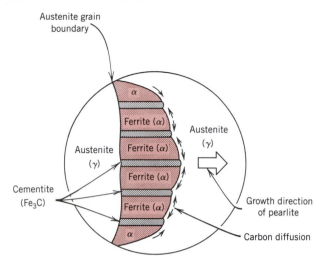

Figure 10.3 Schematic representation of the formation of pearlite from austenite; direction of carbon diffusion indicated by arrows.

Temperature plays an important role in the rate of the austenite-to-pearlite transformation. The temperature dependence for an iron–carbon alloy of eutectoid composition is indicated in Figure 10.4, which plots S-shaped curves of the percentage transformation versus the logarithm of time at three different temperatures. For each curve, data were collected after rapidly cooling a specimen composed of 100% austenite to the temperature indicated; that temperature was maintained constant throughout the course of the reaction.

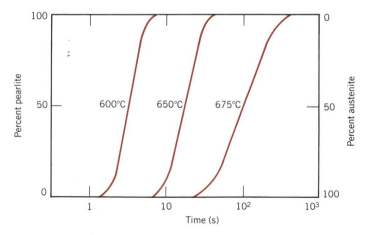

Figure 10.4 For an iron–carbon alloy of eutectoid composition (0.77 wt% C), isothermal fraction reacted versus the logarithm of time for the austenite-to-pearlite transformation.

A more convenient way of representing both the time and temperature dependence of this transformation is in the bottom portion of Figure 10.5. Here, the vertical and horizontal axes are, respectively, temperature and the logarithm of time. Two solid curves are plotted; one represents the time required at each temperature for the initiation or start of the transformation; the other is for the transformation conclusion. The dashed curve corresponds to 50% of transformation completion. These curves were generated from a series of plots of the percentage transformation versus the logarithm of time taken over a range of temperatures. The S-shaped curve [for 675°C (1247°F)], in the upper portion of Figure 10.5, illustrates how the data transfer is made.

In interpreting this diagram, note first that the eutectoid temperature [727°C (1341°F)] is indicated by a horizontal line; at temperatures above the eutectoid and for all times, only austenite will exist, as indicated in the figure. The austenite-to-pearlite transformation will occur only if an alloy is supercooled to below the eutectoid; as indicated by the curves, the time necessary for the transformation to begin

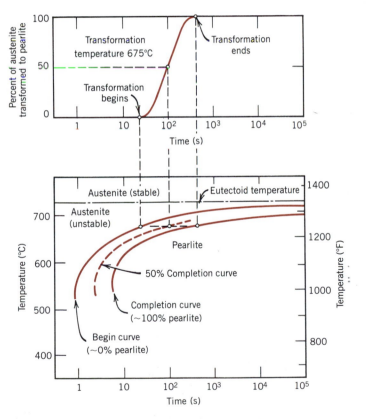

Figure 10.5 Demonstration of how an isothermal transformation diagram (bottom) is generated from percent transformation-versus-logarithm of time measurements (top). (Adapted from H. Boyer, Editor, *Atlas of Isothermal Transformation and Cooling Transformation Diagrams,* American Society for Metals, 1977, p. 369.)

and then end depends on temperature. The start and finish curves are nearly parallel, and they approach the eutectoid line asymptotically. To the left of the transformation start curve, only austenite (which is unstable) will be present, whereas to the right of the finish curve, only pearlite will exist. In between, the austenite is in the process of transforming to pearlite, and thus both microconstituents will be present.

According to Equation 10.2, the transformation rate at some particular temperature is inversely proportional to the time required for the reaction to proceed to 50% completion (to the dashed line in Figure 10.5). That is, the shorter this time, the higher the rate. Thus from Figure 10.5, at temperatures just below the eutectoid (corresponding to just a slight degree of undercooling) very long times (on the order of 10^5 s) are required for the 50% transformation, and therefore the reaction rate is very slow. The transformation rate increases with decreasing temperature such that at 540°C (1000°F) only about 3 s is required for the reaction to go to 50% completion.

This rate–temperature behavior is in apparent contradiction of Equation 10.3, which stipulates that rate *increases* with increasing temperature. The reason for this disparity is that over this range of temperatures (i.e., 540 to 727°C), the transformation rate is controlled by the rate of pearlite nucleation, and nucleation rate decreases with rising temperature (i.e., less supercooling). This behavior may be explained by Equation 10.3, wherein the activation energy Q is a function of, and increases with, increasing temperature. We shall find that at lower temperatures, the austenite decomposition transformation is diffusion-controlled and that the rate behavior is as predicted by Equation 10.3, with a temperature-independent activation energy.

Several constraints are imposed on using diagrams like Figure 10.5. First, this particular plot is valid only for an iron–carbon alloy of eutectoid composition; for other compositions, the curves will have different configurations. In addition, these plots are accurate only for transformations in which the temperature of the alloy is held constant throughout the duration of the reaction. Conditions of constant temperature are termed *isothermal;* thus plots such as Figure 10.5 are referred to as **isothermal transformation diagrams,** or sometimes as *time–temperature–transformation* (or *T–T–T*) plots.

An actual isothermal heat treatment curve (*ABCD*) is superimposed on the isothermal transformation diagram for a eutectoid iron–carbon alloy in Figure 10.6. Very rapid cooling of austenite to a temperature is indicated by the near-vertical line *AB*, and the isothermal treatment at this temperature is represented by the horizontal segment *BCD*. Of course, time increases from left to right along this line. The transformation of austenite to pearlite begins at the intersection, point *C* (after approximately 3.5 s), and has reached completion by about 15 s, corresponding to point *D*. Figure 10.6 also shows schematic microstructures at various times during the progression of the reaction.

The thickness ratio of the ferrite and cementite layers in pearlite is approximately 8 to 1. However, the absolute layer thickness depends on the temperature at which the isothermal transformation is allowed to occur. At temperatures just below the eutectoid, relatively thick layers of both the α-ferrite and Fe_3C phases are produced; this microstructure is called **coarse pearlite,** and the region at which it forms is indicated to the right of the completion curve on Figure 10.6. At these temperatures, diffusion rates are relatively high, such that during the transformation illustrated in Figure 10.3 carbon atoms can diffuse relatively long distances, which results in the formation of

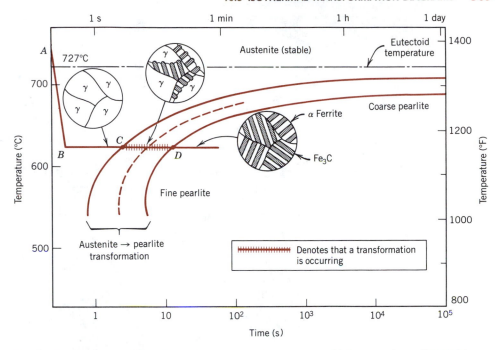

Figure 10.6 Isothermal transformation diagram for an eutectoid iron–carbon alloy, with superimposed isothermal heat treatment curve (*ABCD*). Microstructures before, during, and after the austenite-to-pearlite transformation are shown. (Adapted from H. Boyer, Editor, *Atlas of Isothermal Transformation and Cooling Transformation Diagrams,* American Society for Metals, 1977, p. 28.)

thick lamellae. With decreasing temperature, the layers become progressively thinner inasmuch as the carbon diffusion rate decreases. The thin-layered structure produced in the vicinity of 540°C is termed **fine pearlite;** this is also indicated in Figure 10.6. To be discussed in Section 10.7 is the dependence of mechanical properties on lamellar thickness. Photomicrographs of coarse and fine pearlite for a eutectoid composition are shown in Figure 10.7.

For iron–carbon alloys of other compositions, a proeutectoid phase (either ferrite or cementite) will coexist with pearlite, as discussed in Section 9.14. Thus additional curves corresponding to a proeutectoid transformation also must be included on the isothermal transformation diagram. A portion of one such diagram for a 1.13 wt% C alloy is shown in Figure 10.8.

Bainite

In addition to pearlite, other microconstituents that are products of the austenitic transformation exist; one of these is called **bainite.** The microstructure of bainite consists of ferrite and cementite phases, and thus diffusional processes are involved in its formation. Bainite forms as needles or plates, depending on the temperature of

Figure 10.7 Photomicrographs of (*a*) coarse pearlite and (*b*) fine pearlite. 3000×. (From K. M. Ralls, et al., *Introduction to Materials Science and Engineering*, p. 361. Copyright © 1976 by John Wiley & Sons, New York. Reprinted by permission of John Wiley & Sons, Inc.)

the transformation; the microstructural details of bainite are so fine that their resolution is possible only using electron microscopy. Figure 10.9 is an electron micrograph which shows a needle of bainite (positioned diagonally from lower left to upper right); it is composed of a ferrite matrix and elongated particles of Fe_3C. The phase that surrounds the needle is martensite, the topic to which a subsequent section is addressed.

The time–temperature dependence of the bainite transformation may also be represented on the isothermal transformation diagram. It occurs at temperatures below those at which pearlite forms; begin-, end-, and half-reaction curves are just extensions of those for the pearlitic transformation, as shown in Figure 10.10, the isothermal transformation diagram for an iron–carbon alloy of eutectoid composition, which has been extended to lower temperatures. All three curves are C-shaped and have a "nose" at point N, where the rate of transformation is a maximum. As may be noted, whereas pearlite forms above the nose—that is, over the temperature range of about 540 to 727°C (1000 to 1341°F)—for isothermal treatments at temperatures between about 215 and 540°C (420 and 1000°F), bainite is the transformation product. The pearlitic and bainitic transformations are really competitive with each other; and once some portion of an alloy has transformed to either pearlite or bainite, transformation to the other microconstituent is not possible without reheating to form austenite.

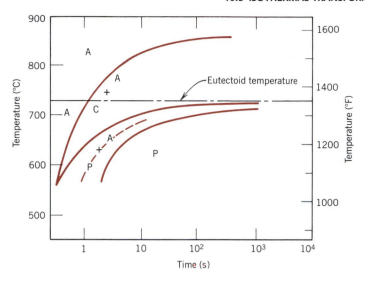

Figure 10.8 Isothermal transformation diagram for 1.13 wt% C iron–carbon alloy: A, austenite; C, proeutectoid cementite; P, pearlite. (Adapted from H. Boyer, Editor, *Atlas of Isothermal Transformation and Cooling Transformation Diagrams,* American Society for Metals, 1977, p. 33.)

In passing, it should be mentioned that the kinetics of the bainite transformation (below the nose in Figure 10.10) obey Equation 10.3; that is, rate ($1/t_{0.5}$, Equation 10.2) increases exponentially with rising temperature. Furthermore, the kinetics of many solid-state transformations are represented by this characteristic C-shaped curve (Figure 10.10).

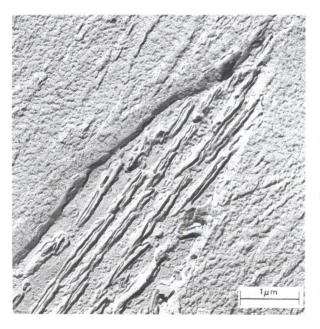

Figure 10.9 Replica transmission electron micrograph showing the structure of bainite. A needle of bainite passes from lower left to upper right-hand corners, which consists of elongated particles of Fe_3C within a ferrite matrix. The phase surrounding the bainite needle is martensite. (Reproduced with permission from *Metals Handbook,* Vol. 8, 8th edition, *Metallography, Structures and Phase Diagrams,* American Society for Metals, Materials Park, OH, 1973.)

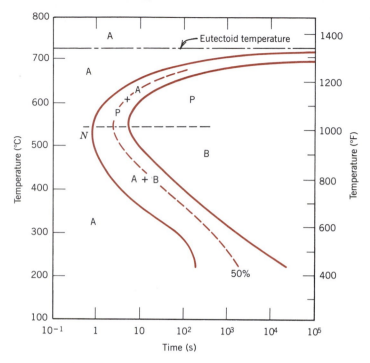

Figure 10.10 Isothermal transformation diagram for an iron–carbon alloy of eutectoid composition, including austenite-to-pearlite (A–P) and austenite-to-bainite (A–B) transformations. (Adapted from H. Boyer, Editor, *Atlas of Isothermal Transformation and Cooling Transformation Diagrams*, American Society for Metals, 1977, p. 28.)

Spheroidite

If a steel alloy having either pearlitic or bainitic microstructures is heated to, and left at, a temperature below the eutectoid for a sufficiently long period of time—for example, at about 700°C (1300°F) for between 18 and 24 h—yet another microstructure will form. It is called **spheroidite** (Figure 10.11). Instead of the alternating ferrite and cementite lamellae (pearlite), or the elongated Fe_3C particles in a ferrite matrix (bainite), the Fe_3C phase appears as spherelike particles embedded in a continuous α phase matrix. This transformation has occurred by additional carbon diffusion with no change in the compositions or relative amounts of ferrite and cementite phases. The driving force for this transformation is the reduction in α–Fe_3C phase boundary area. The kinetics of spheroidite formation are not included on isothermal transformation diagrams.

Martensite

Yet another microconstituent or phase called **martensite** is formed when austenitized iron–carbon alloys are rapidly cooled (or quenched) to a relatively low temperature

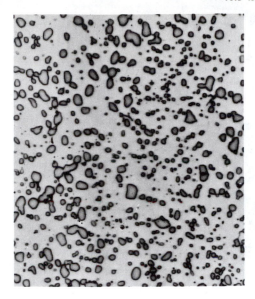

Figure 10.11 Photomicrograph of a steel having a spheroidite microstructure. The small particles are cementite; the continuous phase is α-ferrite. 1000×. (Copyright 1971 by United States Steel Corporation.)

(in the vicinity of the ambient). Martensite is a nonequilibrium single-phase structure that results from a diffusionless transformation of austenite. It may be thought of as a transformation product that is competitive with pearlite and bainite. The martensitic transformation occurs when the quenching rate is rapid enough to prevent carbon diffusion. Any diffusion whatsoever will result in the formation of ferrite and cementite phases.

The martensitic transformation is not well understood. However, large numbers of atoms experience cooperative movements, in that there is only a slight displacement of each atom relative to its neighbors. This occurs in such a way that the FCC austenite experiences a polymorphic transformation to a body-centered tetragonal (BCT) martensite. A unit cell of this crystal structure (Figure 10.12) is simply a body-centered cube that has been elongated along one of its dimensions; this structure is distinctly different from that for BCC ferrite. All the carbon atoms remain as interstitial impurities in martensite; as such, they constitute a supersaturated solid

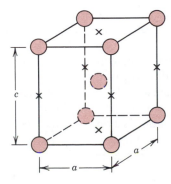

Figure 10.12 The body-centered tetragonal unit cell for martensitic steel showing iron atoms (circles) and sites that may be occupied by carbon atoms (crosses). For this tetragonal unit cell, $c > a$.

Figure 10.13 Photomicrograph showing the martensitic microstructure. The needle-shaped grains are the martensite phase, and the white regions are austenite that failed to transform during the rapid quench. 1220×. (Photomicrograph courtesy of United States Steel Corporation.)

solution that is capable of rapidly transforming to other structures if heated to temperatures at which diffusion rates become appreciable. Many steels, however, retain their martensitic structure almost indefinitely at room temperature.

The martensitic transformation is not, however, unique to iron–carbon alloys. It is found in other systems and is characterized, in part, by the diffusionless transformation.

Since the martensitic transformation does not involve diffusion, it occurs almost instantaneously; the martensite grains nucleate and grow at a very rapid rate—the velocity of sound within the austenite matrix. Thus the martensitic transformation rate, for all practical purposes, is time independent.

Martensite grains take on a platelike or needlelike appearance, as indicated in Figure 10.13. The white phase in the micrograph is most probably austenite (retained austenite) that did not transform during the rapid quench. As has already been mentioned, martensite as well as other microconstituents (e.g., pearlite) can coexist.

Being a nonequilibrium phase, martensite does not appear on the iron–iron carbide phase diagram (Figure 9.20). The austenite-to-martensite transformation is, however, represented on the isothermal transformation diagram. Since the martensitic transformation is diffusionless and instantaneous, it is not depicted in this diagram like the pearlitic and bainitic reactions. The beginning of this transformation is represented by a horizontal line designated M(start) (Figure 10.14). Two other horizontal and dashed lines, labeled M(50%) and M(90%), indicate percentages of the austenite-to-martensite transformation. The temperatures at which these lines are located vary with alloy composition but, nevertheless, must be relatively low because carbon diffusion must be virtually nonexistent. The horizontal and linear character of these lines indicates that the martensitic transformation is independent of time; it is a function only of the temperature to which the alloy is quenched or rapidly cooled. A transformation of this type is termed an **athermal transformation.**

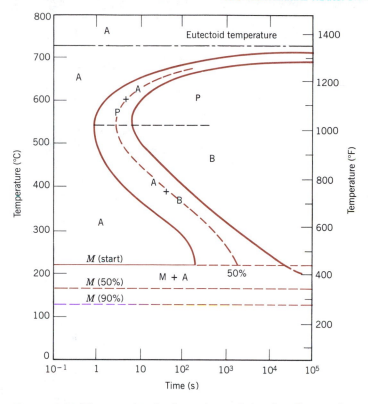

Figure 10.14 The complete isothermal transformation diagram for an iron–carbon alloy of eutectoid composition: A, austenite; B, bainite; M, martensite; P, pearlite.

Consider an alloy of eutectoid composition that is very rapidly cooled from a temperature above 727°C (1341°F) to, say, 165°C (330°F). From the isothermal transformation diagram (Figure 10.14) it may be noted that 50% of the austenite will immediately transform to martensite; and as long as this temperature is maintained, there will be no further transformation.

The presence of alloying elements other than carbon (e.g., Cr, Ni, Mo, and W) may cause significant changes in the positions and shape of the curves in the isothermal transformation diagrams. These include (1) shifting to longer times the nose of the austenite-to-pearlite transformation (and also a proeutectoid phase nose, if such exists), and (2) the formation of a separate bainite nose. These alterations may be observed by comparing Figures 10.14 and 10.15, which are isothermal transformation diagrams for carbon and alloy steels, respectively.

Steels in which carbon is the prime alloying element are termed **plain carbon steels,** whereas **alloy steels** contain appreciable concentrations of other elements, including those cited in the preceding paragraph. Chapter 12 tells more about the classification and properties of ferrous alloys.

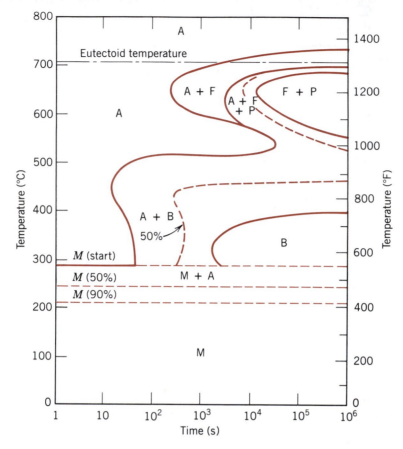

Figure 10.15 Isothermal transformation diagram for an alloy steel (type 4340): A, austenite; B, bainite; P, pearlite; M, martensite. (Adapted from H. Boyer, Editor, *Atlas of Isothermal Transformation and Cooling Transformation Diagrams,* American Society for Metals, 1977, p. 181.)

EXAMPLE PROBLEM 10.1

Using the isothermal transformation diagram for an iron–carbon alloy of eutectoid composition (Figure 10.14), specify the nature of the final microstructure (in terms of microconstituents present and approximate percentages) of a small specimen that has been subjected to the following time–temperature treatments. In each case assume that the specimen begins at 760°C (1400°F) and that it has been held at this temperature long enough to have achieved a complete and homogeneous austenitic structure.

(a) Rapidly cool to 350°C (660°F), hold for 10^4 s, and quench to room temperature.

(b) Rapidly cool to 250°C (480°F), hold for 100 s, and quench to room temperature.

(c) Rapidly cool to 650°C (1200°F), hold for 20 s, rapidly cool to 400°C (750°F), hold for 10^3 s, and quench to room temperature.

SOLUTION

The time–temperature paths for all three treatments are shown in Figure 10.16. In each case the initial cooling is rapid enough to prevent any transformation from occurring.

(a) At 350°C austenite isothermally transforms to bainite; this reaction begins after about 10 s and reaches completion at about 500 s elapsed time. Therefore, by 10^4 s, as stipulated in the problem, 100% of the specimen is bainite, and no further transformation is possible, even though the final quenching line passes through the martensite region of the diagram.

(b) In this case it takes about 150 s at 250°C for the bainite transformation to begin, so that at 100 s the specimen is still 100% austenite. As the specimen is cooled through the martensite region, beginning at about 215°C, progressively

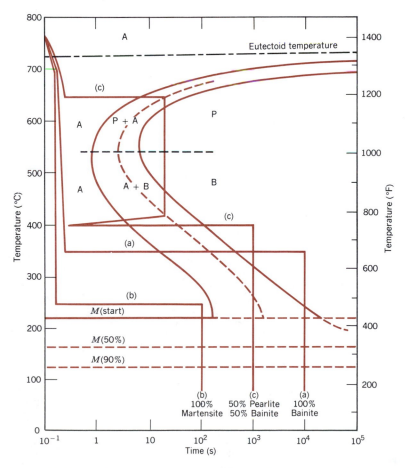

Figure 10.16 Isothermal transformation diagram for an iron–carbon alloy of eutectoid composition and the isothermal heat treatments (a), (b), and (c) in Example Problem 10.1.

more of the austenite instantaneously transforms to martensite. This transformation is complete by the time room temperature is reached, such that the final microstructure is 100% martensite.

(c) For the isothermal line at 650°C, pearlite begins to form after about 7 s; by the time 20 s has elapsed, only approximately 50% of the specimen has transformed to pearlite. The rapid cool to 400°C is indicated by the vertical line; during this cooling, very little, if any, remaining austenite will transform to either pearlite or bainite, even though the cooling line passes through pearlite and bainite regions of the diagram. At 400°C, we begin timing at essentially zero time (as indicated in Figure 10.16); thus by the time 10^3 s has elapsed, all of the remaining 50% austenite will have completely transformed to bainite. Upon quenching to room temperature, any further transformation is not possible inasmuch as no austenite remains; and so the final microstructure at room temperature consists of 50% pearlite and 50% bainite.

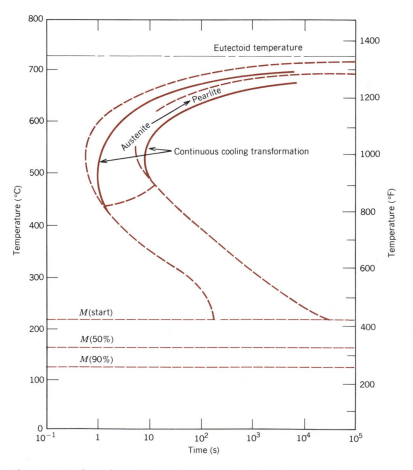

Figure 10.17 Superimposition of isothermal and continuous cooling transformation diagrams for a eutectoid iron–carbon alloy. (Adapted from H. Boyer, Editor, *Atlas of Isothermal Transformation and Cooling Transformation Diagrams,* American Society for Metals, 1977, p. 376.)

10.6 CONTINUOUS COOLING TRANSFORMATION DIAGRAMS

Isothermal heat treatments are not the most practical to conduct because an alloy must be rapidly cooled to and maintained at an elevated temperature from a higher temperature above the eutectoid. Most heat treatments for steels involve the continuous cooling of a specimen to room temperature. An isothermal transformation diagram is valid only for conditions of constant temperature, which must be modified for transformations that occur as the temperature is constantly changing. For continuous cooling, the time required for a reaction to begin and end is delayed. Thus the isothermal curves are shifted to longer times and lower temperatures, as indicated in Figure 10.17 for an iron–carbon alloy of eutectoid composition. A plot containing such modified beginning and ending reaction curves is termed a **continuous cooling transformation (CCT) diagram.** Some control may be maintained over the rate of temperature change depending on the cooling environment. Two cooling curves corresponding to moderately fast and slow rates are superimposed and labeled in Figure 10.18, again for a eutectoid steel. The transformation starts after a time period corresponding to the intersection of the cooling curve with the beginning reaction curve

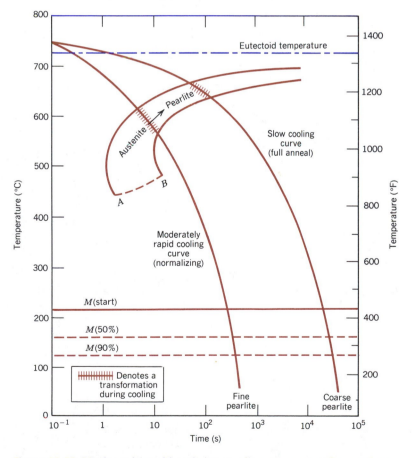

Figure 10.18 Moderately rapid and slow cooling curves superimposed on a continuous cooling transformation diagram for a eutectoid iron–carbon alloy.

and concludes upon crossing the completion transformation curve. The microstructural products for the moderately rapid and slow cooling rate curves in Figure 10.18 are fine and coarse pearlite, respectively.

Normally, bainite will not form when an alloy of eutectoid composition or, for that matter, any plain carbon steel is continuously cooled to room temperature. This is because all the austenite will have transformed to pearlite by the time the bainite transformation has become possible. Thus the region representing the austenite–pearlite transformation terminates just below the nose (Figure 10.18) as indicated by the curve *AB*. For any cooling curve passing through *AB* in Figure 10.18, the transformation ceases at the point of intersection; with continued cooling, the unreacted austenite begins transforming to martensite upon crossing the *M*(start) line.

With regard to the representation of the martensitic transformation, the *M*(start), *M*(50%), and *M*(90%) lines occur at identical temperatures for both isothermal and continuous cooling transformation diagrams. This may be verified for an iron–carbon alloy of eutectoid composition by comparison of Figures 10.14 and 10.17.

For the continuous cooling of a steel alloy, there exists a critical quenching rate, which represents the minimum rate of quenching that will produce a totally martensitic structure. This critical cooling rate, when included on the continuous transformation

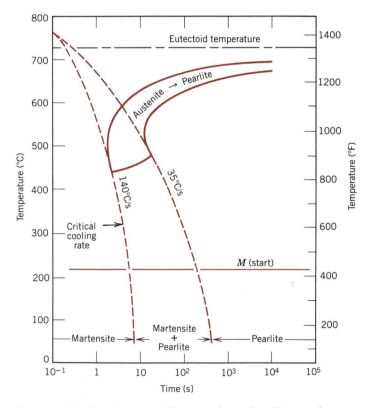

Figure 10.19 Continuous cooling transformation diagram for a eutectoid iron–carbon alloy and superimposed cooling curves, demonstrating the dependence of the final microstructure on the transformations that occur during cooling.

diagram, will just miss the nose at which the pearlite transformation begins, as illustrated in Figure 10.19. As the figure also shows, only martensite will exist for quenching rates greater than the critical; in addition, there will be a range of rates over which both pearlite and martensite are produced. Finally, a totally pearlitic structure develops for low cooling rates.

Carbon and other alloying elements also shift the pearlite (as well as the proeutectoid phase) and bainite noses to longer times, thus decreasing the critical cooling rate. In fact, one of the reasons for alloying steels is to facilitate the formation of martensite so that totally martensitic structures can develop in relatively thick cross sections. Figure 10.20 shows the continuous cooling transformation diagram for the same alloy steel the isothermal transformation diagram for which is presented in Figure 10.15. The presence of the bainite nose accounts for the possibility of formation of bainite for a continuous cooling heat treatment. Several cooling curves superimposed on Figure 10.20 indicate the critical cooling rate, and also how the transformation behavior and final microstructure are influenced by the rate of cooling.

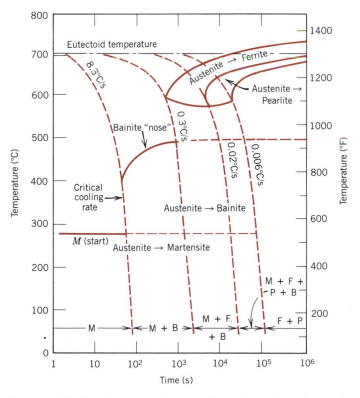

Figure 10.20 Continuous cooling transformation diagram for an alloy steel (type 4340) and several superimposed cooling curves demonstrating dependence of the final microstructure of this alloy on the transformations that occur during cooling. (Adapted from H. E. McGannon, Editor, *The Making, Shaping and Treating of Steel,* 9th edition, United States Steel Corporation, Pittsburgh, 1971, p. 1096.)

Interestingly enough, the critical cooling rate is diminished even by the presence of carbon. In fact, iron–carbon alloys containing less than about 0.25 wt% carbon are not normally heat treated to form martensite because quenching rates too rapid to be practical are required. Other alloying elements that are particularly effective in rendering steels heat treatable are chromium, nickel, molybdenum, manganese, silicon, and tungsten; however, these elements must be in solid solution with the austenite at the time of quenching.

In summary, isothermal and continuous transformation diagrams are, in a sense, phase diagrams in which the parameter of time is introduced. Each is experimentally determined for an alloy of specified composition, the variables being temperature and time. These diagrams allow prediction of the microstructure after some time period for constant temperature and continuous cooling heat treatments, respectively.

10.7 MECHANICAL BEHAVIOR OF IRON–CARBON ALLOYS

We shall now discuss the mechanical behavior of iron–carbon alloys having the microstructures discussed heretofore, namely, fine and coarse pearlite, spheroidite, bainite, and martensite. For all but martensite, two phases are present (i.e., ferrite and cementite); and so an opportunity to explore several mechanical property–microstructure relationships that exist for these alloys is provided.

Pearlite

Cementite is much harder but more brittle than ferrite. Thus increasing the fraction of Fe_3C in a steel alloy while holding other microstructural elements constant will result in a harder and stronger material. This is demonstrated in Figure 10.21a, in which the tensile and yield strengths as well as the Brinell hardness number are plotted as a function of the weight percent carbon (or equivalently as the percent of Fe_3C) for steels that are composed of fine pearlite. All three parameters increase with increasing carbon concentration. Inasmuch as cementite is more brittle, increasing its content will result in a decrease in both ductility and toughness (or impact energy). These effects are shown in Figure 10.21b for the same fine pearlitic steels.

The layer thickness of each of the ferrite and cementite phases in the microstructure also influences the mechanical behavior of the material. Fine pearlite is harder and stronger than coarse pearlite, as demonstrated in Figure 10.22a, which plots hardness versus the carbon concentration.

The reasons for this behavior relate to phenomena that occur at the α–Fe_3C phase boundaries. First, there is a large degree of adherence between the two phases across a boundary. Therefore, the strong and rigid cementite phase severely restricts deformation of the softer ferrite phase in the regions adjacent to the boundary; thus the cementite may be said to reinforce the ferrite. The degree of this reinforcement is substantially higher in fine pearlite because of the greater phase boundary area per unit volume of material. In addition, phase boundaries serve as barriers to dislocation motion in much the same way as grain boundaries (Section 7.8). For fine pearlite there are more boundaries through which a dislocation must pass during plastic deformation. Thus the greater reinforcement and restriction of dislocation motion in fine pearlite account for its greater hardness and strength.

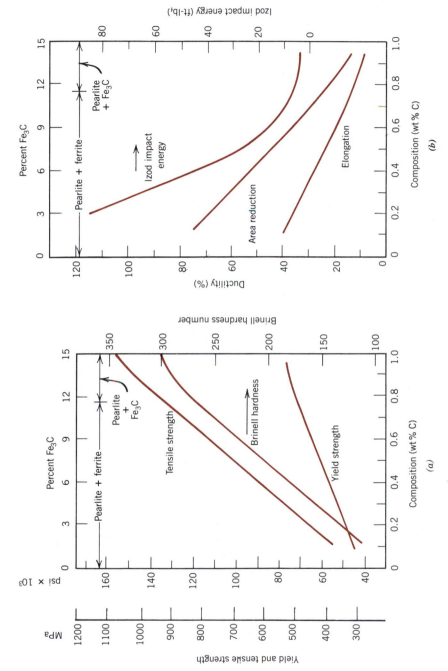

Figure 10.21 (a) Yield strength, tensile strength, and Brinell hardness versus carbon concentration for plain carbon steels having microstructures consisting of fine pearlite. (b) Ductility (%EL and %AR) and Izod impact energy versus carbon concentration for plain carbon steels having microstructures consisting of fine pearlite. (Data taken from *Metals Handbook: Heat Treating*, Vol. 4, 9th edition, V. Masseria, Managing Editor, American Society for Metals, 1981, p. 9.)

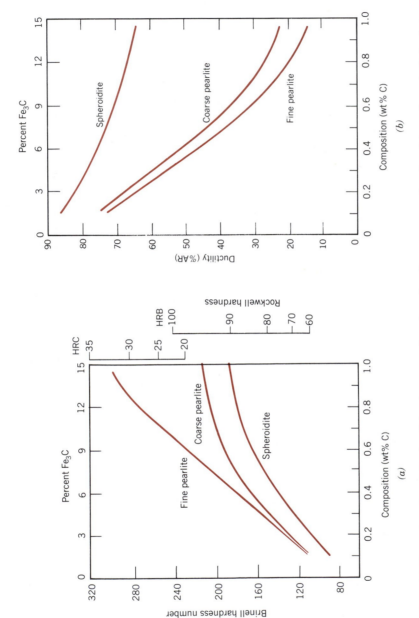

Figure 10.22 (a) Brinell and Rockwell hardness as a function of carbon concentration for plain carbon steels having fine and coarse pearlite as well as spheroidite microstructures. (b) Ductility (%AR) as a function of carbon concentration for plain carbon steels having fine and coarse pearlite as well as spheroidite microstructures. (Data taken from *Metals Handbook: Heat Treating*, Vol. 4, 9th edition, V. Masseria, Managing Editor, American Society for Metals, 1981, pp. 9 and 17.)

Coarse pearlite is more ductile than fine pearlite, as illustrated in Figure 10.22*b*, which plots percent area reduction versus carbon concentration for both microstructure types. This behavior results from the greater restriction to plastic deformation of the fine pearlite.

Spheroidite

Other elements of the microstructure relate to the shape and distribution of the phases. In this respect, the cementite phase has distinctly different shapes and arrangements in the pearlite and spheroidite microstructures (Figures 10.7 and 10.11). Alloys containing pearlitic microstructures have greater strength and hardness than do those with spheroidite. This is demonstrated in Figure 10.22*a*, which compares the hardness as a function of the weight percent carbon for spheroidite with both the other pearlite structure types. This behavior is again explained in terms of reinforcement at, and impedence to, dislocation motion across the ferrite–cementite boundaries as discussed above. There is less boundary area per unit volume in spheroidite, and consequently plastic deformation is not nearly as constrained, which gives rise to a relatively soft and weak material. In fact, of all steel alloys, those that are softest and weakest have a spheroidite microstructure.

As would be expected, spheroidized steels are extremely ductile, much more than either fine or coarse pearlite (Figure 10.22*b*). In addition, they are notably tough because any crack can encounter only a very small fraction of the brittle cementite particles as it propagates through the ductile ferrite matrix.

Bainite

Because bainitic steels have a finer structure (i.e., smaller Fe_3C particles in the ferrite matrix), they are generally stronger and harder than pearlitic ones; yet they exhibit a desirable combination of strength and ductility.

Martensite

Of the various microstructures that may be produced for a given steel alloy, martensite is the hardest and strongest and, in addition, the most brittle; it has, in fact, negligible ductility. Its hardness is dependent on the carbon content, up to about 0.6 wt% as demonstrated in Figure 10.23, which plots the hardness of martensite and fine pearlite as a function of weight percent carbon. In contrast to pearlitic steels, strength and hardness of martensite are not thought to be related to microstructure. Rather, these properties are attributed to the effectiveness of the interstitial carbon atoms in hindering dislocation motion (as a solid-solution effect, Section 7.9), and to the relatively few slip systems (along which dislocations move) for the BCT structure.

Austenite is slightly denser than martensite, and therefore, during the phase transformation upon quenching, there is a net volume increase. Consequently, relatively large pieces that are rapidly quenched may crack as a result of internal stresses; this becomes a problem especially when the carbon content is greater than about 0.5 wt%.

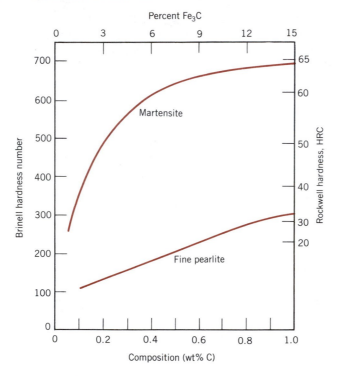

Figure 10.23 Hardness as a function of carbon concentration for plain carbon martensitic and fine pearlitic steels. (Adapted from Dr. Edgar C. Bain, *Functions of the Alloying Elements in Steel,* American Society for Metals, 1939, p. 36.)

10.8 TEMPERED MARTENSITE

In the as-quenched state, martensite, in addition to being very hard, is so brittle that it cannot be used for most applications; also, any internal stresses that may have been introduced during quenching have a weakening effect. The ductility and toughness of martensite may be enhanced and these internal stresses relieved by a heat treatment known as *tempering.*

Tempering is accomplished by heating a martensitic steel to a temperature below the eutectoid for a specified time period. Normally, tempering is carried out at temperatures between 250 and 650°C (480 and 1200°F); internal stresses, however, may be relieved at temperatures as low as 200°C (390°F). This tempering heat treatment allows, by diffusional processes, the formation of **tempered martensite,** according to the reaction

$$\text{martensite (BCT, single phase)} \longrightarrow \text{tempered martensite } (\alpha + \text{Fe}_3\text{C phases})$$

$$(10.4)$$

where the single-phase BCT martensite, which is supersaturated with carbon, transforms to the tempered martensite, composed of the stable ferrite and cementite phases, as indicated on the iron–iron carbide phase diagram.

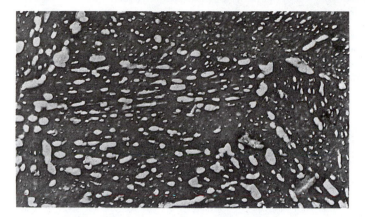

Figure 10.24 Electron micrograph of tempered martensite. Tempering was carried out at 594°C (1100°F). The small particles are the cementite phase; the matrix phase is α-ferrite. 9300×. (Copyright 1971 by United States Steel Corporation.)

The microstructure of tempered martensite consists of extremely small and uniformly dispersed cementite particles embedded within a continuous ferrite matrix. This is similar to the microstructure of spheroidite except that the cementite particles are much, much smaller. An electron micrograph showing the microstructure of tempered martensite at a very high magnification is presented in Figure 10.24.

Tempered martensite may be nearly as hard and strong as martensite, but with substantially enhanced ductility and toughness. The hardness and strength may be explained by the large ferrite–cementite phase boundary area per unit volume that exists for the very fine and numerous cementite particles. Again, the hard cementite phase reinforces the ferrite matrix along the boundaries, and these boundaries also act as barriers to dislocation motion during plastic deformation. The continuous ferrite phase is also very ductile and relatively tough, which accounts for the improvement of these two properties for tempered martensite.

The size of the cementite particles influences the mechanical behavior of tempered martensite; increasing the particle size decreases the ferrite–cementite phase boundary area and, consequently, results in a softer and weaker material yet one that is tougher and more ductile. Furthermore, the tempering heat treatment determines the size of the cementite particles. Heat treatment variables are temperature and time, and most treatments are constant-temperature processes. Since carbon diffusion is involved in the martensite-tempered martensite transformation, increasing the temperature will accelerate diffusion, the rate of cementite particle growth, and, subsequently, the rate of softening. The dependence of tensile and yield strength and ductility on tempering temperature for an alloy steel is shown in Figure 10.25. Before tempering, the material was quenched in oil to produce the martensitic structure; the tempering time at each temperature was 1 h. This type of tempering data is ordinarily provided by the steel manufacturer.

The time dependence of hardness at several different temperatures is presented in Figure 10.26 for a water-quenched steel of eutectoid composition; the time scale

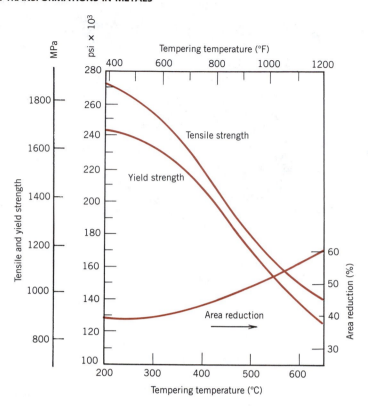

Figure 10.25 Tensile and yield strengths and ductility (%AR) versus tempering temperature for an oil-quenched alloy steel (type 4340). (Adapted from figure furnished courtesy Republic Steel Corporation.)

is logarithmic. With increasing time the hardness decreases, which corresponds to the growth and coalescence of the cementite particles. At temperatures approaching the eutectoid [700°C (1300°F)] and after several hours, the microstructure will have become spheroiditic (Figure 10.11), with large cementite spheroids embedded within the continuous ferrite phase. Correspondingly, overtempered martensite is relatively soft and ductile.

Temper Embrittlement

The tempering of some steels may result in a reduction of toughness as measured by impact tests (Section 8.6); this is termed *temper embrittlement*. The phenomenon occurs when the steel is tempered at a temperature above about 575°C (1070°F) followed by slow cooling to room temperature, or when tempering is carried out at between approximately 375 and 575°C (700 and 1070°F). Steel alloys that are susceptible to temper embrittlement have been found to contain appreciable concentrations of the alloying elements manganese, nickel, or chromium, and, in addition, one or more of antimony, phosphorus, arsenic, and tin as impurities in relatively low concentrations.

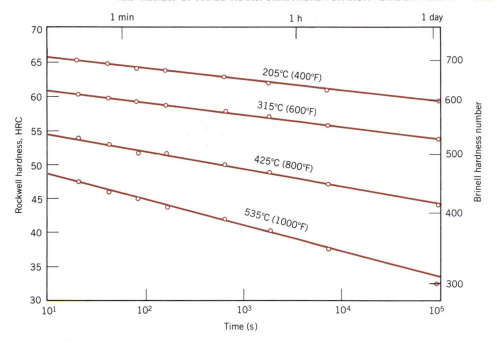

Figure 10.26 Hardness versus tempering time for a water-quenched eutectoid plain carbon (1080) steel. (Adapted from Dr. Edgar C. Bain, *Functions of the Alloying Elements in Steel,* American Society for Metals, 1939, p. 233.)

The presence of these alloying elements and impurities shifts the ductile-to-brittle transition to significantly higher temperatures; the ambient temperature thus lies below this transition in the brittle regime. It has been observed that crack propagation of these embrittled materials is intergranular; that is, the fracture path is along the grain boundaries of the precursor austenite phase. Furthermore, alloy and impurity elements have been found to preferentially segregate in these regions.

Temper embrittlement may be avoided by (1) compositional control; and/or (2) tempering above 575°C or below 375°C, followed by quenching to room temperature. Furthermore, the toughness of steels that have been embrittled may be improved significantly by heating to about 600°C (1100°F) and then rapidly cooling to below 300°C (570°F).

10.9 REVIEW OF PHASE TRANSFORMATIONS FOR IRON–CARBON ALLOYS

In this chapter several different microstructures which may be produced in iron–carbon alloys depending on heat treatment have been discussed. Figure 10.27 summarizes the transformation paths that produce these various microstructures. Here, it is assumed that pearlite, bainite, and martensite result from continuous cooling treatments; furthermore, the formation of bainite is only possible for alloy steels (not plain carbon ones) as outlined above.

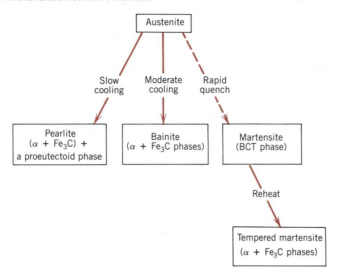

Figure 10.27 Possible transformations involving the decomposition of austenite. Solid arrows, transformations involving diffusion; dashed arrow, diffusionless transformation.

SUMMARY

The themes of this chapter have been phase transformations in metals—modifications in the phase structure or microstructure—and how they affect mechanical properties. Some transformations involve diffusional phenomena, which means that their progress is time dependent. For these, some of the basic kinetic concepts were explored, including the relation between degree of reaction completion and time, the notion of transformation rate, and how rate depends on temperature.

As a practical matter, phase diagrams are severely restricted relative to transformations in multiphase alloys, because they provide no information as to phase transformation rates. The element of time is incorporated into both isothermal transformation and continuous cooling transformation diagrams; transformation progress as a function of temperature and elapsed time is expressed for a specific alloy at constant temperature and for continuous cooling treatments, respectively. Diagrams of both types were presented for iron–carbon steel alloys, and their utility with regard to the prediction of microstructural products was discussed.

Several microconstituents are possible for steels, the formation of which depends on composition and heat treatment. These microconstituents include fine and coarse pearlite, and bainite, which are composed of ferrite and cementite phases and result from the decomposition of austenite via diffusional processes. A spheroidite microstructure (also consisting of ferrite and cementite phases) may be produced when a steel specimen composed of any of the preceding microstructures is heat treated at a temperature just below the eutectoid. The mechanical characteristics of pearlitic, bainitic, and spheroiditic steels were compared and also explained in terms of their microconstituents.

Martensite, yet another transformation product in steels, results when austenite is cooled very rapidly. It is a metastable and single-phase structure that may be produced in steels by a diffusionless and almost instantaneous transformation of austenite. Transformation progress is dependent on temperature rather than time, and may be represented on both isothermal and continuous cooling transformation diagrams. Furthermore, alloying element additions retard the formation rate of pearlite and bainite, thus rendering the martensitic transformation more competitive. Mechanically, martensite is extremely hard; applicability, however, is limited by its brittleness. A tempering heat treatment increases the ductility at some sacrifice of strength and hardness. During tempering, martensite transforms to tempered martensite, which consists of the equilibrium ferrite and cementite phases. Embrittlement of some steel alloys results when specific alloying and impurity elements are present, and upon tempering within a definite temperature range.

IMPORTANT TERMS AND CONCEPTS

Alloy steel

Athermal
 transformation

Bainite

Coarse pearlite

Continuous cooling
 transformation
 diagram

Fine pearlite

Isothermal
 transformation
 diagram

Kinetics

Martensite

Nucleation

Phase transformation

Plain carbon steel

Spheroidite

Supercooling

Superheating

Tempered martensite

Thermally activated
 transformation

Transformation rate

REFERENCES

ATKINS, M., *Atlas of Continuous Cooling Transformation Diagrams for Engineering Steels,* British Steel Corporation, Sheffield, England, 1980.

Atlas of Isothermal Transformation and Cooling Transformation Diagrams, American Society for Metals, Metals Park, OH, 1977.

BROPHY, J. H., R. M. ROSE, and J. WULFF, *The Structure and Properties of Materials,* Vol. II, *Thermodynamics of Structure,* John Wiley & Sons, New York, 1964.

PECKNER, D. (Editor), *The Strengthening of Metals,* Reinhold Publishing Corp., New York, 1964.

PORTER, D. A. and K. E. EASTERLING, *Phase Transformations in Metals and Alloys,* Van Nostrand Reinhold (International) Co. Ltd., Workingham, Berkshire, England, 1981.

SHEWMON, P. G., *Transformations in Metals,* McGraw-Hill Book Co., New York, 1969.

QUESTIONS AND PROBLEMS

10.1 Name the two stages involved in the formation of particles of a new phase. Briefly describe each.

10.2 For some transformation having kinetics that obey the Avrami equation (Equation 10.1), the parameter n is known to have a value of 1.7. If, after 100 s, the reaction is 50% complete, how long (total time) will it take the transformation to go to 99% completion?

10.3 Compute the rate of some reaction that obeys Avrami kinetics, assuming that the constants n and k have values of 2.0 and 5×10^{-4}, respectively, for time expressed in seconds.

10.4 It is known that the kinetics of recrystallization for some alloy obey the Avrami equation and that the value of n in the exponential is 5.0. If, at some temperature, the fraction recrystallized is 0.30 after 100 min, determine the rate of recrystallization at this temperature.

10.5 The kinetics of the austenite-to-pearlite transformation obey the Avrami relationship. Using the fraction transformed–time data given below, determine the total time required for 95% of the austenite to transform to pearlite:

Fraction Transformed	Time (s)
0.2	280
0.6	425

10.6 Below, the fraction recrystallized–time data for the recrystallization at 350°C of a previously deformed aluminum are tabulated. Assuming that the kinetics of this process obey the Avrami relationship, determine the fraction recrystallized after a total time of 116.8 min.

Fraction Recrystallized	Time (min)
0.30	95.2
0.80	126.6

10.7 **(a)** From the curves shown in Figure 10.2 and using Equation 10.2, determine the rate of recrystallization for pure copper at the several temperatures. **(b)** Make a plot of ln (rate) versus the reciprocal of temperature (in K^{-1}), and determine the activation energy for this recrystallization process. (See Section 5.5.) **(c)** By extrapolation, estimate the length of time required for 50% recrystallization at room temperature, 20°C (293 K).

10.8 In terms of heat treatment and the development of microstructure, what are two major limitations of the iron–iron carbide phase diagram?

10.9 **(a)** Briefly describe the phenomena of superheating and supercooling. **(b)** Why do they occur?

10.10 Suppose that a steel of eutectoid composition is cooled to 550°C (1020°F) from 760°C (1400°F) in less than 0.5 s and held at this temperature. **(a)** How long will it take for the austenite-to-pearlite reaction to go to 50% completion? To 100% completion? **(b)** Estimate the hardness of the alloy that has completely transformed to pearlite.

10.11 Briefly explain why the reaction rate for the austenite-to-pearlite transformation, as determined from Figure 10.6 and utilizing Equation 10.2, decreases with increasing temperature, in apparent contradiction with Equation 10.3.

10.12 Briefly cite the differences between pearlite, bainite, and spheroidite relative to microstructure and mechanical properties.

10.13 What is the driving force for the formation of spheroidite?

10.14 Using the isothermal transformation diagram for an iron–carbon alloy of eutectoid composition (Figure 10.14), specify the nature of the final microstructure (in terms of microconstituents present and approximate percentages of each) of a small specimen that has been subjected to the following time–temperature treatments. In each case assume that the specimen begins at 760°C (1400°F) and that it has been held at this temperature long enough to have achieved a complete and homogeneous austenitic structure.

(a) Cool rapidly to 700°C (1290°F), hold for 10^4 s, then quench to room temperature.

(b) Reheat the specimen in part a to 700°C (1290°F) for 20 h.

(c) Rapidly cool to 600°C (1110°F), hold for 4 s, rapidly cool to 450°C (840°F), hold for 10 s, then quench to room temperature.

(d) Cool rapidly to 400°C (750°F), hold for 2 s, then quench to room temperature.

(e) Cool rapidly to 400°C (750°F), hold for 20 s, then quench to room temperature.

(f) Cool rapidly to 400°C (750°F), hold for 200 s, then quench to room temperature.

(g) Rapidly cool to 575°C (1065°F), hold for 20 s, rapidly cool to 350°C (660°F), hold for 100 s, then quench to room temperature.

(h) Rapidly cool to 250°C (480°F), hold for 100 s, then quench to room temperature in water. Reheat to 315°C (600°F) for 1 h and slowly cool to room temperature.

10.15 Make a copy of the isothermal transformation diagram for an iron–carbon alloy of eutectoid composition (Figure 10.14) and then sketch and label on this diagram time–temperature paths to produce the following microstructures:

(a) 100% coarse pearlite.

(b) 50% martensite and 50% austenite.

(c) 50% coarse pearlite, 25% bainite, and 25% martensite.

10.16 Using the isothermal transformation diagram for a 1.13 wt% C steel alloy (Figure 10.28), determine the final microstructure (in terms of just the microconstituents present) of a small specimen that has been subjected to the following time–temperature treatments. In each case assume that the specimen begins at 920°C (1690°F), and that it has been held at this temperature long enough to have achieved a complete and homogeneous austenitic structure.

(a) Rapidly cool to 750°C (1380°F), hold for 100 s, then quench to room temperature.

(b) Rapidly cool to 650°C (1200°F), hold for 3 s, then quench to room temperature.

(c) Rapidly cool to 550°C (1020°F), hold for 10 s, then quench to room temperature.

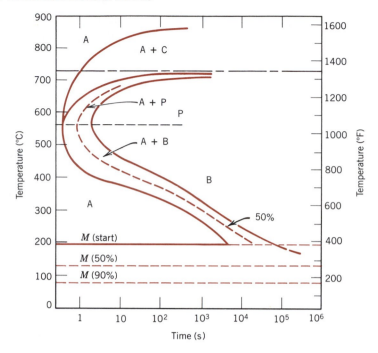

Figure 10.28 Isothermal transformation diagram for a 1.13 wt% C iron–carbon alloy: A, austenite; B, bainite; C, proeutectoid cementite; M, martensite; P, pearlite. (Adapted from H. Boyer, Editor, *Atlas of Isothermal Transformation and Cooling Transformation Diagrams,* American Society for Metals, 1977, p. 33.)

(d) Rapidly cool to 700°C (1290°F), hold at this temperature for 10^5 s, then quench to room temperature.

(e) Rapidly cool to 650°C (1200°F), hold at this temperature for 3 s, rapidly cool to 400°C (750°F), hold for 25 s, then quench to room temperature.

(f) Rapidly cool to 300°C (570°F), hold for 100 s, then quench to room temperature.

(g) Rapidly cool to 300°C (570°F), hold for 1000 s, then quench to room temperature.

(h) Rapidly cool to 600°C (1110°F), hold at this temperature for 7 s, rapidly cool to 450°C (840°F), hold at this temperature for 4 s, then quench to room temperature.

10.17 For parts c, d, f, g, and h of Problem 10.16, determine the approximate percentages of the microconstituents that form.

10.18 Make a copy of the isothermal transformation diagram for a 1.13 wt% C iron–carbon alloy (Figure 10.28), and then sketch and label on this diagram the time–temperature paths to produce the following microstructures:

(a) 6.1% proeutectoid cementite and 93.9% coarse pearlite.

(b) 50% fine pearlite and 50% bainite.

(c) 100% martensite.

(d) 100% tempered martensite.

10.19 Name the microstructural products of eutectoid iron–carbon alloy (0.77 wt% C) specimens that are first completely transformed to austenite, then cooled to room temperature at the following rates: **(a)** 200°C/s, **(b)** 100°C/s, and **(c)** 20°C/s.

10.20 Figure 10.29 shows the continuous cooling transformation diagram for a 0.35 wt% C iron–carbon alloy. Make a copy of this figure and then sketch and label continuous cooling curves to yield the following microstructures:
(a) Fine pearlite and proeutectoid ferrite.
(b) Martensite.
(c) Martensite and proeutectoid ferrite.
(d) Coarse pearlite and proeutectoid ferrite.
(e) Martensite, fine pearlite, and proeutectoid ferrite.

10.21 Cite two major differences between martensitic and pearlitic transformations.

10.22 Cite two important differences between continuous cooling transformation diagrams for plain carbon and alloy steels.

10.23 Briefly explain why there is no bainite transformation region on the continuous cooling transformation diagram for an iron–carbon alloy of eutectoid composition.

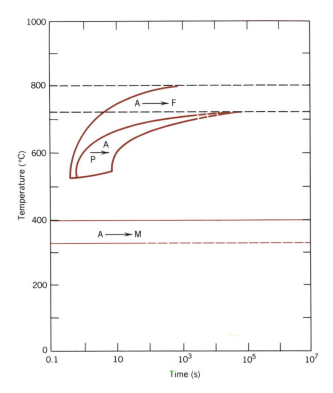

Figure 10.29 Continuous cooling transformation diagram for a 0.35 wt% C iron–carbon alloy.

10.24 Name the microstructural products of 4340 alloy steel specimens that are first completely transformed to austenite, then cooled to room temperature at the following rates: **(a)** 10°C/s, **(b)** 1°C/s, **(c)** 0.1°C/s, and **(d)** 0.01°C/s.

10.25 Briefly describe the simplest continuous cooling heat treatment procedure that would be used in converting a 4340 steel from one microstructure to another.
(a) (Martensite + ferrite + bainite) to (martensite + ferrite + pearlite + bainite).
(b) (Martensite + ferrite + bainite) to spheroidite.
(c) (Martensite + bainite + ferrite) to tempered martensite.

10.26 On the basis of diffusion considerations, explain why fine pearlite forms for the moderate cooling of austenite through the eutectoid temperature, whereas coarse pearlite is the product for relatively slow cooling rates.

10.27 **(a)** Which is the more stable, the pearlitic or the spheroiditic microstructure? **(b)** Why?

10.28 Briefly explain why fine pearlite is harder and stronger than coarse pearlite, which in turn is harder and stronger than spheroidite.

10.29 Cite two reasons why martensite is so hard and brittle.

10.30 Rank the following iron–carbon alloys and associated microstructures from the highest to the lowest tensile strength: **(a)** 0.25 wt% C with spheroidite, **(b)** 0.25 wt% C with coarse pearlite, **(c)** 0.6 wt% C with fine pearlite, and **(d)** 0.6 wt% C with coarse pearlite. Justify this ranking.

10.31 Briefly explain why the hardness of tempered martensite diminishes with tempering time (at constant temperature) and with increasing temperature (at constant tempering time).

10.32 Briefly describe the simplest heat treatment procedure that would be used in converting a 0.77 wt% C steel from one microstructure to the other, as follows:
(a) Martensite to spheroidite.
(b) Spheroidite to martensite.
(c) Bainite to pearlite.
(d) Pearlite to bainite.
(e) Spheroidite to pearlite.
(f) Pearlite to spheroidite.
(g) Tempered martensite to martensite.
(h) Bainite to spheroidite.

10.33 **(a)** Briefly describe the microstructural difference between spheroidite and tempered martensite. **(b)** Explain why tempered martensite is much harder and stronger.

10.34 Estimate the Rockwell hardnesses for specimens of an iron–carbon alloy of eutectoid composition that have been subjected to the heat treatments described in parts b, d, g, and h of Problem 10.14.

10.35 Estimate the Brinell hardnesses for specimens of a 1.13 wt% C iron–carbon alloy that have been subjected to the heat treatments described in parts d, f, and h of Problem 10.16.

10.36 Determine the approximate tensile strengths for specimens of a eutectoid iron–carbon alloy that have experienced the heat treatments described in parts a and c of Problem 10.19.

10.37 Is it possible to produce an iron–carbon alloy of eutectoid composition that has a minimum hardness of 200 HB and a minimum ductility of 25%AR? If so, describe the continuous cooling heat treatment to which the alloy would be subjected to achieve these properties. If it is not possible, explain why.

10.38 Is it possible to produce an iron–carbon alloy that has a minimum tensile strength of 90,000 psi (620 MPa) and a minimum ductility of 50%AR? If so, what will be its composition and microstructure (coarse and fine pearlites and spheroidite are alternatives)? If this is not possible explain why.

10.39 It is desired to produce an iron–carbon alloy that has a minimum hardness of 200 HB and a minimum ductility of 35%AR. Is such an alloy possible? If so, what will be its composition and microstructure (coarse and fine pearlites and spheroidite are alternatives)? If this is not possible, explain why.

10.40 For a eutectoid steel, describe isothermal heat treatments that would be required to yield specimens having the following Brinell hardnesses: **(a)** 180 HB, **(b)** 220 HB, and **(c)** 500 HB.

10.41 **(a)** For a 1080 steel that has been water quenched, estimate the tempering time at 425°C (800°F) to achieve a hardness of 50 HRC. **(b)** What will be the tempering time at 315°C (600°F) necessary to attain the same hardness?

10.42 An alloy steel (4340) is to be used in an application requiring a minimum tensile strength of 200,000 psi (1380 MPa) and a minimum ductility of 43%AR. Oil quenching followed by tempering is to be used. Briefly describe the tempering heat treatment.

10.43 Is it possible to produce an oil-quenched and tempered 4340 steel that has a minimum yield strength of 180,000 psi (1240 MPa) and a ductility of at least 50%AR? If this is possible, describe the tempering heat treatment. If it is not possible, explain why.

CHAPTER 11

THERMAL PROCESSING
OF METAL ALLOYS

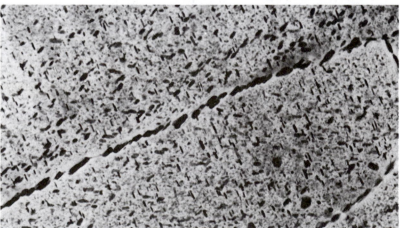

Top: A Boeing 767 airplane in flight. (Photograph courtesy of the Boeing Commercial Airplane Company.) Bottom: A transmission electron micrograph showing the microstructure of the aluminum alloy that is used for the upper wing skins, parts of the internal wing structures, and selected areas of the fuselage of the Boeing 767 above. This is a 7150-T651 alloy (6.2Zn, 2.3Cu, 2.3Mg, 0.12Zr, the balance Al) that has been precipitation hardened. The light matrix phase in the micrograph is an aluminum solid solution. The majority of the small plate-shaped dark precipitate particles are a transition η' phase, the remainder being the equilibrium η ($MgZn_2$) phase. Note that grain boundaries are "decorated" by some of these particles. $80,475 \times$. (Electron micrograph courtesy of G. H. Narayanan and A. G. Miller, Boeing Commercial Airplane Company.)

11.1 INTRODUCTION

Earlier chapters have discussed a number of phenomena that occur in metals and alloys at elevated temperatures, for example, recrystallization and the decomposition of austenite. These are effective in altering the mechanical characteristics when appropriate heat treatments or thermal processes are employed. In fact, the use of heat treatments on commercial alloys is an exceedingly common practice. Therefore, we consider the details of some of these processes, including annealing procedures, the heat treating of steels, and precipitation hardening.

ANNEALING PROCESSES

The term **annealing** refers to a heat treatment in which a material is exposed to an elevated temperature for an extended time period and then slowly cooled. Ordinarily, annealing is carried out to (1) relieve stresses; (2) increase softness, ductility, and toughness; and/or (3) produce a specific microstructure. A variety of annealing heat treatments are possible; they are characterized by the changes that are induced, which many times are microstructural and are responsible for the alteration of the mechanical properties.

Any annealing process consists of three stages: (1) heating to the desired temperature, (2) holding or "soaking" at that temperature, and (3) cooling, usually to room temperature. Time is an important parameter in these procedures. During heating and cooling, there exist temperature gradients between the outside and interior portions of the piece; their magnitudes depend on the size and geometry of the piece. If the rate of temperature change is too great, temperature gradients and internal stresses may be induced that may lead to warping or even cracking. Also, the actual annealing time must be long enough to allow for any necessary transformation reactions. Annealing temperature is also an important consideration; annealing may be accelerated by increasing the temperature, since diffusional processes are normally involved.

11.2 PROCESS ANNEALING

Process annealing is a heat treatment that is used to negate the effects of cold work, that is, to soften and increase the ductility of a previously strain-hardened metal. It is commonly utilized during fabrication procedures that require extensive plastic deformation, to allow a continuation of deformation without fracture or excessive energy consumption. Recovery and recrystallization processes are allowed to occur. Ordinarily a fine-grained microstructure is desired, and therefore, the heat treatment is terminated before appreciable grain growth has occurred. Surface oxidation or scaling may be prevented or minimized by annealing at a relatively low temperature (but above the recrystallization temperature), or in a nonoxidizing atmosphere.

11.3 STRESS RELIEF

Internal residual stresses may develop in metal pieces in response to the following: (1) plastic deformation processes such as machining and grinding; (2) nonuniform cooling of a piece that was processed or fabricated at an elevated temperature, such as a weld or a casting; and **(3)** a phase transformation that is induced upon cooling wherein parent and product phases have different densities. Distortion and warpage may result if these residual stresses are not removed. They may be eliminated by a **stress relief** annealing heat treatment in which the piece is heated to the recommended temperature, held there long enough to attain a uniform temperature, and finally cooled to room temperature in air. The annealing temperature is ordinarily a relatively low one such that effects resulting from cold working and other heat treatments are not affected.

11.4 ANNEALING OF FERROUS ALLOYS

Several different annealing procedures are employed to enhance the properties of steel alloys. However, before they are discussed, some comment relative to the labeling of phase boundaries is necessary. Figure 11.1 shows the portion of the iron–iron carbide phase diagram in the vicinity of the eutectoid. The horizontal line at the eutectoid temperature, conventionally labeled A_1, is termed the **lower critical temperature,** below which, under equilibrium conditions, all austenite will have transformed into ferrite and cementite phases. The phase boundaries denoted as A_3 and A_{cm} represent the **upper critical temperature** lines, for hypoeutectoid and hypereutectoid steels, respectively. For temperatures and compositions above these boundaries, only the austenite phase will prevail. As explained in Section 9.15, other alloying elements will shift the eutectoid and the positions of these phase boundary lines.

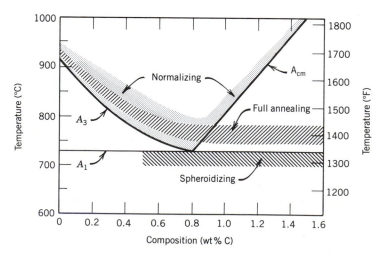

Figure 11.1 The iron–iron carbide phase diagram in the vicinity of the eutectoid, indicating heat treating temperature ranges for plain carbon steels. (Adapted from *Metals Handbook,* T. Lyman, Editor, American Society for Metals, 1948, p. 661.)

Normalizing

Steels that have been plastically deformed by, for example, a rolling operation, consist of grains of pearlite (and most likely a proeutectoid phase), which are irregularly shaped and relatively large, but varying substantially in size. An annealing heat treatment called **normalizing** is used to refine the grains (i.e., to decrease the average grain size) and produce a more uniform and desirable size distribution; fine-grained pearlitic steels are tougher than coarse-grained ones. Normalizing is accomplished by heating at approximately 55 to 85°C (100 to 150°F) above the upper critical temperature, which is, of course, dependent on composition, as indicated in Figure 11.1. After sufficient time has been allowed for the alloy to completely transform to austenite—a procedure termed **austenitizing**—the treatment is terminated by cooling in air. A normalizing cooling curve is superimposed on the continuous cooling transformation diagram (Figure 10.18).

Full Anneal

A heat treatment known as a **full anneal** is often utilized in low- and medium-carbon steels that will be machined or will experience extensive plastic deformation during a forming operation. The alloy is austenitized by heating to 15 to 40°C (30 to 70°F) above the A_3 or A_1 lines as indicated in Figure 11.1 until equilibrium is achieved. The alloy is then furnace cooled; that is, the heat-treating furnace is turned off and both furnace and steel cool to room temperature at the same rate, which takes several hours. The microstructural product of this anneal is coarse pearlite (in addition to any proeutectoid phase) that is relatively soft and ductile. The full-anneal cooling procedure (also shown in Figure 10.18) is time consuming; however, a microstructure having small grains and a uniform grain structure results.

Spheroidizing

Medium- and high-carbon steels having a microstructure containing even coarse pearlite may still be too hard to conveniently machine or plastically deform. These steels, and in fact any steel, may be heat treated or annealed to develop the spheroidite structure as described in Section 10.5. Spheroidized steels have a maximum softness and ductility and are easily machined or deformed. The **spheroidizing** heat treatment consists of heating the alloy at a temperature just below the eutectoid [line A_1 in Figure 11.1, or at about 700°C (1300°F)] in the $\alpha + Fe_3C$ region of the phase diagram. If the precursor microstructure contains pearlite, spheroidizing times will ordinarily range between 15 and 25 h. During this annealing there is a coalescence of the Fe_3C to form the spheroid particles (see page 301).

Still other annealing treatments are possible. For example, glasses are annealed, as outlined in Section 14.4, to remove residual internal stresses that render the material excessively weak. In addition, microstructural alterations and the attendant modification of mechanical properties of cast irons, as discussed in Section 12.6, result from what are in a sense annealing treatments.

HEAT TREATMENT OF STEELS

Conventional heat treatment procedures for producing martensitic steels ordinarily involve continuous and rapid cooling of an austenitized specimen in some type of quenching medium, such as water, oil, or air. The optimum properties of a steel that has been quenched and then tempered can be realized only if, during the quenching heat treatment, the specimen has been converted to a high content of martensite; the formation of any pearlite and/or bainite will result in other than the best combination of mechanical characteristics. During the quenching treatment, it is impossible to cool the specimen at a uniform rate throughout—the surface will always cool more rapidly than interior regions. Therefore, the austenite will transform over a range of temperatures, yielding a possible variation of microstructure and properties with position within a specimen.

The successful heat treating of steels to produce a predominantly martensitic microstructure throughout the cross section depends mainly on three factors: (1) the composition of the alloy, (2) the type and character of the quenching medium, and (3) the size and shape of the specimen. The influence of each of these factors is now addressed.

11.5 HARDENABILITY

The influence of alloy composition on the ability of a steel alloy to transform to martensite for a particular quenching treatment is related to a parameter called **hardenability.** For every different steel alloy there is a specific relationship between the mechanical properties and the cooling rate. "Hardenability" is a term that is used to describe the ability of an alloy to be hardened by the formation of martensite as a result of a given heat treatment. Hardenability is not "hardness"; rather, hardness measurements are utilized to determine the extent of a martensitic transformation in the interior of a specimen. A steel alloy that has a high hardenability is one that hardens, or forms martensite, not only at the surface but to a large degree throughout the entire interior; in other words, hardenability is a measure of the depth to which a specific alloy may be hardened.

The Jominy End-Quench Test

One standard procedure that is widely utilized to determine hardenability is the **Jominy end-quench test.** With this procedure, except for alloy composition, all factors that may influence the depth to which a piece hardens (i.e., specimen size and shape, and quenching treatment) are maintained constant. A cylindrical specimen 1 in. (25 mm) in diameter and 4 in. (100 mm) long is austenitized at a prescribed temperature for a prescribed time. After removal from the furnace, it is quickly mounted in a fixture as diagrammed in Figure 11.2a. The lower end is quenched by a jet of water of specified flow rate and temperature. Thus the cooling rate is a maximum at the quenched end and diminishes with position from this point along the length of the specimen. After the piece has cooled to room temperature, shallow flats 0.015 in. (0.4 mm) deep are ground along the specimen length and Rockwell hardness measurements are made for the first 2 in. (50 mm) along each flat (Figure 11.2b); for the

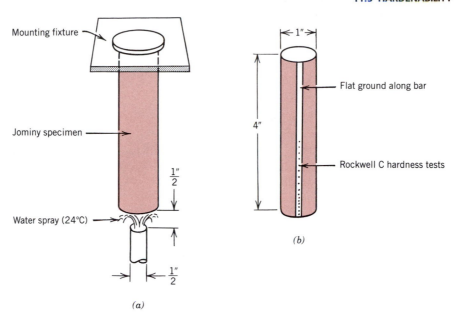

Figure 11.2 Schematic diagram of Jominy end-quench specimen (*a*) mounted during quenching and (*b*) after hardness testing from the quenched end along a ground flat. (Adapted from A. G. Guy, *Essentials of Materials Science*. Copyright 1978 by McGraw-Hill Book Company, New York.)

first $\frac{1}{2}$ in., hardness readings are taken at $\frac{1}{16}$-in. (1.6 mm) intervals, and for the remaining $1\frac{1}{2}$ in., every $\frac{1}{8}$ in. (3.2 mm). A hardenability curve is produced when hardness is plotted as a function of position from the quenched end.

Hardenability Curves

A typical hardenability curve is represented in Figure 11.3. The quenched end is cooled most rapidly and exhibits the maximum hardness; 100% martensite is the product at this position for most steels. Cooling rate decreases with distance from the quenched end, and the hardness also decreases, as indicated in the figure. With diminishing cooling rate more time is allowed for carbon diffusion and the formation of a greater proportion of the softer pearlite, which may be mixed with martensite and bainite.

Hardenability is a qualitative measure of the rate at which the hardness drops off with distance from the quenched end. A steel that is highly hardenable will retain large hardness values for relatively long distances; a low hardenable one will not. Also, each steel alloy has its own unique hardenability curve.

Sometimes, it is convenient to relate hardness to a cooling rate rather than to the location from the quenched end of a standard Jominy specimen. Cooling rate [taken at 700°C (1300°F)] is ordinarily shown on the upper horizontal axis of a hardenability diagram; this scale is included with the hardenability plots presented here. This correlation between position and cooling rate is the same for plain carbon and many alloy steels because the rate of heat transfer is nearly independent of

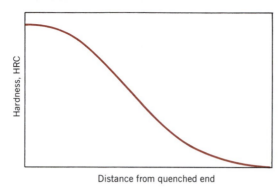

Figure 11.3 Typical hardenability plot of Rockwell C hardness as a function of distance from the quenched end.

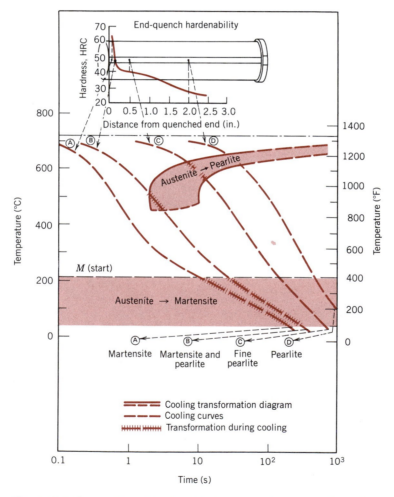

Figure 11.4 Correlation of hardenability and continuous cooling information for an iron–carbon alloy of eutectoid composition. (Adapted from H. Boyer, Editor, *Atlas of Isothermal Transformation and Cooling Transformation Diagrams,* American Society for Metals, 1977, p. 376.)

composition. On occasion, cooling rate or position from the quenched end is specified in terms of Jominy distance, one Jominy distance unit being $\frac{1}{16}$ in. (1.6 mm).

A correlation may be drawn between position along the Jominy specimen and continuous cooling transformations. For example, Figure 11.4 is a continuous transformation diagram for a eutectoid iron–carbon alloy onto which is superimposed the cooling curves at four different Jominy positions, and corresponding microstructures that result for each. The hardenability curve for this alloy is also included.

The hardenability curves for five different steel alloys all having 0.40 wt% C, yet differing amounts of other alloying elements, are shown in Figure 11.5. One specimen is a plain carbon steel (1040); the other four (4140, 4340, 5140, and 8640) are alloy steels. The compositions of the four alloy steels are included with the figure. The significance of the alloy designation numbers (e.g., 1040) is explained in Chapter 12. Several details are worth noting from this figure. First, all five alloys have identical hardnesses at the quenched end (57 HRC); this hardness is a function of carbon content only, which is the same for all these alloys.

Probably the most significant feature of these curves is shape, which relates to hardenability. The hardenability of the plain carbon 1040 steel is low because the hardness drops off precipitously (to about 30 HRC) after a relatively short Jominy distance ($\frac{1}{4}$th in., 6.4 mm). By way of contrast, the decreases in hardness for the other four alloy steels are distinctly more gradual. For example, at a Jominy distance of 2 in. (50 mm), the hardnesses of the 4340 and 8640 alloys are approximately 50 and

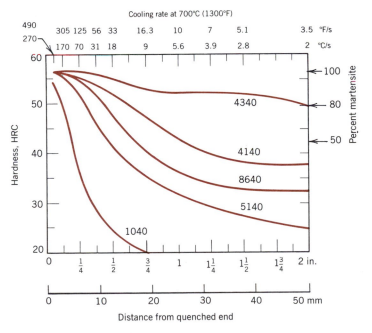

Figure 11.5 Hardenability curves for five different steel alloys, each containing 0.4 wt% C. Approximate alloy compositions (wt%) are as follows: 4340–1.85 Ni, 0.80 Cr, and 0.25 Mo; 4140–1.0 Cr and 0.20 Mo; 8640–0.55 Ni, 0.50 Cr, and 0.20 Mo; 5140–0.85 Cr; 1040 is an unalloyed steel. (Adapted from figure furnished courtesy Republic Steel Corporation.)

32 HRC, respectively; thus of these two alloys, the 4340 is more hardenable. A water-quenched specimen of the 1040 plain carbon steel would harden only to a shallow depth below the surface, whereas for the other four alloy steels the high quenched hardness would persist to a much greater depth.

The hardness profiles in Figure 11.5 are indicative of the influence of cooling rate on the microstructure. At the quenched end, where the quenching rate is approximately 600°C/s (1100°F/s), 100% martensite is present for all five alloys. For cooling rates less than about 70°C/s (125°F/s) or Jominy distances greater than about $\frac{1}{4}$ in. (6.4 mm), the microstructure of the 1040 steel is predominantly pearlitic, with some proeutectoid ferrite. However, the microstructures of the four alloy steels consists primarily of a mixture of martensite and bainite; bainite content increases with decreasing cooling rate.

This disparity in hardenability behavior for the five alloys in Figure 11.5 is explained by the presence of nickel, chromium, and molybdenum in the alloy steels. These alloying elements delay the austenite-to-pearlite and/or bainite reactions, as explained above; this permits more martensite to form for a particular cooling rate, yielding a greater hardness. The right-hand axis of Figure 11.5 shows the approximate fraction of martensite that is present at various hardnesses for these alloys.

The hardenability curves also depend on carbon content. This effect is demonstrated in Figure 11.6 for a series of alloy steels in which only the concentration of carbon is varied. The hardness at any Jominy position increases with the concentration of carbon.

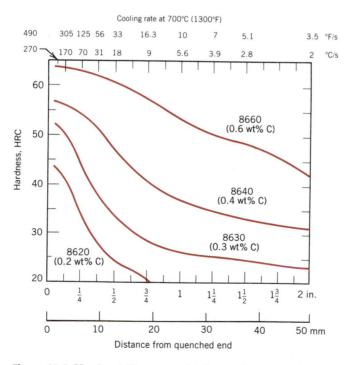

Figure 11.6 Hardenability curves for four 8600 series alloys of indicated carbon content. (Adapted from figure furnished courtesy Republic Steel Corporation.)

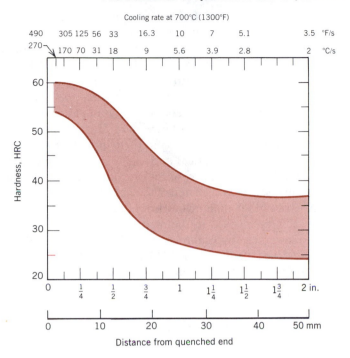

Figure 11.7 The hardenability band for an 8640 steel indicating maximum and minimum limits. (Adapted from figure furnished courtesy Republic Steel Corporation.)

Also, during the industrial production of steel, there is always a slight, unavoidable variation in composition and average grain size from one batch to another. This results in some scatter in measured hardenability data, which frequently are plotted as a band representing the maximum and minimum values that would be expected for the particular alloy. Such a hardenability band is plotted in Figure 11.7 for an 8640 steel. An H following the designation specification for an alloy (e.g., 8640H) indicates that the composition and characteristics of the alloy are such that its hardenability curve will lie within a specified band.

11.6 INFLUENCE OF QUENCHING MEDIUM, SPECIMEN SIZE, AND GEOMETRY

The preceding treatment of hardenability discussed the influence of both alloy composition and cooling or quenching rate on the hardness. The cooling rate of a specimen depends on the rate of heat energy extraction, which is a function of the characteristics of the quenching medium in contact with the specimen surface, as well as the specimen size and geometry.

"Severity of quench" is a term often used to indicate the rate of cooling; the more rapid the quench, the more severe the quench. Of the three most common quenching media—water, oil, and air—water produces the most severe quench, followed by oil, which is more effective than air. The degree of agitation of each medium also influences the rate of heat removal. Increasing the velocity of the quenching medium across the specimen surface enhances the quenching effectiveness. Oil quenches are suitable for the heat treating of many alloy steels. In fact, for higher-carbon steels, a water quench is too severe because cracking and warping may be produced. Air cooling of austenitized plain carbon steels ordinarily produces an almost totally pearlitic structure.

During the quenching of a steel specimen, heat energy must be transported to the surface before it can be dissipated into the quenching medium. As a consequence, the cooling rate within and throughout the interior of a steel structure varies with position and depends on the geometry and size. Figures 11.8*a* and 11.8*b* show the quenching rate [at 700°C (1300°F)] as a function of diameter for cylindrical bars at four radial positions (surface, three-quarters radius, midradius, and center). Quenching is in mildly agitated water (Figure 11.8*a*) and oil (Figure 11.8*b*); cooling rate is also expressed as equivalent Jominy distance, since these data are often used in conjunction with hardenability curves. Diagrams similar to those in Figure 11.8 have also been generated for geometries other than cylindrical (e.g., flat plates).

One utility of such diagrams is in the prediction of the hardness traverse along the cross section of a specimen. For example, Figure 11.9*a* compares the radial hardness distributions for cylindrical plain carbon (1040) and alloy (4140) steel specimens; both have a diameter of 2 in. (50 mm) and are water quenched. The difference in hard-

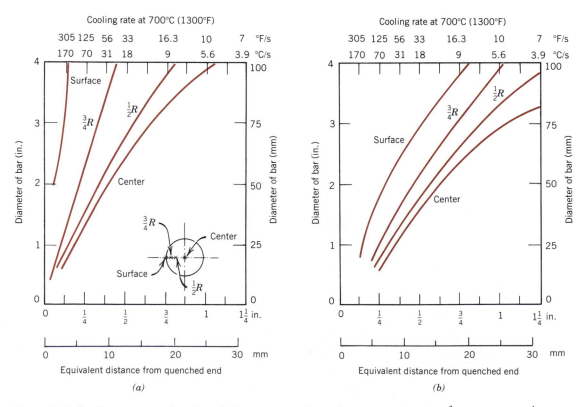

Figure 11.8 Cooling rate as a function of diameter at surface, three-quarters radius ($\frac{3}{4}R$), midradius ($\frac{1}{2}R$), and center positions for cylindrical bars quenched in mildly agitated (*a*) water and (*b*) oil. Equivalent Jominy positions are included along the bottom axis. (Adapted from *Metals Handbook: Properties and Selection, Irons and Steels,* Vol. 1, 9th edition, B. Bardes, Editor, American Society for Metals, 1978, p. 492.)

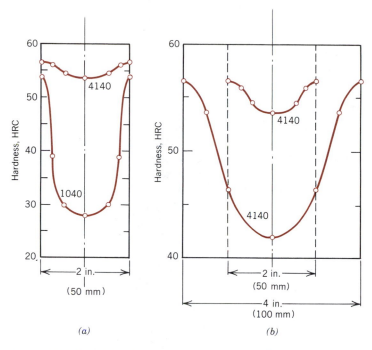

Figure 11.9 Radial hardness profiles for (*a*) 2 in. (50 mm) diameter cylindrical 1040 and 4140 steel specimens quenched in mildly agitated water, and (*b*) 2 and 4 in. (50 and 100 mm) diameter cylindrical specimens of 4140 steel quenched in mildly agitated water.

enability is evident from these two profiles. Specimen diameter also influences the hardness distribution, as demonstrated in Figure 11.9*b*, which plots the hardness profiles for water-quenched 4140 cylinders 2 and 4 in. (50 and 100 mm) in diameter. Example Problem 11.1 illustrates how these hardness profiles are determined.

As far as specimen shape is concerned, since the heat energy is dissipated to the quenching medium at the specimen surface, the rate of cooling for a particular quenching treatment depends on the ratio of surface area to the mass of the specimen. The larger this ratio, the more rapid will be the cooling rate and, consequently, the deeper the hardening effect. Irregular shapes with edges and corners have larger surface-to-mass ratios than regular and rounded shapes (e.g., spheres and cylinders) and are thus more amenable to hardening by quenching.

There are a multitude of steels that are responsive to a martensitic heat treatment, and one of the most important criteria in the selection process is hardenability. Hardenability curves, when utilized in conjunction with plots such as those in Figure 11.8 for various quenching media, may be used to ascertain the suitability of a specific steel alloy for a particular application. Or, conversely, the appropriateness of a quenching procedure for an alloy may be determined. For parts that are to be involved in relatively high stress applications, a minimum of 80% martensite must be produced throughout the interior as a consequence of the quenching procedure. Only a 50% minimum is required for moderately stressed parts.

EXAMPLE PROBLEM 11.1

Determine the radial hardness profile for a 2 in. (50 mm) diameter cylindrical specimen of 1040 steel that has been quenched in moderately agitated water.

SOLUTION

First, evaluate the cooling rate (in terms of the Jominy end-quench distance) at center, surface, mid-, and three-quarter radial positions of the cylindrical specimen. This is accomplished using the cooling rate-versus-bar diameter plot for the appropriate quenching medium, in this case, Figure 11.8a. Then, convert the cooling rate at each of these radial positions into a hardness value from a hardenability plot for the particular alloy. Finally, determine the hardness profile by plotting the hardness as a function of radial position.

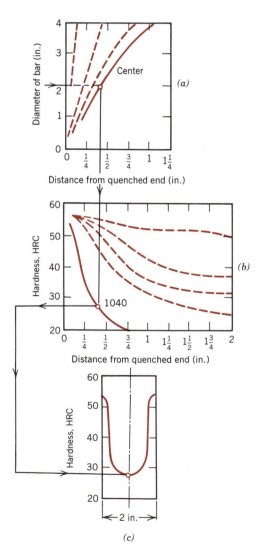

Figure 11.10 Use of hardenability data in the generation of hardness profiles. (a) The cooling rate at the center of a water-quenched 2 in. (50 mm) diameter specimen is determined. (b) The cooling rate is converted into an HRC hardness for a 1040 steel. (c) The Rockwell hardness is plotted on the radial hardness profile.

This procedure is demonstrated in Figure 11.10, for the center position. Note that for a water-quenched cylinder of 2 in. diameter, the cooling rate at the center is equivalent to that approximately $\frac{3}{8}$ in. from the Jominy specimen quenched end (Figure 11.10a). This corresponds to a hardness of about 28 HRC, as noted from the hardenability plot for the 1040 steel alloy (Figure 11.10b). Finally, this data point is plotted on the hardness profile in Figure 11.10c.

Surface, midradius, and three-quarter radius hardnesses would be determined in a similar manner. The complete profile has been included, and the data that were used are tabulated below.

Radial Position	Equivalent Distance from Quenched End (in.)	Hardness HRC
Center	$\frac{3}{8}$	28
Midradius	$\frac{5}{16}$	30
Three-quarters radius	$\frac{3}{16}$	39
Surface	$\frac{1}{16}$	54

PRECIPITATION HARDENING

The strength and hardness of some metal alloys may be enhanced by the formation of extremely small uniformly dispersed particles of a second phase within the original phase matrix; this must be accomplished by appropriate heat treatments. The process is called **precipitation hardening** because the small particles of the new phase are termed "precipitates." "Age hardening" is also used to designate this procedure because the strength develops with time, or as the alloy ages. Examples of alloys that are hardened by precipitation treatments include aluminum–copper, aluminum–silicon, copper–beryllium, copper–tin, and magnesium–aluminum; some ferrous alloys are also precipitation hardenable.

Precipitation hardening and the treating of steel to form tempered martensite are totally different phenomena, even though the heat treatment procedures are similar; therefore, the processes should not be confused. The principal difference lies in the mechanisms by which hardening and strengthening are achieved. These should become apparent as precipitation hardening is explained.

11.7 HEAT TREATMENTS

Inasmuch as precipitation hardening results from the development of particles of a new phase, an explanation of the heat treatment procedure is facilitated by use of a phase diagram. Even though, in practice, many precipitation-hardenable alloys contain two or more alloying elements, the discussion is simplified by reference to a binary system. The phase diagram must be of the form shown for the hypothetical A–B system in Figure 11.11.

Two requisite features must be displayed by the phase diagrams of alloy systems for precipitation hardening: an appreciable maximum solubility of one component in the other, on the order of several percent; and a solubility limit that rapidly decreases in concentration of the major component with temperature reduction. Both these

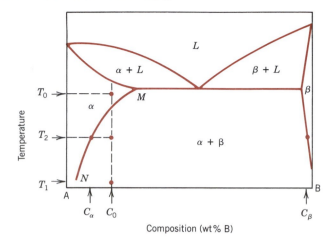

Figure 11.11 Hypothetical phase diagram for a precipitation hardenable alloy of composition C_0.

conditions are satisfied by this hypothetical phase diagram (Figure 11.11). The maximum solubility corresponds to the composition at point M. In addition, the solubility limit boundary between the α and $\alpha + \beta$ phase fields diminishes from this maximum concentration to a very low B content in A at point N. Furthermore, the composition of a precipitation-hardenable alloy must be less than the maximum solubility. These conditions are necessary but *not* sufficient for precipitation hardening to occur in an alloy system. An additional requirement is discussed below.

Solution Heat Treating

Precipitation hardening is accomplished by two different heat treatments. The first is a **solution heat treatment** in which all solute atoms are dissolved to form a single-phase solid solution. Consider an alloy of composition C_0 in Figure 11.11. The treatment consists of heating the alloy to a temperature within the α phase field—say T_0—and waiting until all the β phase that may have been present is completely dissolved. At this point, the alloy consists only of an α phase of composition C_0. This procedure is followed by rapid cooling or quenching to temperature T_1, which for many alloys is room temperature, to the extent that any diffusion and the accompanying formation of any of the β phase is prevented. Thus, a nonequilibrium situation exists in which only the α phase solid solution supersaturated with B atoms is present at T_1; in this state the alloy is relatively soft and weak. Furthermore, for most alloys diffusion rates at T_1 are extremely slow, such that the single α phase is retained at this temperature for relatively long periods.

Precipitation Heat Treating

For the second or **precipitation heat treatment,** the supersaturated α solid solution is ordinarily heated to an intermediate temperature T_2 (Figure 11.11) within the $\alpha + \beta$ two-phase region, at which temperature diffusion rates become appreciable. The β

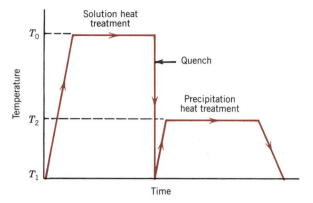

Figure 11.12 Schematic temperature-versus-time plot showing both solution and precipitation heat treatments for precipitation hardening.

precipitate phase begins to form as finely dispersed particles of composition C_β, which process is sometimes termed "aging." After the appropriate aging time at T_2, the alloy is cooled to room temperature; normally, this cooling rate is not an important consideration. Both solution and precipitation heat treatments are represented on the temperature-versus-time plot, Figure 11.12. The character of these β particles, and subsequently the strength and hardness of the alloy, depend on both the precipitation temperature T_2 and the aging time at this temperature. For some alloys, aging occurs spontaneously at room temperature over extended time periods.

The dependence of the growth of the precipitate β particles on time and temperature under isothermal heat treatment conditions may be represented by C-shaped curves similar to those in Figure 10.10 for the eutectoid transformation in steels. However, it is more useful and convenient to present the data as tensile strength or yield strength at room temperature as a function of the logarithm of aging time, at constant temperature T_2. The behavior for a typical precipitation-hardenable alloy is represented in Figure 11.13. With increasing time, the strength increases, reaches

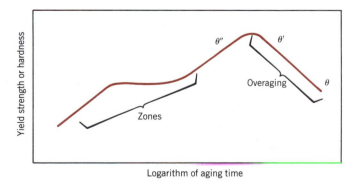

Figure 11.13 Schematic diagram showing tensile strength and hardness as a function of the logarithm of aging time at constant temperature during the precipitation heat treatment.

a maximum, and finally diminishes. This reduction in strength and hardness that occurs after long time periods is known as **overaging.** The influence of temperature is incorporated by the superposition, on a single plot, of curves at a variety of temperatures.

11.8 MECHANISM OF HARDENING

Precipitation hardening is commonly employed with high-strength aluminum alloys. Although a large number of these alloys have different proportions and combinations of alloying elements, the mechanism of hardening has perhaps been studied most extensively for the aluminum–copper alloys. Figure 11.14 presents the aluminum-rich portion of the aluminum–copper phase diagram. The α phase is a substitutional solid solution of copper in aluminum, whereas the intermetallic compound $CuAl_2$ is designated the θ phase. For an aluminum–copper alloy of, say, composition 96 wt% Al– 4 wt% Cu, in the development of this equilibrium θ phase during the precipitation heat treatment, several transition phases are first formed in a specific sequence. The mechanical properties are influenced by the character of the particles of these transition phases. During the initial hardening stage (at short times, Figure 11.13), copper atoms cluster together in very small and thin discs that are only one or two atoms thick and approximately 25 atoms in diameter; these form at countless positions within the α phase. The clusters, sometimes called zones, are so small that they are really not regarded as distinct precipitate particles. However, with time and the subsequent

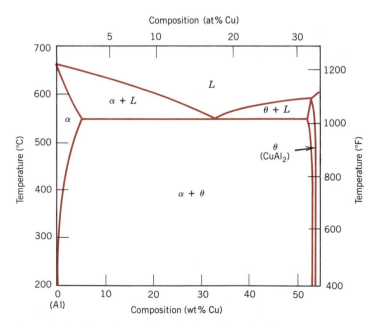

Figure 11.14 The aluminum-rich side of the aluminum–copper phase diagram. (Adapted from *Metals Handbook: Metallography, Structures and Phase Diagrams,* Vol. 8, 8th edition, ASM Handbook Committee, T. Lyman, Editor, American Society for Metals, 1973, p. 259.)

diffusion of copper atoms, zones become particles as they increase in size. These precipitate particles then pass through two transition phases (denoted as θ'' and θ'), before the formation of the equilibrium θ phase (Figure 11.15c). Transition phase particles for a precipitation-hardened 7150 aluminum alloy are shown in the electron micrograph on page 338.

The strengthening and hardening effects shown in Figure 11.13 result from the innumerable particles of these transition and metastable phases. As noted in the figure, maximum strength coincides with the formation of the θ'' phase, which may be preserved upon cooling the alloy to room temperature. Overaging results from continued particle growth and the development of θ' and θ phases.

The strengthening process is accelerated as the temperature is increased. This is demonstrated in Figure 11.16a, a plot of yield strength versus the logarithm of time for a 2014 aluminum alloy at several different precipitation temperatures. Ideally, temperature and time for the precipitation heat treatment should be designed to produce a hardness or strength in the vicinity of the maximum. Associated with an increase in strength is a reduction in ductility. This is demonstrated in Figure 11.16b for the same 2014 aluminum alloy at the several temperatures.

Not all alloys that satisfy the aforementioned conditions relative to composition and phase diagram configuration are amenable to precipitation hardening. In addition, lattice strains must be established at the precipitate–matrix interface. For aluminum–copper alloys, there is a distortion of the crystal lattice structure around and within the vicinity of particles of these transition phases (Figure 11.15b). During plastic deformation, dislocation motions are effectively impeded as a result of these distortions, and, consequently, the alloy becomes harder and stronger. As the θ phase forms, the resultant overaging (softening and weakening) is explained by a reduction in the resistance to slip that is offered by these precipitate particles.

Alloys that experience appreciable precipitation hardening at room temperature and after relatively short time periods must be quenched to and stored under refrigerated conditions. Several aluminum alloys that are used for rivets exhibit this

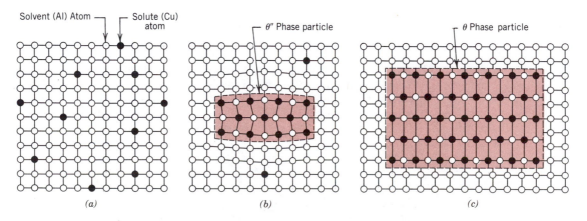

Figure 11.15 Schematic depiction of several stages in the formation of the equilibrium precipitate (θ) phase. (a) A supersaturated α solid solution. (b) A transition, θ'', precipitate phase. (c) The equilibrium θ phase, within the α matrix phase. Actual phase particle sizes are much larger than shown here.

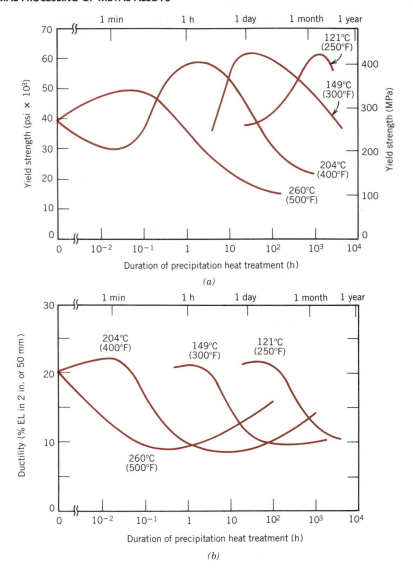

Figure 11.16 The precipitation hardening characteristics of a 2014 aluminum alloy (0.9 wt% Si, 4.4 wt% Cu, 0.8 wt% Mn, 0.5 wt% Mg) at four different aging temperatures: (*a*) yield strength, and (*b*) ductility (%EL). (Adapted from *Metals Handbook: Properties and Selection: Nonferrous Alloys and Pure Metals,* Vol. 2, 9th edition, H. Baker, Managing Editor, American Society for Metals, 1979, p. 41.)

behavior. They are driven while still soft, then allowed to age harden at the normal ambient temperature. This is termed **natural aging; artificial aging** is carried out at elevated temperatures.

11.9 MISCELLANEOUS CONSIDERATIONS

The combined effects of strain hardening and precipitation hardening may be employed in high-strength alloys. The order of these hardening procedures is important in the production of alloys having the optimum combination of mechanical properties. Normally, the alloy is solution heat treated and then quenched. This is followed by cold working and finally by the precipitation hardening heat treatment. In the final treatment, little strength loss is sustained as a result of recrystallization. If the alloy is precipitation hardened before cold working, more energy must be expended in its deformation; in addition, cracking may also result because of the reduction in ductility that accompanies the precipitation hardening.

Most precipitation-hardened alloys are limited in their maximum service temperatures. Exposure to temperatures at which aging occurs may lead to a loss of strength due to overaging.

SUMMARY

Some of the heat treatments that are used to fashion the mechanical properties of metal alloys were discussed in this chapter. The exposure to an elevated temperature for an extended time period followed by cooling to room temperature at a relatively slow rate is termed annealing; several specific annealing treatments were discussed briefly. During process annealing, a cold-worked piece is rendered softer yet more ductile as a consequence of recrystallization. Internal residual stresses that have been introduced are eliminated during a stress relief anneal. For ferrous alloys, normalizing is used to refine and improve the grain structure. Fabrication characteristics may also be enhanced by full anneal and spheroidizing treatments that produce microstructures consisting of coarse pearlite and spheroidite, respectively.

For high-strength steels, the best combination of mechanical characteristics may be realized if a predominantly martensitic microstructure is developed over the entire cross section; this is converted to tempered martensite during a tempering heat treatment. Hardenability is a parameter used to ascertain the influence of composition on the susceptibility to the formation of a predominantly martensitic structure for some specific heat treatment. Determination of hardenability is accomplished by the standard Jominy end-quench test, from which hardenability curves are generated.

Other factors also influence the extent to which martensite will form. Of the common quenching media, water is the most efficient, followed by oil and air, in that order. The relationships between cooling rate and specimen size and geometry for a specific quenching medium frequently are expressed on empirical charts; two were introduced for cylindrical specimens. These may be used in conjunction with hardenability data to generate cross-sectional hardness profiles.

Some alloys are amenable to precipitation hardening, that is, to strengthening by the formation of very small particles of a second, or precipitate, phase. Control of particle size, and subsequently the strength, is accomplished by two heat treatments. For the second or precipitation treatment at constant temperature, strength increases with time to a maximum and decreases during overaging. This process is accelerated with rising temperature. The strengthening phenomenon is explained in terms of an increased resistance to dislocation motion by lattice strains, which are established in the vicinity of these microscopically small precipitate particles.

IMPORTANT TERMS AND CONCEPTS

Annealing	Natural aging	Process annealing
Artificial aging	Normalizing	Solution heat treatment
Austenitizing	Overaging	Spheroidizing
Full annealing	Precipitation hardening	Stress relief
Hardenability	Precipitation heat	Upper critical
Jominy end-quench test	treatment	temperature
Lower critical		
temperature		

REFERENCES

GROSSMANN, M. A. and E. C. BAIN, *Principles of Heat Treatment,* American Society for Metals, Metals Park, Ohio, 1964.

KRAUSS, G., *Principles of Heat Treatment of Steel,* American Society for Metals, Metals Park, Ohio, 1980.

Metals Handbook, 9th edition, Vol. 4, *Heat Treating,* American Society for Metals, Metals Park, Ohio, 1981.

QUESTIONS AND PROBLEMS

11.1   In your own words describe the following heat treatment procedures for steels and, for each, the intended final microstructure: full annealing, normalizing, quenching, and tempering.

11.2   Cite three sources of internal residual stresses in metal components. What are two possible adverse consequences of these stresses?

11.3   Give the temperature range over which it is desirable to austenitize each of the following iron–carbon alloys during a normalizing heat treatment: **(a)** 0.20 wt% C, **(b)** 0.77 wt% C, and **(c)** 0.95 wt% C.

11.4   Give the temperature range over which it is possible to austenitize each of the following iron–carbon alloys during a full anneal heat treatment: **(a)** 0.25 wt% C, **(b)** 0.45 wt% C, **(c)** 0.85 wt% C, and **(d)** 1.10 wt% C.

11.5 What is the purpose of a spheroidizing heat treatment? On what classes of alloys is it normally used?

11.6 Briefly explain the difference between hardness and hardenability.

11.7 What influence does the presence of alloying elements (other than carbon) have on the shape of a hardenability curve? Briefly explain this effect.

11.8 How would you expect a decrease in the austenite grain size to affect the hardenability of a steel alloy? Why?

11.9 (a) Name the three factors that influence the degree to which martensite is formed throughout the cross section of a steel specimen. (b) For each, tell how the extent of martensite formation may be increased.

11.10 Name two thermal properties of a liquid medium that will influence its quenching effectiveness.

11.11 Construct radial hardness profiles for the following:
(a) A 2 in. (50 mm) diameter cylindrical specimen of an 8640 steel alloy that has been quenched in moderately agitated oil.
(b) A 3 in. (75 mm) diameter cylindrical specimen of a 5140 steel alloy that has been quenched in moderately agitated oil.
(c) A $3\frac{1}{2}$ in. (90 mm) diameter cylindrical specimen of an 8630 steel alloy that has been quenched in moderately agitated water.
(d) A 4 in. (100 mm) diameter cylindrical specimen of an 8660 steel alloy that has been quenched in moderately agitated water.

11.12 A cylindrical piece of steel 1.0 in. (25 mm) in diameter is to be quenched in moderately agitated oil. Surface and center hardnesses must be at least 55 and 50 HRC, respectively. Which of the following alloys will satisfy these requirements: 1040, 5140, 4340, 4140, and 8640? Justify your choices.

11.13 A cylindrical piece of steel $2\frac{1}{4}$ in. (57 mm) in diameter is to be austenitized and quenched such that a minimum hardness of 45 HRC is to be produced throughout the entire piece. Of the alloys 8660, 8640, 8630, and 8620, which will qualify if the quenching medium is (a) moderately agitated water, and (b) moderately agitated oil? Justify your choice(s).

11.14 A cylindrical piece of steel $1\frac{1}{2}$ in. (38 mm) in diameter is to be austenitized and quenched such that a microstructure consisting of at least 80% martensite will be produced throughout the entire piece. Of the alloys 4340, 4140, 8640, 5140, and 1040, which will qualify if the quenching medium is (a) moderately agitated oil and (b) moderately agitated water? Justify your choices.

11.15 A cylindrical piece of steel 2 in. (50 mm) in diameter is to be quenched in moderately agitated water. Surface and center hardnesses must be at least 50 and 40 HRC, respectively. Which of the following alloys will satisfy these requirements: 1040, 5140, 4340, 4140, 8620, 8630, 8640, and 8660? Justify your choices.

11.16 A cylindrical piece of 4140 steel is to be austenitized and quenched in moderately agitated oil. If the microstructure is to consist of at least 50% martensite throughout the entire piece, what is the maximum allowable diameter? Justify your answer.

11.17 Compare the effectiveness of quenching in moderately agitated water and oil by graphing, on a single plot, radial hardness profiles for a 3 in. (75 mm) diameter cylindrical specimen of an 8640 steel that has been quenched in both media.

11.18 A cylindrical piece of 8660 steel is to be austenitized and quenched in moderately agitated oil. If the hardness at the surface of the piece must be at least 58 HRC, what is the maximum allowable diameter? Justify your answer.

11.19 Copper-rich copper–beryllium alloys are precipitation hardenable. After consulting the portion of the phase diagram (Figure 11.17), do the following:
  (a) Specify the range of compositions over which these alloys may be precipitation hardened.
  (b) Briefly describe the heat treatment procedures (in terms of temperatures) that would be used to precipitation harden an alloy having a composition of your choosing, yet lying within the range given for part a.

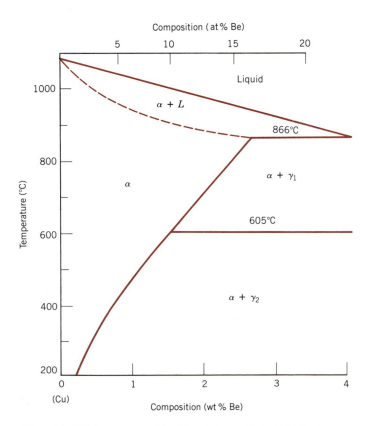

Figure 11.17 The copper-rich side of the copper–beryllium phase diagram. (Adapted from *Metals Handbook: Metallography, Structures and Phase Diagrams,* Vol. 8, 8th edition, ASM Handbook Committee, T. Lyman, Editor, American Society for Metals, 1973, p. 271.)

11.20 Compare precipitation hardening (Sections 11.7 and 11.8) and the hardening of steel by quenching and tempering (Sections 10.5 and 10.8) with regard to

(a) The total heat treatment procedure.

(b) The microstructures that develop.

(c) How the mechanical properties change during the several heat treatment stages.

11.21 A solution heat-treated 2014 aluminum alloy is to be precipitation hardened to have a minimum yield strength of 50,000 psi (345 MPa) and a ductility of at least 12%EL. Specify a practical precipitation heat treatment in terms of temperature and time that would give these mechanical characteristics.

11.22 Is it possible to produce a precipitation-hardened 2014 aluminum alloy having a minimum yield strength of 55,000 psi (380 MPa) and a ductility of at least 15%EL? If so, specify the precipitation heat treatment. If it is not possible, explain why.

11.23 What is the principal difference between natural and artificial aging processes?

CHAPTER 12

METAL ALLOYS

Top: A sectioned four cylinder, gray iron automobile engine block. Bottom: A scanning electron micrograph of a gray iron similar to that of which the engine block is made. The graphite flake structure is evident in this micrograph. (Reprinted with permission of the Iron Castings Society, Des Plaines, IL.)

12.1 INTRODUCTION

It is appropriate to conclude the treatment of metallic materials by discussing some of the important engineering alloys in terms of their compositions, properties, applications, and fabrication techniques. We draw on a number of concepts and phenomena that have been developed in preceding chapters, including mechanical properties, phase diagrams, and various strengthening mechanisms. The first section is devoted to a brief exploration of several techniques by which metals are fabricated. Metal alloys, by virtue of composition, are often grouped into two classes—ferrous and nonferrous. Ferrous alloys, those in which iron is the principal constituent, include steels and cast irons. These alloys and their characteristics are treated in the second portion of this chapter. The nonferrous ones—all the alloys that are not iron based— are the final topic of discussion.

Many times a materials problem is really one of selecting that material which has the right combination of characteristics for a specific application. Therefore, the persons who are involved in the decision making should have some knowledge of the available options. This extremely abbreviated presentation provides an overview of some of the commercial alloys, their general properties, and their limitations.

FABRICATION OF METALS

On occasion, the suitability of a material for an application is dictated by the ease of producing a desired shape and the cost involved. Metal fabrication techniques are the methods by which metals and alloys are formed or manufactured into useful products. They are preceded by refining, alloying, and often heat-treating processes that produce alloys with the desired characteristics.

The classifications of fabrication techniques include various metal-forming methods, casting, powder metallurgy, welding, and machining; invariably, two or more of them must be used before a piece is finished. The methods chosen depend on several factors; the most important are the properties of the metal, the size and shape of the finished piece, and, of course, cost. The metal fabrication techniques discussed in this chapter are classified according to the scheme illustrated in Figure 12.1.

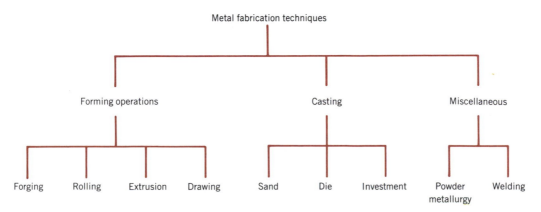

Figure 12.1 Classification scheme of metal fabrication techniques discussed in this chapter.

12.2 FORMING OPERATIONS

Forming operations are those in which the shape of a metal piece is changed by plastic deformation; for example, forging, rolling, extrusion, and drawing are common forming techniques. Of course the deformation must be induced by an external force or stress, the magnitude of which must exceed the yield strength of the material. Most metallic materials are especially amenable to these procedures, being at least moderately ductile and capable of some permanent deformation without cracking or fracturing.

When deformation is achieved at a temperature above that at which recrystallization occurs, the process is termed **hot working** (Section 7.12); otherwise, it is cold working. With most of the forming techniques, both hot- and cold-working procedures are possible. For hot-working operations, large deformations are possible, which may be successively repeated because the metal remains soft and ductile. Also, deformation energy requirements are less than for cold working. However, most metals experience some surface oxidation, which results in material loss and a poor final surface finish. **Cold working** produces an increase in strength with the attendant decrease in ductility, since the metal strain hardens; advantages over hot working include a higher quality surface finish, better mechanical properties and a greater variety of them, and closer dimensional control of the finished piece. On occasion, the total deformation is accomplished in a series of steps in which the piece is successively cold worked a small amount and then process annealed (Section 11.2); however, this is an expensive and inconvenient procedure.

A very brief description of the forming techniques is illustrated schematically in Figure 12.2.

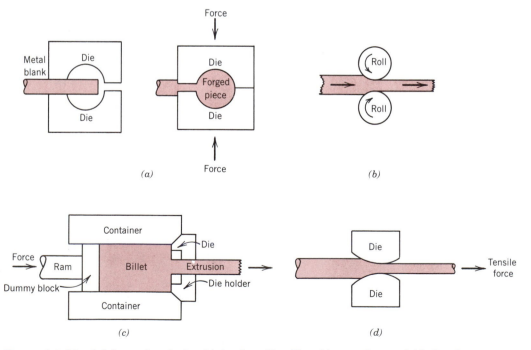

Figure 12.2 Metal deformation during (a) forging, (b) rolling, (c) extrusion, and (d) drawing.

Forging

Forging is accomplished by hammering on a single piece of metal. A force is brought to bear on two die halves having the finished shape such that the metal is deformed in the cavity between them. Forged pieces have outstanding grain structures and the best combination of mechanical properties. Wrenches, railroad wheels, and automotive crankshafts are typical of articles formed using this technique.

Rolling

Rolling, the most widely used deformation process, consists of passing a piece of metal between two rolls; a reduction in thickness results from compressive stresses exerted by the rolls. Cold rolling may be used in the production of sheet, strip, and foil with high quality surface finish. Circular shapes as well as I beams and railroad rails are fabricated using grooved rolls.

Extrusion

For **extrusion,** a bar of metal is forced through a die orifice by a compressive force that is applied to a ram; the extruded piece that emerges has the desired shape and a reduced cross-sectional area. Extrusion products include rods and tubing that have rather complicated cross-sectional geometries; seamless tubing may also be extruded.

Drawing

Drawing is the pulling of a metal piece through a die having a tapered bore by means of a tensile force that is applied on the exit side. A reduction in cross section results, with a corresponding increase in length. The total drawing operation may consist of a number of dies in a series sequence. Rod, wire, and tubing products are commonly fabricated in this way.

12.3 CASTING

Casting is a fabrication process whereby a totally molten metal is poured into a mold cavity having the desired shape; upon solidification, the metal assumes the shape of the mold but experiences some shrinkage. Casting techniques are employed when (1) the finished shape is so large or complicated that any other method would be impractical, (2) quality and strength are not important considerations (i.e., it is accepted that the inevitable existence of internal defects and a less than desirable grain structure lead to poor mechanical characteristics), (3) a particular alloy is so low in ductility that forming by either hot or cold working would be difficult, and (4) in comparison to other fabrication processes, casting is the most economical. Furthermore, the final step in the refining of even ductile metals may involve a casting process. A number of different casting techniques are commonly employed, including sand, die, and investment casting. Only a cursory treatment of each of these is offered.

Sand Casting

With sand casting, probably the most common method, ordinary sand is used as the mold material. A two-piece mold is formed by packing sand around a pattern that has the shape of the intended casting. Furthermore, a *gating system* is usually incorporated into the mold to expedite the flow of molten metal into the cavity and to minimize internal casting defects. Sand-cast parts include automotive cylinder blocks, fire hydrants, and large pipe fittings.

Die Casting

In die casting, the liquid metal is forced into a mold under pressure and at a relatively high velocity, and allowed to solidify with the pressure maintained. A two-piece permanent steel mold or die is employed; when clamped together, the two pieces form the desired shape. When complete solidification has been achieved, the die pieces are opened and the cast piece is ejected. Rapid casting rates are possible, making this an inexpensive method; furthermore, a single set of dies may be used for thousands of castings. However, this technique lends itself only to relatively small pieces and to alloys of zinc, aluminum, and magnesium, which have low melting temperatures.

Investment Casting

For investment (sometimes called lost-wax) casting, the pattern is made from a wax or plastic that has a low melting temperature. Around the pattern is poured a fluid slurry, which sets up to form a solid mold or investment; plaster of paris is usually used. The mold is then heated, such that the pattern melts and is burned out, leaving behind a mold cavity having the desired shape. This technique is employed when high dimensional accuracy, reproduction of fine detail, and an excellent finish are required— for example, in jewelry and dental crowns and inlays. Also, blades for gas turbines and jet engine impellers are investment cast.

12.4 MISCELLANEOUS TECHNIQUES

Powder Metallurgy

Yet another fabrication technique involves the compaction of powdered metal, followed by a heat treatment to produce a more dense piece. The process is appropriately called **powder metallurgy,** frequently designated as P/M. Powder metallurgy makes it possible to produce a virtually nonporous piece having properties almost equivalent to the fully dense parent material. This method is especially suitable for metals having low ductilities, since only small plastic deformation of the powder particles need occur. Metals having high melting temperatures are difficult to melt and cast, and fabrication is expedited using P/M. Furthermore, parts that require very close dimensional tolerances (e.g., bushings and gears) may be economically produced using this technique.

Welding

In a sense, welding may be considered to be a fabrication technique. In **welding,** two or more metal parts are joined to form a single piece when one-part fabrication is expensive or inconvenient. Both similar and dissimilar metals may be welded. The joining bond is metallurgical (involving some diffusion) rather than just mechanical, as with riveting and bolting. A variety of welding methods exist, including arc and gas welding, as well as brazing and soldering.

FERROUS ALLOYS

Ferrous alloys—those of which iron is the prime constituent—are produced in larger quantities than any other metal type. They are especially important as engineering construction materials. Their widespread use is accounted for by three factors: (1) iron-containing compounds exist in abundant quantities within the earth's crust; (2) metallic iron and steel alloys may be produced using relatively economical extraction, refining, alloying, and fabrication techniques; and (3) ferrous alloys are extremely versatile, in that they may be tailored to have a wide range of mechanical and physical properties. The principal disadvantage of many ferrous alloys is their susceptibility to corrosion. This chapter discusses compositions, microstructures, and properties of a number of different classes of steels and cast irons. A taxonomic classification scheme for the various ferrous alloys is presented in Figure 12.3.

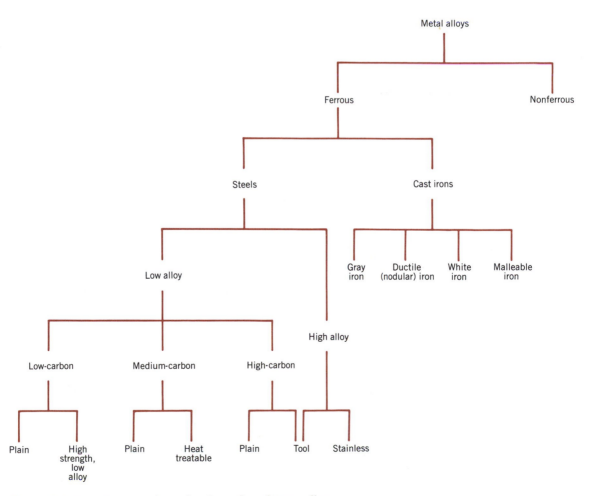

Figure 12.3 Classification scheme for the various ferrous alloys.

12.5 STEELS

Steels are iron–carbon alloys that may contain appreciable concentrations of other alloying elements; there are thousands of alloys that have different compositions and/or heat treatments. The mechanical properties are sensitive to the content of carbon, which is normally less than 1.0 wt%. Some of the more common steels are classified according to carbon concentration, namely, into low-, medium-, and high-carbon types. Subclasses also exist within each group according to the concentration of other alloying elements. **Plain carbon steels** contain only residual concentrations of impurities other than carbon. For **alloy steels,** alloying elements are intentionally added in specific concentrations.

Low-Carbon Steels

Of all the different steels, those produced in the greatest quantities fall within the low-carbon classification. These generally contain less than about 0.25 wt% C and are unresponsive to heat treatments intended to form martensite; strengthening is accomplished by cold work. Microstructures consist of ferrite and pearlite constituents. As a consequence, these alloys are relatively soft and weak, but have outstanding ductility and toughness; in addition, they are machinable, weldable, and, of all steels, are the least expensive to produce. Typical applications include automobile body components, structural shapes (I beams, channel and angle iron), and sheets that are used in pipelines, buildings, bridges, and tin cans. Tables 12.1a and 12.1b, respectively, present the compositions and mechanical properties of several plain low-carbon steels. They typically have a yield strength of 40,000 psi (275 MPa), tensile strengths between 60,000 and 80,000 psi (415 and 550 MPa), and a ductility of 25%EL.

TABLE 12.1a Compositions of Five Plain Low-Carbon Steels and Three High-Strength, Low-Alloy Steels

Designation[a]		Composition (wt%)[b]		
AISI/SAE or ASTM Number	UNS Number	C	Mn	Other
Plain Low-Carbon Steels				
1010	G10100	0.10	0.45	
1020	G10200	0.20	0.45	
A36	K02600	0.29	1.00	0.20 Cu (min)
A516 Grade 70	K02700	0.31	1.00	0.25 Si
High-Strength, Low-Alloy Steels				
A440	K12810	0.28	1.35	0.30 Si (max), 0.20 Cu (min)
A633 Grade E	K12002	0.22	1.35	0.30 Si, 0.08 V, 0.02 N, 0.03 Nb
A656 Grade 1	K11804	0.18	1.60	0.60 Si, 0.1 V, 0.20 Al, 0.015 N

[a] The codes used by the American Iron and Steel Institute (AISI), the Society of Automotive Engineers (SAE), and the American Society for Testing and Materials (ASTM), and in the Uniform Numbering System (UNS) are explained in the text.

[b] Also a maximum of 0.04 wt% P, 0.05 wt% S, and 0.30 wt% Si (unless indicated otherwise).

Source: Adapted from *Metals Handbook: Properties and Selection, Irons and Steels,* Vol. 1, 9th edition, B. Bardes (Editor), American Society for Metals, 1978, pp. 185, 407.

TABLE 12.1b Mechanical Characteristics of Hot-Rolled Material and Typical Applications for Various Plain Low-Carbon and High-Strength, Low-Alloy Steels

AISI/SAE or ASTM Number	Tensile Strength [psi × 10^3 (MPa)]	Yield Strength [psi × 10^3 (MPa)]	Ductility (%EL in 2 in.)	Typical Applications
Plain Low-Carbon Steels				
1010	47 (325)	26 (180)	28	Automobile panels, nails, and wire
1020	55 (380)	30 (205)	25	Pipe; structural and sheet steel
A36	58 (400)	32 (220)	23	Structural (bridges and buildings)
A516 Grade 70	70 (485)	38 (260)	21	Low temperature pressure vessels
High-Strength, Low-Alloy Steels				
A440	63 (435)	42 (290)	21	Structures that are bolted or riveted
A633 Grade E	75 (520)	55 (380)	23	Structures used at low ambient temperatures
A656 Grade 1	95 (655)	80 (552)	15	Truck frames and railway cars

Source: Adapted from *Metals Handbook: Properties and Selection, Irons and Steels,* Vol. 1, 9th edition, B. Bardes (Editor), American Society for Metals, 1978, pp. 190, 192, 405, 406.

Another group of low carbon alloys are the **high-strength, low-alloy (HSLA) steels.** They contain other alloying elements such as copper, vanadium, nickel, and molybdenum in combined concentrations as high as 10 wt%, and possess higher strengths than the plain low-carbon steels. Most may be strengthened by heat treatment, yielding tensile strengths in excess of 70,000 psi (480 MPa); in addition, they are ductile, formable, and machinable. Several are listed in Table 12.1. In normal atmospheres, the HSLA steels are more resistant to corrosion than the plain carbon steels, which they have replaced in many applications where structural strength is critical (e.g., bridges, towers, support columns in high-rise buildings, and pressure vessels).

Medium-Carbon Steels

The medium-carbon steels have carbon concentrations between about 0.25 and 0.60 wt%. These alloys may be heat treated by austenitizing, quenching, and then tempering to improve their mechanical properties. They are most often utilized in the tempered condition, having microstructures of tempered martensite. The plain medium-carbon steels have low hardenabilities and can be successfully heat treated only in very thin sections and with very rapid quenching rates. Additions of chromium, nickel, and molybdenum improve the capacity of these alloys to be heat treated (Section 11.5), giving rise to a variety of strength–ductility combinations. These heat-treated alloys are stronger than the low-carbon steels, but at a sacrifice of ductility and toughness. Applications include railway wheels and tracks, gears, crankshafts, and other machine parts and high-strength structural components calling for a combination of high strength, wear resistance, and toughness.

The compositions of several of these alloyed medium-carbon steels are presented in Table 12.2a. Some comment is in order regarding the designation schemes which are also included. The Society of Automotive Engineers (SAE), the American Iron and Steel Institute (AISI) and the American Society for Testing and Materials (ASTM) are responsible for the classification and specification of steels as well as other alloys. The AISI/SAE designation for these steels is a four-digit number: the first two digits indicate the alloy content; the last two, the carbon concentration. For plain carbon steels, the first two digits are 1 and 0; alloy steels are designated by other initial two-digit combinations (e.g., 13, 41, 43). The third and fourth digits represent the weight percent carbon multiplied by 100. For example, a 1060 steel is a plain carbon steel containing 0.60 wt% C.

A unified numbering system (UNS) is used for uniformly indexing both ferrous and nonferrous alloys. Each UNS number consists of a single-letter prefix followed by a five-digit number. The letter is indicative of the family of metals to which an alloy belongs. The UNS designation for these alloys begins with a G, followed by the AISI/SAE number; the fifth digit is a zero. Table 12.2b contains the mechanical characteristics and typical applications of several of these steels, which have been quenched and tempered.

High-Carbon Steels

The high-carbon steels, normally having carbon contents between 0.60 and 1.4 wt%, are the hardest, strongest, and yet least ductile of the carbon steels. They are almost

TABLE 12.2a AISI/SAE and UNS Designation Systems and Composition Ranges for Plain Carbon Steel and Various Low-Alloy Steels

AISI/SAE Designation[a]	UNS Designation	Composition Ranges (wt% of alloying elements in addition to C)[b]			
		Ni	Cr	Mo	Other
10xx, Plain carbon	G10xx0				
11xx, Free machining	G11xx0				0.08–0.33S
12xx, Free machining	G12xx0				0.10–0.35S, 0.04–0.12P
13xx	G13xx0				1.60–1.90Mn
40xx	G40xx0			0.20–0.30	
41xx	G41xx0		0.80–1.10	0.15–0.25	
43xx	G43xx0	1.65–2.00	0.40–0.90	0.20–0.30	
46xx	G46xx0	0.70–2.00		0.15–0.30	
48xx	G48xx0	3.25–3.75		0.20–0.30	
51xx	G51xx0		0.70–1.10		
61xx	G61xx0		0.50–1.10		0.10–0.15V
86xx	G86xx0	0.40–0.70	0.40–0.60	0.15–0.25	
92xx	G92xx0				1.80–2.20Si

[a] The carbon concentration, in weight percent times 100, is inserted in the place of "xx" for each specific steel.
[b] Except for 13xx alloys, manganese concentration is less than 1.00 wt%.
Except for 12xx alloys, phosphorus concentration is less than 0.35 wt%.
Except for 11xx and 12xx alloys, sulfur concentration is less than 0.04 wt%.
Except for 92xx alloys, silicon concentration varies between 0.15 and 0.35 wt%.

TABLE 12.2b Typical Applications and Mechanical Property Ranges for Oil-Quenched and Tempered Plain Carbon and Alloy Steels

AISI Number	UNS Number	Tensile Strength [psi × 10³ (MPa)]	Yield Strength [psi × 10³ (MPa)]	Ductility (%EL in 2 in.)	Typical Applications
		Plain Carbon Steels			
1040	G10400	88–113 (605–780)	62–85 (430–585)	33–19	Crankshafts, bolts
1080[a]	G10800	116–190 (800–1310)	70–142 (480–980)	24–13	Chisels, hammers
1095[a]	G10950	110–186 (760–1280)	74–120 (510–830)	26–10	Knives, hacksaw blades
		Alloy Steels			
4063	G40630	114–345 (786–2380)	103–257 (710–1770)	24–4	Springs, hand tools
4340	G43400	142–284 (980–1960)	130–228 (895–1570)	21–11	Bushings, aircraft tubing
6150	G61500	118–315 (815–2170)	108–270 (745–1860)	22–7	Shafts, pistons, gears

[a] Classified as high-carbon steels.

always used in a hardened and tempered condition, and, as such, are especially wear resistant and capable of holding a sharp cutting edge. The tool and die steels are high-carbon alloys, usually containing chromium, vanadium, tungsten, and molybdenum. These alloying elements combine with carbon to form very hard and wear-resistant carbide compounds (e.g., $Cr_{23}C_6$, V_4C_3, and WC). Some tool steel compositions and their applications are listed in Table 12.3. These steels are utilized

TABLE 12.3 Designations, Compositions, and Applications for Six Tool Steels

AISI Number	UNS Number	C	W	Mo	Cr	V	Other	Typical Applications
W1	T72301	0.6–1.4						Blacksmith tools, woodworking tools
S1	T41901	0.50	2.50		1.50			Pipe cutters, concrete drills
O1	T31501	0.90	0.50		0.50		1.00Mn	Shear blades, cutting tools
A2	T30102	1.00		1.00	5.00			Punches, embossing dies
D2	T30402	1.50		1.00	12.00	1.00		Cutlery, drawing dies
M1	T11301	0.85	1.50	8.50	4.00	1.00		Drills, saws; lathe and planer tools

[a] The balance of the composition is iron.
Source: *Metal Progress 1982 Materials and Processing Databook.* Copyright © 1982 American Society for Metals.

as cutting tools and dies for forming and shaping materials, as well as in knives, razors, hacksaw blades, springs, and high-strength wire.

Stainless Steels

The stainless steels are highly resistant to corrosion (rusting) in a variety of environments, especially the ambient atmosphere. Their predominant alloying element is chromium; a concentration of at least 11 wt% Cr is required. Corrosion resistance may also be enhanced by nickel and molybdenum additions.

Stainless steels are divided into three classes on the basis of the predominant phase constituent of the microstructure—martensitic, ferritic, or austenitic. Table 12.4 lists several stainless steels, by class, along with composition, typical mechanical properties, and applications. A wide range of mechanical properties combined with excellent resistance to corrosion make stainless steels very versatile in their applicability.

Martensitic stainless steels are capable of being heat treated in such a way that martensite is the prime microconstituent. Additions of alloying elements in significant concentrations produce dramatic alterations in the iron–iron carbide phase diagram (Figure 9.20). For austenitic stainless steels, the austenite (or γ) phase field is extended to room temperature. Ferritic stainless steels are composed of the α-ferrite (BCC) phase. Austenitic and ferritic stainless steels are hardened and strengthened by cold work because they are not heat treatable. The austenitic stainless steels are the most corrosion resistant because of the high chromium contents and also the nickel additions; and they are produced in the largest quantities. Both martensitic and ferritic stainless steels are magnetic; the austenitic stainlesses are not.

Some stainless steels are frequently used at elevated temperatures and in severe environments because they resist oxidation and maintain their mechanical integrity under such conditions; the upper temperature limit in oxidizing atmospheres is about 1000°C (1800°F). Equipment employing these steels includes gas turbines, high-temperature steam boilers, heat-treating furnaces, aircraft, missiles, and nuclear power generating units. Also included in Table 12.4 is one ultrahigh-strength stainless steel (17-7 PH), which is unusually strong and corrosion resistant. Strengthening is accomplished by precipitation-hardening heat treatments (Section 11.7).

12.6 CAST IRONS

Generically, cast irons are a class of ferrous alloys with carbon contents above 2.1 wt%; in practicality, however, most cast irons contain between 3.0 and 4.5 wt% C and, in addition, other alloying elements. A reexamination of the iron–iron carbide phase diagram (Figure 9.20) reveals that alloys within this composition range become completely liquid at temperatures between approximately 1150 and 1300°C (2100 and 2350°F), which is considerably lower than for steels. Thus they are easily melted and amenable to casting. Furthermore, some cast irons are very brittle, and casting is the most convenient fabrication technique.

Cementite (Fe_3C) is a metastable compound, and under some circumstances it can be made to dissociate or decompose to form α-ferrite and graphite, according to the reaction

$$Fe_3C \longrightarrow 3Fe\ (\alpha) + C\ (graphite) \tag{12.1}$$

TABLE 12.4 Designations, Compositions, Mechanical Properties, and Typical Applications for Austenitic, Ferritic, Martensitic, and Precipitation-Hardenable Stainless Steels

AISI Number	UNS Number	Composition (wt%)[a]				Condition[b]	Mechanical Properties			Typical Applications
		C	Cr	Ni	Other		Tensile Strength [psi × 10³ (MPa)]	Yield Strength [psi × 10³ (MPa)]	Ductility (%EL in 2 in.)	
Ferritic										
409	S40900	0.08	11		1.0Mn, 0.75Ti	Annealed	65 (448)	35 (240)	25	Automotive exhaust
446	S44600	0.20	25		1.5Mn	Annealed	80 (552)	50 (345)	20	Valves (high temperature), glass molds
Austenitic										
304	S30400	0.08	19	9	2.0Mn	Annealed	85 (586)	35 (240)	55	Food processing
316L	S31603	0.03	17	12	2.0Mn, 2.5Mo	Annealed	80 (552)	35 (240)	50	Welding construction
Martensitic										
410	S41000	0.15	12.5		1.0Mn	Annealed	70 (483)	40 (275)	30	Rifle barrels, cutlery
						Q and T	140 (965)	100 (690)	23	
440A	S44002	0.70	17		1.0Mn, 0.75Mo	Annealed	105 (724)	60 (414)	20	Cutlery, surgical tools
						Q and T	260 (1790)	240 (1655)	5	
Precipitation Hardenable										
17-7PH	S17700	0.09	17	7	1.0Mn, 1.0Al	Solution treated	130 (897)	40 (275)	35	Knives, springs
						Precipitation hardened	215 (1480)	195 (1345)	9	

[a] The balance of the composition is iron.

[b] "Q and T" denotes quenched and tempered.

Source: Adapted from *Metal Progress 1982 Materials and Processing Databook.* Copyright © 1982 American Society for Metals.

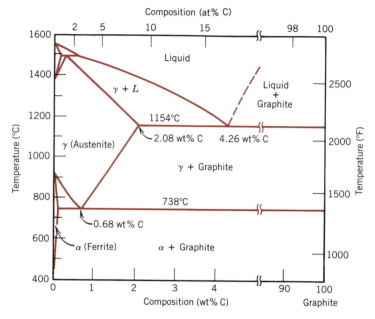

Figure 12.4 The true equilibrium iron–carbon phase diagram with graphite instead of cementite as a stable phase. (Adapted from *Metal Progress 1982 Materials and Processing Databook.* Copyright © 1982 American Society for Metals.)

Thus the true equilibrium diagram for iron and carbon is not that presented in Figure 9.20, but rather as shown in Figure 12.4. The two diagrams are virtually identical on the iron-rich side (e.g., eutectic and eutectoid temperatures for the Fe–Fe₃C system are 1148 and 727°C, respectively, as compared to 1154 and 738°C for Fe–C); however, Figure 12.4 extends to 100 wt% carbon such that graphite is the carbon-rich phase, instead of cementite at 6.7 wt% C (Figure 9.20).

This tendency to form graphite is regulated by the composition and rate of cooling. Graphite formation is promoted by the presence of silicon in concentrations greater than about 1 wt%. Also, slower cooling rates during solidification favor graphitization (the formation of graphite). For most cast irons, the carbon exists as graphite, and both microstructure and mechanical behavior depend on composition and heat treatment. The most common cast iron types are gray, nodular, white, and malleable.

Gray Iron

The carbon and silicon contents of **gray cast irons** vary between 2.5 to 4.0 and 1.0 to 3.0 wt%, respectively. For most of these cast irons, the graphite exists in the form of flakes (similar to corn flakes), which are normally surrounded by an α-ferrite or pearlite matrix; the microstructure of a typical gray iron is shown in Figure 12.5a and on page 362. Because of these graphite flakes, a fractured surface takes on a gray appearance, hence its name.

Mechanically, gray iron is comparatively weak and brittle in tension as a consequence of its microstructure; the tips of the graphite flakes are sharp and pointed,

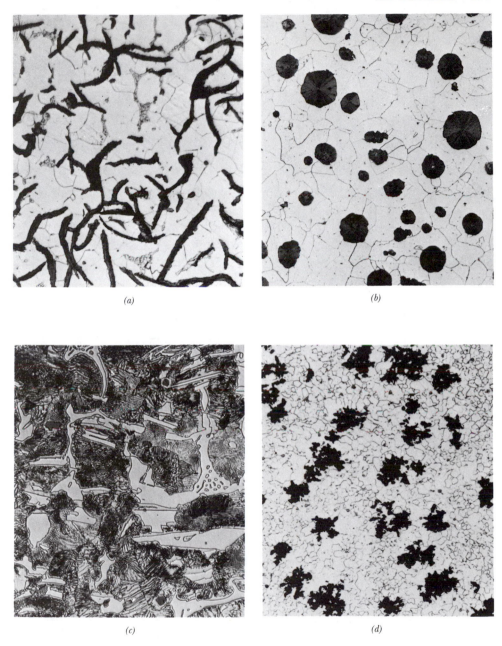

(a)

(b)

(c)

(d)

Figure 12.5 Optical photomicrographs of various cast irons. (*a*) Gray iron: the dark graphite flakes are embedded in an α-ferrite matrix. 500×. (Courtesy of C. H. Brady, National Bureau of Standards, Washington, D.C.) (*b*) Nodular (ductile) iron: the dark graphite nodules are surrounded by an α-ferrite matrix. 200×. (Courtesy of C. H. Brady and L. C. Smith, National Bureau of Standards, Washington, D.C.) (*c*) White iron: the light cementite regions are surrounded by pearlite, which has the ferrite–cementite layered structure. 400×. (Courtesy of Amcast Industrial Corporation.) (*d*) Malleable iron: dark graphite rosettes (temper carbon) in an α-ferrite matrix. 150×. (Reprinted with permission of the Iron Castings Society, Des Plaines, IL.)

TABLE 12.5 Designations, Minimum Mechanical Properties, Approximate Compositions, and Typical Applications for Various Gray, Nodular, and Malleable Cast Irons

| Grade | UNS Number | Composition (wt%)[a] | | | Matrix Structure | Mechanical Properties | | | Typical Applications |
		C	Si	Other		Tensile Strength [psi × 10³ (MPa)]	Yield Strength [psi × 10³ (MPa)]	Ductility (%EL in 2 in.)	
Gray Iron									
SAE G2500	F10005	3.3	2.2	0.7Mn	Pearlite + ferrite	25 (173)	—	—	Engine blocks, brake drums
SAE G4000	F10008	3.2	2.0	0.8Mn	Pearlite + ferrite	40 (276)	—	—	Engine cylinders and pistons
Ductile (Nodular) Iron									
ASTM A536 60-40-18	F32800	3.5–3.8	2.0–2.8	0.05Mg, <0.20Ni, <0.10Mo	Ferrite	60 (414)	40 (276)	18	Valve and pump bodies
100-70-03	F34800				Pearlite	100 (690)	70 (483)	3	High strength gears
120-90-02	F36200				Tempered martensite	120 (828)	90 (621)	2	Gears, rollers
Malleable Iron									
32510	F22200	2.3–2.7	1.0–1.75	<0.55Mn	Ferrite	50 (345)	32 (224)	10	General engineering service at room and elevated temperatures
45006		2.4–2.7	1.25–1.55	<0.55Mn	Ferrite + pearlite	65 (448)	45 (310)	6	

[a] The balance of the composition is iron.
Source: Adapted from *Metals Handbook: Properties and Selection: Irons and Steels*, Vol. I, 9th edition, Bardes, B. ed., American Society for Metals, 1978.

and may serve as points of stress concentration when an external tensile stress is applied. Strength and ductility are much higher under compressive loads. Typical mechanical properties and compositions of several of the common gray cast irons are listed in Table 12.5. Gray irons do have some desirable characteristics and, in fact, are utilized extensively. They are very effective in damping vibrational energy; this is represented in Figure 12.6, which compares the relative damping capacities of steel and gray iron. Base structures for machines and heavy equipment that are exposed to vibrations are frequently constructed of this material. In addition, gray irons exhibit a high resistance to wear. Furthermore, in the molten state they have a high fluidity at casting temperature, which permits casting pieces having intricate shapes; also, casting shrinkage is low. Finally, and perhaps most important, gray cast irons are the least expensive of all metallic materials.

Gray irons having microstructures different from that shown in Figure 12.5a may be generated by adjustment of composition and/or by using an appropriate treatment. For example, lowering the silicon content or increasing the cooling rate may prevent the complete dissociation of cementite to form graphite (Equation 12.1). Under these circumstances the microstructure consists of graphite flakes embedded in a pearlite matrix. Figure 12.7 compares schematically the several cast iron microstructures obtained by varying the composition and heat treatment.

Ductile (or Nodular) Iron

Adding a small amount of magnesium and/or cerium to the gray iron before casting produces a distinctly different microstructure and set of mechanical properties. Graphite still forms, but as nodules or spherelike particles instead of flakes. The resulting alloy is called **nodular** or **ductile iron,** and a typical microstructure is shown in Figure 12.5b. The matrix phase surrounding these particles is either pearlite or ferrite, depending on heat treatment (Figure 12.7); it is normally pearlite for an as-cast piece. However, a heat treatment for several hours at about 700°C (1300°F) will yield a ferrite matrix as in this photomicrograph. Castings are stronger and much more ductile than gray iron, as a comparison of their mechanical properties in Table 12.5 shows. In fact, ductile iron has mechanical characteristics approaching those of steel. For example, ferritic ductile iron has tensile strengths ranging between 55,000 and 70,000 psi (380 and 480 MPa), and ductilities (as percent elongation) from 10 to 20%. Typical applications for this material include valves, pump bodies, crankshafts, gears, and other automotive and machine components.

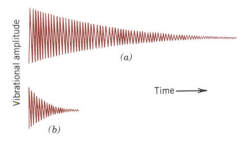

Figure 12.6 Comparison of the relative vibrational damping capacities of (a) steel and (b) gray cast iron. (From *Metals Engineering Quarterly,* February 1961. Copyright 1961 American Society for Metals.)

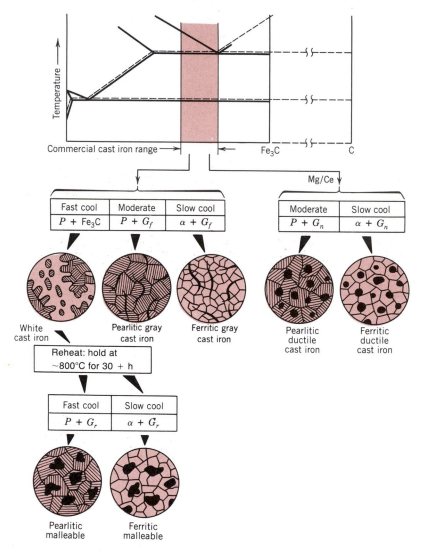

Figure 12.7 From the iron–carbon phase diagram, composition ranges for commercial cast irons. Also shown are microstructures that result from a variety of heat treatments. G_f, flake graphite; G_r, graphite rosettes; G_n, graphite nodules; P, pearlite; α, ferrite. (Adapted from W. G. Moffatt, G. W. Pearsall, and J. Wulff, *The Structure and Properties of Materials,* Vol. 1, *Structure,* p. 195. Copyright © 1964 by John Wiley & Sons, New York. Reprinted by permission of John Wiley & Sons, Inc.)

White Iron and Malleable Iron

For low silicon cast irons (containing less than 1.0 wt% Si) and rapid cooling rates, most of the carbon exists as cementite instead of graphite, as indicated in Figure 12.7. A fracture surface of this alloy has a white appearance, and thus it is termed **white cast iron.** An optical photomicrograph showing the microstructure of white iron is presented in Figure 12.5c. Thick sections may have only a surface layer of white iron that was "chilled" during the casting process; gray iron forms at interior regions, which cool more slowly. As a consequence of large amounts of the cementite phase, white iron is extremely hard but also very brittle, to the point of being virtually unmachinable. Its use is limited to applications that necessitate a very hard and wear-resistant surface, and without a high degree of ductility—for example, as rollers in rolling mills. Generally, white iron is used as an intermediary in the production of yet another cast iron, **malleable iron.**

Heating white iron at temperatures between 800 and 900°C (1470 and 1650°F) for a prolonged time period and in a neutral atmosphere (to prevent oxidation) causes a decomposition of the cementite, forming graphite, which exists in the form of clusters or rosettes surrounded by a ferrite or pearlite matrix, depending on cooling rate, as indicated in Figure 12.7. A photomicrograph of a ferritic malleable iron is presented in Figure 12.5d. The microstructure is similar to that for nodular iron (Figure 12.5b), which accounts for relatively high strength and appreciable ductility or malleability. Some typical mechanical characteristics are also listed in Table 12.5. Representative applications include connecting rods, transmission gears, and differential cases for the automotive industry, and also flanges, pipe fittings, and valve parts for railroad, marine, and other heavy-duty services.

NONFERROUS ALLOYS

Steel and other ferrous alloys are consumed in exceedingly large quantities because they have such a wide range of mechanical properties, may be fabricated with relative ease, and are economical to produce. However, they have some distinct limitations, chiefly: (1) a relatively high density, (2) a comparatively low electrical conductivity, and (3) an inherent susceptibility to corrosion in some common environments. Thus for many applications it is advantageous or even necessary to utilize other alloys having more suitable property combinations. Alloy systems are classified either according to the base metal or according to some specific characteristic that a group of alloys share. This chapter discusses the following metal and alloy systems: copper, aluminum, magnesium, and titanium alloys, the refractory metals, the superalloys, the noble metals, and miscellaneous alloys, including those that have nickel, lead, tin, and zinc as base metals.

On occasion, a distinction is made between cast and wrought alloys. Alloys that are so brittle that forming or shaping by appreciable deformation is not possible ordinarily are cast; these are classified as *cast alloys*. On the other hand, those that are amenable to mechanical deformation are termed **wrought alloys.**

In addition, the heat treatability of an alloy system is mentioned frequently. "Heat treatable" designates an alloy whose mechanical strength is improved by precipitation hardening or a martensitic transformation (normally the former), both of which involve specific heat-treating procedures.

12.7 COPPER AND ITS ALLOYS

Copper and copper-based alloys, possessing a desirable combination of physical properties, have been utilized in quite a variety of applications since antiquity. Unalloyed copper is so soft and ductile that it is difficult to machine; also, it has an almost unlimited capacity to be cold-worked. Furthermore, it is highly resistant to corrosion in diverse environments including the ambient atmosphere, seawater, and some industrial chemicals. The mechanical and corrosion-resistance properties of copper may be improved by alloying. Most copper alloys cannot be hardened or strengthened by heat-treating procedures; consequently, cold working and/or solid solution alloying must be utilized to improve these mechanical properties.

The most common copper alloys are the **brasses** for which zinc, as a substitutional impurity, is the predominant alloying element. As may be observed for the copper–zinc phase diagram (Figure 9.15), the α phase is stable for concentrations up to approximately 35 wt% Zn. This phase has an FCC crystal structure, and α brasses are relatively soft, ductile, and easily cold worked. Brass alloys having a higher zinc content contain both α and β' phases at room temperature. The β' phase has a BCC crystal structure and is harder and stronger than the α phase; consequently, $\alpha + \beta'$ alloys are generally hot worked.

Some of the common brasses are yellow, naval, and cartridge brass, muntz metal, and gilding metal. The compositions, properties, and typical uses of several of these alloys are listed in Table 12.6. Some of the common uses for brass alloys include costume jewelry, cartridge casings, automotive radiators, musical instruments, and coins.

The **bronzes** are alloys of copper and several other elements, including tin, aluminum, silicon, and nickel. These alloys are somewhat stronger than the brasses, yet they still have a high degree of corrosion resistance. Table 12.6 contains several of the bronze alloys, their compositions, properties, and applications. Generally they are utilized when, in addition to corrosion resistance, good tensile properties are required.

A recent generation of high strength copper alloys are the beryllium coppers. They possess a remarkable combination of properties: tensile strengths as high as 200,000 psi (1400 MPa), excellent electrical and corrosion properties, and wear resistance when properly lubricated; they may be cast, hot worked, or cold worked. High strengths are attained by precipitation-hardening heat treatments. These alloys are costly because of the beryllium additions, which range between 1.0 and 2.5 wt%. Applications include jet aircraft landing gear bearings and bushings, springs, and surgical and dental instruments. One of these alloys (C17200) is included in Table 12.6.

12.8 ALUMINUM AND ITS ALLOYS

Aluminum and its alloys are characterized by a relatively low density (2.7 g/cm^3 as compared to 7.9 g/cm^3 for steel), high electrical and thermal conductivities, and a resistance to corrosion in some common environments, including the ambient atmosphere. Many of these alloys are easily formed by virtue of high ductility; this is evidenced by the thin aluminum foil sheet into which the relatively pure material may be rolled. Since aluminum has an FCC crystal structure, its ductility is retained even at very low temperatures. The chief limitation of aluminum is its low melting temperature [660°C (1220°F)], which restricts the maximum temperature at which it can be used.

TABLE 12.6 Compositions, Mechanical Properties, and Typical Applications for Eight Copper Alloys

Alloy Name	UNS Number	Composition (wt%)				Condition	Mechanical Properties			Typical Applications
		Cu	Zn	Sn	Other		Tensile Strength [psi × 10³ (MPa)]	Yield Strength [psi × 10³ (MPa)]	Ductility (%EL in 2 in.)	
Wrought Alloys										
Electrolytic tough pitch	C11000	99.9			0.04O	Annealed	32 (220)	10 (69)	55	Roofing, rivets, radiators
Beryllium–copper	C17200	97.9			1.9Be, 0.2Co	Annealed	68 (470)	25 (172)	48	Springs, diaphragms
						Precipitation hardened	165 (1140)	145 (1000)	7	
Cartridge brass	C26000	70	30			Annealed	44 (303)	11 (76)	66	Ammunition components
Phosphor bronze, 5% A	C51000	95		5	Trace P	Annealed	47 (324)	19 (131)	64	Bellows, welding rods
Copper–nickel, 30%	C71500	70			30Ni	Annealed	54 (372)	20 (138)	45	Saltwater piping
Cast Alloys										
Leaded yellow brass	C85400	67	29	1	3Pb	As cast	34 (234)	12 (83)	35	Battery clamps, fittings
Tin bronze	C90500	88	2	10		As cast	45 (310)	22 (152)	25	Bearings, bushings
Aluminum bronze	C95400	85			4Fe, 11Al	As cast	85 (586)	35 (241)	18	Gears, valve seats

Source: *Metal Progress 1980 Databook.* Copyright © 1980 American Society for Metals.

TABLE 12.7 Compositions, Mechanical Properties, and Typical Applications for Eight Common Aluminum Alloys

Aluminum Association Number	UNS Number	Composition (wt%)[a]				Condition	Mechanical Properties			Typical Applications
		Cu	Mg	Mn	Other		Tensile Strength [psi × 10³ (MPa)]	Yield Strength [psi × 10³ (MPa)]	Ductility (%EL in 2 in.)	
Wrought, Nonheat-Treatable Alloys										
1100	A91100	0.12				Annealed	13 (90)	5 (34)	35	Sheet metalwork
3003	A93003	0.12		1.2		Annealed	16 (110)	6 (42)	30	Cooking utensils
5052	A95052		2.5		0.25Cr	Annealed	28 (195)	13 (90)	25	Bus, truck uses
Wrought, Heat-Treatable Alloys										
2014	A92014	4.4	0.5	0.8	0.8Si	Heat treated	70 (485)	60 (415)	13	General structures
6061	A96061	0.3	1.0		0.6Si, 0.2Cr	Heat treated	45 (310)	40 (275)	12	Trucks, towers, furniture
7075	A97075	1.6	2.5		5.6Zn, 0.23Cr	Heat treated	83 (570)	73 (505)	11	Aircraft structural parts
Cast, Heat-Treatable Alloys										
295.0	A02950	4.5			1.1Si	Heat treated	36 (250)	24 (165)	5	Crankcases, aircraft wheels
356.0	A03560		0.3		7.0Si	Heat treated	33 (230)	24 (165)	4	Water-cooled cylinder blocks

[a] The balance of the composition is aluminum.

Source: Adapted from *Metals Handbook: Properties and Selection: Nonferrous Alloys and Pure Metals*, Vol. 2, 9th edition, H. Baker (Managing Editor), American Society for Metals, 1979.

The mechanical strength of aluminum may be enhanced by cold work and by alloying; however, both processes tend to diminish resistance to corrosion. Principal alloying elements include copper, magnesium, silicon, manganese, and zinc. Nonheat-treatable alloys consist of a single phase, for which an increase in strength is achieved by solid solution strengthening. Others are rendered heat treatable (capable of being precipitation hardened) as a result of alloying.

Generally, aluminum alloys are classified as either cast or wrought. Composition for both types is designated by a four-digit number that indicates the principal impurities, and in some cases, the purity level. For cast alloys, a decimal point is located between the last two digits. After these digits is a hyphen and the basic **temper designation**—a letter and possibly a one- to three-digit number, which indicates the mechanical and/or heat treatment to which the alloy has been subjected. For example, F, H, and O represent, respectively, the as-fabricated, strain-hardened, and annealed states; T3 means that the alloy was solution heat treated, cold worked, and then naturally aged (age hardened). A solution heat treatment followed by artificial aging is indicated by T6. The composition, properties, and applications of several wrought and cast alloys are contained in Table 12.7. Some of the more common applications of aluminum alloys include aircraft structural parts, soda pop cans, bus bodies, and automotive parts (engine blocks, pistons, and manifolds).

Recent attention has been given to alloys of aluminum and other low-density metals (e.g., Mg and Ti) as engineering materials for transportation, to effect reductions in fuel consumption. An important characteristic of these materials is **specific strength,** which is quantified by the tensile strength–specific gravity ratio. Even though an alloy of one of these metals may have a tensile strength that is inferior to a more dense material (such as steel), on a weight basis it will be able to sustain a larger load.

12.9 MAGNESIUM AND ITS ALLOYS

Perhaps the most outstanding characteristic of magnesium is its density, 1.7 g/cm^3, which is the lowest of all the structural metals; therefore, its alloys are used where light weight is an important consideration (e.g., in aircraft components). Magnesium has an HCP crystal structure, is relatively soft, and has a low elastic modulus: 6.5×10^6 psi (45×10^3 MPa). At room temperature magnesium and its alloys are difficult to deform; in fact, only small degrees of cold work may be imposed without annealing. Consequently, most fabrication is by casting or hot working at temperatures between 200 and 350°C (400 and 650°F). Magnesium, like aluminum, has a moderately low melting temperature [651°C (1204°F)]. Chemically, magnesium alloys are relatively unstable and especially susceptible to corrosion in marine environments. On the other hand, corrosion or oxidation resistance is reasonably good in the normal atmosphere. Fine magnesium powder ignites easily when heated in air; consequently, care should be exercised when handling it in this state.

These alloys are also classified as either cast or wrought, and some of them are heat treatable. Aluminum, zinc, manganese, and some of the rare earths are the major alloying elements. A composition–temper designation scheme similar to that for aluminum alloys is also used. Table 12.8 lists several common magnesium alloys, their compositions, properties, and applications. These alloys are used in aircraft and missile applications, as well as in luggage and automobile wheels.

TABLE 12.8 Compositions, Mechanical Properties, and Typical Applications for Six Common Magnesium Alloys

ASTM Number	UNS Number	Composition (wt%)[a]				Condition	Mechanical Properties			Typical Applications
		Al	Mn	Zn	Other		Tensile Strength [psi × 10³ (MPa)]	Yield Strength [psi × 10³ (MPa)]	Ductility (%EL in 2 in.)	
Wrought Alloys										
AZ80A	M11800	8.5	0.12	0.5		As extruded	49 (340)	36 (250)	11	Highly stressed extrusions
HM31A	M13312		1.20		3.0Th	Artificially aged	37 (255)	26 (179)	4	Missile and aircraft use to 425°C
ZK60A	M16600			5.5	0.45Zr	Artificially aged	51 (350)	41 (285)	11	Forgings of maximum strength for aircraft
Cast Alloys										
AZ92A	M11920	9.0	0.10	2.0		As cast	25 (170)	14 (97)	2	Pressure-tight castings
EZ33A	M12330			2.6	3.2 Rare earths, 0.7Zr	Artificially aged	23 (160)	16 (110)	3	Pressure-tight castings for use between 175 and 250°C
AZ91A	M11910	9.0	0.13	0.7		As cast	33 (230)	24 (165)	3	Parts for cars, lawnmowers, luggage

[a] The balance of the composition is magnesium.

Source: Adapted from *Metals Handbook: Properties and Selection: Nonferrous Alloys and Pure Metals*, Vol. 2, 9th edition, H. Baker (Managing Editor), American Society for Metals, 1979.

12.10 TITANIUM AND ITS ALLOYS

Titanium and its alloys are relatively new engineering materials that possess an extraordinary combination of properties. The pure metal has a relatively low density (4.5 g/cm^3), a high melting point [1668°C (3035°F)], and an elastic modulus of 15.5 × 10^6 psi (107 × 10^3 MPa). Titanium alloys are extremely strong; room temperature tensile strengths as high as 200,000 psi (1400 MPa) are attainable, yielding remarkable specific strengths. Furthermore, the alloys are highly ductile and easily forged and machined.

The major limitation of titanium is its chemical reactivity with other materials at elevated temperatures. This property has necessitated the development of nonconventional refining, melting, and casting techniques; consequently, titanium alloys are quite expensive. In spite of this high temperature reactivity, the corrosion resistance of titanium alloys at normal temperatures is unusually high; they are virtually immune to air, marine, and a variety of industrial environments. Table 12.9 presents several titanium alloys along with their typical properties and applications. They are commonly utilized in airplane structures, space vehicles, and in the petroleum and chemical industries.

12.11 THE REFRACTORY METALS

Metals that have extremely high melting temperatures are classified as the refractory metals. Included in this group are niobium (Nb), molybdenum (Mo), tungsten (W), and tantalum (Ta). Melting temperatures range between 2468°C (4474°F) for niobium to 3410°C (6170°F), the highest melting temperature of any metal, for tungsten. Interatomic bonding in these metals is extremely strong, which accounts for the melting temperatures, and, in addition, large elastic moduli and high strengths and hardnesses, at ambient as well as elevated temperatures. The applications of these metals are varied. For example, tantalum and molybdenum are alloyed with stainless steel to improve its corrosion resistance. Molybdenum alloys are utilized for extrusion dies and structural parts in space vehicles; incandescent light filaments, x-ray tubes, and welding electrodes employ tungsten alloys. Tantalum is immune to chemical attack by virtually all environments at temperatures below 150°C, and is frequently used in applications requiring such a corrosion-resistant material.

12.12 THE SUPERALLOYS

The superalloys have superlative combinations of properties. Most are used in aircraft turbine components, which must withstand exposure to severely oxidizing environments and high temperatures for reasonable time periods. Mechanical integrity under these conditions is critical; in this regard, density is an important consideration because centrifugal stresses are diminished in rotating members when the density is reduced. These materials are classified according to the predominant metal in the alloy, which may be cobalt, nickel, or iron. Other alloying elements include the refractory metals (Nb, Mo, W, Ta), chromium, and titanium. In addition to turbine applications, these alloys are utilized in nuclear reactors and petrochemical equipment.

TABLE 12.9 Compositions, Mechanical Properties, and Typical Applications for Four Common Titanium Alloys

Alloy Type	UNS Number	Composition (wt%)	Condition	Mechanical Properties			Typical Applications
				Tensile Strength [psi × 10³ (MPa)]	Yield Strength [psi × 10³ (MPa)]	Ductility (%EL in 2 in.)	
Commercially pure	R50550	99.1Ti	Annealed	75 (517)	65 (448)	25	Chemical, marine, aircraft parts
α	R54521	5Al, 2.5Sn, balance Ti	Annealed	125 (862)	117 (807)	16	Aircraft engine compressor blades
α–β	R56401	6Al, 4V, balance Ti	Annealed	144 (993)	134 (924)	14	Rocket motor cases
β	R58010	13V, 11Cr, 3Al, balance Ti	Precipitation hardened	177 (1220)	170 (1172)	8	High strength fasteners

Source: Adapted from *Metal Progress 1978 Databook*. Copyright © 1978 American Society for Metals.

12.13 THE NOBLE METALS

The noble or precious metals are a group of eight elements that have some physical characteristics in common. They are highly resistant to oxidation and corrosion (hence the "noble"), expensive (precious), and also characteristically soft, ductile, and heat resistant. The noble metals are silver, gold, platinum, palladium, rhodium, ruthenium, iridium, and osmium; the first three are most common and are used extensively in jewelry. Silver and gold may be strengthened by solid solution alloying with copper; sterling silver is a silver–copper alloy containing approximately 7.5 wt% Cu. Alloys of both silver and gold are employed as dental restoration materials; also, some electrical contacts on printed circuit boards are of gold. Platinum is used for chemical laboratory equipment, as a catalyst (especially in the manufacture of gasoline), and in thermocouples to measure elevated temperatures.

12.14 MISCELLANEOUS NONFERROUS ALLOYS

The discussion above covers the vast majority of nonferrous alloys; however, a number of others are found in a variety of engineering applications, and a brief exposure of these is worthwhile.

Nickel and its alloys are highly resistant to corrosion in many environments, especially those that are basic (alkaline). Nickel is often coated or plated on some metals that are susceptible to corrosion as a protective measure. Monel, a nickel-based alloy containing approximately 65 wt% Ni and 28 wt% Cu (the balance iron), has very high strength and is extremely corrosion resistant; it is used in pumps, valves, and other components that are in contact with some acid and petroleum solutions. As already mentioned, nickel is one of the principal alloying elements in stainless steels, and one of the major constituents in the superalloys.

Lead, tin, and their alloys find some use as engineering materials. Both are mechanically soft and weak, have low melting temperatures, are quite resistant to many corrosion environments, and have recrystallization temperatures below room temperature. Many common solders are lead–tin alloys, which have low melting temperatures. Applications for lead and its alloys include x-ray shields, storage batteries, and some plumbing. The primary use of tin is as a very thin coating on the inside of plain carbon steel cans (tin cans) that are used for food containers; this coating inhibits chemical reactions between the steel and the food products.

Unalloyed zinc also is a relatively soft metal having a low melting temperature and a subambient recrystallization temperature. Chemically, it is reactive in a number of common environments and, therefore, susceptible to corrosion. Galvanized steel is just plain carbon steel that has been coated with a thin zinc layer; the zinc preferentially corrodes and protects the steel (Section 18.9). Typical applications of galvanized steel are familiar (sheet metal, fences, screen, screws, etc.). Common applications of zinc alloys include padlocks, automotive parts (door handles and grilles), and office equipment.

SUMMARY

This chapter began with a discussion of various fabrication techniques that may be applied to metallic materials. Forming operations are those in which a metal piece is shaped by plastic deformation. When deformation is carried out above the recrystallization temperature it is termed hot working; otherwise, it is cold working. Forging, rolling, extrusion, and drawing are four of the more common forming techniques. Depending on the properties and shape of the finished piece, casting may be the most desirable and economical fabrication process; sand, die, and investment casting methods were also treated. Additional fabrication procedures, including powder metallurgy and welding, may be utilized alone or in combination with other methods.

With regard to composition, metals and alloys are classified as either ferrous or nonferrous. Ferrous alloys (steels and cast irons) are those in which iron is the prime constituent. Most steels contain less than 1.0 wt% C, and, in addition, other alloying elements, which render them susceptible to heat treatment (and an enhancement of mechanical properties) and/or more corrosion resistant. Plain low-carbon steels and high-strength low-alloy, medium-carbon, tool, and stainless steels are the most common types.

Cast irons contain a higher carbon content, normally between 3.0 and 4.5 wt% C, and other alloying elements, notably silicon. For these materials, most of the carbon exists in graphite form rather than combined with iron as cementite. Gray, ductile (or nodular), and malleable irons are the three most widely used cast irons; the latter two are reasonably ductile.

All other alloys fall within the nonferrous category, which is further subdivided according to base metal or some distinctive characteristic that is shared by a group of alloys. The composition, typical properties, and applications of copper, aluminum, magnesium, titanium, nickel, lead, tin, and zinc alloys, as well as the refractory metals, the superalloys, and the noble metals were discussed.

IMPORTANT TERMS AND CONCEPTS

Alloy steel

Brass

Bronze

Cast iron

Cold working

Drawing

Ductile (nodular) iron

Extrusion

Ferrous alloy

Forging

Gray cast iron

High-strength, low-alloy (HSLA) steel

Hot working

Malleable cast iron

Nonferrous alloy

Plain carbon steel

Powder metallurgy (P/M)

Rolling

Specific strength

Temper designation

Welding

White cast iron

Wrought alloy

REFERENCES

BRICK, R. M., A. W. PENSE, and R. B. GORDON, *Structure and Properties of Engineering Materials,* 4th edition, McGraw-Hill Book Company, New York, 1977.

DIETER, G. E., *Mechanical Metallurgy,* 3rd edition, McGraw-Hill Book Company, New York, 1986. Chapters 15–21 provide an excellent discussion of various metal-forming techniques.

Metals Handbook, 9th edition, Vol. I, *Properties and Selection: Irons and Steels,* American Society for Metals, Metals Park, OH, 1978.

Metals Handbook, 9th edition, Vol. 2, *Properties and Selection: Nonferrous Alloys and Pure Metals,* American Society for Metals, Metals Park, OH, 1979.

Metals Handbook, 9th edition, Vol. 3, *Properties and Selection: Stainless Steels, Tool Materials and Special-Purpose Metals,* American Society for Metals, Metals Park, OH, 1980.

Metals Handbook, 9th edition, Vol. 6, *Welding, Brazing, and Soldering,* American Society for Metals, Metals Park, OH, 1983.

Metals Handbook, 9th edition, Vol. 14, *Forming and Forging,* ASM International, Metals Park, OH, 1988.

Metals Handbook, 9th edition, Vol. 15, *Casting,* ASM International, Metals Park, OH, 1988.

Metals and Alloys in the Unified Numbering System, 5th edition, Society of Automotive Engineers, and American Society for Testing and Materials, Warrendale, PA, 1989.

Walton, C. F. and T. F. Opar (Editors), *Iron Castings Handbook,* Iron Castings Society, Des Plains, IL, 1981.

Welding Handbook, 7th edition, American Welding Society, Miami, FL, 1976. In five volumes.

QUESTIONS AND PROBLEMS

12.1   Cite advantages and disadvantages of hot working and cold working.

12.2   **(a)** Cite advantages of forming metals by extrusion as opposed to rolling. **(b)** Cite some disadvantages.

12.3   List four situations in which casting is the preferred fabrication technique.

12.4   Compare sand, die, and investment casting techniques.

12.5   **(a)** Cite some advantages of powder metallurgy over casting. **(b)** Cite some disadvantages.

12.6   What are the principal differences between welding, brazing, and soldering? You may need to consult another reference.

12.7   Describe one problem that might exist with a steel weld that was cooled very rapidly.

12.8   **(a)** List the four classifications of steels. **(b)** For each, briefly describe the properties and typical applications.

12.9   **(a)** Cite three reasons why ferrous alloys are used so extensively. **(b)** Cite three characteristics of ferrous alloys that limit their utilization.

12.10 Briefly explain why ferritic and austenitic stainless steels are not heat treatable.

12.11 What is the function of alloying elements in tool steels?

12.12 Compute the volume percent of graphite V_{Gr} in a 3.5 wt% C cast iron, assuming that all the carbon exists as the graphite phase. Assume densities of 7.9 and 2.3 g/cm^3 for ferrite and graphite, respectively.

12.13 On the basis of microstructure, briefly explain why gray iron is brittle and weak in tension.

12.14 Compare gray and malleable cast irons with respect to **(a)** composition and heat treatment, **(b)** microstructure, and **(c)** mechanical characteristics.

12.15 It is possible to produce cast irons that consist of a martensite matrix in which graphite is embedded in either flake, nodule, or rosette form. Briefly describe the treatment necessary to produce each of these three microstructures.

12.16 Compare white and nodular cast irons with respect to **(a)** composition and heat treatment, **(b)** microstructure, and **(c)** mechanical characteristics.

12.17 Is it possible to produce malleable cast iron in pieces having large cross-sectional dimensions? Why or why not?

12.18 What is the principal difference between wrought and cast alloys?

12.19 What is the main difference between a brass and a bronze?

12.20 Why must rivets of a 2017 aluminum alloy be refrigerated before they are used?

12.21 Explain why, under some circumstances, it is not advisable to weld a structure that is fabricated with a 3003 aluminum alloy.

12.22 What is the chief difference between heat-treatable and nonheat-treatable alloys?

12.23 Of the following alloys, pick the one(s) that may be strengthened by heat treatment, cold work, or both: tin, 4340 steel, 5052 aluminum, C17200 beryllium–copper, 409 stainless steel, EZ33A magnesium, and cartridge brass.

12.24 A structural member 10 in. (250 mm) long must be able to support a load of 10,000 lb$_f$ (44,400 N) without experiencing any plastic deformation. Given the data below for brass, steel, and aluminum, which alloy will meet these criteria at the minimum weight?

Alloy	Yield Strength (psi [MPa])	Density (g/cm^3)
Brass	60,000 (415)	8.5
Steel	125,000 (860)	7.9
Aluminum	40,000 (275)	2.7

12.25 Give the distinctive features, limitations, and applications of the following alloy groups: titanium alloys, refractory metals, superalloys, and noble metals.

12.26 Discuss whether it would be advisable to hot work or cold work the following metals or alloys on the basis of melting temperature, oxidation resistance, yield strength, and degree of brittleness: tin, tungsten, aluminum alloys, magnesium alloys, and a 4140 steel.

12.27 Below is a list of metals and alloys:

Plain carbon steel	Magnesium
Brass	Zinc
Gray cast iron	Tool steel
Platinum	Aluminum
Stainless steel	Tungsten
Titanium alloy	

Select from this list the one metal or alloy that is best suited for each of the following applications, and cite at least one reason for your choice:

(a) The base for a milling machine.
(b) The walls of a steam boiler.
(c) High-speed aircraft.
(d) Drill bit.
(e) Cryogenic (i.e., very low temperature) container.
(f) As a pyrotechnic (i.e., in flares and fireworks).
(g) High-temperature furnace elements to be used in oxidizing atmospheres.

CHAPTER 13

STRUCTURES AND PROPERTIES OF CERAMICS

Electron micrograph of kaolinite crystals. They are in the form of hexagonal plates, some of which are stacked on top of one another. 21,000×. (Photograph courtesy of Georgia Kaolin Co., Inc.)

13.1 INTRODUCTION

Ceramic materials were discussed briefly in Chapter 1, which noted that they are inorganic and nonmetallic materials. Most ceramics are compounds between metallic and nonmetallic elements for which the interatomic bonds are either totally ionic or predominantly ionic but having some covalent character. The term "ceramic" comes from the Greek word *keramikos,* which means "burnt stuff," indicating that desirable properties of these materials are normally achieved through a high-temperature heat treatment process called firing.

Up until the past 40 or so years, the most important materials in this class were termed the "traditional ceramics," those for which the primary raw material is clay; products considered to be traditional ceramics are china, porcelain, bricks, tiles, and, in addition, glasses and high-temperature ceramics. Of late, significant progress has been made in understanding the fundamental character of these materials and of the phenomena that occur in them that are responsible for their unique properties. Consequently, a new generation of these materials has evolved, and the term "ceramic" has taken on a much broader meaning. To one degree or another, these new materials have a rather dramatic effect on our lives; electronic, computer, communication, aerospace, and a host of other industries rely on their use.

This chapter discusses the types of crystal structure and atomic point defect that are found in ceramic materials and, in addition, some of their mechanical characteristics. Applications and fabrication techniques for this class of materials are treated in the next chapter.

CERAMIC STRUCTURES

Since the atomic bonding in ceramic materials is either partially or totally ionic, most ceramic crystal structures may be thought of as being composed of electrically charged ions instead of atoms. The metallic ions, or **cations,** are positively charged, since they have given up their valence electrons to the nonmetallic ions, or **anions,** which are negatively charged. Because ceramics are composed of at least two elements, and often more, their crystal structures are generally more complex than those of metals.

13.2 CRYSTAL STRUCTURES

Two characteristics of the component ions in crystalline ceramic materials influence the crystal structure: the magnitude of the electrical charge on each of the component ions, and the relative sizes of the cations and anions. With regard to the first characteristic, the crystal must be electrically neutral; that is, all the cation positive charges must be balanced by an equal number of anion negative charges. The chemical formula of a compound indicates the ratio of cations to anions, or the composition that achieves this charge balance. For example, in calcium fluoride, each calcium ion has a $+2$ charge (Ca^{2+}), and associated with each fluorine ion is a single negative charge (F^-). Thus there must be twice as many F^- as Ca^{2+} ions, which is reflected in the chemical formula CaF_2.

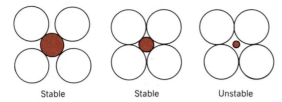

Figure 13.1 Stable and unstable anion–cation coordination configurations. Open circles represent anions; closed circles denote cations.

Stable Stable Unstable

TABLE 13.1 Coordination Numbers and Geometries for Various Cation–Anion Radius Ratios (r_C/r_A)

Coordination Number	Cation–Anion Radius Ratio	Coordination Geometry
2	<0.155	
3	0.155–0.225	
4	0.225–0.414	
6	0.414–0.732	
8	0.732–1.0	

Source: W. D. Kingery, H. K. Bowen, and D. R. Uhlmann, *Introduction to Ceramics,* 2nd edition. Copyright © 1976 by John Wiley & Sons, New York. Reprinted by permission of John Wiley & Sons, Inc.

The second criterion involves the sizes or ionic radii of the cations and anions, r_C and r_A, respectively. Because the metallic elements give up electrons when ionized, cations are ordinarily smaller than anions, and, consequently, the ratio r_C/r_A is less than unity. Each cation prefers to have as many nearest-neighbor anions as possible. The anions also desire a maximum number of cation nearest neighbors.

Stable ceramic crystal structures form when those anions surrounding a cation are all in contact with that cation, as illustrated in Figure 13.1. The coordination number (i.e., number of anion nearest neighbors for a cation) is related to the cation–anion radius ratio. For a specific coordination number, there is a critical or minimum r_C/r_A ratio for which this cation–anion contact is established (Figure 13.1), which ratio may be determined from pure geometrical considerations (see Example Problem 13.1).

The coordination numbers and nearest-neighbor geometries for various r_C/r_A ratios are presented in Table 13.1. For r_C/r_A ratios less than 0.155, the very small cation is bonded to two anions in a linear manner. If r_C/r_A has a value between 0.155 and 0.225, the coordination number for the cation is 3. This means each cation is surrounded by three anions in the form of a planar equilateral triangle, with the cation located in the center. The coordination number is 4 for r_C/r_A between 0.225 and 0.414; the cation is located at the center of a tetrahedron, with anions at each of the four corners. For r_C/r_A between 0.414 and 0.732, the cation may be thought of as being situated at the center of an octahedron surrounded by six anions, one at each corner, as also shown in the table. The coordination number is 8 for r_C/r_A between 0.732 and 1.0, with anions at all corners of a cube and a cation positioned at the center. For a radius ratio greater than unity, the coordination number is 12. The most common coordination numbers for ceramic materials are 4, 6, and 8. Table 13.2 gives the ionic radii for several anions and cations that are common to ceramic materials.

TABLE 13.2 Ionic Radii for Several Cations and Anions (for a Coordination Number of 6)

Cation	Ionic Radius (nm)	Anion	Ionic Radius (nm)
Al^{3+}	0.053	Br^-	0.196
Ba^{2+}	0.136	Cl^-	0.181
Ca^{2+}	0.100	F^-	0.133
Cs^+	0.170	I^-	0.220
Fe^{2+}	0.077	O^{2-}	0.140
Fe^{3+}	0.069	S^{2-}	0.184
K^+	0.138		
Mg^{2+}	0.072		
Mn^{2+}	0.067		
Na^+	0.102		
Ni^{2+}	0.069		
Si^{4+}	0.040		
Ti^{4+}	0.061		

Show that the minimum cation-to-anion radius ratio for the coordination number 3 is 0.155.

SOLUTION

For this coordination, the small cation is surrounded by three anions to form an equilateral triangle as shown below—triangle ABC; the centers of all four ions are coplanar.

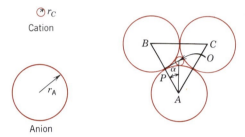

This boils down to a relatively simple plane trigonometry problem. Consideration of the right triangle APO makes it clear that the side lengths are related to the anion and cation radii r_A and r_C as

$$\overline{AP} = r_A$$

and

$$\overline{AO} = r_A + r_C$$

Furthermore, the side length ratio $\overline{AP}/\overline{AO}$ is a function of the angle α as

$$\frac{\overline{AP}}{\overline{AO}} = \cos \alpha$$

The magnitude of α is $30°$, since line $\overline{AO}$ bisects the $60°$ angle BAC. Thus

$$\frac{\overline{AP}}{\overline{AO}} = \frac{r_A}{r_A + r_C} = \cos 30° = \frac{\sqrt{3}}{2}$$

Or, solving for the cation–anion radius ratio,

$$\frac{r_C}{r_A} = \frac{1 - \sqrt{3}/2}{\sqrt{3}/2} = 0.155$$

AX Type Crystal Structures

Some of the common ceramic materials are those in which there are equal numbers of cations and anions. These are often referred to as AX compounds, where A denotes the cation and X the anion. There are several different crystal structures for AX compounds; each is normally named after a common material that assumes the particular structure.

Rock Salt Structure.   Perhaps the most common AX crystal structure is the *sodium chloride* (NaCl) or *rock salt* type. The coordination number for both cations and

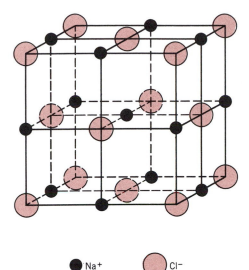

Figure 13.2 A unit cell for the rock salt, or sodium chloride (NaCl), crystal structure.

anions is 6, and therefore the cation–anion radius ratio is between approximately 0.414 and 0.732. A unit cell for this crystal structure (Figure 13.2) is generated from an FCC arrangement of anions with one cation situated at the cube center and one at the center of each of the 12 cube edges. An equivalent crystal structure results from a face-centered arrangement of cations. Thus the rock salt crystal structure may be thought of as two interpenetrating FCC lattices, one composed of the cations, the other of anions. Some of the common ceramic materials that form with this crystal structure are NaCl, MgO, MnS, LiF, and FeO.

Cesium Chloride Structure.   Figure 13.3 shows a unit cell for the *cesium chloride* (CsCl) crystal structure; the coordination number is 8 for both ion types. The anions are located at each of the corners of a cube, whereas at the cube center is a single

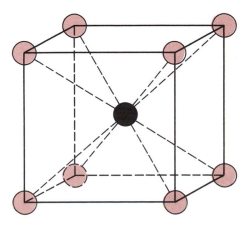

Figure 13.3 A unit cell for the cesium chloride (CsCl) crystal structure.

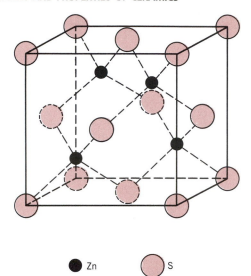

● Zn   ◯ S

Figure 13.4 A unit cell for the zinc blende (ZnS) crystal structure.

cation. Interchange of anions with cations, and vice versa, produces the same crystal structure. This is *not* a BCC crystal structure because ions of two different kinds are involved.

Zinc Blende Structure.   A third AX structure is one in which the coordination number is 4; that is, all ions are tetrahedrally coordinated. This is called the *zinc blende,* or *sphalerite,* structure, after the mineralogical term for zinc sulfide (ZnS). A unit cell is presented in Figure 13.4; all corner and face positions of the cubic cell are occupied by S atoms, while the Zn atoms fill interior tetrahedral positions. An equivalent structure results if Zn and S atom positions are reversed. Thus each Zn atom is bonded to four S atoms, and vice versa. Most often the atomic bonding is highly covalent in compounds exhibiting this crystal structure, which include ZnS, ZnTe, and SiC.

Diamond Cubic Structure.   For diamond, the crystal structure is a variant of the zinc blende, in which carbon atoms occupy all positions (both Zn and S), as indicated in Figure 13.5. Thus each carbon bonds to four other carbons, and these bonds are totally covalent. This is appropriately called the *diamond cubic* crystal structure, which is also found for other Group IVA elements in the periodic table [e.g., germanium and gray tin, below 13°C (55°F)].

Structure of Graphite.   Another polymorph of carbon is graphite, which has a crystal structure (Figure 13.6) distinctly different from that of diamond. This structure is composed of layers of hexagonally arranged carbon atoms; within the layers, each carbon atom is bonded to three coplanar neighbor atoms by strong covalent bonds. Interlayer bonds are weaker and of the van der Waals type. Graphite is more stable than diamond at ambient temperature and pressure. Furthermore, carbon may also exist in a noncrystalline, or amorphous, state.

Graphite is sometimes considered to be a ceramic material even though it is an element, contrary to the normal ceramic definition (Section 1.3). As a consequence

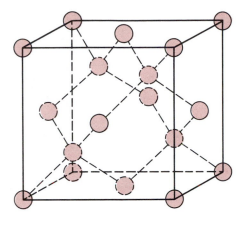

 C

Figure 13.5 A unit cell for the diamond cubic crystal structure.

of the weak van der Waals bonds, interplanar cleavage is facile, which gives rise to graphite's excellent lubricative properties. The electrical conductivity is relatively high in crystallographic directions parallel to the hexagonal sheets.

A$_m$X$_p$ Type Crystal Structures

If the charges on the cations and anions are not the same, a compound can exist with the chemical formula A$_m$X$_p$, where m and/or $p \neq 1$. An example would be AX$_2$, for which a common crystal structure is found in *fluorite* (CaF$_2$). The ionic radii ratio r_C/r_A for CaF$_2$ is about 0.8 which, according to Table 13.1, gives a coordination number of 8. Calcium ions are positioned at the centers of cubes, with fluorine ions at the corners. The chemical formula shows that there are only half as many Ca^{2+}

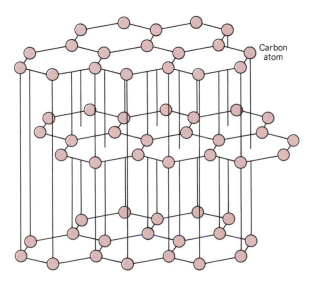

Carbon atom

Figure 13.6 The structure of graphite.

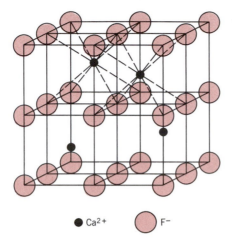

● Ca²⁺   ○ F⁻

Figure 13.7 A unit cell for the fluorite (CaF$_2$) crystal structure.

ions as F⁻ ions, and therefore the crystal structure would be similar to CsCl (Figure 13.3), except that only half the center cube positions are occupied by Ca^{2+} ions. One unit cell consists of eight cubes, as indicated in Figure 13.7. Other compounds that have this crystal structure include UO_2, PuO_2, and ThO_2.

A$_m$B$_n$X$_p$ Type Crystal Structures

It is also possible for ceramic compounds to have more than one type of cation; for two types of cations (represented by A and B), their chemical formula may be designated as A$_m$B$_n$X$_p$. Barium titanate ($BaTiO_3$), having both Ba^{2+} and Ti^{4+} cations, falls into this classification. This material has a *perovskite crystal structure* and rather interesting electromechanical properties to be discussed later. At temperatures above 120°C (248°F), the crystal structure is cubic. A unit cell of this structure is shown in

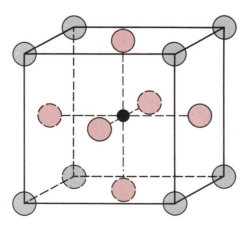

● Ti⁴⁺   ○ Ba²⁺   ○ O²⁻

Figure 13.8 A unit cell for the perovskite crystal structure.

TABLE 13.3 Summary of Some Common Ceramic Crystal Structures

Structure Name	Structure Type	Anion Packing	Coordination Numbers		Examples
			Cation	Anion	
Rock salt (sodium chloride)	AX	FCC	6	6	NaCl, MgO, FeO
Cesium chloride	AX	Simple cubic	8	8	CsCl
Zinc blende (sphalerite)	AX	FCC	4	4	ZnS, SiC
Fluorite	AX_2	Simple cubic	8	4	CaF_2, UO_2, ThO_2
Perovskite	ABX_3	FCC	12(A) 6(B)	6	$BaTiO_3$, $SrZrO_3$, $SrSnO_3$
Spinel	AB_2X_4	FCC	4(A) 6(B)	4	$MgAl_2O_4$, $FeAl_2O_4$

Source: W. D. Kingery, H. K. Bowen, and D. R. Uhlmann, *Introduction to Ceramics*, 2nd edition. Copyright © 1976 by John Wiley & Sons, New York. Reprinted by permission of John Wiley & Sons, Inc.

Figure 13.8; Ba^{2+} ions are situated at all eight corners of the cube and a single Ti^{4+} is at the cube center, with O^{2-} ions located at the center of each of the six faces.

Table 13.3 summarizes the rock salt, cesium chloride, zinc blende, fluorite, and perovskite crystal structures in terms of cation–anion ratios and coordination numbers, and gives examples for each. Of course, many other ceramic crystal structures are possible.

Crystal Structures from the Close Packing of Anions

It may be recalled (Section 3.11) that for metals, close-packed planes of atoms stacked on one another generate both FCC and HCP crystal structures. Similarly, a number of ceramic crystal structures may be considered in terms of close-packed planes of ions, as well as by unit cells. Ordinarily, the close-packed planes are composed of the large anions. As these planes are stacked atop each other, small interstitial sites are created between them in which the cations may reside.

These interstitial positions exist in two different types, as illustrated in Figure 13.9. Four atoms (three in one plane, and a single one in the adjacent plane) surround one type, labeled T in the figure; this is termed a **tetrahedral position,** since straight lines drawn from the centers of the surrounding spheres form a four-sided tetrahedron. The other site type, denoted as O in Figure 13.9, involves six ion spheres, three in each of the two planes. Because an octahedron is produced by joining these six sphere centers, this site is called an **octahedral position.** Thus the coordination numbers for cations filling tetrahedral and octahedral positions are 4 and 6, respectively. Furthermore, for each of these anion spheres, one octahedral and two tetrahedral positions will exist.

Ceramic crystal structures of this type depend on two factors: (1) the stacking of the close-packed anion layers (both FCC and HCP arrangements are possible, which correspond to *ABCABC*... and *ABABAB*... sequences, respectively), and (2) the manner in which the interstitial sites are filled with cations. For example, consider

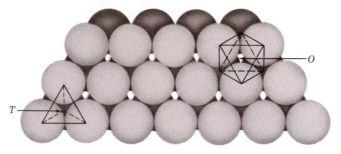

Figure 13.9 The stacking of one plane of close-packed spheres (anions) on top of another; tetrahedral and octahedral positions between the planes are designated by T and O, respectively. (From W. G. Moffatt, G. W. Pearsall, and J. Wulff, *The Structure and Properties of Materials*, Vol. 1, *Structure*. Copyright © 1964 by John Wiley & Sons, New York. Reprinted by permission of John Wiley & Sons, Inc.)

the rock salt crystal structure discussed above. The unit cell has cubic symmetry, and each cation (Na$^+$ ion) has six Cl$^-$ ion nearest neighbors, as may be verified from Figure 13.2. That is, the Na$^+$ ion at the center has as nearest neighbors the six Cl$^-$ ions that reside at the centers of each of the cube faces. The crystal structure, having cubic symmetry, may be considered in terms of an FCC array of close-packed planes of anions, and all planes are of the {111} type. The cations reside in octahedral positions because they have as nearest neighbors six anions. Furthermore, all octahedral positions are filled, since there is a single octahedral site per anion, and the ratio of anions to cations is 1:1. For this crystal structure, the relationship between

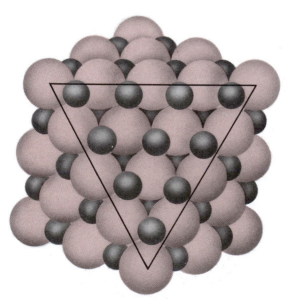

Figure 13.10 A section of the rock salt crystal structure from which a corner has been removed. The exposed plane of anions (light spheres inside the triangle) is a {111}-type plane; the cations (dark spheres) occupy the interstitial octahedral positions.

the unit cell and close-packed anion plane stacking schemes is illustrated in Figure 13.10.

Other, but not all, ceramic crystal structures may be treated in a similar manner; included are the zinc blende and perovskite structures. The *spinel structure* is one of the $A_m B_n X_p$ types, which is found for magnesium aluminate or spinel $(MgAl_2O_4)$. With this structure, the O^{2-} ions form an FCC lattice, whereas Mg^{2+} ions fill tetrahedral sites and Al^{3+} reside in octahedral positions. Magnetic ceramics, or ferrites, have a crystal structure that is a slight variant of this spinel structure; and the magnetic characteristics are affected by the occupancy of tetrahedral and octahedral positions (see Section 21.5).

EXAMPLE PROBLEM 13.2.

On the basis of ionic radii, what crystal structure would you predict for FeO?

SOLUTION
First, note that FeO is an AX-type compound. Next, determine the cation–anion radius ratio, which from Table 13.2 is

$$\frac{r_{Fe^{2+}}}{r_{O^{2-}}} = \frac{0.077 \text{ nm}}{0.140 \text{ nm}} = 0.550$$

This value lies between 0.414 and 0.732, and, therefore, from Table 13.1 the coordination number for the Fe^{2+} ion is 6; this is also the coordination number of O^{2-}, since there are equal numbers of cations and anions. The predicted crystal structure will be sodium chloride, which is the AX crystal structure having a coordination number of 6, as given in Table 13.3.

Ceramic Density Computations

It is possible to compute the theoretical density of a crystalline ceramic material from unit cell data in a manner similar to that described in Section 3.5 for metals. In this case the density ρ may be determined using a modified form of Equation 3.5, as follows:

$$\rho = \frac{n'(\sum A_C + \sum A_A)}{V_C N_A} \tag{13.1}$$

where

n' = the number of formula units[1] within the unit cell
$\sum A_C$ = the sum of the atomic weights of all cations in the formula unit
$\sum A_A$ = the sum of the atomic weights of all anions in the formula unit
V_C = the unit cell volume
N_A = Avogadro's number, 6.023×10^{23} formula units/mol.

---

[1] By "formula unit" we mean all the ions that are included in the chemical formula unit. For example, for $BaTiO_3$, a formula unit consists of one barium ion, a titanium ion, and three oxygen ions.

EXAMPLE PROBLEM 13.3

On the basis of crystal structure, compute the theoretical density for sodium chloride. How does this compare with its measured density?

SOLUTION

The density may be determined using Equation 13.1, where n', the number of NaCl units per unit cell, is 4 because both sodium and chloride ions form FCC lattices. Furthermore,

$$\sum A_C = A_{Na} = 22.99 \text{ g/mol}$$

$$\sum A_A = A_{Cl} = 35.45 \text{ g/mol}$$

Since the unit cell is cubic, $V_C = a^3$, a being the unit cell edge length. For the face of the cubic unit cell shown here,

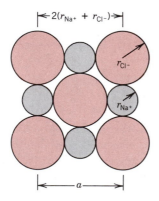

$$a = 2r_{Na^+} + 2r_{Cl^-}$$

r_{Na^+} and r_{Cl^-} being the sodium and chloride ionic radii, given in Table 13.2 as 0.102 and 0.181 nm, respectively.

Thus

$$V_C = a^3 = (2r_{Na^+} + 2r_{Cl^-})^3$$

And finally,

$$\rho = \frac{n'(A_{Na} + A_{Cl})}{(2r_{Na^+} + 2r_{Cl^-})^3 N_A}$$

$$= \frac{4(22.99 + 35.45)}{[2(0.102 \times 10^{-7}) + 2(0.181 \times 10^{-7})]^3(6.023 \times 10^{23})}$$

$$= 2.14 \text{ g/cm}^3$$

This compares very favorably with the experimental value of 2.16 g/cm³.

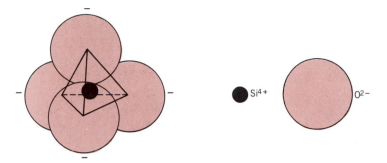

Figure 13.11 A silicon–oxygen (SiO_4^{4-}) tetrahedron.

13.3 SILICATE CERAMICS

Silicates are materials composed primarily of silicon and oxygen, the two most abundant elements in the earth's crust; consequently, the bulk of soils, rocks, clays, and sand come under the silicate classification. Rather than characterizing the crystal structures of these materials in terms of unit cells, it is more convenient to use various arrangements of an SiO_4^{4-} tetrahedron (Figure 13.11). Each atom of silicon is bonded to four oxygen atoms, which are situated at the corners of the tetrahedron; the silicon atom is positioned at the center. Since this is the basic unit of the silicates, it is often treated as a negatively charged entity.

Often the silicates are not considered to be ionic because there is a significant covalent character to the interatomic Si–O bonds, which are directional and relatively strong. Regardless of the character of the Si–O bond, there is a -4 charge associated with every SiO_4^{4-} tetrahedron, since each of the four oxygen atoms requires an extra electron to achieve a stable electronic structure. Various silicate structures arise from the different ways in which the SiO_4^{4-} units can be combined into one-, two-, and three-dimensional arrangements.

Silica

Chemically, the most simple silicate material is silicon dioxide, or silica (SiO_2). Structurally, it is a three-dimensional network that is generated when every corner oxygen atom in each tetrahedron is shared by adjacent tetrahedra. Thus the material is electrically neutral and all atoms have stable electronic structures. Under these circumstances the ratio of Si to O atoms is 1:2, as indicated by the chemical formula.

If these tetrahedra are arrayed in a regular and ordered manner, a crystalline structure is formed. There are three primary polymorphic crystalline forms of silica: quartz, cristobalite (Figure 13.12), and tridymite. Their structures are relatively complicated, and comparatively open; that is, the atoms are not closely packed together. As a consequence, these crystalline silicas have relatively low densities; for example, at room temperature quartz has a density of only 2.65 g/cm^3. The strength of the Si–O interatomic bonds is reflected in a relatively high melting temperature, 1710°C (3110°F).

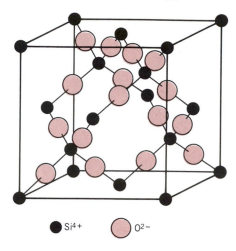

Si⁴⁺   O²⁻

Figure 13.12 The arrangement of silicon and oxygen atoms in a unit cell of cristobalite, a polymorph of SiO_2.

Silica Glasses

Silica can also be made to exist as a noncrystalline solid or glass, having a high degree of atomic randomness, which is characteristic of the liquid; such a material is called *fused silica,* or *vitreous silica.* As with crystalline silica, the SiO_4^{4-} tetrahedron is the basic unit; beyond this structure, considerable disorder exists. The structures for crystalline and noncrystalline silica are compared schematically in Figure 3.21. Other

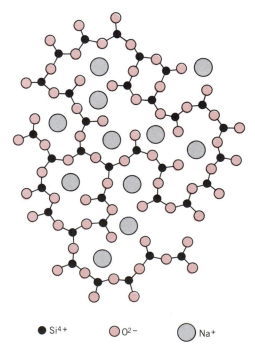

Si⁴⁺   O²⁻   Na⁺

Figure 13.13 Schematic representation of ion positions in a sodium–silicate glass.

oxides (e.g., B_2O_3 and GeO_2) may also form glassy structures, in that crystallization occurs with some difficulty upon cooling from the liquid.

The common inorganic glasses that are used for containers, windows, and so on are silica glasses to which have been added other oxides, such as CaO and Na_2O. The cations from these other oxides (Na^+ and Ca^{2+}) fit into the network and modify it to the degree that the formation of a glass structure is more probable than a crystalline one. Figure 13.13 is a schematic representation of the structure of a sodium–silicate glass.

The Silicates

For the various silicate minerals, one, two, or three of the corner oxygen atoms of the SiO_4^{4-} tetrahedra are shared by other tetrahedra to form some rather complex structures. Some of these, represented in Figure 13.14, have formulas SiO_4^{4-}, $Si_2O_7^{6-}$, $Si_3O_9^{6-}$, and so on; single-chain structures are also possible, as in Figure 13.14e. Positively charged cations such as Ca^{2+}, Mg^{2+}, and Al^{3+} serve two roles. First, they compensate the negative charges from the SiO_4^{4-} units so that charge neutrality is achieved; and second, these cations ionically bond the SiO_4^{4-} tetrahedra together.

Simple Silicates.   Of these silicates, the most structurally simple ones involve isolated tetrahedra (Figure 13.14a). For example, forsterite (Mg_2SiO_4) has the equivalent of two Mg^{2+} ions associated with each tetrahedron in such a way that every Mg^{2+} ion has six oxygen nearest neighbors.

The $Si_2O_7^{6-}$ ion is formed when two tetrahedra share a common oxygen atom (Figure 13.14b). Akermanite ($Ca_2MgSi_2O_7$) is a mineral having the equivalent of two Ca^{2+} ions and one Mg^{2+} ion bonded to each $Si_2O_7^{6-}$ unit.

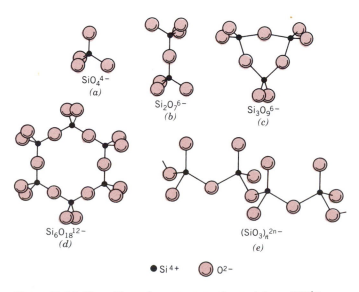

SiO_4^{4-}
(a)

$Si_2O_7^{6-}$
(b)

$Si_3O_9^{6-}$
(c)

$Si_6O_{18}^{12-}$
(d)

$(SiO_3)_n^{2n-}$
(e)

● Si^{4+} ◯ O^{2-}

Figure 13.14 Five silicate ion structures formed from SiO_4^{4-} tetrahedra.

Layered Silicates.   A two-dimensional sheet or layered structure can also be produced by the sharing of three oxygen ions in each of the tetrahedra (Figure 13.15); for this structure the repeating unit formula may be represented by $(Si_2O_5)^{2-}$. The net negative charge is associated with the unbonded oxygen atoms projecting out of the plane of the page. Electroneutrality is ordinarily established by a second planar sheet structure having an excess of cations, which bond to these unbonded oxygen atoms from the Si_2O_5 sheet. Such materials are called the sheet or layered silicates, and their basic structure is characteristic of the clays and other minerals.

One of the most common clay minerals, kaolinite, has a relatively simple two-layer silicate sheet structure. Kaolinite clay has the formula $Al_2(Si_2O_5)(OH)_4$ in which the silica tetrahedral layer, represented by $(Si_2O_5)^{2-}$, is made electrically neutral by an adjacent $Al_2(OH)_4^{2+}$ layer. A single sheet of this structure is shown in Figure 13.16, which is exploded in the vertical direction to provide a better perspective of the ion positions; the two distinct layers are indicated in the figure. The midplane of anions consists of O^{2-} ions from the $(Si_2O_5)^{2-}$ layer, as well as OH^- ions that are a part of the $Al_2(OH)_4^{2+}$ layer. Whereas the bonding within this two-layered sheet is strong and intermediate ionic–covalent, adjacent sheets are only loosely bound to one another by weak van der Waals forces.

A crystal of kaolinite is made of a series of these double layers or sheets stacked parallel to each other, which form small flat plates typically less than 1 μm in diameter and nearly hexagonal. On page 392 is an electron micrograph of kaolinite crystals at a high magnification, showing the hexagonal crystal plates some of which are piled one on top of the other.

These silicate sheet structures are not confined to the clays; other minerals also in this group are talc $[Mg_3(Si_2O_5)_2(OH)_2]$ and the micas [e.g., muscovite,

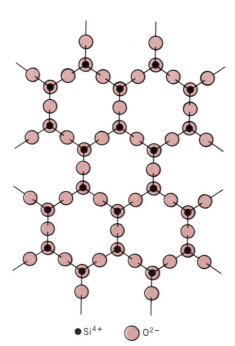

● Si^{4+} ◉ O^{2-}

Figure 13.15 Schematic representation of the two-dimensional silicate sheet structure having a repeat unit formula of $(Si_2O_5)^{2-}$.

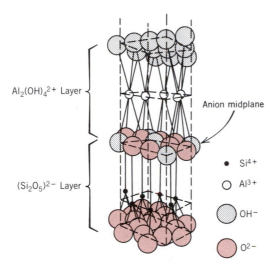

Al$_2$(OH)$_4^{2+}$ Layer

Anion midplane

(Si$_2$O$_5$)$^{2-}$ Layer

• Si^{4+}

○ Al^{3+}

◍ OH$^-$

⬤ O^{2-}

Figure 13.16 The structure of kaolinite clay. (Adapted from W. E. Hauth, "Crystal Chemistry of Ceramics," *American Ceramic Society Bulletin,* Vol. 30, No. 4, 1951, p. 140.)

KAl$_3$Si$_3$O$_{10}$(OH)$_2$], which are important ceramic raw materials. As might be deduced from the chemical formulas, the structures for some silicates are among the most complex of all the inorganic materials.

13.4 IMPERFECTIONS IN CERAMICS

Atomic Point Defects

Atomic defects involving host atoms may exist in ceramic compounds. As with metals, both vacancies and interstitials are possible; however, since ceramic materials contain ions of at least two kinds, defects for each ion type may occur. For example, in NaCl, Na interstitials and vacancies and Cl interstitials and vacancies may exist. It is highly improbable that there would be appreciable concentrations of anion interstitials. The anion is relatively large, and to fit into a small interstitial position, substantial strains on the surrounding ions must be introduced. Anion and cation vacancies and a cation interstitial are represented in Figure 13.17.

The expression **defect structure** is often used to designate the types and concentrations of atomic defects in ceramics. Because the atoms exist as charged ions, when defect structures are considered, conditions of electroneutrality must be maintained. **Electroneutrality** is the state that exists when there are equal numbers of positive and negative charges from the ions. As a consequence, defects in ceramics do not occur alone. One such type of defect involves a cation–vacancy and a cation–interstitial pair. This is called a **Frenkel defect** (Figure 13.18). It might be thought of as being formed by a cation leaving its normal position and moving into an interstitial site. There is no change in charge because the cation maintains the same positive charge as an interstitial.

Another type of defect found in AX materials is a cation vacancy–anion vacancy pair known as a **Schottky defect,** also schematically diagrammed in Figure 13.18. This defect might be thought of as being created by removing one cation and one anion from the interior of the crystal and then placing them both at an external surface.

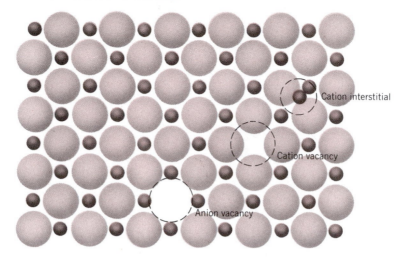

Figure 13.17 Schematic representations of cation and anion vacancies and a cation interstitial. (From W. G. Moffatt, G. W. Pearsall, and J. Wulff, *The Structure and Properties of Materials,* Vol. 1, *Structure,* p. 78. Copyright © 1964 by John Wiley & Sons, New York. Reprinted by permission of John Wiley & Sons, Inc.)

Since both cations and anions have the same charge, and since for every anion vacancy there exists a cation vacancy, the charge neutrality of the crystal is maintained.

The ratio of cations to anions is not altered by the formation of either a Frenkel or a Schottky defect. If no other defects are present, the material is said to be stoichiometric. **Stoichiometry** may be defined as a state for ionic compounds wherein

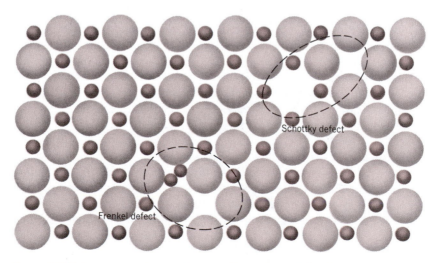

Figure 13.18 Schematic diagram showing Frenkel and Schottky defects in ionic solids. (From W. G. Moffatt, G. W. Pearsall, and J. Wulff, *The Structure and Properties of Materials,* Vol. 1, *Structure,* p. 78. Copyright © 1964 by John Wiley & Sons, New York. Reprinted by permission of John Wiley & Sons, Inc.)

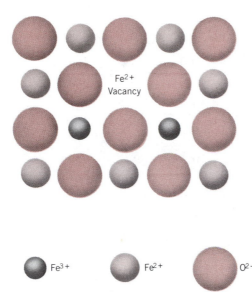

Fe^{2+}
Vacancy

Fe^{3+} Fe^{2+} O^{2-}

Figure 13.19 Schematic representation of an Fe^{2+} vacancy in FeO that results from the formation of two Fe^{3+} ions.

there is the exact ratio of cations to anions as predicted by the chemical formula. For example, NaCl is stoichiometric if the ratio of Na^+ ions to Cl^- ions is exactly 1:1. A ceramic is *nonstoichiometric* if there is any deviation from this exact ratio.

Nonstoichiometry may occur for some ceramic materials in which two valence (or ionic) states exist for one of the ion types. Iron oxide (wüstite, FeO), is one such material, for the iron can be present in both Fe^{2+} and Fe^{3+} states; the number of each of these ion types depends on temperature and the ambient oxygen pressure. The formation of an Fe^{3+} ion disrupts the electroneutrality of the crystal by introducing an excess $+1$ charge, which must be offset by some type of defect. This may be accomplished by the formation of one Fe^{2+} vacancy (or the removal of two positive charges) for every two Fe^{3+} ions that are formed (Figure 13.19). The crystal is no longer stoichiometric because there is one more O ion than Fe ion; however, the crystal remains electrically neutral. This phenomenon is fairly common in iron oxide, and, in fact, its chemical formula is often written as $Fe_{1-x}O$ (where x is some small and variable fraction substantially less than unity) to indicate a condition of nonstoichiometry with a deficiency of Fe.

Impurities in Ceramics

Impurity atoms can form solid solutions in ceramic materials much as they do in metals. Solid solutions of both substitutional and interstitial types are possible. For interstitial, the ionic radius of the impurity must be relatively small in comparison to the anion. Since there are both anions and cations, a substitutional impurity will substitute for the host ion to which it is most similar in an electrical sense: if the impurity atom normally forms a cation in a ceramic material, it most probably will substitute for a host cation. For example, in sodium chloride, impurity Ca^{2+} and O^{2-} ions would most likely substitute for Na^+ and Cl^- ions, respectively. Schematic

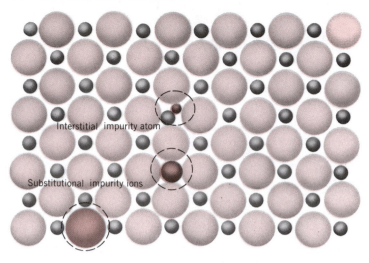

Figure 13.20 Schematic representation of interstitial, anion-substitutional, and cation-substitutional impurity atoms in an ionic compound. (Adapted from W. G. Moffatt, G. W. Pearsall, and J. Wulff, *The Structure and Properties of Materials,* Vol. 1, *Structure,* p. 78. Copyright © 1964 by John Wiley & Sons, New York. Reprinted by permission of John Wiley & Sons, Inc.)

representations for cation and anion substitutional as well as interstitial impurities are shown in Figure 13.20. To achieve any appreciable solid solubility of substituting impurity atoms, the ionic size and charge must be very nearly the same as those of one of the host ions. For an impurity ion having a charge different from the host ion for which it substitutes, the crystal must compensate for this difference in charge so that electroneutrality is maintained with the solid. One way this is accomplished is by the formation of lattice defects—vacancies or interstitials of both ion types, as discussed above.

EXAMPLE PROBLEM 13.4.

If electroneutrality is to be preserved, what point defects are possible in NaCl when a Ca^{2+} substitutes for an Na^+ ion? How many of these defects exist for every Ca^{2+} ion?

SOLUTION
Replacement of an Na^+ by a Ca^{2+} ion introduces one extra positive charge. Electroneutrality is maintained when either a single positive charge is eliminated or another single negative charge is added. Removal of a positive charge is accomplished by the formation of one Na^+ vacancy. Alternatively, a Cl^- interstitial will supply an additional negative charge, negating the effect of each Ca^{2+} ion. However, as mentioned above, the formation of this defect is highly unlikely.

13.5 CERAMIC PHASE DIAGRAMS

Phase diagrams have been experimentally determined for a large number of ceramic systems. For binary or two-component phase diagrams, it is frequently the case that the two components are compounds that share a common element, often oxygen. These diagrams may have configurations similar to metal–metal systems, and they are interpreted in the same way. For a review of the interpretation of phase diagrams, the reader is referred to Section 9.6.

The Al_2O_3–Cr_2O_3 System

One of the relatively simple ceramic phase diagrams is that found for the aluminum oxide–chromium oxide system, Figure 13.21. This diagram has the same form as the isomorphous copper–nickel phase diagram (Figure 9.2a), consisting of single liquid and single solid phase regions separated by a two-phase solid–liquid region having the shape of a blade. The Al_2O_3–Cr_2O_3 solid solution is a substitutional one in which Al^{3+} substitutes for Cr^{3+}, and vice versa. It exists for all compositions below the melting point of Al_2O_3 inasmuch as both aluminum and chromium ions have the same charge as well as similar radii (0.053 and 0.062 nm, respectively). Furthermore, both Al_2O_3 and Cr_2O_3 have the same crystal structure.

The MgO–Al_2O_3 System

The phase diagram for the magnesium oxide–aluminum oxide system (Figure 13.22) is similar in many respects to the lead–magnesium diagram (Figure 9.16). There exists an intermediate phase, or better, a compound called *spinel,* which has the chemical formula $MgAl_2O_4$ (or MgO–Al_2O_3). Even though spinel is a distinct compound [of composition 50 mol% Al_2O_3–50 mol% MgO (73 wt% Al_2O_3–27 wt% MgO)], it is represented on the phase diagram as a single-phase field rather than as a vertical line, as for Mg_2Pb (Figure 9.16); that is, there is a range of compositions over which spinel is a stable compound. Thus spinel is nonstoichiometric for other than the 50 mol% Al_2O_3–50 mol% MgO composition. Furthermore, there is limited solubility of Al_2O_3 in MgO below about 1600°C (2900°F) at the left-hand extremity of Figure 13.22, which is due primarily to the differences in charge and radii of the Mg^{2+} and Al^{3+} ions (0.072 versus 0.053 nm). For the same reasons, MgO is virtually insoluble in Al_2O_3, as evidenced by a lack of a terminal solid solution on the right-hand side of the phase diagram. Also, two eutectics are found, one on either side of the spinel phase field, and stoichiometric spinel melts congruently at about 2100°C (3800°F).

The ZrO_2–CaO System

Another important binary ceramic system is that for zirconium oxide (zirconia) and calcium oxide (calcia); a portion of this phase diagram is shown in Figure 13.23. The horizontal axis extends only to about 31 wt% CaO (50 mol% CaO), at which composition the compound $ZrCaO_3$ forms. It is worth noting that eutectic (2280°C and 24.5 wt% CaO), eutectoid (940°C and 6.3 wt% CaO), and peritectic (2560°C and 7.7 wt% CaO) reactions are found for this system.

It may also be noted from Figure 13.23 that phases having three different crystal structures exist in this system, namely, tetragonal, monoclinic, and cubic. Pure ZrO_2

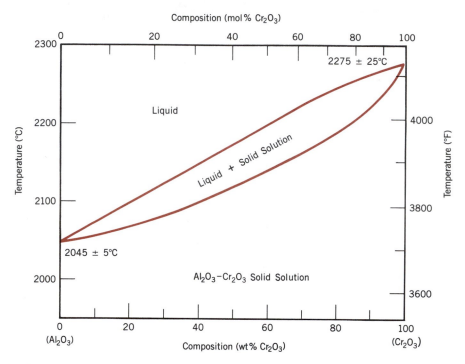

Figure 13.21 The aluminum oxide–chromium oxide phase diagram. (Adapted from E. N. Bunting, "Phase Equilibria in the System Cr_2O_3–Al_2O_3," *Bur. Standards J. Research,* **6,** 1931, p. 948.)

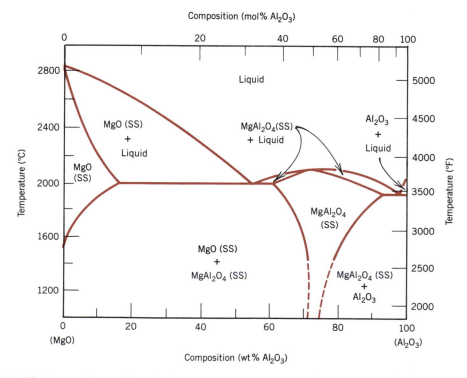

Figure 13.22 The magnesium oxide–aluminum oxide phase diagram; *SS* denotes solid solution. (Adapted from A. M. Alper, R. N. McNally, P. H. Ribbe, and R. C. Doman, "The System MgO–$MgAl_2O_4$," *J. Am. Ceram. Soc.,* **45,** 1962, p. 264; and D. M. Roy, R. Roy, and E. F. Osborn, "The System MgO–Al_2O_3–H_2O and Influence of Carbonate and Nitrate Ions on the Phase Equilibria," *Am. J. Sci.,* **251,** 1953, p. 341.)

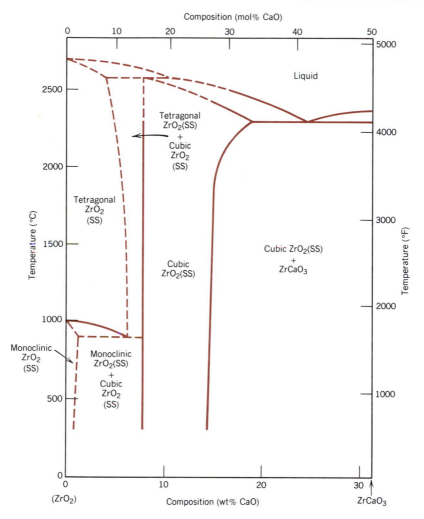

Figure 13.23 A portion of the zirconia–calcia phase diagram; *SS* denotes solid solution. (Adapted from "Stabilization of Zirconia with Calcia and Magnesia," by P. Duwez, F. Odell and F. H. Brown, Jr., *J. Am. Ceram. Soc.,* **35** [5], 109, (1952). Reprinted by permission of the American Ceramic Society.)

experiences a tetragonal-to-monoclinic phase transformation at 1000°C (1832°F). A relatively large volume change accompanies this transformation, resulting in the formation of cracks, which render a ceramic ware useless. This problem is overcome by "stabilizing" the zirconia by adding between 8 and 14 wt% CaO to form the cubic phase. Upon either heating or cooling, this cubic phase experiences no such phase transition, and thus the crack resistance is improved significantly.

The SiO₂–Al₂O₃ System

Commercially, the silica–alumina system is an important one since the principal constituents of many ceramic refractories are these two materials. Figure 13.24 shows

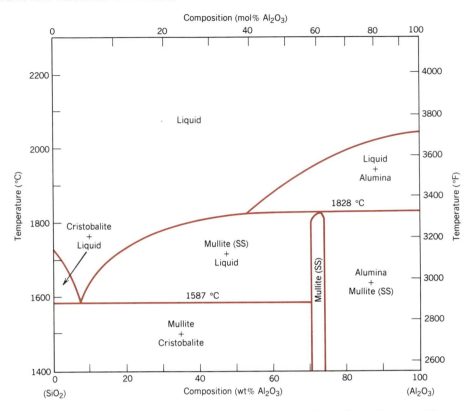

Figure 13.24 The silica–alumina phase diagram. (Adapted from I. A. Aksay and J. A. Pask, *Science,* **183,** 70, 1974. Copyright 1974 by the AAAS.)

the SiO_2–Al_2O_3 phase diagram. The polymorphic form of silica that is stable at these temperatures is termed *cristobalite,* the unit cell for which is shown in Figure 13.12. Silica and alumina are not mutually soluble in one another, which is evidenced by the absence of terminal solid solutions at both extremities of the phase diagram. Also, it may be noted that the intermediate compound *mullite,* $3Al_2O_3$–$2SiO_2$, forms at a composition of about 72 wt% Al_2O_3, which melts incongruently at 1828°C (3322°F). A single eutectic exists at 1587°C (2890°F) and 7.7 wt% Al_2O_3. In Sections 14.10 and 14.11 refractory ceramic materials, the prime constituents for which are silica and alumina, are discussed.

MECHANICAL PROPERTIES

Ceramic materials are somewhat limited in applicability by their mechanical properties, which in many respects are inferior to those of metals. The principal drawback is a disposition to catastrophic fracture in a brittle manner with very little energy absorption.

13.6 BRITTLE FRACTURE OF CERAMICS

At room temperature, both crystalline and noncrystalline ceramics almost always fracture before any plastic deformation can occur in response to an applied tensile load. The topics of brittle fracture and fracture mechanics, as discussed previously in Sections 8.4 and 8.5, also relate to the fracture of ceramic materials; they will be reviewed briefly in this context.

The brittle fracture process consists of the formation and propagation of cracks through the cross section of material in a direction perpendicular to the applied load. Crack growth in crystalline ceramics is usually through the grains (i.e., transgranular) and along specific crystallographic (or cleavage) planes, planes of high atomic density.

The measured fracture strengths of ceramic materials are substantially lower than predicted by theory from interatomic bonding forces. This may be explained by very small and omnipresent flaws in the material which serve as stress raisers—points at which the magnitude of an applied tensile stress is amplified. The degree of stress amplification depends on crack length and tip radius of curvature according to Equation 8.1, being greatest for long and pointed flaws. These stress raisers may be minute surface or interior cracks (microcracks), internal pores, and grain corners, which are virtually impossible to eliminate or control. For example, even moisture and contaminants in the atmosphere can introduce surface cracks in freshly drawn glass fibers; these cracks deleteriously affect the strength. A stress concentration at a flaw tip can cause a crack to form, which may propagate until the eventual failure.

The measure of a ceramic material's ability to resist fracture when a crack is present is specified in terms of fracture toughness. The plane strain fracture toughness K_{Ic}, as discussed in Section 8.5, is defined according to the expression

$$K_{Ic} = Y\sigma\sqrt{\pi a} \tag{13.2}$$

where Y is a dimensionless parameter which is a function of both specimen and crack geometries, σ is the applied stress, and a is the length of a surface crack or half of the length of an internal crack. Crack propagation will not occur as long as the right-hand side of Equation 13.2 is less than the plane strain fracture toughness of the material. Plane strain fracture toughness values for ceramic materials are smaller than for metals; typically they are below 9000 psi$\sqrt{in.}$ (10 MPa$\sqrt{m}$). Values of K_{Ic} for several ceramic materials are included in Table 8.1.

Under some circumstances, fracture of ceramic materials will occur by the slow propagation of cracks, when stresses are static in nature, and the right-hand side of Equation 13.2 is less than K_{Ic}. This phenomenon is called *static fatigue,* or *delayed fracture;* use of the term "fatigue" is somewhat misleading inasmuch as fracture may occur in the absence of cyclic stresses (metal fatigue was discussed in Chapter 8). It has been observed that this type of fracture is especially sensitive to environmental conditions, specifically when moisture is present in the atmosphere. Relative to mechanism, a stress–corrosion process probably occurs at the crack tips; that is, the combination of an applied tensile stress and material dissolution leads to a sharpening and lengthening of the cracks until, ultimately, one crack grows to a size capable of rapid propagation according to Equation 8.3. Furthermore, the duration of stress application preceeding fracture diminishes with increasing stress. Consequently, when specifying the static fatigue strength, the time of stress application should also be stipulated. Silicate glasses are especially susceptible to this type of fracture; it has also been observed in other ceramic materials to include porcelain, portland cement, high-alumina ceramics, barium titanate, and silicon nitride.

There is usually considerable variation and scatter in the fracture strength for many specimens of a specific brittle ceramic material. A distribution of fracture strengths for portland cement is shown in Figure 13.25. This phenomenon may be explained by the dependence of fracture strength on the probability of the existence of a flaw that is capable of initiating a crack. This probability varies from specimen to specimen of the same material and depends on fabrication technique and any subsequent treatment. Specimen size or volume also influences fracture strength; the larger the specimen, the greater this flaw existence probability, and the lower the fracture strength.

For compressive stresses, there is no stress amplification associated with any existent flaws. For this reason, brittle ceramics display much higher strengths in compression than in tension, and they are generally utilized when load conditions are compressive. Also, the fracture strength of a brittle ceramic may be enhanced dramatically by imposing residual compressive stresses at its surface. One way this may be accomplished is by thermal tempering (see Section 14.4).

Statistical theories have been developed which in conjunction with experimental data are used to determine the risk of fracture for a given material; a discussion of these is beyond the scope of the present treatment. However, due to the dispersion in the measured fracture strengths of brittle ceramic materials, the use of average values and factors of safety as discussed in Section 6.12 are not normally employed for design purposes.

As a consequence of the fabrication process, some ceramic systems retain residual porosity, which normally exists as small, nearly spherical, isolated pores. Porosity is

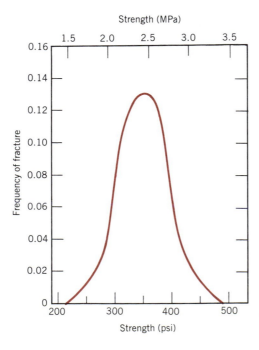

Figure 13.25 The frequency distribution of observed fracture strengths for a portland cement. (From W. Weibull, *Ing. Vetensk. Akad.*, Proc. 151, No. 153, 1939.)

usually deleterious to the fracture strength of these materials for two reasons: (1) pores reduce the cross-sectional area across which a load is applied; and (2) they also act as stress concentrators—for an isolated spherical pore, an applied tensile stress is amplified by a factor of 2. The influence of porosity is rather dramatic; experimentally it has been observed that the strength of many ceramics decreases exponentially with increasing porosity.

13.7 STRESS–STRAIN BEHAVIOR

Modulus of Rupture

The stress–strain behavior of brittle ceramics is not usually ascertained by a tensile test as outlined in Section 6.2, for two reasons. First, it is difficult to prepare and test specimens having the required geometry; and second, there is a significant difference in results obtained from tests conducted in compressive and in tensile modes. Therefore, a more suitable transverse bending test is most frequently employed, in which a rod specimen having either a circular or rectangular cross section is bent until fracture using a three- or four-point loading technique; the three-point loading scheme is illustrated in Figure 13.26. At the point of loading, the top surface of the specimen is placed in a state of compression, whereas the bottom surface is in tension. Stress is computed from the specimen thickness, the bending moment, and the moment of inertia of the cross section; these parameters are noted in Figure 13.26 for rectangular and circular cross sections.

The maximum stress, or stress at fracture using this bend test, is known as the **modulus of rupture,** or the *bend strength,* an important mechanical parameter for brittle ceramics. For a rectangular cross section, the modulus of rupture σ_{mr} is equal

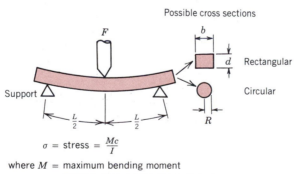

Possible cross sections

$$\sigma = \text{stress} = \frac{Mc}{I}$$

where M = maximum bending moment
c = distance from center of specimen to outer fibers
I = moment of inertia of cross section
F = applied load

	M	c	I	σ
Rectangular	$\frac{FL}{4}$	$\frac{d}{2}$	$\frac{bd^3}{12}$	$\frac{3FL}{2bd^2}$
Circular	$\frac{FL}{4}$	R	$\frac{\pi R^4}{4}$	$\frac{FL}{\pi R^3}$

Figure 13.26 A three-point loading scheme for measuring the stress–strain behavior and modulus of rupture of brittle ceramics, including expressions for computing stress for rectangular and circular cross sections.

TABLE 13.4 Tabulation of Modulus of Rupture (Bend Strength) and Modulus of Elasticity for Eight Common Ceramic Materials

Material	Modulus of Rupture		Modulus of Elasticity	
	psi $\times$ 10^3	MPa	psi $\times$ 10^6	MPa $\times$ 10^4
Titanium carbide[a] (TiC)	160	1100	45	31
Aluminum oxide[a] (Al_2O_3)	30–50	200–345	53	37
Beryllium oxide[a] (BeO)	20–40	140–275	45	31
Silicon carbide[a] (SiC)	25	170	68	47
Magnesium oxide[a] (MgO)	15	105	30	21
Spinel[a] ($MgAl_2O_4$)	13	90	35	24
Fused silica	16	110	11	7.5
Glass	10	70	10	7

[a] Sintered and containing approximately 5% porosity.
Source: W. D. Kingery, H. K. Bowen, and D. R. Uhlmann, *Introduction to Ceramics*, 2nd edition. Copyright © by John Wiley & Sons, New York. Reprinted by permission of John Wiley & Sons, Inc.

to

$$\sigma_{mr} = \frac{3F_f L}{2bd^2} \tag{13.3a}$$

where F_f is the load at fracture, L is the distance between support points, and the other parameters are as indicated in Figure 13.26. When the cross section is circular, then

$$\sigma_{mr} = \frac{F_f L}{\pi R^3} \tag{13.3b}$$

R being the specimen radius.

Characteristic modulus of rupture values for several ceramic materials are given in Table 13.4. Since, during bending, a specimen is subjected to both compressive and tensile stresses, the magnitude of its modulus of rupture is greater than the tensile fracture strength. Furthermore, σ_{mr} will depend on specimen size; as explained in the preceding section, with increasing specimen volume (under stress) there is an increase in flaw severity and, consequently, a decrease in modulus of rupture.

Elastic Behavior

The elastic stress–strain behavior for ceramic materials using these transverse bending tests is similar to the tensile test results for metals: a linear relationship exists between stress and strain. Figure 13.27 compares the stress–strain behavior to fracture for aluminum oxide (alumina) and glass. Again, the slope in the elastic region is the modulus of elasticity; the range of moduli of elasticity for ceramic materials is between about 10×10^6 and 70×10^6 psi (7×10^4 and 50×10^4 MPa), being slightly higher than for metals. Table 13.4 lists values for several ceramic materials.

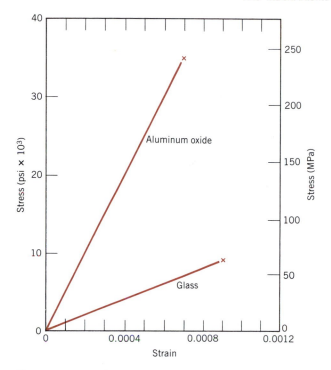

Figure 13.27 Typical stress–strain behavior to fracture for aluminum oxide and glass.

13.8 MECHANISMS OF PLASTIC DEFORMATION

Although at room temperature most ceramic materials suffer fracture before the onset of plastic deformation, a brief exploration into the possible mechanisms is worthwhile. Plastic deformation is different for crystalline and noncrystalline ceramics; however, each is discussed.

Crystalline Ceramics

For crystalline ceramics, plastic deformation occurs, as with metals, by the motion of dislocations (Chapter 7). One reason for the hardness and brittleness of these materials is the difficulty of slip (or dislocation motion). There are very few slip systems (crystallographic planes and directions within those planes) along which dislocations may move. This is a consequence of the electrically charged nature of the ions. For slip in some directions, ions of like charge are brought into close proximity to one another; because of electrostatic repulsion, this mode of slip is very restricted. This is not a problem in metals, since all atoms are electrically neutral.

Noncrystalline Ceramics

Plastic deformation does not occur by dislocation motion for noncrystalline ceramics because there is no regular atomic structure. Rather, these materials deform by *viscous*

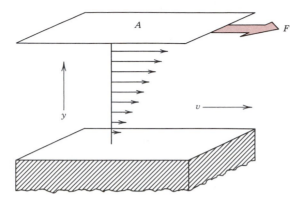

Figure 13.28 Representation of the viscous flow of a liquid or fluid glass in response to an applied shear force.

flow, the same manner in which liquids deform; the rate of deformation is proportional to the applied stress. In response to an applied shear stress, atoms or ions slide past one another by the breaking and reforming of interatomic bonds. However, there is no prescribed manner or direction in which this occurs, as with dislocations. Viscous flow on a macroscopic scale is demonstrated in Figure 13.28.

The characteristic property for viscous flow, **viscosity,** is a measure of a noncrystalline material's resistance to deformation. For viscous flow in a liquid, which originates from shear stresses imposed by two flat and parallel plates, the viscosity η is the ratio of the applied shear stress τ and the change in velocity dv with distance dy in a direction perpendicular to and away from the plates, or

$$\eta = \frac{\tau}{dv/dy} = \frac{F/A}{dv/dy} \qquad (13.4)$$

This scheme is represented in Figure 13.28.

The units for viscosity are poises (P) and pascal-seconds (Pa-s); $1\,P = 1$ dyne-s/cm^2, and $1\,\text{Pa-s} = 1\,\text{N-s/m}^2$. Conversion from one system of units to the other is according to

$$10\,P = 1\,\text{Pa-s}$$

Liquids have relatively low viscosities; for example, the viscosity of water at room temperature is about 10^{-3} Pa-s. On the other hand, glasses have extremely large viscosities at ambient temperatures, which is accounted for by strong interatomic bonding. As the temperature is raised, the magnitude of the bonding is diminished, the sliding motion or flow of the atoms or ions is facilitated, and subsequently there is an attendant decrease in viscosity. A discussion of the temperature dependence of viscosity for glasses is deferred to Section 14.2.

13.9 MISCELLANEOUS MECHANICAL CONSIDERATIONS

Influence of Porosity

The residual porosity that exists in many ceramic pieces as a result of fabrication processes has a deleterious influence on the elastic properties. For example, the magnitude of the modulus of elasticity E decreases with volume fraction porosity P

TABLE 13.5 Approximate Knoop Hardness
(100 g load) for Seven
Ceramic Materials

Material	Approximate Knoop Hardness
Diamond (carbon)	7000
Boron carbide (B_4C)	2800
Silicon carbide (SiC)	2500
Tungsten carbide (WC)	2100
Aluminum oxide (Al_2O_3)	2100
Quartz (SiO_2)	800
Glass	550

according to

$$E = E_0(1 - 1.9P + 0.9P^2) \tag{13.5}$$

where E_0 is the modulus of elasticity of the nonporous material.

Hardness

One beneficial mechanical property of ceramics is their hardness, which is often utilized when an abrasive or grinding action is required; in fact, the hardest known materials are ceramics. A listing of a number of different ceramic materials according to Knoop hardness is contained in Table 13.5. Only ceramics having Knoop hardnesses of about 1000 or greater are utilized for their abrasive characteristics.

Creep

Often ceramic materials experience creep deformation as a result of exposure to stresses (usually compressive) at elevated temperatures. In general, the time–deformation creep behavior of ceramics is similar to that for metals (Section 8.13); however, creep occurs at higher temperatures in ceramics. High-temperature compressive creep tests are conducted on ceramic materials to ascertain creep deformation as a function of temperature and stress level.

SUMMARY

Both crystalline and noncrystalline states are possible for ceramics. Since these materials are composed of electrically charged cations and anions, crystal structure is determined by the charge magnitude and the radius of each kind of ion. Some of the simpler crystal structures are described in terms of unit cells; several of these were discussed (rock salt, cesium chloride, zinc blende, diamond cubic, graphite, fluorite, perovskite, and spinel structures).

For the silicates, structure is more conveniently represented by means of interconnecting SiO_4^{4-} tetrahedra. Relatively complex structures may result when other cations (e.g., Ca^{2+}, Mg^{2+}, Al^{3+}) and anions (e.g., OH^-) are added. The structures of silica (SiO_2), silica glass, and several of the simple and layered silicates were presented.

With regard to atomic point defects, interstitials and vacancies for each anion and cation type are possible. These imperfections often occur in pairs as Frenkel and Schottky defects to ensure that crystal electroneutrality is maintained. Addition of impurity atoms may result in the formation of substitutional or interstitial solid solutions. Any charge imbalance created by the impurity ions may be compensated by the generation of host ion vacancies or interstitials.

Phase diagrams for the Al_2O_3–Cr_2O_3, MgO–Al_2O_3, ZrO_2–CaO, and SiO_2–Al_2O_3 systems were discussed. These diagrams are especially useful in assessing the high-temperature performance of ceramic materials.

At room temperature, virtually all ceramics are brittle. Microcracks, the presence of which is very difficult to control, result in amplification of applied tensile stresses and account for relatively low fracture strengths (moduli of rupture). This amplification does not occur with compressive loads, and, consequently, ceramics are stronger in compression. Representative strengths of ceramic materials are determined by performing transverse bending tests to fracture.

Any plastic deformation of crystalline ceramics is a result of dislocation motion; the brittleness of these materials is, in part, explained by the limited number of operable slip systems. The mode of plastic deformation for noncrystalline materials is by viscous flow; a material's resistance to deformation is expressed as viscosity. At room temperature, the viscosity of many noncrystalline ceramics is extremely high.

In addition to their inherent brittleness, ceramic materials are distinctively hard. Also, since these materials are frequently utilized at elevated temperatures and under applied loads, creep characteristics are important.

IMPORTANT TERMS AND CONCEPTS

Anion	Frenkel defect	Stoichiometry
Cation	Modulus of rupture	Tetrahedral position
Defect structure	Octahedral position	Viscosity
Electroneutrality	Schottky defect	

REFERENCES

BERGERON, C. G. and S. H. RISBUD, *Introduction to Phase Equilibria in Ceramics,* American Ceramic Society, Columbus, OH, 1984.

BUDWORTH, D. W., *An Introduction to Ceramic Science,* Pergamon Press, Oxford, 1970.

CHARLES, R. J., "The Nature of Glasses," *Scientific American,* Vol. 217, No. 3, September 1967, pp. 126–136.

GILMAN, J. J., "The Nature of Ceramics," *Scientific American,* Vol. 217, No. 3, September 1967, pp. 112–124.

HAUTH, W. E., "Crystal Chemistry in Ceramics," *American Ceramic Society Bulletin,* Vol. 30, 1951: No. 1, pp. 5–7; No. 2, pp. 47–49; No. 3, pp. 76–77; No. 4, pp. 137–142; No. 5, pp. 165–167; No. 6, pp. 203–205. A good overview of silicate structures.

KINGERY, W. D., H. K. BOWEN, and D. R. UHLMANN, *Introduction to Ceramics,* 2nd edition, John Wiley & Sons, New York, 1976. Chapters 1–4, 14, and 15.

LEVIN, E. M., C. R. ROBBINS, and H. F. McMURDIE (Editors), *Phase Diagrams for Ceramists,* Vol. I, American Ceramic Society, Columbus, OH, 1964. Also supplementary Volumes II, III, IV, V and VI, published in 1969, 1973, 1981, 1983 and 1987, respectively.

NORTON, F. H., *Elements of Ceramics,* 2nd edition, Addison-Wesley Publishing Company, Reading, MA, 1974. Chapters 2 and 23.

VAN VLACK, L. H., *Physical Ceramics for Engineers,* Addison-Wesley Publishing Company, Reading, MA, 1964. Chapters 1–4 and 6–8.

QUESTIONS AND PROBLEMS

13.1 Distinguish between a cation and an anion. Which is normally larger? Why?

13.2 For a ceramic compound, what are the two characteristics of the component ions that determine the crystal structure?

13.3 Show that the minimum cation-to-anion radius ratio for a coordination number of 4 is 0.225.

13.4 Show that the minimum cation-to-anion radius ratio for a coordination number of 6 is 0.414. *Hint:* Use the NaCl crystal structure (Figure 13.2), and assume that anions and cations are just touching along cube edges and across face diagonals.

13.5 Demonstrate that the minimum cation-to-anion radius ratio for a coordination number of 8 is 0.732.

13.6 On the basis of ionic charge and ionic radii, predict the crystal structures for the following materials: **(a)** CsI, **(b)** NiO, **(c)** KI, and **(d)** NiS. Justify your selections.

13.7 Which of the cations in Table 13.2 would you predict to form fluorides having the cesium chloride crystal structure? Justify your choices.

13.8 Compute the atomic packing factor for the rock salt crystal structure in which $r_C/r_A = 0.414$.

13.9 Table 13.2 gives the ionic radii for K^+ and O^{2-} as 0.138 and 0.140 nm, respectively. What would be the coordination number for each O^{2-} ion? Briefly describe the resulting crystal structure for K_2O. Explain why this is called the antifluorite structure.

13.10 The zinc blende crystal structure is one that may be generated from close-packed planes of anions. **(a)** Will the stacking sequence for this structure be FCC or HCP? Why? **(b)** Will cations fill tetrahedral or octahedral positions? Why? **(c)** What fraction of the positions will be occupied?

13.11 The corundum crystal structure, found for Al_2O_3, consists of an HCP arrangement of O^{2-} ions; the Al^{3+} ions occupy octahedral positions. **(a)** What fraction of the available octahedral positions are filled with Al^{3+} ions? **(b)**

Sketch two close-packed O^{2-} planes stacked in an AB sequence, and note octahedral positions that will be filled with the Al^{3+} ions.

13.12 Beryllium oxide (BeO) may form a crystal structure that consists of an HCP arrangement of O^{2-} ions. If the ionic radius of Be^{2+} is 0.035 nm, then
   (a) Which type of interstitial site will the Be^{2+} ions occupy?
   (b) What fraction of these available interstitial sites will be occupied by Be^{2+} ions?

13.13 Iron titanate, $FeTiO_3$, forms in the ilmenite crystal structure which consists of an HCP arrangement of O^{2-} ions.
   (a) Which type of interstitial site will the Fe^{2+} ions occupy? Why?
   (b) Which type of interstitial site will the Ti^{4+} ions occupy? Why?
   (c) What fraction of the total tetrahedral sites will be occupied?
   (d) What fraction of the total octahedral sites will be occupied?

13.14 Calculate the density of FeO, given that it has the rock salt crystal structure.

13.15 Magnesium oxide has the rock salt crystal structure and a density of 3.58 g/cm^3. **(a)** Determine the unit cell edge length. **(b)** How does this result compare with the edge length as determined from the radii in Table 13.2, assuming that the Mg^{2+} and O^{2-} ions just touch each other along the edges?

13.16 Compute the theoretical density of diamond given that the C—C distance and bond angle are 0.154 nm and 109.5°, respectively. How does this value compare with the measured density?

13.17 Compute the theoretical density of ZnS given that the Zn—S distance and bond angle are 0.234 nm and 109.5°, respectively. How does this value compare with the measured density?

13.18 Cadmium sulfide (CdS) has a cubic unit cell, and from x-ray diffraction data it is known that the cell edge length is 0.582 nm. If the measured density is 4.82 g/cm^3, how many Cd^{2+} and S^{2-} ions are there per unit cell?

13.19 **(a)** Using the ionic radii in Table 13.2, compute the density of CsCl. *Hint:* Use a modification of the result of Problem 3.4. **(b)** The measured density is 3.99 g/cm^3. How do you explain the slight discrepancy between your calculated value and the measured one?

13.20 From the data in Table 13.2, compute the density of CaF_2, which has the fluorite structure.

13.21 A hypothetical AX type of ceramic material is known to have a density of 2.10 g/cm^3 and a unit cell of cubic symmetry with a cell edge length of 0.57 nm. The atomic weights of the A and X elements are 28.5 and 30.0 g/mol, respectively. On the basis of this information, which of the following crystal structures is (are) possible for this material: sodium chloride, cesium chloride, or zinc blende? Justify your choice(s).

13.22 The unit cell for Fe_3O_4 ($FeO \cdot Fe_2O_3$) has cubic symmetry with a unit cell edge length of 0.839 nm. If the density of this material is 5.24 g/cm^3, compute its atomic packing factor. For this computation you will need to use ionic radii listed in Table 13.2.

13.23 The unit cell for Al_2O_3 has hexagonal symmetry with lattice parameters $a = 0.4759$ nm and $c = 1.2989$ nm. If the density of this material is 3.99 g/cm^3, calculate its atomic packing factor. For this computation use ionic radii listed in Table 13.2.

13.24 Compute the atomic packing factor for the diamond cubic crystal structure (Figure 13.5). Assume that bonding atoms touch one another, that the angle between adjacent bonds is 109.5°, and that each atom internal to the unit cell is positioned $a/4$ of the distance away from the two nearest cell faces (a is the unit cell edge length).

13.25 Compute the atomic packing factor for cesium chloride using the ionic radii in Table 13.2 and assuming that the ions touch along the cube diagonals.

13.26 For each of the following crystal structures, represent the indicated plane in the manner of Figures 3.9 and 3.10, showing both anions and cations: (a) (100) plane for the rock salt crystal structure, (b) (110) plane for the cesium chloride crystal structure, (c) (111) plane for the zinc blende crystal structure, and (d) (110) plane for the perovskite crystal structure.

13.27 In terms of bonding, explain why silicate materials have relatively low densities.

13.28 Determine the angle between covalent bonds in an SiO_4^{4-} tetrahedron.

13.29 When kaolinite clay $[Al_2(Si_2O_5)(OH)_4]$ is heated to a sufficiently high temperature, chemical water is driven off.
(a) Under these circumstances what is the composition of the remaining product?
(b) What are the liquidus and solidus temperatures of this material?

13.30 Would you expect Frenkel defects for anions to exist in ionic ceramics in relatively large concentrations? Why or why not?

13.31 In your own words, briefly define the term "stoichiometric."

13.32 Let us assume that there exists a hypothetical ceramic material which is composed of A^{2+} and X^{2-} ions. Suppose that a small fraction of the X^{2-} ions may exist as X^-.
(a) Under these conditions name one defect that you would expect to form in order to maintain charge neutrality.
(b) How many of these X^- ions are required for each defect?
(c) How would you express the chemical formula for this nonstoichiometric material?

13.33 (a) Suppose that Li_2O is added as an impurity to CaO. If the Li^+ substitutes for Ca^{2+}, what kind of vacancies would you expect to form? How many of these vacancies are created for every Li^+ added? (b) Suppose that $CaCl_2$ is added as an impurity to CaO. If the Cl^+ substitutes for O^{2-}, what kind of vacancies would you expect to form? How many of the vacancies are created for every Cl^- added?

13.34 What point defects are possible for MgO as an impurity in Al_2O_3? How many Mg^{2+} ions must be added to form each of these defects?

13.35 (a) What minimum concentration of CaO is necessary to stabilize ZrO_2? (b) For the ZrO_2–CaO system (Figure 13.23), write eutectic, eutectoid, and peritectic reactions for cooling.

13.36 From Figure 13.22, the phase diagram for the MgO–Al_2O_3 system, it may be noted that the spinel solid solution exists over a range of compositions, which means that it is nonstoichiometric at compositions other than 50 mol% MgO–50 mol% Al_2O_3.

(a) The maximum nonstoichiometry on the Al_2O_3-rich side of the spinel phase field exists at about 1900°C (3450°F) corresponding to approximately 80 mol% (93 wt%) Al_2O_3. Determine the type of vacancy defect that is produced and the percentage of vacancies that exist at this composition.

(b) The maximum nonstoichiometry on the MgO-rich side of the spinel phase field exists at about 2000°C (3630°F) corresponding to approximately 38 mol% (61 wt%) Al_2O_3. Determine the type of vacancy defect that is produced and the percentage of vacancies that exist at this composition.

13.37 Briefly explain (a) why there may be significant scatter in the fracture strength for some given ceramic material, and (b) why fracture strength increases with decreasing specimen size.

13.38 The tensile strength of brittle materials may be determined using a variation of Equation 8.1. Compute the critical crack tip radius for an Al_2O_3 specimen that experiences tensile fracture at an applied stress of 40,000 psi (275 MPa). Assume a critical surface crack length of 2×10^{-3} mm and a theoretical fracture strength of $E/10$, where E is the modulus of elasticity.

13.39 The fracture strength of glass may be increased by etching away a thin surface layer. It is believed that the etching may alter surface crack geometry (i.e., reduce crack length and increase the tip radius). Compute the ratio of the original and etched crack tip radii for a fourfold increase in fracture strength if half the crack length is removed.

13.40 A three-point bending test is performed on a glass specimen having a rectangular cross section of height d 0.2 in. (5 mm) and width b 0.4 in. (10 mm); the distance between support points is 1.75 in. (44.5 mm). (a) Compute the modulus of rupture if the load at fracture is 65 lb$_f$ (290 N). (b) The point of maximum deflection Δy occurs at the center of the specimen and is described by

$$\Delta y = \frac{FL^3}{48EI}$$

where E is the modulus of elasticity and I the cross-sectional moment of inertia. Compute Δy at a load of 60 lb$_f$ (266 N).

13.41 A circular specimen of MgO is loaded using a three-point bending mode. Compute the minimum possible radius of the specimen, given that the applied load is 4000 lb$_f$ (17,800 N), the modulus of rupture is 35,000 psi (240 MPa), and the separation between load points is 1.5 in. (38 mm).

13.42 It is necessary to select a ceramic material to be stressed using a three-point loading scheme (Figure 13.26). The specimen must have a circular cross section and a radius of 0.15 in. (3.8 mm), and must not experience fracture nor a deflection of more than 8.5×10^{-4} in. (0.021 mm) at its center when a load of 100 lb$_f$ (445 N) is applied. If the distance between support points is 2 in. (50.8 mm), which of the materials in Table 13.4 are candidates? The magnitude of the center-point deflection may be computed using the equation supplied in Problem 13.40.

13.43 A three-point bending test was performed on an aluminum oxide specimen having a circular cross section of radius 0.20 in. (5.0 mm); the specimen fractured at a load of 675 lb$_f$ (3000 N) when the distance between the support points was 1.6 in. (40 mm). Another test is to be performed on a specimen of this same material, but one that has a square cross section of 0.6 in. (15 mm) length on each edge. At what load would you expect this specimen to fracture if the support point separation is maintained at 1.6 in. (40 mm)?

13.44 **(a)** A three-point transverse bending test is conducted on a cylindrical specimen of aluminum oxide having a reported modulus of rupture of 300 MPa (43,500 psi). If the specimen radius is 5.0 mm (0.20 in.) and the support point separation distance is 15.0 mm (0.60 in.), predict whether or not you would expect the specimen to fracture when a load of 7500 N (1690 lb$_f$) is applied? Justify your prediction. **(b)** Would you be 100% certain of the prediction in part a? Why or why not?

13.45 Cite one reason why ceramic materials are, in general, harder yet more brittle than metals.

13.46 Using the data in Table 13.4, compute the modulus of elasticity for nonporous MgO.

13.47 At what volume percent porosity will the modulus of elasticity for Al$_2$O$_3$ be 39×10^6 psi (27×10^4 MPa)? Assume that the data in Table 13.4 is for Al$_2$O$_3$ having 5% porosity.

CHAPTER 14

APPLICATIONS AND PROCESSING OF CERAMICS

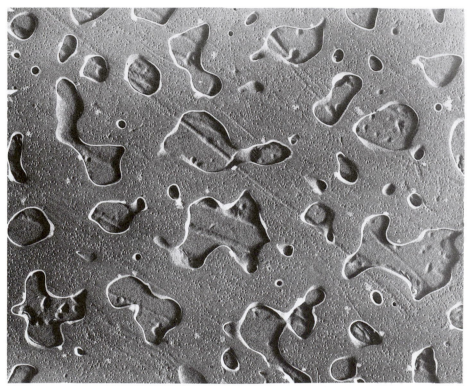

A scanning electron micrograph showing the microstructure of a glass–ceramic from which commercial dinnerware is made. A borosilicate phase was etched out, leaving a phase-separated glass phase matrix. 40,000 ×. (Photograph courtesy of D. G. Pickles, Corning Glass Works Research Laboratory, Corning, NY.)

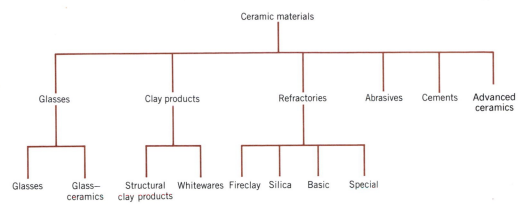

Figure 14.1 Classification of ceramic materials on the basis of application.

14.1 INTRODUCTION

The preceding discussion of the properties of materials has demonstrated that there is a significant disparity between the physical characteristics of metals and ceramics. Consequently, these materials are utilized in totally different kinds of applications and, in this regard, tend to complement each other, and also the polymers. Most ceramic materials fall into an application–classification scheme that includes the following groups: glasses, structural clay products, whitewares, refractories, abrasives, cements, and the newly developed advanced ceramics. Figure 14.1 presents a taxonomy of these several types; some discussion is devoted to each in this chapter.

One chief concern in the application of ceramic materials is the method of fabrication. Many of the metal-forming operations discussed in Chapter 12 rely on casting and/or techniques that involve some form of plastic deformation. Since ceramic materials have relatively high melting temperatures, casting them is normally impractical. Furthermore, in most instances the brittleness of these materials precludes deformation. Some ceramic pieces are formed from powders (or particulate collections) that must ultimately be dried and fired. Glass shapes are formed at elevated temperatures from a fluid mass that becomes very viscous upon cooling. Cements are shaped by placing into forms a fluid paste that hardens and assumes a permanent set by virtue of chemical reactions. A taxonomical scheme for the several types of ceramic-forming techniques is presented in Figure 14.2. Each type is discussed with the group of ceramics with which it is most frequently employed.

GLASSES

The glasses are a familiar group of ceramics; containers, windows, lenses, and fiberglass represent typical applications. As already mentioned, they are noncrystalline silicates containing other oxides, notably CaO, Na_2O, K_2O, and Al_2O_3, which influence the glass properties. A typical soda–lime glass consists of approximately 70 wt% SiO_2, the balance being mainly Na_2O (soda) and CaO (lime). The compositions of several

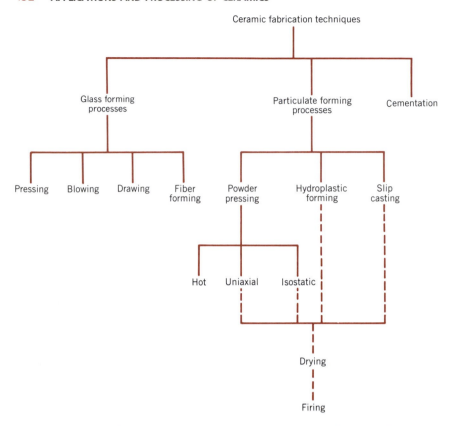

Figure 14.2 A classification scheme for the ceramic-forming techniques discussed in this chapter.

common glass materials are contained in Table 14.1. Possibly the two prime assets of these materials are their optical transparency and the relative ease with which they may be fabricated.

14.2 GLASS PROPERTIES

Before we discuss specific glass-forming techniques, some of the temperature-sensitive properties of glass materials must be presented. Glassy, or noncrystalline, materials do not solidify in the same sense as do those that are crystalline. Upon cooling, a glass becomes more and more viscous in a continuous manner with decreasing temperature; there is no definite temperature at which the liquid transforms to a solid as with crystalline materials. In fact, one of the distinctions between crystalline and noncrystalline materials lies in the dependence of specific volume (or volume per unit weight—the reciprocal of density) on temperature, as illustrated in Figure 14.3. For crystalline materials, there is a discontinuous decrease in volume at the melting temperature T_m. However, for glassy materials, volume decreases continuously with temperature reduction; a slight decrease in slope of the curve occurs at what is called

TABLE 14.1 Compositions and Characteristics of Some of the Common
Commercial Glasses

| Glass Type | Composition (wt%) | | | | | | Characteristics and Applications |
	SiO$_2$	Na$_2$O	CaO	Al$_2$O$_3$	B$_2$O$_3$	Other	
Fused silica	>99.5						High melting temperature, very low coefficient of expansion (shock resistant)
96% Silica (Vycor)	96				4		Thermally shock and chemically resistant— laboratory ware
Borosilicate (Pyrex)	81	3.5		2.5	13		Thermally shock and chemically resistant— ovenware
Container (soda–lime)	74	16	5	1		4MgO	Low melting temperature, easily worked, also durable
Fiberglass	55		16	15	10	4MgO	Easily drawn into fibers— glass–resin composites
Optical flint	54	1				37PbO, 8K$_2$O	High density and high index of refraction— optical lenses
Glass–ceramic	70			18		4.5TiO$_2$, 2.5Li$_2$O	Easily fabricated; strong; resists thermal shock— ovenware

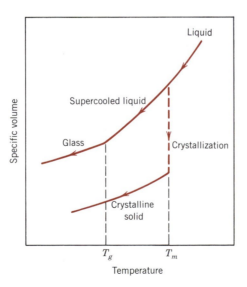

Figure 14.3 Contrast of specific volume-versus-temperature behavior of crystalline and noncrystalline materials. Crystalline materials solidify at the melting temperature T_m. Characteristic of the noncrystalline state is the glass transition temperature T_g.

the **glass transition temperature,** or *fictive* temperature, T_g. Below this temperature, the material is considered to be a glass; above, it is first a supercooled liquid, and finally a liquid.

Also important in glass-forming operations are the viscosity–temperature characteristics of the glass. Figure 14.4 plots the logarithm of viscosity versus the temperature for fused silica, high silica, borosilicate, and soda–lime glasses. On the viscosity scale several specific points that are important in the fabrication and processing of glasses are labeled:

1. The **melting point** corresponds to the temperature at which the viscosity is 10 Pa-s (100 P); the glass is fluid enough to be considered a liquid.

2. The **working point** represents the temperature at which the viscosity is 10^3 Pa-s (10^4 P); the glass is easily deformed at this viscosity.

3. The **softening point,** the temperature at which the viscosity is 4×10^6 Pa-s (4×10^7 P), is the maximum temperature at which a glass piece may be handled without causing significant dimensional alterations.

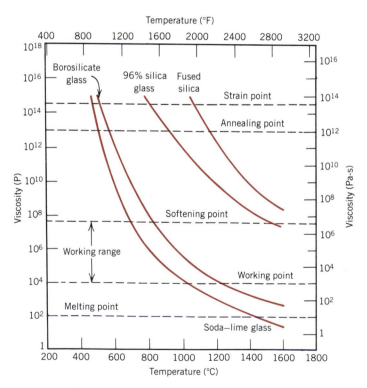

Figure 14.4 Logarithm of viscosity versus temperature for fused silica and several silica glasses. (From E. B. Shand, *Engineering Glass,* Modern Materials, Vol. 6, Academic Press, New York, 1968, p. 262.)

4. The **annealing point** is the temperature at which the viscosity is 10^{12} Pa-s (10^{13} P); at this temperature atomic diffusion is sufficiently rapid that any residual stresses may be removed within about 15 min.

5. The **strain point** corresponds to the temperature at which the viscosity becomes 3×10^{13} Pa-s (3×10^{14} P); for temperatures below the strain point, fracture will occur before the onset of plastic deformation. The glass transition temperature will be above the strain point.

Most glass-forming operations are carried out within the working range—between the working and softening temperatures.

Of course, the temperature at which each of these points occurs depends on glass composition. For example, the softening point for soda–lime and 96% silica glasses from Figure 14.4 are about 700 and 1550°C (1300 and 2825°F), respectively. That is, forming operations may be carried out at significantly lower temperatures for the soda–lime glass. The formability of a glass is tailored to a large degree by its composition.

14.3 GLASS FORMING

Glass is produced by heating the raw materials to an elevated temperature above which melting occurs. Most commercial glasses are of the silica–soda–lime variety; the silica is usually supplied as common quartz sand, whereas Na_2O and CaO are added as soda ash (Na_2CO_3) and limestone ($CaCO_3$). For most applications, especially when optical transparency is important, it is essential that the glass product be homogeneous and pore free. Homogeneity is achieved by complete melting and mixing of the raw ingredients. Porosity results from small gas bubbles that are produced; these must be absorbed into the melt or otherwise eliminated, which requires proper adjustment of the viscosity of the molten material.

Four different forming methods are used to fabricate glass products: pressing, blowing, drawing, and fiber forming. Pressing is used in the fabrication of relatively thick-walled pieces such as plates and dishes. The glass piece is formed by pressure application in a graphite-coated cast iron mold having the desired shape; the mold is ordinarily heated to ensure an even surface.

Although some glass blowing is done by hand, especially for art objects, the process has been completely automated for the production of glass jars, bottles, and light bulbs. The several steps involved in one such technique are illustrated in Figure 14.5. From a raw gob of glass, a *parison,* or temporary shape, is formed by mechanical pressing in a mold. This piece is inserted into a finishing or blow mold and forced to conform to the mold contours by the pressure created from a blast of air.

Drawing is used to form long glass pieces such as sheet, rod, tubing, and fibers, which have a constant cross section. One process by which sheet glass is formed is illustrated in Figure 14.6; it may also be fabricated by hot rolling. Flatness and the surface finish may be improved significantly by floating the sheet on a bath of molten tin at an elevated temperature; the piece is slowly cooled and subsequently heat treated by annealing.

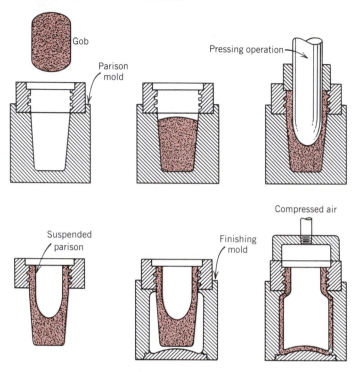

Figure 14.5 The press-and-blow technique for producing a glass bottle. (Adapted from C. J. Phillips, *Glass: The Miracle Maker*. Reproduced by permission of Pitman Publishing Ltd., London.)

Continuous glass fibers are formed in a rather sophisticated drawing operation. The molten glass is contained in a platinum heating chamber. Fibers are formed by drawing the molten glass through many small orifices at the chamber base. The glass viscosity, which is critical, is controlled by chamber and orifice temperatures.

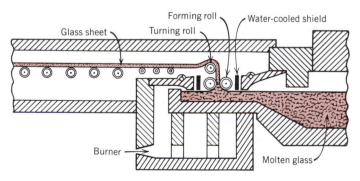

Figure 14.6 A process for the continuous drawing of sheet glass. (From W. D. Kingery, *Introduction to Ceramics*. Copyright © 1960 by John Wiley & Sons, New York. Reprinted by permission of John Wiley & Sons, Inc.)

14.4 HEAT TREATING GLASSES

Annealing

When a ceramic material is cooled from an elevated temperature, internal stresses, called thermal stresses, may be introduced as a result of the difference in cooling rate and thermal contraction between the surface and interior regions. These thermal stresses are important in brittle ceramics, especially glasses, since they may weaken the material or, in extreme cases, lead to fracture, which is termed **thermal shock** (see Section 20.5). Normally, attempts are made to avoid thermal stresses, which may be accomplished by cooling the piece at a sufficiently slow rate. Once such stresses have been introduced, however, elimination, or at least a reduction in their magnitude, is possible by an annealing heat treatment in which the glassware is heated to the annealing point, then slowly cooled to room temperature.

Glass Tempering

The strength of a glass piece may be enhanced by intentionally inducing compressive residual surface stresses. This can be accomplished by a heat treatment procedure called **thermal tempering.** With this technique, the glassware is heated to a temperature above the glass transition region yet below the softening point. It is then cooled to room temperature in a jet of air or, in some cases, an oil bath. The residual stresses arise from differences in cooling rates for surface and interior regions. Initially, the surface cools more rapidly and, once having dropped to a temperature below the strain point, becomes rigid. At this time, the interior, having cooled less rapidly, is at a higher temperature (above the strain point) and, therefore, is still plastic. With continued cooling, the interior attempts to contract to a greater degree than the now rigid exterior will allow. Thus the inside tends to draw in the outside, or to impose inward radial stresses. As a consequence, after the glass piece has cooled to room temperature, it sustains compressive stresses on the surface, with tensile stresses at interior regions. The room-temperature stress distribution over a cross section of a glass plate is represented schematically in Figure 14.7.

The failure of ceramic materials almost always results from a crack that is initiated at the surface by an applied tensile stress. To cause fracture of a tempered glass piece, the magnitude of an externally applied tensile stress must be great enough to first overcome the residual compressive surface stress and, in addition, to stress the surface in tension sufficiently to initiate a crack, which may then propagate. For an untempered glass, a crack will be introduced at a lower external stress level, and, consequently, the fracture strength will be smaller.

Tempered glass is used for applications in which high strength is important; these include large doors, automobile windshields, and eyeglass lenses.

14.5 GLASS–CERAMICS

Most inorganic glasses can be made to transform from a noncrystalline state to one that is crystalline by the proper high-temperature heat treatment. This process, called **devitrification,** ordinarily is avoided because devitrified glass, being polycrystalline, is not transparent. Also, stresses may be introduced as a result of volume changes that attend the transformation, yielding a relatively weak product material.

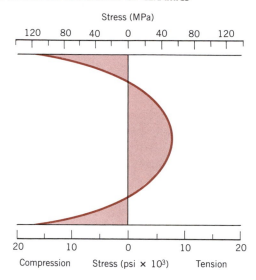

Figure 14.7 Room temperature residual stress distribution over the cross section of a tempered glass plate. (From W. D. Kingery, H. K. Bowen, and D. R. Uhlmann, *Introduction to Ceramics,* 2nd edition. Copyright © 1976 by John Wiley & Sons, New York. Reprinted by permission of John Wiley & Sons, Inc.)

However, for some glasses this devitrification transformation may be managed to the extent that a fine-grained material is produced that is free of these residual stresses; such is often called a **glass–ceramic.** A nucleating agent (frequently titanium dioxide) must be added to induce the crystallization or devitrification process. Desirable characteristics of glass–ceramics include a low coefficient of thermal expansion, such that the glass–ceramic ware will not experience thermal shock; in addition relatively high mechanical strengths and thermal conductivities are achieved. Possibly the most attractive attribute of this class of materials is the ease with which they may be fabricated; conventional glass-forming techniques may be used conveniently in the mass production of nearly pore-free ware.

Glass–ceramics are manufactured commercially under the trade names of Pyroceram, Cer-Vit, and Hercuvit. The most common uses for these materials are as ovenware and tableware, primarily because of their excellent resistance to thermal shock and their high thermal conductivity. They also serve as insulators and as substrates for printed circuit boards. A typical glass–ceramic is also included in Table 14.1, and the microstructure of a commercial material is shown on page 430.

CLAY PRODUCTS

One of the most widely used ceramic raw materials is clay. This inexpensive ingredient, found naturally in great abundance, often is used as mined without any upgrading of quality. Another reason for its popularity lies in the ease with which clay products may be formed; when mixed in the proper proportions, clay and water form a plastic mass that is very amenable to shaping. The formed piece is dried to remove some of the moisture, after which it is fired at an elevated temperature to improve its mechanical strength.

Most of the clay-based products fall within two broad classifications: the **structural clay products** and the **whitewares.** Structural clay products include building bricks, tiles, and sewer pipes—applications in which structural integrity is important. The whiteware ceramics become white after the high-temperature **firing.** Included in this group are porcelain, pottery, tableware, china, and plumbing fixtures (sanitary ware). In addition to clay, many of these products also contain nonplastic ingredients, which influence the changes that take place during the drying and firing processes, and the characteristics of the finished piece.

14.6 THE CHARACTERISTICS OF CLAY

The clay minerals play two very important roles in ceramic bodies. First, when water is added, they become very plastic, a condition termed hydroplasticity. This property is very important in forming operations, as discussed below. In addition, clay fuses or melts over a range of temperatures; thus a dense and strong ceramic piece may be produced during firing without complete melting such that the desired shape is maintained. This fusion temperature range, of course, depends on the composition of the clay.

Clays are aluminosilicates, being composed of alumina (Al_2O_3) and silica (SiO_2), that contain chemically bound water. They have a broad range of physical characteristics, chemical compositions, and structures; common impurities include compounds (usually oxides) of barium, calcium, sodium, potassium, and iron, and also some organic matter. Crystal structures for the clay minerals are relatively complicated; however, one prevailing characteristic is a layered structure. The most common clay minerals that are of interest have what is called the kaolinite structure. Kaolinite clay [$Al_2(Si_2O_5)(OH)_4$] has the crystal structure shown in Figure 13.16. When water is added, the water molecules fit in between these layered sheets and form a thin film around the clay particles. The particles are thus free to move over one another, which accounts for the resulting plasticity of the water–clay mixture.

14.7 COMPOSITIONS OF CLAY PRODUCTS

In addition to clay, many of these products (in particular the whitewares) also contain some nonplastic ingredients; the nonclay minerals include flint, or finely ground quartz, and a flux such as feldspar. The quartz is used primarily as a filler material, being inexpensive, relatively hard, and chemically unreactive. It experiences little change during high-temperature heat treatment because it has a high melting temperature; when melted, however, quartz has the ability to form a glass.

When mixed with clay, a flux forms a glass that has a relatively low melting point. The feldspars are some of the more common fluxing agents; they are a group of aluminosilicate materials that contain K^+, Na^+, and Ca^{2+} ions.

As would be expected, the changes that take place during drying and firing processes, and also the characteristics of the finished piece, are influenced by the proportions of these three constituents: clay, quartz, and flux. A typical porcelain might contain approximately 50% clay, 25% quartz, and 25% feldspar.

14.8 FABRICATION TECHNIQUES

The as-mined raw materials usually have to go through a milling or grinding operation in which particle size is reduced; this is followed by screening or sizing to yield a powdered product having a desired range of particle sizes. For multicomponent systems, powders must be thoroughly mixed with water and perhaps other ingredients to give flow characteristics that are compatible with the particular forming technique. The formed piece must have sufficient mechanical strength to remain intact during transporting, drying, and firing operations. Two common shaping techniques are utilized for forming clay-based compositions: **hydroplastic forming** and **slip casting.**

Hydroplastic Forming

As mentioned above, clay minerals, when mixed with water, become highly plastic and pliable and may be molded without cracking; however, they have extremely low yield strengths. The consistency (water–clay ratio) of the hydroplastic mass must give a yield strength sufficient to permit a formed ware to maintain its shape during handling and drying.

The most common hydroplastic forming technique is extrusion, in which a stiff plastic ceramic mass is forced through a die orifice having the desired cross-sectional geometry; it is similar to the extrusion of metals (Figure 12.2c). Brick, pipe, ceramic blocks, and tiles are all commonly fabricated using hydroplastic forming. Usually the plastic ceramic is forced through the die by means of a motor-driven auger, and often air is removed in a vacuum chamber to enhance the density. Hollow internal columns in the extruded piece (e.g., building brick) are formed by inserts situated within the die.

Slip Casting

Another forming process used for clay-based compositions is slip casting. A slip is a suspension of clay and/or other nonplastic materials in water. When poured into a porous mold (commonly made of plaster of paris), water from the slip is absorbed into the mold, leaving behind a solid layer on the mold wall the thickness of which depends on the time. This process may be continued until the entire mold cavity becomes solid (solid casting), as demonstrated in Figure 14.8a. Or it may be terminated when the solid shell wall reaches the desired thickness, by inverting the mold and pouring out the excess slip; this is termed drain casting (Figure 14.8b). As the cast piece dries and shrinks, it will pull away (or release) from the mold wall; at this time the mold may be disassembled and the cast piece removed.

The nature of the slip is extremely important; it must have a high specific gravity, and yet be very fluid and pourable. These characteristics depend on the solid-to-water ratio and other agents that are added. A satisfactory casting rate is an essential requirement. In addition, the cast piece must be free of bubbles, and it must have a low drying shrinkage and a relatively high strength.

The properties of the mold itself influence the quality of the casting. Normally, plaster of paris, which is economical, relatively easy to fabricate into intricate shapes, and reusable, is used as the mold material. Most molds are multipiece items that must be assembled before casting. Also, the mold porosity may be varied to control the casting rate. The rather complex ceramic shapes that may be produced by means of slip casting include sanitary lavatory ware, art objects, and specialized scientific laboratory ware such as ceramic tubes.

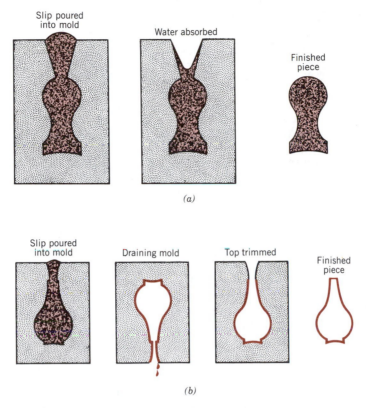

Figure 14.8 The steps in (*a*) solid and (*b*) drain slip casting using a plaster of paris mold. (From W. D. Kingery, *Introduction to Ceramics.* Copyright © 1960 by John Wiley & Sons, New York. Reprinted by permission of John Wiley & Sons, Inc.)

14.9 DRYING AND FIRING

A ceramic piece that has been formed hydroplastically or by slip casting retains significant porosity and insufficient strength for most practical applications. In addition, it may still contain some liquid (e.g., water), which was added to assist in the forming operation. This liquid is removed in a drying process; density and strength are enhanced as a result of a high-temperature heat treatment or firing procedure. A body that has been formed and dried but not fired is termed **green.** Drying and firing techniques are critical inasmuch as defects that ordinarily render the ware useless (e.g., warpage, distortion, and cracks) may be introduced during the operation. These defects normally result from stresses that are set up from nonuniform shrinkage.

Drying

As a clay-based ceramic body dries, it also experiences some shrinkage. In the early stages of drying the clay particles are virtually surrounded by and separated from one another by a thin film of water. As drying progresses and water is removed, the

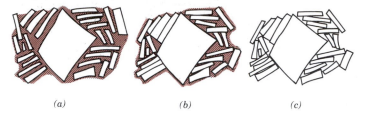

(a)   *(b)*   *(c)*

Figure 14.9 Several stages in the removal of water from between clay particles during the drying process. (*a*) Wet body. (*b*) Partially dry body. (*c*) Completely dry body. (From W. D. Kingery, *Introduction to Ceramics*. Copyright © 1960 by John Wiley & Sons, New York. Reprinted by permission of John Wiley & Sons, Inc.)

interparticle separation decreases, which is manifested as shrinkage (Figure 14.9). During drying it is critical to control the rate of water removal. Drying at interior regions of a body is accomplished by the diffusion of water molecules to the surface where evaporation occurs. If the rate of evaporation is greater than the rate of diffusion, the surface will dry (and as a consequence shrink) more rapidly than the interior, with a high probability of the formation of the aforementioned defects. The rate of surface evaporation should be diminished to, at most, the rate of water diffusion; evaporation rate may be controlled by temperature, humidity, and the rate of airflow.

Other factors also influence shrinkage. One of these is body thickness; nonuniform shrinkage and defect formation are more pronounced in thick pieces than in thin ones. Water content of the formed body is also critical: the greater the water content, the more extensive the shrinkage. Consequently, the water content is ordinarily kept as low as possible. Clay particle size also has an influence; shrinkage is enhanced as the particle size is decreased. To minimize shrinkage, the size of the particles may be increased, or nonplastic materials having relatively large particles may be added to the clay.

Firing

After drying, a body is usually fired at a temperature between 900 and 1400°C (1650 and 2550°F); the firing temperature depends on the composition and desired properties of the finished piece. During the firing operation, the density is further increased (with an attendant decrease in porosity) and the mechanical strength is enhanced.

When clay-based materials are heated to elevated temperatures, some rather complex and involved reactions occur. One of these is **vitrification,** the gradual formation of a liquid glass that flows into and fills some of the pore volume. The degree of vitrification depends on firing temperature and time, as well as the composition of the body. The temperature at which the liquid phase forms is lowered by the addition of fluxing agents such as feldspar. This fused phase flows around the remaining unmelted particles and fills in the pores as a result of surface tension forces (or capillary action); shrinkage also accompanies this process. Upon cooling, this fused phase forms a glassy matrix that results in a dense, strong body. Thus the final microstructure consists of the vitrified phase, any unreacted quartz particles, and some porosity.

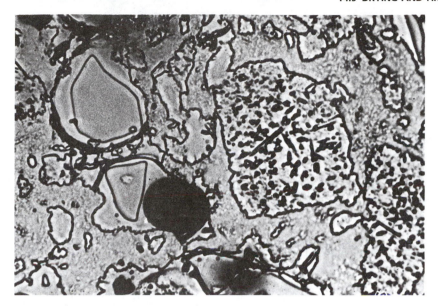

Figure 14.10 Photomicrograph of a sintered porcelain specimen (etched 10 s, 0°C, 4% HF) showing quartz grains (large light particles), partially dissolved feldspar grains (small indistinct particles), a glassy (clay) matrix, and dark pores. 400×. (From W. D. Kingery, H. K. Bowen, and D. R. Uhlmann, *Introduction to Ceramics,* 2nd edition, p. 536. Copyright © 1976 by John Wiley & Sons, New York. Reprinted by permission of John Wiley & Sons, Inc.)

Figure 14.10 is a photomicrograph of a fired porcelain in which may be seen these microstructural elements.

The degree of vitrification, of course, controls the room-temperature properties of the ceramic ware; strength, durability, and density are all enhanced as it increases. The firing temperature determines the extent to which vitrification occurs; that is, vitrification increases as the firing temperature is raised. Building bricks are ordinarily fired around 900°C (1650°F) and are relatively porous. On the other hand, firing of highly vitrified porcelain, which borders on being optically translucent, takes place at much higher temperatures. Complete vitrification is avoided during firing, since a body becomes too soft and will collapse.

REFRACTORIES

Another important class of ceramics that are utilized in large tonnages are the refractory ceramics. The salient properties of these materials include the capacity to withstand high temperatures without melting or decomposing, and the capacity to remain unreactive and inert when exposed to severe environments. In addition, the ability to provide thermal insulation is often an important consideration. Refractory materials are marketed in a variety of forms, but bricks are the most common. Typical

TABLE 14.2 Compositions of Five Common Ceramic Refractory Materials

Refractory Type	Composition (wt%)							Apparent Porosity (%)
	Al_2O_3	SiO_2	MgO	Cr_2O_3	Fe_2O_3	CaO	TiO_2	
Fireclay	25–45	70–50	0–1		0–1	0–1	1–2	10–25
High-alumina fireclay	90–50	10–45	0–1		0–1	0–1	1–4	18–25
Silica	0.2	96.3	0.6			2.2		25
Periclase	1.0	3.0	90.0	0.3	3.0	2.5		22
Periclase–chrome ore	9.0	5.0	73.0	8.2	2.0	2.2		21

Source: From W. D. Kingery, H. K. Bowen, and D. R. Uhlmann, *Introduction to Ceramics,* 2nd edition. Copyright © 1976 by John Wiley & Sons, New York. Reprinted by permission of John Wiley & Sons, Inc.

applications include furnace linings for metal refining, glass manufacturing, metallurgical heat treatment, and power generation.

Of course, the performance of a refractory ceramic, to a large degree, depends on its composition. On this basis, there are several classifications, namely, fireclay, silica, basic, and special refractories. Compositions for a number of commercial refractories are listed in Table 14.2. For many commercial materials, the raw ingredients consist of both large (or grog) particles and fine particles, which may have different compositions. Upon firing, the fine particles normally are involved in the formation of a bonding phase, which is responsible for the increased strength of the brick; this phase may be predominantly either glassy or crystalline. The service temperature is normally below that at which the refractory piece was fired.

Porosity is one microstructural variable that must be controlled to produce a suitable refractory brick. Strength, load-bearing capacity, and resistance to attack by corrosive materials all increase with porosity reduction. At the same time, thermal insulation characteristics and resistance to thermal shock are diminished. Of course, the optimum porosity depends on the conditions of service.

14.10 FIRECLAY REFRACTORIES

The primary ingredients for the fireclay refractories are high-purity fireclays, alumina and silica mixtures, usually containing between 25 and 45 wt% alumina. According to the SiO_2–Al_2O_3 phase diagram, Figure 13.24, over this composition range the highest temperature possible without the formation of a liquid phase is 1587°C (2890°F). Below this temperature the equilibrium phases present are mullite and silica (cristobalite). During refractory service use the presence of a small amount of a liquid phase may be allowable without compromising mechanical integrity. Above 1587°C the fraction of liquid phase present will depend on refractory composition. Upgrading the alumina content will increase the maximum service temperature, allowing for the formation of a small amount of liquid.

Fireclay bricks are used principally in furnace construction, to confine hot atmospheres, and to thermally insulate structural members from excessive temperatures. For fireclay brick, strength is not ordinarily an important consideration, because support of structural loads is usually not required. Some control is normally maintained over the dimensional accuracy and stability of the finished product.

14.11 SILICA REFRACTORIES

The prime ingredient for silica refractories, sometimes termed acid refractories, is silica. These materials, well known for their high-temperature load-bearing capacity, are commonly used in the arched roofs of steel- and glass-making furnaces; for these applications, temperatures as high as 1650°C (3000°F) may be realized. Under these conditions some small portion of the brick will actually exist as a liquid. The presence of even small concentrations of alumina has an adverse influence on the performance of these refractories, which may be explained by the silica–alumina phase diagram, Figure 13.24. Since the eutectic composition (7.7 wt% Al_2O_3) is very near to the silica extremity of the phase diagram, even small additions of Al_2O_3 lower the liquidus temperature significantly, which means that substantial amounts of liquid may be present at temperatures in excess of 1600°C (2910°F). Thus the alumina content should be held to a minimum, normally to between 0.2 and 1.0 wt%.

These refractory materials are also resistant to slags that are rich in silica (called acid slags) and are often used as containment vessels for them. On the other hand, they are readily attacked by slags composed of a high proportion of CaO and/or MgO (basic slags), and contact with these oxide materials should be avoided.

14.12 BASIC REFRACTORIES

The refractories that are rich in periclase, or magnesia (MgO), are termed basic; they may also contain calcium, chromium, and iron compounds. The presence of silica is deleterious to their high-temperature performance. Basic refractories are especially resistant to attack by slags containing high concentrations of MgO and CaO, and find extensive use in some steel-making open hearth furnaces.

14.13 SPECIAL REFRACTORIES

There are yet other ceramic materials that are used for rather specialized refractory applications. Some of these are relatively high-purity oxide materials, many of which may be produced with very little porosity. Included in this group are alumina, silica, magnesia, beryllia (BeO), zirconia (ZrO_2), and mullite ($3Al_2O_3–2SiO_2$). Others include carbide compounds, in addition to carbon and graphite. Silicon carbide (SiC) has been used for electrical resistance heating elements, as a crucible material, and in internal furnace components. Carbon and graphite are very refractory, but find limited application because they are susceptible to oxidation at temperatures in excess of about 800°C (1470°F). As would be expected, these specialized refractories are relatively expensive.

OTHER APPLICATIONS AND PROCESSING METHODS

14.14 ABRASIVES

Abrasive ceramics are used to wear, grind, or cut away other material, which necessarily is softer. Therefore, the prime requisite for this group of materials is hardness or wear resistance; in addition, a high degree of toughness is essential to ensure that the abrasive particles do not easily fracture. Furthermore, high temperatures may be produced from abrasive frictional forces, so some refractoriness is also desirable.

Diamonds, both natural and synthetic, are utilized as abrasives; however, they are relatively expensive. The more common ceramic abrasives include silicon carbide, tungsten carbide (WC), aluminum oxide (or corundum), and silica sand.

Figure 14.11 Photomicrographs of (a) silicon carbide (50×) and (b) aluminum oxide (100×) bonded ceramic abrasives. In both photomicrographs the light regions are the abrasive (SiC or Al_2O_3) grains; the gray and dark areas are the bonding phase and porosity, respectively. (From W. D. Kingery, H. K. Bowen, and D. R. Uhlmann, *Introduction to Ceramics,* 2nd edition, p. 568. Copyright © 1976 by John Wiley & Sons. Reprinted by permission of John Wiley & Sons, Inc.)

Abrasives are used in several forms—bonded to grinding wheels, as coated abrasives, and as loose grains. In the first case, the abrasive particles are bonded to a wheel by means of a glassy ceramic or an organic resin. The surface structure should contain some porosity; a continual flow of air currents or liquid coolants within the pores that surround the refractory grains will prevent excessive heating. Figure 14.11 shows the microstructures of two bonded abrasives, revealing abrasive grains, the bonding phase, and pores.

Coated abrasives are those in which an abrasive powder is coated on some type of paper or cloth material; sandpaper is probably the most familiar example. Wood, metals, ceramics, and plastics are all frequently ground and polished using this form of abrasive.

Grinding, lapping, and polishing wheels often employ loose abrasive grains that are delivered in some type of oil- or water-based vehicle. Diamonds, corundum, silicon carbide, and rouge (an iron oxide) are used in loose form over a variety of grain size ranges.

14.15 POWDER PRESSING

Several ceramic-forming techniques have already been discussed relative to the fabrication of glass and clay products. Another important and commonly used method that warrants a brief treatment is powder pressing. Powder pressing, the ceramic analogue to powder metallurgy, is used to fabricate both clay and nonclay compositions, including electronic and magnetic ceramics as well as some refractory brick products. In essence, a powdered mass, usually containing a small amount of water or other binder, is compacted into the desired shape, by pressure. The degree of compaction is maximized and fraction of void space is minimized by using coarse and fine particles mixed in appropriate proportions. There is no plastic deformation of the particles during compaction, as there may be with metal powders. One function of the binder is to lubricate the powder particles as they move past one another in the compaction process.

There are three basic powder pressing procedures: uniaxial, isostatic (or hydrostatic), and hot pressing. For uniaxial pressing, the powder is compacted in a metal die by pressure that is applied in a single direction. The formed piece takes on the configuration of die and platens through which the pressure is applied. This method is confined to shapes that are relatively simple; however, production rates are high and the process is inexpensive. The steps involved in one technique are illustrated in Figure 14.12.

For isostatic pressing, the powdered material is contained in a rubber envelope and the pressure is applied by a fluid, isostatically (i.e., it has the same magnitude in all directions). More complicated shapes are possible than with uniaxial pressing; however, the isostatic technique is more time consuming and expensive.

For both uniaxial and isostatic procedures, a firing operation is required after the pressing operation. During firing the formed piece shrinks, and experiences a reduction of porosity and an improvement in mechanical integrity. These changes occur by the coalescence of the powder particles into a more dense mass in a process termed **sintering.** The mechanism of sintering is schematically illustrated in Figure 14.13. After pressing, many of the powder particles touch one another (Figure 14.13a). During the initial sintering stage, necks form along the contact region between adjacent

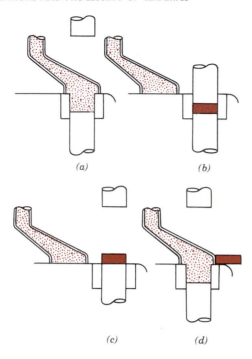

Figure 14.12 Schematic representation of the steps in uniaxial powder pressing. (*a*) The die cavity is filled with powder. (*b*) The powder is compacted by means of pressure applied to the top die. (*c*) The compacted piece is ejected by rising action of the bottom punch. (*d*) The fill shoe pushes away the compacted piece, and the fill step is repeated. (From W. D. Kingery, Editor, *Ceramic Fabrication Processes,* MIT Press. Copyright © 1958 by the Massachusetts Institute of Technology.)

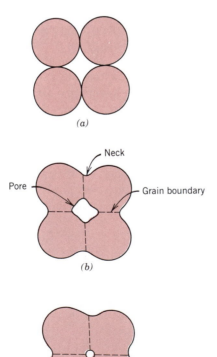

Figure 14.13 For a powder compact, microstructural changes that occur during firing. (*a*) Powder particles after pressing. (*b*) Particle coalescence and pore formation as sintering begins. (*c*) As sintering proceeds, the pores change size and shape.

particles; in addition, a grain boundary forms within each neck, and every interstice between particles becomes a pore (Figure 14.13b). As sintering progresses, the pores become smaller and more spherical in shape (Figure 14.13c). Scanning electron micrographs of a sintered alumina material are shown in Figure 14.14. The driving force for sintering is the reduction in total particle surface area; surface energies are larger in magnitude than grain boundary energies. Sintering is carried out below the melting

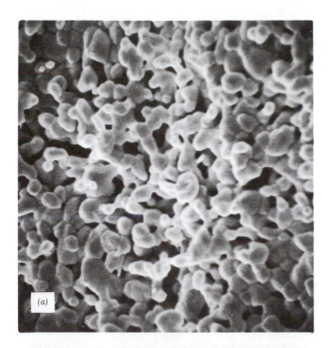

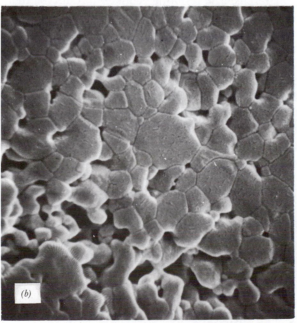

Figure 14.14 Scanning electron micrographs of an aluminum oxide powder compact sintered at 1700°C for (a) $2\frac{1}{2}$ min and (b) 6 min. As may be noted, with increasing sintering time there is a decrease in total porosity and an increase in grain size. 5000×. (From W. D. Kingery, H. K. Bowen, and D.R. Uhlmann, *Introduction to Ceramics*, 2nd edition, p. 483. Copyright © 1976 by John Wiley & Sons, New York. Reprinted by permission of John Wiley & Sons, Inc.)

temperature so that a liquid phase is normally not present. Mass transport necessary to affect the changes shown in Figure 14.13 is accomplished by atomic diffusion from the bulk particles to the neck regions.

With hot pressing, the powder pressing and heat treatment are performed simultaneously—the powder aggregate is compacted at an elevated temperature. This procedure is used for materials that do not form a liquid phase except at very high and impractical temperatures; in addition, it is utilized when high densities without appreciable grain growth are desired. This is an expensive fabrication technique that has some limitations. It is costly in terms of time, since both mold and die must be heated and cooled during each cycle. In addition, the mold is usually expensive to fabricate and ordinarily has a short lifetime.

14.16 CEMENTS

Several familiar ceramic materials are classified as inorganic cements: cement, plaster of paris, and lime, which, as a group, are produced in extremely large quantities. The characteristic feature of these materials is that when mixed with water, they form a paste that subsequently sets and hardens. This trait is especially useful in that solid and rigid structures having just about any shape may be expeditiously formed. Also, some of these materials act as a bonding phase that chemically binds particulate aggregates into a single cohesive structure. Under these circumstances the role of the cement is similar to that of the glassy bonding phase that forms when clay products and some refractory bricks are fired. One important difference, however, is that the cementitious bond develops at room temperature.

Of this group of materials, portland cement is consumed in the largest tonnages. It is produced by grinding and intimately mixing clay and lime-bearing minerals in the proper proportions, and then heating the mixture to about 1400°C (2550°F) in a rotary kiln; this process, sometimes called **calcination,** produces physical and chemical changes in the raw materials. The resulting "clinker" product is then ground into a very fine powder to which is added a small amount of gypsum ($CaSO_4$–$2H_2O$) to retard the setting process. This product is portland cement. The properties of portland cement, including setting time and final strength, to a large degree depend on its composition.

Several different constituents are found in portland cement, the principal ones being tricalcium silicate ($3CaO$–SiO_2) and dicalcium silicate ($2CaO$–SiO_2). The setting and hardening of this material results from relatively complicated hydration reactions that occur between the various cement constituents and the water that is added. For example, one hydration reaction involving dicalcium silicate is as follows:

$$2CaO\text{–}SiO_2 + xH_2O = 2CaO\text{–}SiO_2\text{–}xH_2O \tag{14.1}$$

where x is variable and depends on how much water is available. These hydrated products are in the form of complex gels or crystalline substances that form the cementitious bond. Hydration reactions begin just as soon as water is added to the cement. These are first manifested as setting (i.e., the stiffening of the once-plastic paste), which takes place soon after mixing, usually within several hours. Hardening of the mass follows as a result of further hydration, a relatively slow process that may continue for as long as several years. It should be emphasized that the process by which cement hardens is not one of drying, but rather, of hydration in which water actually participates in a chemical bonding reaction.

Portland cement is termed a hydraulic cement because its hardness develops by chemical reactions with water. It is used primarily in mortar and concrete to bind into a cohesive mass, aggregates of inert particles (sand and/or gravel); these are considered to be composite materials (see Section 17.2). Other cement materials, such as lime, are nonhydraulic; that is, compounds other than water (e.g., CO_2) are involved in the hardening reaction.

14.17 ADVANCED CERAMICS

Although the traditional ceramics discussed previously account for the bulk of the production, the development of new and what are termed "advanced ceramics" has begun and will continue to establish a prominent niche in our advanced technologies. In particular, electrical, magnetic, and optical properties and property combinations unique to ceramics have been exploited in a host of new products; some of these are discussed in Chapters 19, 21, and 22. Furthermore, advanced ceramics are, or have the potential to be, utilized in internal combustion and turbine engines, as armor plate, in electronic packaging, as cutting tools, and for energy conversion, storage, and generation. Some of these will now be discussed.

Heat Engine Applications

Advanced ceramic materials are just beginning to be used in automobile internal combustion engines. The principal advantages of these new materials over the conventional metal alloys include: the ability to withstand higher operating temperatures, thereby increasing fuel efficiency; excellent wear and corrosion resistance; lower frictional losses; the ability to operate without a cooling system; and lower densities, which result in decreased engine weights. Such engines are still in the developmental stage; however, ceramic engine blocks as well as valves, cylinder liners, pistons, bearings, and other components have been demonstrated in the United States and Japan. Furthermore, research is also being conducted on automobile gas turbine engines that employ ceramic rotors, stators, regenerators, and combustion housings.

On the basis of their desirable physical and chemical characteristics mentioned above, advanced ceramic materials will, at some future time, most certainly be utilized in jet aircraft engines. Of particular significance are the relatively low densities of these materials, which will result in turbine blades that are lighter than their superalloy counterparts, and which will also lead to other light-weight components. Materials presently under consideration for use in ceramic heat engines include silicon nitride (Si_3N_4), silicon carbide (SiC), and zirconia (ZrO_2).

The wear resistance and/or high-temperature deterioration characteristics of some metal heat engine parts currently in use have been improved significantly by using ceramic surface coatings. For example, superalloy aircraft turbine engine components are protected by a zirconia thermal barrier coating. The successful performance of these coatings is dependent on the establishment of a strong bond between the ceramic and its metal substrate which must be maintained during thermal cycling.

The chief drawback to the use of ceramics in heat engines is their disposition to brittle and catastrophic failure, due to their relatively low fracture toughnesses. Techniques are currently being developed to enhance the toughness characteristics of these materials. One particularly interesting and promising toughening technique employs a phase transformation to arrest the propagation of cracks, and is aptly termed

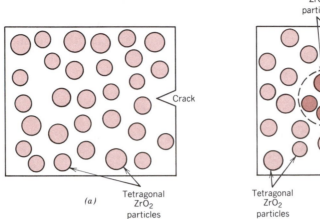
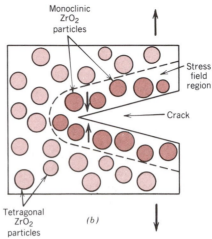

Figure 14.15 Schematic demonstration of transformation toughening. (*a*) A crack prior to inducement of the ZrO_2 particle phase transformation. (*b*) Crack arrestment due to the stress-induced phase transformation.

transformation toughening. Small particles of partially stabilized ZrO_2 are dispersed within the structural material under consideration (which may include ZrO_2 itself). Typically, CaO, MgO, Y_2O_3 and CeO are used as stabilizers. Partial stabilization allows retention of the metastable tetragonal phase at ambient conditions rather than the stable monoclinic phase. In the near stress field of a propagating crack, metastably retained tetragonal particles undergo transformation to the stable monoclinic phase. Accompanying this transformation is a slight particle volume increase, and the net result is that compressive stresses are established on the crack surface that tend to pinch the crack shut, thereby arresting its growth. This process is demonstrated schematically in Figure 14.15.

Fracture toughness and strength may also be improved by addition of another ceramic phase, which is usually in fiber or whisker form; such materials are called *ceramic–matrix composites*. The fibers or whiskers may inhibit crack propagation by the deflection of crack tips, or by pulling the fibers/whiskers out of the matrix and/ or by causing a redistribution of stresses in regions adjacent to the crack tips.

Furthermore, improved material processing techniques are necessary so as to produce materials that have specific microstructures and, therefore, uniform and reliable mechanical and corrosion-resistant characteristics at elevated temperatures. Several key issues relative to processing include starting ceramic powder character-ization (i.e., purity, particle size, and size distribution); powder synthesis and processing; improved densification techniques; and property (e.g., mechanical, physical, and chemical) evaluation. In addition, it is important to develop a methodology whereby more reliable service life predictions may be made for these materials.

Ceramic Armor

Some of the new advanced ceramics are being used in armor systems to protect military personnel and vehicles from ballistic projectiles. The chief consideration in these applications is the weight of the protective material required to thwart projectile impact. On a per weight basis, some ceramic materials are highly armor efficient.

Most ceramic armor systems consist of one or more outer ceramic facing plates which are combined with a ductile and softer backing sheet. Upon impact, the facing plates must be hard enough to fracture the high-velocity projectile, which impact also causes brittle fracture of the plate itself. Ceramic armor materials include alumina (Al_2O_3), boron carbide (B_4C), silicon carbide (SiC), and titanium diboride (TiB_2).

The armor backing must absorb the remaining projectile kinetic energy by deformation and, in addition, restrain continued penetration of projectile and ceramic fragments. Aluminum and laminates of synthetic fibers embedded in a plastic matrix are commonly used.

Electronic Packaging

The electronics industry is continually looking for new materials to keep up with its ever changing technologies. Of particular interest is the packaging of integrated circuits (ICs). The ICs are mounted on a substrate material which must be electrically insulating, have appropriate dielectric characteristics (i.e., low dielectric constant), as well as dissipate heat generated by electrical currents that pass through the electronic components (i.e., be thermally conductive). As the IC electronic components become packed closer together, this dissipation of heat becomes an increasingly more critical consideration.

Aluminum oxide has been the standard-bearer substrate material; its chief limitation, however, is a relatively low thermal conductivity. As a general rule, materials that are poor electrical conductors are also poor thermal conductors, and vice versa. Exceptions (i.e., electrical insulators and thermal conductors) are high-purity ceramic materials that have simple crystal structures; these include boron nitride (BN), silicon carbide (SiC), and aluminum nitride (AlN). Currently, the most promising substrate alternative is AlN, which has a thermal conductivity a factor of 10 better than that for alumina. Furthermore, the thermal expansion of AlN is very close to the silicon IC chips for which it serves as a substrate.

SUMMARY

This chapter discussed the various types of ceramic materials and, for each, the principal fabrication methods.

Since glasses are formed at elevated temperatures, the temperature–viscosity behavior is an important consideration. Melting, working, softening, annealing, and strain points represent temperatures that correspond to specific viscosity values. Knowledge of these points is important in the fabrication and processing of a glass of given composition. Four of the more common glass-forming techniques—pressing, blowing, drawing, and fiber forming—were discussed briefly. After fabrication, glasses may be annealed and/or tempered to improve mechanical characteristics. Glass–ceramics are initially fabricated as a glass, then crystallized or devitrified.

Clay is the principal component of the whitewares and structural clay products. Other ingredients may be added, such as feldspar and quartz, which influence the changes that occur during firing. Two fabrication techniques that are frequently utilized are hydroplastic forming and slip casting. After forming, a body must be first dried and then fired at an elevated temperature to reduce porosity and enhance strength. Shrinkage that is excessive or too rapid may result in cracking and/or warping, and a worthless ware. Densification during firing is accomplished by vitrification, the formation of a glassy bonding phase.

The materials that are employed at elevated temperatures and often in reactive environments are the refractory ceramics; on occasion, their ability to thermally insulate is also utilized. On the basis of composition and application, the four main subdivisions are fireclay, silica, basic, and special.

The abrasive ceramics, being hard and tough, are utilized to cut, grind, and polish other softer materials. Diamond, silicon carbide, tungsten carbide, corundum, and silica sand are the most common examples. The abrasives may be employed in the form of loose grains, bonded to an abrasive wheel, or coated on paper or a fabric.

Some ceramic pieces are formed by powder compaction; uniaxial, isostatic, and hot-pressing techniques are possible.

When mixed with water, inorganic cements form a paste that is capable of assuming just about any desired shape. Subsequent setting or hardening is a result of chemical reactions involving the cement particles and occurs at the ambient temperature. For hydraulic cements, of which portland cement is the most common, the chemical reaction is one of hydration.

Many of our modern technologies utilize and will continue to utilize advanced ceramics because of their unique mechanical, chemical, electrical, magnetic, and optical properties and property combinations. Advanced materials characterization, processing, and reliability techniques need to be developed to make these materials cost effective.

IMPORTANT TERMS AND CONCEPTS

Abrasive	Green ceramic body	Strain point (glass)
Annealing point (glass)	Hydroplastic forming	Structural clay product
Calcination	Melting point (glass)	Thermal shock
Cement	Refractory (ceramic)	Thermal tempering
Devitrification	Sintering	Vitrification
Firing	Slip casting	Whiteware
Glass–ceramic	Softening point (glass)	Working point (glass)
Glass transition temperature		

REFERENCES

Coes, L., Jr., *Abrasives,* Springer-Verlag, New York, 1971.

Kingery, W. D., H. K. Bowen, and D. R. Uhlmann, *Introduction to Ceramics,* 2nd edition, John Wiley & Sons, New York, 1976. Chapters 1, 10, 11, and 16.

KINGERY, W. D. (Editor), *Ceramic Fabrication Processes,* Technology Press of Massachussetts Institute of Technology, and John Wiley & Sons, New York, 1958.

LEA, F. M., *The Chemistry of Cement and Concrete,* 3rd edition, Edward Arnold, London, 1970.

NORTON, F. H., *Elements of Ceramics,* 2nd edition, Addison-Wesley Publishing Company, Reading, MA, 1974. Chapters 3–23.

REED, J. S., *Introduction to the Principles of Ceramic Processing,* John Wiley & Sons, New York, 1988.

RICHERSON, D. W., *Modern Ceramic Engineering,* Marcel Dekker, New York, 1982.

TOOLEY, F. V. (Editor), *The Handbook of Glass Manufacture,* Books for Industry, and *Glass Industry Magazine,* New York, 1974. In two volumes.

VAN VLACK, L. H., *Physical Ceramics for Engineers,* Addison-Wesley Publishing Company, Reading, MA, 1964. Chapters 8, 13, and 14.

QUESTIONS AND PROBLEMS

14.1 Cite the two desirable characteristics of glasses.

14.2 Soda and lime are added to a glass batch in the form of soda ash (Na_2CO_3) and limestone ($CaCO_3$). During heating, these two ingredients decompose to give off carbon dioxide (CO_2), the resulting products being soda and lime. Compute the weight of soda ash and limestone that must be added to 100 lb_m of quartz (SiO_2) to yield a glass of composition 75 wt% SiO_2, 15 wt% Na_2O, and 10 wt% CaO.

14.3 What is the distinction between glass transition temperature and melting temperature?

14.4 On the basis of mechanical characteristics associated with the behaviors shown in Figure 14.3, explain why glass may be drawn into fibers whereas crystalline aluminum oxide may not.

14.5 Compare the temperatures at which soda–lime, borosilicate, 96% silica, and fused silica may be annealed.

14.6 Compare the softening points for 96% silica, borosilicate, and soda–lime glasses.

14.7 The viscosity η of a glass varies with temperature according to the relationship

$$\eta = A \exp\left(\frac{Q_{vis}}{RT}\right)$$

where Q_{vis} is the energy of activation for viscous flow, A is a temperature-independent constant, and R and T are, respectively, the gas constant and the absolute temperature. A plot of $\ln \eta$ verses $1/T$ should be nearly linear, and with a slope of Q_{vis}/R. Using the data in Figure 14.4, **(a)** make such a plot for the borosilicate glass, and **(b)** determine the activation energy between temperatures of 500 and 900°C.

14.8 For many viscous materials, the viscosity η may be defined in terms of the expression

$$\eta = \frac{\sigma}{d\epsilon/dt}$$

where σ and $d\epsilon/dt$ are, respectively, the tensile stress and the strain rate. A cylindrical specimen of a soda–lime glass of diameter 5 mm (0.2 in.) and length 100 mm (4 in.) is subjected to a tensile force of 1 N (0.224 lb$_f$) along its axis. If its deformation is to be less than 1 mm (0.04 in.) over a week's time, using Figure 14.4, determine the maximum temperature to which the specimen may be heated.

14.9 **(a)** Explain why residual thermal stresses are introduced into a glass piece when it is cooled. **(b)** Are thermal stresses introduced upon heating? Why or why not? **(c)** How does the thickness of a glass ware affect the magnitude of the thermal stresses? Why?

14.10 Borosilicate glasses and fused silica are resistant to thermal shock. Why is this so?

14.11 In your own words, briefly describe what happens as a glass piece is thermally tempered.

14.12 Glass pieces may also be strengthened by chemical tempering. With this procedure, the glass surface is put in a state of compression by exchanging some of the cations near the surface with other cations having a larger diameter. Suggest one type of cation which, by replacing Na$^+$, will induce chemical tempering in a soda–lime glass.

14.13 **(a)** What is devitrification? **(b)** Cite two properties that may be improved by devitrification and two that may be impaired.

14.14 Briefly explain why glass–ceramics are not transparent. You may want to consult Chapter 22.

14.15 Cite the two desirable characteristics of clay minerals relative to fabrication processes.

14.16 From a molecular perspective, briefly explain the mechanism by which clay minerals become hydroplastic when water is added.

14.17 Thick ceramic wares are more likely to crack upon drying than thin wares. Why is this so?

14.18 Explain why a clay, once having been fired at an elevated temperature, loses its hydroplasticity.

14.19 **(a)** What are the three main components of a whiteware ceramic such as porcelain? **(b)** What role does each component play in the forming and firing procedures?

14.20 **(a)** Why is it so important to control the rate of drying of a ceramic body that has been hydroplastically formed or slip cast? **(b)** Cite three factors that influence the rate of drying, and explain how each affects the rate.

14.21 Cite one reason why drying shrinkage is greater for slip cast or hydroplastic products that have smaller clay particles.

14.22 **(a)** Name three factors that influence the degree to which vitrification occurs in clay-based ceramic wares. **(b)** Explain how density, firing distortion, strength, corrosion resistance, and thermal conductivity are affected by the extent of vitrification.

14.23 For refractory ceramic materials, cite three characteristics that improve with and two characteristics that are adversely affected by increasing porosity.

14.24 Find the maximum temperature to which the following two magnesia–alumina refractory materials may be heated before a liquid phase will appear. **(a)** A spinel-bonded magnesia material of composition 88.5 wt% MgO–11.5 wt% Al_2O_3. **(b)** A magnesia–alumina spinel of composition 25 wt% MgO–75 wt% Al_2O_3. Consult Figure 13.22.

14.25 Upon consideration of the SiO_2–Al_2O_3 phase diagram, Figure 13.24, for each pair of compositions listed below, which would you judge to be the more desirable refractory? Justify your choices:
(a) 99.8 wt% SiO_2–0.2 wt% Al_2O_3 and 99.0 wt% SiO_2–1.0 wt% Al_2O_3.
(b) 70 wt% Al_2O_3–30 wt% SiO_2 and 74 wt% Al_2O_3–26 wt% SiO_2.
(c) 90 wt% Al_2O_3–10 wt% SiO_2 and 95 wt% Al_2O_3–5 wt% SiO_2.

14.26 Compute the mass fractions of liquid in the following fireclay refractory materials at 1600°C (2910°F):
(a) 25 wt% Al_2O_3–75 wt% SiO_2.
(b) 45 wt% Al_2O_3–55 wt% SiO_2.

14.27 **(a)** For the SiO_2–Al_2O_3 system, what is the maximum temperature that is possible without the formation of a liquid phase? At what composition or over what range of compositions will this maximum temperature be achieved? **(b)** For the MgO–Al_2O_3 system, what is the maximum temperature that is possible without the formation of a liquid phase? At what composition or over what range of compositions will this maximum temperature be achieved?

14.28 For a 99.8 wt% SiO_2–0.2 wt% Al_2O_3 refractory, at what temperature will the same mass fraction of a liquid phase exist as for a 99.0 wt% SiO_2–1.0 wt% Al_2O_3 material at 1600°C (2910°F)?

14.29 Some ceramic materials are fabricated by hot isostatic pressing. Cite some of the limitations and difficulties associated with this technique.

14.30 Compare the manner in which the aggregate particles become bonded together in clay-based mixtures during firing and in cements during setting.

14.31 Explain why it is important to grind cement into a fine powder.

CHAPTER 15

POLYMER STRUCTURES

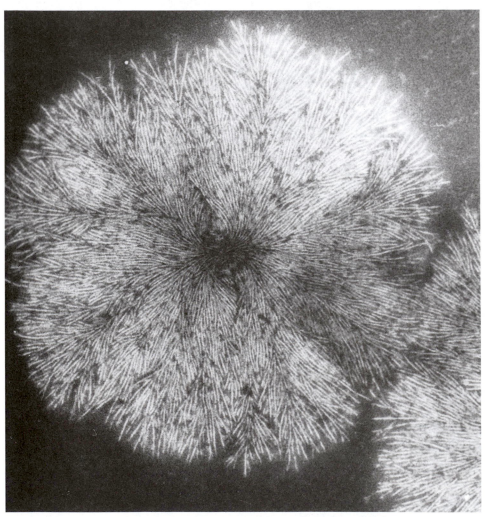

Transmission electron micrograph showing the spherulite structure in a natural rubber specimen. Chain-folded lamellar crystallites approximately 10 nm thick extend in radial directions from the center; they appear as white lines in the micrograph. $30,000\times$. (Photograph supplied by P. J. Phillips. First published in R. Bartnikas and R. M. Eichhorn, *Engineering Dielectrics,* Vol. IIA, *Electrical Properties of Solid Insulating Materials: Molecular Structure and Electrical Behavior.* Copyright ASTM, 1916 Race Street, Philadelphia, PA 19103. Reprinted with permission.)

15.1 INTRODUCTION

Naturally occurring polymers—those derived from plants and animals—have been used for many centuries; these materials include wood, rubber, cotton, wool, leather, and silk. Other natural polymers such as proteins, enzymes, starches, and cellulose are important in biological and physiological processes in plants and animals. Since the early 1900s, modern scientific research tools have made possible the determination of the molecular structures of this group of materials, and the development of numerous polymers, which are synthesized from small organic molecules. Many of our useful plastics, rubbers, and fiber materials are synthetic polymers. In fact, since the conclusion of World War II, the field of materials has been virtually revolutionized by the advent of synthetic polymers. The synthetics can be produced inexpensively, and their properties may be managed to the degree that many are superior to their natural counterparts. In some applications metal and wood parts have been replaced by plastics, which have satisfactory properties and may be produced at a lower cost.

As with metals and ceramics, the properties of polymers are intricately related to the structural elements of the material. This chapter explores molecular and crystal structures of polymers; Chapter 16 discusses the relationships between structure and some of the physical and chemical properties, along with typical applications and forming methods.

15.2 HYDROCARBON MOLECULES

Since most polymers are organic in origin, we briefly review some of the basic concepts relating to the structure of their molecules. First, many organic materials are *hydrocarbons;* that is, they are composed of hydrogen and carbon. Furthermore, the intramolecular bonds are covalent. Each carbon atom has four electrons that may participate in covalent bonding, whereas every hydrogen atom has only one bonding electron. A single covalent bond exists when each of the two bonding atoms contributes one electron, as represented schematically in Figure 2.10 for a molecule of methane (CH_4). A bond between two carbon atoms may involve the sharing of two pairs of electrons; this is termed a *double bond.* For example, in ethylene, which has the chemical formula C_2H_4, the two carbon atoms are doubly bonded together, and each is also singly bonded to two hydrogen atoms, as represented by the structural formula

$$
\begin{array}{ccc}
\text{H} & & \text{H} \\
| & & | \\
\text{C} & = & \text{C} \\
| & & | \\
\text{H} & & \text{H}
\end{array}
$$

where — and = denote single and double covalent bonds, respectively. In rare cases triple bonds exist, as in acetylene, C_2H_2:

$$ \text{H}-\text{C}\equiv\text{C}-\text{H} $$

Molecules that have double and triple covalent bonds are termed **unsaturated;** that is, each carbon atom is not bonded to the maximum (or four) other atoms. For an unsaturated molecule, a double bond may be thought of as consisting of two single bonds. A transfer in position around the carbon atom of one of these single bonds permits the addition of another atom or group of atoms to the original molecule. Of course, in a **saturated** hydrocarbon, all bonds are single ones (and saturated), and no new atoms may be joined without the removal of others that are already bonded.

Some of the simple hydrocarbons belong to the paraffin family; the chainlike paraffin molecules include methane (CH_4), ethane (C_2H_6), propane (C_3H_8), and butane (C_4H_{10}). Compositions and molecular structures for paraffin molecules are contained in Table 15.1. The covalent bonds in each molecule are strong, but only weak hydrogen and van der Waals bonds exist between molecules, and thus these hydrocarbons have relatively low melting and boiling points. However, melting and boiling temperatures rise with increasing molecular weight.

Hydrocarbon compounds with the same composition may have different atomic arrangements, a phenomenon termed **isomerism.** For example, there are two isomers for butane; normal butane has the structure

$$
\begin{array}{c}
\quad\;\; H \quad H \quad H \quad H \\[-2pt]
\quad\;\; | \qquad | \qquad | \qquad | \\[-2pt]
H - C - C - C - C - H \\[-2pt]
\quad\;\; | \qquad | \qquad | \qquad | \\[-2pt]
\quad\;\; H \quad H \quad H \quad H
\end{array}
$$

whereas a molecule of isobutane is represented as follows:

$$
\begin{array}{c}
\qquad\qquad H \\[-2pt]
\qquad\qquad | \\[-2pt]
\qquad H - C - H \\[-2pt]
\qquad H \quad\; | \quad\; H \\[-2pt]
\qquad | \qquad | \qquad | \\[-2pt]
H - C - C - C - H \\[-2pt]
\qquad | \qquad | \qquad | \\[-2pt]
\qquad H \quad H \quad H
\end{array}
$$

TABLE 15.1 Compositions and Molecular Structures for Some of the Paraffin Compounds: C_nH_{2n+2}

Name	Composition	Structure
Methane	CH_4	$\begin{array}{c} H \\ | \\ H-C-H \\ | \\ H \end{array}$
Ethane	C_2H_6	$\begin{array}{c} H \quad H \\ | \quad\; | \\ H-C-C-H \\ | \quad\; | \\ H \quad H \end{array}$
Propane	C_3H_8	$\begin{array}{c} H \quad H \quad H \\ | \quad\; | \quad\; | \\ H-C-C-C-H \\ | \quad\; | \quad\; | \\ H \quad H \quad H \end{array}$
Butane	C_4H_{10}	.
Pentane	C_5H_{12}	.
Hexane	C_6H_{14}	.

TABLE 15.2 Some Common Hydrocarbon Groups

Family	Characteristic Unit	Representative Compound	
Alcohols	R—OH	$\begin{array}{c} H \\ | \\ H-C-OH \\ | \\ H \end{array}$	Methyl alcohol
Ethers	R—O—R'	$\begin{array}{c} H \quad\quad H \\ | \quad\quad | \\ H-C-O-C-H \\ | \quad\quad | \\ H \quad\quad H \end{array}$	Dimethyl ether
Acids	$R-C{\Large\diagdown}{}^{OH}_{O}$	$\begin{array}{c} H \\ | \\ H-C-C{\Large\diagdown}{}^{OH}_{O} \\ | \\ H \end{array}$	Acetic acid
Aldehydes	$\begin{array}{c} R \\ \diagdown \\ \diagup \\ H \end{array}C=O$	$\begin{array}{c} H \\ \diagdown \\ \diagup \\ H \end{array}C=O$	Formaldehyde
Aromatic hydrocarbons	⬡ R [a]	⬡ OH	Phenol

[a] The simplified structure ⬡ denotes the benzene ring,

$$\begin{array}{c} H \quad\quad\quad H \\ \diagdown \; C \quad C \; \diagup \\ C \quad\quad\quad C \\ \| \quad\quad\quad \| \\ C \quad\quad\quad C \\ \diagup \quad\quad\quad \diagdown \\ H \quad C \quad\quad C \quad H \\ | \\ H \end{array}$$

Some of the physical properties of hydrocarbons will depend on the isomeric state; for example, the boiling temperatures for normal butane and isobutane are -0.5 and $-12.3°C$ (31.1 and 9.9°F), respectively.

There are numerous other hydrocarbons, many of which are involved in organic polymer structures. Several of the more common hydrocarbon groups are presented in Table 15.2, where R and R' represent organic radicals—groups of atoms that remain as a single unit and maintain their identity during chemical reactions. Examples of organic radicals include the CH_3, C_2H_5, and C_6H_5 (methyl, ethyl, and benzyl) groups.

15.3 POLYMER MOLECULES

The molecules in polymers are gigantic in comparison to the hydrocarbon molecules heretofore discussed; because of their size they are often referred to as **macromolecules.**

Within each molecule, the atoms are bound together by covalent interatomic bonds. For most polymers, these molecules are in the form of long and flexible chains, the backbone of which is a string of carbon atoms; many times each carbon atom singly bonds to two adjacent carbons atoms on either side, represented schematically in two dimensions as follows:

$$-\overset{|}{\underset{|}{C}}-\overset{|}{\underset{|}{C}}-\overset{|}{\underset{|}{C}}-\overset{|}{\underset{|}{C}}-\overset{|}{\underset{|}{C}}-\overset{|}{\underset{|}{C}}-\overset{|}{\underset{|}{C}}-$$

Each of the two remaining valence electrons for every carbon atom may be involved in side-bonding with atoms or radicals that are positioned adjacent to the chain. Of course, both chain and side double bonds are also possible.

These long molecules are composed of structural entities called **mer** units, which are successively repeated along the chain. "Mer" originates from the Greek word *meros,* which means part; a single mer is called a **monomer;** the term **polymer** was coined to mean many mers. "Mer" denotes the repeat unit in a polymer chain, whereas, "monomer" is used in the context of a molecule consisting of a single mer.

15.4 THE CHEMISTRY OF POLYMER MOLECULES

Consider again the hydrocarbon ethylene (C_2H_4), which is a gas at ambient temperature and pressure, and has the following molecular structure:

$$\overset{\textstyle H \quad H}{\underset{\textstyle H \quad H}{\overset{|}{\underset{|}{C}}=\overset{|}{\underset{|}{C}}}}$$

If the ethylene gas is subjected catalytically to appropriate conditions of temperature and pressure, it will transform to polyethylene (PE), which is a solid polymeric material. This process begins when an active mer is formed by the reaction between an initiator or catalyst species (R·) and the ethylene mer unit, as follows:

$$R\cdot + \overset{\textstyle H \quad H}{\underset{\textstyle H \quad H}{\overset{|}{\underset{|}{C}}=\overset{|}{\underset{|}{C}}}} \longrightarrow R-\overset{\textstyle H \quad H}{\underset{\textstyle H \quad H}{\overset{|}{\underset{|}{C}}-\overset{|}{\underset{|}{C}}}}\cdot \qquad (15.1)$$

The polymer chain then forms by the sequential addition of polyethylene monomer units to this active initiator-mer center. The active site, or unpaired electron (denoted by ·), is transferred to each successive end monomer as it is linked to the chain. This may be represented schematically as follows:

$$R-\overset{\textstyle H \quad H}{\underset{\textstyle H \quad H}{\overset{|}{\underset{|}{C}}-\overset{|}{\underset{|}{C}}}}\cdot + \overset{\textstyle H \quad H}{\underset{\textstyle H \quad H}{\overset{|}{\underset{|}{C}}=\overset{|}{\underset{|}{C}}}} \longrightarrow R-\overset{\textstyle H \quad H \quad H \quad H}{\underset{\textstyle H \quad H \quad H \quad H}{\overset{|}{\underset{|}{C}}-\overset{|}{\underset{|}{C}}-\overset{|}{\underset{|}{C}}-\overset{|}{\underset{|}{C}}}}\cdot \qquad (15.2)$$

The final result, after the addition of many ethylene monomer units, is the polyethylene molecule, a portion of which is shown in Figure 15.1a. This representation is not

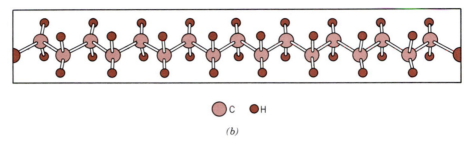

Figure 15.1 For polyethylene, (a) a schematic representation of mer and chain structures, and (b) a perspective of the molecule, indicating the zigzag backbone structure.

strictly correct in that the angle between the singly bonded carbon atoms is not 180°
as shown, but rather close to 109°. A more accurate three-dimensional model is one
in which the carbon atoms form a zigzag pattern (Figure 15.1b), the C—C bond
length being 0.154 nm. In this discussion, depiction of polymer molecules is frequently
simplified using the linear chain model.

If all the hydrogen atoms in polyethylene are replaced by fluorine, the resulting
polymer is *polytetrafluoroethylene* (PTFE); its mer and chain structures are shown in
Figure 15.2a. Polytetrafluoroethylene (having the trade name Teflon) belongs to a
family of polymers called the fluorocarbons.

Polyvinyl chloride (PVC), another common polymer, has a structure that is a
slight variant of that for polyethylene, in which every fourth hydrogen is replaced
with a Cl atom. Furthermore, substitution of the CH_3 group

$$H—C—H$$
$$|$$
$$H$$

for each Cl atom in PVC yields *polypropylene* (PP). Polyvinyl chloride and polypropylene chain structures are also represented in Figure 15.2. Table 15.3 lists mer
structures for some of the more common polymers; as may be noted, some of them,
for example, nylon, polyester, and polycarbonate, are relatively complex.

When all the repeating units along a chain are of the same type, the resulting
polymer is called a **homopolymer.** There is no restriction in polymer synthesis that
prevents the formation of compounds other than homopolymers; and, in fact, chains
may be composed of two or more different mer units, in what are termed **copolymers**
(see Section 15.9).

The mer units discussed thus far have two active bonds that may covalently bond
with other mers, as indicated above for ethylene; such a mer is termed **bifunctional;**
that is, it may bond with two other units in forming the two-dimensional chainlike

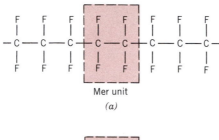

Mer unit

(a)

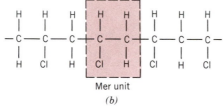

Mer unit

(b)

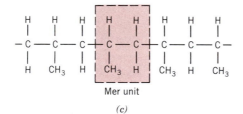

Mer unit

(c)

Figure 15.2 Mer and chain structures for (a) polytetrafluoroethylene, (b) polyvinyl chloride, and (c) polypropylene.

TABLE 15.3 A Listing of Mer Structures for 10 of the More Common Polymeric Materials

Polymer	Repeating (Mer) Structure
Polyethylene (PE)	H H | | —C—C— | | H H
Polyvinyl chloride (PVC)	H H | | —C—C— | | H Cl
Polytetrafluoroethylene (PTFE)	F F | | —C—C— | | F F
Polypropylene (PP)	H H | | —C—C— | | H CH$_3$

TABLE 15.3 (*continued*)

Polymer	Repeating (Mer) Structure
Polystyrene (PS)	
Polymethyl methacrylate (PMMA)	
Phenol-formaldehyde (Bakelite)	
Polyhexamethylene adipamide (nylon 6,6)	
Polyethylene terephthalate (PET, a polyester)	
Polycarbonate	

molecular structure. However, other mers, such as phenol-formaldehyde (Table 15.3), are **trifunctional;** they have three active bonds, from which a three-dimensional molecular network structure results, as discussed later in this chapter.

15.5 MOLECULAR WEIGHT

Extremely large molecular weights are to be found in polymers with very long chains. During the polymerization process in which these large macromolecules are synthesized from smaller molecules, not all polymer chains will grow to the same length; this results in a distribution of chain lengths or molecular weights. Ordinarily, an

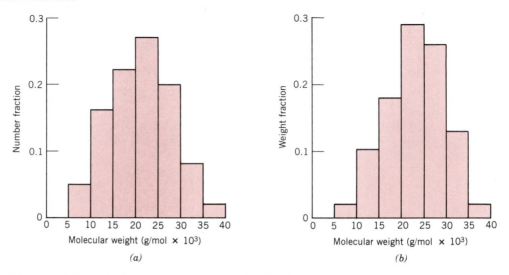

Figure 15.3 Hypothetical polymer molecule size distributions on the basis of (a) number and (b) weight fractions of molecules.

average molecular weight is specified, which may be determined by the measurement of various physical properties such as viscosity and osmotic pressure.

There are several ways of defining average molecular weight. The number-average molecular weight $\overline{M}_n$ is obtained by dividing the chains into a series of size ranges and then determining the number fraction of chains within each size range (Figure 15.3a). This number-average molecular weight is expressed as

$$\overline{M}_n = \sum x_i M_i \tag{15.3a}$$

where M_i represents the mean (middle) molecular weight of size range i, and x_i is the fraction of the total number of chains within the corresponding size range.

A weight-average molecular weight $\overline{M}_w$ is based on the weight fraction of molecules within the various size ranges (Figure 15.3b). It is calculated according to

$$\overline{M}_w = \sum w_i M_i \tag{15.3b}$$

where, again, M_i is the mean molecular weight within a size range, whereas w_i denotes the weight fraction of molecules within the same size interval. Computations for both number-average and weight-average molecular weights are carried out in Example Problem 15.1. A typical molecular weight distribution along with these molecular weight averages are shown in Figure 15.4.

An alternate way of expressing average chain size of a polymer is as the **degree of polymerization** n, which represents the average number of mer units in a chain. Both number-average (n_n) and weight-average (n_w) degrees of polymerization are possible, as follows:

$$n_n = \frac{\overline{M}_n}{\overline{m}} \tag{15.4a}$$

$$n_w = \frac{\overline{M}_w}{\overline{m}} \tag{15.4b}$$

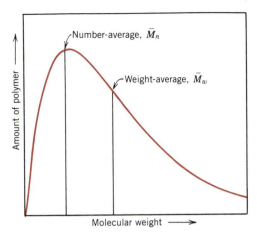

Figure 15.4 Distribution of molecular weights for a typical polymer.

where $\overline{M}_n$ and $\overline{M}_w$ are the number-average and weight-average molecular weights as defined above, while $\overline{m}$ is the mer molecular weight. For a copolymer (having two or more different mer units), $\overline{m}$ is determined from

$$\overline{m} = \sum f_j m_j \tag{15.5}$$

In this expression, f_j and m_j are, respectively, the chain fraction and molecular weight of mer j.

EXAMPLE PROBLEM 15.1

Assume that the molecular weight distributions shown in Figure 15.3 are for polyvinyl choride. For this material, compute (a) the number-average molecular weight; (b) the number-average degree of polymerization; and (c) the weight-average molecular weight.

SOLUTION

(a) The data necessary for this computation, as taken from Figure 15.3a, are presented in Table 15.4a. According to Equation 15.3a, summation of all the $x_i M_i$ products (from the right-hand column) yields the number-average molecular weight, which in this case is 21,150 g/mol.

(b) To determine the number-average degree of polymerization (Equation 15.4a), it first becomes necessary to compute the mer molecular weight. For PVC, each mer consists of two carbon atoms, three hydrogen atoms, and a single chlorine atom (Table 15.3). Furthermore, the atomic weights of C, H, and Cl are, respectively, 12.01, 1.01, and 35.45 g/mol. Thus for PVC

$$\overline{m} = 2(12.01 \text{ g/mol}) + 3(1.01 \text{ g/mol}) + 35.45 \text{ g/mol}$$
$$= 62.50 \text{ g/mol}$$

and

$$n_n = \frac{\overline{M}_n}{\overline{m}} = \frac{21{,}150 \text{ g/mol}}{62.50 \text{ g/mol}} = 338$$

TABLE 15.4a Data Used for Number-Average Molecular Weight Computations in Example Problem 15.1

Molecular Weight Range (g/mol)	Mean M_i (g/mol)	x_i	$x_i M_i$
5,000–10,000	7,500	0.05	375
10,000–15,000	12,500	0.16	2000
15,000–20,000	17,500	0.22	3850
20,000–25,000	22,500	0.27	6075
25,000–30,000	27,500	0.20	5500
30,000–35,000	32,500	0.08	2600
35,000–40,000	37,500	0.02	750
			$\overline{M}_n = 21,150$

TABLE 15.4b Data Used for Weight-Average Molecular Weight Computations in Example Problem 15.1

Molecular Weight Range (g/mol)	Mean M_i (g/mol)	w_i	$w_i M_i$
5,000–10,000	7,500	0.02	150
10,000–15,000	12,500	0.10	1250
15,000–20,000	17,500	0.18	3150
20,000–25,000	22,500	0.29	6525
25,000–30,000	27,500	0.26	7150
30,000–35,000	32,500	0.13	4225
35,000–40,000	37,500	0.02	750
			$\overline{M}_w = 23,200$

(c) Table 15.4b shows the data for the weight-average molecular weight, as taken from Figure 15.3b. The $w_i M_i$ products for the several size intervals are tabulated in the right-hand column. The sum of these products (Equation 15.3b) yields a value of 23,200 g/mol for $\overline{M}_w$.

Various polymer characteristics are affected by the magnitude of the molecular weight. One of these is the melting or softening temperature; melting temperature is raised with increasing molecular weight (for $\overline{M}$ up to about 100,000 g/mol). At room temperature, polymers with very short chains (having molecular weights on the order of 100 g/mol) exist as liquids or gases. Those with molecular weights of approximately 1000 g/mol are waxy solids (such as paraffin wax) and soft resins. Solid polymers (sometimes termed *high polymers*), which are of prime interest here, commonly have molecular weights ranging between 10,000 and several million g/mol. Melting temperature rises with an increase in intermolecular forces. In general, as the average chain length increases, so also does the total degree of bonding between the molecules; the intermolecular bonds are generally van der Waals and/or hydrogen.

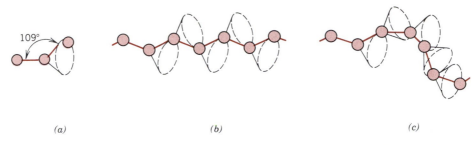

Figure 15.5 Schematic representations of how polymer chain shape is influenced by the positioning of backbone carbon atoms (solid circles). For (*a*), the right-most atom may lie anywhere on the dashed circle and still subtend a 109° angle with the bond between the other two atoms. Straight and twisted chain segments are generated when the backbone atoms are situated as in (*b*) and (*c*), respectively. (Adapted from Donald R. Askeland, *The Science and Engineering of Materials, 2nd ed.* [Boston: PWS-KENT Publishing Company, 1989] p. 517. Copyright © by Wadsworth, Inc. Reprinted by permission of PWS-KENT Publishing Company, a division of Wadsworth, Inc.)

15.6 MOLECULAR SHAPE

There is no reason to suppose that polymer chain molecules are strictly straight, straight in the sense that the zigzag arrangement of the backbone atoms (Figure 15.1*b*) is disregarded. Single chain bonds are capable of rotation and bending in three dimensions. Consider the chain atoms in Figure 15.5*a*; a third carbon atom may lie at any point on the cone of revolution and still subtend about a 109° angle with the bond between the other two atoms. A straight chain segment results when successive chain atoms are positioned as in Figure 15.5*b*. On the other hand, chain bending and twisting are possible when there is a rotation of the chain atoms into other positions, as illustrated in Figure 15.5*c*.[1] Thus a single chain molecule composed of many chain atoms might assume a shape similar to that represented schematically in Figure 15.6, having a multitude of bends, twists, and kinks.[2] Also indicated in this figure is the end-to-end distance of the polymer chain *r*; this distance is much smaller than the total chain length.

Some polymers consist of large numbers of molecular chains, each of which may bend, coil, and kink in the manner of Figure 15.6. This leads to extensive intertwining and entanglement of neighboring chain molecules, a situation similar to that of a fishing line that has experienced backlash from a fishing reel. These random coils and molecular entanglements are responsible for a number of important characteristics of polymers, to include the large elastic extensions displayed by the rubber materials.

Some of the mechanical and thermal characteristics of polymers are a function of the ability of chain segments to experience rotation in response to applied stresses or thermal vibrations. Rotational flexibility is dependent on mer structure and chemistry. For example, the region of a chain segment that has a double bond (C=C) is

---

[1] For some polymers, rotation of carbon backbone atoms within the cone may be hindered by bulky side group elements on neighboring chains.

[2] The term *conformation* is often used in reference to the physical outline of a molecule, or molecular shape, that can only be altered by rotation of chain atoms about single bonds.

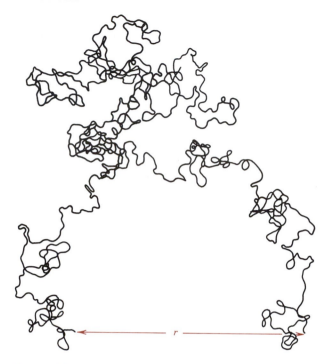

Figure 15.6 Schematic representation of a single polymer chain molecule that has numerous random kinks and coils produced by chain bond rotations. (From L. R. G. Treloar, *The Physics of Rubber Elasticity,* 2nd edition, Oxford University Press, Oxford, 1958, p. 47.)

rotationally rigid. Also, introduction of a bulky or large side group of atoms restricts rotational movement. For example, polystyrene molecules, which have a benzene ring side group (Table 15.3), are more resistant to rotational motion than are polyethylene chains.

15.7 MOLECULAR STRUCTURE

The physical characteristics of a polymer depend not only on its molecular weight and shape, but also on differences in the structure of the molecular chains. Modern polymer synthesis techniques permit considerable control over various structural possibilities. This section discusses several molecular structures including linear, branched, crosslinked, and network, in addition to various isomeric configurations.

Linear Polymers

Linear polymers are those in which the mer units are joined together end to end in single chains. These long chains are flexible and may be thought of as a mass of spaghetti, as represented schematically in Figure 15.7a, where each circle represents a mer unit. For linear polymers, there may be extensive van der Waals bonding between the chains. Some of the common polymers that form with linear structures

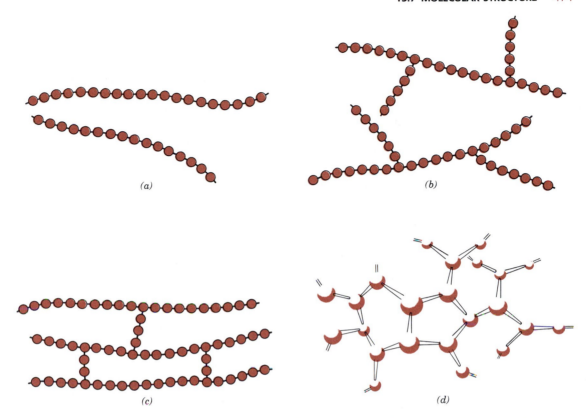

Figure 15.7 Schematic representations of (*a*) linear, (*b*) branched, (*c*) crosslinked, and (*d*) network (three-dimensional) molecular structures. Circles designate individual mer units.

are polyethylene, polyvinyl chloride, polystyrene, polymethyl methacrylate, nylon, and the fluorocarbons.

Branched Polymers

Polymers may be synthesized in which side-branch chains are connected to the main ones, as indicated schematically in Figure 15.7*b*; these are fittingly called **branched polymers.** The branches, considered to be part of the main-chain molecule, result from side reactions that occur during the synthesis of the polymer. The chain packing efficiency is reduced with the formation of side branches, which results in a lowering of the polymer density.

Crosslinked Polymers

In **crosslinked polymers,** adjacent linear chains are joined one to another at various positions by covalent bonds, as represented in Figure 15.7*c*. The process of crosslinking is achieved either during synthesis or by a nonreversible chemical reaction that is usually carried out at an elevated temperature. Often, this crosslinking is accomplished by additive atoms or molecules that are covalently bonded to the chains. Many of the rubber elastic materials are crosslinked; in rubbers, this is called vulcanization, a process described in Section 16.14.

Network Polymers

Trifunctional mer units, having three active covalent bonds, form three-dimensional networks (Figure 15.7d) instead of the linear chain framework assumed by bifunctional mers. Polymers composed of trifunctional units are termed **network polymers.** Actually, a polymer that is highly crosslinked may be classified as a network polymer. These materials have distinctive mechanical and thermal properties; the epoxies and phenol-formaldehyde belong to this group.

It should be pointed out that polymers are not usually of only one distinctive structural type. For example, a predominantly linear polymer might have some limited branching and crosslinking.

15.8 MOLECULAR CONFIGURATIONS

For polymers having more than one side atom or group of atoms bonded to the main chain, the regularity and symmetry of the side group arrangement can significantly influence the properties. Consider the mer unit

$$
\begin{array}{c}
\text{H} \quad \text{H} \\
| \quad\quad | \\
-\text{C}-\text{C}- \\
| \quad\quad | \\
\text{H} \quad \text{\textcircled{R}}
\end{array}
$$

in which R represents an atom or side group other than hydrogen (e.g., Cl, CH_3). One arrangement is possible when the R side groups of successive mer units bond to alternate carbon atoms as follows:

$$
\begin{array}{c}
\text{H} \quad \text{H} \quad \text{H} \quad \text{H} \\
| \quad\quad | \quad\quad | \quad\quad | \\
-\text{C}-\text{C}-\text{C}-\text{C}- \\
| \quad\quad | \quad\quad | \quad\quad | \\
\text{H} \quad \text{\textcircled{R}} \quad \text{H} \quad \text{R}
\end{array}
\qquad (15.6a)
$$

This is designated as a head-to-tail configuration.[3] Its complement, the head-to-head configuration, occurs when R groups bond to adjacent chain atoms:

$$
\begin{array}{c}
\text{H} \quad \text{H} \quad \text{H} \quad \text{H} \\
| \quad\quad | \quad\quad | \quad\quad | \\
-\text{C}-\text{C}-\text{C}-\text{C}- \\
| \quad\quad | \quad\quad | \quad\quad | \\
\text{H} \quad \text{\textcircled{R}} \quad \text{\textcircled{R}} \quad \text{H}
\end{array}
\qquad (15.6b)
$$

In most polymers, the head-to-tail configuration predominates; there is often a polar repulsion that occurs between R groups for the head-to-head configuration.

Isomerism (Section 15.2) is also found in polymer molecules, wherein different atomic configurations are possible for the same composition. Two isomeric subclasses, stereoisomerism and geometrical isomerism, are topics of discussion in the succeeding sections.

---

[3] The term *configuration* is used in reference to arrangements of units along the axis of the chain, or atom positions that are not alterable except by the breaking and then reforming of primary bonds.

Stereoisomerism

Stereoisomerism denotes the situation in which atoms are linked together in the same order (head-to-tail) but differ in their spatial arrangement. For one stereoisomer, all the R groups are situated on the same side of the chain as follows:

$$
\begin{array}{cccccccc}
H & H & H & H & H & H & H & H \\
| & | & | & | & | & | & | & | \\
-C- & C- & C- & C- & C- & C- & C- & C- \\
| & | & | & | & | & | & | & | \\
H & \text{\textcircled{R}} & H & \text{\textcircled{R}} & H & \text{\textcircled{R}} & H & \text{\textcircled{R}}
\end{array}
\qquad (15.7)
$$

This is called an **isotactic configuration.**

In a **syndiotactic configuration,** the R groups alternate sides of the chain:

$$
\begin{array}{cccccccc}
H & H & H & \text{\textcircled{R}} & H & H & H & \text{\textcircled{R}} \\
| & | & | & | & | & | & | & | \\
-C- & C- & C- & C- & C- & C- & C- & C- \\
| & | & | & | & | & | & | & | \\
H & \text{\textcircled{R}} & H & H & H & \text{\textcircled{R}} & H & H
\end{array}
\qquad (15.8)
$$

And for random positioning,

$$
\begin{array}{cccccccc}
H & H & H & H & H & \text{\textcircled{R}} & H & H \\
| & | & | & | & | & | & | & | \\
-C- & C- & C- & C- & C- & C- & C- & C- \\
| & | & | & | & | & | & | & | \\
H & \text{\textcircled{R}} & H & \text{\textcircled{R}} & H & H & H & \text{\textcircled{R}}
\end{array}
\qquad (15.9)
$$

the term **atactic configuration** is used.

Conversion from one stereoisomer to another (e.g., isotactic to syndiotactic) is not possible by a simple rotation about single chain bonds; these bonds must first be severed, and then, after the appropriate rotation, they are reformed.

In reality, a specific polymer does not exhibit just one of these configurations; the predominant form depends on the method of synthesis.

Geometrical Isomerism

Other important chain configurations, or geometrical isomers, are possible within mer units having a double bond between chain carbon atoms. Bonded to each of the carbon atoms participating in the double bond is a single side-bonded atom or radical, which may be situated on one side of the chain or its opposite. Consider the isoprene mer having the structure

$$
\begin{array}{ccc}
\text{\textcircled{CH}_3} & & \text{\textcircled{H}} \\
\diagdown & & \diagup \\
& C=C & \\
\diagup & & \diagdown \\
-CH_2 & & CH_2-
\end{array}
$$

in which the CH_3 group and the H atom are positioned on the same side of the chain. This is termed a **cis** structure, and the resulting polymer, cis-isoprene, is natural rubber. For the alternative

$$
\begin{array}{ccc}
\text{\textcircled{CH}_3} & & CH_2- \\
\diagdown & & \diagup \\
& C=C & \\
\diagup & & \diagdown \\
-CH_2 & & \text{\textcircled{H}}
\end{array}
$$

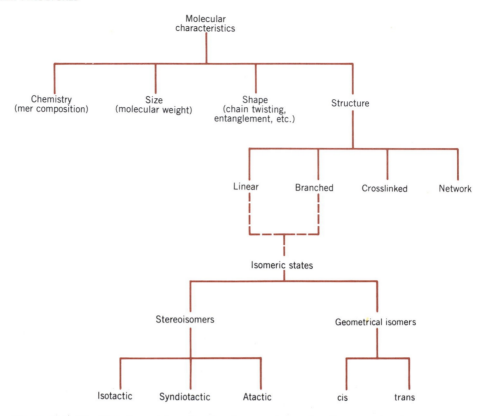

Figure 15.8 Classification scheme for the characteristics of polymer molecules.

the **trans** structure, the CH_3 and H reside on opposite chain sides. Trans-isoprene, sometimes called gutta percha, has properties that are distinctly different from natural rubber as a result of this configurational alteration. Conversion of trans to cis, or vice versa, is not possible by a simple chain bond rotation because the chain double bond is extremely rigid.

By way of summary of the preceding sections, polymer molecules may be characterized in terms of their size, shape, and structure. Molecular size is specified in terms of molecular weight (or degree of polymerization). Molecular shape relates to the degree of chain twisting, coiling, and bending. Molecular structure depends on the manner in which structural units are joined together. Linear, branched, crosslinked, and network structures are all possible, in addition to several isomeric configurations (isotactic, syndiotactic, atactic, cis, and trans). These molecular characteristics are presented in the taxonomic chart, Figure 15.8. It should be noted that some of the structural elements are not mutually exclusive of one another, and, in fact, it may be necessary to specify molecular structure in terms of more than one. For example, a linear polymer may also be isotactic.

15.9 COPOLYMERS

Polymer chemists and scientists are continually searching for new materials that can be easily and economically synthesized and fabricated, with improved properties or better property combinations than are offered by the homopolymers heretofore discussed. One group of these materials are the copolymers.

Consider a copolymer that is composed of two mer units as represented by ○ and ●. Depending on the polymerization process and the relative fractions of these mer types, different sequencing arrangements along the polymer chains are possible. For one, as depicted in Figure 15.9a, the two different units are randomly dispersed along the chain in what is termed a **random copolymer.** For an **alternating copolymer,** as the name suggests, the two mer units alternate chain positions, as illustrated in Figure 15.9b. A **block copolymer** is one in which identical mers are clustered in blocks along the chain (Figure 15.9c). And, finally, homopolymer side branches of one type may be grafted to homopolymer main chains that are composed of a different mer; such a material is termed a **graft copolymer** (Figure 15.9d).

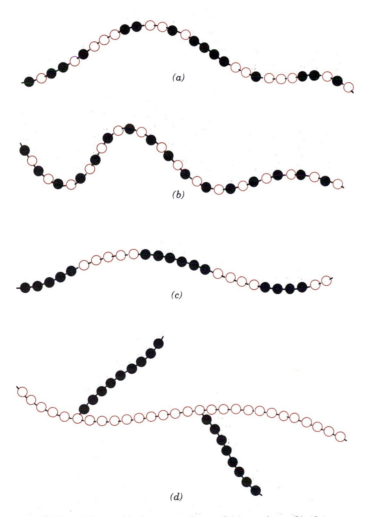

Figure 15.9 Schematic representations of (a) random, (b) alternating, (c) block, and (d) graft copolymers. The two different mer types are designated by dark and light circles.

TABLE 15.5 Mers that Are Employed in Copolymer Rubbers

Mer Name	Mer Structure
Acrylonitrile	$\begin{array}{cc} H & H \\ | & | \\ -C & -C- \\ | & | \\ H & C\equiv N \end{array}$
Styrene	$\begin{array}{cc} H & H \\ | & | \\ -C & -C- \\ | \\ H \end{array}$ (with benzene ring)
Butadiene	$\begin{array}{cccc} H & H & H & H \\ | & | & | & | \\ -C & -C & =C & -C- \\ | & & & | \\ H & & & H \end{array}$
Chloroprene	$\begin{array}{cccc} H & Cl & H & H \\ | & | & | & | \\ -C & -C & =C & -C- \\ | & & & | \\ H & & & H \end{array}$
cis-Isoprene	$\begin{array}{cccc} H & CH_3 & H & H \\ | & | & | & | \\ -C & -C & =C & -C- \\ | & & & | \\ H & & & H \end{array}$
Isobutylene	$\begin{array}{cc} H & CH_3 \\ | & | \\ -C & -C- \\ | & | \\ H & CH_3 \end{array}$
Dimethylsiloxane	$\begin{array}{c} CH_3 \\ | \\ -Si-O- \\ | \\ CH_3 \end{array}$

Synthetic rubbers, discussed in Section 16.14, are often copolymers; mer structures that are employed in some of these rubbers are contained in Table 15.5. Styrene-butadiene rubber (SBR) is a common random copolymer from which automobile tires are made. Nitrile rubber (NBR) is another random copolymer composed of acrylonitrile and butadiene. It is also highly elastic and, in addition, resistant to swelling in organic solvents; gasoline hoses are made of NBR.

15.10 POLYMER CRYSTALLINITY

The crystalline state may exist in polymeric materials. However, since it involves molecules instead of just atoms or ions, as with metals and ceramics, the atomic arrangements will be more complex for polymers. We think of polymer crystallinity as the packing of molecular chains so as to produce an ordered atomic array. Crystal structures may be specified in terms of unit cells, which are often quite complex. For example, Figure 15.10 shows the unit cell for polyethylene and its relationship to the molecular chain structure; this unit cell has orthorhombic geometry (Table 3.2). Of course, the chain molecules also extend beyond the unit cell shown in the figure.

Molecular substances having small molecules (e.g., water and methane) are normally either totally crystalline (as solids) or totally amorphous (as liquids). As a consequence of their size and often complexity, polymer molecules are often only partially crystalline (or semicrystalline), having crystalline regions dispersed within the remaining amorphous material. Any chain disorder or misalignment will result in an amorphous region, a condition that is fairly common, since twisting, kinking, and coiling of the chains prevent the strict ordering of every segment of every chain. Other structural effects are also influential in determining the extent of crystallinity, as discussed below.

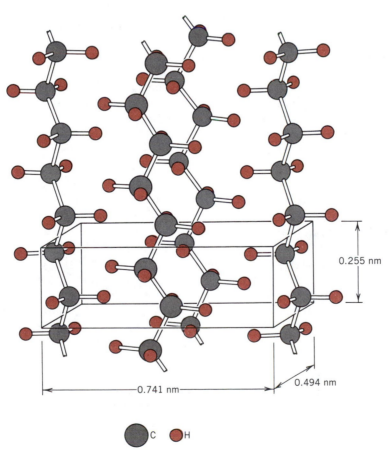

Figure 15.10 Arrangement of molecular chains in a unit cell for poly-ethylene. (Adapted from C. W. Bunn, *Chemical Crystallography*, Oxford University Press, Oxford, 1945, p. 233.)

The degree of crystallinity may range from completely amorphous to almost entirely (up to about 95%) crystalline; by way of contrast, metal specimens are almost always entirely crystalline, whereas many ceramics are either totally crystalline or totally noncrystalline. Semicrystalline polymers are, in a sense, analogous to two-phase metal alloys, discussed previously.

The density of a crystalline polymer will be greater than an amorphous one of the same material and molecular weight, since the chains are more closely packed together for the crystalline structure. The degree of crystallinity by weight may be determined from accurate density measurements, according to

$$\% \text{ crystallinity} = \frac{\rho_c(\rho_s - \rho_a)}{\rho_s(\rho_c - \rho_a)} \times 100 \tag{15.10}$$

where ρ_s is the density of a specimen for which the percent crystallinity is to be determined, ρ_a is the density of the totally amorphous polymer, and ρ_c is the density of the perfectly crystalline polymer. The values of ρ_a and ρ_c must be measured by other experimental means.

The degree of crystallinity of a polymer depends on the rate of cooling during solidification as well as on the chain configuration. During crystallization upon cooling through the melting temperature, the chains, which are highly random and entangled in the viscous liquid, must assume an ordered configuration. For this to occur, sufficient time must be allowed for the chains to move and align themselves.

The molecular chemistry as well as chain configuration also influence the ability of a polymer to crystallize. Crystallization is not favored in polymers that are composed of chemically complex mer structures (e.g., polyisoprene). On the other hand, crystallization is not easily prevented in chemically simple polymers such as polyethylene and polytetrafluoroethylene, even for very rapid cooling rates.

For linear polymers, crystallization is easily accomplished because there are virtually no restrictions to prevent chain alignment. Any side branches interfere with crystallization, such that branched polymers never are highly crystalline; in fact, excessive branching may prevent any crystallization whatsoever. Network polymers are almost totally amorphous, whereas various degrees of crystallinity are possible for those that are crosslinked. With regard to the stereoisomers, atactic polymers are difficult to crystallize; however isotactic and syndiotactic polymers crystallize much more easily because the regularity of the geometry of the side groups facilitates the process of fitting together adjacent chains. Also, the bulkier or larger the side-bonded groups of atoms, the less tendency there is for crystallization.

For copolymers, as a general rule, the more irregular and random the mer arrangements, the greater is the tendency for the development of noncrystallinity. For alternating and block copolymers there is some likelihood of crystallization. On the other hand, random and graft copolymers are normally amorphous.

To some extent, the physical properties of polymeric materials are influenced by the degree of crystallinity. Crystalline polymers are usually stronger and more resistant to dissolution and softening by heat. Some of these properties are discussed in subsequent chapters.

15.11   POLYMER CRYSTALS

We shall now briefly discuss some of the models that have been proposed to describe the spatial arrangement of molecular chains in polymer crystals. One early model, accepted for many years, is the *fringed-micelle* model (Figure 15.11). It was proposed that a semicrystalline polymer consists of small crystalline regions (**crystallites,** or micelles), each having a precise alignment, which are embedded within an amorphous matrix composed of randomly oriented molecules. Thus a single chain molecule might pass through several crystallites as well as the intervening amorphous regions.

More recently, investigations centered on polymer single crystals grown from dilute solutions. These crystals are regularly shaped, thin platelets (or lamellae), approximately 10 to 20 nm thick, and on the order of 10 μm long. Frequently, these platelets will form a multilayered structure, like that shown in the electron micrograph of a single crystal of polyethylene, Figure 15.12. It is theorized that the molecular chains within each platelet fold back and forth on themselves, with folds occurring at the faces; this structure, aptly termed the **chain-folded model,** is illustrated schematically in Figure 15.13. Each platelet will consist of a number of molecules; however, the average chain length will be much greater than the thickness of the platelet.

Many bulk polymers that are crystallized from a melt form **spherulites.** As implied by the name, each spherulite may grow to be spherical in shape; one of them, as found in natural rubber, is shown in the transmission electron micrograph, page 458. The spherulite consists of an aggregate of ribbonlike chain-folded crystallites (lamellae) approximately 10 nm thick that radiate from the center outward. In this electron

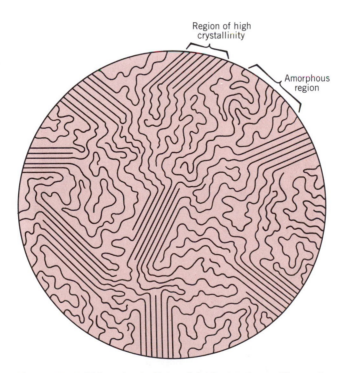

Figure 15.11 Fringed-micelle model of a semicrystalline polymer, showing both crystalline and amorphous regions. (From H. W. Hayden, W. G. Moffatt, and J. Wulff, *The Structure and Properties of Materials,* Vol. III, *Mechanical Behavior.* Copyright © 1965 by John Wiley & Sons, New York. Reprinted by permission of John Wiley & Sons, Inc.)

Figure 15.12 Electron micrograph of a polyethylene single crystal. 20,000×. (From A. Keller, R. H. Doremus, B. W. Roberts, and D. Turnbull, Editors, *Growth and Perfection of Crystals.* General Electric Company and John Wiley & Sons, Inc., 1958, p. 498.)

micrograph these lamellae appear as thin white lines. The detailed structure of a spherulite is illustrated schematically in Figure 15.14; shown here are the individual chain-folded lamellar crystals that are separated by amorphous material. Tie-chain molecules that act as connecting links between adjacent lamellae pass through these amorphous regions.

As the crystallization of a spherulitic structure nears completion, the extremities of adjacent spherulites begin to impinge on one another, forming more or less planar boundaries; prior to this time, they maintain their spherical shape. These boundaries are evident in Figure 15.15, which is a photomicrograph of polyethylene using cross-polarized light. A characteristic Maltese cross pattern appears within each spherulite.

~ 10 nm

Figure 15.13 The chain-folded structure for a plate-shaped polymer crystallite.

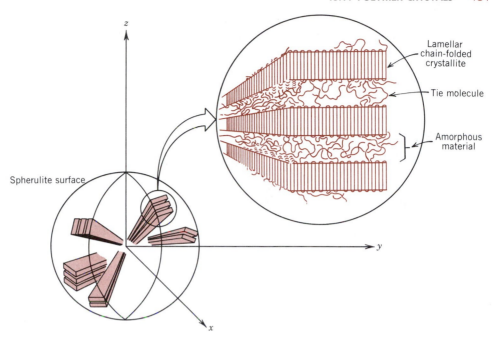

Figure 15.14 Schematic representation of the detailed structure of a spherulite. (From R. H. Boyd and J. Coburn.)

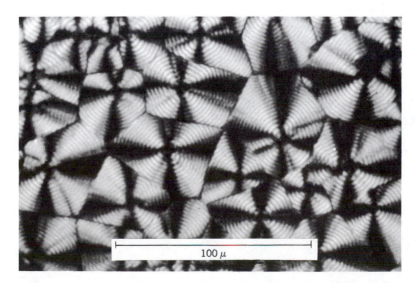

Figure 15.15 A transmission photomicrograph (using cross-polarized light) showing the spherulite structure of polyethylene. Linear boundaries form between adjacent spherulites, and within each spherulite appears a Maltese cross. 525×. (Courtesy F. P. Price, General Electric Company.)

Spherulites are considered to be the polymer analogue of grains in polycrystalline metals and ceramics. However, as discussed above, each spherulite is really composed of many different lamellar crystals and, in addition, some amorphous material. Polyethylene, polypropylene, polyvinyl chloride, polytetrafluoroethylene, and nylon form a spherulitic structure when they crystallize from a melt.

SUMMARY

Most polymeric materials are composed of very large molecules—chains of carbon atoms, to which are side-bonded various atoms or radicals. These macromolecules may be thought of as being composed of mers, smaller structural entities, which are repeated along the chain. Mer structures of some of the chemically simple polymers (e.g., polyethylene, polytetrafluoroethylene, polyvinyl chloride, and polypropylene) were presented.

Molecular weights for high polymers may be in excess of a million. Since all molecules are not of the same size, there is a distribution of molecular weights. Molecular weight is often expressed in terms of number and weight averages. Chain length may also be specified by degree of polymerization, the number of mer units per average molecule.

Several molecular characteristics that have an influence on the properties of polymers were discussed. Molecular entanglements occur when the chains assume twisted, coiled, and kinked shapes or contours. With regard to molecular structure, linear, branched, cross-linked, and network structures are possible, in addition to isotactic, syndiotactic, and atactic stereoisomers, and the cis and trans geometrical isomers. The copolymers include random, alternating, block, and graft types.

When the packing of molecular chains is such as to produce an ordered atomic arrangement, the condition of crystallinity is said to exist. In addition to being entirely amorphous, polymers may also exhibit virtually total and also partial crystallinity; for the latter case, crystalline regions are interdispersed within amorphous areas. Crystallinity is facilitated for polymers that are chemically simple and that have regular and symmetrical chain structures.

Polymer single crystals as thin platelets and having chain-folded structures may be grown from dilute solutions. Many semicrystalline polymers form spherulites; each spherulite consists of a collection of ribbonlike chain-folded lamellar crystallites that radiate outward from its center.

IMPORTANT TERMS AND CONCEPTS

Alternating copolymer

Atactic structure

Bifunctional mer

Block copolymer

Branched polymer

Chain-folded model

Cis (structure)

Copolymer

Crosslinked polymer

Crystallite

Degree of polymerization

Graft copolymer

Homopolymer

Isomerism

Isotactic configuration	Monomer	Stereoisomerism
Linear polymer	Network polymer	Syndiotactic
Macromolecule	Polymer	configuration
Mer	Polymer crystallinity	Trans (structure)
Molecular chemistry	Random copolymer	Trifunctional mer
Molecular structure	Saturated	Unsaturated
Molecular weight	Spherulite	

REFERENCES

BOVEY, F. A., and F. H. WINSLOW (Editors), *Macromolecules: An Introduction to Polymer Science,* Academic Press, New York, 1979.

COWIE, J. M. G., *Polymers: Chemistry and Physics of Modern Materials,* International Textbook Company Ltd., Glasgow, 1973.

MEARES, P., *Polymers: Structure and Bulk Properties,* D. Van Nostrand Company, Ltd., London, 1965.

RODRIGUEZ, F., *Principles of Polymer Systems,* 2nd edition, McGraw-Hill Book Co., New York, 1982.

SCHULTZ, J., *Polymer Materials Science,* Prentice-Hall, Englewood Cliffs, NJ, 1974.

YOUNG, R. J., *Introduction to Polymers,* Chapman and Hall, London, 1981.

WILLIAMS, D. J., *Polymer Science and Engineering,* Prentice-Hall, Inc., Englewood Cliffs, NJ, 1971.

QUESTIONS AND PROBLEMS

15.1 Differentiate between polymorphism and isomerism.

15.2 On the basis of the structures presented in this chapter, sketch mer structures for the following polymers: **(a)** polyvinyl fluoride, **(b)** polychlorotrifluoroethylene, and **(c)** polyvinyl alcohol.

15.3 Compute mer molecular weights for the following: **(a)** polytetrafluoroethylene, **(b)** polymethyl methacrylate, **(c)** nylon 6,6, and **(d)** polyethylene terephthalate.

15.4 The number-average molecular weight of a polypropylene is 1,000,000 g/mol. Compute the number-average degree of polymerization.

15.5 **(a)** Compute the mer molecular weight of polystyrene. **(b)** Compute the weight-average molecular weight for a polystyrene for which the weight-average degree of polymerization is 25,000.

15.6 Below, molecular weight data for a polypropylene material are tabulated. Compute **(a)** the number-average molecular weight, **(b)** the weight-average molecular weight, **(c)** the number-average degree of polymerization, and **(d)** the weight-average degree of polymerization.

Molecular Weight Range (g/mol)	x_i	w_i
8,000–16,000	0.05	0.02
16,000–24,000	0.16	0.10
24,000–32,000	0.24	0.20
32,000–40,000	0.28	0.30
40,000–48,000	0.20	0.27
48,000–56,000	0.07	0.11

15.7 Below, molecular weight data for some polymer are tabulated. Compute **(a)** the number-average molecular weight, and **(b)** the weight-average molecular weight. **(c)** If it is known that this material's number-average degree of polymerization is 477, which one of the polymers listed in Table 15.3 is this polymer? Why? **(d)** What is this material's weight-average degree of polymerization?

Molecular Weight Range (g/mol)	x_i	w_i
8,000–20,000	0.05	0.02
20,000–32,000	0.15	0.08
32,000–44,000	0.21	0.17
44,000–56,000	0.28	0.29
56,000–68,000	0.18	0.23
68,000–80,000	0.10	0.16
80,000–92,000	0.03	0.05

15.8 Is it possible to have a polyvinyl chloride homopolymer with the following molecular weight data and a number-average degree of polymerization of 1120? Why or why not?

Molecular Weight Range (g/mol)	w_i	x_i
8,000–20,000	0.02	0.05
20,000–32,000	0.08	0.15
32,000–44,000	0.17	0.21
44,000–56,000	0.29	0.28
56,000–68,000	0.23	0.18
68,000–80,000	0.16	0.10
80,000–92,000	0.05	0.03

15.9 High-density polyethylene may be chlorinated by inducing the random substitution of chlorine atoms for hydrogen. **(a)** Determine the concentration of Cl (in wt%) that must be added if this substitution occurs for 5% of all the original hydrogen atoms. **(b)** In what ways does this chlorinated polyethylene differ from polyvinyl chloride?

15.10 What is the difference between *configuration* and *conformation* in relation to polymer chains?

15.11 For a linear polymer molecule, the total chain length L depends on the bond length between chain atoms d, the total number of bonds in the molecule N, and the angle between adjacent backbone chain atoms θ, as follows:

$$L = Nd \sin\left(\frac{\theta}{2}\right) \tag{15.11}$$

Furthermore, the average end-to-end distance for a series of polymer molecules r in Figure 15.6 is equal to

$$r = d\sqrt{N} \tag{15.12}$$

A linear polyethylene has a number-average molecular weight of 300,000 g/mol; compute average values of L and r for this material.

15.12 Using the definitions for total chain molecule length L (Equation 15.11) and average chain end-to-end distance r (Equation 15.12), for a linear polytetrafluoroethylene determine **(a)** the number-average molecular weight for $L = 2000$ nm; and **(b)** the number-average molecular weight for $r = 15$ nm.

15.13 Sketch portions of a linear polypropylene molecule that are **(a)** syndiotactic, **(b)** atactic, and **(c)** isotactic.

15.14 Sketch cis and trans mer structures for **(a)** butadiene, and **(b)** chloroprene.

15.15 Sketch the mer structure for each of the following alternating copolymers: **(a)** poly(ethylene-propylene), **(b)** poly(butadiene-styrene), and **(c)** poly(isobutylene-isoprene).

15.16 The number-average molecular weight of a styrene-butadiene alternating copolymer is 1,350,000 g/mol; determine the average number of styrene and butadiene mer units per molecule.

15.17 Calculate the number-average molecular weight of a random nitrile rubber (acrylonitrile-butadiene copolymer) in which the fraction of butadiene mers is 0.30; assume that this concentration corresponds to a number-average degree of polymerization of 2000.

15.18 An alternating copolymer is known to have a number-average molecular weight of 100,000 g/mol and a number-average degree of polymerization of 2210. If one of the mers is ethylene, which of styrene, propylene, tetrafluoroethylene, and vinyl chloride is the other mer? Why?

15.19 **(a)** Determine the ratio of butadiene to acrylonitrile mers in a copolymer having a weight-average molecular weight of 250,000 g/mol and a weight-average degree of polymerization of 4640. **(b)** Which type(s) of copolymer(s) will this copolymer be of the following possibilities: random, alternating, graft, and block? Why?

15.20 Crosslinked copolymers consisting of 60 wt% ethylene and 40 wt% propylene may have elastic properties similar to those for natural rubber. For a copolymer of this composition, determine the fraction of both mer types.

15.21 A random poly(isobutylene-isoprene) copolymer has a weight-average molecular weight of 200,000 g/mol and a weight-average degree of polymerization of 3000. Compute the fraction of isobutylene and isoprene mers in this copolymer.

15.22 **(a)** Compare the crystalline state in metals and polymers. **(b)** Compare the noncrystalline state as it applies to polymers and ceramic glasses.

15.23 Explain briefly why the tendency of a polymer to crystallize decreases with increasing molecular weight.

15.24 For each of the following pairs of polymers, decide which is more likely to crystallize, and cite the reasons for your choice:
(a) Linear and isotactic polyvinyl chloride; linear and atactic polypropylene.
(b) Linear and syndiotactic polypropylene; crosslinked *cis*-isoprene.
(c) Network phenol-formaldehyde; linear and isotactic polystyrene.
(d) Lightly branched polytetrafluoroethylene; linear polyethylene.

15.25 Compute the density of totally crystalline polyethylene. The orthorhombic unit cell for polyethylene is shown in Figure 15.10; also, the equivalent of two ethylene mer units is contained within each unit cell.

15.26 The density of totally crystalline nylon 6,6 at room temperature is 1.213 g/cm³. Also, at room temperature the unit cell for this material is triclinic with lattice parameters

$a = 0.497$ nm	$\alpha = 48.4°$
$b = 0.547$ nm	$\beta = 76.6°$
$c = 1.729$ nm	$\gamma = 62.5°$

If the volume of a triclinic unit cell V_{tri} is a function of these lattice parameters as

$$V_{tri} = abc \sqrt{1 - \cos^2\alpha - \cos^2\beta - \cos^2\gamma + 2 \cos \alpha \cos \beta \cos \gamma}$$

determine the number of mer units associated with each unit cell.

15.27 The density and associated percent crystallinity for two polyethylene materials are as follows:

ρ(g/cm³)	Crystallinity (%)
0.965	76.8
0.925	46.4

(a) Compute the densities of totally crystalline and totally amorphous polyethylene.
(b) Determine the percent crystallinity of a specimen having a density of 0.950 g/cm³.

15.28 The density and associated percent crystallinity for two polypropylene materials are as follows:

ρ(g/cm³)	Crystallinity (%)
0.904	62.8
0.895	54.4

(a) Compute the densities of totally crystalline and totally amorphous polypropylene.
(b) Determine the density of a specimen having 74.6% crystallinity.

CHAPTER 16

CHARACTERISTICS, APPLICATIONS, AND PROCESSING OF POLYMERS

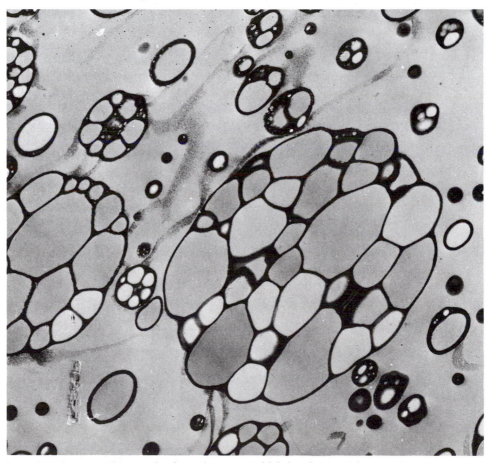

Scanning electron micrograph of a polystyrene which has been made more impact resistant by the addition of a rubber phase. The continuous matrix phase (gray) is polystyrene; the fine dispersed structure consists of particles of both polystyrene and rubber (white). 15,000 ×. (From E. R. Wagner and L. M. Robeson, *Rubber Chemistry and Technology*, **43,** 1129, 1970. Reprinted with permission.)

16.1 INTRODUCTION

This chapter discusses some of the characteristics important to polymeric materials and, in addition, the various types and processing techniques.

MECHANICAL AND THERMOMECHANICAL CHARACTERISTICS

16.2 STRESS–STRAIN BEHAVIOR

The mechanical properties of polymers are specified with many of the same parameters that are used for metals, that is, modulus of elasticity, and tensile, impact, and fatigue strengths. For many polymeric materials, the simple stress–strain test is employed for the characterization of some of these mechanical parameters. The mechanical characteristics of polymers, for the most part, are highly sensitive to the rate of deformation (strain rate), the temperature, and the chemical nature of the environment (the presence of water, oxygen, organic solvents, etc.). Some modifications of the testing techniques and specimen configurations used for metals (Chapter 6) are necessary with polymers, especially for the highly elastic materials, such as rubbers.

Three typically different types of stress–strain behavior are found for polymeric materials, as represented in Figure 16.1. Curve *A* illustrates the stress–strain character for a brittle polymer, inasmuch as it fractures while deforming elastically. The behavior for the plastic material, curve *B*, is similar to that found for many metallic materials; the initial deformation is elastic, which is followed by yielding and a region of plastic deformation. Finally, the deformation displayed by curve *C* is totally elastic; this rubberlike elasticity (large recoverable strains produced at low stress levels) is displayed by a class of polymers termed the **elastomers.**

Modulus of elasticity, tensile strength, and ductility, in percent elongation, are determined for polymers in the same manner as for metals (Section 6.6). Table 16.1

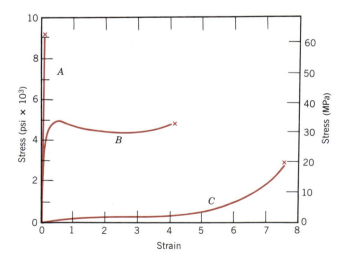

Figure 16.1 The stress–strain behavior for brittle (curve *A*), plastic (curve *B*), and highly elastic (elastomeric) (curve *C*) polymers.

TABLE 16.1 Room Temperature Mechanical Characteristics of Some of the More Common Polymers

Material	Specific Gravity	Tensile Modulus [psi × 10⁵ (MPa × 10²)]	Tensile Strength		Elongation at Break (%)	Impact Strength (ft-lb$_f$/in.)[a]
			[psi × 10³	(MPa)]		
Polyethylene (low density)	0.917–0.932	0.25–0.41 (1.7–2.8)	1.2–4.6	(8.3–31.7)	100–650	No break
Polyethylene (high density)	0.952–0.965	1.55–1.58 (10.6–10.9)	3.2–4.5	(22–31)	10–1200	0.4–4.0
Polyvinyl chloride	1.30–1.58	3.5–6.0 (24–41)	5.9–7.5	(41–52)	40–80	0.4–22
Polytetrafluoroethylene	2.14–2.20	0.58–0.80 (4.0–5.5)	2.0–5.0	(14–34)	200–400	3
Polypropylene	0.90–0.91	1.6–2.3 (11–16)	4.5–6.0	(31–41)	100–600	0.4–1.2
Polystyrene	1.04–1.05	3.3–4.8 (23–33)	5.2–7.5	(36–52)	1.2–2.5	0.35–0.45
Polymethyl methacrylate	1.17–1.20	3.3–4.7 (22–32)	7.0–11.0	(48–76)	2–10	0.3–0.6
Phenol-formaldehyde	1.24–1.32	4.0–7.0 (28–48)	5.0–9.0	(34–62)	1.5–2.0	0.24–4.0
Nylon 6,6	1.13–1.15	2.3–5.5 (16–38)	11.0–13.7	(76–94)	15–300	0.55–2.1
Polyester (PET)	1.29–1.40	4.0–6.0 (28–41)	7.0–10.5	(48–72)	30–300	0.25–0.70
Polycarbonate	1.20	3.5 (24.0)	9.5	(66)	110	16

[a] $\frac{1}{8}$ in. (3.2 mm) thick specimen.

Source: *Modern Plastics Encyclopedia* 1988. Copyright 1987, McGraw-Hill Inc. Reprinted with permission.

gives these mechanical properties for several polymeric materials. Polymers are, in many respects, mechanically dissimilar to metals. For example, the modulus for highly elastic polymeric materials may be as low as 10^3 psi (7 MPa), but may run as high as 0.6×10^6 psi (4×10^3 MPa) for some of the very stiff polymers; modulus values for metals are much larger and range between 7×10^6 and 60×10^6 psi (48×10^3 to 410×10^3 MPa). Maximum tensile strengths for polymers are on the order of 15,000 psi (100 MPa)—for some metal alloys, 600,000 psi (4100 MPa). And, whereas metals rarely elongate plastically to more than 100%, some highly elastic polymers may experience elongations to as much as 1000%.

In addition, the mechanical characteristics of polymers are much more sensitive to temperature changes within the vicinity of room temperature. Consider the stress–strain behavior for polymethyl methacrylate (Plexiglas) at several temperatures between 4 and 60°C (40 and 140°F) (Figure 16.2). Several features of this figure are worth noting, as follows: increasing the temperature produces (1) a decrease in elastic modulus, (2) a reduction in tensile strength, and (3) an enhancement of ductility—at 4°C (40°F) the material is totally brittle, whereas considerable plastic deformation is realized at both 50 and 60°C (122 and 140°F).

The influence of strain rate on the mechanical behavior may also be important. In general, decreasing the rate of deformation has the same influence on the stress–strain characteristics as increasing the temperature; that is, the material becomes softer and more ductile.

An understanding of deformation mechanisms of polymers is important in order for us to be able to manage the mechanical characteristics of these materials. In this

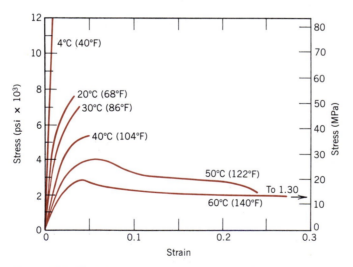

Figure 16.2 The influence of temperature on the stress–strain characteristics of polymethyl methacrylate. (From T. S. Carswell and H. K. Nason, "Effect of Environmental Conditions on the Mechanical Properties of Organic Plastics," *Symposium on Plastics,* American Society for Testing Materials, Philadelphia, 1944. Copyright, ASTM, 1916 Race Street, Philadelphia, PA 19103. Reprinted with permission.)

regard, two different deformation models deserve our attention. One of these involves plastic deformation that occurs in semicrystalline polymers, which is the topic of the succeeding section. The strength of these materials is often an important consideration. On the other hand, elastomers are utilized on the basis of their unusual elastic properties. The deformation mechanism of these polymers is treated in Section 16.7.

16.3 DEFORMATION OF SEMICRYSTALLINE POLYMERS

Mechanism

Many semicrystalline polymers in bulk form will have the spherulitic structure described in Section 15.11. By way of review, let us repeat here that each spherulite consists of numerous chain-folded ribbons, or lamellae, that radiate outward from the center. Separating these lamellae are areas of amorphous material (Figure 15.14); adjacent lamellae are connected by tie chains that pass through these amorphous regions.

The mechanism of plastic deformation is best described by the interactions between lamellar and intervening amorphous regions in response to an applied tensile load. This process occurs in several stages, which are schematically diagrammed in Figure 16.3. Two adjacent chain-folded lamellae and the interlamellar amorphous material, prior to deformation, are shown in Figure 16.3a. During the initial stage of deformation (Figure 16.3b), the lamellar ribbons simply slide past one another as the tie chains within the amorphous regions become extended. Continued deformation in the second stage occurs by the tilting of the lamellae so that the chain folds become aligned with the tensile axis (Figure 16.3c). Next, crystalline block segments separate from the lamellae, which segments remain attached to one another by tie chains (Figure 16.3d). In the final stage (Figure 16.3e), the blocks and tie chains become oriented in the direction of the tensile axis. Thus appreciable tensile deformation of semicrystalline polymers produces a highly oriented structure. Of course, during this process the spherulites also experience changes in shape.

The mechanical characteristics of semicrystalline polymers are subject to modification. An increase in strength results whenever any restraint is imposed on the process illustrated in Figure 16.3. For example, increasing the degree of crosslinking will inhibit relative chain motion and thus strengthen the polymer and make it more brittle. Crosslinking may be promoted by irradiation; when a polymer specimen is exposed to certain types of radiation, side chain bonds are broken and become sites for crosslinkages to form.

Even though secondary intermolecular (e.g., van der Waals) bonds are much weaker than the primary covalent ones, they are, nevertheless, effective in inhibiting relative chain motion. In fact, the mechanical properties of polymers are highly dependent on the magnitude of these weak intermolecular forces. For a specific polymer, the degree of crystallinity can have a rather significant influence on the mechanical properties of polymers, since it affects the extent of this intermolecular secondary bonding. For crystalline regions in which molecular chains are closely packed in an ordered and parallel arrangement, extensive secondary bonding ordinarily exists between adjacent chain segments. This secondary bonding is much less prevalent in amorphous regions, by virtue of the chain misalignment. Thus increasing the crystallinity of a particular polymer generally enhances its mechanical properties. The

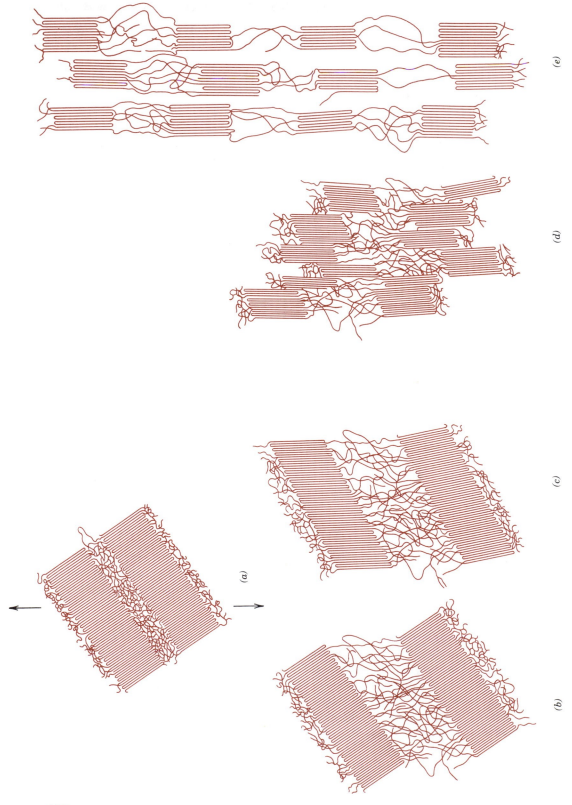

Figure 16.3 Stages in deformation of a semicrystalline polymer. (*a*) Two adjacent chain-folded lamellae and interlamellar amorphous material before deformation. (*b*) Elongation of amorphous tie chains during the first stage of deformation. (*c*) Tilting of lamellar chain folds during the second stage. (*d*) Separation of crystalline block segments during the third stage. (*e*) Orientation of block segments and tie chains with the tensile axis in the final deformation stage. (From Jerold M. Schultz, *Polymer Materials Science*, copyright © 1974, pp. 500–501. Reprinted by permission of Prentice-Hall, Inc., Englewood Cliffs, NJ.)

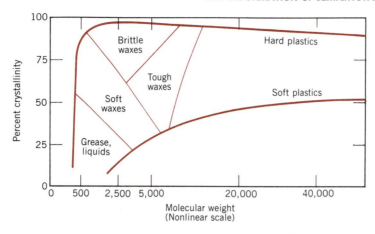

Figure 16.4 The influence of degree of crystallinity and molecular weight on the physical characteristics of polyethylene. (From R. B. Richards, "Polyethylene—Structure, Crystallinity and Properties," *J. Appl. Chem.*, **1**, 370, 1951.)

influence of chain chemistry and structure (branching, stereoisomerism, etc.) on degree of crystallinity was discussed in Chapter 15.

Other molecular chain characteristics, including molecular weight, influence the mechanical behavior. For relatively low molecular weight polymers, the mechanical strength increases with molecular weight. The physical state of polyethylene as a function of percent crystallinity and molecular weight is illustrated in Figure 16.4.

On a commercial basis, one of the most important techniques used to improve mechanical strength is by predeforming the polymer, so that it becomes oriented, having the structure depicted in Figure 16.3e. This process might be thought analogous to the strain hardening of metals. Predeformation by drawing is commonly used to strengthen fiber materials.

Macroscopic Deformation

Some aspects of the macroscopic deformation of semicrystalline polymers deserve our attention. The tensile stress–strain curve for a semicrystalline material, which was initially unoriented, is shown in Figure 16.5; also included in the figure are schematic representations of specimen profile at various stages of deformation. Both upper and lower yield points are evident on the curve, which are followed by a near horizontal region. At the upper yield point, a small neck forms within the gauge section of the specimen. Within this neck, the chains become oriented, which leads to localized strengthening. Consequently, there is a resistance to continued deformation at this point, and specimen elongation proceeds by the propagation of this neck region along the gauge length; the chain orientation phenomenon accompanies this neck extension. This tensile behavior may be contrasted to that found for ductile metals (Section 6.6), wherein once a neck has formed, all subsequent deformation is confined to within the neck region.

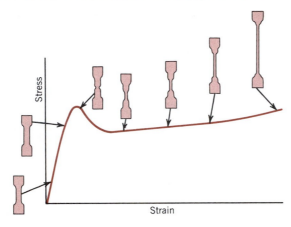

Figure 16.5 Schematic tensile stress–strain curve for a semicrystalline polymer. Specimen contours at several stages of deformation are included. (From Jerold M. Schultz, *Polymer Materials Science,* copyright © 1974, p. 488. Reprinted by permission of Prentice-Hall, Inc., Englewood Cliffs, NJ.)

16.4 MELTING AND GLASS TRANSITION PHENOMENA

Inasmuch as the mechanical properties of polymers are sensitive to temperature changes, the several succeeding sections are devoted to discussions of thermal and thermomechanical characteristics of these materials. We begin with a treatment of melting and glass transition phenomena.

The temperatures at which melting and/or the glass transition occur for a polymer are determined in the same manner as for ceramic materials—from a plot of specific volume versus temperature. Figure 16.6 is such a plot, wherein curves A and C, for amorphous and crystalline polymers, respectively, have the same configurations as their ceramic counterparts (Figure 14.3). For the crystalline material, there is a discontinuous change in specific volume at the melting temperature T_m. The curve for the totally amorphous material is continuous but, upon cooling, experiences a slight decrease in slope at the **glass transition temperature** T_g. Below T_g the material is considered to be an amorphous solid; above T_g, it is a rubbery solid and then a viscous liquid.

The behavior is intermediate between these extremes for a semicrystalline polymer (curve B), in that both melting and glass transition temperatures are observed; T_m

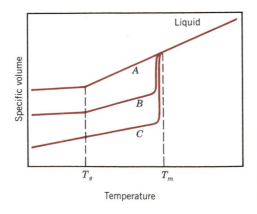

Figure 16.6 Specific volume versus temperature, upon cooling from the liquid melt, for totally amorphous (curve A), semicrystalline (curve B), and crystalline (curve C) polymers.

and T_g are properties of the respective crystalline and amorphous phases. As a rule, T_g is normally on the order of $\frac{2}{3}T_m$, for absolute units of temperature.

Factors That Influence the Melting and Glass Transition Temperatures

The melting of a crystalline polymer corresponds to the transformation of a solid material, having an ordered structure of aligned molecular chains, to a viscous liquid in which the structure is highly random. At low temperatures, the atoms vibrate with small amplitudes and relatively independently of one another; consequently, large numbers of secondary or noncovalent bonds form between adjacent chains. With increasing temperature, however, the vibrations increase in magnitude and eventually become coordinated to the degree that translational chain motions are produced, which involve many chain atoms at elevated temperatures. At the melting temperature, these chain motions become energetic enough to sever large numbers of the secondary bonds and to produce the highly disordered molecular structure. The magnitude of the melting temperature of a crystalline or partially crystalline polymer depends on the structural elements that influence the ability of chains to form van der Waals and/or hydrogen bonds (degree of branching, molecular weight, etc.). For example, the influence of branching is to decrease the chain packing efficiency, and the ability of chains to align and bond. Consequently, the melting temperature is diminished as the degree of branching is increased. Or, increasing the molecular weight (or chain length) raises the melting temperature. Chain ends are free to move in response to the vibrational motions. As chain length is increased, the number of chain ends is reduced. This results in an increase in energy, which is necessary to stimulate sufficient vibrational motion to cause melting; this additional energy must be supplied by a higher melting temperature. The melting temperatures of a number of polymers are contained in Table 16.2.

For amorphous polymers and upon heating, the glass transition corresponds to the transformation from a rigid material to one that has rubberlike characteristics. The glass transition temperature also depends on some of the structural components, which influence the ability of molecular chains to vibrate and rotate as the temperature

TABLE 16.2 Melting and Glass Transition Temperatures for Some of the More Common Polymeric Materials

Material	Glass Transition Temperature [°C (°F)]	Melting Temperature [°C (°F)]
Polyethylene (low density)	−110 (−166)	115 (239)
Polyethylene (high density)	−90 (−130)	137 (279)
Polyvinyl chloride	105 (221)	212 (414)
Polytetrafluoroethylene	−90 (−130)	327 (621)
Polypropylene	−20 (−4)	175 (347)
Polystyrene	100 (212)	
Nylon 6,6	57 (135)	265 (509)
Polyester (PET)	73 (163)	265 (509)
Polycarbonate	150 (302)	

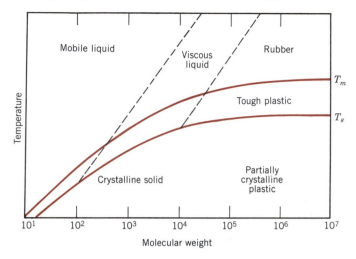

Figure 16.7 Dependence of polymer properties as well as melting and glass transition temperatures on molecular weight. (From F. W. Billmeyer, Jr., *Textbook of Polymer Science,* 3rd edition. Copyright © 1984 by John Wiley & Sons, New York. Reprinted by permission of John Wiley & Sons, Inc.)

is raised. Chain flexibility probably has the most pronounced influence. The more rigid (or less flexible) a chain, the less likely it is to experience rotational motion with increasing temperature, and the higher the magnitude of T_g. Chain flexibility may be decreased by the introduction to the polymer chain of bulky or large groups of atoms, which restrict molecular rotation. Furthermore, if crosslinks between chains are introduced, molecular motion will also be restricted, with a subsequent increase in T_g. Table 16.2 gives the glass transition temperatures for several polymers.

The dependence of T_g and T_m on molecular weight is demonstrated in Figure 16.7. Both temperatures rise, then becoming independent of the molecular weight with its increase. Figure 16.7 also illustrates the influence of temperature and molecular weight on the nature of the polymer. At relatively high temperatures, a fluid liquid exists for low molecular weights; whereas for intermediate and high molecular weights, viscous liquid and rubbery states, respectively, prevail. Solid crystalline polymers are formed at low temperatures and low molecular weights, which develop more amorphous character as the molecular weight increases.

16.5 THERMOPLASTIC AND THERMOSETTING POLYMERS

One classification scheme of polymeric materials is according to the mechanical response at elevated temperatures. *Thermoplasts* (or **thermoplastic polymers**) and *thermosets* (or **thermosetting polymers**) are the two subdivisions. Thermoplasts soften when heated (and eventually liquefy) and harden when cooled—processes that are totally reversible and may be repeated. These materials are normally fabricated by the simultaneous application of heat and pressure. On a molecular level, as the temperature is raised, secondary bonding forces are diminished (by increased molecular motion) so that the relative movement of adjacent chains is facilitated when a stress is applied.

Irreversible degradation results when the temperature of a molten thermoplastic polymer is raised to the point at which molecular vibrations become violent enough to break the primary covalent bonds. In addition, thermoplasts are relatively soft and ductile. Most linear polymers and those having some branched structures with flexible chains are thermoplastic.

Thermosetting polymers become permanently hard when heat is applied and do not soften upon subsequent heating. During the initial heat treatment, covalent cross-links are formed between adjacent molecular chains; these bonds anchor the chains together to resist the vibrational and rotational chain motions at high temperatures. Crosslinking is usually extensive, in that 10 to 50% of the chain mer units are cross-linked. Only heating to excessive temperatures will cause severance of these crosslink bonds and polymer degradation. Thermoset polymers are generally harder, stronger, and more brittle, than thermoplastics, and have better dimensional stability. Most of the crosslinked and network polymers, which include vulcanized rubbers, epoxies, and phenolic and polyester resins, are thermosetting.

16.6 VISCOELASTICITY

We know that an amorphous polymer may behave like a glass at low temperatures, a rubbery solid at intermediate temperatures (above the glass transition temperature), and a viscous liquid as the temperature is further raised. For relatively small deformations, the mechanical behavior at low temperatures may be elastic; that is, in conformity to Hooke's law, $\sigma = E\epsilon$. At the highest temperatures, viscous, or liquid-like behavior, prevails. For intermediate temperatures is found a rubbery solid that exhibits the combined mechanical characteristics of these two extremes; the condition is termed **viscoelasticity.**

Elastic deformation is instantaneous, which means that total deformation (or strain) occurs the instant the stress is applied or released (i.e., the strain is independent of time). In addition, upon release of the external stress, the deformation is totally recovered—the specimen assumes its original dimensions. This behavior is represented in Figure 16.8b as strain versus time for the instantaneous load–time curve, shown in Figure 16.8a.

By way of contrast, for totally viscous behavior, deformation or strain is not instantaneous; that is, in response to an applied stress, deformation is delayed or dependent on time. Also, this deformation is not reversible or completely recovered after the stress is released. This phenomenon is demonstrated in Figure 16.8d.

For the intermediate viscoelastic behavior, the imposition of a stress in the manner of Figure 16.8a results in an instantaneous elastic strain, which is followed by a viscous, time-dependent strain, a form of anelasticity (Section 6.4); this behavior is illustrated in Figure 16.8c.

A familiar example of these viscoelastic extremes is found in a silicone polymer that is sold as a novelty and known by some as "silly putty." When rolled into a ball and dropped onto a horizontal surface, it bounces elastically—the rate of deformation during the bounce is very rapid. On the other hand, if pulled in tension with a gradually increasing applied stress, the material elongates or flows like a highly viscous liquid. For this and other viscoelastic materials, the rate of strain determines whether the deformation is elastic or viscous.

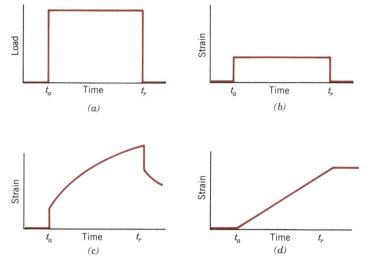

Figure 16.8 (*a*) Load versus time, where load is applied instantaneously at time t_a and released at t_r. For the load–time cycle in (*a*), the strain-versus-time responses are for totally elastic (*b*), viscoelastic (*c*), and viscous (*d*) behavior.

Viscoelastic Relaxation Modulus

The viscoelastic behavior of polymeric materials is dependent on both time and temperature; several experimental techniques may be used to measure and quantify this behavior. *Stress relaxation* measurements represent one possibility. With these tests, a specimen is initially strained rapidly in tension to a predetermined and relatively low strain level. The stress necessary to maintain this strain is measured as a function of time, while temperature is held constant. Stress is found to decrease with time due to molecular relaxation processes that take place within the polymer. We may define a **relaxation modulus** $E_r(t)$, a time-dependent elastic modulus for viscoelastic polymers, as

$$E_r(t) = \frac{\sigma(t)}{\epsilon_0} \tag{16.1}$$

where $\sigma(t)$ is the measured time-dependent stress and ϵ_0 is the strain level, which is maintained constant.

Furthermore, the magnitude of the relaxation modulus is a function of temperature; and to more fully characterize the viscoelastic behavior of a polymer, isothermal stress relaxation measurements must be conducted over a range of temperatures. Figure 16.9 is a schematic log $E_r(t)$-versus-log time plot for a polymer that exhibits viscoelastic behavior; included are several curves generated at a variety of temperatures. Worth noting from this figure are (1) the decrease of $E_r(t)$ with time (corresponding to the decay of stress, Equation 16.1), and (2) the displacement of the curves to lower $E_r(t)$ levels with increasing temperature.

To represent the influence of temperature, data points are taken at a specific time from the log $E_r(t)$-versus-log time plot—for example, t_1 in Figure 16.9—and then

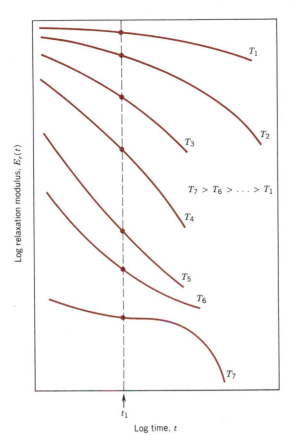

Figure 16.9 Schematic plot of logarithm of relaxation modulus versus logarithm of time for a viscoelastic polymer; isothermal curves are generated at temperatures T_1 through T_7. The temperature dependence of the relaxation modulus is represented as log $E_r(t_1)$ versus temperature.

cross-plotted as log $E_r(t_1)$ versus temperature. Figure 16.10 is such a plot for an amorphous (atactic) polystyrene; in this case, t_1 was arbitrarily taken 10 s after the load application. Several distinct regions may be noted on the curve shown in this figure. For the first, at the lowest temperatures, in the glassy region, the material is rigid and brittle, and the value of $E_r(10)$ is that of the elastic modulus, which initially is virtually independent of temperature. Over this temperature range, the strain–time characteristics are as represented in Figure 16.8b. On a molecular level, the long molecular chains are essentially frozen in position at these temperatures.

As the temperature is increased, $E_r(10)$ drops abruptly by about a factor of 10^3 within a 20°C (35°F) temperature span; this is sometimes called the leathery, or glass transition, region, and T_g lies near the upper temperature extremity; for polystyrene (Figure 16.10), $T_g = 100$°C (212°F). Within this temperature region, a polymer specimen will be leathery; that is, deformation will be time dependent and not totally recoverable on release of an applied load, characteristics depicted in Figure 16.8c. Over these temperatures, the atomic vibrations are such that the molecules begin to experience coordinated chain motions.

Within the rubbery plateau temperature region (Figure 16.10), the material deforms in a rubbery manner; here, both elastic and viscous components are present, and deformation is easy to produce because the relaxation modulus is relatively low.

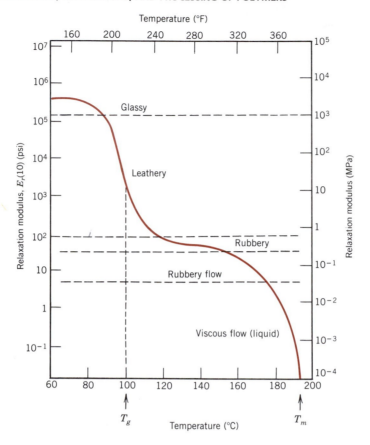

Figure 16.10 Logarithm of the relaxation modulus versus temperature for amorphous polystyrene, showing the five different regions of viscoelastic behavior. (From A. V. Tobolsky, *Properties and Structures of Polymers*. Copyright © 1960 by John Wiley & Sons, New York. Reprinted by permission of John Wiley & Sons, Inc.)

The final two high temperature regions are rubbery flow and viscous flow. Upon heating through these temperatures, the material experiences a gradual transition to a soft rubbery state, and finally to a viscous liquid. Within the viscous flow region, the modulus decreases dramatically with increasing temperature; and, again, the strain–time behavior is as represented in Figure 16.8d. From a molecular standpoint, chain motion intensifies so greatly that for viscous flow, the chain segments experience vibration and rotational motion quite independently of one another. At these temperatures, any deformation is entirely viscous.

Normally, the deformation behavior of a viscous polymer is specified in terms of viscosity, a measure of a material's resistance to flow by shear forces. Viscosity is discussed for the inorganic glasses in Section 13.8.

The rate of stress application also influences the viscoelastic characteristics. Increasing the loading rate has the same influence as lowering temperature.

The log $E_r(10)$-versus-temperature behavior for polystyrene materials having several molecular configurations is plotted in Figure 16.11. The curve for the amorphous material (curve C) is the same as in Figure 16.10. For a lightly crosslinked atactic polystyrene (curve B), the rubbery region forms a plateau that extends to the temperature at which the polymer decomposes. For increased crosslinking, the magnitude of the plateau $E_r(10)$ value will also increase. Rubber or elastomeric materials display this type of behavior and are ordinarily utilized at temperatures within this plateau range.

Also shown in Figure 16.11 is the temperature dependence for an almost totally crystalline isotactic polystyrene (curve A). The decrease in $E_r(10)$ at T_g is much less pronounced than the other polystyrene materials, and the relaxation modulus is maintained at a relatively high value with increasing temperature until its melting temperature T_m is approached. From Figure 16.11, the melting temperature of this isotactic polystyrene is about 240°C (460°F).

Viscoelastic Creep

Many polymeric materials are susceptible to time-dependent deformation when the stress level is maintained constant; such deformation is termed *viscoelastic creep*. This type of deformation may be significant even at room temperature and under modest stresses which lie below the yield strength of the material. For example, automobile tires may develop flat spots on their contact surfaces when the automobile is parked for prolonged time periods. Creep tests on polymers are conducted in the same manner as for metals (Chapter 8); that is, a stress (normally tensile) is applied instantaneously, which is maintained at a constant level while strain is measured as a function of time.

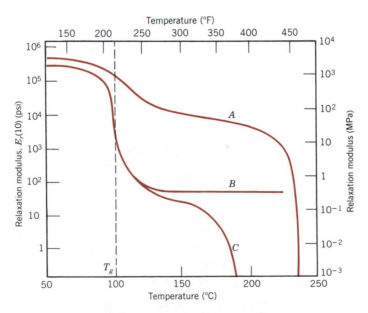

Figure 16.11 Logarithm of the relaxation modulus versus temperature for crystalline isotactic (curve A), lightly crosslinked atactic (curve B), and amorphous (curve C) polystyrene. (From A. V. Tobolsky, *Properties and Structures of Polymers*. Copyright © 1960 by John Wiley & Sons, New York. Reprinted by permission of John Wiley & Sons, Inc.)

Furthermore, the tests are performed under isothermal conditions. Creep results are represented as a time-dependent *creep modulus* $E_c(t)$, defined by

$$E_c(t) = \frac{\sigma_0}{\epsilon(t)} \tag{16.2}$$

wherein σ_0 is the constant applied stress and $\epsilon(t)$ is the time-dependent strain. The creep modulus is also temperature sensitive and diminishes with increasing temperature.

With regard to the influence of molecular structure on the creep characteristics, as a general rule the susceptibility to creep decreases [i.e., $E_c(t)$ increases] as the degree of crystallinity increases.

16.7 DEFORMATION OF ELASTOMERS

One of the fascinating properties of the elastomeric materials is their rubberlike elasticity. That is, they have the ability to be deformed to quite large deformations, and then elastically spring back to their original form. This behavior was probably first observed in natural rubber; however, the past few years have brought about the synthesis of a large number of elastomers with a wide variety of properties. Typical stress–strain characteristics of elastomeric materials are displayed in Figure 16.1, curve *C*. Their moduli of elasticity are quite small and, furthermore, vary with strain since the stress–strain curve is nonlinear.

In an unstressed state, an elastomer will be amorphous and composed of molecular chains that are highly twisted, kinked, and coiled. Elastic deformation, upon application of a tensile load, is simply the partial uncoiling, untwisting, and straightening, and the resultant elongation of the chains in the stress direction, a phenomenon represented in Figure 16.12. Upon release of the stress, the chains spring back to their prestressed conformations, and the macroscopic piece returns to its original shape.

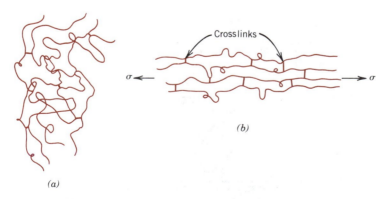

Figure 16.12 Schematic representation of crosslinked polymer chain molecules (*a*) in an unstressed state and (*b*) during elastic deformation in response to an applied tensile stress. (Adapted from Z. D. Jastrzebski, *The Nature and Properties of Engineering Materials,* 3rd edition. Copyright © 1987 by John Wiley & Sons, New York. Reprinted by permission of John Wiley & Sons, Inc.)

The driving force for elastic deformation is a thermodynamic parameter called *entropy,* which is a measure of the degree of disorder within a system; entropy increases with increasing disorder. As an elastomer is stretched and the chains straighten and become more aligned, the system becomes more ordered. From this state, the entropy increases if the chains return to their original kinked and coiled contours. Two intriguing phenomena result from this entropic effect. First, when stretched, an elastomer experiences a rise in temperature; second, the modulus of elasticity increases with increasing temperature, which is opposite to the behavior found in other materials (see Figure 6.7).

Several criteria must be met in order for a polymer to be elastomeric: (1) It must not easily crystallize; elastomeric materials are amorphous. (2) Chain bond rotations must be relatively free in order for the coiled chains to readily respond to an applied force. (3) For elastomers to experience relatively large elastic deformations, the onset of plastic deformation must be delayed. Restricting the motions of chains past one another by crosslinking accomplishes this objective. The crosslinks act as anchor points between the chains and prevent chain slippage from occurring; the role of crosslinks in the deformation process is illustrated in Figure 16.12. Crosslinking in many elastomers is carried out in a process called vulcanization, to be discussed in Section 16.14. (4) Finally, the elastomer must be above its glass transition temperature. The lowest temperature at which rubberlike behavior persists is T_g (Figure 16.10), which for many of the common elastomers is between -50 and $-90°C$ (-60 and $-130°F$). Below its glass transition temperature, an elastomer becomes brittle such that its stress–strain behavior resembles curve *A* in Figure 16.1.

16.8 FRACTURE OF POLYMERS

The fracture strengths of polymeric materials are low relative to those of metals and ceramics. As a general rule, the mode of fracture in thermosetting polymers is brittle. In simple terms, associated with the fracture process is the formation of cracks at regions where there is a localized stress concentration (i.e., scratches, notches, and sharp flaws). Covalent bonds in the network or crosslinked structure are severed during fracture.

For thermoplastic polymers, both ductile and brittle modes are possible, and many of these materials are capable of experiencing a ductile-to-brittle transition. Factors that favor brittle fracture are a reduction in temperature, an increase in strain rate, the presence of a sharp notch, increased specimen thickness, and, in addition, a modification of the polymer structure (chemical, molecular, and/or microstructure). Glassy thermoplastics are brittle at relatively low temperatures; as the temperature is raised, they become ductile in the vicinity of their glass transition temperatures and experience plastic yielding prior to fracture. This behavior is demonstrated by the stress–strain characteristics of polymethyl methacrylate in Figure 16.2. At 4°C, PMMA is totally brittle, whereas at 60°C it becomes extremely ductile.

One phenomenon that is involved in the fracture of some glassy thermoplastic polymers is *crazing.* Crazes form at highly stressed regions associated with scratches, flaws, dust particles, and molecular inhomogeneities; they normally propagate perpendicular to the tensile stress axis. Associated with crazes are regions of very localized yielding, which lead to the formation of fibrils (regions wherein the molecular chains are oriented) and also interspersed small voids (microvoids) that are interconnected;

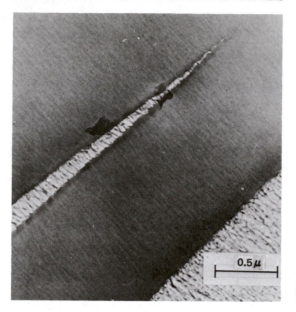

Figure 16.13 Photomicrograph of a craze in polyphenylene oxide. (From R. P. Kambour and R. E. Robertson, "The Mechanical Properties of Plastics," in *Polymer Science, A Materials Science Handbook*, A. D. Jenkins, editor. Reprinted with permission of Elsevier Science Publishers.)

typical craze thicknesses are 5 μm or less. Figure 16.13 is a photomicrograph in which a craze is shown.

Unlike cracks, crazes are capable of supporting loads across their faces. The loads supported will be less than in the uncrazed, uncracked material. If an applied tensile load is sufficient, cracks form along crazes by the breakdown of the fibrillar structure and expansion of the voids, which is followed by crack tip extension through the craze.

Principles of fracture mechanics developed in Section 8.5 also apply to brittle and quasi-brittle polymers; the susceptibility of these materials to fracture when a crack is present may be expressed in terms of the plane strain fracture toughness. The magnitude of K_{Ic} will depend on characteristics of the polymer (i.e., molecular weight, percent crystallinity, etc.) as well as temperature, strain rate, and the external environment. Representative values of K_{Ic} for several polymers are included in Table 8.1.

16.9 MISCELLANEOUS CHARACTERISTICS

Impact Strength

The degree of resistance of a polymeric material to impact loading of a notched piece may be of concern in some applications. Izod or Charpy tests are ordinarily conducted to assess impact strength. As with metals, polymers may exhibit ductile or brittle fracture under impact loading conditions, depending on the temperature, specimen size, strain rate, and mode of loading, as discussed in the preceding section. Crystalline and amorphous polymers are brittle at low temperatures, and both have relatively low impact strengths. However, they experience a ductile-to-brittle transition over a relatively narrow temperature range, similar to that shown for a steel in Figure 8.14. Of course, impact strength undergoes a gradual decrease at still higher temperatures as the polymer begins to soften. Ordinarily, the two impact characteristics most sought after are a high impact strength at the ambient temperature, and a ductile-to-brittle transition temperature that lies below room temperature.

Fatigue

Polymers may experience fatigue failure under conditions of cyclic loading. As with metals, fatigue occurs at stress levels that are low relative to the yield strength. Fatigue testing in polymers has not been nearly as extensive as with metals; however, fatigue data are plotted in the same manner for both types of material, and the resulting curves have the same general shape. Fatigue curves for nylon and polymethyl methacrylate are shown in Figure 16.14, as stress versus the number of cycles to failure (on a logarithmic scale). Some polymers have a fatigue limit (a stress level at which the stress at failure becomes independent of the number of cycles); others do not appear to have such a limit. As would be expected, fatigue strengths and fatigue limits for polymeric materials are much lower than for metals.

Tear Strength and Hardness

Other mechanical properties that are sometimes influential in the suitability of a polymer for some particular application include tear resistance and hardness. The ability to resist tearing is an important property of some plastics, especially those used for thin films in packaging. *Tear strength,* the mechanical parameter that is measured, is the energy required to tear apart a cut specimen that has a standard geometry. The magnitude of tensile and tear strengths are related.

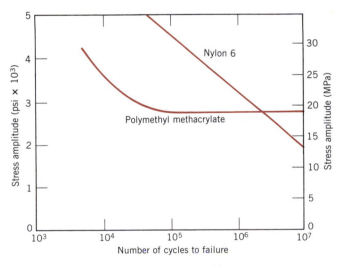

Figure 16.14 Fatigue curves (stress amplitude versus the number of cycles to failure) for nylon 6 and polymethyl methacrylate. The PMMA shows a fatigue limit, whereas the nylon does not. (From M. N. Riddell, G. P. Koo, and J. L. O'Toole, "Fatigue Mechanisms of Thermoplastics," *Polymer Eng. Sci.,* **6,** 363, 1966.)

As with metals, hardness represents a material's resistance to scratching, penetration, marring, and so on. Most hardness tests are conducted by penetration techniques similar to those described for metals in Section 6.10.

POLYMER APPLICATIONS AND PROCESSING

The large macromolecules of the commercially useful polymers must be synthesized from substances having smaller molecules in a process termed polymerization. Furthermore, the properties of a polymer may be modified and enhanced by the inclusion of additive materials. Finally, a finished piece having a desired shape must be fashioned during a forming operation. This section treats polymerization processes and the various forms of additives; specific forming procedures are discussed according to polymer type.

16.10 POLYMERIZATION

The synthesis of the large molecular weight polymers is termed *polymerization;* it is simply the process by which monomer units are joined over and over, to generate each of the constituent giant molecules. Most generally, the raw materials for synthetic polymers are derived from coal and petroleum products, which are composed of molecules having low molecular weights. The reactions by which polymerization occurs are grouped into two general classifications—addition and condensation—according to the reaction mechanism, as discussed below.

Addition Polymerization

Addition polymerization (sometimes called chain reaction polymerization) is a process by which bifunctional monomer units are attached one at a time in chainlike fashion to form a linear macromolecule; the composition of the resultant product molecule is an exact multiple for that of the original reactant monomer.

Three distinct stages—initiation, propagation, and termination—are involved in addition polymerization. During the initiation step, an active center capable of propagation is formed by a reaction between an initiator (or catalyst) species and the monomer unit. This process has already been demonstrated for polyethylene (Equation 15.1), which is repeated as follows:

$$\text{R} \cdot + \begin{matrix} \text{H} & \text{H} \\ | & | \\ \text{C} = \text{C} \\ | & | \\ \text{H} & \text{H} \end{matrix} \longrightarrow \text{R} - \begin{matrix} \text{H} & \text{H} \\ | & | \\ \text{C} - \text{C} \\ | & | \\ \text{H} & \text{H} \end{matrix} \qquad (16.3)$$

Again, R· represents the active initiator, and · is an unpaired electron.

Propagation involves the linear growth of the molecule as monomer units become attached to one another in succession to produce the chain molecule, which is represented, again for polyethylene, as follows:

$$R-\underset{\underset{H}{|}}{\overset{\overset{H}{|}}{C}}-\underset{\underset{H}{|}}{\overset{\overset{H}{|}}{C}}\cdot + \underset{\underset{H}{|}}{\overset{\overset{H}{|}}{C}}=\underset{\underset{H}{|}}{\overset{\overset{H}{|}}{C}} \longrightarrow R-\underset{\underset{H}{|}}{\overset{\overset{H}{|}}{C}}-\underset{\underset{H}{|}}{\overset{\overset{H}{|}}{C}}-\underset{\underset{H}{|}}{\overset{\overset{H}{|}}{C}}-\underset{\underset{H}{|}}{\overset{\overset{H}{|}}{C}}\cdot \qquad (16.4)$$

Chain growth is relatively rapid; the period required to grow a molecule consisting of, say, 1000 mer units is on the order of 10^{-2} to 10^{-3} s.

Propagation may end or terminate in different ways. First, the active ends of two propagating chains may react or link together to form a nonreactive molecule, as follows:

$$(16.5)$$

thus terminating the growth of each chain. Or, an active chain end may react with an initiator or other chemical species having a single active bond, as follows:

$$(16.6)$$

with the resultant cessation of chain growth.

Molecular weight is governed by the relative rates of initiation, propagation, and termination. Ordinarily, they are controlled to ensure the production of a polymer having the desired degree of polymerization. Since termination is a somewhat random process and does not occur at the same point for each molecule, a variety of chain lengths is achieved, which accounts for a distribution of molecular weights.

Addition polymerization is used in the synthesis of polyethylene, polypropylene, polyvinyl chloride, and polystyrene, as well as many of the copolymers.

Condensation Polymerization

Condensation (or step reaction) **polymerization** is the formation of polymers by stepwise intermolecular chemical reactions that normally involve more than one monomer species; there is usually a small molecular weight by-product such as water, which is eliminated. No reactant species has the chemical formula of the mer repeat unit, and the intermolecular reaction occurs every time a mer repeat unit is formed. For example, consider the formation of a polyester from the reaction between ethylene glycol and adipic acid; the intermolecular reaction is as follows:

$$+ \ H_2O \quad (16.7)$$

This stepwise process is successively repeated, producing, in this case, a linear molecule. The chemistry of the specific reaction is not important, but rather, the condensation polymerization mechanism.

Reaction times for condensation are generally longer than for addition polymerization. To produce large molecular weight materials, it is essential that the reaction times be sufficiently long and the conversion of the monomer reactants complete. As with addition polymerization, various chain lengths are produced, yielding a molecular weight distribution.

Condensation reactions often produce trifunctional monomers capable of forming crosslinked and network polymers. The thermosetting polyesters and phenol-formaldehyde, the nylons, and the polycarbonates are produced by condensation polymerization. Some polymers, such as nylon, may be polymerized by either technique.

16.11 POLYMER ADDITIVES

Most of the properties of polymers discussed earlier in this chapter are intrinsic ones—that is, characteristic of or fundamental to the specific polymer. Some of these properties are related to and controlled by the molecular structure. Many times, however, it is necessary to modify the mechanical, chemical, and physical properties to a much greater degree than is possible by the simple alteration of this fundamental molecular structure. Foreign substances called *additives* are intentionally introduced to enhance or modify many of these properties, and thus render a polymer more serviceable. Typical additives include filler materials, plasticizers, stabilizers, colorants, and flame retardants.

Fillers

Filler materials are most often added to polymers to improve tensile and compressive strengths, abrasion resistance, toughness, dimensional and thermal stability, and other properties. Materials used as particulate fillers include wood flour (finely powdered sawdust), silica flour and sand, glass, clay, talc, limestone, and even some synthetic polymers. Particle sizes range all the way from 10 nm to macroscopic dimensions. Because these inexpensive materials replace some volume of the more expensive polymer, the cost of the final product is reduced.

Plasticizers

The flexibility, ductility, and toughness of polymers may be improved with the aid of additives called **plasticizers.** Their presence also produces reductions in hardness and stiffness. Plasticizers are generally liquids having low vapor pressures and low molecular weights. The small plasticizer molecules occupy positions between the large polymer chains, effectively increasing the interchain distance with a reduction in the secondary intermolecular bonding. Plasticizers are commonly used in polymers that are intrinsically brittle at room temperature, such as polyvinyl chloride and some of the acetate copolymers. In effect, the plasticizer lowers the glass transition temperature, so that at ambient conditions the polymers may be used in applications requiring some degree of pliability and ductility. These applications include thin sheets or films, tubing, raincoats, and curtains.

Stabilizers

Some polymeric materials, under normal environmental conditions, are subject to rapid deterioration, generally in terms of mechanical integrity. Most often, this deterioration is a result of exposure to light, in particular ultraviolet radiation, and also oxidation (Section 18.12). Ultraviolet radiation interacts with, and causes a severance of some of the covalent bonds along the molecular chain which may also result in some crosslinking. Oxidation deterioration is a consequence of the chemical interaction between oxygen atoms and the polymer molecules. Additives that counteract these deteriorative processes are called **stabilizers.**

Colorants

Colorants impart a specific color to a polymer; they may be added in the form of dyes or pigments. The molecules in a dye actually dissolve and become part of the molecular structure of the polymer. Pigments are filler materials that do not dissolve, but remain as a separate phase; normally they have a small particle size, are transparent, and have a refractive index near to that of the parent polymer. Others may impart opacity as well as color to the polymer.

Flame Retardants

The flammability of polymeric materials is a major concern, especially in the manufacture of textiles and children's toys. Most polymers are flammable in their pure form; exceptions include those containing significant contents of chlorine and/or fluorine, such as polyvinyl chloride and polytetrafluoroethylene. The flammability resistance of the remaining combustible polymers may be enhanced by additives called **flame retardants.** These retardants may function by interfering with the combustion process through the gas phase, or by initiating a chemical reaction that causes a cooling of the combustion region and a cessation of burning.

16.12 POLYMER TYPES

There are many different polymeric materials that are familiar to us and find a wide variety of applications. These include plastics, elastomers (or rubbers), fibers, coatings, adhesives, foams, and films. Depending on its properties, a particular polymer may be used in two or more of these application categories. For example, a plastic, if crosslinked and utilized above its glass transition temperature, may make a satisfactory elastomer. Or, a fiber material may be used as a plastic if it is not drawn into filaments. This portion of the chapter includes a brief discussion of each of these types of polymer. In addition, for each, some of the common fabrication methods are noted.

16.13 PLASTICS

Characteristics and Applications

Possibly the largest number of different polymeric materials come under the plastic classification. Polyethylene, polypropylene, polyvinyl chloride, polystyrene, and the fluorocarbons, epoxies, phenolics, and polyesters may all be classified as **plastics.** They have a wide variety of combinations of properties. Some plastics are very rigid and brittle; others are flexible, exhibiting both elastic and plastic deformations when stressed, and sometimes experiencing considerable deformation before fracture.

Polymers falling within this classification may have any degree of crystallinity, and all molecular structures and configurations (linear, branching, isotactic, etc.) are possible. Plastic materials may be either thermoplastic or thermosetting; in fact, this is the manner in which they are usually subclassified. The trade names, characteristics, and typical applications for a number of plastics are given in Table 16.3.

Several plastics exhibit especially outstanding properties. For applications in which optical transparency is critical, polystyrene and polymethyl methacrylate are especially well suited; however, it is imperative that the material be highly amorphous. The fluorocarbons have a low coefficient of friction and are extremely resistant to attack by a host of chemicals, even at relatively high temperatures. They are utilized as coatings on nonstick cookware, in bearings and bushings, and for high-temperature electronic components.

Forming Techniques

Quite a variety of different techniques are employed in the forming of polymeric materials. The method used for a specific polymer depends on several factors: (1) whether the material is thermoplastic or thermosetting; (2) if thermoplastic, the temperature at which it softens; (3) the atmospheric stability of the material being formed; and (4) the geometry and size of the finished product. There are numerous similarities between some of these techniques and those utilized for fabricating metals and ceramics.

Fabrication of polymeric materials normally occurs at elevated temperatures and often by the application of pressure. Thermoplastics are formed above their glass transition temperature, and an applied pressure must be maintained as the piece is cooled to temperatures below T_g so that the formed article will retain its shape while still soft and in a plastic state. One significant economic benefit of using thermoplastics is that they may be recycled; scrap thermoplastic pieces may be remelted and reformed into new shapes.

Fabrication of thermosetting polymers is ordinarily accomplished in two stages. First comes the preparation of a linear polymer (sometimes called a prepolymer) as a liquid, having a low molecular weight. This material is converted into the final hard

TABLE 16.3 Trade Names, Characteristics, and Typical Applications for a Number of Plastic Materials

Material Type	Trade Names	Major Application Characteristics	Typical Applications
Thermoplastics			
Acrylonitrile-butadiene-styrene (ABS)	Marbon, Cycolac, Lustran, Abson	Outstanding strength and toughness; resistant to heat distortion; good electrical properties; flammable and soluble in some organic solvents	Refrigerator linings, lawn and garden equipment, toys, highway safety devices
Acrylics (polymethyl methacrylate)	Lucite, Plexiglas	Outstanding light transmission and resistance to weathering; only fair mechanical properties	Lenses, transparent aircraft enclosures, drafting equipment, outdoor signs
Fluorocarbons (PTFE or TFE)	Teflon TFE, Halon TFE	Chemically inert in almost all environments; excellent electrical properties; low coefficient of friction; may be used to 260°C (500°F); relatively weak and poor cold-flow properties	Anticorrosive seals, chemical pipes and valves, bearings, antiadhesive coatings, high temperature electronic parts
Nylons	Zytel, Plaskon	Good mechanical strength, abrasion resistance, and toughness; low coefficient of friction; absorbs water and some other liquids	Bearings, gears, cams, bushings, handles, and jacketing for wires and cables
Polycarbonates	Merlon, Lexan	Dimensionally stable; low water absorption; transparent; very good impact resistance and ductility; chemical resistance not outstanding	Safety helmets, lenses, light globes, base for photographic film
Polyethylene	Alathon, Petrothene, Hi-fax	Chemically resistant and electrically insulating; tough and relatively low coefficient of friction; low strength and poor resistance to weathering	Flexible bottles, toys, tumblers, battery parts, ice trays, film wrapping materials
Polypropylene	Pro-fax, Tenite, Moplen	Resistant to heat distortion; excellent electrical properties and fatigue strength; chemically inert; relatively inexpensive; poor resistance to UV light	Sterilizable bottles, packaging film, TV cabinets, luggage

(Continued)

TABLE 16.3 (*continued*)

Material Type	Trade Names	Major Application Characteristics	Typical Applications
Thermoplastics			
Polystyrene	Styron, Lustrex, Rexolite	Excellent electrical properties and optical clarity; good thermal and dimensional stability; relatively inexpensive	Wall tile, battery cases, toys, indoor lighting panels, appliance housings
Vinyls	PVC, Pliovic, Saran, Tygon	Good low-cost, general-purpose materials; ordinarily rigid, but may be made flexible with plasticizers; often copolymerized; susceptible to heat distortion	Floor coverings, pipe, electrical wire insulation, garden hose, phonograph records
Polyester (PET)	Mylar, Celanar, Dacron	One of the toughest of plastic films; excellent fatigue and tear strength, and resistance to humidity, acids, greases, oils, and solvents	Magnetic recording tapes, clothing, automotive tire cords
Thermosetting Polymers			
Epoxies	Epon, Epi-rez, Araldite	Excellent combination of mechanical properties and corrosion resistance; dimensionally stable; good adhesion; relatively inexpensive; good electrical properties	Electrical moldings, sinks, adhesives, protective coatings, used with fiberglass laminates
Phenolics	Bakelite, Durez, Resinox	Excellent thermal stability to over 150°C (300°F); may be compounded with a large number of resins, fillers, etc.; inexpensive	Motor housings, telephones, auto distributors, electrical fixtures
Polyesters	Selectron, Laminac, Paraplex	Excellent electrical properties and low cost; can be formulated for room- or high-temperature use; often fiber reinforced	Helmets, fiberglass boats, auto body components, chairs, fans
Silicones	DC resins	Excellent electrical properties; chemically inert, but susceptible to attack by steam; outstanding heat resistance; relatively expensive	Laminates, terminal strips, high-temperature insulation

Source: Adapted from C. A. Harper (Editor), *Handbook of Plastics and Elastomers.* Copyright © 1975 by McGraw-Hill Book Company. Reproduced with permission.

and stiff product during the second stage, which is normally carried out in a mold having the desired shape. This second stage, termed "curing," may occur during heating and/or by the addition of catalysts, and often under pressure. During curing, chemical and structural changes occur on a molecular level: a crosslinked or a network structure forms. After curing, thermoset polymers may be removed from a mold while still hot, since they are now dimensionally stable. Thermosets cannot be recycled, do not melt, are usable at higher temperatures than thermoplastics, and are more chemically inert.

Molding is the most common method for forming plastic polymers. The several molding techniques used include compression, transfer, blow, injection, and extrusion molding. For each, a finely pelletized or granulized plastic is forced, at an elevated temperature and by pressure, to flow into, fill, and assume the shape of a mold cavity.

Compression and Transfer Molding. For compression molding, the appropriate amounts of thoroughly mixed polymer and necessary additives are placed between male and female mold members, as illustrated in Figure 16.15. Both mold pieces are heated; however, only one is movable. The mold is closed, and heat and pressure are applied, causing the plastic material to become viscous and conform to the mold shape. Before molding, raw materials may be mixed and cold pressed into a disc, which is called a preform. Preheating of the preform reduces molding time and pressure, extends the die lifetime, and produces a more uniform finished piece. This molding technique lends itself to the fabrication of both thermoplastic and thermosetting polymers; however, its use with thermoplastics is more time consuming and expensive.

In transfer molding, a variation of compression molding, the solid ingredients are first melted in a heated transfer chamber. As the molten material is injected into the mold chamber, the pressure is distributed more uniformly over all surfaces. This process is used with thermosetting polymers and for pieces having complex geometries.

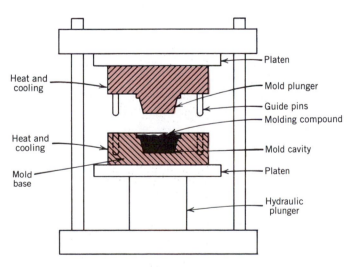

Figure 16.15 Schematic diagram of a compression molding apparatus. (From F. W. Billmeyer, Jr., *Textbook of Polymer Science,* 3rd edition. Copyright © 1984 by John Wiley & Sons, New York. Reprinted by permission of John Wiley & Sons, Inc.)

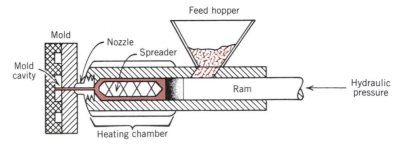

Figure 16.16 Schematic diagram of an injection molding apparatus. (Adapted from F. W. Billmeyer, Jr., *Textbook of Polymer Science,* 2nd edition. Copyright © 1971 by John Wiley & Sons, New York. Reprinted by permission of John Wiley & Sons, Inc.)

Injection Molding.   Injection molding, the polymer analogue of die casting for metals, is the most widely used technique for fabricating thermoplastic materials. A schematic cross section of the apparatus used is illustrated in Figure 16.16. The correct amount of pelletized material is fed from a loading hopper into a cylinder by the motion of a plunger or ram. This charge is pushed forward into a heating chamber, at which point the thermoplastic material melts to form a viscous liquid. Next, the molten plastic is impelled, again by ram motion, through a nozzle into the enclosed mold cavity; pressure is maintained until the molding has solidified. Finally, the mold is opened, the piece is ejected, the mold is closed, and the entire cycle is repeated. Probably the most outstanding feature of this technique is the speed with which pieces may be produced. For thermoplastics, solidification of the injected charge is almost immediate; consequently, cycle times for this process are short (commonly within the range of 10 to 30 s). Thermosetting polymers may also be injection molded; curing takes place while the material is under pressure in a heated mold, which results in longer cycle times than for thermoplastics.

Extrusion.   The extrusion process is simply injection molding of a viscous thermoplastic through an open-ended die, similar to the extrusion of metals (Figure 12.2c). A mechanical screw or auger propels through a chamber the pelletized material, which is successively compacted, melted, and formed into a continuous charge of viscous fluid. Extrusion takes place as this molten mass is forced through a die orifice. Solidification of the extruded length is expedited by blowers or a water spray just before passing onto a moving conveyor. The technique is especially adapted to producing continuous lengths having constant cross-sectional geometries, for example, rods, tubes, hose channels, sheets, and filaments.

Blow Molding.   The blow molding process for the fabrication of plastic containers is similar to that used for blowing glass bottles, as represented in Figure 14.5. First, a parison, or length of polymer tubing is extruded. While still in a semimolten state, the parison is placed in a two-piece mold having the desired container configuration. The hollow piece is formed by blowing air or steam under pressure into the parison, forcing the tube walls to conform to the contours of the mold. Of course the temperature and viscosity of the parison must be carefully regulated.

Casting.   Like metals and ceramics, polymeric materials may be cast, as when a molten plastic material is poured into a mold and allowed to solidify. Both thermoplastic and thermosetting plastics may be cast. For thermoplastics, solidification occurs upon cooling from the molten state; however, for thermosets, hardening is a consequence of the actual polymerization or curing process, which is usually carried out at an elevated temperature.

16.14 ELASTOMERS

The characteristics of and deformation mechanism for elastomers were treated previously (Section 16.7). The present discussion, therefore, focuses on the processing and types of elastomeric materials.

Vulcanization

One requisite characteristic for elastomeric behavior is that the molecular structure be lightly crosslinked. The crosslinking process in elastomers is called **vulcanization,** which is achieved by a nonreversible chemical reaction, ordinarily carried out at an elevated temperature. In most vulcanizing reactions, sulfur compounds are added to the heated elastomer; sulfur atoms bond with adjacent chains and crosslink them. Sulfur bridge crosslinks are formed in polyisoprene according to the following reaction:

$$
\begin{array}{cc}
\begin{array}{l}
\ \ \ \ \text{H CH}_3\text{ H H} \\
\ \ \ \ |\ \ \ \ |\ \ \ \ |\ \ \ |\ \\
-\text{C}-\text{C}=\text{C}-\text{C}- \\
\ \ \ \ |\ \ \ \ \ \ \ \ \ \ \ \ | \\
\ \ \ \ \text{H}\ \ \ \ \ \ \ \ \ \ \text{H} \\
\\
\ \ \ \ \text{H}\ \ \ \ \ \ \ \ \ \ \text{H} \\
\ \ \ \ |\ \ \ \ \ \ \ \ \ \ \ \ | \\
-\text{C}-\text{C}=\text{C}-\text{C}- \\
\ \ \ \ |\ \ \ \ |\ \ \ \ |\ \ \ | \\
\ \ \ \ \text{H CH}_3\text{ H H}
\end{array}
&
\begin{array}{l}
\ \ \ \ \text{H CH}_3\text{ H H} \\
\ \ \ \ |\ \ \ \ |\ \ \ \ |\ \ \ |\ \\
-\text{C}-\text{C}-\text{C}-\text{C}- \\
\ \ \ \ |\ \ \ \ |\ \ \ \ |\ \ \ | \\
\ \ \ \ \text{H}\ \ \ \ \text{S}\ \ \ \text{S}\ \text{H} \\
\ \ \ \ \ \ \ \ \ \ |\ \ \ \ | \\
-\text{C}-\text{C}-\text{C}-\text{C}- \\
\ \ \ \ |\ \ \ \ |\ \ \ \ |\ \ \ | \\
\ \ \ \ \text{H CH}_3\text{ H H}
\end{array}
\end{array}
$$

$$+ 2S \longrightarrow \qquad\qquad\qquad (16.8)$$

Unvulcanized rubber is soft and tacky, and has poor resistance to abrasion. Modulus of elasticity, tensile strength, and resistance to degradation by oxidation are all enhanced by vulcanization. The magnitude of the modulus of elasticity is directly proportional to the density of the crosslinks. Stress–strain curves for vulcanized and unvulcanized natural rubber are presented in Figure 16.17. To produce a rubber that is capable of large extensions without rupture of the primary chain bonds, there must be relatively few crosslinks, and these must be widely separated. Useful rubbers result when about 1 to 5 parts (by weight) of sulfur is added to 100 parts of rubber. Increasing the sulfur content further hardens the rubber and also reduces its extensibility. Also, since they are crosslinked, elastomeric materials are thermosetting in nature.

Elastomeric Types

Table 16.4 lists properties and applications of common elastomers; these properties are typical and, of course, depend on the degree of vulcanization and on whether any reinforcement is used. Natural rubber is still utilized to a large degree because it has

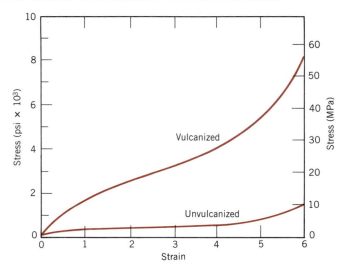

Figure 16.17 Stress–strain curves to 600% elongation for unvulcanized and vulcanized natural rubber.

an outstanding combination of desirable properties. However, the most important synthetic elastomer is SBR, which is used predominantly in automobile tires, reinforced with carbon black. NBR, which is highly resistant to degradation and swelling, is another common synthetic elastomer.

For many applications (e.g., automobile tires), the mechanical properties of even vulcanized rubbers are not satisfactory in terms of tensile strength, abrasion and tear resistance, and stiffness. These characteristics may be further improved by additives such as carbon black (Section 17.2). Furthermore, the techniques used in the actual fabrication of rubber parts are essentially the same as those discussed for plastics as described above, that is, compression molding, extrusion, and so on.

Finally, some mention should be made of the silicone rubbers. For these materials, the backbone carbon chain is replaced by a chain that alternates silicon and oxygen atoms:

$$\begin{array}{c} R \\ | \\ -Si-O- \\ | \\ R' \end{array}$$

where R and R′ represent side-bonded atoms such as hydrogen or groups of atoms such as CH_3. For example, polydimethylsiloxane has the mer structure

$$\begin{array}{c} CH_3 \\ | \\ -Si-O- \\ | \\ CH_3 \end{array}$$

Of course, as elastomers, these materials are crosslinked.

TABLE 16.4 Tabulation of Important Characteristics and Typical Applications for Five Commercial Elastomers

Chemical Type	Trade (Common) Name	Elongation (%)	Useful Temperature Range [°C (°F)]	Major Application Characteristics	Typical Applications
Natural polyisoprene	Natural Rubber (NR)	500–700	−55 to 120 (−65 to 250)	Excellent physical properties; good resistance to cutting, gouging, and abrasion; low heat, ozone, and oil resistance; good electrical properties	Pneumatic tires and tubes; heels and soles; gaskets
Styrene-butadiene copolymer	GRS, Buna S (SBR)	450–500	−60 to 120 (−75 to 250)	Good physical properties; excellent abrasion resistance; not oil, ozone, or weather resistant; electrical properties good, but not outstanding	Same as natural rubber
Acrylonitrile-butadiene copolymer	Buna A, Nitrile (NBR)	400–600	−50 to 150 (−60 to 300)	Excellent resistance to vegetable, animal, and petroleum oils; poor low temperature properties; electrical properties not outstanding	Gasoline, chemical, and oil hose; seals and O-rings; heels and soles
Chloroprene	Neoprene (CR)	100–800	−50 to 105 (−60 to 225)	Excellent ozone, heat, and weathering resistance; good oil resistance; excellent flame resistance; not as good in electrical applications as natural rubber	Wire and cable; chem. tank linings; belts, hoses, seals, and gaskets
Polysiloxane	Silicone (SIL)	600	−90 to 250 (−130 to 480)	Excellent resistance to high and low temperatures; low strength; excellent electrical properties	High- and low-temperature insulation; seals, diaphragms; tubing for food and medical uses

Sources: Adapted from: C. A. Harper (Editor), *Handbook of Plastics and Elastomers.* Copyright © 1975 by McGraw-Hill Book Company, reproduced with permission; and Materials Engineering's *Materials Selector*, copyright Penton/IPC.

The silicone elastomers possess a high degree of flexibility at low temperatures [to $-90°C$ $(-130°F)$] and yet are stable to temperatures as high as $250°C$ $(480°F)$. In addition, they are resistant to weathering and lubricating oils. A further attractive characteristic is that some silicone rubbers vulcanize at room temperature (RTV rubbers).

16.15 FIBERS

Characteristics and Applications

The fiber polymers are capable of being drawn into long filaments having at least a $100:1$ length-to-diameter ratio. Most commerical fiber polymers are utilized in the textile industry, being woven or knit into cloth or fabric. To be useful as a textile material, a fiber polymer must meet a host of rather restrictive physical and chemical properties. While in use, fibers may be subjected to a variety of mechanical deformations—stretching, twisting, shearing, and abrasion. Consequently, they must have a high tensile strength (over a relatively wide temperature range) and a high modulus of elasticity, as well as abrasion resistance. These properties are governed by the chemistry of the polymer chains and also by the fiber drawing process.

The molecular weight of fiber materials should be relatively high. Also, since the tensile strength increases with degree of crystallinity, the structure and configuration of the chains should allow the production of a highly crystalline polymer; that translates into a requirement for linear and unbranched chains that are symmetrical and have regularly repeating mer units.

Convenience in washing and maintaining clothing depends primarily on the thermal properties of the fiber polymer, that is, its melting and glass transition temperatures. Furthermore, fiber polymers must exhibit chemical stability to a rather extensive variety of environments, including acids, bases, bleaches, dry cleaning solvents, and sunlight. In addition, they must be relatively nonflammable and amenable to drying.

Forming Techniques

The process by which fibers are formed from bulk polymer material is termed **spinning.** Most often, fibers are spun from the molten state in a process called melt spinning. The material to be spun is first heated until it forms a relatively viscous liquid. Next, it is pumped down through a plate called a spinnerette, which contains numerous small, round holes. As the molten material passes through each of these orifices, a single fiber is formed, which solidifies almost immediately upon passing into the air.

The crystallinity of a spun fiber will depend on its rate of cooling during spinning. The strength of fibers is improved by a postforming process called **drawing.** Drawing is simply the mechanical elongation of a fiber in the direction of its axis. During this process the molecular chains become oriented in the direction of drawing (Section 16.3), such that the tensile strength, modulus of elasticity, and toughness are improved. Although the mechanical strength of a drawn fiber is improved in this axial direction, strength is reduced in a transverse or radial direction. However, since fibers are normally stressed only along the axis, this strength differential is not critical. The cross section of drawn fibers is nearly circular, and the properties are uniform throughout the cross section.

16.16 MISCELLANEOUS APPLICATIONS

Coatings

Coatings are frequently applied to the surface of materials to serve one or more of the following functions: (1) to protect the item from the environment that may produce corrosive or deteriorative reactions; (2) to improve the item's appearance; and (3) to provide electrical insulation. Many of the ingredients in coating materials are polymers, the majority of which are organic in origin. These organic coatings fall into several different classifications, as follows: paint, varnish, enamel, lacquer, and shellac.

Adhesives

An **adhesive** is a substance used to join together the surfaces of two solid materials (termed "adherends") to produce a joint with a high shear strength. The bonding forces between the adhesive and adherend surfaces are thought to be electrostatic, similar to the secondary bonding forces between the molecular chains in thermoplastic polymers. Even though the inherent strength of the adhesive may be much less than that of the adherend materials, nevertheless, a strong joint may be produced if the adhesive layer is thin and continuous. If a good joint is formed, the adherend material may fracture or rupture before the adhesive.

Polymeric materials that fall within the classifications of thermoplastics, thermosetting resins, elastomeric compounds, and natural adhesives (animal glue, casein, starch, and rosin) may serve adhesive functions. Polymer adhesives may be used to join a large variety of material combinations: metal–metal, metal–plastic, metal–ceramic, and so on. The primary drawback is the service temperature limitation. Organic polymers maintain their mechanical integrity only at relatively low temperatures, and strength decreases rapidly with increasing temperature.

Films

Within relatively recent times, polymeric materials have found widespread use in the form of thin *films*. Films having thicknesses between 0.001 and 0.005 in. (0.025 and 0.125 mm) are fabricated and used extensively as bags for packaging food products and other merchandise, as textile products, and a host of other uses. Important characteristics of the materials produced and used as films include low density, a high degree of flexibility, high tensile and tear strengths, resistance to attack by moisture and other chemicals, and low permeability to some gases, especially water vapor. Some of the polymers that meet these criteria and are manufactured in film form are polyethylene, polypropylene, cellophane, and cellulose acetate.

There are several forming methods. Many films are simply extruded through a thin die slit; this may be followed by a rolling operation that serves to reduce thickness and improve strength. Alternately, film may be blown: continuous tubing is extruded through an annular die; then, by maintaining a carefully controlled positive gas pressure inside the tube, wall thickness may be continuously reduced to produce a thin cylindrical film, which may be cut and laid flat. Some of the newer films are produced by coextrusion; that is, multilayers of more than one polymer type are extruded simultaneously.

Foams

Very porous plastic materials are produced in a process called foaming. Both thermoplastic and thermosetting materials may be foamed by including in the batch a blowing agent that upon heating, decomposes with the liberation of a gas. Gas bubbles are generated throughout the now-fluid mass, which remain as pores upon cooling and give rise to a spongelike structure. The same effect is produced by bubbling an inert gas through a material while it is in a molten state. Some of the commonly foamed polymers are polyurethane, rubber, polystyrene, and polyvinyl chloride. **Foams** are commonly used as cushions in automobiles and furniture as well as in packaging and thermal insulation.

SUMMARY

On the basis of stress–strain behavior, polymers fall within three general classifications: brittle, plastic, and highly elastic. These materials are neither as strong nor as stiff as metals, and their mechanical properties are sensitive to changes in temperature.

The mechanism of plastic deformation for semicrystalline polymers having the spherulitic structure was presented. Tensile deformation is thought to occur in several stages as both amorphous tie chains and chain-folded block segments (which separate from the ribbonlike lamellae) become oriented with the tensile axis. The strength of these materials may be enhanced by radiation-induced crosslinking, and by both increased crystallinity and average molecular weight.

Melting and glass transition temperatures are important parameters relative to the temperature range over which a particular polymer may be utilized and processed. The magnitudes of T_m and T_g depend on the ability of molecules to resist chain motions whereby coordinated atomic vibrations arise with increasing temperature. The same molecular factors that affect mechanical strength and stiffness influence the melting and glass transition temperatures.

With regard to mechanical behavior at elevated temperatures, polymers are classified as either thermoplastic or thermosetting. The former soften when heated and harden when cooled; this cycle is reversible and repeatable. In contrast, thermosets, once having hardened, will not soften upon heating.

Viscoelastic mechanical behavior, being intermediate between totally elastic and totally viscous, is displayed by a number of polymeric materials. It is characterized by the relaxation modulus, a time-dependent modulus of elasticity. The magnitude of the relaxation modulus is very sensitive to temperature; critical to the in-service temperature range for elastomers is this temperature dependence.

Large elastic extensions are possible for the elastomeric materials that are amorphous and lightly crosslinked. Deformation corresponds to the unkinking and uncoiling of chains in response to an applied tensile stress. Crosslinking is often achieved during a vulcanization process. Many of the elastomers are copolymers, whereas the silicone elastomers are really inorganic materials.

Fracture strengths of polymeric materials are low relative to metals and ceramics. Both brittle and ductile fracture modes are possible, and some thermoplastic materials experience a ductile-to-brittle transition with a lowering of temperature, an increase in strain rate, and/or an alteration of specimen thickness or geometry. In some glassy thermoplastics, the crack formation process may be preceded by crazing.

Synthesis of large molecular weight polymers is attained by polymerization, of which there are two types: addition and condensation. The various properties of polymers may be further modified by using additives; these include fillers, plasticizers, stabilizers, colorants, and flame retardants.

The plastic materials are perhaps the most widely used group of polymers. Fabrication is usually accomplished by plastic deformation at an elevated temperature, using at least one of several different molding techniques; casting is also possible.

Many polymeric materials may be spun into fibers, which are used primarily in textiles. Mechanical, thermal, and chemical characteristics of these materials are especially critical. Some fibers are spun from a viscous melt, after which they are plastically elongated during a drawing operation, which improves the mechanical strength.

Other miscellaneous applications that employ polymers include coatings, adhesives, films, and foams.

IMPORTANT TERMS AND CONCEPTS

Addition polymerization

Adhesive

Colorant

Condensation
  polymerization

Drawing

Elastomer

Fiber

Filler

Flame retardant

Foam

Glass transition
  temperature

Molding

Plasticizer

Plastics

Relaxation modulus

Spinning

Stabilizer

Thermoplastic polymer

Thermosetting polymer

Viscoelasticity

Vulcanization

REFERENCES

BILLMEYER, F. W., JR., *Textbook of Polymer Science,* 3rd edition, Wiley-Interscience, New York, 1984.

Engineered Materials Handbook, Vol. 2, *Engineering Plastics,* ASM International, Metals Park, OH, 1988.

HARPER, C. A. (Editor), *Handbook of Plastics and Elastomers,* McGraw-Hill Book Company, New York, 1975.

McCLINTOCK, F. A. and A. S. ARGON, *Mechanical Behavior of Materials,* Addison-Wesley Publishing Company, Reading, MA, 1966.

MEARES, P., *Polymers: Structure and Bulk Properties,* D. Van Nostrand Company, Ltd., London, 1965.

MONCRIEFF, R. W., *Man-Made Fibres,* Newnes-Butterworths, London, 1975.

MOORE, G. R. and D. E. KLINE, *Properties and Processing of Polymers for Engineers,* Prentice-Hall, Inc., Englewood Cliffs, NJ, 1984.

NIELSEN, L. E., *Mechanical Properties of Polymers and Composites,* Vol. 1, Marcel Dekker, New York, 1974.

ROSEN, S. L., *Fundamental Principles of Polymeric Materials,* John Wiley & Sons, New York, 1982.

Tobolosky, A. V., *Properties and Structure of Polymers,* John Wiley & Sons, New York, 1960. Advanced treatment.

Williams, D. J., *Polymer Science and Engineering,* Prentice-Hall, Englewood Cliffs, NJ, 1971.

QUESTIONS AND PROBLEMS

16.1   From the stress–strain data for polymethyl methacrylate shown in Figure 16.2, determine the modulus of elasticity and tensile strength at room temperature [20°C (68°F)], and compare these values with those given in Table 16.1.

16.2   In your own words, describe the mechanisms by which **(a)** semicrystalline polymers plastically deform and **(b)** elastomers elastically deform.

16.3   Briefly explain how each of the following influences the mechanical strength of a semicrystalline polymer and why: **(a)** molecular weight; **(b)** degree of crystallinity; and **(c)** extent of crosslinking.

16.4   Normal butane and isobutane have boiling temperatures of -0.5 and $-12.3°C$ (31.1 and 9.9°F), respectively. Briefly explain this behavior on the basis of their molecular structures, as presented in Section 15.2.

16.5   For each of the following pairs of polymers, decide which is more likely to have the greater tensile strength, and then cite the reasons for your choice:
(a) Lightly crosslinked polyethylene; network phenol-formaldehyde.
(b) Lightly crosslinked polyvinyl chloride; branched polyvinyl chloride.
(c) 95% crystalline and linear PTFE having a number-average molecular weight of 650,000 g/mol; 80% crystalline and linear PTFE having a number-average molecular weight of 500,000 g/mol.
(d) Atactic polypropylene having a weight-average molecular weight of 750,000 g/mol; isotactic polypropylene having a weight-average molecular weight of 750,000 g/mol.
(e) Alternating styrene–butadiene copolymer; graft styrene–butadiene copolymer.

16.6   Would you expect the tensile strength of polychlorotrifluoroethylene to be greater than, the same as, or less than that of a polytetrafluoroethylene specimen having the same molecular weight and degree of crystallinity? Why?

16.7   For each of the following pairs of polymers, plot and label schematic stress–strain curves on the same graph (i.e., make separate plots for parts a, b, and c).
(a) Polyisoprene having a number-average molecular weight of 100,000 g/mol and 10% of available sites crosslinked; polyisoprene having a number-average molecular weight of 100,000 g/mol and 20% of available sites crosslinked.
(b) Syndiotactic polypropylene having a weight-average molecular weight of 100,000 g/mol; atactic polypropylene having a weight-average molecular weight of 75,000 g/mol.
(c) Branched polyethylene having a number-average molecular weight of 90,000 g/mol; heavily crosslinked polyethylene having a number-average molecular weight of 90,000 g/mol.

16.8 When citing the ductility as percent elongation for semicrystalline polymers, it is not necessary to specify the specimen gauge length, as is the case with metals. Why is this so?

16.9 **(a)** In your own words describe, from a molecular perspective, the melting phenomenon for crystalline polymers.
(b) Briefly explain why, for relatively low molecular weights, the melting temperature of a polymer increases with increasing molecular weight.
(c) In your own words describe, from a molecular perspective, the glass transition phenomenon for amorphous polymers.

16.10 **(a)** Make a comparison of the glass transition temperatures for polyethylene (low density), polypropylene, and polystyrene (Table 16.2). **(b)** Explain these relative values on the basis of molecular structure and chemistry.

16.11 For each of the following pairs of polymers, plot and label schematic specific volume-versus-temperature curves on the same graph (i.e., make separate plots for parts a, b, and c).
(a) Linear polyethylene with a weight-average molecular weight of 75,000 g/mol; branched polyethylene with a weight-average molecular weight of 50,000 g/mol.
(b) Spherulitic polyvinyl chloride, 50% crystallinity, and degree of polymerization 5000; spherulitic polypropylene, 50% crystallinity, and degree of polymerization 10,000.
(c) Totally amorphous polystyrene having a degree of polymerization of 7000; totally amorphous polypropylene having a degree of polymerization of 7000.

16.12 For each of the following pairs of polymers, decide which polymer is more likely to have the greater melting temperature, and cite the reasons for your choice:
(a) Branched polyethylene having a number-average molecular weight of 850,000 g/mol; linear polyethylene having a number-average molecular weight of 850,000 g/mol.
(b) Polytetrafluoroethylene having a density of 2.14 g/cm^3 and a weight-average molecular weight of 600,000 g/mol; PTFE having a density of 2.20 g/cm^3 and a weight-average molecular weight of 600,000 g/mol.
(c) Linear and syndiotactic polyvinyl chloride having a number-average molecular weight of 500,000 g/mol; linear polyethylene having a number-average molecular weight of 225,000 g/mol.
(d) Linear and syndiotactic polypropylene having a weight-average molecular weight of 750,000 g/mol; linear and atactic polypropylene having a weight-average molecular weight of 500,000 g/mol.

16.13 Make a schematic plot showing how the modulus of elasticity of an amorphous polymer depends on the glass transition temperature. Assume that molecular weight is held constant.

16.14 Name the following polymer(s) that would be suitable for the fabrication of cups to contain hot coffee: polyethylene, polypropylene, polyvinyl chloride, PET polyester, and polycarbonate. Why?

16.15 Of those polymers listed in Table 16.2, which polymer(s) would be best suited for use as ice cube trays? Why?

16.16 Make comparisons of thermoplastic and thermosetting polymers **(a)** on the basis of mechanical characteristics upon heating, and **(b)** according to possible molecular structures.

16.17 Some of the polyesters may be either thermoplastic or thermosetting. Suggest one reason for this.

16.18 **(a)** Is it possible to grind up and reuse phenol-formaldehyde? Why or why not? **(b)** Is it possible to grind up and reuse polypropylene? Why or why not?

16.19 In your own words, briefly describe the phenomenon of viscoelasticity.

16.20 For some viscoelastic polymers that are subjected to stress relaxation tests, the stress decays with time according to

$$\sigma(t) = \sigma(0) \exp\left(-\frac{t}{\tau}\right) \qquad (16.9)$$

where $\sigma(t)$ and $\sigma(0)$ represent the time-dependent and initial (i.e., time = 0) stresses, respectively, and t and τ denote elapsed time and the relaxation time; τ is a time-independent constant characteristic of the material. A specimen of some viscoelastic polymer the stress relaxation of which obeys Equation 16.9 was suddenly pulled in tension to a measured strain of 0.5; the stress necessary to maintain this constant strain was measured as a function of time. Determine $E_r(10)$ for this material if the initial stress level was 500 psi (3.5 MPa) which dropped to 70 psi (0.5 MPa) after 30 s.

16.21 In Figure 16.18 the logarithm of $E_r(t)$ versus the logarithm of time is plotted

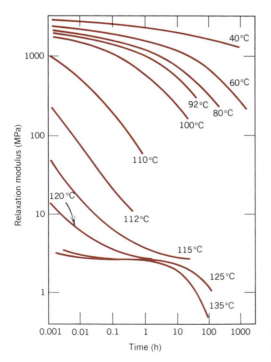

Figure 16.18 Logarithm of relaxation modulus versus logarithm of time for polymethyl methacrylate between 40 and 135°C. [From J. R. McLoughlin and A. V. Tobolsky, *J. Colloid Sci.*, **7**, 555 (1952). Reprinted with permission.]

for PMMA at a variety of temperatures. Make a plot of log $E_r(10)$ versus temperature and then estimate the T_g.

16.22 On the same plot, superimpose and label schematic log $E_r(10)$–temperature curves for two elastomeric materials, one having a higher degree of crosslinking than the other.

16.23 On the basis of the curves in Figure 16.8, sketch schematic strain–time plots for the following polystyrene materials at the specified temperatures:
(a) Crystalline at 70°C.
(b) Amorphous at 180°C.
(c) Crosslinked at 180°C.
(d) Amorphous at 100°C.

16.24 **(a)** Contrast the manner in which stress relaxation and viscoelastic creep tests are conducted. **(b)** For each of these tests, cite the experimental parameter of interest and how it is determined.

16.25 Make two schematic plots of the logarithm of relaxation modulus versus temperature for an amorphous polymer (curve C in Figure 16.11).
(a) On one of these plots demonstrate how the behavior changes with increasing molecular weight.
(b) On the other plot, indicate the change in behavior with increasing crosslinking.

16.26 For thermoplastic polymers, cite five factors that favor brittle fracture.

16.27 **(a)** Compare the fatigue limits for PMMA (Figure 16.14) and the steel alloy for which fatigue data are given in Problem 8.32. **(b)** Compare the fatigue strengths at 10^6 cycles for nylon 6 (Figure 16.14) and 2014-T6 aluminum (Figure 8.38).

16.28 Cite the primary differences between addition and condensation polymerization techniques.

16.29 Cite whether the molecular weight of a polymer that is synthesized by addition polymerization is relatively high, medium, or relatively low for the following situations:
(a) Rapid initiation, slow propagation, and rapid termination.
(b) Slow initiation, rapid propagation, and slow termination.
(c) Rapid initiation, rapid propagation, and slow termination.
(d) Slow initiation, slow propagation, and rapid termination.

16.30 **(a)** How much ethylene glycol must be added to 20.0 kg adipic acid to produce a linear chain structure of polyester according to Equation 16.7? **(b)** What is the mass of the resulting polyester?

16.31 Nylon 6,6 may be formed by means of a condensation polymerization reaction in which hexamethylene diamine $[NH_2-(CH_2)_6-NH_2]$ and adipic acid react with one another with the formation of water as a by-product. Write out this reaction in the manner of Equation 16.7.

16.32 It is desired to produce nylon 6,6 by condensation polymerization using hexamethylene diamine and adipic acid as described in Problem 16.31. What masses of these two components are necessary to yield 20 kg of completely linear nylon 6,6?

16.33 (a) Why must the vapor pressure of a plasticizer be relatively low? **(b)** How will the crystallinity of a polymer be affected by the addition of a plasticizer? Why? **(c)** Is it possible for a crosslinked polymer to be plasticized? Why or why not? **(d)** How does the addition of a plasticizer influence the tensile strength of a polymer? Why?

16.34 What is the distinction between dye and pigment colorants?

16.35 Cite four factors that determine what fabrication technique is used to form polymeric materials.

16.36 Explain why molding is more time consuming for thermoplastics than for thermosets.

16.37 Contrast compression, injection, and transfer molding techniques that are used to form plastic materials.

16.38 Ten kg polychloroprene is vulcanized with 0.72 kg sulfur. What fraction of the possible crosslink sites is bonded to sulfur, assuming a single sulfur atom participates in each bond?

16.39 Compute the weight percent sulfur that must be added to completely crosslink an alternating styrene-butadiene copolymer.

16.40 The vulcanization of polyisoprene is accomplished with sulfur atoms according to Equation 16.8. If 32 wt% sulfur is combined with polyisoprene, how many crosslinks will there be per isoprene mer?

16.41 (a) For the vulcanization of polyisoprene, how much sulfur (wt%) must be combined to ensure that 5% of possible sites will be crosslinked? **(b)** Is it possible to have more combined sulfur atoms than there are crosslink sites for polyisoprene? Explain.

16.42 It is desired that some rubber component in its final form be vulcanized. Should vulcanization be carried out prior or subsequent to the forming operation? Why?

16.43 List the two molecular characteristics that are essential for elastomers.

16.44 Which of the following would you expect to be elastomers and which thermosetting polymers at room temperature? Justify each choice.
(a) Linear and crystalline polyethylene.
(b) Phenol-formaldehyde having a network molecular structure.
(c) Heavily crosslinked polyisoprene having a glass transition temperature of 50°C (122°F).
(d) Lightly crosslinked polyisoprene having a glass transition temperature of −60°C (−76°F).
(e) Linear and partially amorphous polyvinyl chloride.

16.45 In terms of molecular structure, explain why phenol-formaldehyde (Bakelite) will not be an elastomer.

16.46 Demonstrate, in a manner similar to Equation 16.8, how vulcanization may occur in a butadiene rubber.

16.47 During the winter months the temperature in some parts of Alaska may go as low as −55°C (−65°F). Of the elastomers natural isoprene, styrene-butadiene, acrylonitrile-butadiene, chloroprene, and polysiloxane, which would be suitable for automobile tires under these conditions? Why?

16.48 Briefly explain the difference in molecular chemistry between silicone polymers and other polymeric materials.

16.49 Silicone polymers may be prepared to exist as liquids at room temperature. Cite differences in molecular structure between them and the silicone elastomers.

16.50 Why must fiber materials that are melt spun and then drawn be thermoplastic? Cite two reasons.

16.51 List two important characteristics for polymers that are to be used in fiber applications.

16.52 Cite five important characteristics for polymers that are to be used in thin film applications.

16.53 Which of the following polyethylene thin films would have the better mechanical characteristics: (1) formed by blowing, or (2) formed by extrusion and then rolled? Why?

CHAPTER 17

COMPOSITES

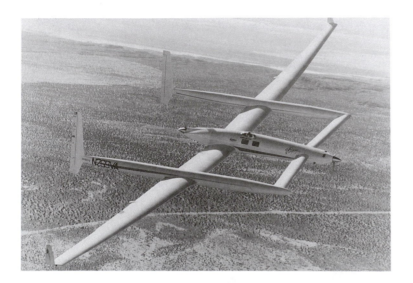

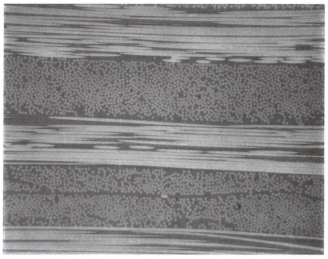

Top: The Voyager aircraft, which made a 25,000 mile nonstop flight around the world without midair refueling. High-strength, low-density structural members of the Voyager are constructed from a series of crossplies consisting of graphite fibers that are aligned and embedded within an epoxy matrix. (Photograph courtesy of Hercules Inc.) Bottom: This photomicrograph shows a section taken through one of these crossplies. In the top, center, and bottom ply regions the fibers are aligned parallel to the plane of the page. The fibers run approximately perpendicular to the page plane in the other two plies, and their circular cross sections may be discerned. 100 ×. (Photograph courtesy of S. Y. Gweon and W. D. Bascom.)

17.1 INTRODUCTION

Many of our modern technologies require materials with unusual combinations of properties that cannot be met by the conventional metal alloys, ceramics, and polymeric materials. This is especially true for materials that are needed for aerospace, underwater, and transportation applications. For example, aircraft engineers are increasingly searching for structural materials that have low densities, are strong, stiff, and abrasion and impact resistant, and are not easily corroded. This is a rather formidable combination of characteristics. Frequently, strong materials are relatively dense; also, increasing the strength or stiffness generally results in a decrease in impact strength.

Material property combinations and ranges have been, and are yet being, extended by the development of composite materials. Generally speaking, a composite is considered to be any multiphase material that exhibits a significant proportion of the properties of both constituent phases such that a better combination of properties is realized. According to this **principle of combined action,** better property combinations are fashioned by the judicious combination of two or more distinct materials. Property trade-offs are also made for many composites.

Composites of sorts have already been discussed; these include multiphase metal alloys, ceramics, and polymers. For example, pearlitic steels (Section 9.14) have a microstructure consisting of alternating layers of α-ferrite and cementite (Figure 9.23). The ferrite phase is soft and ductile, whereas cementite is hard and very brittle. The combined mechanical characteristics of the pearlite (reasonably high ductility and strength) are superior to those of either of the constituent phases. There are also a number of composites that occur in nature. For example, wood consists of strong and flexible cellulose fibers surrounded and held together by a stiffer material called lignin. Also, bone is a composite of the strong yet soft protein collagen and the hard, brittle mineral apatite.

A composite, in the present context, is a multiphase material that is artificially made, as opposed to one that occurs or forms naturally. In addition, the constituent phases must be chemically dissimilar and separated by a distinct interface. Thus most metallic alloys and many ceramics do not fit this definition because their multiple phases are formed as a consequence of natural phenomena.

In designing composite materials, scientists and engineers have ingeniously combined various metals, ceramics, and polymers to produce a new generation of extraordinary materials. Most composites have been created to improve combinations of mechanical characteristics such as stiffness, toughness, and ambient and high-temperature strength.

Many composite materials are composed of just two phases; one is termed the **matrix,** which is continuous and surrounds the other phase, often called the **dispersed phase.** The properties of composites are a function of the properties of the constituent phases, their relative amounts, and the geometry of the dispersed phase. "Dispersed phase geometry" in this context means the shape of the particles and the particle size, distribution, and orientation; these characteristics are represented in Figure 17.1.

One simple scheme for the classification of composite materials is shown in Figure 17.2, which consists of three main divisions—particle-reinforced, fiber-reinforced, and structural composites; also, at least two subdivisions exist for each. The dispersed phase for particle-reinforced composites is equiaxed (i.e., particle dimensions are approximately the same in all directions); for fiber-reinforced composites, the dispersed

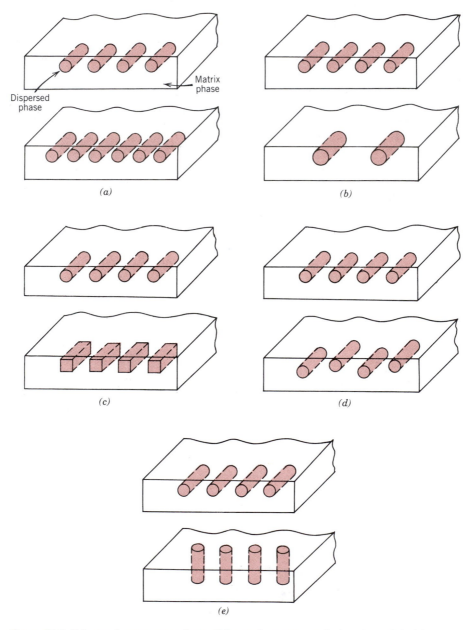

Figure 17.1 Schematic representations of the various geometrical and spatial character-istics of particles of the dispersed phase that may influence the properties of composites: (a) concentration, (b) size, (c) shape, (d) distribution, and (e) orientation. (From Richard A. Flinn and Paul K. Trojan, *Engineering Materials and Their Applications,* Third Edition. Copyright © 1986 by Houghton Mifflin Company. Used with permission.)

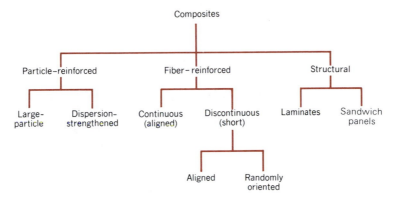

Figure 17.2 A classification scheme for the various composite types discussed in this chapter.

phase has the geometry of a fiber (i.e., a large length-to-diameter ratio). Structural composites are combinations of composites and homogeneous materials. The discussion of the remainder of this chapter will be organized according to this classification scheme.

PARTICLE-REINFORCED COMPOSITES

As noted in Figure 17.2, **large-particle** and **dispersion-strengthened composites** are the two subclassifications of particle-reinforced composites. The distinction between these is based upon reinforcement or strengthening mechanism. The term "large" is used to indicate that particle–matrix interactions cannot be treated on the atomic or molecular level; rather, continuum mechanics is used. For most of these composites, the particulate phase is harder and stiffer than the matrix. These reinforcing particles tend to restrain movement of the matrix phase in the vicinity of each particle. In essence, the matrix transfers some of the applied stress to the particles, which bear a fraction of the load. The degree of reinforcement or improvement of mechanical behavior depends on strong bonding at the matrix–particle interface.

For dispersion-strengthened composites, particles are normally much smaller, having diameters between 0.01 and 0.1 μm (10 and 100 nm). Particle–matrix interactions that lead to strengthening occur on the atomic or molecular level. The mechanism of strengthening is similar to that for precipitation hardening discussed in Section 11.8. Whereas the matrix bears the major portion of an applied load, the small dispersed particles hinder or impede the motion of dislocations. Thus plastic deformation is restricted such that yield and tensile strengths, as well as hardness, improve.

17.2 LARGE-PARTICLE COMPOSITES

Some polymeric materials to which fillers have been added (Section 16.11) are really large-particle composites. Again, the fillers modify or improve the properties of the material and/or replace some of the polymer volume with a less expensive material—the filler.

Another familiar large-particle composite is concrete, being composed of cement (the matrix), and sand and gravel (the particulates). Concrete is the discussion topic of the succeeding section.

Particles can have quite a variety of geometries, but they should be of approximately the same dimension in all directions (equiaxed). For effective reinforcement, the particles should be small and evenly distributed throughout the matrix. Furthermore, the volume fraction of the two phases influences the behavior; mechanical properties are enhanced with increasing particulate content. Two mathematical expressions have been formulated for the dependence of the elastic modulus on the volume fraction of the constituent phases for a two-phase composite. These **rule of mixture** equations predict that the elastic modulus should fall between an upper bound represented by

$$E_c = E_m V_m + E_p V_p \tag{17.1}$$

and a lower bound, or limit,

$$E_c = \frac{E_m E_p}{V_m E_p + V_p E_m} \tag{17.2}$$

In these expressions, E and V denote the elastic modulus and volume fraction, respectively, whereas the subscripts c, m, and p represent composite, matrix, and particulate phases. Figure 17.3 plots upper- and lower-bound E_c-versus-V_p curves for a copper–tungsten composite, in which tungsten is the particulate phase; experimental

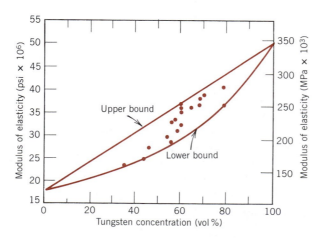

Figure 17.3 Modulus of elasticity versus volume percent tungsten for a composite of tungsten particles dispersed within a copper matrix. Upper and lower bounds are according to Equations 17.1 and 17.2; experimental data points are included. (From R. H. Krock, *ASTM Proceedings,* Vol. 63, 1963. Copyright ASTM, 1916 Race Street, Philadelphia, PA 19103. Reprinted with permission.)

data points fall between the two curves. Equations analogous to 17.1 and 17.2 for fiber-reinforced composites are derived in Section 17.5.

Large-particle composites are utilized with all three material types (metals, polymers, and ceramics). The **cermets** are examples of ceramic–metal composites. The most common cermet is the cemented carbide, which is composed of extremely hard particles of a refractory carbide ceramic such as tungsten carbide (WC) or titanium carbide (TiC), embedded in a matrix of a metal such as cobalt or nickel. These composites are utilized extensively as cutting tools for hardened steels. The hard carbide particles provide the cutting surface but, being extremely brittle, are not themselves capable of withstanding the cutting stresses. Toughness is enhanced by their inclusion in the ductile metal matrix, which isolates the carbide particles from one another and prevents particle-to-particle crack propagation. Both matrix and particulate phases are quite refractory, to withstand the high temperatures generated by the cutting action on materials that are extremely hard. No single material could possibly provide the combination of properties possessed by a cermet. Relatively large volume fractions of the particulate phase may be utilized, often exceeding 90 vol%; thus the abrasive action of the composite is maximized. A photomicrograph of a WC–Co cemented carbide is shown in Figure 17.4.

Both elastomers and plastics are frequently reinforced with various particulate materials. Our use of many of the modern rubbers would be severely restricted without reinforcing particulate materials such as carbon black. Carbon black consists of very small and essentially spherical particles of carbon, produced by the combustion of natural gas or oil in an atmosphere that has only a limited air supply. When added to vulcanized rubber, this extremely inexpensive material enhances tensile strength, toughness, and tear and abrasion resistance. Automobile tires contain on the order of 15 to 30 vol% of carbon black. For the carbon black to provide significant reinforcement, the particle size must be extremely small, with diameters between 20 and

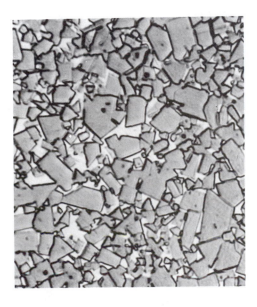

Figure 17.4 Photomicrograph of a WC–Co cemented carbide. Light areas are the cobalt matrix; dark regions, the particles of tungsten carbide. 100×. (Courtesy of Carboloy Systems Department, General Electric Company.)

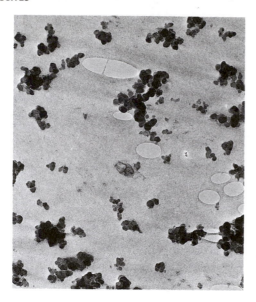

Figure 17.5 Electron micrograph showing the spherical reinforcing carbon black particles in a synthetic rubber tire tread compound. The areas resembling water marks are tiny air pockets in the rubber. 80,000×. (Courtesy of Goodyear Tire & Rubber Company.)

50 nm; also, the particles must be evenly distributed throughout the rubber and must form a strong adhesive bond with the rubber matrix. Particle reinforcement using other materials (e.g., silica) is much less effective because this special interaction between the rubber molecules and particle surfaces does not exist. Figure 17.5 is an electron micrograph of a carbon black-reinforced rubber.

Concrete

Concrete is a common large-particle composite in which both matrix and dispersed phases are ceramic materials. Since the terms "concrete" and "cement" are sometimes incorrectly interchanged, perhaps it is appropriate to make a distinction between them. In a broad sense, concrete implies a composite material consisting of an aggregate of particles that are bound together in a solid body by some type of binding medium, that is, a cement. The two most familiar concretes are those made with portland and asphaltic cements, where the aggregate is gravel and sand. Asphaltic concrete is widely used primarily as a paving material, whereas portland cement concrete is employed extensively as a structural building material. Only the latter is treated in this discussion.

Portland Cement Concrete.   The ingredients for this concrete are portland cement, a fine aggregate (sand), a coarse aggregate (gravel), and water. The process by which portland cement is produced and the mechanism of setting and hardening were discussed very briefly in Section 14.16. The aggregate particles act as a filler material to reduce the overall cost of the concrete product because they are cheap, whereas cement is relatively expensive. To achieve the optimum strength and workability of a concrete mixture, the ingredients must be added in the correct proportions. Dense packing of the aggregate and good interfacial contact are achieved by having particles of two different sizes; the fine particles of sand should fill the void spaces between the gravel particles. Ordinarily these aggregates comprise between 60 and 80% of the total volume. The amount of cement–water paste should be sufficient to coat all the sand and gravel particles, otherwise the cementitious bond will be incomplete. Furthermore, all the constituents should be thoroughly mixed. Complete bonding between cement and the aggregate particles is contingent upon the addition of the correct quantity of water. Too little water leads to incomplete bonding, and too much results in excessive porosity; in either case the final strength is less than the optimum.

The character of the aggregate particles is an important consideration. In particular, the size distribution of the aggregates influences the amount of cement–water paste required. Also, the surfaces should be clean and free from clay and silt, which prevent the formation of a sound bond at the particle surface.

Portland cement concrete is a major material of construction, primarily because it can be poured in place and hardens at room temperature, and even when submerged in water. However, as a structural material, there are some limitations and disadvantages. Like most ceramics, portland cement concrete is relatively weak and extremely brittle; its tensile strength is approximately 10 to 15 times smaller than its compressive strength. Also, large concrete structures can experience considerable thermal expansion and contraction with temperature fluctuations. In addition, water penetrates into external pores, which can cause severe cracking in cold weather as a consequence of freeze–thaw cycles. Most of these inadequacies may be eliminated or at least improved by reinforcement and/or the incorporation of additives.

Reinforced Concrete.   The strength of portland cement concrete may be increased by additional reinforcement. This is usually accomplished by means of steel rods, wires, bars (rebar), or mesh, which are embedded into the fresh and uncured concrete. Thus the reinforcement renders the hardened structure capable of supporting greater tensile, compressive, and shear stresses. Even if cracks develop in the concrete, considerable reinforcement is maintained.

Steel serves as a suitable reinforcement material because its coefficient of thermal expansion is nearly the same as that of concrete. In addition, steel is not rapidly corroded in the cement environment, and a relatively strong adhesive bond is formed between it and the cured concrete. This adhesion may be enhanced by the incorporation of contours into the surface of the steel member, which permits a greater degree of mechanical interlocking.

Portland cement concrete may also be reinforced by mixing into the fresh concrete fibers of a high-modulus material such as glass, steel, nylon, and polyethylene. Care must be exercised in utilizing this type of reinforcement, since some fiber materials experience rapid deterioration when exposed to the cement environment.

Still another reinforcement technique for strengthening concrete involves the introduction of residual compressive stresses into the structural member; the resulting material is called **prestressed concrete.** This method utilizes one characteristic of brittle ceramics—namely, that they are stronger in compression than in tension. Thus, to fracture a prestressed concrete member, the magnitude of the precompressive stress must be exceeded by an applied tensile stress.

In one such prestressing technique high-strength steel wires are positioned inside the empty molds and stretched with a high tensile force, which is maintained constant. After the concrete has been placed and allowed to harden, the tension is released. As the wires contract, they put the structure in a state of compression because the stress is transmitted to the concrete via the concrete–wire bond that is formed.

Another technique is also utilized in which stresses are applied after the concrete hardens; it is appropriately called *posttensioning*. Sheet metal or rubber tubes are situated inside and pass through the concrete forms, around which the concrete is cast. After the cement has hardened, steel wires are fed through the resulting holes, and tension is applied to the wires by means of jacks attached and abutted to the faces of the structure. Again, a compressive stress is imposed on the concrete piece, this time by the jacks. Finally, the empty spaces inside the tubing are filled with a grout to protect the wire from corrosion.

Concrete that is prestressed should be of a high quality, having a low shrinkage and a low creep rate. Prestressed concretes, usually prefabricated, are commonly used for highway and railway bridges.

17.3 DISPERSION-STRENGTHENED COMPOSITES

Metals and metal alloys may be strengthened and hardened by the uniform dispersion of several volume percent of fine particles of a very hard and inert material. The dispersed phase may be metallic or nonmetallic; oxide materials are often used. Again, the strengthening mechanism involves interactions between the particles and dislocations within the matrix, as with precipitation hardening. The dispersion strengthening effect is not as pronounced as with precipitation hardening; however, the strengthening is retained at elevated temperatures and for extended time periods because the dispersed particles are chosen to be unreactive with the matrix phase. For precipitation-hardened alloys, the increase in strength may disappear upon heat treatment as a consequence of precipitate growth or dissolution of the precipitate phase.

The high-temperature strength of nickel alloys may be enhanced significantly by the addition of about 3 vol% of thoria (ThO_2) as finely dispersed particles; this material is known as thoria-dispersed (or TD) nickel. The same effect is produced in the aluminum–aluminum oxide system. A very thin and adherent alumina coating is caused to form on the surface of extremely small (0.1 to 0.2 μm thick) flakes of aluminum, which are dispersed within an aluminum metal matrix; this material is termed sintered aluminum powder (SAP).

FIBER-REINFORCED COMPOSITES

Technologically, the most important composites are those in which the dispersed phase is in the form of a fiber. Design goals of fiber-reinforced composites often include high strength and/or stiffness on a weight basis. These characteristics are expressed in terms of **specific strength** and **specific modulus** parameters, which correspond, respectively, to the ratios of tensile strength to specific gravity and modulus of elasticity to specific gravity. Fiber-reinforced composites with exceptionally high specific strengths and moduli have been produced that utilize low-density fiber and matrix materials.

As noted in Figure 17.2, fiber-reinforced composites are subclassified by fiber length. For short fiber, the fibers are too short to produce a significant improvement in strength.

17.4 INFLUENCE OF FIBER LENGTH

The mechanical characteristics of a fiber-reinforced composite depend not only on the properties of the fiber, but also on the degree to which an applied load is transmitted to the fibers by the matrix phase. Important to the extent of this load transmittance is the magnitude of the interfacial bond between the fiber and matrix phases. Under an applied stress, this fiber–matrix bond ceases at the fiber ends, yielding a matrix deformation pattern as shown schematically in Figure 17.6; in other words, there is no load transmittance from the matrix at each fiber extremity.

Some critical fiber length is necessary for effective strengthening and stiffening of the composite material. This critical length l_c is dependent on the fiber diameter d and its ultimate (or tensile) strength σ_f, and on the fiber–matrix bond strength (or the shear yield strength of the matrix) τ_c according to

$$l_c = \frac{\sigma_f d}{\tau_c} \tag{17.3}$$

For a number of glass and carbon fiber–matrix combinations, this critical length is on the order of 1 mm, which ranges between 20 and 150 times the fiber diameter.

When a stress equal to σ_f is applied to a fiber having just this critical length, the stress–position profile shown in Figure 17.7a results; that is, the maximum fiber load is achieved only at the axial center of the fiber. As fiber length l increases, the fiber reinforcement becomes more effective; this is demonstrated in Figure 17.7b, a stress–axial position profile for $l > l_c$ when the applied stress is equal to the fiber strength. Figure 17.7c shows the stress–position profile for $l < l_c$.

Fibers for which $l \gg l_c$ (normally $l > 15 l_c$) are termed *continuous; discontinuous* or *short fibers* have lengths shorter than this. For discontinuous fibers of lengths significantly less than l_c, the matrix deforms around the fiber such that there is virtually no stress transference and little reinforcement by the fiber. These are essentially the particulate composites as described above. To affect a significant improvement in strength of the composite, the fibers must be continuous.

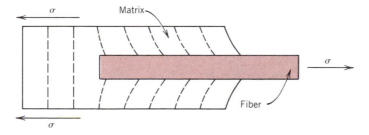

Figure 17.6 The deformation pattern in the matrix surrounding a fiber that is subjected to an applied tensile load.

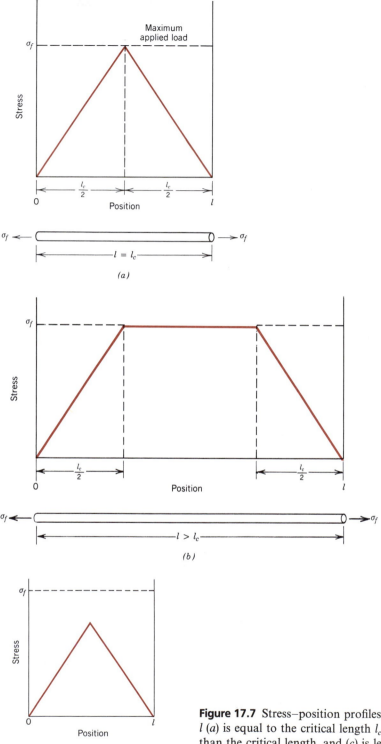

Figure 17.7 Stress–position profiles when fiber length l (a) is equal to the critical length l_c, (b) is greater than the critical length, and (c) is less than the critical length for a fiber-reinforced composite that is subjected to a tensile stress equal to the fiber tensile strength σ_f.

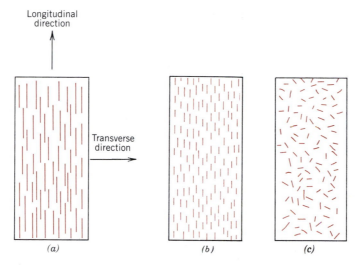

Figure 17.8 Schematic representations of (*a*) continuous and aligned, (*b*) discontinuous and aligned, and (*c*) discontinuous and randomly oriented fiber-reinforced composites.

17.5 INFLUENCE OF FIBER ORIENTATION AND CONCENTRATION

The arrangement or orientation of the fibers relative to one another, the fiber concentration, and the distribution all have a significant influence on the strength and other properties of fiber-reinforced composites. With respect to orientation, two extremes are possible: (1) a parallel alignment of the longitudinal axis of the fibers in a single direction, and (2) a totally random alignment. Continuous fibers are normally aligned (Figure 17.8*a*), whereas discontinuous fibers may be aligned (Figure 17.8*b*), randomly oriented (Figure 17.8*c*), or partially oriented. Better overall composite properties are realized when the fiber distribution is uniform.

Continuous and Aligned Fiber Composites

Longitudinal Loading. The properties of a composite having its fibers aligned are highly anisotropic, that is, dependent on the direction in which they are measured. Let us first consider the deformation of this type of composite in which a stress is applied along the direction of alignment, the **longitudinal direction,** as indicated in Figure 17.8*a*. Assume also that the fiber–matrix interfacial bond is very good, such that deformation of both matrix and fibers is the same (an *isostrain* situation). Under these conditions, the total load sustained by the composite F_c is equal to the loads carried by the matrix phase F_m and the fiber phase F_f, or

$$F_c = F_m + F_f \tag{17.4}$$

From the definition of stress, Equation 6.1, $F = \sigma A$; and thus expressions for F_c, F_m, and F_f in terms of their respective stresses (σ_c, σ_m, and σ_f) and cross-sectional areas (A_c, A_m, and A_f) are possible. Substitution of these into Equation 17.4 yields

$$\sigma_c A_c = \sigma_m A_m + \sigma_f A_f \tag{17.5}$$

and then, dividing through by the total cross-sectional area of the composite, A_c, we have

$$\sigma_c = \sigma_m \frac{A_m}{A_c} + \sigma_f \frac{A_f}{A_c} \tag{17.6}$$

where A_m/A_c and A_f/A_c are the area fractions of the matrix and fiber phases, respectively. If the composite, matrix, and fiber phase lengths are all equal, A_m/A_c is equivalent to the volume fraction of the matrix, V_m; and likewise for the fibers, $V_f = A_f/A_c$. Equation 17.6 now becomes

$$\sigma_c = \sigma_m V_m + \sigma_f V_f \tag{17.7}$$

The previous assumption of an isostrain state means that

$$\epsilon_c = \epsilon_m = \epsilon_f \tag{17.8}$$

and when each term in Equation 17.7 is divided by its respective strain,

$$\frac{\sigma_c}{\epsilon_c} = \frac{\sigma_m}{\epsilon_m} V_m + \frac{\sigma_f}{\epsilon_f} V_f \tag{17.9}$$

Furthermore, if composite, matrix, and fiber deformations are all elastic, then $\sigma_c/\epsilon_c = E_c$, $\sigma_m/\epsilon_m = E_m$, and $\sigma_f/\epsilon_f = E_f$, the E's being the moduli of elasticity for the respective phases. Substitution into Equation 17.9 yields

$$E_c = E_m V_m + E_f V_f \tag{17.10a}$$

or

$$E_c = E_m(1 - V_f) + E_f V_f \tag{17.10b}$$

since the composite consists of only matrix and fiber phases; that is, $V_m + V_f = 1$.

Thus the modulus of elasticity of a continuous and aligned fiber-reinforced composite *in the direction of alignment* is equal to the volume-fraction weighted average of the moduli of elasticity of the fiber and matrix phases. Other properties, including tensile strength, also have this dependence on volume fractions. Equation 17.10a is the fiber analogue of Equation 17.1, the upper bound for particle-reinforced composites.

It can also be shown, for longitudinal loading, that the ratio of the load carried by the fibers to that carried by the matrix is

$$\frac{F_f}{F_m} = \frac{E_f V_f}{E_m V_m} \tag{17.11}$$

The demonstration is left as a homework problem.

EXAMPLE PROBLEM 17.1

A continuous and aligned glass-reinforced composite consists of 40 vol% of glass fibers having a modulus of elasticity of 10×10^6 psi (69×10^3 MPa) and 60 vol% of a polyester resin that, when hardened, displays a modulus of 5.0×10^5 psi (3.4×10^3 MPa).

(a) Compute the modulus of elasticity of this composite in the longitudinal direction.

(b) If the cross-sectional area is 0.4 in.^{2} (258 mm^2) and a stress of 7000 psi (48.3 MPa) is applied in this longitudinal direction, compute the magnitude of the load carried by each of the fiber and matrix phases.

(c) Determine the strain that is sustained by each phase when the stress in part b is applied.

(d) Assuming tensile strengths of 500,000 and 10,000 psi (3.5×10^3 and 69 MPa), respectively, for glass fibers and polyester resin, determine the longitudinal tensile strength of this fiber composite.

SOLUTION

(a) The modulus of elasticity of the composite is calculated using Equation 17.10a:

$$E_c = (5.0 \times 10^5 \text{ psi})(0.6) + (10 \times 10^6 \text{ psi})(0.4)$$
$$= 4.3 \times 10^6 \text{ psi } (3 \times 10^4 \text{ MPa})$$

(b) To solve this portion of the problem, first find the ratio of fiber load to matrix load, using Equation 17.11; thus

$$\frac{F_f}{F_m} = \frac{(10 \times 10^6 \text{ psi})(0.4)}{(5 \times 10^5 \text{ psi})(0.6)} = 13.3$$

or $F_f = 13.3 F_m$.

In addition, the total force sustained by the composite F_c may be computed from the applied stress σ and total composite cross-sectional area A_c according to

$$F_c = A_c \sigma = (0.4 \text{ in.}^2)(7000 \text{ psi})$$
$$= 2800 \text{ lb}_f \ (12{,}400 \text{ N})$$

However, this total load is just the sum of the loads carried by fiber and matrix phases, that is,

$$F_c = F_f + F_m = 2800 \text{ lb}_f \ (12{,}400 \text{ N})$$

Substitution for F_f from the above yields

$$13.3 F_m + F_m = 2800 \text{ lb}_f$$

or

$$F_m = 195 \text{ lb}_f \ (870 \text{ N})$$

whereas

$$F_f = F_c - F_m = 2800 \text{ lb}_f - 195 \text{ lb}_f = 2605 \text{ lb}_f \ (11{,}530 \text{ N})$$

Thus the fiber phase supports the vast majority of the applied load.

(c) The stress for both fiber and matrix phases must first be calculated. Then, by using the elastic modulus for each (from part a), the strain values may be determined.

For stress calculations, phase cross-sectional areas are necessary:

$$A_m = V_m A_c = (0.6)(0.4 \text{ in.}^2) = 0.24 \text{ in.}^2 \ (155 \text{ mm}^2)$$

and

$$A_f = V_f A_c = (0.4)(0.4 \text{ in.}^2) = 0.16 \text{ in.}^2 \, (103 \text{ mm}^2)$$

Thus

$$\sigma_m = \frac{F_m}{A_m} = \frac{195 \text{ lb}_f}{0.24 \text{ in.}^2} = 812.5 \text{ psi} \, (5.6 \text{ MPa})$$

$$\sigma_f = \frac{F_f}{A_f} = \frac{2605 \text{ lb}_f}{0.16 \text{ in.}^2} = 16{,}280 \text{ psi} \, (112.3 \text{ MPa})$$

Finally, strains are computed as

$$\epsilon_m = \frac{\sigma_m}{E_m} = \frac{812.5 \text{ psi}}{5 \times 10^5 \text{ psi}} = 1.63 \times 10^{-3}$$

$$\epsilon_f = \frac{\sigma_f}{E_f} = \frac{16{,}280 \text{ psi}}{10 \times 10^6 \text{ psi}} = 1.63 \times 10^{-3}$$

Therefore, strains for both matrix and fiber phases are identical, which they should be, according to Equation 17.8 in the previous development.

(d) For tensile strength *TS*, we write

$$(TS)_c = (TS)_m V_m + (TS)_f V_f \tag{17.12}$$

which, for this particular composite (in the longitudinal direction), is

$$(TS)_c = (10{,}000 \text{ psi})(0.60) + (500{,}000 \text{ psi})(0.40)$$
$$= 206{,}000 \text{ psi} \, (1{,}420 \text{ MPa})$$

Transverse Loading.   A continuous and oriented fiber composite may be loaded in the **transverse direction;** that is, the load is applied at a 90° angle to the direction of fiber alignment as shown in Figure 17.8a. For this situation the stress σ to which the composite as well as both phases are exposed is the same, or

$$\sigma_c = \sigma_m = \sigma_f = \sigma \tag{17.13}$$

this is termed an *isostress* state. Also, the strain or deformation of the entire composite ϵ_c is

$$\epsilon_c = \epsilon_m V_m + \epsilon_f V_f \tag{17.14}$$

but, since $\epsilon = \sigma/E$,

$$\frac{\sigma}{E_c} = \frac{\sigma}{E_m} V_m + \frac{\sigma}{E_f} V_f \tag{17.15}$$

Or, dividing through by σ yields

$$\frac{1}{E_c} = \frac{V_m}{E_m} + \frac{V_f}{E_f} \tag{17.16}$$

which reduces to

$$E_c = \frac{E_m E_f}{V_m E_f + V_f E_m} = \frac{E_m E_f}{(1 - V_f)E_f + V_f E_m} \tag{17.17}$$

Equation 17.17 is analogous to the lower-bound expression for particulate composites, Equation 17.2.

Compute the elastic modulus of the composite material described in Example Problem 17.1, but assume that the stress is applied perpendicular to the direction of fiber alignment.

SOLUTION
According to Equation 17.17,

$$E_c = \frac{(5 \times 10^5 \text{ psi})(10 \times 10^6 \text{ psi})}{(0.6)(10 \times 10^6 \text{ psi}) + (0.4)(5 \times 10^5 \text{ psi})}$$

$$= 8.1 \times 10^5 \text{ psi } (5.6 \times 10^3 \text{ MPa})$$

This value for E_c is slightly greater than that of the matrix phase but, from Example Problem 17.1a, only approximately one fifth of the modulus of elasticity along the fiber direction, which indicates the degree of anisotropy of continuous and oriented fiber composites.

Discontinuous and Aligned Fiber Composites

Even though reinforcement efficiency is lower for discontinuous than for continuous fibers, discontinuous and aligned fiber composites (Figure 17.8b) are becoming increasingly more important in the commercial market. Chopped glass fibers are used most extensively; however, carbon and aramid discontinuous fibers are also employed. These short fiber composites can be produced having moduli of elasticity and tensile strengths that approach 90% and 50%, respectively, of their continuous fiber counterparts.

For a discontinuous and aligned fiber composite having a uniform distribution of fibers and in which $l > l_c$, the longitudinal strength $(TS)_c$ is given by the relationship

$$(TS)_c = (TS)_f V_f \left(1 - \frac{l_c}{2l}\right) + (TS)'_m (1 - V_f) \tag{17.18}$$

where $(TS)_f$ and $(TS)'_m$ represent, respectively, the fracture strength of the fiber and the stress in the matrix when the composite fails.

If the fiber length is less than critical $(l < l_c)$, then the longitudinal strength is given by

$$(TS)_c = \frac{l\tau_c}{d} V_f + (TS)'_m (1 - V_f) \tag{17.19}$$

where d is the fiber diameter.

Discontinuous and Randomly Oriented Fiber Composites

Normally, when the fiber orientation is random, short and discontinuous fibers are used; reinforcement of this type is schematically demonstrated in Figure 17.8c. Under these circumstances, a "rule-of-mixtures" expression for the elastic modulus similar

TABLE 17.1 Properties of Unreinforced and Reinforced Polycarbonates with Randomly Oriented Glass Fibers

Property	Unreinforced	Fiber Reinforcement (vol%)		
		20	30	40
Specific gravity	1.19–1.22	1.35	1.43	1.52
Tensile strength [psi × 10^3 (MPa)]	8.5–9.0 (59–62)	16 (110)	19 (131)	23 (159)
Modulus of elasticity [psi × 10^3 (MPa)]	325–340 (2240–2345)	860 (5930)	1250 (8620)	1680 (11,590)
Elongation (%)	90–115	4–6	3–5	3–5
Impact strength, notched Izod (lb/in.)	12–16	2.0	2.0	2.5

Source: Adapted from Materials Engineering's *Materials Selector,* copyright Penton/IPC.

to Equation 17.10a may be utilized, as follows:

$$E_c = KE_fV_f + E_mV_m \qquad (17.20)$$

In this expression K is a fiber efficiency parameter, which depends on V_f and the E_f/E_m ratio. Of course, its magnitude will be less than unity, usually in the range 0.1 to 0.6. Thus for random fiber reinforcement (as with oriented), the modulus increases in some proportion of the volume fraction of fiber. Table 17.1, which gives some of the mechanical properties of unreinforced and reinforced polycarbonates for discontinuous and randomly oriented glass fibers, provides an idea of the magnitude of the reinforcement that is possible.

By way of summary, then, aligned fibrous composites are inherently anisotropic, in that the maximum strength and reinforcement are achieved along the alignment (longitudinal) direction. In the transverse direction, fiber reinforcement is virtually nonexistent: fracture usually occurs at relatively low tensile stresses. For other stress orientations, composite strength lies between these extremes. The efficiency of fiber reinforcement for several situations is presented in Table 17.2; this efficiency is taken to be unity for an oriented fiber composite in the alignment direction, and zero perpendicular to it.

When multidirectional stresses are imposed within a single plane, aligned layers that are fastened together one on top of another at different orientations are frequently utilized. These are termed *laminar composites,* which are discussed in Section 17.13.

Applications involving totally multidirectional applied stresses normally use discontinuous fibers, which are randomly oriented in the matrix material. Table 17.2 shows that the reinforcement efficiency is only one fifth that of an aligned composite in the longitudinal direction; however, the mechanical characteristics are isotropic.

Consideration of orientation and fiber length for a particular composite will depend on the level and nature of the applied stress as well as fabrication cost. Production rates for short-fiber composites (both aligned and randomly oriented) are

TABLE 17.2 Reinforcement Efficiency of Fiber-Reinforced Composites for Several Fiber Orientations and at Various Directions of Stress Application

Fiber Orientation	Stress Direction	Reinforcement Efficiency
All fibers parallel	Parallel to fibers	1
	Perpendicular to fibers	0
Fibers randomly and uniformly distributed within a specific plane	Any direction in the plane of the fibers	$\frac{3}{8}$
Fibers randomly and uniformly distributed within three dimensions in space	Any direction	$\frac{1}{5}$

Source: H. Krenchel, *Fibre Reinforcement*, Copenhagen: Akademisk Forlag, 1964 [33].

rapid, and intricate shapes can be formed which are not possible with continuous fiber reinforcement. Furthermore, fabrication costs are considerably lower than for continuous and aligned; fabrication techniques applied to short-fiber composite materials include compression, injection, and extrusion molding, which are described for unreinforced polymers in Section 16.13.

17.6 THE FIBER PHASE

An important characteristic of most materials, especially brittle ones, is that a small-diameter fiber is much stronger than the bulk material. As discussed in Section 13.6, the probability of the presence of a critical surface flaw that can lead to fracture diminishes with decreasing specimen volume, and this feature is used to advantage in the fiber-reinforced composites. Also, the materials used for reinforcing fibers have high tensile strengths.

On the basis of diameter and character, fibers are grouped into three different classifications: *whiskers, fibers,* and *wires.* **Whiskers** are very thin single crystals that have extremely large length-to-diameter ratios. As a consequence of their small size, they have a high degree of crystalline perfection and are virtually flaw-free, which accounts for their exceptionally high strengths; they are the strongest known materials. In spite of these high strengths, whiskers are not utilized extensively as a reinforcement medium because they are extremely expensive. Moreover, it is difficult and often impractical to incorporate whiskers into a matrix. Whisker materials include graphite, silicon carbide, silicon nitride, and aluminum oxide; some mechanical characteristics of these materials are given in Table 17.3.

Materials that are classified as **fibers** are either polycrystalline or amorphous and have small diameters; fibrous materials are generally either polymers or ceramics (e.g., the polymer aramids, glass, carbon, boron, aluminum oxide, and silicon carbide). Table 17.3 also presents some data on a few materials that are used in fiber form.

Fine wires have relatively large diameters; typical materials include steel, molybdenum, and tungsten. Wires are utilized as a radial steel reinforcement in automobile tires, in filament-wound rocket casings, and in wire-wound high-pressure hoses.

TABLE 17.3 Characteristics of Several Fiber-Reinforcement Materials

Material	Specific Gravity	Tensile Strength psi × 10^6 (MPa × 10^3)	Specific Strength (psi × 10^6)	Modulus of Elasticity psi × 10^6 (MPa × 10^3)	Specific Modulus (psi × 10^6)
Whiskers					
Graphite	2.2	3 (20)	1.36	100 (690)	45.5
Silicon carbide	3.2	3 (20)	0.94	70 (480)	22
Silicon nitride	3.2	2 (14)	0.63	55 (380)	17.2
Aluminum oxide	3.9	2–4 (14–28)	0.5–1.0	60–80 (415–550)	15.4–20.5
Fibers					
Aramid (Kevlar 49)	1.4	0.5 (3.5)	0.36	19 (124)	13.5
E-Glass	2.5	0.5 (3.5)	0.20	10.5 (72)	4.2
Carbon[a]	1.8	0.25–0.80 (1.5–5.5)	0.18–0.57	22–73 (150–500)	15.7–52.1
Aluminum oxide	3.2	0.3 (2.1)	0.09	25 (170)	7.8
Silicon carbide	3.0	0.50 (3.9)	0.17	62 (425)	20.7
Metallic Wires					
High-carbon steel	7.8	0.6 (4.1)	0.08	30 (210)	3.9
Molybdenum	10.2	0.2 (1.4)	0.02	52 (360)	5.1
Tungsten	19.3	0.62 (4.3)	0.03	58 (400)	3.0

[a] The term "carbon" instead of "graphite" is used to denote these fibers, since they are composed of crystalline graphite regions, and also of noncrystalline material and areas of crystal misalignment.

17.7 THE MATRIX PHASE

The matrix phase of fiber composites serves several functions. First, it binds the fibers together and acts as the medium by which an externally applied stress is transmitted and distributed to the fibers; only a very small proportion of an applied load is sustained by the matrix phase. Furthermore, the matrix material should be ductile. In addition, the elastic modulus of the fiber should be much higher than that of the matrix. The second function of the matrix is to protect the individual fibers from surface damage as a result of mechanical abrasion or chemical reactions with the environment. Such interactions may introduce surface flaws capable of forming cracks, which may lead to failure at low tensile stress levels. Finally, the matrix separates the fibers and, by virtue of its relative softness and plasticity, prevents the propagation of brittle cracks from fiber to fiber, which could result in catastrophic failure; in other words, the matrix phase serves as a barrier to crack propagation. Even though some of the individual fibers fail, total composite fracture will not occur until large numbers of adjacent fibers, once having failed, form a cluster of critical size.

It is essential that adhesive bonding forces between fiber and matrix be high to minimize fiber pull-out. In fact, bonding strength is an important consideration in the choice of the matrix–fiber combination. The ultimate strength of the composite depends to a large degree on the magnitude of this bond; adequate bonding is essential to maximize the stress transmittance from the weak matrix to the strong fibers.

In general, only metals and polymers are used as matrix materials because some ductility is desirable. Metals that are frequently fiber reinforced include aluminum and copper. However, polymers are used as matrix materials in the greatest diversity of composite applications, as well as in the largest quantities, in light of their properties and ease of fabrication. Common matrix polymers include nearly all commercial thermoplastic and thermosetting polymers.

17.8 FIBERGLASS-REINFORCED COMPOSITES

Fiberglass is simply a composite consisting of glass fibers, either continuous or discontinuous, contained within a plastic matrix; this type of composite is produced in the largest quantities. The composition of the glass that is most commonly drawn into fibers (sometimes referred to as E-glass) is contained in Table 14.1. Glass is popular as a fiber reinforcement material for several reasons:

1. It is easily drawn into high-strength fibers from the molten state.
2. It is readily available and may be fabricated into a glass-reinforced plastic economically using a wide variety of composite-manufacturing techniques.
3. As a fiber, it is relatively strong, and when embedded in a plastic matrix, it produces a composite having a very high specific strength.
4. When coupled with the various plastics, it possesses a chemical inertness that renders the composite useful in a variety of corrosive environments.

The surface characteristics of glass fibers are extremely important because even minute surface flaws can deleteriously affect the tensile properties, as discussed in Section 13.6. Surface flaws are easily introduced by rubbing or abrading the surface with another hard material. Also, glass surfaces that have been exposed to the normal atmosphere for even short time periods generally have a weakened surface layer that interferes with bonding to the matrix. Newly drawn fibers are normally coated during drawing with a material that protects the surface from damage and undesirable interactions and, in addition, promotes a better bond between the fiber and matrix.

A large number of different plastic materials are utilized for the matrix in fiberglasses, the polyesters being the most common. Some of the relatively new commercial fiber-reinforced composites utilize glass fibers in a nylon matrix; these materials are extremely strong and highly impact resistant.

There are several limitations to this group of materials. In spite of having high strengths, they are not very stiff and do not display the rigidity that is necessary for some applications (e.g., as structural members for airplanes and bridges). Most fiberglass materials are limited to service temperatures below 200°C (400°F); at higher temperatures most polymers begin to flow or to deteriorate. Service temperatures may be extended to approximately 300°C (575°F) by using high-purity fused silica for the fibers and high-temperature polymers such as the polyimide resins.

Many fiberglass applications are familiar: automotive and marine bodies, plastic pipes, storage containers, and industrial floorings. The transportation industries are utilizing increasing amounts of glass fiber-reinforced plastics in an effort to decrease vehicle weight and boost fuel efficiencies. A host of new applications are being used or currently investigated by the automotive industry.

17.9 MISCELLANEOUS FIBER-REINFORCED PLASTIC MATRIX COMPOSITES

Plastics impregnated with other fiber materials—carbon being the most common—also form composites. Table 17.3 indicates that carbon in fiber form has a much higher specific modulus than glass. It also has a better resistance to temperatures and corrosive chemicals, but is more expensive and has only limited short-fiber utilization. The aircraft industry is currently implementing carbon-reinforced composites as structural components of their new aircraft as a weight-saving measure. They hope to be able to manufacture new structures which are 20 to 30% lighter than those fabricated from sheet metal parts.

Various plastics impregnated with boron fibers have been utilized to some degree. For example, some helicopter rotor blades are constructed using boron fibers in an epoxy resin.

A new generation of high-strength polymeric aramid fibers is beginning to be selected for composites used in lightweight structural components, such as aerospace, aircraft, marine, and sporting equipment. Extensive research is currently in progress on the possibility of using other fiber materials such as silicon carbide (SiC) and silicon nitride (Si_3N_4) in plastic matrices.

Also included with these advanced materials are carbon–carbon composites, which are composed of carbon fibers embedded within carbonized resin matrices; these are designed principally for high-temperature aerospace applications.

17.10 METAL MATRIX–FIBER COMPOSITES

Since most metals are ductile, they may also be utilized as the matrix phase in fiber-reinforced composites; in practice, however, their application in this regard is somewhat limited. Recently, there have been developed a number of continuous-fiber composites in which alloys of aluminum, magnesium, copper, and titanium have served as the matrix phase, which may be reinforced with carbon, silicon carbide, boron, borsic as well as metal fibers. Fiber concentrations normally range between 20 and 50 vol%. Probably the most common fiber-reinforced metal is the borsic–aluminum composite. The continuous borsic fiber is prepared by the vapor deposition of a layer of boron on a thin (10 μm diameter) tungsten wire. This fiber is subsequently given a thin coating of silicon carbide to retard undesirable reactions between the boron and aluminum.

These fiber-reinforced metals may be utilized at higher temperatures than the polymer composites. In addition, high specific strengths and high specific moduli are possible because the densities of these base metals are relatively low. This combination of properties makes these materials especially attractive for use in some aerospace and new engine applications.

The high-temperature creep and rupture properties of some of the superalloys (Ni- and Co-based alloys) may be enhanced by fiber reinforcement using refractory metals such as tungsten. Excellent high-temperature oxidation resistance and impact strength are also maintained. Designs incorporating these composites permit higher operating temperatures and better efficiencies for turbine engines.

17.11 HYBRID COMPOSITES

A relatively new fiber-reinforced composite is the **hybrid,** which is obtained by using two or more different kinds of fibers in a single matrix; hybrids have a better all-around combination of properties than composites containing only a single fiber type. A variety of fiber combinations and matrix materials are used, but in the most common system, both carbon and glass fibers are incorporated into a polymeric resin. The carbon fibers are strong and relatively stiff and provide a low-density reinforcement; however, they are expensive. Glass fibers are inexpensive and lack the stiffness of carbon. The glass–carbon hybrid is stronger and tougher, has a higher impact resistance, and may be produced at a lower cost than either of the comparable all-carbon or all-glass reinforced plastics.

There are a number of ways in which the two different fibers may be combined, which will ultimately affect the overall properties. For example, the fibers may all be aligned and intimately mixed with one another; or laminations may be constructed consisting of layers, each of which consists of a single fiber type, alternating one with another. In virtually all hybrids the properties are anisotropic.

When hybrid composites are stressed in tension, failure is usually noncatastrophic (i.e., does not occur suddenly). The carbon fibers are the first to fail, at which time the load is transferred to the glass fibers. Upon failure of the glass fibers, the matrix phase must sustain the applied load. Eventual composite failure concurs with that of the matrix phase.

Principal applications for hybrid composites are lightweight land, water, and air transport structural components, sporting goods, and lightweight orthopedic components.

17.12 PROCESSING OF FIBER-REINFORCED COMPOSITES

To fabricate continuous fiber-reinforced plastics that meet design specifications, the fibers should be uniformly distributed within the plastic matrix and, in most instances, all oriented in virtually the same direction. In this section newly developed techniques (pultrusion, filament winding, and prepreg production processes) by which useful products of these materials are manufactured will be discussed.

Pultrusion

Pultrusion is used for the manufacture of components having continuous lengths and a constant cross-sectional shape (i.e., rods, tubes, beams, etc.). With this technique,

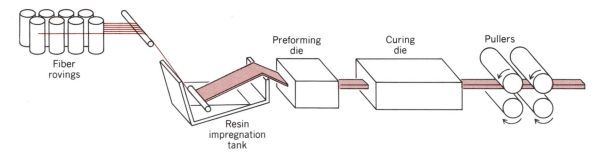

Figure 17.9 Schematic diagram showing the pultrusion process.

illustrated schematically in Figure 17.9, continuous fiber *rovings,* or *tows,*[1] are first impregnated with a thermosetting resin; these are then pulled through a steel die which preforms to the desired shape and also establishes the resin/fiber ratio. The stock then passes through a curing die which is precision machined so as to impart the final shape; this die is also heated in order to initiate curing of the resin matrix. A pulling device draws the stock through the dies and also determines the production speed. Tubes and hollow sections are made possible by using center mandrels or inserted hollow cores. Principal reinforcements are glass, carbon, and aramid fibers, normally added in concentrations between 40 and 70 vol%. Commonly used matrix materials include polyesters, vinyl esters, and epoxy resins.

Pultrusion is a continuous process which is easily automated; production rates are relatively high, making it very cost effective. Furthermore, a wide variety of shapes are possible, and there is really no practical limit to the length of stock that may be manufactured.

Prepreg Production Processes

Prepreg is the composite industry's term for continuous fiber reinforcement preimpregnated with a polymer resin that is only partially cured. This material is delivered in tape form to the manufacturer, who then directly molds and fully cures the product without having to add any resin. This is probably the composite material form most widely used for structural applications.

The prepregging process, represented schematically for thermoset polymers in Figure 17.10, begins by collimating a series of spool-wound continuous fiber tows. These tows are then sandwiched and pressed between sheets of release and carrier paper using heated rollers, a process termed "calendering." The release paper sheet has been coated with a thin film of heated resin solution of relatively low viscosity so as to provide for its thorough impregnation of the fibers. A "doctor knife" metal blade spreads the resin into a film of uniform thickness and width. The final prepreg product—the thin tape consisting of continuous and aligned fibers embedded in a partially cured resin—is prepared for packaging by winding onto a cardboard core. As shown in Figure 17.10, the release paper sheet is removed as the impregnated tape

---

[1] A roving, or tow, is a loose and untwisted bundle of continuous fibers that are drawn together as parallel strands.

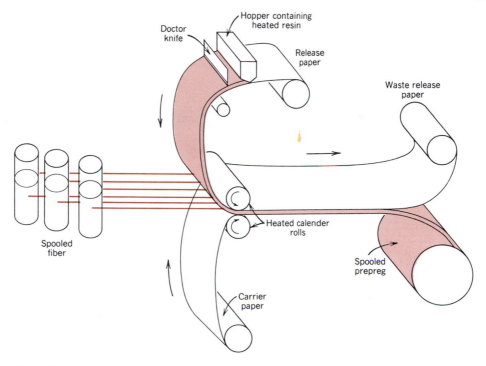

Figure 17.10 Schematic diagram illustrating the production of prepreg tape using thermoset polymers.

is spooled. Typical tape thicknesses range between 3×10^{-3} and 10^{-2} in. (0.08 and 0.25 mm), tape widths between 1 and 60 in. (25 and 1525 mm), whereas resin content usually lies between about 35 and 45 vol%.

At room temperature the thermoset matrix undergoes curing reactions; therefore, the prepreg is stored at 0°C (32°F) or lower. Also, the time in use at room temperature (or "out-time") must be minimized. If properly handled, thermoset prepregs have a lifetime of at least six months and usually longer.

Both thermoplastic and thermosetting resins are utilized; carbon, glass, and aramid fibers are the common reinforcements.

Actual fabrication begins with the "lay-up"—laying of the prepreg tape onto a tooled surface. Normally a number of plies are laid up (after removal from the carrier backing paper) to provide the desired thickness. The lay-up arrangement may be unidirectional, but more often the fiber orientation is alternated to produce a cross-ply or angle-ply laminate. Final curing is accomplished by the simultaneous application of heat and pressure.

The lay-up procedure may be carried out entirely by hand (hand lay-up), wherein the operator both cuts the lengths of tape and then positions them in the desired orientation on the tooled surface. Alternately, tape patterns may be machine cut, which are then hand laid. Fabrication costs can be further reduced by automation of prepreg lay-up and other manufacturing procedures (e.g., filament winding, as

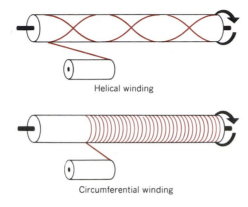

Helical winding

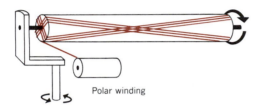

Circumferential winding

Polar winding

Figure 17.11 Schematic representations of helical, circumferential, and polar filament winding techniques. (From N. L. Hancox, editor, *Fibre Composite Hybrid Materials*.)

discussed below), which virtually eliminate the need for hand labor. These automated methods are essential for many applications of composite materials to be cost effective.

Filament Winding

Filament winding is a process by which continuous reinforcing fibers are accurately positioned in a predetermined pattern to form a hollow (usually cylindrical) shape. The fibers, either as individual strands or as tows, are first fed through a resin bath and then continuously wound onto a mandrel, usually using automated winding equipment (Figure 17.11). After the appropriate number of layers have been applied, curing is carried out either in an oven or at room temperature, after which the mandrel is removed. As an alternative, narrow and thin prepregs (i.e., tow pregs) 10 mm or less in width may be filament wound.

Various winding patterns are possible (i.e., circumferential, helical, and polar) to give the desired mechanical characteristics. Filament-wound parts have very high strength-to-weight ratios. Also, a high degree of control over winding uniformity and orientation is afforded with this technique. Furthermore, when automated, the process is most economically attractive. Common filament-wound structures include rocket motor casings, storage tanks and pipes, and pressure vessels.

Manufacturing techniques are now being used to produce a wide variety of structural shapes that are not necessarily limited to surfaces of revolution (e.g., I-beams). This technology is advancing very rapidly because it is very cost effective.

STRUCTURAL COMPOSITES

A **structural composite** is normally composed of both homogeneous and composite materials the properties of which depend, not only on the properties of the constituent materials but also on the geometrical design of the various structural elements. Laminar composites and sandwich panels are two of the most common structural composites; only a relatively superficial examination is offered here for them.

17.13 LAMINAR COMPOSITES

A **laminar composite** is composed of two-dimensional sheets or panels that have a preferred high-strength direction such as is found in wood and continuous and aligned fiber-reinforced plastics. The layers are stacked and subsequently cemented together such that the orientation of the high-strength direction varies with each successive layer (Figure 17.12). For example, adjacent wood sheets in plywood are aligned with the grain direction at right angles to each other. Laminations may also be constructed

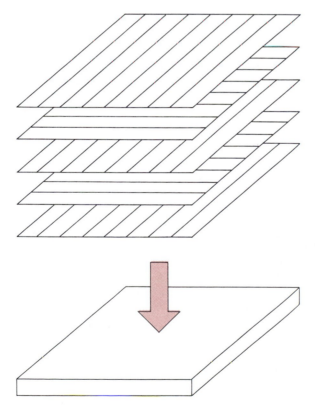

Figure 17.12 The stacking of successive oriented, fiber-reinforced layers for a laminar composite.

using fabric material such as cotton, paper, or woven glass fibers embedded in a plastic matrix. Thus a laminar composite has relatively high strength in a number of directions in the two-dimensional plane; however, the strength in any given direction is, of course, lower than it would be if all the fibers were oriented in that direction. One example of a relatively complex laminated structure is the modern ski.

17.14 SANDWICH PANELS

Sandwich panels, considered to be a class of structural composites, consist of two strong outer sheets, or faces, separated by a layer of less-dense material, or core, which has lower stiffness and lower strength. The faces bear most of the in-plane loading, and also any transverse bending stresses. Typical face materials include aluminum alloys, fiber-reinforced plastics, titanium, steel, and plywood.

Structurally, the core serves two functions. First, it separates the faces and resists deformations perpendicular to the face plane. Secondly, it provides a certain degree of shear rigidity along planes which are perpendicular to the faces. Various materials and structures are utilized for cores, including foamed polymers, synthetic rubbers, inorganic cements, as well as balsa wood.

Another popular core consists of a "honeycomb" structure—thin foils that have been formed into interlocking hexagonal cells, with axes oriented perpendicular to the face planes. The material of which the honeycomb is made may be similar to the face material. Figure 17.13 shows a cutaway view of a honeycomb core sandwich panel.

Sandwich panels are found in a wide variety of applications; they include roofs, floors, and walls of buildings; and, in aircraft, for wings, fuselage, and tailplane skins.

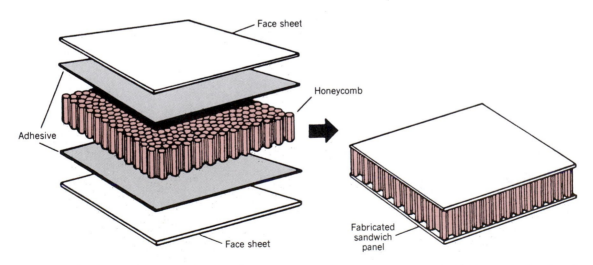

Figure 17.13 Schematic diagram showing the construction of a honeycomb core sandwich panel. (Reprinted with permission from *Engineered Materials Handbook,* Vol. 1, *Composites,* ASM International, Metals Park, OH, 1987.)

SUMMARY

Composites are artificially produced multiphase materials having a desirable combination of the best properties of the constituent phases. Usually, one phase (the matrix) is continuous and completely surrounds the other (the dispersed phase). In this discussion composites were classified as particle-reinforced, fiber-reinforced, and structural composites.

Large-particle and dispersion-strengthened composites fall within the particle-reinforced classification. For dispersion strengthening, improved strength is achieved by extremely small particles of the dispersed phase, which inhibit dislocation motion; that is, the strengthening mechanism involves interactions that may be treated on the atomic level. The particle size is normally greater with large-particle composites, whose mechanical characteristics are enhanced by reinforcement action.

Concrete, a type of large-particle composite, consists of an aggregate of particles bonded together with cement. In the case of portland cement concrete, the aggregate consists of sand and gravel; the cementitious bond develops as a result of chemical reactions between the portland cement and water. The mechanical strength of this concrete may be improved by reinforcement methods (e.g., embedment into the fresh concrete of steel rods, wires, etc.). Additional reinforcement is possible by the imposition of residual compressive stresses using prestressing and posttensioning techniques.

Of the several composite types, the potential for reinforcement efficiency is greatest for those that are fiber reinforced. With these composites an applied load is transmitted and distributed among the fibers via the matrix phase, which should be at least moderately ductile. Significant reinforcement is possible only if the matrix–fiber bond is strong. On the basis of diameter, fiber reinforcements are classified as whiskers, fibers, or wires. Since reinforcement discontinues at the fiber extremities, reinforcement efficiency depends on fiber length. For each fiber–matrix combination, there exists some critical length; the length of continuous fibers greatly exceeds this critical value, whereas shorter fibers are discontinuous.

Fiber arrangement is also crucial relative to composite characteristics. The mechanical properties of continuous and aligned fiber composites are highly anisotropic. In the alignment direction, reinforcement and strength are a maximum; perpendicular to the alignment, they are a minimum. Composite rule-of-mixture expressions for both longitudinal and transverse orientations were also developed.

For short and discontinuous fibrous composites, the fibers may be either aligned or randomly oriented. Significant strengths and stiffnesses are possible for aligned short-fiber composites in the longitudinal direction. Despite some limitations on reinforcement efficiency, the properties of randomly oriented short-fiber composites are isotropic.

The objective of many fiber-reinforced composites is a high specific strength and/or specific modulus. Therefore, matrix materials are commonly polymers or metals having relatively low densities. The most common composite of this type is fiberglass—glass fibers embedded in a polymer matrix, often polyester. Carbon fibers are also frequently used. Better overall property combinations are possible with hybrid composites, those containing at least two different fiber types.

Several composite processing techniques have been developed which provide for a uniform fiber distribution and a high degree of alignment. With pultrusion, components of continuous length and constant cross section are formed as resin-impregnated fiber tows are pulled through a die. Composites utilized for many structural applications are commonly prepared using a lay-up operation (either hand or automated), wherein prepreg tape plies are laid down on a tooled surface, which are subsequently fully cured by the simultaneous application of heat and pressure. Some hollow structures may be fabricated using automated filament winding procedures, whereby resin-coated strands or tows or prepreg tape are continuously wound onto a mandrel, followed by a curing operation.

Two general kinds of structural composites were discussed: the laminar composites and sandwich panels. The properties of laminar composites are virtually isotropic in a two-dimensional plane. This is made possible with several sheets of a highly anisotropic composite, which are cemented onto one another such that the high-strength direction is varied with each successive layer. Sandwich panels consist of two strong and stiff sheet faces that are separated by a core material or structure. These structures combine relatively high strengths and stiffnesses with low densities.

IMPORTANT TERMS AND CONCEPTS

Cermet

Concrete

Dispersed phase

Dispersion-strengthened
  composite

Fiber

Fiber-reinforced
  composite

Hybrid composite

Laminar composite

Large-particle
  composite

Longitudinal direction

Matrix phase

Prepreg

Prestressed concrete

Principle of combined
  action

Reinforced concrete

Rule of mixtures

Sandwich panel

Specific modulus

Specific strength

Structural composite

Transverse direction

Whisker

REFERENCES

BROUTMAN, L. J. and R. H. KROCK, *Modern Composite Materials,* Addison-Wesley Publishing Co., Reading, MA, 1967.

Engineered Materials Handbook, Vol. 1, *Composites,* ASM International, Metals Park, OH, 1987.

HULL, D., *An Introduction to Composite Materials,* Cambridge University Press, Cambridge, 1981.

KELLY, A. and N. H. MACMILLAN, *Strong Solids,* 3rd edition, Clarendon Press, Oxford, 1986.

RICHARDSON, M. O. W. (Editor), *Polymer Engineering Composites,* Applied Science Publishers, Ltd., London, 1977.

VINSON, J. R. and T. W. CHOU, *Composite Materials and Their Use in Structures,* Applied Science Publishers, Ltd., London, 1975.

WEETON, J. W. (Editor), *Engineers' Guide to Composite Materials,* American Society for Metals, Metals Park, OH, 1986.

QUESTIONS AND PROBLEMS

17.1 Cite the general difference in strengthening mechanism between large-particle and dispersion-strengthened particle-reinforced composites.

17.2 Cite one similarity and two differences between precipitation hardening and dispersion strengthening.

17.3 The mechanical properties of cobalt may be improved by incorporating fine particles of tungsten carbide (WC). Given that the moduli of elasticity of these materials are, respectively, 30×10^6 psi (2.0×10^5 MPa) and 102×10^6 psi (7.0×10^5 MPa), plot modulus of elasticity versus the volume percent of WC in Co from 0 to 100 vol%, using both upper- and lower-bound expressions.

17.4 Estimate the maximum and minimum thermal conductivity values for a cermet that contains 65 vol% titanium carbide (TiC) particles in a nickel matrix. Assume thermal conductivities of 27 W/m-K (0.065 cal/s-cm-K) and 67 W/m-K (0.16 cal/s-cm-K) for TiC and Ni, respectively.

17.5 A large-particle composite consisting of tungsten particles within a copper matrix is to be prepared. If the volume fractions of tungsten and copper are 0.60 and 0.40, respectively, estimate the upper limit for the specific stiffness of this composite given the data below.

	Specific Gravity	Modulus of Elasticity (MPa)
Copper	8.9	11.0×10^4
Tungsten	19.3	40.7×10^4

17.6 **(a)** What is the distinction between matrix and dispersed phases in a composite material? **(b)** Contrast the mechanical characteristics of matrix and dispersed phases for fiber-reinforced composites.

17.7 **(a)** What is the distinction between cement and concrete? **(b)** Cite three important limitations that restrict the use of concrete as a structural material. **(c)** Briefly explain three techniques that are utilized to strengthen concrete by reinforcement.

17.8 For a fiber-reinforced composite, **(a)** list three functions of the matrix phase; **(b)** compare the desired mechanical characteristics of matrix and fiber phases; and **(c)** cite two reasons why there must be a strong bond between fiber and matrix at their interface.

17.9 For some glass fiber–epoxy matrix combination, the critical fiber length–fiber diameter ratio is 40. Using the data in Table 17.3, determine the fiber–matrix bond strength.

17.10 **(a)** For a fiber-reinforced composite, the efficiency of reinforcement η is dependent on fiber length l according to

$$\eta = \frac{l - 2x}{l}$$

where x represents the length of the fiber at each end that does not contribute to the load transfer. Make a plot of η versus l to $l = 2$ in. (50 mm), assuming that $x = 0.05$ in. (1.25 mm). **(b)** What length is required for a 0.9 efficiency of reinforcement?

17.11 It is desired to produce an aligned and continuous fiber-reinforced epoxy composite having a maximum of 50 vol% fibers; in addition, minimum longitudinal modulus of elasticity and tensile strength values must be 7×10^6 and 200,000 psi (4.8×10^4 and 1.38×10^3 MPa), respectively. Of glass, carbon, and aramid fiber materials, which are possible candidates, and why? Assume that the epoxy has a modulus of elasticity of 4.5×10^5 psi (3.1×10^3 MPa) and a tensile strength of 10,000 psi (69 MPa). The fiber data are contained in Table 17.3; for carbon, assume average values of the extremes.

17.12 A continuous and aligned fiber-reinforced composite is to be produced consisting of 30 vol% aramid fibers and 70 vol% of a polycarbonate matrix; mechanical characteristics of these two materials are as follows:

	Modulus of Elasticity [psi (MPa)]	Tensile Strength [psi (MPa)]
Aramid fiber	19×10^6 (1.3×10^5)	500,000 (3500)
Polycarbonate	3.5×10^5 (2.4×10^3)	8000 (55)

For this composite, compute **(a)** the longitudinal tensile strength, and **(b)** the longitudinal modulus of elasticity.

17.13 It is desired to produce a continuous and oriented carbon fiber-reinforced epoxy having a modulus of elasticity of at least 12×10^6 psi (8.3×10^4 MPa) in the direction of fiber alignment. The maximum permissible specific gravity is 1.40. Given the data below, is such a composite possible? Why or why not? Assume that composite specific gravity may be determined using a relationship similar to Equation 17.10a.

	Specific Gravity	Modulus of Elasticity [psi (MPa)]
Carbon	1.80	37×10^6 (2.6×10^5)
Epoxy	1.25	3.5×10^5 (2.4×10^3)

17.14 It is desired to fabricate a continuous and aligned glass fiber-reinforced polyester having a tensile strength of at least 180,000 psi (1250 MPa) in the longitudinal direction. The maximum possible specific gravity is 1.80. Using the data below, determine if such a composite is possible. Justify your decision.

	Specific Gravity	Tensile Strength [psi (MPa)]
Glass	2.50	5×10^5 (3.5×10^3)
Polyester	1.35	7.25×10^3 (50)

17.15 Is it possible to produce a continuous and oriented aramid fiber–epoxy matrix composite having longitudinal and transverse moduli of elasticity of 5×10^6 psi $(3.5 \times 10^4$ MPa) and 7.5×10^5 psi (5170 MPa), respectively? Why or why not? Assume that the modulus of elasticity of the epoxy is 4.93×10^5 psi $(3.4 \times 10^3$ MPa).

17.16 **(a)** Calculate and compare the specific strength of the fiberglass-reinforced polycarbonate containing 40 vol% fibers in Table 17.1, and the following metal alloys: an A656 grade 1 HSLA steel (Table 12.1b), an annealed C26000 cartridge brass (Table 12.6), an annealed 5052 aluminum alloy (Table 12.7), and an artificially aged HM31A magnesium alloy (Table 12.8). **(b)** Compare the specific modulus of the same fiber-reinforced polycarbonate with aluminum, brass, magnesium, steel, and molybdenum, using the data in Tables C.1 and C.2, Appendix C.

17.17 For a continuous and oriented fiber-reinforced composite, the moduli of elasticity in the longitudinal and transverse directions are 2.80×10^6 and 5.26×10^5 psi $(1.97 \times 10^4$ and 3.63×10^3 MPa), respectively. If the volume fraction of fibers is 0.25, determine the moduli of elasticity of fiber and matrix phases.

17.18 **(a)** Verify that Equation 17.11, the expression for the fiber load–matrix load ratio (F_f/F_m) is valid. **(b)** What is the F_f/F_c ratio in terms of E_f, E_m, and V_f?

17.19 In an aligned and continuous carbon fiber-reinforced nylon 6,6 composite, the fibers are to carry 95% of a load applied in the longitudinal direction. **(a)** Using the data provided, determine the volume fraction of fibers that will be required. **(b)** What will be the tensile strength of this composite?

	Modulus of Elasticity [psi (MPa)]	Tensile Strength [psi (MPa)]
Carbon	37×10^6 (2.6×10^5)	2.5×10^5 (1.7×10^3)
Nylon 6,6	4×10^5 (2.8×10^3)	1.1×10^4 (76)

17.20 Assume that the composite described in Problem 17.12 has a cross-sectional area of 0.5 in.^{2} (320 mm^2) and is subjected to a longitudinal load of 10,000 lb$_f$ (44,500 N). **(a)** Calculate the fiber–matrix load ratio. **(b)** Calculate the actual loads carried by both fiber and matrix phases. **(c)** Compute the magnitude of the stress on each of the fiber and matrix phases. **(d)** What strain is experienced by the composite?

17.21 A continuous and aligned fiber-reinforced composite having a cross-sectional area of 1.5 in.^{2} (970 mm^2) is subjected to an external tensile load. If the stresses sustained by the fiber and matrix phases are 31,100 psi (215 MPa) and 780 psi (5.38 MPa), respectively, the force sustained by the fiber phase is 17,265 lb$_f$ (76,800 N), and the total longitudinal composite strain is 1.56×10^{-3}, determine **(a)** the force sustained by the matrix phase; **(b)** the modulus of elasticity of the composite material in the longitudinal direction; and **(c)** the moduli of elasticity for the fiber and matrix phases.

17.22 Compute the longitudinal strength of an aligned carbon fiber–epoxy matrix composite having a 0.20 volume fraction of fibers, assuming the following: (1) an average fiber diameter of 6×10^{-3} mm $(2.4 \times 10^{-4}$ in.); (2) an average fiber length of 8.0 mm (0.31 in.); (3) a fiber fracture strength of 4.5×10^3 MPa (6.5×10^5 psi); (4) a fiber–matrix bond strength of 75 MPa (10,900 psi); (5) a matrix stress at composite failure of 6.0 MPa (870 psi); and (6) a matrix tensile strength of 60 MPa (8,700 psi).

17.23 It is desired to produce an aligned carbon fiber–epoxy matrix composite having a longitudinal tensile strength of 100,000 psi (690 MPa). Calculate the volume fraction of fibers necessary if (1) the average fiber diameter and length are 3.9×10^{-4} in. (0.01 mm) and 2×10^{-2} in. (0.5 mm), respectively; (2) the fiber fracture strength is 5.8×10^5 psi (4×10^3 MPa); (3) the fiber–matrix bond strength is 7250 psi (50 MPa); and (4) the matrix stress at composite failure is 1000 psi (7.0 MPa).

17.24 Compute the longitudinal tensile strength of an aligned glass fiber–epoxy matrix composite in which the average fiber diameter and length are 5.9×10^{-4} in. (0.015 mm) and 0.08 in. (2.0 mm), respectively, and the volume fraction of fibers is 0.25. Assume that (1) the fiber–matrix bond strength is 14,500 psi (100 MPa); (2) the fracture strength of the fibers is 5×10^5 psi (3.5×10^3 MPa); and (3) the matrix stress at composite failure is 800 psi (5.5 MPa).

17.25 It is necessary to fabricate an aligned and discontinuous glass fiber–epoxy matrix composite having a longitudinal tensile strength of 1200 MPa (175,000 psi) using a 0.35 volume fraction of fibers. Compute the required fiber fracture strength assuming that the average fiber diameter and length are 0.015 mm (5.9×10^{-4} in.) and 5.0 mm (0.20 in.), respectively. The fiber–matrix bond strength is 80 MPa (11,600 psi), and the matrix stress at composite failure is 6.55 MPa (950 psi).

17.26 **(a)** From the moduli of elasticity data in Table 17.1 for glass fiber-reinforced polycarbonate composites, determine the value of the fiber efficiency parameter for each of 20, 30, and 40 vol% fibers. **(b)** Estimate the modulus of elasticity for 50 vol% glass fibers.

17.27 Cite one desirable characteristic and one less desirable characteristic for each of (1) discontinuous-oriented, and (2) discontinuous-random fiber-reinforced composites.

17.28 **(a)** List four reasons why glass fibers are most commonly used for reinforcement. **(b)** Why is the surface perfection of glass fibers so important? **(c)** What measures are taken to protect the surface of glass fibers?

17.29 Cite the distinction between carbon and graphite.

17.30 **(a)** Cite several reasons why fiberglass-reinforced composites are utilized extensively. **(b)** Cite several limitations of this type of composite.

17.31 **(a)** What is a hybrid composite? **(b)** List two important advantages of hybrid composites over normal fiber composites.

17.32 **(a)** Write an expression for the modulus of elasticity for a hybrid composite in which all fibers of both types are oriented in the same direction. **(b)** Using this expression, compute the longitudinal modulus of elasticity of a hybrid composite consisting of carbon and glass fibers in volume fractions of 0.3 and 0.4, respectively, within a polyester resin matrix [$E_m = 6 \times 10^5$ psi (4×10^3 MPa)].

17.33 Derive a generalized expression analogous to Equation 17.17 for the transverse modulus of elasticity of an aligned hybrid composite consisting of two types of continuous fibers.

17.34 Briefly describe pultrusion, filament winding, and prepreg production fabrication processes; cite the advantages and disadvantages of each.

17.35 Briefly describe laminar composites. What is the prime reason for fabricating these materials?

17.36 **(a)** Briefly describe sandwich panels. **(b)** What is the prime reason for fabricating these structural composites? **(c)** What are the functions of the faces and the core?

CHAPTER 18

CORROSION AND DEGRADATION OF MATERIALS

Photograph showing a bar of steel that has been bent into a "horseshoe" shape using a nut-and-bolt assembly. While immersed in seawater, stress corrosion cracks formed along the bend at those regions where the tensile stresses are the greatest. (Photograph courtesy of F. L. LaQue. From F. L. LaQue, *Marine Corrosion, Causes and Prevention.* Copyright 1975 by John Wiley & Sons, Inc. Reprinted by permission of John Wiley & Sons, Inc.)

18.1 INTRODUCTION

To one degree or another, most materials experience some type of interaction with a large number of diverse environments. Often, such interactions impair a material's usefulness as a result of the deterioration of its mechanical properties (e.g., ductility and strength), other physical properties, or appearance. Occasionally, to the chagrin of a design engineer, the degradation behavior of a material for some application is ignored, with adverse consequences.

Deteriorative mechanisms are different for the three material types. In metals, there is actual material loss either by dissolution (**corrosion**) or by the formation of nonmetallic scale or film (*oxidation*). Ceramic materials are relatively resistant to deterioration, which usually occurs at elevated temperatures or in rather extreme environments; the process is frequently also called corrosion. For polymers, mechanisms and consequences differ from those for metals and ceramics, and the term **degradation** is most frequently used. Polymers may dissolve when exposed to a liquid solvent, or they may absorb the solvent and swell; also, electromagnetic radiation (primarily ultraviolet) and heat may cause alterations in their molecular structure.

The deterioration of each of these material types is discussed in this chapter, with special regard to mechanism, resistance to attack by various environments, and measures to prevent or reduce degradation.

CORROSION OF METALS

Corrosion is defined as the destructive and unintentional attack of a metal; it is electrochemical and ordinarily begins at the surface. The problem of metallic corrosion is one of significant proportions; in economic terms, it has been estimated that approximately 5% of an industrialized nation's income is spent on corrosion prevention and the maintenance or replacement of products lost or contaminated as a result of corrosion reactions. The consequences of corrosion are all too common. Familiar examples include the rusting of automotive body panels and radiator and exhaust components.

Corrosion processes are occasionally used to advantage. For example, etching procedures, as discussed in Section 4.9, make use of the selective chemical reactivity of grain boundaries or various microstructural constituents. Also, the current developed in dry-cell batteries is a result of corrosion processes.

18.2 ELECTROCHEMICAL CONSIDERATIONS

For metallic materials, the corrosion process is normally electrochemical, that is, a chemical reaction in which there is transfer of electrons from one chemical species to another. Metal atoms characteristically lose or give up electrons in what is called an **oxidation** reaction. For example, the hypothetical metal M, which has a valence of n (or n valence electrons) may experience oxidation according to the reaction

$$M \longrightarrow M^{n+} + ne^-$$ (18.1)

in which M becomes an $n+$ positively charged ion and in the process loses its n valence electrons; e^- is used to symbolize an electron. Examples in which metals oxidize are

$$Fe \longrightarrow Fe^{2+} + 2e^- \tag{18.2a}$$

$$Al \longrightarrow Al^{3+} + 3e^- \tag{18.2b}$$

The site at which oxidation takes place is called the **anode;** oxidation is sometimes called an anodic reaction.

The electrons generated from each metal atom that is oxidized must be transferred to and become a part of another chemical species in what is termed a **reduction** reaction. For example, some metals undergo corrosion in acid solutions, which have a high concentration of hydrogen (H^+) ions; the H^+ ions are reduced as follows:

$$2H^+ + 2e^- \longrightarrow H_2 \tag{18.3}$$

and hydrogen gas (H_2) is evolved.

Other reduction reactions are possible, depending on the nature of solution to which the metal is exposed. For an acid solution having dissolved oxygen, reduction according to

$$O_2 + 4H^+ + 4e^- \longrightarrow 2H_2O \tag{18.4}$$

will probably occur. Or, for a neutral or basic aqueous solution in which oxygen is also dissolved,

$$O_2 + 2H_2O + 4e^- \longrightarrow 4(OH^-) \tag{18.5}$$

Any metal ions present in the solution may also be reduced; for ions that can exist in more than one valence state (multivalent ions), reduction may occur by

$$M^{n+} + e^- \longrightarrow M^{(n-1)+} \tag{18.6}$$

in which the metal ion decreases its valence state by accepting an electron. Or, a metal may be totally reduced from an ionic to a neutral metallic state according to

$$M^{n+} + ne^- \longrightarrow M \tag{18.7}$$

That position at which reduction occurs is called the **cathode.** Of course, it is possible for two or more of the reduction reactions above to occur simultaneously.

An overall electrochemical reaction must consist of at least one oxidation and one reduction reaction, and will be the sum of them; often oxidation and reduction reactions are termed *half-reactions*. There can be no net electric charge accumulation from the electrons and ions; that is, the total rate of oxidation must equal the total rate of reduction, or all electrons generated through oxidation must be consumed by reduction.

For example, consider zinc metal immersed in an acid solution containing H^+ ions. At some regions on the metal surface, zinc will experience oxidation or corrosion as illustrated in Figure 18.1:

$$Zn \longrightarrow Zn^{2+} + 2e^- \tag{18.8}$$

Since zinc is a metal, and therefore a good electrical conductor, these electrons may be transferred to an adjacent region at which the H^+ ions are reduced according to

$$2H^+ + 2e^- \longrightarrow H_2 \text{ (gas)} \tag{18.9}$$

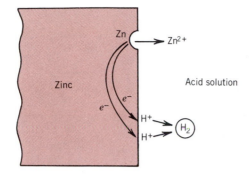

Zn

Zinc

Zn → Zn²⁺

Acid solution

e^-

H⁺
H⁺ → H₂

Figure 18.1 The electrochemical reactions associated with the corrosion of zinc in an acid solution. (From M. G. Fontana, *Corrosion Engineering*, 3rd edition. Copyright © 1986 by McGraw-Hill Book Company. Reproduced with permission.)

If no other oxidation or reduction reactions occur, the total electrochemical reaction is just the sum of reactions 18.8 and 18.9, or

$$\begin{aligned}
\text{Zn} &\longrightarrow \text{Zn}^{2+} + 2e^- \\
2\text{H}^+ + 2e^- &\longrightarrow \text{H}_2 \text{ (gas)} \\
\hline
\text{Zn} + 2\text{H}^+ &\longrightarrow \text{Zn}^{2+} + \text{H}_2 \text{ (gas)}
\end{aligned}$$
(18.10)

Another example is the oxidation or rusting of iron in water, which contains dissolved oxygen. This process occurs in two steps; in the first, Fe is oxidized to Fe^{2+} [as Fe(OH)_2],

$$\text{Fe} + \tfrac{1}{2}\text{O}_2 + \text{H}_2\text{O} \longrightarrow \text{Fe}^{2+} + 2\text{OH}^- \longrightarrow \text{Fe(OH)}_2 \qquad (18.11)$$

and, in the second stage, to Fe^{3+} [as Fe(OH)_3] according to

$$2\text{Fe(OH)}_2 + \tfrac{1}{2}\text{O}_2 + \text{H}_2\text{O} \longrightarrow 2\text{Fe(OH)}_3 \qquad (18.12)$$

The compound Fe(OH)_3 is the all-too-familiar rust.

As a consequence of oxidation, the metal ions may either go into the corroding solution as ions (reaction 18.8), or they may form an insoluble compound with non-metallic elements as in reaction 18.12.

Electrode Potentials

Not all metallic materials oxidize to form ions with the same degree of ease. Consider the electrochemical cell shown in Figure 18.2. On the left-hand side is a piece of pure iron immersed in a solution containing Fe^{2+} ions of $1M$ concentration.[1] The other side of the cell consists of a pure copper electrode in a $1M$ solution of Cu^{2+} ions. The cell halves are separated by a membrane, which limits the mixing of the two solutions. If the iron and copper electrodes are connected electrically, reduction will occur for copper at the expense of the oxidation of iron, as follows:

$$\text{Cu}^{2+} + \text{Fe} \longrightarrow \text{Cu} + \text{Fe}^{2+} \qquad (18.13)$$

or Cu^{2+} ions will deposit (electrodeposit) as metallic copper on the copper electrode, while iron dissolves (corrodes) on the other side of the cell and goes into solution as

---

[1] Concentration of liquid solutions is often expressed in terms of **molarity**, M, the number of moles of solute per million cubic millimeters (10^6 mm³, or 1000 cm³) of solution.

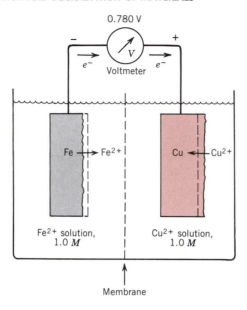

Figure 18.2 An electrochemical cell consisting of iron and copper electrodes, each of which is immersed in a $1M$ solution of its ion. Iron corrodes while copper electrodeposits.

Fe^{2+} ions. Thus the two half-cell reactions are represented by the relations

$$Fe \longrightarrow Fe^{2+} + 2e^- \qquad \text{(18.14a)}$$

$$Cu^{2+} + 2e^- \longrightarrow Cu \qquad \text{(18.14b)}$$

When a current passes through the external circuit, electrons generated from the oxidation of iron flow to the copper cell in order that Cu^{2+} be reduced. In addition, there will be some net ion motion from each cell to the other across the membrane. This is called a *galvanic couple*—two metals electrically connected in a liquid **electrolyte** wherein one metal becomes an anode and corrodes, while the other acts as a cathode.

An electric potential or voltage will exist between the two cell halves, and its magnitude can be determined if a voltmeter is connected in the external circuit. A potential of 0.780 V results for a copper–iron galvanic cell when the temperature is 25°C (77°F).

Now consider another galvanic couple consisting of the same iron half-cell connected to a metal zinc electrode that is immersed in a $1M$ solution of Zn^{2+} ions (Figure 18.3). In this case the zinc is the anode and corrodes, whereas the Fe now becomes the cathode. The electrochemical reaction is thus

$$Fe^{2+} + Zn \longrightarrow Fe + Zn^{2+} \qquad \text{(18.15)}$$

The potential associated with this cell reaction is 0.323 V.

Thus various electrode pairs have different voltages; the magnitude of such a voltage may be thought of as representing the driving force for the electrochemical oxidation–reduction reaction. Consequently, metallic materials may be rated as to their tendency to experience oxidation when coupled to other metals in solutions of their respective ions. A half-cell similar to those described above [i.e., a pure metal electrode immersed in a $1M$ solution of its ions and at 25°C (77°F)] is termed a **standard half-cell**.

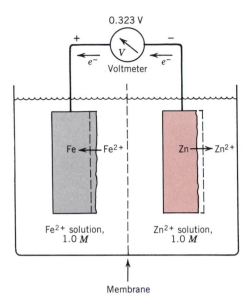

Figure 18.3 An electrochemical cell consisting of iron and zinc electrodes, each of which is immersed in a 1*M* solution of its ion. The iron electrodeposits while the zinc corrodes.

The Standard Emf Series

These measured cell voltages represent only differences in electrical potential, and thus it is convenient to establish a reference point, or reference cell, to which other cell halves may be compared. This reference cell, arbitrarily chosen, is the standard hydrogen electrode (Figure 18.4). It consists of an inert platinum electrode in a 1*M* solution of H^+ ions, saturated with hydrogen gas that is bubbled through the solution

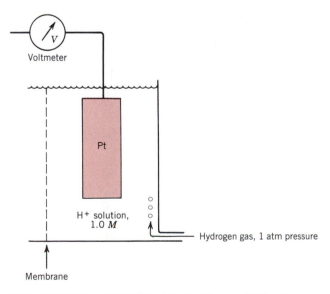

Figure 18.4 The standard hydrogen reference half-cell.

at a pressure of 1 atm and a temperature of 25°C (77°F). The platinum itself does not take part in the electrochemical reaction; it acts only as a surface on which hydrogen atoms may be oxidized or hydrogen ions may be reduced. The **electromotive force (emf) series** (Table 18.1) is generated by coupling to the standard hydrogen electrode, standard half-cells for various metals and ranking them according to measured voltage. Table 18.1 represents the corrosion tendencies for the several metals; those at the top (i.e., gold and platinum) are noble, or chemically inert. Moving down the table, the metals become increasingly more active, that is, more susceptible to oxidation. Sodium and potassium have the highest reactivities.

The voltages in Table 18.1 are for the half-reactions as *reduction reactions,* with the electrons on the left-hand side of the chemical equation; for oxidation, the direction of the reaction is reversed and the sign of the voltage changed.

Consider the generalized reactions involving the oxidation of metal M_1 and the reduction of metal M_2, as

$$M_1 \longrightarrow M_1^{n+} + ne^- \qquad -V_1^0 \qquad (18.16a)$$

$$M_2^{n+} + ne^- \longrightarrow M_2 \qquad +V_2^0 \qquad (18.16b)$$

where the V^0's are the standard potentials as taken from the standard emf series. Since metal M_1 is oxidized, the sign of V_1^0 is opposite to that as it appears in Table 18.1. Addition of Equations 18.16a and 18.16b yields

$$M_1 + M_2^{n+} \longrightarrow M_1^{n+} + M_2 \qquad (18.17)$$

and the overall cell potential ΔV^0 is

$$\Delta V^0 = V_2^0 - V_1^0 \qquad (18.18)$$

TABLE 18.1 The Standard emf Series

	Electrode Reaction	Standard Electrode Potential, V^0 (V)
↑	$Au^{3+} + 3e^- \longrightarrow Au$	+1.420
	$O_2 + 4H^+ + 4e^- \longrightarrow 2H_2O$	+1.229
	$Pt^{2+} + 2e^- \longrightarrow Pt$	~ +1.2
	$Ag^+ + e^- \longrightarrow Ag$	+0.800
Increasingly inert	$Fe^{3+} + e^- \longrightarrow Fe^{2+}$	+0.771
(cathodic)	$O_2 + 2H_2O + 4e^- \longrightarrow 4(OH^-)$	+0.401
	$Cu^{2+} + 2e^- \longrightarrow Cu$	+0.340
	$2H^+ + 2e^- \longrightarrow H_2$	0.000
	$Pb^{2+} + 2e^- \longrightarrow Pb$	−0.126
	$Sn^{2+} + 2e^- \longrightarrow Sn$	−0.136
	$Ni^{2+} + 2e^- \longrightarrow Ni$	−0.250
	$Co^{2+} + 2e^- \longrightarrow Co$	−0.277
	$Cd^{2+} + 2e^- \longrightarrow Cd$	−0.403
	$Fe^{2+} + 2e^- \longrightarrow Fe$	−0.440
Increasingly active	$Cr^{3+} + 3e^- \longrightarrow Cr$	−0.744
(anodic)	$Zn^{2+} + 2e^- \longrightarrow Zn$	−0.763
	$Al^{3+} + 3e^- \longrightarrow Al$	−1.662
↓	$Mg^{2+} + 2e^- \longrightarrow Mg$	−2.363
	$Na^+ + e^- \longrightarrow Na$	−2.714
	$K^+ + e^- \longrightarrow K$	−2.924

For this reaction to occur spontaneously, ΔV^0 must be positive; if it is negative, the spontaneous cell direction is just the reverse of Equation 18.17. When standard half-cells are coupled together, the metal that lies lower in Table 18.1 will experience oxidation, while the higher one will be reduced.

Influence of Concentration and Temperature on Cell Potential

The emf series applies to highly idealized electrochemical cells (i.e., pure metals in $1M$ solutions of their ions, at 25°C). Altering temperature or solution concentration or using alloy electrodes instead of pure metals will change the cell potential, and, in some cases, the spontaneous reaction direction may be reversed.

Consider again the electrochemical reaction described by Equation 18.17. If M_1 and M_2 electrodes are pure metals, the cell potential depends on the absolute temperature T and the molar ion concentrations $[M_1^{n+}]$ and $[M_2^{n+}]$ according to the Nernst equation:

$$\Delta V = (V_2^0 - V_1^0) - \frac{RT}{n\mathscr{F}} \ln \frac{[M_1^{n+}]}{[M_2^{n+}]} \qquad (18.19)$$

Where R is the gas constant, n is the number of electrons participating in either of the half-cell reactions, and $\mathscr{F}$ is the Faraday constant, 96,500 C/mol—the magnitude of charge per mole (6.023×10^{23}) of electrons. At 25°C (about room temperature),

$$\Delta V = (V_2^0 - V_1^0) - \frac{0.0592}{n} \log \frac{[M_1^{n+}]}{[M_2^{n+}]} \qquad (18.20)$$

to give ΔV in volts. Again, for reaction spontaneity, ΔV must be positive. As expected, for $1M$ concentrations of both ion types (that is, $[M_1^{n+}] = [M_2^{n+}] = 1$), Equation 18.19 simplifies to Equation 18.18.

EXAMPLE PROBLEM 18.1

One half of an electrochemical cell consists of a pure nickel electrode in a solution of Ni^{2+} ions; the other is a cadmium electrode immersed in a Cd^{2+} solution.

(a) If the cell is a standard one, write the spontaneous overall reaction and calculate the voltage that is generated.

(b) Compute the cell potential at 25°C if the Cd^{2+} and Ni^{2+} concentrations are 0.5 and $10^{-3}M$, respectively. Is the spontaneous reaction direction still the same as for the standard cell?

SOLUTION

(a) The cadmium electrode will be oxidized and nickel reduced because cadmium is lower in the emf series; thus the spontaneous reactions will be

$$\begin{array}{rl} Cd &\longrightarrow Cd^{2+} + 2e^- \\ \underline{Ni^{2+} + 2e^- \longrightarrow Ni} & \\ Ni^{2+} + Cd &\longrightarrow Ni + Cd^{2+} \end{array} \qquad (18.21)$$

From Table 18.1, the half-cell potentials for cadmium and nickel are, respectively, -0.403 and -0.250 V. Therefore, from Equation 18.18,

$$\Delta V = V_{Ni}^0 - V_{Cd}^0 = -0.250 \text{ V} - (-0.403 \text{ V}) = +0.153 \text{ V}$$

(b) For this portion of the problem, Equation 18.20 must be utilized, since the half-cell solution concentrations are no longer $1M$. At this point it is necessary to make a calculated guess as to which metal species is oxidized (or reduced). This choice will either be affirmed or refuted on the basis of the sign of ΔV at the conclusion of the computation. For the sake of argument, let us assume that in contrast to part a, nickel is oxidized and cadmium reduced according to

$$Cd^{2+} + Ni \longrightarrow Cd + Ni^{2+} \qquad (18.22)$$

Thus

$$\Delta V = (V^0_{Cd} - V^0_{Ni}) - \frac{RT}{n\mathscr{F}} \ln \frac{[Ni^{2+}]}{[Cd^{2+}]}$$

$$= -0.403 \text{ V} - (-0.250 \text{ V}) - \frac{0.0592}{2} \log \left(\frac{10^{-3}}{0.50}\right)$$

$$= -0.073 \text{ V}$$

Since ΔV is negative, the spontaneous reaction direction is the opposite to that of Equation 18.22, or the same as that of the standard cell (Equation 18.21).

The Galvanic Series

Even though Table 18.1 was generated under highly idealized conditions and has limited utility, it nevertheless indicates the relative reactivities of the metals. A more realistic and practical ranking, however, is provided by the **galvanic series,** Table 18.2. This represents the relative reactivities of a number of metals and commercial alloys in seawater. The alloys near the top are cathodic and unreactive, whereas those at the bottom are most anodic; no voltages are provided. Comparison of the standard emf and the galvanic series reveals a high degree of correspondence between the relative positions of the pure base metals.

Most metals and alloys are subject to oxidation or corrosion to one degree or another in a wide variety of environments; that is, they are more stable in an ionic state than as metals. In thermodynamic terms, there is a net decrease in free energy in going from metallic to oxidized states. Consequently, essentially all metals occur in nature as compounds—for example, oxides, hydroxides, carbonates, silicates, sulfides, and sulfates. Two notable exceptions are the noble metals gold and platinum. For them, oxidation in most environments is not favorable, and, therefore, they may exist in nature in the metallic state.

18.3 CORROSION RATES

The half-cell potentials listed in Table 18.1 are thermodynamic parameters that relate to systems at equilibrium. For example, for the discussions pertaining to Figures 18.2 and 18.3, it was tacitly assumed that there was no current flow through the external circuit. Real corroding systems are not at equilibrium; there will be a flow of electrons from anode to cathode (corresponding to the short-circuiting of the electrochemical cells in Figures 18.2 and 18.3), which means that the half-cell potential parameters (Table 18.1) cannot be applied.

Furthermore, these half-cell potentials represent the magnitude of a driving force, or the tendency for the occurrence of the particular half-cell reaction. However, it

TABLE 18.2 The Galvanic Series

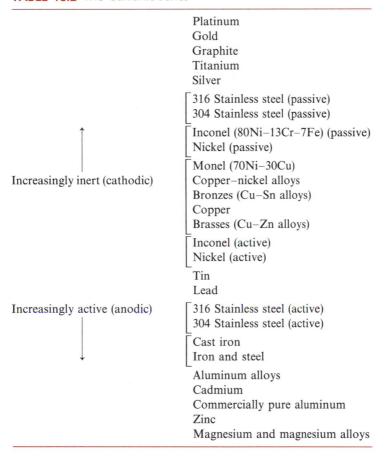

	Platinum
	Gold
	Graphite
	Titanium
	Silver
	316 Stainless steel (passive)
	304 Stainless steel (passive)
	Inconel (80Ni–13Cr–7Fe) (passive)
	Nickel (passive)
Increasingly inert (cathodic)	Monel (70Ni–30Cu)
	Copper–nickel alloys
	Bronzes (Cu–Sn alloys)
	Copper
	Brasses (Cu–Zn alloys)
	Inconel (active)
	Nickel (active)
	Tin
	Lead
Increasingly active (anodic)	316 Stainless steel (active)
	304 Stainless steel (active)
	Cast iron
	Iron and steel
	Aluminum alloys
	Cadmium
	Commercially pure aluminum
	Zinc
	Magnesium and magnesium alloys

Source: M. G. Fontana, *Corrosion Engineering,* 3rd edition. Copyright 1986 by McGraw-Hill Book Company. Reprinted with permission.

should be noted that although these potentials may be used to determine spontaneous reaction directions, they provide no information as to corrosion rates. That is, even though a ΔV potential computed for a specific corrosion situation using Equation 18.20 is a relatively large positive number, the reaction may occur at only an insignificantly slow rate. From an engineering perspective, we are interested in predicting the rates at which systems corrode; this requires the utilization of other parameters, as discussed below.

The corrosion rate, or the rate of material removal as a consequence of the chemical action, is an important corrosion parameter. This may be expressed as the **corrosion penetration rate (CPR),** or the thickness loss of material per unit of time. The formula for this calculation is

$$\text{CPR} = \frac{KW}{\rho At} \tag{18.23}$$

where W is the weight loss after exposure time t; ρ and A represent the density and exposed specimen area, respectively, and K is a constant, its magnitude depending on the system of units used. The CPR is conveniently expressed in terms of either mils per year (mpy) or millimeters per year (mm/yr). In the first case, $K = 534$ to give CPR in mpy (where 1 mil = 0.001 in.), and W, ρ, A, and t are specified in units of milligrams, grams per cubic centimeter, square inches, and hours, respectively. In the second case, $K = 87.6$ for mm/yr, and units for the other parameters are the same as for mils per year, except that A is given in square centimeters. For most applications a corrosion penetration rate less than about 20 mpy (0.50 mm/yr) is acceptable.

Inasmuch as there is an electrical current associated with electrochemical corrosion reactions, we can also express corrosion rate in terms of this current, or more specifically, current density—that is, the current per unit surface area of material corroding—which is designated i. The rate r, in units of mol/m²-s, is determined using the expression

$$r = \frac{i}{n\mathscr{F}} \tag{18.24}$$

where, again, n is the number of electrons associated with the ionization of each metal atom, and $\mathscr{F}$ is 96,500 C/mol.

18.4 PREDICTION OF CORROSION RATES

Polarization

Consider the standard Zn/H_2 electrochemical cell shown in Figure 18.5, which has been short-circuited such that oxidation of zinc and reduction of hydrogen will occur at their respective electrode surfaces. The potentials of the two electrodes will not be at the values determined from Table 18.1 because the system is now a nonequilibrium one. The displacement of each electrode potential from its equilibrium value is termed **polarization,** and the magnitude of this displacement is the *overvoltage,* normally represented by the symbol η. Overvoltage is expressed in terms of plus or minus volts (or millivolts) relative to the equilibrium potential. For example, suppose that the zinc electrode in Figure 18.5 has a potential of -0.621 V after it has been connected to the platinum electrode. The equilibrium potential is -0.763 V (Table 18.1), and, therefore,

$$\eta = -0.621 \text{ V} - (-0.763 \text{ V}) = +0.142 \text{ V}$$

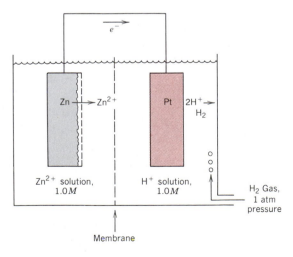

Figure 18.5 Electrochemical cell consisting of standard zinc and hydrogen electrodes that has been short-circuited.

There are two types of polarization—activation and concentration—the mechanisms of which will now be discussed since they control the rate of electrochemical reactions.

Activation Polarization.   All electrochemical reactions consist of a sequence of steps that occur in series at the interface between the metal electrode and the electrolyte solution. **Activation polarization** refers to the condition wherein the reaction rate is controlled by the one step in the series that occurs at the slowest rate. The term "activation" is applied to this type of polarization because an activation energy barrier is associated with this slowest, rate-limiting step.

To illustrate, let us consider the reduction of hydrogen ions to form bubbles of hydrogen gas on the surface of a zinc electrode (Figure 18.6). It is conceivable that this reaction could proceed by the following step sequence:

1. Adsorption of H^+ ions from the solution onto the zinc surface.
2. Electron transfer from the zinc to form a hydrogen atom,
$$H^+ + e^- \longrightarrow H$$
3. Combining of two hydrogen atoms to form a molecule of hydrogen,
$$2H \longrightarrow H_2$$
4. The coalescence of many hydrogen molecules to form a bubble.

The slowest of these steps determines the rate of the overall reaction.

For activation polarization, the relationship between overvoltage η_a and current density i is

$$\eta_a = \pm \beta \log \frac{i}{i_0} \qquad (18.25)$$

where β and i_0 are constants for the particular half-cell. The parameter i_0 is termed the *exchange current density*, which deserves a brief explanation. Equilibrium for some particular half-cell reaction is really a dynamic state on the atomic level. That is, oxidation and reduction processes are occurring, but both at the same rate, so that there is no net reaction. For example, for the standard hydrogen cell (Figure 18.4)

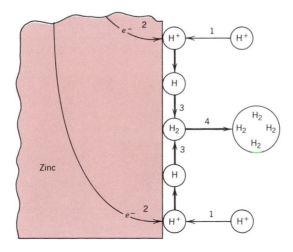

Figure 18.6 Schematic representation of possible steps in the hydrogen reduction reaction, the rate of which is controlled by activation polarization. (From M. G. Fontana, *Corrosion Engineering*, 3rd edition. Copyright © 1986 by McGraw-Hill Book Company. Reproduced with permission.)

reduction of hydrogen ions in solution will take place at the surface of the platinum electrode according to

$$2H^+ + 2e^- \longrightarrow H_2$$

with a corresponding rate r_{red}. Similarly, hydrogen gas in the solution will experience oxidation as

$$H_2 \longrightarrow 2H^+ + 2e^-$$

at rate r_{oxid}. Equilibrium exists when

$$r_{red} = r_{oxid}$$

This exchange current density is just the current density from Equation 18.24 at equilibrium, or

$$r_{red} = r_{oxid} = \frac{i_0}{n\mathscr{F}} \qquad (18.26)$$

Use of the term "current density" for i_0 is a little misleading inasmuch as there is no net current. Furthermore, the value for i_0 is determined experimentally and will vary from system to system.

According to Equation 18.25, when overvoltage is plotted as a function of the logarithm of current density, straight line segments result; these are shown in Figure 18.7 for the hydrogen electrode. The line segment with a slope of $+\beta$ corresponds to the oxidation half-reaction, whereas the line with a $-\beta$ slope is for reduction. Also worth noting is that both line segments originate at i_0 (H_2/H^+), the exchange current density, and at zero overvoltage, since at this point the system is at equilibrium and there is no net reaction.

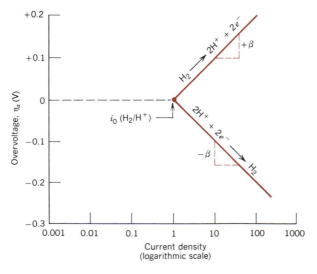

Figure 18.7 For a hydrogen electrode, plot of activation polarization overvoltage versus logarithm of current density for both oxidation and reduction reactions. (Adapted from M. G. Fontana, *Corrosion Engineering,* 3rd edition. Copyright © 1986 by McGraw-Hill Book Company. Reproduced with permission.)

Concentration Polarization.   **Concentration polarization** exists when the reaction rate is limited by diffusion in the solution. For example, consider again the hydrogen evolution reduction reaction. When the reaction rate is low and/or the concentration of H^+ is high, there is always an adequate supply of hydrogen ions available in the solution at the region near the electrode interface (Figure 18.8a). On the other hand, at high rates and/or low H^+ concentrations, a depletion zone may be formed in the

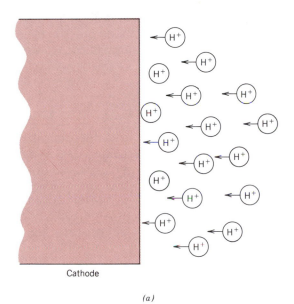

Cathode

(a)

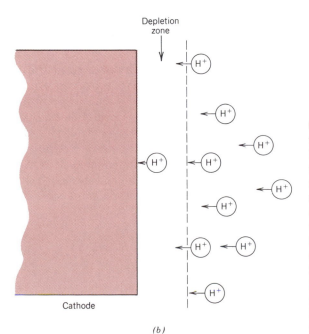

Depletion zone

Cathode

(b)

Figure 18.8 For hydrogen reduction, schematic representations of H^+ distribution in the vicinity of the cathode for (a) low reaction rates and/or high concentrations, and (b) for high reaction rates and/or low concentrations wherein a depletion zone is formed which gives rise to concentration polarization. (Adapted from M. G. Fontana, *Corrosion Engineering,* 3rd edition. Copyright © 1986 by McGraw-Hill Book Company. Reproduced with permission.)

vicinity of the interface, inasmuch as the H^+ ions are not replenished at a rate sufficient to keep up with the reaction (Figure 18.8b). Thus diffusion of H^+ to the interface is rate controlling, and the system is said to be concentration polarized. Concentration polarization generally occurs only for reduction reactions because for oxidation, there is virtually an unlimited supply of metal atoms at the corroding electrode interface.

Concentration polarization data are also normally plotted as overvoltage versus the logarithm of current density; such a plot is represented schematically in Figure 18.9a.[2] It may be noted from this figure that overvoltage is independent of current density until i approaches i_L; at this point η_c decreases abruptly in magnitude.

Both concentration and activation polarization are possible for reduction reactions. Under these circumstances, the total overvoltage is just the sum of both overvoltage contributions. Figure 18.9b shows such a schematic η-versus-log i plot.

Corrosion Rates from Polarization Data

Let us now apply the concepts developed above to the determination of corrosion rates. Two types of systems will be discussed. In the first case, both oxidation and reduction reactions are rate limited by activation polarization. In the second case, concentration polarization controls the reduction reaction, whereas only activation polarization is important for oxidation. Case one will be illustrated by considering the corrosion of zinc immersed in an acid solution (see Figure 18.1). The reduction of H^+ ions to form H_2 gas bubbles occurs at the surface of the zinc according to

$$2H^+ + 2e^- \longrightarrow H_2 \tag{18.3}$$

and the zinc oxidizes as

$$Zn \longrightarrow Zn^{2+} + 2e^- \tag{18.8}$$

No net charge accumulation may result from these two reactions; that is, all electrons generated by reaction 18.8 must be consumed by reaction 18.3, which is to say that rates of oxidation and reduction must be equal.

Activation polarization for both reactions is expressed graphically in Figure 18.10 as cell potential referenced to the standard hydrogen electrode (not overvoltage) versus the logarithm of current density. The potentials of the uncoupled hydrogen and zinc half-cells, $V(H^+/H_2)$ and $V(Zn/Zn^{2+})$, respectively, are indicated, along with their respective exchange current densities, $i_0(H^+/H_2)$ and $i_0(Zn/Zn^{2+})$. Straight line segments are shown for hydrogen reduction and zinc oxidation. Upon immersion, both hydrogen and zinc experience activation polarization along their respective lines. Also, oxidation and reduction rates must be equal as explained above, which is only possible at the intersection of the two line segments; this intersection occurs at the corrosion

---

[2] The mathematical expression relating concentration polarization overvoltage η_c and current density i is

$$\eta_c = \frac{2.3RT}{n\mathscr{F}} \log\left(1 - \frac{i}{i_L}\right) \tag{18.27}$$

where R and T are the gas constant and absolute temperature, respectively, n and $\mathscr{F}$ have the same meanings as above, and i_L is the limiting diffusion current density.

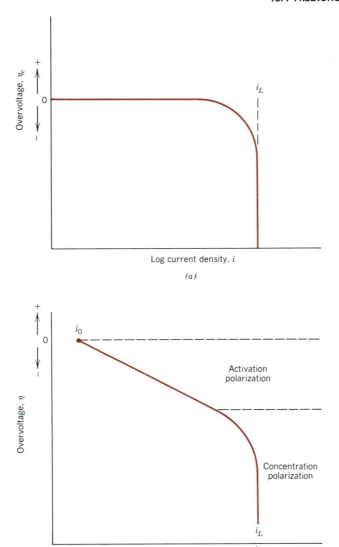

Figure 18.9 For reduction reactions, schematic plots of overvoltage versus logarithm of current density for (a) concentration polarization, and (b) combined activation-concentration polarization.

potential, designated V_C, and the corrosion current density i_C. The corrosion rate of zinc (which also corresponds to the rate of hydrogen evolution) may thus be computed by insertion of this i_C value into Equation 18.24.

The second corrosion case (combined activation and concentration polarization for hydrogen reduction and activation polarization for oxidation of metal M) is treated

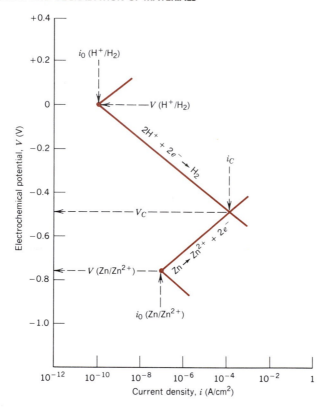

Figure 18.10 Electrode kinetic behavior of zinc in an acid solution; both oxidation and reduction reactions are rate limited by activation polarization. (Adapted from M. G. Fontana, *Corrosion Engineering,* 3rd edition. Copyright © 1986 by McGraw-Hill Book Company. Reproduced with permission.)

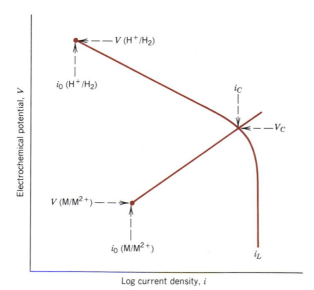

Figure 18.11 Schematic electrode kinetic behavior for metal M; the reduction reaction is under combined activation–concentration polarization control.

in a like manner. Figure 18.11 shows both polarization curves; as above, corrosion potential and corrosion current density correspond to the point at which the oxidation and reduction lines intersect.

18.5 PASSIVITY

Some normally active metals and alloys exist that, under particular environmental conditions, lose their chemical reactivity and become extremely inert. This phenomenon, termed **passivity,** is displayed by chromium, iron, nickel, titanium, and many of their alloys. It is felt that this passive behavior results from the formation of a highly adherent and very thin oxide film on the metal surface, which serves as a protective barrier to further corrosion. Stainless steels are highly resistant to corrosion in a rather wide variety of atmospheres as a result of passivation. They contain at least 11% chromium which, as a solid-solution alloying element in iron, minimizes the formation of rust; instead, a protective surface film forms. Aluminum is highly corrosion resistant in many environments because it also passivates. If damaged, the protective film normally reforms very rapidly. However, a change in the character of the environment (e.g., alteration in the concentration of the active corrosive species) may cause a passivated material to revert to an active state. Subsequent damage to a preexisting passive film could result in a substantial increase in corrosion rate, by as much as 100,000 times.

This passivation phenomenon may be explained in terms of polarization potential–log current density curves discussed in the preceding section. The polarization curve for a metal that passivates will have the general shape shown in Figure 18.12. At relatively low potential values, within the "active" region the behavior is linear as

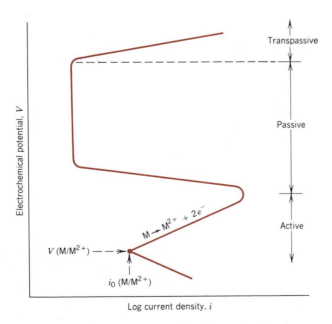

Figure 18.12 Schematic polarization curve for a metal that displays an active–passive transition.

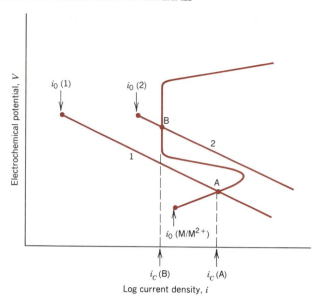

Figure 18.13 Demonstration of how an active–passive metal can exhibit both active and passive corrosion behavior.

it is for normal metals. With increasing potential, the current density suddenly decreases to a very low value which remains independent of potential; this is termed the "passive" region. Finally, at even higher potential values, the current density again increases with potential in the "transpassive" region.

Figure 18.13 illustrates how a metal can experience both active and passive behavior depending on the corrosion environment. Included in this figure is the S-shaped oxidation polarization curve for an active–passive metal M and, in addition, reduction polarization curves for two different solutions, which are labeled 1 and 2. Curve 1 intersects the oxidation polarization curve in the active region at point A, yielding a corrosion current density i_C (A). The intersection of curve 2 at point B, is in the passive region and at current density i_C (B). The corrosion rate of metal M in solution 1 is greater than in solution 2 since i_C (A) is greater than i_C (B) and rate is proportional to current density according to Equation 18.24. This difference in corrosion rate between the two solutions may be significant—several orders of magnitude—when one considers that the current density scale in Figure 18.13 is scaled logarithmically.

18.6 ENVIRONMENTAL EFFECTS

The variables in the corrosion environment, which include fluid velocity, temperature, and composition, can have a decided influence on the corrosion properties of the materials that are in contact with it. In most instances, increasing fluid velocity enhances the rate of corrosion due to erosive effects, as discussed later in the chapter. The rates of most chemical reactions rise with increasing temperature; this also holds for the great majority of corrosion situations. Increasing the concentration of the corrosive species (e.g., H^+ ions in acids) in many situations produces a more rapid rate of corrosion. However, for materials capable of passivation, raising the corrosive content may result in an active-to-passive inversion, with a considerable reduction in corrosion.

Cold working or plastically deforming ductile metals is used to increase their strength; however, a cold-worked metal is more susceptible to corrosion than the same material in an annealed state. For example, deformation processes are used to shape the head and point of a nail; consequently, these positions are anodic with respect to the shank region. Thus differential cold working on a structure should be a consideration when a corrosive environment may be encountered during service.

18.7 FORMS OF CORROSION

It is convenient to classify corrosion according to the manner in which it is manifest. Metallic corrosion is sometimes classified into eight forms: uniform, galvanic, crevice, pitting, intergranular, selective leaching, erosion–corrosion, and stress corrosion. The causes and means of prevention of each of these forms are discussed briefly.

Uniform Attack

Uniform attack is a form of electrochemical corrosion that occurs with equivalent intensity over the entire exposed surface and often leaves behind a scale or deposit. In a microscopic sense, the oxidation and reduction reactions occur randomly over the surface. Some familiar examples include general rusting of steel and iron and the tarnishing of silverware. This is probably the most common form of corrosion. It is also the least objectionable because it can be predicted and designed for with relative ease.

Galvanic Corrosion

Galvanic corrosion occurs when two metals or alloys having different compositions are electrically coupled while exposed to an electrolyte. This is the type of corrosion or dissolution that was described in Section 18.2. The less noble or more reactive metal in the particular environment will experience corrosion; the more inert metal, the cathode, will be protected from corrosion. For example, steel screws corrode when in contact with brass in a marine environment; or if copper and steel tubing are joined in a domestic water heater, the steel will corrode in the vicinity of the junction. Depending on the nature of the solution, one or more of the reduction reactions, Equations 18.3 through 18.7, will occur at the surface of the cathode material. Figure 18.14 shows galvanic corrosion.

Again, the galvanic series (Table 18.2) indicates the relative reactivities, in seawater, of a number of metals and alloys. When two alloys are coupled in seawater, the one lower in the series will experience corrosion. Some of the alloys in the table are grouped in brackets. Generally the base metal is the same for these bracketed alloys, and there is little danger of corrosion if alloys within a single bracket are coupled. It is also worth noting from this series that some alloys are listed twice (e.g., nickel and the stainless steels), in both active and passive states.

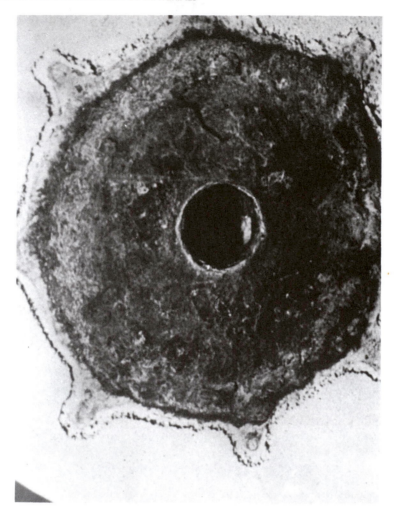

Figure 18.14 Galvanic corrosion of a magnesium shell that was cast around a steel core. (Photograph courtesy of LaQue Center for Corrosion Technology, Inc.)

The rate of galvanic attack depends on the relative anode-to-cathode surface areas that are exposed to the electrolyte, and rate is related directly to the cathode–anode area ratio; that is, for a given cathode area, a smaller anode will corrode more rapidly than a larger one.

A number of measures may be taken to significantly reduce the effects of galvanic corrosion. These include the following:

1. If coupling of dissimilar metals is necessary, choose two that are close together in the galvanic series.

2. Avoid an unfavorable anode-to-cathode surface area ratio; use an anode area as large as possible.

3. Electrically insulate dissimilar metals from each other.

4. Electrically connect a third, anodic metal to the other two; this is a form of **cathodic protection,** discussed presently.

Crevice Corrosion

Electrochemical corrosion may also occur as a consequence of concentration differences of ions or dissolved gases in the electrolyte solution, and between two regions of the same metal piece. For such a *concentration cell,* corrosion occurs in the locale that has the lower concentration. A good example of this type of corrosion occurs in crevices and recesses or under deposits of dirt or corrosion products where the solution becomes stagnant and there is localized depletion of dissolved oxygen. Corrosion preferentially occurring at these positions is called **crevice corrosion** (Figure 18.15). The crevice must be wide enough for the solution to penetrate, yet narrow enough for stagnancy; usually the width is several thousandths of an inch.

The proposed mechanism for crevice corrosion is illustrated in Figure 18.16. After oxygen has been depleted within the crevice, oxidation of the metal occurs at this position according to Equation 18.1. Electrons from this electrochemical reaction are conducted through the metal to adjacent external regions, where they are consumed by reduction—most probably reaction 18.5. In many aqueous environments, the solution within the crevice has been found to develop high concentrations of H^+ and Cl^- ions, which are especially corrosive. Many alloys that passivate are susceptible to crevice corrosion because protective films are often destroyed by the H^+ and Cl^- ions.

Crevice corrosion may be prevented by using welded instead of riveted or bolted joints, using nonabsorbing gaskets when possible, removing accumulated deposits frequently, and designing containment vessels to avoid stagnant areas and ensure complete drainage.

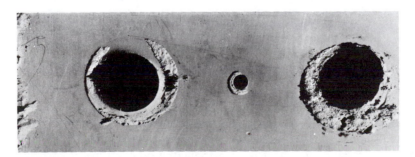

Figure 18.15 On this plate, which was immersed in seawater, crevice corrosion has occurred at the regions that were covered by washers. (Photograph courtesy of LaQue Center for Corrosion Technology, Inc.)

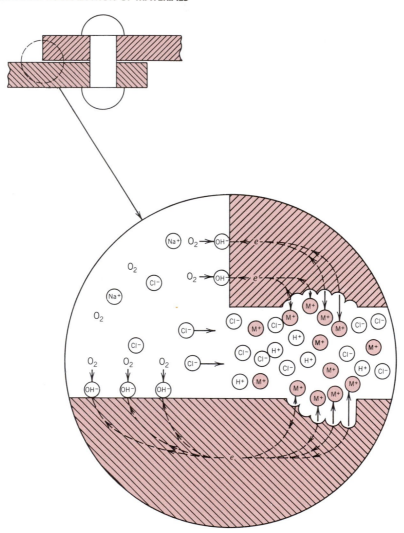

Figure 18.16 Schematic illustration of the mechanism of crevice corrosion between two riveted sheets. (From M. G. Fontana, *Corrosion Engineering,* 3rd edition. Copyright © 1986 by McGraw-Hill Book Company. Reproduced with permission.)

Pitting

Pitting is another form of very localized corrosion attack in which small pits or holes form. They ordinarily penetrate from the top of a horizontal surface downward in a nearly vertical direction. It is an extremely insidious type of corrosion, often going undetected and with very little material loss until failure occurs. An example of pitting corrosion is shown in Figure 18.17.

The mechanism for pitting is probably the same as for crevice corrosion in that oxidation occurs within the pit itself, with complementary reduction at the surface.

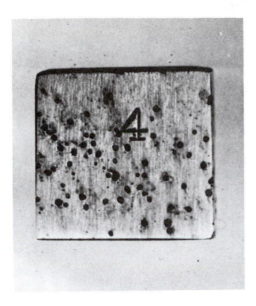

Figure 18.17 The pitting of a 304 stainless steel plate by an acid-chloride solution. (Photograph courtesy of Mars G. Fontana. From M. G. Fontana, *Corrosion Engineering,* 3rd edition. Copyright © 1986 by McGraw-Hill Book Company. Reproduced with permission.)

It is supposed that gravity causes the pits to grow downward, the solution at the pit tip becoming more concentrated and dense as pit growth progresses. A pit may be initiated by a localized surface defect such as a scratch or a slight variation in composition. In fact, it has been observed that specimens having polished surfaces display a greater resistance to pitting corrosion. Stainless steels are somewhat susceptible to this form of corrosion; however, alloying with about 2% molybdenum enhances their resistance significantly.

Intergranular Corrosion

As the name suggests, **intergranular corrosion** occurs preferentially along grain boundaries for some alloys and in specific environments. The net result is that a macroscopic specimen disintegrates along its grain boundaries. This type of corrosion is especially prevalent in some stainless steels. When heated to temperatures between 500 and 800°C (950 and 1450°F) for sufficiently long time periods, these alloys become sensitized to intergranular attack. It is believed that this heat treatment permits the formation of small precipitate particles of chromium carbide ($Cr_{23}C_6$) by reaction between the chromium and carbon in the stainless steel. These particles form along the grain boundaries, as illustrated in Figure 18.18. Both the chromium and the carbon must diffuse to the grain boundaries to form the precipitates, which leaves a chromium-depleted zone adjacent to the grain boundary. Consequently, this grain boundary region is now highly susceptible to corrosion.

Intergranular corrosion is an especially severe problem in the welding of stainless steels, when it is often termed **weld decay.** Figure 18.19 shows this type of intergranular corrosion.

Stainless steels may be protected from intergranular corrosion by the following measures: (1) subjecting the sensitized material to high-temperature heat treatment

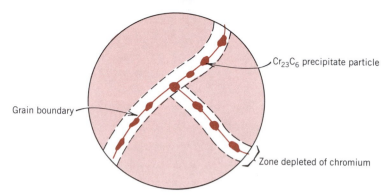

Figure 18.18 Schematic illustration of chromium carbide particles that have precipitated along grain boundaries in stainless steel, and the attendant zones of chromium depletion.

in which all the chromium carbide particles are redissolved, (2) lowering the carbon content below 0.03 wt% C so that carbide formation is minimal, and (3) alloying the stainless steel with another metal such as niobium or titanium, which has a greater tendency to form carbides than does chromium so that the Cr remains in solid solution.

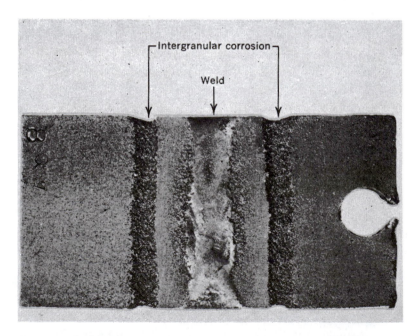

Figure 18.19 Weld decay in a stainless steel. The regions along which the grooves have formed were sensitized as the weld cooled. (From H. H. Uhlig and R. W. Revie, *Corrosion and Corrosion Control,* 3rd edition, Fig. 2, p. 307. Copyright © 1985 by John Wiley & Sons, Inc. Reprinted by permission of John Wiley & Sons, Inc.)

Selective Leaching

Selective leaching is found in solid solution alloys and occurs when one element or constituent is preferentially removed as a consequence of corrosion processes. The most common example is the dezincification of brass, in which zinc is selectively leached from a copper–zinc brass alloy. The mechanical properties of the alloy are significantly impaired, since only a porous mass of copper remains in the region that has been dezincified. In addition, the material changes from yellow to a red or copper color. Selective leaching may also occur with other alloy systems in which aluminum, iron, cobalt, chromium, and other elements are vulnerable to preferential removal.

Erosion–Corrosion

Erosion–corrosion arises from the combined action of chemical attack and mechanical abrasion or wear as a consequence of fluid motion. Virtually all metal alloys, to one degree or another, are susceptible to erosion–corrosion. It is especially harmful to alloys that passivate by forming a protective surface film; the abrasive action may erode away the film, leaving exposed a bare metal surface. If the coating is not capable of continuously and rapidly reforming as a protective barrier, corrosion may be severe. Relatively soft metals such as copper and lead are also sensitive to this form of attack. Usually it can be identified by surface grooves and waves having contours that are characteristic of the flow of the fluid.

The nature of the fluid can have a dramatic influence on the corrosion behavior. Increasing fluid velocity normally enhances the rate of corrosion. Also, a solution is more erosive when bubbles and suspended particulate solids are present.

Erosion–corrosion is commonly found in piping, especially at bends, elbows, and abrupt changes in pipe diameter—positions where the fluid changes direction or flow suddenly becomes turbulent. Propellers, turbine blades, valves, and pumps are also susceptible to this form of corrosion. Figure 18.20 illustrates the impingement failure of an elbow fitting.

One of the best ways to reduce erosion–corrosion is to change the design to eliminate fluid turbulence and impingement effects. Other materials may also be

Figure 18.20 Impingement failure of an elbow that was part of a steam condensate line. (Photograph courtesy of Mars G. Fontana. From M. G. Fontana, *Corrosion Engineering*, 3rd edition. Copyright © 1986 by McGraw-Hill Book Company. Reproduced with permission.)

utilized that inherently resist erosion. Furthermore, removal of particulates and bubbles from the solution will lessen its ability to erode.

Stress Corrosion

Stress corrosion, sometimes termed stress corrosion cracking, results from the combined action of an applied tensile stress and a corrosive environment; both influences are necessary. In fact, some materials that are virtually inert in a particular corrosive medium become susceptible to this form of corrosion when a stress is applied. Small cracks form and then propagate in a direction perpendicular to the stress (see page 562), with the result that failure may eventually occur. Failure behavior is characteristic of that for a brittle material, even though the metal alloy is intrinsically ductile. Furthermore, cracks may form at relatively low stress levels, significantly below the tensile strength. Most alloys are susceptible to stress corrosion in specific environments, especially at moderate stress levels. For example, most stainless steels stress corrode in solutions containing chloride ions, whereas brasses are especially vulnerable when exposed to ammonia. Figure 18.21 is a photomicrograph in which an example of intergranular stress corrosion cracking in brass is shown.

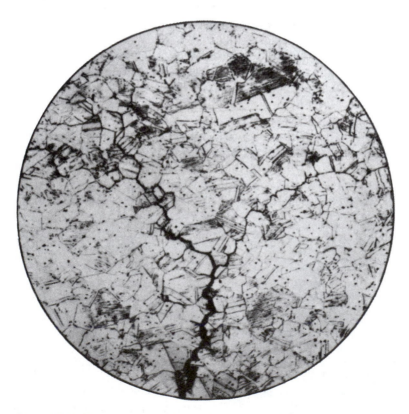

Figure 18.21 Photomicrograph showing intergranular stress corrosion cracking in brass. (From H. H. Uhlig and R. W. Revie, *Corrosion and Corrosion Control,* 3rd edition, Fig. 5, p. 335. Copyright 1985 by John Wiley & Sons, Inc. Reprinted by permission of John Wiley & Sons, Inc.)

The stress that produces stress corrosion cracking need not be externally applied; it may be a residual one that results from rapid temperature changes and uneven contraction, or for two-phase alloys in which each phase has a different coefficient of expansion. Also, gaseous and solid corrosion products that are entrapped internally can give rise to internal stresses.

Probably the best measure to take in reducing or totally eliminating stress corrosion is to lower the magnitude of the stress. This may be accomplished by reducing the external load or increasing the cross-sectional area perpendicular to the applied stress. Furthermore, an appropriate heat treatment may be used to anneal out any residual thermal stresses.

18.8 CORROSION ENVIRONMENTS

Corrosive environments include the atmosphere, aqueous solutions, soils, acids, bases, inorganic solvents, molten salts, liquid metals, and last but not least, the human body. On a tonnage basis, atmospheric corrosion accounts for the greatest losses. Moisture containing dissolved oxygen is the primary corrosive agent, but other substances, including sulfur compounds and sodium chloride, may also contribute. This is especially true of marine atmospheres, which are highly corrosive because of the presence of sodium chloride. Dilute sulfuric acid solutions (acid rain) in industrial environments can also cause corrosion problems. Metals commonly used for atmospheric applications include alloys of aluminum and copper, and galvanized steel.

Water environments can also have a variety of compositions and corrosion characteristics. Fresh water normally contains dissolved oxygen, as well as other minerals several of which account for hardness. Seawater contains approximately 3.5% salt (predominantly sodium chloride), as well as some minerals and organic matter. Seawater is generally more corrosive than fresh water, frequently producing pitting and crevice corrosion. Cast iron, steel, aluminum, copper, brass, and some stainless steels are generally suitable for freshwater use, whereas titanium, brass, some bronzes, copper–nickel alloys, and nickel–chromium–molybdenum alloys are highly corrosion resistant in seawater.

Soils have a wide range of compositions and susceptibilities to corrosion. Compositional variables include moisture, oxygen, salt content, alkalinity, and acidity, as well as the presence of various forms of bacteria. Cast iron and plain carbon steels, both with and without protective surface coatings, are found most economical for underground structures.

Because there are so many acids, bases, and organic solvents, no attempt is made to discuss these solutions. Good references are available that treat these topics in detail.

18.9 CORROSION PREVENTION

Some corrosion prevention methods were treated relative to the eight forms of corrosion; however, only the measures specific to each of the various corrosion types were discussed. Now, some more general techniques are presented; these include material selection, environmental alteration, design, coatings, and cathodic protection.

Perhaps the most common and easiest way of preventing corrosion is through the judicious selection of materials once the corrosion environment has been characterized. Standard corrosion references are helpful in this respect. Here, cost may be a significant factor. It is not always economically feasible to employ the material that renders the optimum corrosion resistance; sometimes, either another alloy and/or some other measure must be used.

Changing the character of the environment, if possible, may also significantly influence corrosion. Lowering the fluid temperature and/or velocity usually produces a reduction in the rate at which corrosion occurs. Many times increasing or decreasing the concentration of some species in the solution will have a positive effect; for example, the metal may experience passivation.

Inhibitors are substances that, when added in relatively low concentrations to the environment, decrease its corrosiveness. Of course, the specific inhibitor depends both on the alloy and on the corrosive environment. There are several mechanisms that may account for the effectiveness of inhibitors. Some react with and virtually eliminate a chemically active species in the solution (such as dissolved oxygen). Other inhibitor molecules attach themselves to the corroding surface and interfere with either the oxidation or the reduction reaction, or form a very thin protective coating. Inhibitors are normally used in closed systems such as automobile radiators and steam boilers.

Several aspects of design consideration have already been discussed, especially with regard to galvanic and crevice corrosion, and erosion–corrosion. In addition, the design should allow for complete drainage in the case of a shutdown, and easy washing. Since dissolved oxygen may enhance the corrosivity of many solutions, the design should, if possible, include provision for the exclusion of air.

Physical barriers to corrosion are applied on surfaces in the form of films and coatings. A large diversity of metallic and nonmetallic coating materials are available. It is essential that the coating maintain a high degree of surface adhesion, which undoubtedly requires some preapplication surface treatment. In most cases, the coating must be virtually nonreactive in the corrosive environment and resistant to mechanical damage that exposes the bare metal to the corrosive environment. All three material types—metals, ceramics, and polymers—are used as coatings for metals.

Cathodic Protection

One of the most effective means of corrosion prevention is **cathodic protection;** it can be used for all eight different forms of corrosion as discussed above, and may, in some situations, completely stop corrosion. Again, oxidation or corrosion of a metal M occurs by the generalized reaction

$$M \longrightarrow M^{n+} + ne^- \tag{18.1}$$

Cathodic protection simply involves supplying, from an external source, electrons to the metal to be protected, making it a cathode; the reaction above is thus forced in the reverse (or reduction) direction.

One cathodic protection technique employs a galvanic couple: the metal to be protected is electrically connected to another metal that is more reactive in the particular environment. The latter experiences oxidation, and, upon giving up electrons, protects the first metal from corrosion. The oxidized metal is often called a **sacrificial anode,** and magnesium and zinc are commonly used as such because they lie at the anodic end of the galvanic series. This form of galvanic protection, for structures buried in the ground, is illustrated in Figure 18.22a.

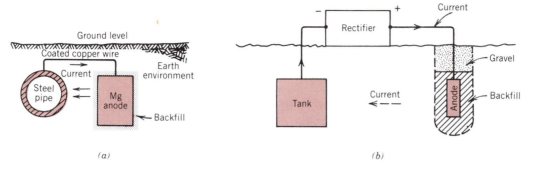

(a)   (b)

Figure 18.22 Cathodic protection of (a) an underground pipeline using a magnesium sacrificial anode, and (b) an underground tank using an impressed current. (From M. G. Fontana, *Corrosion Engineering*, 3rd edition. Copyright © 1986 by McGraw-Hill Book Company. Reproduced with permission.)

The process of *galvanizing* is simply one in which a layer of zinc is applied to the surface of steel by hot dipping. In the atmosphere and most aqueous environments, zinc is anodic to and will thus cathodically protect the steel if there is any surface damage (Figure 18.23). Any corrosion of the zinc coating will proceed at an extremely slow rate because the ratio of the anode-to-cathode surface area is quite large.

For another method of cathodic protection, the source of electrons is an impressed current from an external dc power source, as represented in Figure 18.22b for an underground tank. The negative terminal of the power source is connected to the structure to be protected. The other terminal is joined to an inert anode (often graphite), which is, in this case, buried in the soil; high-conductivity backfill material provides good electrical contact between the anode and surrounding soil. A current path exists between the cathode and anode through the intervening soil, completing the electrical circuit. Cathodic protection is especially useful in preventing corrosion of water heaters, underground tanks and pipes, and marine equipment.

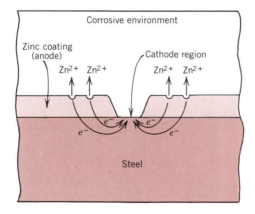

Figure 18.23 Galvanic protection of steel as provided by a coating of zinc.

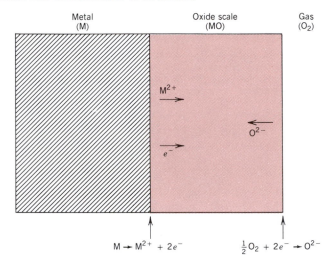

Figure 18.24 Schematic representation of processes that are involved in the gaseous oxidation at a metal surface.

18.10 OXIDATION

The discussion of Section 18.2 treated the corrosion of metallic materials in terms of electrochemical reactions that take place in aqueous solutions. In addition, oxidation of metal alloys is also possible in gaseous atmospheres, normally air, wherein an oxide layer or scale forms on the surface of the metal. This phenomenon is frequently termed *scaling, tarnishing,* or *dry corrosion.* In this section possible mechanisms for this type of corrosion, the types of oxide layers that can form, and the kinetics of oxide formation will be discussed.

Mechanisms

As with aqueous corrosion, the process of oxide layer formation is an electrochemical one, which may be expressed, for divalent metal M, by the following reaction[3]:

$$M + \tfrac{1}{2}O_2 \longrightarrow MO \tag{18.29}$$

Furthermore, the above reaction consists of oxidation and reduction half-reactions. The former, with the formation of metal ions,

$$M \longrightarrow M^{2+} + 2e^- \tag{18.30}$$

occurs at the metal–scale interface. The reduction half-reaction produces oxygen ions as follows:

$$\tfrac{1}{2}O_2 + 2e^- \longrightarrow O^{2-} \tag{18.31}$$

---

[3] For other than divalent metals, this reaction may be expressed as

$$aM + \frac{b}{2}O_2 \longrightarrow M_aO_b \tag{18.28}$$

and takes place at the scale–gas interface. A schematic representation of this metal–scale–gas system is shown in Figure 18.24.

For the oxide layer to increase in thickness via Equation 18.29, it is necessary that electrons be conducted to the scale–gas interface, at which point the reduction reaction occurs; in addition, M^{2+} ions must diffuse away from the metal–scale interface, and/or O^{2-} ions must diffuse toward this same interface (Figure 18.24).[4] Thus the oxide scale serves as both an electrolyte through which ions diffuse and as an electrical circuit for the passage of electrons. Furthermore, the scale may protect the metal from rapid oxidation when it acts as a barrier to ionic diffusion and/or electrical conduction; most metal oxides are highly electrically insulative.

Scale Types

Rate of oxidation (i.e., the rate of film thickness increase) and the tendency of the film to protect the metal from further oxidation are related to the relative volumes of the oxide and metal. The ratio of these volumes, termed the **Pilling-Bedworth ratio,** may be determined from the following expression[5]:

$$\text{P–B ratio} = \frac{A_O \rho_M}{A_M \rho_O} \tag{18.32}$$

where A_O is the molecular (or formula) weight of the oxide, A_M is the atomic weight of the metal, and ρ_O and ρ_M are the oxide and metal densities, respectively. For metals having P–B ratios less than unity, the oxide film tends to be porous and unprotective because it is insufficient to fully cover the metal surface. If the ratio is greater than unity, compressive stresses result in the film as it forms. For a ratio greater than 2–3, the oxide coating may crack and flake off, continually exposing a fresh and unprotected metal surface. The ideal P–B ratio for the formation of a protective oxide film is unity. Table 18.3 presents P–B ratios both for metals that form protective coatings and those that do not. It may be noted from these data that protective coatings normally form for metals having P–B ratios between 1 and 2, whereas nonprotective ones usually result when this ratio is less than 1 or greater than about 2. In addition to the P–B ratio, other factors also influence the oxidation resistance imparted by the film; these include a high degree of adherence between film and metal, comparable coefficients of thermal expansion for metal and oxide, and, for the oxide, a relatively high melting point and good high-temperature plasticity.

Several techniques are available for improving the oxidation resistance of a metal. One involves application of a protective surface coating of another material which both adheres well to the metal and also is itself resistant to oxidation. In some instances, the addition of alloying elements will form a more adherent and protective oxide scale by virtue of producing a more favorable Pilling–Bedworth ratio and/or improving other scale characteristics.

---

[4] Alternatively, electron holes (Section 19.10) and vacancies may diffuse instead of electrons and ions.

[5] For other than divalent metals, Equation 18.32 becomes

$$\text{P–B ratio} = \frac{A_O \rho_M}{a A_M \rho_O} \tag{18.33}$$

where a is the coefficient of the metal species for the overall oxidation reaction described by Equation 18.28.

TABLE 18.3 Pilling–Bedworth Ratios for a Number of Metals

Protective		Nonprotective	
Be	1.59	Li	0.57
Cu	1.68	Na	0.57
Al	1.28	K	0.45
Si	2.27	Ag	1.59
Cr	1.99	Cd	1.21
Mn	1.79	Ti	1.95
Fe	1.77	Mo	3.40
Co	1.99	Nb	2.61
Ni	1.52	Sb	2.35
Pd	1.60	W	3.40
Pb	1.40	Ta	2.33
Ce	1.16	U	3.05

Source: B. Chalmers, *Physical Metallurgy.* Copyright © 1959 by John Wiley & Sons, New York. Reprinted by permission of John Wiley & Sons, Inc.

Kinetics

One of the primary concerns relative to metal oxidation is the rate at which the reaction progresses. Inasmuch as the oxide scale reaction product normally remains on the surface, the rate of reaction may be determined by measuring the weight gain per unit area of surface as a function of time.

When the oxide that forms is nonporous and adheres to the metal surface, the rate of layer growth is controlled by ionic diffusion. A *parabolic* relationship exists between the weight gain per unit area W and the time t as follows:

$$W^2 = K_1 t + K_2 \tag{18.34}$$

where K_1 and K_2 are time-independent constants. This weight gain–time behavior is plotted schematically in Figure 18.25. The oxidations of iron, copper, and cobalt follow this rate expression.

The oxidation of metals for which the scale is porous or flakes off (i.e., for P–B ratios less than about 1 or greater than about 2), the oxidation rate expression is *linear;* that is,

$$W = K_3 t \tag{18.35}$$

where K_3 is a constant. Under these circumstances oxygen is always available for reaction with an unprotected metal surface because the oxide does not act as a reaction barrier. Sodium, potassium, and tantalum oxidize according to this rate expression and, incidently, have P–B ratios significantly different than unity (Table 18.3). Linear growth rate kinetics is also represented in Figure 18.25.

Still a third reaction rate law has been observed for very thin oxide layers [generally less than 100 nm (1000 Å)] that form at relatively low temperatures. The dependence of weight gain on time is *logarithmic* and takes the form

$$W = K_4 \log(K_5 t + K_6) \tag{18.36}$$

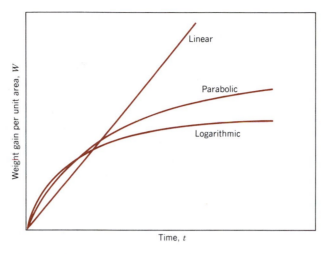

Figure 18.25 Oxidation film growth curves for linear, parabolic, and logarithmic rate laws.

Again, the K's are constants. This oxidation behavior, also shown in Figure 18.25, has been observed for aluminum, iron, and copper at near-ambient temperatures.

CORROSION OF CERAMIC MATERIALS

Ceramic materials, being compounds between metallic and nonmetallic elements, may be thought of as having already been corroded. Thus they are exceedingly immune to corrosion by almost all environments, especially at room temperature. Corrosion of ceramic materials generally involves simple chemical dissolution, in contrast to the electrochemical processes found in metals, as described above.

Ceramic materials are frequently utilized because of their resistance to corrosion. Glass is often used to contain liquids for this reason. Refractory ceramics must not only withstand high temperatures and provide thermal insulation but, in many instances, must resist high-temperature attack by molten metals, salts, slags, and glasses. Some of the new technology schemes for converting energy from one form to another that is more useful require relatively high temperatures, corrosive atmospheres, and pressures above the ambient. Ceramic materials are much better suited to withstand most of these environments for reasonable time periods than are metals.

DEGRADATION OF POLYMERS

Polymeric materials also experience deterioration by means of environmental interactions. However, an undesirable interaction is specified as degradation rather than corrosion because the processes are basically dissimilar. Whereas most metallic corrosion reactions are electrochemical, by contrast, polymeric degradation is physiochemical; that is, it involves physical as well as chemical phenomena. Furthermore, a wide variety of reactions and adverse consequences are possible for polymer degradation. Polymers may deteriorate by swelling and dissolution. Covalent bond rupture, as a result of heat energy, chemical reactions, and radiation is also possible, ordinarily with an attendant reduction in mechanical integrity. It should also be mentioned that because of the chemical complexity of polymers, their degradation mechanisms are not well understood.

To briefly cite a couple of examples of polymer degradation, polyethylene, if exposed to high temperatures in an oxygen atmosphere, suffers an impairment of its mechanical properties by becoming brittle. Or, the utility of polyvinyl chloride may be limited because this material may become colored when exposed to high temperatures, although such environments do not affect its mechanical characteristics.

18.11 SWELLING AND DISSOLUTION

When polymers are exposed to liquids, the main forms of degradation are swelling and dissolution. With swelling, the liquid or solute diffuses into and is absorbed within the polymer; the small solute molecules fit into and occupy positions among the polymer molecules. This forces the macromolecules apart such that the specimen expands or swells. Furthermore, this increase in chain separation results in a reduction of the secondary intermolecular bonding forces; as a consequence, the material becomes softer and more ductile. The liquid solute also lowers the glass transition temperature and, if depressed below the ambient temperature, will cause a once strong material to become rubbery and weak.

Swelling may be considered to be a partial dissolution process in which there is only limited solubility of the polymer in the solvent. Dissolution, which occurs when

TABLE 18.4 Resistance to Degradation by Various Environments for Selected Plastic Materials[a]

Material	Acids Weak	Acids Strong	Alkalis Weak	Alkalis Strong	Organic Solvents	Water Absorption (%/24 h)	Oxygen and Ozone
Thermoplastics							
Fluorocarbons	N	N	N	N	N	0.0	N
Polymethyl methacrylate	R	AO	R	A	A	0.2	R
Nylon	G	A	R	R	R	1.5	SA
Polyethylene (low density)	R	AO	R	R	G	0.15	A
Polyethylene (high density)	R	AO	R	R	G	0.1	A
Polypropylene	R	AO	R	R	R	<0.01	A
Polystyrene	R	AO	R	R	A	0.04	SA
Polyvinyl chloride	R	R	R	R	A	0.10	R
Thermosetting Polymers							
Epoxy	R	SA	R	R	G	0.1	SA
Phenolics	SA	A	SA	A	SA	0.6	—
Polyesters	SA	A	A	A	SA	0.2	A

[a] N = none; R = generally resistant; SA = slight attack; A = attacked; AO = attacked by oxidizing acids; G = good.
Source: M. G. Fontana and N. D. Greene, *Corrosion Engineering,* 2nd edition, McGraw-Hill Book Company, New York, 1978.

TABLE 18.5 Resistance to Degradation by Various Environments for Five Elastomeric Materials[a]

Material	Dilute Acids	Oxidizing Agents	Alkalies	Petroleum Oils and Greases	Water and Antifreezes	Weather and Ozone
Polyisoprene (natural rubber)	G	P	F	P	G	F
Chloroprene (neoprene)	G	P	G	G	F	E
Nitrile	G	P	F	E	E	F
Styrene–butadiene	G	P	F	P	G	F
Silicon rubber	—	—	—	G	F	E

[a] E = excellent; G = good; F = fair; P = poor.
Source: M. G. Fontana and N. D. Greene, *Corrosion Engineering,* 2nd edition, McGraw-Hill Book Company, New York, 1978.

the polymer is completely soluble, may be thought of as just a continuation of swelling. As a rule of thumb, the greater the similarity of chemical structure between the solvent and polymer, the greater the likelihood of swelling and/or dissolution. For example, many hydrocarbon rubbers readily absorb hydrocarbon liquids such as gasoline. The responses of selected polymeric materials to organic solvents are contained in Tables 18.4 and 18.5.

Swelling and dissolution traits also are affected by temperature as well as characteristics of the molecular structure. In general, increasing molecular weight, increasing degree of crosslinking and crystallinity, and decreasing temperature result in a reduction of these deteriorative processes.

Resistance to attack by acidic and alkaline solutions is, in general, much better for polymers than for metals. A qualitative comparison of the behavior of various polymers in these solutions is also presented in Tables 18.4 and 18.5. Materials that exhibit outstanding resistance to attack by both solution types include the fluorocarbons and polyvinyl chloride.

18.12 BOND RUPTURE

Polymers may also experience degradation by a process termed **scission**—the severance or rupture of molecular chain bonds. This causes a separation of chain segments at the point of scission and a reduction in the molecular weight. As previously discussed (Chapter 16), several properties of polymeric materials, including mechanical strength and resistance to chemical attack, depend on molecular weight. Consequently, some of the physical and chemical properties of polymers may be adversely affected by this form of degradation. Bond rupture may result from exposure to radiation or to heat, and from chemical reaction.

Radiation Effects

Certain types of radiation (electron beams, x-rays, β- and γ-rays, and ultraviolet radiation) possess sufficient energy to penetrate a polymer specimen and interact with the constituent atoms or their electrons. One such reaction is *ionization,* in which the radiation removes an orbital electron from a specific atom, converting that atom into a positively charged ion. As a consequence, one of the covalent bonds associated with the specific atom is broken, and there is a rearrangement of atoms or groups of atoms at that point. This bond breaking leads to either scission or crosslinking at the ionization site, depending on the chemical structure of the polymer and also on the dose of radiation. Stabilizers (Section 16.11) may be added to protect polymers from ultraviolet damage.

Not all consequences of radiation exposure are deleterious. Crosslinking may be induced by irradiation to improve the mechanical behavior and degradation characteristics. For example, γ-radiation is used commercially to crosslink polyethylene to enhance its resistance to softening and flow at elevated temperatures; indeed, this process may be carried out on products that have already been fabricated.

Chemical Reaction Effects

Oxygen, ozone, and other substances can cause or accelerate chain scission as a result of chemical reaction. This effect is especially prevalent in vulcanized rubbers that have doubly bonded carbon atoms along the backbone molecular chains, and which are exposed to ozone (O_3), an atmospheric pollutant. One such scission reaction may be represented by

$$-R-\underset{\underset{H}{|}}{C}=\underset{\underset{H}{|}}{C}-R'- + O_3 \longrightarrow -R-\underset{\underset{H}{|}}{C}=O + O=\underset{\underset{H}{|}}{C}-R'- + O\cdot \quad (18.37)$$

where the chain is severed at the point of the double bond; R and R' represent groups of atoms bonded to the chain that are unaffected during the reaction. Ordinarily, if the rubber is in an unstressed state, a film will form on the surface, protecting the bulk material from any further reaction. However, when these materials are subjected to tensile stresses, cracks and crevices form and grow in a direction perpendicular to the stress; eventually, rupture of the material may occur. Apparently these cracks result from large numbers of ozone-induced scissions. The elastomers in Table 18.5 are rated as to their resistance to degradation by exposure to ozone.

Thermal Effects

Thermal degradation corresponds to the scission of molecular chains at elevated temperatures; as a consequence, some polymers undergo chemical reactions in which gaseous species are produced. These reactions are evidenced by a weight loss of material; a polymer's thermal stability is a measure of its resilience to this decomposition. Thermal stability is related primarily to the magnitude of the bonding energies between the various atomic constituents of the polymer: higher bonding energies result in more thermally stable materials. For example, the magnitude of the C—F bond is greater than that of the C—H bond, which in turn is greater than that of the C—Cl bond. The fluorocarbons, having C—F bonds, are among the most thermally resistant polymeric materials and may be utilized at relatively high temperatures.

18.13 WEATHERING

Many polymeric materials serve in applications that require exposure to outdoor conditions. Any resultant degradation is termed *weathering*, which may, in fact, be a combination of several different processes. Under these conditions deterioration is primarily a result of oxidation, which is initiated by ultraviolet radiation from the sun. Some polymers such as nylon and cellulose are also susceptible to water absorption, which produces a reduction in their hardness and stiffness. Resistance to weathering among the various polymers is quite diverse. The fluorocarbons are virtually inert under these conditions; but some materials, including polyvinyl chloride and polystyrene, are susceptible to weathering.

SUMMARY

Metallic corrosion is ordinarily electrochemical, involving both oxidation and reduction reactions. Oxidation is the loss of the metal atoms' valence electrons; the resulting metal ions may either go into the corroding solution or form an insoluble compound. During reduction, these electrons are transferred to at least one other chemical species. The character of the corrosion environment dictates which of several possible reduction reactions will occur.

Not all metals oxidize with the same degree of ease, which is demonstrated with a galvanic couple; when in an electrolyte, one metal (the anode) will corrode, whereas a reduction reaction will occur at the other metal (the cathode). The magnitude of the electric potential that is established between anode and cathode is indicative of the driving force for the corrosion reaction.

The standard emf and galvanic series are simply rankings of metallic materials on the basis of their tendency to corrode when coupled to other metals. For the standard emf series, ranking is based on the magnitude of the voltage generated when the standard cell of a metal is coupled to the standard hydrogen electrode at 25°C (77°F). The galvanic series consists of the relative reactivities of metals and alloys in seawater.

The half-cell potentials in the standard emf series are thermodynamic parameters that are valid only at equilibrium; corroding systems are not in equilibrium. Furthermore, the magnitudes of these potentials provide no indication as to the rates at which corrosion reactions occur.

The rate of corrosion may be expressed as corrosion penetration rate, that is, the thickness loss of material per unit of time. Mils per year and millimeters per year are the common units for this parameter. Alternatively, rate is proportional to the current density associated with the electrochemical reaction.

Corroding systems will experience polarization, which is the displacement of each electrode potential from its equilibrium value; the magnitude of this displacement is termed the overvoltage. The corrosion rate of a reaction is limited by polarization, of which there are two types—activation and concentration. Polarization data are plotted as potential versus the logarithm of current density. The corrosion rate for a particular reaction may be computed using the current density associated with the intersection point of oxidation and reduction polarization curves.

A number of metals and alloys passivate, or lose their chemical reactivity, under some environmental circumstances. This phenomenon is thought to involve the formation of a thin protective oxide film. Stainless steels and aluminum alloys exhibit this type of behavior. The active-to-passive behavior may be explained by the alloy's S-shaped electrochemical potential-versus-log current density curve. Intersections with reduction polarization curves in active and passive regions correspond, respectively, to high and low corrosion rates.

Metallic corrosion is sometimes classified into eight different forms: uniform attack, galvanic corrosion, crevice corrosion, pitting, intergranular corrosion, selective leaching, erosion–corrosion, and stress corrosion.

The measures that may be taken to prevent, or at least reduce, corrosion include material selection, environmental alteration, the use of inhibitors, design changes, application of coatings, and cathodic protection.

Oxidation of metallic materials by electrochemical action is also possible in dry, gaseous atmospheres. An oxide film forms on the surface which may act as a barrier to further oxidation if the volumes of metal and oxide film are similar, that is, if the Pilling–Bedworth ratio is near unity. The kinetics of film formation may follow parabolic, linear, or logarithmic rate laws.

Ceramic materials, being inherently corrosion resistant, are frequently utilized at elevated temperatures and/or in extremely corrosive environments.

Polymeric materials deteriorate by noncorrosive processes. Upon exposure to liquids, they may experience degradation by swelling or dissolution. With swelling, solute molecules actually fit into the molecular structure. Scission, or the severance of molecular chain bonds, may be induced by radiation, chemical reactions, or heat. This results in a reduction of molecular weight and a deterioration of the physical and chemical properties of the polymer.

IMPORTANT TERMS AND CONCEPTS

Activation polarization	Electrolyte	Pilling–Bedworth ratio
Anode	Electromotive force	Pitting
Cathode	(emf) series	Polarization
Cathodic protection	Erosion–corrosion	Reduction
Concentration	Galvanic corrosion	Sacrificial anode
polarization	Galvanic series	Scission
Corrosion	Inhibitor	Selective leaching
Corrosion penetration	Intergranular corrosion	Standard half-cell
rate	Molarity	Stress corrosion
Crevice corrosion	Oxidation	Weld decay
Degradation	Passivity	

REFERENCES

EVANS, U. R., *An Introduction to Metallic Corrosion,* 3rd edition, Edward Arnold, Baltimore, 1981.

FONTANA, M. G., *Corrosion Engineering,* 3rd edition, McGraw-Hill Book Company, New York, 1986.

Metals Handbook, 9th edition, Vol. 13, *Corrosion,* ASM International, Metals Park, OH, 1987.

POURBAIX, M., *Lectures on Electrochemical Corrosion,* Plenum Press, New York, 1973.

REICH, L. and S. S. STIVALA, *Elements of Polymer Degradation,* McGraw-Hill Book Company, New York, 1971.

SCHREIR, L. L. (Editor), *Corrosion:* Vol. 1, *Metal/Environment Reactions;* Vol. 2, *Corrosion Control,* Newnes-Butterworths, London, 1976.

SCHWEITZER, P. A., *Corrosion Resistance Tables,* 2nd edition, Marcel Dekker, New York, 1986.

UHLIG, H. H. and R. W. REVIE, *Corrosion and Corrosion Control,* 3rd edition, John Wiley & Sons, New York, 1985.

QUESTIONS AND PROBLEMS

18.1 (a) Briefly explain the difference between oxidation and reduction electrochemical reactions. (b) Which reaction occurs at the anode, and which at the cathode?

18.2 (a) Write the possible oxidation and reduction half-reactions that occur when magnesium is immersed in each of the following solutions: (i) HCl, (ii) an HCl solution containing dissolved oxygen, (iii) an HCl solution containing dissolved oxygen and, in addition, Fe^{2+} ions. (b) In which of these solutions would you expect the magnesium to oxidize most rapidly? Why?

18.3 Would you expect iron to corrode in water of high purity? Why or why not?

18.4 Demonstrate that (a) the value of $\mathscr{F}$ in Equation 18.19 is 96,500 C/mol, and (b) at 25°C (298 K),

$$\frac{RT}{n\mathscr{F}} \ln x = \frac{0.0592}{n} \log x$$

18.5 (a) Compute the voltage at 25°C of an electrochemical cell consisting of pure cadmium immersed in a $2 \times 10^{-3} M$ solution of Cd^{2+} ions, and pure iron in a $0.4 M$ solution of Fe^{2+} ions. (b) Write the spontaneous electrochemical reaction.

18.6 A Zn/Zn^{2+} concentration cell is constructed in which both electrodes are pure zinc. The Zn^{2+} concentration for one cell half is $1.0 M$, for the other, $10^{-2} M$. Is a voltage generated between the two cell halves? If so, what is its magnitude and which electrode will be oxidized? If no voltage is produced, explain this result.

18.7 An electrochemical cell is composed of pure copper and pure lead electrodes immersed in solutions of their respective divalent ions. For a $0.6 M$ concentration of Cu^{2+}, the lead electrode is oxidized yielding a cell potential of 0.507 V. Calculate the concentration of Pb^{2+} ions if the temperature is 25°C.

18.8 An electrochemical cell is constructed such that on one side a pure Zn electrode is in contact with a solution containing Zn^{2+} ions at a concentration of $10^{-2} M$. The other cell half consists of a pure Pb electrode that is immersed in a solution of Pb^{2+} ions having a concentration of $10^{-4} M$. At what temperature will the potential between the two electrodes be $+0.568$ V?

18.9 Modify Equation 18.19 for the case in which metals M_1 and M_2 are alloys.

18.10 (a) From the galvanic series (Table 18.2), cite three metals or alloys that may be used to galvanically protect 304 stainless steel in the active state. (b) Sometimes galvanic corrosion is prevented by making an electrical contact between both metals in the couple and a third metal that is anodic to these other two. Using the galvanic series, name one metal that could be used to protect a copper–aluminum galvanic couple.

18.11 Demonstrate that the constant K in Equation 18.23 will have values of 534 and 87.6 for the CPR in units of mpy and mm/yr, respectively.

18.12 A piece of corroded steel plate was found in a submerged ocean vessel. It was estimated that the original area of the plate was 40 in.^{2} and that approximately 2.6 kg had corroded away during the submersion. Assuming a corrosion penetration rate of 200 mpy for this alloy in seawater, estimate the time of submersion in years. The density of steel is 7.9 g/cm^3.

18.13 A thick steel sheet of area 100 in.^{2} is exposed to air near the ocean. After a one-year period it was found to experience a weight loss of 485 g due to corrosion. To what rate of corrosion, in both mpy and mm/yr, does this correspond?

18.14 **(a)** Demonstrate that the CPR is related to the corrosion current density i (A/cm^2) through the expression

$$\mathrm{CPR} = \frac{KAi}{n\rho} \qquad (18.38)$$

where K is a constant, A is the atomic weight of the metal experiencing corrosion, n is the number of electrons associated with the ionization of each metal atom, and ρ is the density of the metal. **(b)** Calculate the value of the constant K for the CPR in mpy and i in μA/cm^2 (10^{-6} A/cm^2).

18.15 Using the results of Problem 18.14, compute the corrosion penetration rate, in mpy, for the corrosion of iron in HCl (to form Fe^{2+} ions) if the corrosion current density is 8×10^{-5} A/cm^2.

18.16 **(a)** Cite the major differences between activation and concentration polarizations. **(b)** Under what conditions is activation polarization rate controlling? **(c)** Under what conditions is concentration polarization rate controlling?

18.17 **(a)** Describe the phenomenon of dynamic equilibrium as it applies to oxidation and reduction electrochemical reactions. **(b)** What is the exchange current density?

18.18 Briefly explain why concentration polarization is not normally rate controlling for oxidation reactions.

18.19 Nickel experiences corrosion in an acid solution according to the reaction

$$\mathrm{Ni} + 2H^+ \longrightarrow Ni^{2+} + H_2$$

The rates of both oxidation and reduction half-reactions are controlled by activation polarization. **(a)** Compute the rate of oxidation of Ni (in mol/cm^2-s) given the following activation polarization data:

For Nickel	For Hydrogen
$V_{(Ni/Ni^{2+})} = -0.25$ V	$V_{(H^+/H_2)} = 0$ V
$i_0 = 10^{-8}$ A/cm^2	$i_0 = 6 \times 10^{-7}$ A/cm^2
$\beta = +0.12$	$\beta = -0.10$

(b) Compute the value of the corrosion potential.

18.20 The corrosion rate is to be determined for some divalent metal M in a solution containing hydrogen ions. The following corrosion data are known about the metal and solution:

For Metal M	For Hydrogen
$V_{(M/M^{2+})} = -0.90$ V	$V_{(H^+/H_2)} = 0$ V
$i_0 = 10^{-12}$ A/cm^2	$i_0 = 10^{-10}$ A/cm^2
$\beta = +0.10$	$\beta = -0.15$

(a) Assuming that activation polarization controls both oxidation and reduction reactions, determine the rate of corrosion of metal M (in mol/cm^2-s). (b) Compute the corrosion potential for this reaction.

18.21 The influence of increasing solution velocity on the overvoltage-versus-log current density behavior for a solution that experiences combined activation–concentration polarization is indicated in Figure 18.26. On the basis of this behavior, make a schematic plot of corrosion rate versus solution velocity for the oxidation of a metal; assume that the oxidation reaction is controlled by activation polarization.

18.22 Briefly describe the phenomenon of passivity. Name two common types of alloy that passivate.

18.23 Why does chromium in stainless steels make them more corrosion resistant in many environments than plain carbon steels?

18.24 For each form of corrosion, other than uniform, do the following: (a) describe why, where, and the conditions under which the corrosion occurs; and (b) cite three measures that may be taken to prevent or control it.

18.25 Cite two examples of the beneficial use of galvanic corrosion.

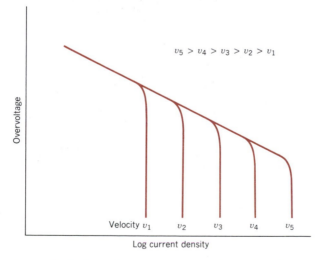

Figure 18.26 Plot of overvoltage versus logarithm of current density for a solution that experiences combined activation–concentration polarization at various solution velocities.

18.26 Briefly explain why cold-worked metals are more susceptible to corrosion than noncold-worked metals.

18.27 Briefly explain why, for a small anode-to-cathode area ratio, the corrosion rate will be higher than for a large ratio.

18.28 For a concentration cell, briefly explain why corrosion occurs at that region having the lower concentration.

18.29 Is Equation 18.23 equally valid for uniform corrosion and pitting? Why or why not?

18.30 **(a)** What are inhibitors? **(b)** What possible mechanisms account for their effectiveness?

18.31 Briefly describe the two techniques that are used for galvanic protection.

18.32 Tin cans are made of a steel the inside of which is coated with a thin layer of tin. The tin protects the steel from corrosion by food products in the same manner as zinc protects steel from atmospheric corrosion. Briefly explain how this cathodic protection of tin cans is possible since tin is electrochemically less active than steel in the galvanic series (Table 18.2).

18.33 A brine solution is used as a cooling medium in a steel heat exchanger. The brine is circulated within the heat exchanger and contains some dissolved oxygen. Suggest three methods, other than cathodic protection, for reducing corrosion of the steel by the brine. Explain the rationale for each suggestion.

18.34 For each of the metals listed below, compute the Pilling–Bedworth ratio. Also, on the basis of this value, specify whether or not you would expect the oxide scale that forms on the surface to be protective, and then justify your decision. Density data for both the metal and its oxide are also tabulated.

Metal	Metal Density (g/cm^3)	Metal Oxide	Oxide Density (g/cm^3)
Mg	1.74	MgO	3.58
V	6.11	V$_2$O$_5$	3.36
Zn	7.13	ZnO	5.61

18.35 According to Table 18.3, the oxide coating that forms on silver should be nonprotective, and yet Ag does not oxidize appreciably at room temperature and in air. How do you explain this apparent discrepancy?

18.36 Below, weight gain–time data for the oxidation of nickel at an elevated temperature are tabulated.

W (mg/cm^2)	Time (min)
0.527	10
0.857	30
1.526	100

(a) Determine whether the oxidation kinetics obey a linear, parabolic, or logarithmic rate expression.

(b) If the answer to part a is either linear or parabolic, determine W after a total time of 600 min.

18.37 Below, weight gain–time data for the oxidation of some metal at an elevated temperature are tabulated.

W (mg/cm^2)	Time (min)
1.10	50
1.34	200
1.67	1000

(a) Determine whether the oxidation kinetics obey a linear, parabolic, or logarithmic rate expression.

(b) If the answer to part a is linear or parabolic, determine W after a time of 5000 min.

18.38 Below, weight gain–time data for the oxidation of some metal at an elevated temperature are tabulated.

W (mg/cm^2)	Time (min)
1.54	10
23.24	150
95.37	620

(a) Determine whether the oxidation kinetics obey a linear, parabolic, or logarithmic rate expression.

(b) If the answer to part a is linear or parabolic, determine W after a time of 1200 min.

18.39 From a molecular perspective, explain why increasing crosslinking and crystallinity of a polymeric material will enhance its resistance to swelling and dissolution. Would you expect crosslinking or crystallinity to have the greater influence? Justify your choice.

18.40 List three differences between the corrosion of metals and **(a)** the corrosion of ceramics and **(b)** the degradation of polymers.

CHAPTER 19

ELECTRICAL PROPERTIES

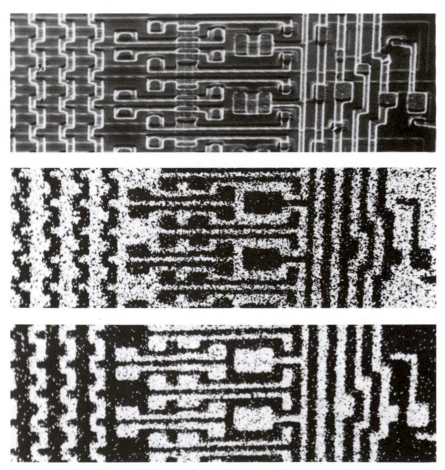

It was noted in Section 4.9 that an image is generated on a scanning electron micrograph as a beam of electrons scans the surface of the specimen being examined. The electrons in this beam cause some of the specimen surface atoms to emit x-rays; the energy of an x-ray photon depends on the particular atom from which it radiates. It is possible to selectively filter out all but the x-rays emitted from one kind of atom. When projected on a cathode ray tube, small white dots are produced indicating the locations of the particular atom type; thus a "dot map" of the image is generated.

Top: Scanning electron micrograph of an integrated circuit.

Center: A silicon dot map for the integrated circuit above, showing regions where silicon atoms are concentrated. Doped silicon is the semiconducting material from which integrated circuit elements are made.

Bottom: An aluminum dot map. Metallic aluminum is an electrical conductor, and, as such, wires the circuit elements together. Approximately 200 ×.

19.1 INTRODUCTION

The prime objective of this chapter is to explore the electrical properties of materials, that is, their responses to an applied electric field. We begin with the phenomenon of electrical conduction: the parameters by which it is expressed, the mechanism of conduction by electrons, and how the electron energy band structure of a material influences its ability to conduct. These principles are extended to metals, semiconductors, and insulators. Particular attention is given to the characteristics of semiconductors, and then to semiconducting devices. Also treated are the dielectric characteristics of insulating materials. The final sections are devoted to the peculiar phenomena of ferroelectricity and piezoelectricity.

ELECTRICAL CONDUCTION

19.2 OHM'S LAW

One of the most important electrical characteristics of a solid material is the ease with which it transmits an electric current. **Ohm's law** relates the current I—or time rate of charge passage—to the applied voltage V as follows:

$$V = IR \tag{19.1}$$

where R is the resistance of the material through which the current is passing. The units for V, I, and R are, respectively, volts (J/C), amperes (C/s), and ohms (V/A). The value of R is influenced by specimen configuration, and for many materials is independent of current. The **resistivity** ρ is independent of specimen geometry but related to R through the expression

$$\rho = \frac{RA}{l} \tag{19.2}$$

where l is the distance between the two points at which the voltage is measured, and A is the cross-sectional area perpendicular to the direction of the current. The units for ρ are ohm-meters (Ω-m). From the expression for Ohm's law and Equation 19.2,

$$\rho = \frac{VA}{Il} \tag{19.3}$$

Figure 19.1 is a schematic diagram of an experimental arrangement for measuring electrical resistivity.

19.3 ELECTRICAL CONDUCTIVITY

Sometimes, **electrical conductivity** σ is used to specify the electrical character of a material. It is simply the reciprocal of the resistivity, or

$$\sigma = \frac{1}{\rho} \tag{19.4}$$

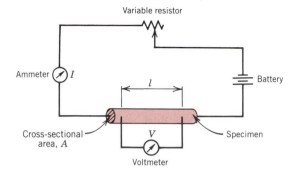

Figure 19.1 Schematic representation of the apparatus used to measure electrical resistivity.

and is indicative of the ease with which a material is capable of conducting an electric current. The units for σ are reciprocal ohm-meters $[(\Omega\text{-m})^{-1}$, or mho/m$]$. The following discussions on electrical properties use both resistivity and conductivity.

In addition to Equation 19.1, Ohm's law may be expressed as

$$J = \sigma \mathscr{E} \tag{19.5}$$

in which J is the current density, the current per unit of specimen area I/A, and $\mathscr{E}$ is the electric field intensity, or the voltage difference between two points divided by the distance separating them, that is,

$$\mathscr{E} = \frac{V}{l} \tag{19.6}$$

The demonstration of the equivalence of the two Ohm's law expressions (Equations 19.1 and 19.5) is left as a homework exercise.

Solid materials exhibit an amazing range of electrical conductivities, extending over 27 orders of magnitude; probably no other physical property experiences this breadth of variation. In fact, one way of classifying solid materials is according to the ease with which they conduct an electric current; within this classification scheme there are three groupings: *conductors, semiconductors,* and *insulators.* **Metals** are good conductors, typically having conductivities on the order of 10^7 $(\Omega\text{-m})^{-1}$. At the other extreme are materials with very low conductivities, ranging between 10^{-10} and 10^{-20} $(\Omega\text{-m})^{-1}$; these are electrical **insulators.** Materials with intermediate conductivities, generally from 10^{-6} to 10^4 $(\Omega\text{-m})^{-1}$, are termed **semiconductors.**

19.4 ELECTRONIC AND IONIC CONDUCTION

An electric current results from the motion of electrically charged particles in response to forces that act on them from an externally applied electric field. Positively charged particles are accelerated in the field direction, negatively charged particles in the direction opposite. Within most solid materials a current arises from the flow of electrons, which is termed *electronic conduction.* In addition, for ionic materials a net motion of charged ions is possible that produces a current; such is termed **ionic conduction.** The present discussion deals with electronic conduction; ionic conduction is treated briefly in Section 19.14.

19.5 ENERGY BAND STRUCTURES IN SOLIDS

In all conductors, semiconductors, and many insulating materials, only electronic conduction exists, and the magnitude of the electrical conductivity is strongly dependent on the number of electrons available to participate in the conduction process. However, not all electrons in every atom will accelerate in the presence of an electric field. The number of electrons available for electrical conduction in a particular material is related to the arrangement of electron states or levels with respect to energy, and then the manner in which these states are occupied by electrons. A thorough exploration of these topics is complicated and involves principles of quantum mechanics that are beyond the scope of this book; the ensuing development omits some concepts and simplifies others.

Concepts relating to electron energy states, their occupancy, and the resulting electron configuration for isolated atoms were discussed in Section 2.3. By way of review, for each individual atom there exist discrete energy levels that may be occupied by electrons, arranged into shells and subshells. Shells are designated by integers (1, 2, 3, etc.), and subshells by letters (s, p, d, and f). For each of s, p, d, and f subshells, there exist, respectively, one, three, five, and seven states. The electrons in most atoms fill just the states having the lowest energies, two electrons of opposite spin per state, in accordance with the Pauli exclusion principle. The electron configuration of an isolated atom represents the arrangement of the electrons within the allowed states.

Let us now make an extrapolation of some of these concepts to solid materials. A solid may be thought of as consisting of a large number, say, N, of atoms initially separated from one another, which are subsequently brought together and bonded to form the ordered atomic arrangement found in the crystalline material. At relatively large separation distances, each atom is independent of all the others and will have the atomic energy levels and electron configuration as if isolated. However, as the atoms come within close proximity of one another, electrons are acted upon, or perturbed, by the electrons and nuclei of adjacent atoms. This influence is such that each distinct atomic state may split into a series of closely spaced electron states in the solid, to form what is termed an **electron energy band.** The extent of splitting depends on interatomic separation (Figure 19.2) and begins with the outermost electron shells, since they are the first to be perturbed as the atoms coalesce. Within each band, the energy states are discrete, yet the difference between adjacent states is exceedingly small. At the equilibrium spacing, band formation may not occur for the electron subshells nearest the nucleus, as illustrated in Figure 19.3b. Furthermore, gaps may exist between adjacent bands, as also indicated in the figure; normally, energies lying within these band gaps are not available for electron occupancy. The conventional way of representing electron band structures in solids is shown in Figure 19.3a.

The number of states within each band will equal the total of all states contributed by the N atoms. For example, an s band will consist of N states, and a p band of $3N$ states. With regard to occupancy, each energy state may accommodate two electrons, which must have oppositely directed spins. Furthermore, bands will contain the electrons that resided in the corresponding levels of the isolated atoms; for example, a 4s energy band in the solid will contain those isolated atom's 4s electrons. Of course, there will be empty bands and, possibly, bands that are only partially filled.

The electrical properties of a solid material are a consequence of its electron band structure, that is, the arrangement of the outermost electron bands and the way in

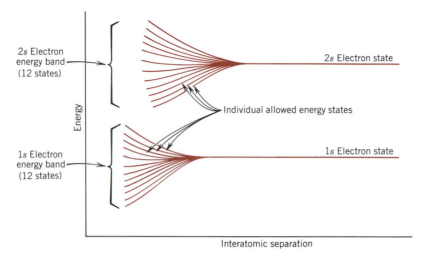

Figure 19.2 Schematic plot of electron energy versus interatomic separation for an aggregate of 12 atoms ($N = 12$). Upon close approach, each of the 1s and 2s atomic states splits to form an electron energy band consisting of 12 states.

which they are filled with electrons. In this regard, the band that contains the highest-energy or valence electrons is termed the **valence band;** the **conduction band** is the next higher energy band, which is, under most circumstances, virtually unoccupied by electrons.

Four different types of band structures are possible at 0 K. In the first (Figure 19.4a), the valence band is only partially filled with electrons. The energy corresponding to the highest filled state at 0 K is called the **Fermi energy** E_f, as indicated. This energy band structure is typified by some metals, in particular those that have a single s valence electron (e.g., copper). Each copper atom has one 4s electron; however, for a solid comprised of N atoms, the 4s band is capable of accommodating $2N$ electrons. Thus only half the available electron positions within this 4s valence band are filled.

For the second band structure, also found in metals (Figure 19.4b), the valence band is full, but it overlaps the conduction band, which, in the absence of any overlapping, would be completely empty. Magnesium has this band structure. Each isolated Mg atom has two 3s valence electrons. However, when a solid is formed, the 3s and 3p bands overlap. In this instance and at 0 K, the Fermi energy is taken as that energy below which, for N atoms, N states are filled, two electrons per state.

The final two band structures are similar; for each, all states in the valence band are completely filled with electrons. However, there is no overlap between this and the empty conduction band; this gives rise to an **energy band gap** in between. For very pure materials, electrons may not have energies within this gap. The difference between the two band structures lies in the magnitude of the energy gap; for materials that are insulators, the band gap is relatively wide (Figure 19.4c), whereas for semiconductors it is narrow (Figure 19.4d). The Fermi energy for these two band structures lies within the band gap—near its center.

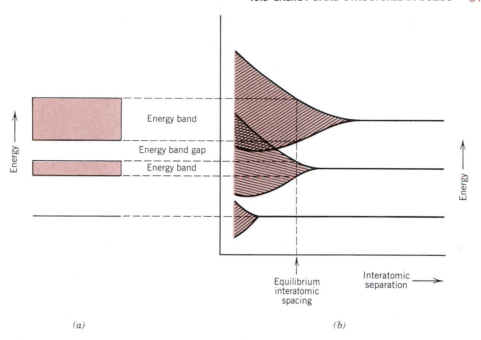

(a)   *(b)*

Figure 19.3 (*a*) The conventional representation of the electron energy band structure for a solid material at the equilibrium interatomic separation. (*b*) Electron energy versus interatomic separation for an aggregate of atoms, illustrating how the energy band structure at the equilibrium separation in (*a*) is generated. (From Z. D. Jastrzebski, *The Nature and Properties of Engineering Materials,* 3rd edition. Copyright © 1987 by John Wiley & Sons, Inc. Reprinted by permission of John Wiley & Sons, Inc.)

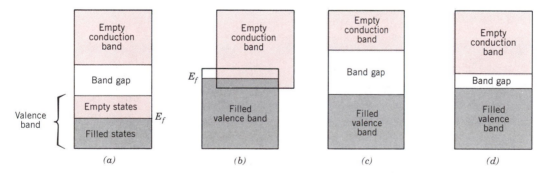

(a)   *(b)*   *(c)*   *(d)*

Figure 19.4 The various possible electron band structures in solids at 0 K. (*a*) The electron band structure found in metals such as copper, in which there are available electron states above and adjacent to filled states, in the same band. (*b*) The electron band structure of metals such as magnesium, wherein there is an overlap of the filled valence band with an empty conduction band. (*c*) The electron band structure characteristic of insulators; the filled valence band is separated from the empty conduction band by a relatively large band gap (>2 eV). (*d*) The electron band structure found in the semiconductors, which is the same as for insulators except that the band gap is relatively narrow (<2 eV).

19.6 CONDUCTION IN TERMS OF BAND AND ATOMIC BONDING MODELS

At this point in the discussion it is vital that another concept be understood, namely, that only electrons with energies greater than the Fermi energy may be acted on and accelerated in the presence of an electric field. These are the electrons that participate in the conduction process, which are termed **free electrons.** Another charged electronic entity called a **hole** is found in semiconductors and insulators. Holes have energies less than E_f and also participate in electronic conduction. As the ensuing discussion reveals, the electrical conductivity is a direct function of the numbers of free electrons and holes. In addition, the distinction between conductors and nonconductors (insulators and semiconductors) lies in the numbers of these free electron and hole charge carriers.

Metals

For an electron to become free, it must be excited or promoted into one of the empty and available energy states above E_f. For metals having either of the band structures shown in Figures 19.4a and 19.4b, there are vacant energy states adjacent to the highest filled state at E_f. Thus very little energy is required to promote electrons into the low-lying empty states, as shown in Figure 19.5. Generally, the energy provided by an electric field is sufficient to excite large numbers of conduction electrons into these conducting states.

For the metallic bonding model discussed in Section 2.6, it was assumed that all the valence electrons have freedom of motion and form an "electron gas," which is uniformly distributed throughout the lattice of ion cores. Even though these electrons are not locally bound to any particular atom, they, nevertheless, must experience some excitation to become conducting electrons that are truly free. Thus although only a fraction are excited, this still gives rise to a relatively large number of free electrons and, consequently, a high conductivity.

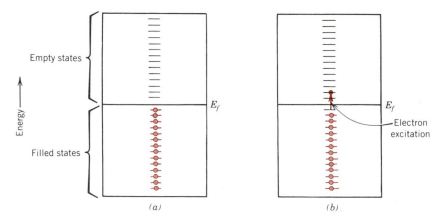

Figure 19.5 For a metal, occupancy of electron states (a) before and (b) after an electron excitation.

Insulators and Semiconductors

For insulators and semiconductors, empty states adjacent to the top of the filled valence band are not available. To become free, therefore, electrons must be promoted across the energy band gap and into empty states at the bottom of the conduction band. This is possible only by supplying to an electron the difference in energy between these two states, which is approximately equal to the band gap energy E_g. This excitation process is demonstrated in Figure 19.6. For many materials this band gap is several electron volts wide. Most often the excitation energy is from a nonelectrical source such as heat or light, usually the former.

The number of electrons excited thermally (by heat energy) into the conduction band depends on the energy band gap width as well as temperature. At a given temperature, the larger the E_g, the lower the probability that a valence electron will be promoted into an energy state within the conduction band; this results in fewer conduction electrons. In other words, the larger the band gap, the lower the electrical conductivity at a given temperature. Thus the distinction between semiconductors and insulators lies in the width of the band gap; for semiconductors it is narrow, whereas for insulating materials it is relatively wide.

Increasing the temperature of either a semiconductor or an insulator results in an increase in the thermal energy that is available for electron excitation. Thus more electrons are promoted into the conduction band, which gives rise to an enhanced conductivity.

The conductivity of insulators and semiconductors may also be viewed from the perspective of atomic bonding models discussed in Section 2.6. For electrically insulating materials, interatomic bonding is ionic or strongly covalent. Thus the valence electrons are tightly bound to or shared with the individual atoms. In other words, these electrons are highly localized and are not in any sense free to wander throughout the crystal. The bonding in semiconductors is covalent (or predominantly covalent)

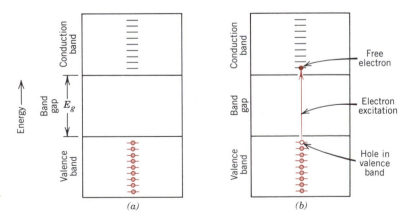

Figure 19.6 For an insulator or semiconductor, occupancy of electron states (a) before and (b) after an electron excitation from the valence band into the conduction band, in which both a free electron and a hole are generated.

and relatively weak, which means that the valence electrons are not as strongly bound to the atoms. Consequently, these electrons are more easily removed by thermal excitation than they are for insulators.

19.7 ELECTRON MOBILITY

When an electric field is applied, a force is brought to bear on the free electrons; as a consequence, they all experience an acceleration in a direction opposite to that of the field, by virtue of their negative charge. According to quantum mechanics, there is no interaction between an accelerating electron and atoms in a perfect crystal lattice of atoms. Under such circumstances all the free electrons should accelerate as long as the electric field is applied, which would give rise to a continuously increasing electric current with time. However, we know that a current reaches a constant value the instant that a field is applied, indicating that there exists what might be termed "frictional forces," which counter this acceleration from the external field. These frictional forces result from the scattering of electrons by imperfections in the crystal lattice, including impurity atoms, vacancies, interstitial atoms, dislocations, and even the thermal vibrations of the atoms themselves. Each scattering event causes an electron to lose kinetic energy and to change its direction of motion, as represented schematically in Figure 19.7. There is, however, some net electron motion in the direction opposite to the field, and this flow of charge is the electric current.

The scattering phenomenon is manifested as a resistance to the passage of an electric current. Several parameters are used to describe the extent of this scattering, these include the *drift velocity* and the **mobility** of an electron. The drift velocity v_d represents the average electron velocity in the direction of the force imposed by the applied field. It is directly proportional to the electric field as follows:

$$v_d = \mu_e \mathscr{E} \tag{19.7}$$

The constant of proportionality μ_e is called the electron mobility, which is an indication of the frequency of scattering events; its units are square meters per volt-second (m^2/V-s).

The conductivity σ of most materials may be expressed as

$$\sigma = n|e|\mu_e \tag{19.8}$$

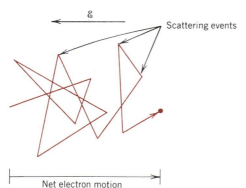

Figure 19.7 Schematic diagram showing the path of an electron that is deflected by scattering events.

where n is the number of free or conducting electrons per unit volume (e.g., per cubic meter), and $|e|$ is the absolute magnitude of the electrical charge on an electron (1.6×10^{-19} C). Thus the electrical conductivity is proportional to both the number of free electrons and the electron mobility.

19.8 ELECTRICAL RESISTIVITY OF METALS

As mentioned previously, most metals are extremely good conductors of electricity; room-temperature conductivities for several of the more common metals are contained in Table 19.1. Again, metals have high conductivities because of the large numbers of free electrons that have been excited into empty states above the Fermi energy. Thus n has a large value in the conductivity expression, Equation 19.8.

At this point it is convenient to discuss conduction in metals in terms of the resistivity, the reciprocal of conductivity; the reason for this switch should become apparent in the ensuing discussion.

Since crystalline defects serve as scattering centers for conduction electrons in metals, increasing their number raises the resistivity (or lowers the conductivity). The concentration of these imperfections depends on temperature, composition, and the degree of cold work of a metal specimen. In fact, it has been observed experimentally that the total resistivity of a metal is the sum of the contributions from thermal vibrations, impurities, and plastic deformation; that is, the scattering mechanisms act independently of one another. This may be represented in mathematical form as follows:

$$\rho_{\text{total}} = \rho_t + \rho_i + \rho_d \tag{19.9}$$

in which ρ_t, ρ_i, and ρ_d represent the individual thermal, impurity, and deformation resistivity contributions, respectively. Equation 19.9 is sometimes known as **Matthiessen's rule.** The influence of each ρ variable on the total resistivity is demonstrated in Figure 19.8, as a plot of resistivity versus temperature for copper and several copper–nickel alloys in annealed and deformed states. The additive nature of the individual resistivity contributions is demonstrated at $-100°C$.

TABLE 19.1 Room Temperature Electrical Conductivities for Eight Common Metals and Alloys

Metal	Electrical Conductivity $[(\Omega\text{-m})^{-1}]$
Silver	6.8×10^7
Copper	6.0×10^7
Gold	4.3×10^7
Aluminum	3.8×10^7
Iron	1.0×10^7
Brass (70 Cu–30 Zn)	1.6×10^7
Plain carbon steel	0.6×10^7
Stainless steel	0.2×10^7

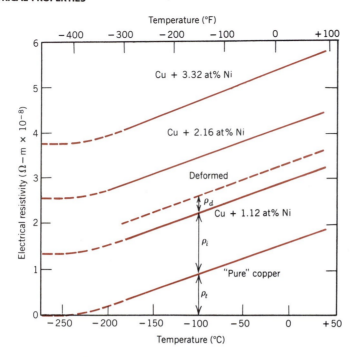

Figure 19.8 The electrical resistivity versus temperature for copper and three copper–nickel alloys, one of which has been deformed. Thermal, impurity, and deformation contributions to the resistivity are indicated at −100°C. [Adapted from J. O. Linde, *Ann. Physik,* **5,** 219 (1932); and C. A. Wert and R. M. Thomson, *Physics of Solids,* 2nd edition, McGraw-Hill Book Company, New York, 1970.]

Influence of Temperature

For the pure metal and all the copper–nickel alloys shown in Figure 19.8, the resistivity rises linearly with temperature above about −200°C. Thus

$$\rho_t = \rho_0 + aT \qquad (19.10)$$

where ρ_0 and a are constants for each particular metal. This dependence of the thermal resistivity component on temperature is due to the increase with temperature in thermal vibrations and other lattice irregularities (e.g., vacancies), which serve as electron-scattering centers.

Influence of Impurities

For additions of a single impurity that forms a solid solution, the impurity resistivity ρ_i is related to the impurity concentration C_i in terms of the atom fraction (at%/100) as follows:

$$\rho_i = AC_i(1 - C_i) \qquad (19.11)$$

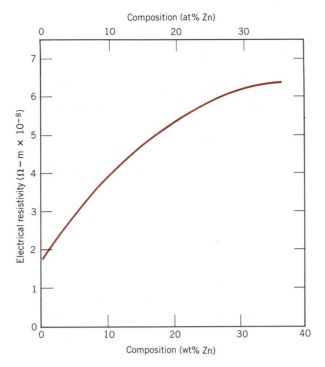

Figure 19.9 Room temperature electrical resistivity versus composition for copper–zinc alloys. (Adapted from *Metals Handbook: Properties and Selection: Nonferrous Alloys and Pure Metals,* Vol. 2, 9th edition, H. Baker, Managing Editor, American Society for Metals, 1979, p. 315.)

where A is a composition-independent constant that is a function of both the impurity and host metals. The influence of zinc impurity additions on the room-temperature resistivity of copper is demonstrated in Figure 19.9, over the composition range of solid solubility for the α phase (Figure 9.15). Again, zinc atoms in copper act as scattering centers, and increasing the concentration of zinc in copper results in an enhancement of resistivity.

For a two-phase alloy consisting of α and β phases, a rule-of-mixture expression may be utilized to approximate the resistivity as follows:

$$\rho_i = \rho_\alpha V_\alpha + \rho_\beta V_\beta \qquad (19.12)$$

where the V's and ρ's represent volume fractions and individual resistivities for the respective phases.

Influence of Plastic Deformation

Plastic deformation also raises the electrical resistivity as a result of increased numbers of electron-scattering dislocations. The effect of deformation on resistivity is also represented in Figure 19.8.

19.9 ELECTRICAL CHARACTERISTICS OF COMMERCIAL ALLOYS

Electrical and other properties of copper render it the most widely used metallic conductor. Oxygen-free high-conductivity (OFHC) copper, having extremely low oxygen and other impurity contents, is produced for many electrical applications. Aluminum, having a conductivity only about one half that of copper, is also frequently used as an electrical conductor. Silver has a higher conductivity than either copper or aluminum; however, its use is restricted on the basis of cost.

On occasion, it is necessary to improve the mechanical strength of a metal alloy without impairing significantly its electrical conductivity. Both solid solution alloying and cold working improve strength at the expense of conductivity, and thus, a tradeoff must be made for these two properties. Most often, strength is enhanced by introducing a second phase that does not have so adverse an effect on conductivity. For example, copper–beryllium alloys are precipitation hardened; but even so, the conductivity is reduced by about a factor of 5 over high-purity copper.

For some applications, such as furnace heating elements, a high electrical resistivity is desirable. The energy loss by electrons that are scattered is dissipated as heat energy. Such materials must have not only a high resistivity, but also a resistance to oxidation at elevated temperatures and, of course, a high melting temperature. Nichrome, a nickel–chromium alloy, is commonly employed in heating elements.

SEMICONDUCTIVITY

The electrical conductivity of the semiconducting materials is not as high as that of the metals; nevertheless, they have some unique electrical characteristics that render them especially useful. The electrical properties of these materials are extremely sensitive to the presence of even minute concentrations of impurities. **Intrinsic semiconductors** are those in which the electrical behavior is based on the electronic structure inherent to the pure material. When the electrical characteristics are dictated by impurity atoms, the semiconductor is said to be **extrinsic.**

19.10 INTRINSIC SEMICONDUCTION

Intrinsic semiconductors are characterized by the electron band structure shown in Figure 19.4d: at 0 K, a completely filled valence band, separated from an empty conduction band by a relatively narrow forbidden band gap, generally less than 2 eV. The two elemental semiconductors are silicon (Si) and germanium (Ge), having band gap energies of approximately 1.1 and 0.7 eV, respectively. Both are found in Group IVA of the periodic table (Figure 2.6) and are covalently bonded. In addition, a host of compound semiconducting materials also display intrinsic behavior. One such group is formed between elements of Groups IIIA and VA, for example, gallium arsenide (GaAs) and indium antimonide (InSb); these are frequently called III–V compounds. The compounds composed of elements of Groups IIB and VIA also display semiconducting behavior; these include cadmium sulfide (CdS) and zinc telluride (ZnTe). As the two elements forming these compounds become more widely separated with respect to their relative positions in the periodic table, the atomic bonding becomes

TABLE 19.2 Band Gap Energies, Electron and Hole Mobilities, and Intrinsic Electrical Conductivities at Room Temperature for Semiconducting Materials

Material	Band Gap (eV)	Electrical Conductivity $[(\Omega\text{-m})^{-1}]$	Electron Mobility $(m^2/V\text{-s})$	Hole Mobility $(m^2/V\text{-s})$
		Elemental		
Si	1.11	4×10^{-4}	0.14	0.05
Ge	0.67	2.2	0.38	0.18
		III–V Compounds		
GaP	2.25	—	0.05	0.002
GaAs	1.42	10^{-6}	0.85	0.45
InSb	0.17	2×10^4	7.7	0.07
		II–VI Compounds		
CdS	2.40	—	0.03	—
ZnTe	2.26	—	0.03	0.01

more ionic and the magnitude of the band gap energy increases—the materials tend to become more insulative. Table 19.2 gives the band gaps for some compound semiconductors.

Concept of a Hole

In intrinsic semiconductors, for every electron excited into the conduction band there is left behind a missing electron in one of the covalent bonds, or in the band scheme, a vacant electron state in the valence band, as shown in Figure 19.6b. Under the influence of an electric field, the position of this missing electron within the crystalline lattice may be thought of as moving by the motion of other valence electrons that repeatedly fill in the incomplete bond (Figure 19.10). This process is expedited by treating a missing electron from the valence band as a positively charged particle called a *hole*. A hole is considered to have a charge that is of the same magnitude as that for an electron, but of opposite sign ($+1.6 \times 10^{-19}$ C). Thus in the presence of an electric field, excited electrons and holes move in opposite directions. Furthermore, in semiconductors both electrons and holes are scattered by lattice imperfections.

Intrinsic Conductivity

Since there are two types of charge carrier (free electrons and holes) in an intrinsic semiconductor, the expression for electrical conduction, Equation 19.8, must be modified to include a term to account for the contribution of the hole current. Therefore, we write

$$\sigma = n|e|\mu_e + p|e|\mu_h \qquad (19.13)$$

where p is the number of holes per cubic meter and μ_h is the hole mobility. The magnitude of μ_h is always less than μ_e for semiconductors. For intrinsic semiconductors,

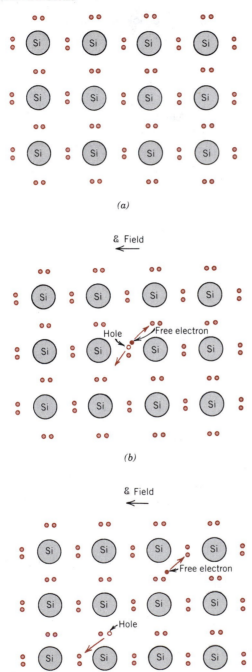

(a)

& Field

(b)

& Field

(c)

Figure 19.10 Electron bonding model of electrical conduction in intrinsic silicon: (a) before excitation; (b) and (c) after excitation (the subsequent free-electron and hole motions in response to an external electric field).

every electron promoted across the band gap leaves behind a hole in the valence band, thus

$$n = p \qquad (19.14)$$

and

$$\sigma = n|e|(\mu_e + \mu_h) = p|e|(\mu_e + \mu_h) \qquad (19.15)$$

The room-temperature intrinsic conductivities and electron and hole mobilities for several semiconducting materials are also presented in Table 19.2.

EXAMPLE PROBLEM 19.1

For intrinsic silicon, the room temperature electrical conductivity is 4×10^{-4} $(\Omega\text{-m})^{-1}$; the electron and hole mobilities are, respectively, 0.14 and 0.048 $m^2/$V-s. Compute the electron and hole concentrations at room temperature.

SOLUTION

Since the material is intrinsic, electron and hole concentrations will be the same, and therefore, from Equation 19.15,

$$n = p = \frac{\sigma}{|e|(\mu_e + \mu_h)}$$

$$= \frac{4 \times 10^{-4} \, (\Omega\text{-m})^{-1}}{(1.6 \times 10^{-19} \, C)(0.14 + 0.048 \, m^2/\text{V-s})}$$

$$= 1.33 \times 10^{16} \, m^{-3}$$

19.11 EXTRINSIC SEMICONDUCTION

Virtually all commercial semiconductors are extrinsic; that is, the electrical behavior is determined by impurities, which, when present in even minute concentrations, introduce excess electrons or holes. For example, an impurity concentration of one atom in 10^{12} is sufficient to render silicon extrinsic at room temperature.

n-Type Extrinsic Semiconduction

To illustrate how extrinsic semiconduction is accomplished, consider again the elemental semiconductor silicon. An Si atom has four electrons, each of which is covalently bonded with one of four adjacent Si atoms. Now, suppose that an impurity atom with a valence of 5 is added as a substitutional impurity; possibilities would include atoms from the Group VA column of the periodic table (e.g., P, As, and Sb). Only four of the five valence electrons of these impurity atoms can participate in the bonding because there are only four possible bonds with neighboring atoms. The extra nonbonding electron is loosely bound to the region around the impurity atom by a weak electrostatic attraction, as illustrated in Figure 19.11a. The binding energy of this electron is relatively small (on the order of 0.01 eV), thus it is easily removed

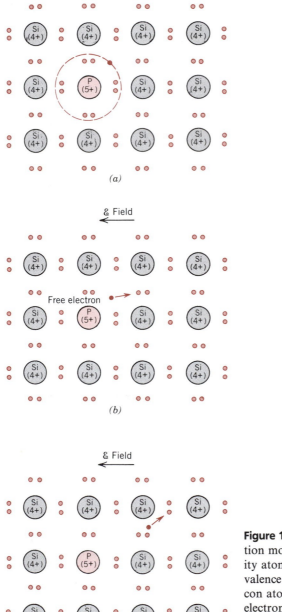

Figure 19.11 Extrinsic *n*-type semiconduction model (electron bonding). (*a*) An impurity atom such as phosphorus, having five valence electrons, may substitute for a silicon atom. This results in an extra bonding electron, which is bound to the impurity atom and orbits it. (*b*) Excitation to form a free electron. (*c*) The motion of this free electron in response to an electric field.

from the impurity atom, in which case it becomes a free or conducting electron (Figures 19.11*b* and 19.11*c*).

The energy state of such an electron may be viewed from the perspective of the electron band model scheme. For each of the loosely bound electrons, there exists a single energy level, or energy state, which is located within the forbidden band gap

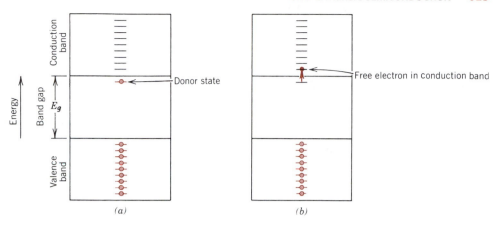

Figure 19.12 (*a*) Electron energy band scheme for a donor impurity level located within the band gap and just below the bottom of the conduction band. (*b*) Excitation from a donor state in which a free electron is generated in the conduction band.

just below the bottom of the conduction band (Figure 19.12*a*). The electron binding energy corresponds to the energy required to excite the electron from one of these impurity states to a state within the conduction band. Each excitation event (Figure 19.12*b*), supplies or donates a single electron to the conduction band; an impurity of this type is aptly termed a *donor*. Since each donor electron is excited from an impurity level, no corresponding hole is created within the valence band.

At room temperature, the thermal energy available is sufficient to excite large numbers of electrons from **donor states;** in addition, some intrinsic valence–conduction band transitions occur, as in Figure 19.6*b*, but to a negligible degree. Thus the number of electrons in the conduction band far exceeds the number of holes in the valence band (or $n \gg p$), and the first term on the right-hand side of Equation 19.13 overwhelms the second; that is,

$$\sigma \cong n|e|\mu_e \qquad (19.16)$$

A material of this type is said to be an n-*type* extrinsic semiconductor. The electrons are *majority carriers* by virtue of their density or concentration; holes, on the other hand, are the *minority charge carriers*. For n-type semiconductors, the Fermi level is shifted upward in the band gap, to within the vicinity of the donor state; its exact position is a function of both temperature and donor concentration.

p-Type Extrinsic Semiconduction

An opposite effect is produced by the addition to silicon or germanium of trivalent substitutional impurities such as aluminum, boron, and gallium from Group IIIA of the periodic table. One of the covalent bonds around each of these atoms is deficient in an electron; such a deficiency may be viewed as a hole that is weakly bound to the impurity atom. This hole may be liberated from the impurity atom by the transfer of an electron from an adjacent bond as illustrated in Figure 19.13. In essence, the electron and the hole exchange positions. A moving hole is considered to be in an excited state and participates in the conduction process, in a manner analogous to an excited donor electron, as described above.

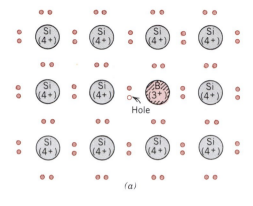

Hole

(a)

& Field

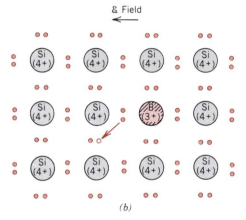

(b)

Figure 19.13 Extrinsic *p*-type semiconduction model (electron bonding). (*a*) An impurity atom such as boron, having three valence electrons, may substitute for a silicon atom. This results in a deficiency of one valence electron, or a hole associated with the impurity atom. (*b*) The motion of this hole in response to an electric field.

Extrinsic excitations, in which holes are generated, may also be represented using the band model. Each impurity atom of this type introduces an energy level within the band gap, above yet very close to the top of the valence band (Figure 19.14*a*). A hole is imagined to be created in the valence band by the thermal excitation of an electron from the valence band into this impurity electron state, as demonstrated in Figure 19.14*b*. With such a transition, only one carrier is produced—a hole in the valence band; a free electron is *not* created in either the impurity level or the conduction band. An impurity of this type is called an *acceptor*, because it is capable of accepting an electron from the valence band, leaving behind a hole. It follows that the energy level within the band gap introduced by this type of impurity is called an **acceptor state**.

For this type of extrinsic conduction, holes are present in much higher concentrations than electrons (i.e., $p \gg n$), and under these circumstances a material is termed p-*type* because positively charged particles are primarily responsible for electrical conduction. Of course, holes are the majority carriers, and electrons are present in minority concentrations. This gives rise to a predominance of the second term on the right-hand side of Equation 19.13, or

$$\sigma \cong p|e|\mu_h \tag{19.17}$$

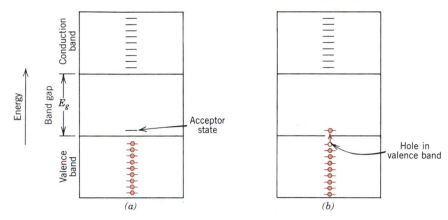

Figure 19.14 (*a*) Energy band scheme for an acceptor impurity level located within the band gap and just above the top of the valence band. (*b*) Excitation of an electron into the acceptor level, leaving behind a hole in the valence band.

For *p*-type semiconductors, the Fermi level is positioned within the band gap and near to the acceptor level.

Extrinsic semiconductors (both *n*- and *p*-type) are produced from materials that are initially of extremely high purity, commonly having total impurity contents on the order of 10^{-7} at%. Controlled concentrations of specific donors or acceptors are then intentionally added, using various techniques. Such an alloying process in semiconducting materials is termed **doping.**

In extrinsic semiconductors, large numbers of charge carriers (either electrons or holes, depending on the impurity type) are created at room temperature, by the available thermal energy. As a consequence, relatively high room-temperature electrical conductivities are obtained in extrinsic semiconductors. Most of these materials are designed for use in electronic devices to be operated at ambient conditions.

EXAMPLE PROBLEM 19.2

Phosphorus is added to high-purity silicon to give a concentration of 10^{23} m^{-3} of charge carriers at room temperature.

(a) Is this material *n*-type or *p*-type?

(b) Calculate the room-temperature conductivity of this material, assuming that electron and hole mobilities are the same as for the intrinsic material.

SOLUTION

(a) Phosphorus is a Group VA element (Figure 2.6) and, therefore, will act as a donor in silicon. Thus the 10^{23} m^{-3} charge carriers will be virtually all electrons. This electron concentration is greater than that for the intrinsic case (1.33×10^{16} m^{-3}, Example Problem 19.1); hence, this material is extrinsically *n*-type.

(b) In this case the conductivity may be determined using Equation 19.16, as follows:

$$\sigma = n|e|\mu_e = (10^{23} \text{ m}^{-3})(1.6 \times 10^{-19} \text{ C})(0.14 \text{ m}^2/\text{V-s})$$
$$= 2240 \ (\Omega\text{-m})^{-1}$$

19.12 THE TEMPERATURE VARIATION OF CONDUCTIVITY AND CARRIER CONCENTRATION

Figure 19.15 plots the logarithm of the electrical conductivity as a function of the logarithm of absolute temperature for intrinsic silicon, and also for silicon that has been doped with 0.0013 and 0.0052 at% boron; again, boron acts as an acceptor in silicon. Worth noting from this figure is that the electrical conductivity in the intrinsic specimen increases dramatically with rising temperature. The numbers of both elec-

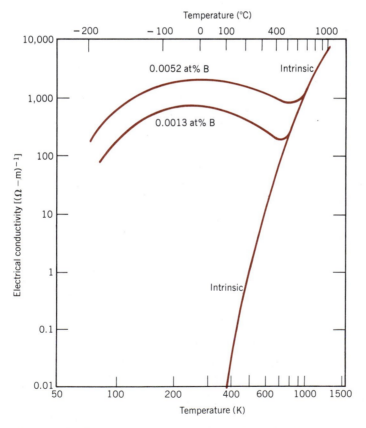

Figure 19.15 The temperature dependence of the electrical conductivity (log–log scales) for intrinsic silicon and boron-doped silicon at two doping levels. [Adapted from G. L. Pearson and J. Bardeen, *Phys. Rev.,* **75**, 865 (1949).]

trons and holes increase with temperature because more thermal energy is available to excite electrons from the valence to the conduction band. Thus both the values of n and p in the intrinsic conductivity expression, Equation 19.15, are enhanced. The magnitudes of electron and hole mobilities decrease slightly with temperature as a result of more effective electron and hole scattering by the thermal vibrations. However, these reductions in μ_e and μ_h by no means offset the increase in n and p, and the net effect of a rise in temperature is to produce a conductivity increase.

Mathematically, the dependence of intrinsic conductivity σ on the absolute temperature T is approximately

$$\ln \sigma \cong C - \frac{E_g}{2kT} \tag{19.18}$$

where C represents a temperature-independent constant and E_g and k are the band gap energy and Boltzmann's constant, respectively. Since the increase of n and p with rising temperature is so much greater than the decrease in μ_e and μ_h, the dependence of carrier concentration on temperature for intrinsic behavior is virtually the same as for the conductivity, or

$$\ln n = \ln p \cong C' - \frac{E_g}{2kT} \tag{19.19}$$

The parameter C' is a constant that is independent of temperature, yet is different from C in Equation 19.18.

In light of Equation 19.19, another method of representing the temperature dependence of the electrical behavior of semiconductors is as the natural logarithm of electron and hole concentrations versus the reciprocal of the absolute temperature. Figure 19.16 is such a plot using data taken from Figure 19.15; and, as may be noted (Figure 19.16), a straight line segment results for the intrinsic material; such a plot expedites the determination of the band gap energy. According to Equation 19.19, the slope of this line segment is equal to $-E_g/2k$, or E_g may be determined as follows:

$$E_g = -2k \left(\frac{\Delta \ln p}{\Delta \left(\frac{1}{T} \right)} \right)$$

$$= -2k \left(\frac{\Delta \ln n}{\Delta \left(\frac{1}{T} \right)} \right) \tag{19.20}$$

This is indicated in the schematic plot of Figure 19.17.

Another important feature of the behavior shown in Figures 19.15 and 19.16 is that at temperatures below about 800 K (527°C), the boron-doped materials are extrinsically p-type; that is, virtually all the carrier holes result from extrinsic excitations—electron transitions from the valence band into the boron acceptor level, which leave behind valence band holes (Figure 19.14). The available thermal energies at the temperatures are sufficient to promote significant numbers of these excitations, yet insufficient to stimulate many electrons from the valence band across the band gap. Thus the extrinsic conductivity far exceeds that of the intrinsic material. For example, at 400 K (127°C) the conductivities for intrinsic silicon and extrinsic

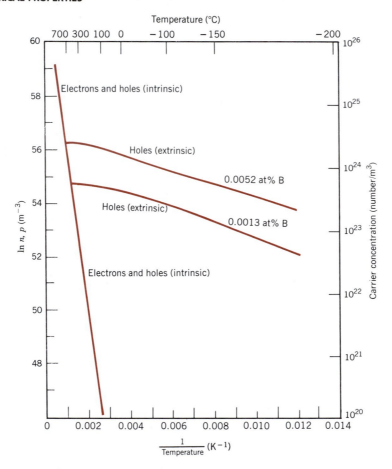

Figure 19.16 The logarithm of carrier (electron and hole) concentration as a function of the reciprocal of the absolute temperature for intrinsic silicon and two boron-doped silicon materials. [Adapted from G. L. Pearson and J. Bardeen, *Phys. Rev.,* **75**, 865 (1949).]

0.0013 at% boron-doped material are approximately 10^{-2} and 600 $(\Omega\text{-m})^{-1}$, respectively (Figure 19.15). This comparison indicates the sensitivity of conductivity to even extremely small concentrations of some impurity elements.

Furthermore, the extrinsic conductivity is also sensitive to temperature, as indicated in Figure 19.15, for both boron-doped materials. Beginning at about 75 K ($-200°$C), the conductivity first increases with temperature, reaches a maximum, and then decreases slightly prior to becoming intrinsic. Or, in terms of carrier (i.e., hole) concentration, Figure 19.16, ln p first increases linearly with decreasing $1/T$ (or increasing temperature). Large numbers of extrinsic excitations are possible even at these relatively low temperatures inasmuch as the acceptor level lies just above the top of the valence band. With further temperature increase ($1/T$ decrease), the hole concentration eventually becomes independent of temperature, Figure 19.16. At this point virtually all of the boron atoms have accepted electrons from the valence band,

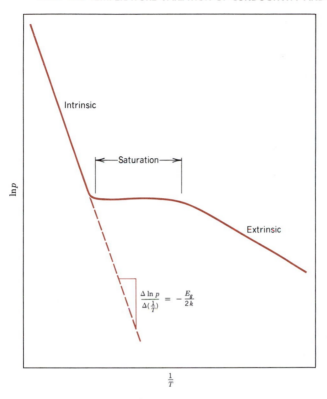

Figure 19.17 Schematic plot of the natural logarithm of hole concentration as a function of the reciprocal of absolute temperature for a p-type semiconductor that exhibits extrinsic, saturation, and intrinsic behavior.

or are said to be *saturated;* this is appropriately termed the *saturation region* (Figure 19.17). (Donor impurities become *exhausted* instead of saturated.) The number of holes in this region is approximately equal to the number of dopant impurity (i.e., boron) atoms.

The decrease of conductivity with increasing temperature within the saturation region for the two extrinsic curves in Figure 19.15 may be explained by the reduction in hole mobility with rising temperature. From the extrinsic conductivity expression, Equation 19.17, both e and p are independent of temperature in this region, and the only temperature dependence comes from the mobility.

Also worth noting from Figures 19.15 and 19.16 is that at about 800 K (527°C), the conductivity of both boron-doped materials becomes intrinsic. At the outset of intrinsic behavior, the number of intrinsic valence band-to-conduction band transitions becomes greater than the number of holes that are extrinsically generated.

A couple of final comments relate to the influence of boron acceptor content on the electrical behavior of silicon. First, the extrinsic and saturation conductivities and hole concentrations are greater for the material with the higher boron content (Figures 19.15 and 19.16), a result not unexpected, since more B atoms are present from which holes may be produced. Also, the intrinsic outset temperature becomes elevated as the dopant content increases.

EXAMPLE PROBLEM 19.3

If the room-temperature [25°C (298 K)] electrical conductivity of intrinsic germanium is 2.2 $(\Omega\text{-m})^{-1}$, estimate its conductivity at 150°C (423 K).

SOLUTION

This problem is solved by employment of Equation 19.18. First, we determine the value of the constant C using the room-temperature data, after which the value at 150°C may be computed. From Table 19.2, the value of E_g for germanium is 0.67 eV, and, therefore,

$$C = \ln \sigma + \frac{E_g}{2kT}$$

$$= \ln(2.2) + \frac{0.67 \text{ eV}}{(2)(8.62 \times 10^{-5} \text{ eV/K})(298 \text{ K})} = 13.83$$

Now, at 150°C (423 K),

$$\ln \sigma = C - \frac{E_g}{2kT}$$

$$= 13.83 - \frac{0.67 \text{ eV}}{(2)(8.62 \times 10^{-5} \text{ eV/K})(423 \text{ K})} = 4.64$$

or

$$\sigma = 103.8 \ (\Omega\text{-m})^{-1}$$

19.13 SEMICONDUCTOR DEVICES

The unique electrical properties of semiconductors permit their use in devices to perform specific electronic functions. Diodes and transistors, which have replaced old-fashioned vacuum tubes, are two familiar examples. Advantages of semiconductor devices (sometimes termed solid state devices) include small size, low power consumption, and no warmup time. Vast numbers of extremely small circuits, each consisting of numerous electronic devices, may be incorporated onto a small silicon "chip." The invention of semiconductor devices, which has given rise to miniaturized circuitry, is responsible for the advent and extremely rapid growth of a host of new industries in the past few years.

The *p–n* Rectifying Junction

A rectifier is an electronic device that allows the current to flow in one direction only; for example, a rectifier transforms an alternating current into direct current. Before the advent of the *p–n* junction semiconductor rectifier, this operation was carried out using the vacuum tube diode. The *p–n* **rectifying junction** is constructed from a single piece of semiconductor which is doped so as to be *n*-type on one side and *p*-type on the other (Figure 19.18*a*). If pieces of *n*- and *p*-type materials are joined together, a poor rectifier results, since the presence of a surface between the two sections renders the device very inefficient. Also, single crystals of semiconducting materials must be used in all devices because electronic phenomena that are deleterious to operation occur at grain boundaries.

Before the application of any potential across the *p–n* specimen, holes will be the dominant carriers on the *p*-side, and electrons will predominate in the *n*-region, as illustrated in Figure 19.18*a*. An external electric potential may be established across a *p–n* junction with two different polarities. When a battery is used, the positive terminal may be connected to the *p*-side, and the negative terminal to the *n*-side; this is referred to as a **forward bias**. The opposite polarity (minus to *p* and plus to *n*) is termed **reverse bias.**

The response of the charge carriers to the application of a forward-biased potential is demonstrated in Figure 19.18*b*. The holes on the *p*-side and the electrons on the *n*-side are attracted to the junction. As electrons and holes encounter one another near the junction, they continuously recombine and annihilate one another, according to

$$\text{electron} + \text{hole} \longrightarrow \text{energy} \tag{19.21}$$

Thus for this bias, large numbers of charge carriers flow across the semiconductor and to the junction, as evidenced by an appreciable current and a low resistivity. The

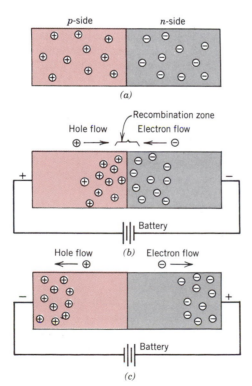

Figure 19.18 For a *p–n* rectifying junction, representations of electron and hole distributions for (*a*) no electrical potential, (*b*) forward bias, and (*c*) reverse bias.

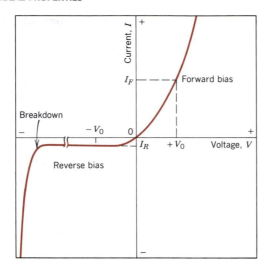

Figure 19.19 The current–voltage characteristics of a *p–n* junction for forward and reverse biases. The phenomenon of breakdown is also shown.

current–voltage characteristics for forward bias are shown on the right-hand half of Figure 19.19.

For reverse bias (Figure 19.18c), both holes and electrons, as majority carriers, are rapidly drawn away from the junction; this separation of positive and negative charges (or polarization) leaves the junction region relatively free of mobile charge carriers. Recombination will not occur to any appreciable extent, so that the junction is now highly insulative. Figure 19.19 also illustrates the current–voltage behavior for reverse bias.

The rectification process in terms of input voltage and output current is demonstrated in Figure 19.20. Whereas voltage varies sinusoidally with time (Figure

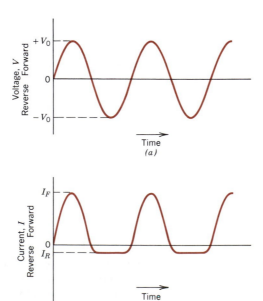

Figure 19.20 (*a*) Voltage versus time for the input to a *p–n* rectifying junction. (*b*) Current versus time, showing rectification of voltage in (*a*) by a *p–n* rectifying junction having the voltage–current characteristics shown in Figure 19.19.

19.20a), maximum current flow for reverse bias voltage I_R is extremely small in comparison to that for forward bias I_F (Figure 19.20b). Furthermore, correspondence between I_F and I_R and the imposed maximum voltage ($\pm V_0$) is noted in Figure 19.19.

At high reverse bias voltages, sometimes on the order of several hundred volts, large numbers of charge carriers (electrons and holes) are generated. This gives rise to a very abrupt increase in current, a phenomenon known as *breakdown,* also shown in Figure 19.19, and discussed in more detail in Section 19.20.

The Transistor

Transistors, which are extremely important semiconducting devices in today's microelectronic circuitry, are capable of two primary types of function. First, they can perform the same operation as their vacuum tube precursor, the triode; that is, they can amplify an electrical signal. In addition, they serve as switching devices in computers for the processing and storage of information. The two major types are the **junction** (or bimodal) **transistor** and the *metal-oxide-semiconductor field-effect transistor* (abbreviated as **MOSFET**).

Junction Transistors. The junction transistor is composed of two *p–n* junctions arranged back to back in either the *n–p–n* or the *p–n–p* configuration; the latter variety is discussed here. Figure 19.21 is a schematic representation of a *p–n–p* junction transistor along with its attendant circuitry. A very thin *n*-type *base* region is sandwiched in between *p*-type *emitter* and *collector* regions. The circuit that includes the emitter–base junction (junction 1) is forward biased, whereas a reverse bias voltage is applied across the base–collector junction (junction 2).

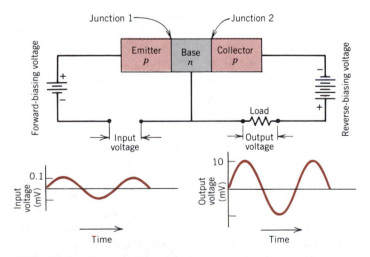

Figure 19.21 Schematic diagram of a *p–n–p* junction transistor and its associated circuitry, including input and output voltage–time characteristics showing voltage amplification. (Adapted from A. G. Guy, *Essentials of Materials Science,* McGraw-Hill Book Company, New York, 1976.)

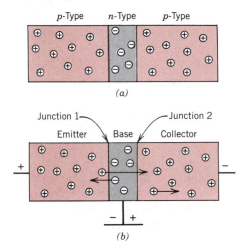

p-Type   *n*-Type   *p*-Type

(a)

Junction 1   Junction 2

Emitter   Base   Collector

+   −

− | +

(b)

Figure 19.22 For a junction transistor (*p–n–p* type), the distributions and directions of electron and hole motion (*a*) when no potential is applied and (*b*) with appropriate bias for voltage amplification.

Figure 19.22 illustrates the mechanics of operation in terms of the motion of charge carriers. Since the emitter is *p*-type and junction 1 is forward biased, large numbers of holes enter the base region. These injected holes are minority carriers in the *n*-type base, and some will combine with the majority electrons. However, if the base is extremely narrow and the semiconducting materials have been properly prepared, most of these holes will be swept through the base without recombination, then across junction 2 and into the *p*-type collector. The holes now become a part of the emitter–collector circuit. A small increase in input voltage within the emitter–base circuit produces a large increase in current across junction 2. This large increase in collector current is also reflected by a large increase in voltage across the load resistor, which is also shown in the circuit (Figure 19.21). Thus a voltage signal that passes through a junction transistor experiences amplification; this effect is also illustrated in Figure 19.21 by the two voltage–time plots.

Similar reasoning applies to the operation of an *n–p–n* transistor, except that electrons instead of holes are injected across the base and into the collector.

The MOSFET. One variety of MOSFET consists of two small islands of *p*-type semiconductor that are created within a substrate of *n*-type silicon, as shown in cross section in Figure 19.23; the islands are joined by a narrow *p*-type channel. Appropriate metal connections (source and drain) are made to these islands; an insulating layer of silicon dioxide is formed by the surface oxidation of the silicon. A final connector (gate) is then fashioned onto the surface of this insulating layer.

The operation of a MOSFET differs from that of the junction transistor in that a single type of charge carrier (either electron or hole) is active. The conductivity of the channel is varied by the presence of an electric field imposed on the gate. For example, imposition of a positive field on the gate will drive charge carriers (in this case holes) out of the channel, thereby reducing the electrical conductivity. Thus a small alteration in the field at the gate will produce a relatively large variation in current between the source and the drain. In some respects, then, the operation of a MOSFET is very similar to that described for the junction transistor. The primary difference is that the gate current is exceedingly small in comparison to the base

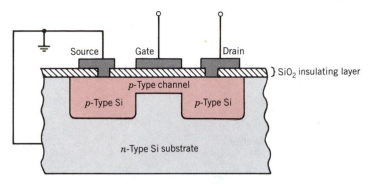

Figure 19.23 Schematic cross-sectional view of a MOSFET transistor.

current of a junction transistor. MOSFETs are, therefore, used where the signal sources to be amplified cannot sustain an appreciable current.

Semiconductors in Computers.   In addition to their ability to amplify an imposed electrical signal, transistors and diodes may also act as switching devices, a feature utilized for arithmetic and logical operations, and also for information storage in computers. Computer numbers and functions are expressed in terms of a binary code (i.e., numbers written to the base 2). Within this framework, numbers are represented by a series of two states (sometimes designated 0 and 1). Now, transistors and diodes within a digital circuit operate as switches that also have two states—on and off, or conducting and nonconducting; "off" corresponds to one binary number state, and "on" to the other. Thus a single number may be represented by a collection of circuit elements containing transistors that are appropriately switched.

Microelectronic Circuitry

During the past few years, the advent of microelectronic circuitry, where thousands of electronic components and circuits are incorporated into a very small space, has revolutionized the field of electronics. This revolution was precipitated, in part, by aerospace technology, which necessitated computers and electronic devices that were small and had low power requirements. As a result of refinement in processing and fabrication techniques, there has been an astonishing depreciation in the cost of integrated circuitry. Consequently, at the time of this writing, personal computers are affordable to a large segment of the population in the United States. Also, the use of **integrated circuits** has become infused into many other facets of our lives—calculators, communications, watches, industrial production and control, and all phases of the electronics industry.

Inexpensive microelectronic circuits are mass produced by using some very ingenious fabrication techniques. The process begins with the growth of relatively large cylindrical single crystals of high-purity silicon from which thin circular wafers are cut. Many microelectronic or integrated circuits, sometimes called "chips," are prepared on a single wafer. A chip is rectangular, typically on the order of $\frac{1}{4}$ in. (6 mm) on a side and contains thousands of circuit elements: diodes, transistors, resistors, and capacitors. One such microprocessor chip is shown in its entirety in Figure 19.24a;

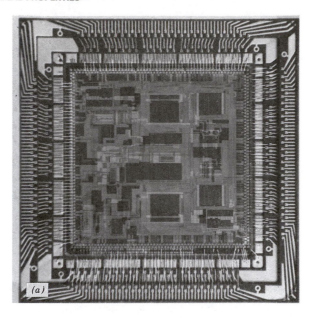

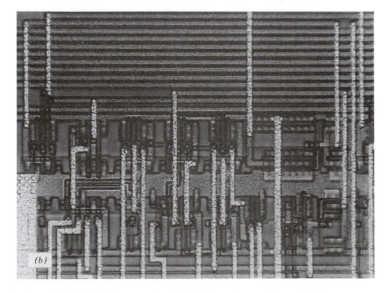

Figure 19.24 (a) Photograph of a 32-bit microprocessor chip which is mounted on a chip carrier. Approximately 3.5 ×. (b) Photograph showing a section of another 32-bit microprocessor chip. Dark and gray regions are diffusion-layer doped silicon. The speckled white regions are an aluminum metal top layer that serves as the wiring for these devices. Approximately 1250 ×. (Both photographs courtesy of Hewlett-Packard.)

more than 400 electrical leads are used to connect this chip to a printed circuit board. Figure 19.24b, an enlarged photograph of a portion of another microprocessor chip, reveals the intricacy of such an integrated circuit. At this time, 2,000,000-component chips are being produced, and even higher memory capabilities will undoubtedly be developed in the future.

Microelectronic circuits consist of many layers that lie within or are stacked on top of the silicon wafer in a precisely detailed pattern. Using photolithographic techniques, for each layer, very small elements are masked in accordance with a microscopic pattern. Circuit elements are constructed by the selective introduction (by diffusion or ion implantation) into unmasked regions to create localized n-type, p-type, high-resistivity, or conductive areas. This procedure is repeated layer by layer until the total integrated circuit has been fabricated, as illustrated in the MOSFET schematic (Figure 19.23). Several elements of an integrated circuit are shown in the scanning electron micrograph on page 606.

ELECTRICAL CONDUCTION IN IONIC CERAMICS AND IN POLYMERS

Most polymers and ionic ceramics are insulating materials at room temperature and, therefore, have electron energy band structures similar to that represented in Figure 19.4c: a filled valence band is separated from an empty conduction band by a relatively large band gap, usually greater than 2 eV. Thus at normal temperatures only very few electrons may be excited across the band gap by the available thermal energy, which accounts for the very small values of conductivity; Table 19.3 gives the room-temperature electrical conductivity of several of these materials. Of course, many

TABLE 19.3 Typical Room-Temperature Electrical Conductivities for 11 Nonmetallic Materials

Material	Electrical Conductivity $[(\Omega\text{-m})^{-1}]$
Graphite	10^5
Ceramics	
Aluminum oxide	10^{-10}–10^{-12}
Porcelain	10^{-10}–10^{-12}
Soda–lime glass	$<10^{-10}$
Mica	10^{-11}–10^{-15}
Polymers	
Phenol-formaldehyde	10^{-9}–10^{-10}
Nylon 6,6	10^{-9}–10^{-12}
Polymethyl methacrylate	$<10^{-12}$
Polyethylene	10^{-13}–10^{-17}
Polystyrene	$<10^{-14}$
Polytetrafluoroethylene	$<10^{-16}$

materials are utilized on the basis of their ability to insulate, and thus a high electrical resistivity is desirable. With rising temperature insulating materials experience an increase in electrical conductivity, which may ultimately be greater than that for semiconductors.

19.14 CONDUCTION IN IONIC MATERIALS

Both cations and anions in ionic materials possess an electric charge and, as a consequence, are capable of migration or diffusion when an electric field is present. Thus an electric current will result from the net movement of these charged ions, which will be present in addition to that due to any electron motion. Of course, anion and cation migrations will be in opposite directions. The total conductivity of an ionic material σ_{total} is thus equal to the sum of both electronic and ionic contributions, as follows:

$$\sigma_{total} = \sigma_{electronic} + \sigma_{ionic} \qquad (19.22)$$

Either contribution may predominate depending on the material, its purity, and, of course, temperature.

A mobility μ_i may be associated with each of the ionic species as follows:

$$\mu_i = \frac{n_i e D_i}{kT} \qquad (19.23)$$

where n_i and D_i represent, respectively, the valence and diffusion coefficient of a particular ion; e, k, and T denote the same parameters as explained earlier in the chapter. Thus the ionic contribution to the total conductivity increases with increasing temperature, as does the electronic component. However, in spite of the two conductivity contributions, most ionic materials remain insulative, even at elevated temperatures.

19.15 ELECTRICAL PROPERTIES OF POLYMERS

Most polymeric materials are poor conductors of electricity (Table 19.3) because of the unavailability of large numbers of free electrons to participate in the conduction process. The mechanism of electrical conduction in these materials is not well understood, but it is felt that conduction in polymers of high purity is electronic.

Conducting Polymers

Within the past several years, polymeric materials have been synthesized that have electrical conductivities on par with metallic conductors; they are appropriately termed *conducting polymers*. Conductivities as high as $1.5 \times 10^7 \; (\Omega\text{-m})^{-1}$ have been achieved in these materials; on a volume basis, this value corresponds to one fourth of the conductivity of copper, or twice its conductivity on the basis of weight.

This phenomenon is observed in a dozen or so polymers, including polyacetylene, polyparaphenylene, polypyrrole, and polyaniline that have been doped with appropriate impurities. As is the case with semiconductors, these polymers may be made either *n*-type (i.e., free-electron dominant) or *p*-type (i.e., hole dominant) depending on the dopant. However, unlike semiconductors, the dopant atoms or molecules do not substitute for or replace any of the polymer atoms.

High-purity polymers have electron band structures characteristic of electrical insulators (Figure 19.4c). The mechanism by which large numbers of free electrons and holes are generated in these conducting polymers is complex and not well understood. In very simple terms, it appears that the dopant atoms lead to the formation of new energy bands that overlap the valence and conduction bands of the intrinsic polymer, giving rise to a partially filled conduction or valence band, and the production at room temperature of a high concentration of free electrons or holes. Orienting the polymer chains, either mechanically (Section 16.3) or magnetically, during synthesis results in a highly anisotropic material having a maximum conductivity along the direction of orientation.

These conducting polymers have the potential to be used in a host of applications inasmuch as they have low densities, are highly flexible, and are easy to produce. Rechargeable batteries are currently being manufactured that employ polymer electrodes; in many respects these are superior to their metallic counterpart batteries. Other possible applications include wiring in aircraft and aerospace components, antistatic coatings for clothing, electromagnetic screening materials, and electronic devices (e.g., transistors and diodes).

DIELECTRIC BEHAVIOR

A **dielectric** material is one that is electrically insulating (nonmetallic) and exhibits or may be made to exhibit an electric dipole structure; that is, there is a separation of positive and negative electrically charged entities on a molecular or atomic level. This concept of an electric dipole was introduced in Section 2.7. As a result of dipole interactions with electric fields, dielectric materials are utilized in capacitors.

19.16 CAPACITANCE

When a voltage is applied across a capacitor, one plate becomes positively charged, the other negatively charged, with the corresponding electric field directed from the positive to the negative. The **capacitance** C is related to the quantity of charge stored on either plate Q by

$$C = \frac{Q}{V} \tag{19.24}$$

where V is the voltage applied across the capacitor. The units of capacitance are coulombs per volt, or farads (F).

Now, consider a parallel-plate capacitor with a vacuum in the region between the plates (Figure 19.25a). The capacitance may be computed from the relationship

$$C = \epsilon_0 \frac{A}{l} \tag{19.25}$$

where A represents the area of the plates and l is the distance between them. The parameter ϵ_0, called the **permittivity** of a vacuum, is a universal constant having the value of 8.85×10^{-12} F/m.

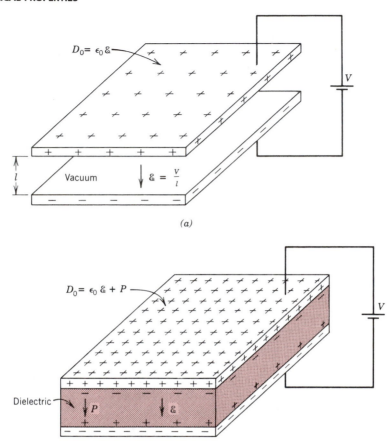

Figure 19.25 A parallel-plate capacitor (a) when a vacuum is present and (b) when a dielectric material is present. (From K. M. Ralls, T. H. Courtney, and J. Wulff, *Introduction to Materials Science and Engineering.* Copyright © 1976 by John Wiley & Sons, Inc. Reprinted by permission of John Wiley & Sons, Inc.)

If a dielectric material is inserted into the region within the plates (Figure 19.25b), then

$$C = \epsilon \frac{A}{l} \tag{19.26}$$

where ϵ is the permittivity of this dielectric medium, which will be greater in magnitude than ϵ_0. The relative permittivity ϵ_r, often called the **dielectric constant,** is equal to the ratio

$$\epsilon_r = \frac{\epsilon}{\epsilon_0} \tag{19.27}$$

TABLE 19.4 Dielectric Constants and Strengths for Some Dielectric Materials

Material	Dielectric Constant		Dielectric Strength (V/mil)[a]
	60 Hz	1 MHz	
Ceramics			
Titanate ceramics	—	15–10,000	50–300
Mica	—	5.4–8.7	1000–2000
Steatite ($MgO–SiO_2$)	—	5.5–7.5	200–350
Soda–lime glass	6.9	6.9	250
Porcelain	6.0	6.0	40–400
Fused silica	4.0	3.8	250
Polymers			
Phenol-formaldehyde	5.3	4.8	300–400
Nylon 6,6	4.0	3.6	400
Polystyrene	2.6	2.6	500–700
Polyethylene	2.3	2.3	450–500
Polytetrafluoroethylene	2.1	2.1	400–500

[a] One mil = 0.001 in. These values of dielectric strength are average ones, the magnitude being dependent on specimen thickness and geometry, as well as the rate of application and duration of the applied electric field.

which is greater than unity and represents the increase in charge storing capacity by insertion of the dielectric medium between the plates. The dielectric constant is one material property that is of prime consideration for capacitor design. The ϵ_r values of a number of dielectric materials are contained in Table 19.4.

19.17 FIELD VECTORS AND POLARIZATION

Perhaps the best approach to an explanation of the phenomenon of capacitance is with the aid of field vectors. To begin, for every electric dipole, there is a separation between a positive and a negative electric charge as demonstrated in Figure 19.26. An electric dipole moment p is associated with each dipole as follows:

$$p = qd \tag{19.28}$$

where q is the magnitude of each dipole charge and d is the distance of separation between them. In reality, a dipole moment is a vector that is directed from the negative to the positive charge, as indicated in Figure 19.26. In the presence of an electric field $\mathscr{E}$, which is also a vector quantity, a force (or torque) will come to bear on an electric dipole to orient it with the applied field; this phenomenon is illustrated in Figure 19.27. The process of dipole alignment is termed **polarization.**

Figure 19.26 Schematic representation of an electric dipole generated by two electric charges (of magnitude q) separated by the distance d; the associated polarization vector p is also shown.

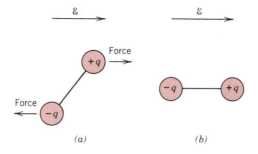

Figure 19.27 (a) Imposed forces (torque) acting on a dipole by an electric field. (b) Final dipole alignment with the field.

Again, returning to the capacitor, the surface charge density D, or quantity of charge per unit area of capacitor plate (C/m^2), is proportional to the electric field. When a vacuum is present, then

$$D_0 = \epsilon_0 \mathscr{E} \qquad (19.29)$$

the constant of proportionality being ϵ_0. Furthermore, an analogous expression exists for the dielectric case, that is,

$$D = \epsilon \mathscr{E} \qquad (19.30)$$

Sometimes, D is also called the **dielectric displacement.**

The increase in capacitance, or dielectric constant, can be explained using a simplified model of polarization within a dielectric material. Consider the capacitor in Figure 19.28a, the vacuum situation, wherein a charge of $+Q_0$ is stored on the top plate, and $-Q_0$ on the bottom one. When a dielectric is introduced and an electric field is applied, the entire solid within the plates becomes polarized (Figure 19.28c). As a result of this polarization, there is a net accumulation of negative charge of magnitude $-Q'$ at the dielectric surface near the positively charged plate, and, in a similar manner, a surplus of $+Q'$ charge at the surface adjacent to the negative plate. For the region of dielectric removed from these surfaces, polarization effects are not important. Thus if each plate and its adjacent dielectric surface are considered to be a single entity, the induced charge from the dielectric ($+Q'$ or $-Q'$) may be thought of as nullifying some of the charge that originally existed on the plate for a vacuum ($-Q_0$ or $+Q_0$). The voltage imposed across the plates is maintained at the vacuum value by increasing the charge at the negative (or bottom) plate by an amount $-Q'$, and the top plate by $+Q'$. Electrons are caused to flow from the positive to the negative plate by the external voltage source such that the proper voltage is reestablished. And so the charge on each plate is now $Q_0 + Q'$, having been increased by an amount Q'.

In the presence of a dielectric, the surface charge density on the plates of a capacitor may also be represented by

$$D = \epsilon_0 \mathscr{E} + P \qquad (19.31)$$

where P is the *polarization,* or the increase in charge density above that for a vacuum because of the presence of the dielectric, or from Figure 19.28c, $P = Q'/A$, where A is the area of each plate. The units of P are the same as for D (C/m^2).

The polarization P may also be thought of as the total dipole moment per unit volume of the dielectric material, or as a polarization electric field within the dielectric

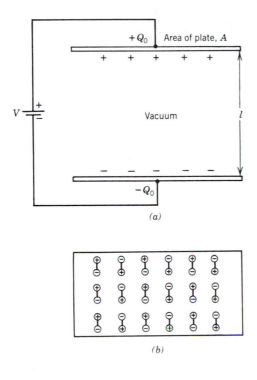

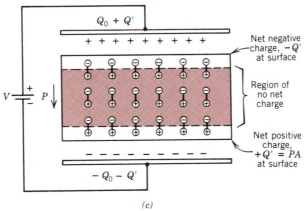

Figure 19.28 Schematic representations of (a) the charge stored on capacitor plates for a vacuum, (b) the dipole arrangement in an unpolarized dielectric, and (c) the increased charge storing capacity resulting from the polarization of a dielectric material. (Adapted from A. G. Guy, *Essentials of Materials Science,* McGraw-Hill Book Company, New York, 1976.)

TABLE 19.5 Primary and Derived Units for Various Electrical Parameters and Field Vectors

Quantity	Symbol	SI Units	
		Derived	Primary
Electric potential	V	volt	$kg\text{-}m^2/s^2\text{-}C$
Electric current	I	ampere	C/s
Electric field strength	$\mathscr{E}$	volt/meter	$kg\text{-}m/s^2\text{-}C$
Resistance	R	ohm	$kg\text{-}m^2/s\text{-}C^2$
Resistivity	ρ	ohm-meter	$kg\text{-}m^3/s\text{-}C^2$
Conductivity	σ	$(\text{ohm-meter})^{-1}$	$s\text{-}C^2/kg\text{-}m^3$
Electric charge	Q	coulomb	C
Capacitance	C	farad	$s^2\text{-}C^2/kg\text{-}m^2$
Permittivity	ϵ	farad/meter	$s^2\text{-}C^2/kg\text{-}m^3$
Dielectric constant	ϵ_r	ratio	ratio
Dielectric displacement	D	farad-volt/m^2	C/m^2
Electric polarization	P	farad-volt/m^2	C/m^2

that results from the mutual alignment of the many atomic or molecular dipoles with the externally applied field $\mathscr{E}$. For many dielectric materials, P is proportional to $\mathscr{E}$ through the relationship

$$P = \epsilon_0(\epsilon_r - 1)\mathscr{E} \tag{19.32}$$

in which case ϵ_r is independent of the magnitude of the electric field.

Table 19.5 lists the several dielectric parameters along with their units.

EXAMPLE PROBLEM 19.4

Consider a parallel-plate capacitor having an area of 1 in.^{2} (6.45×10^{-4} m^2) and a plate separation of 0.08 in. (2×10^{-3} m) across which a potential of 10 V is applied. If a material having a dielectric constant of 6.0 is positioned within the region between the plates, compute

(a) The capacitance.
(b) The magnitude of the charge stored on each plate.
(c) The dielectric displacement D.
(d) The polarization.

SOLUTION

(a) Capacitance is calculated using Equation 19.26; however, the permittivity ϵ of the dielectric medium must first be determined from Equation 19.27 as follows:

$$\epsilon = \epsilon_r\epsilon_0 = (6.0)(8.85 \times 10^{-12} \text{ F/m})$$
$$= 5.31 \times 10^{-11} \text{ F/m}$$

Thus the capacitance is

$$C = \epsilon \frac{A}{l} = (5.31 \times 10^{-11} \text{ F/m})\left(\frac{6.45 \times 10^{-4} \text{ m}^2}{2 \times 10^{-3} \text{ m}}\right)$$

$$= 1.71 \times 10^{-11} \text{ F}$$

(b) Since the capacitance has been determined, the charge stored may be computed using Equation 19.24, according to

$$Q = CV = (1.71 \times 10^{-11} \text{ F})(10 \text{ V}) = 1.71 \times 10^{-10} \text{ C}$$

(c) The dielectric displacement is calculated from Equation 19.30, which yields

$$D = \epsilon \mathcal{E} = \epsilon \frac{V}{l} = \frac{(5.31 \times 10^{-11} \text{ F/m})(10 \text{ V})}{2 \times 10^{-3} \text{ m}}$$

$$= 2.66 \times 10^{-7} \text{ C/m}^2$$

(d) Using Equation 19.31, the polarization may be determined as follows:

$$P = D - \epsilon_0 \mathcal{E} = D - \epsilon_0 \frac{V}{l}$$

$$= 2.66 \times 10^{-7} \text{ C/m}^2 - \frac{(8.85 \times 10^{-12} \text{ F/m})(10 \text{ V})}{2 \times 10^{-3} \text{ m}}$$

$$= 2.22 \times 10^{-7} \text{ C/m}^2$$

19.18 TYPES OF POLARIZATION

Again, polarization is the alignment of permanent or induced atomic or molecular dipole moments with an externally applied electric field. There are three types or sources of polarization: electronic, ionic, and orientation. Dielectric materials ordinarily exhibit at least one of these polarization types depending on the material and also the manner of the external field application.

Electronic Polarization

Electronic polarization may be induced to one degree or another in all atoms. It results from a displacement of the center of the negatively charged electron cloud relative to the positive nucleus of an atom by the electric field (Figure 19.29a). This polarization type is found in all dielectric materials, and, of course, exists only while an electric field is present.

Ionic Polarization

Ionic polarization occurs only in materials that are ionic. An applied field acts to displace cations in one direction and anions in the opposite direction, which gives rise to a net dipole moment. This phenomenon is illustrated in Figure 19.29b. The magnitude of the dipole moment for each ion pair p_i is equal to the product of the

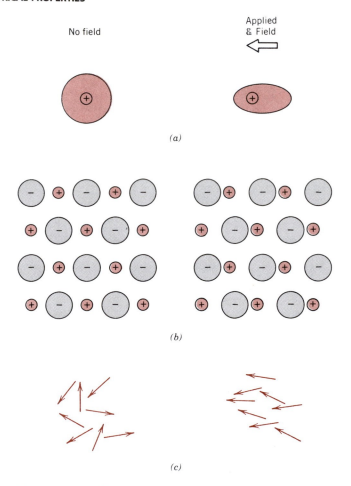

Figure 19.29 (*a*) Electronic polarization that results from the distortion of an atomic electron cloud by an electric field. (*b*) Ionic polarization that results from the relative displacement of electrically charged ions in response to an electric field. (*c*) Response of permanent electric dipoles (arrows) to an applied electric field, producing orientation polarization. (From O. H. Wyatt and D. Dew-Hughes, *Metals, Ceramics and Polymers,* Cambridge University Press, 1974.)

relative displacement d_i and the charge on each ion, or

$$p_i = qd_i \tag{19.33}$$

Orientation Polarization

The third type, **orientation polarization,** is found only in substances that possess permanent dipole moments. Polarization results from a rotation of the permanent moments into the direction of the applied field, as represented in Figure 19.29c. This alignment tendency is counteracted by the thermal vibrations of the atoms, such that polarization decreases with increasing temperature.

The total polarization P of a substance is equal to the sum of the electronic, ionic, and orientation polarizations (P_e, P_i, and P_o, respectively), or

$$P = P_e + P_i + P_o \tag{19.34}$$

It is possible for one or more of these contributions to the total polarization to be either absent or negligible in magnitude relative to the others. For example, ionic polarization will not exist in covalently bonded materials in which no ions are present.

19.19 FREQUENCY DEPENDENCE OF THE DIELECTRIC CONSTANT

In many practical situations the current is alternating (ac); that is, an applied voltage or electric field changes direction with time, as indicated in Figure 19.20a. Now, consider a dielectric material that is subject to polarization by an ac electric field. With each direction reversal, the dipoles attempt to reorient with the field, as illustrated in Figure 19.30, a process requiring some finite time. For each polarization type, some minimum reorientation time exists, which depends on the ease with which the particular dipoles are capable of realignment. A **relaxation frequency** is taken as the reciprocal of this minimum reorientation time.

A dipole cannot keep shifting orientation direction when the frequency of the applied electric field exceeds its relaxation frequency, and therefore, will not make a contribution to the dielectric constant. The dependence of ϵ_r on the field frequency is represented schematically in Figure 19.31 for a dielectric medium that exhibits all three types of polarization; note that the frequency axis is scaled logarithmically. As indicated in Figure 19.31, when a polarization mechanism ceases to function, there is an abrupt drop in the dielectric constant; otherwise, ϵ_r is virtually frequency independent. Table 19.4 gave values of the dielectric constant at 60 Hz and 1 MHz; these provide an indication of this frequency dependence at the low end of the frequency spectrum.

The absorption of electrical energy by a dielectric material that is subjected to an alternating electric field is termed *dielectric loss*. This loss may be important at electric field frequencies in the vicinity of the relaxation frequency for each of the operative dipole types for a specific material. A low dielectric loss is desired at the frequency of utilization.

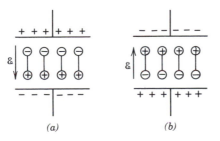

(a)   (b)

Figure 19.30 Dipole orientations for (a) one polarity of an alternating electric field and (b) for the reversed polarity. (Adapted from Richard A. Flinn, Paul K. Trojan, *Engineering Materials and Their Applications*, Third Edition. Copyright © 1986 by Houghton Mifflin Company. Used with permission.)

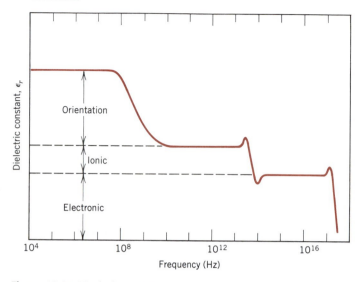

Figure 19.31 Variation of dielectric constant with frequency of an alternating electric field. Electronic, ionic, and orientation polarization contributions to the dielectric constant are indicated.

19.20 DIELECTRIC STRENGTH

When very high electric fields are applied across dielectric materials, large numbers of electrons may suddenly be excited to energies within the conduction band. As a result, the current through the dielectric by the motion of these electrons increases dramatically; sometimes localized melting, burning, or vaporization produces irreversible degradation and perhaps even failure of the material. This phenomenon is known as dielectric breakdown. The **dielectric strength,** sometimes called the breakdown strength, represents the magnitude of an electric field necessary to produce breakdown. Table 19.4 presented dielectric strengths for several materials.

19.21 DIELECTRIC MATERIALS

A number of ceramics and polymers are utilized as insulators and/or in capacitors. Many of the ceramics, including glass, porcelain, steatite, and mica, have dielectric constants within the range of 6 to 10 (Table 19.4). These materials also exhibit a high degree of dimensional stability and mechanical strength. Typical applications include powerline and electrical insulation, switch bases, and light receptacles. The titania (TiO_2) and titanate ceramics, such as barium titanate ($BaTiO_3$), can be made to have extremely high dielectric constants, which render them especially useful for some capacitor applications.

The magnitude of the dielectric constant for most polymers is less than for ceramics, since the latter may exhibit greater dipole moments; ϵ_r values for polymers generally lie between 2 and 5. These materials are commonly utilized for insulation of wires, cables, motors, generators, and so on, and, in addition, for some capacitors.

OTHER ELECTRICAL CHARACTERISTICS OF MATERIALS

Two other relatively important and novel electrical characteristics that are found in some materials deserve brief mention, namely, ferroelectricity and piezoelectricity.

19.22 FERROELECTRICITY

The group of dielectric materials called **ferroelectrics** exhibit spontaneous polarization, that is, polarization in the absence of an electric field. They are the dielectric analogue of ferromagnetic materials, which may display permanent magnetic behavior. There must exist in ferroelectric materials permanent electric dipoles, the origin of which is explained for barium titanate, one of the most common ferroelectrics. The spontaneous polarization is a consequence of the positioning of the Ba^{2+}, Ti^{4+}, and O^{2-} ions within the unit cell, as represented in Figure 19.32. The Ba^{2+} ions are located at the corners of the unit cell, which is of tetragonal symmetry (a cube that has been elongated slightly in one direction). The dipole moment results from the relative displacements of the O^{2-} and Ti^{4+} ions from their symmetrical positions as shown in the side view of the unit cell. The O^{2-} ions are located near, but slightly below, the centers of each of the six faces, whereas the Ti^{4+} ion is displaced upward from the unit cell center. Thus a permanent ionic dipole moment is associated with each unit cell. However, when barium titanate is heated above its *ferroelectric Curie temperature* [120°C (250°F)], the unit cell becomes cubic, and all ions assume symmetric positions

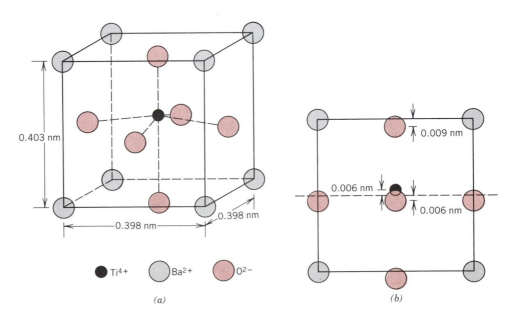

Ti^{4+} Ba^{2+} O^{2-}

0.403 nm   0.398 nm   0.398 nm

0.009 nm   0.006 nm   0.006 nm

(a)   (b)

Figure 19.32 A barium titanate ($BaTiO_3$) unit cell (*a*) in an isometric projection, and (*b*) looking at one face, which shows the displacements of Ti^{4+} and O^{2-} ions from the center of the face.

within the cubic unit cell; the material now has a perovskite crystal structure (Section 13.2), and the ferroelectric behavior ceases.

Spontaneous polarization of this group of materials results as a consequence of interactions between adjacent permanent dipoles wherein they mutually align, all in the same direction. For example, with barium titanate, the relative displacements of O^{2-} and Ti^{4+} ions are in the same direction for all the unit cells within some volume region of the specimen. Other materials display ferroelectricity; these include Rochelle salt ($NaKC_4H_4O_6 \cdot 4H_2O$), potassium dihydrogen phosphate (KH_2PO_4), potassium niobate ($KNbO_3$), and lead zirconate–titanate ($Pb[ZrO_3, TiO_3]$). Ferroelectrics have extremely high dielectric constants at relatively low applied field frequencies; for example, at room temperature, ϵ_r for barium titanate may be as high as 5000. Consequently, capacitors made from these materials can be significantly smaller than capacitors made from other dielectric materials.

19.23 PIEZOELECTRICITY

An unusual property exhibited by a few ceramic materials is piezoelectricity, or, literally, pressure electricity: polarization is induced and an electric field is established across a specimen by the application of external forces. Reversing the sign of an external force (i.e., from tension to compression) reverses the direction of the field. The piezoelectric effect is demonstrated in Figure 19.33.

Piezoelectric materials are utilized in transducers, devices that convert electrical energy into mechanical strains, or vice versa. Familiar applications that employ piezoelectrics include phonograph pickups, microphones, ultrasonic generators, strain gages, and sonar detectors. In a phonograph cartridge, as the stylus traverses the grooves on a record, a pressure variation is imposed on a piezoelectric material located in the cartridge, which is then transformed into an electric signal, and amplified before going to the speaker.

Piezoelectric materials include titanates of barium and lead, lead zirconate ($PbZrO_3$), ammonium dihydrogen phosphate ($NH_4H_2PO_4$), and quartz. This property is characteristic of materials having complicated crystal structures with a low degree of symmetry. The piezoelectric behavior of a polycrystalline specimen may be improved by heating above its Curie temperature and then cooling to room temperature in a strong electric field.

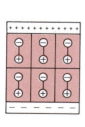 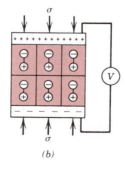

(a)   (b)

Figure 19.33 (a) Dipoles within a piezoelectric material. (b) A voltage is generated when the material is subjected to a compressive stress. (Adapted from L. H. Van Vlack, *Elements of Materials Science and Engineering*, 6th Edition. Copyright © 1989 Addison-Wesley Publishing Co. Reprinted by permission of Addison-Wesley Publishing Co., Inc., Reading, MA.)

SUMMARY

The ease with which a material is capable of transmitting an electric current is expressed in terms of electrical conductivity or its reciprocal, resistivity. On the basis of its conductivity, a solid material may be classified as a metal, a semiconductor, or an insulator.

For most materials, an electric current results from the motion of free electrons, which are accelerated in response to an applied electric field. The number of these free electrons depends on the electron energy band structure of the material. An electron band is just a series of electron states that are closely spaced with respect to energy, and one such band may exist for each electron subshell found in the isolated atom. By "electron energy band structure" is meant the manner in which the outermost bands are arranged relative to one another and then filled with electrons. A distinctive band structure type exists for metals, for semiconductors, and for insulators. An electron becomes free by being excited from a filled state in one band, to an available empty state above the Fermi energy. Relatively small energies are required for electron excitations in metals, giving rise to large numbers of free electrons. Larger energies are required for electron excitations in semiconductors and insulators, which accounts for their lower free electron concentrations and smaller conductivity values.

Free electrons being acted on by an electric field are scattered by imperfections in the crystal lattice. The magnitude of electron mobility is indicative of the frequency of these scattering events. In many materials, the electrical conductivity is proportional to the product of the electron concentration and the mobility.

For metallic materials, electrical resistivity increases with temperature, impurity content, and plastic deformation. The contribution of each to the total resistivity is additive.

Semiconductors may be either elements (Si and Ge) or covalently bonded compounds. With these materials, in addition to free electrons, holes (missing electrons in the valence band), may also participate in the conduction process. On the basis of electrical behavior, semiconductors are classified as either intrinsic or extrinsic. For intrinsic behavior, the electrical properties are inherent to the pure material, and electron and hole concentrations are equal; electrical behavior is dictated by impurities for extrinsic semiconductors. Extrinsic semiconductors may be either n- or p-type depending on whether electrons or holes, respectively, are the predominant charge carriers. Donor impurities introduce excess electrons; acceptor impurities, excess holes.

The electrical conductivity of semiconducting materials is particularly sensitive to impurity type and content, as well as to temperature. The addition of even minute concentrations of some impurities enhances the conductivity drastically. Furthermore, with rising temperature, intrinsic conductivity experiences an exponential increase. Extrinsic conductivity may also increase with temperature.

A number of semiconducting devices employ the unique electrical characteristics of these materials to perform specific electronic functions. Included are the p–n rectifying junction, and junction and MOSFET transistors. Transistors are used for amplification of electrical signals, as well as for switching devices in computer circuitries.

Dielectric materials are electrically insulative, yet susceptible to polarization in the presence of an electric field. This polarization phenomenon accounts for the ability of the dielectrics to increase the charge storing capability of capacitors, the efficiency of which is expressed in terms of a dielectric constant. Polarization results from the inducement by, or orientation with the electric field, of atomic or molecular dipoles; a dipole is said to exist when there is a net spatial separation of positively and negatively charged entities. Possible polarization types include electronic, ionic, and orientation; not all types need be present in a particular dielectric. For alternating electric fields, whether a specific polarization type contributes to the total polarization and dielectric constant depends on frequency; each polarization mechanism ceases to function when the applied field frequency exceeds its relaxation frequency.

This chapter concluded with brief discussions of two other electrical phenomena. Ferroelectric materials are those that may exhibit polarization spontaneously, that is, in the absence of any external electric field. Finally, piezoelectricity is the phenomenon whereby polarization is induced in a material by the imposition of external forces.

IMPORTANT TERMS AND CONCEPTS

Acceptor level	Extrinsic semiconductor	MOSFET
Band gap energy	Fermi energy	Ohm's law
Capacitance	Ferroelectric	Permittivity
Conduction band	Forward bias	Piezoelectric
Dielectric	Free electron	Polarization
Dielectric constant	Hole	Polarization, electronic
Dielectric displacement	Ionic conduction	Polarization, ionic
Dielectric strength	Insulator	Polarization, orientation
Dipole, electric	Integrated circuit	Rectifying junction
Donor level	Intrinsic semiconductor	Relaxation frequency
Doping	Junction transistor	Resistivity, electrical
Electric dipole	Matthiessen's rule	Reverse bias
Electrical conductivity	Metal	Semiconductor
Electrical resistance	Mobility	Valence band
Electron energy band		

REFERENCES

ADLER, R. B., A. C. SMITH, and R. L. LONGINI, *Introduction to Semiconductor Physics,* John Wiley & Sons, New York, 1964.

AZAROFF, L. V. and J. J. BROPHY, *Electronic Processes in Materials,* McGraw-Hill Book Company, New York, 1963. Chapters 6–12.

BUBE, R. H., *Electrons in Solids,* 2nd edition, Academic Press, San Diego, 1988.

BYLANDER, E. G., *Materials for Semiconductor Functions,* Hayden Book Company, New York, 1971. Good fundamental treatment of the physics of semiconductors and various semiconducting devices.

EHRENREICH, H., "The Electrical Properties of Materials," *Scientific American,* Vol. 217, No. 3, September 1967, pp. 194–204.

HUMMEL, R. E., *Electronic Properties of Materials,* Springer-Verlag New York, Inc., New York, 1985.

KINGERY, W. D., H. K. BOWEN, and D. R. UHLMANN, *Introduction to Ceramics,* 2nd edition, John Wiley & Sons, New York, 1976. Chapters 17 and 18.

KITTEL, C., *Introduction to Solid State Physics,* 6th edition, John Wiley & Sons, Inc., New York, 1986. An advanced treatment.

MEINDL, J. D., "Microelectronic Circuit Elements," *Scientific American,* Vol. 237, No. 3, September 1977, pp. 70–81.

NAVON, D. H., *Electronic Materials and Devices,* Houghton Mifflin Company, Boston, 1975. Special emphasis on diodes and bipolar and field effect transistors.

NOYCE, R. N., "Microelectronics," *Scientific American,* Vol. 237, No. 3, September 1977, pp. 62–69.

OLDHAM, W. G., "The Fabrication of Microelectronic Circuits," *Scientific American,* Vol. 237, No. 3, September 1977, pp. 110–128.

ROSE, R. M., L. A. SHEPARD, and J. WULFF, *The Structure and Properties of Materials,* Vol. IV, *Electronic Properties,* John Wiley & Sons, New York, 1966. Chapters 1, 2, 4–8, and 12.

WERT, C. A. and R. M. THOMSON, *Physics of Solids,* 2nd edition, McGraw-Hill Book Company, New York, 1970. Chapters 9 and 11–19.

QUESTIONS AND PROBLEMS

19.1 **(a)** Compute the electrical conductivity of a 0.2 in. (5.1 mm) diameter cylindrical silicon specimen 2 in. (50.8 mm) long in which a current of 0.1 A passes in an axial direction. A voltage of 12.5 V is measured across two probes that are separated by 1.5 in. (38.1 mm). **(b)** Compute the resistance over the entire 2 in. (50.8 mm) of the specimen.

19.2 An aluminum wire 10 m long must experience a voltage drop of less than 1.0 V when a current of 5 A passes through it. Using the data in Table 19.1, compute the minimum diameter of the wire.

19.3 A plain carbon steel wire 3 mm in diameter is to offer a resistance of no more than 20 Ω. Using the data in Table 19.1, compute the maximum wire length.

19.4 Demonstrate that the two Ohm's law expressions, Equations 19.1 and 19.5, are equivalent.

19.5 **(a)** Using the data in Table 19.1, compute the resistance of a copper wire 3 mm (0.12 in.) in diameter and 2 m (78.7 in.) long. **(b)** What would be the current flow if the potential drop across the ends of the wire is 0.05 V? **(c)** What is the current density? **(d)** What is the magnitude of the electric field across the ends of the wire?

19.6 What is the distinction between electronic and ionic conduction?

19.7 How does the electron structure of an isolated atom differ from that of a solid material?

19.8 In terms of electron energy band structure, discuss reasons for the difference in electrical conductivity between metals, semiconductors, and insulators.

19.9 If a metallic material is cooled through its melting temperature at an extremely rapid rate, it will form a noncrystalline solid (i.e., a metallic glass). Will the electrical conductivity of the noncrystalline metal be greater or less than its crystalline counterpart? Why?

19.10 Briefly tell what is meant by the drift velocity and mobility of a free electron.

19.11 **(a)** Calculate the drift velocity of electrons in germanium at room temperature and when the magnitude of the electric field is 1000 V/m. **(b)** Under these circumstances, how long does it take an electron to traverse a 1 in. (25.4 mm) length of crystal?

19.12 An *n*-type semiconductor is known to have an electron concentration of 3×10^{18} m^{-3}. If the electron drift velocity is 100 m/s in an electric field of 500 V/m, calculate the conductivity of this material.

19.13 At room temperature the electrical conductivity and the electron mobility for copper are 6.0×10^7 $(\Omega\text{-m})^{-1}$ and 0.0030 m^2/V-s, respectively. **(a)** Compute the number of free electrons per cubic meter of copper at room temperature. **(b)** What is the number of free electrons per copper atom? Assume a density of 8.9 g/cm^3.

19.14 **(a)** Calculate the number of free electrons per cubic meter for silver assuming that there are 1.3 free electrons per silver atom. The electrical conductivity and density for Ag are 6.8×10^7 $(\Omega\text{-m})^{-1}$ and 10.5 g/cm^3, respectively. **(b)** Now compute the electron mobility for Ag.

19.15 From Figure 19.9, estimate the value of *A* in Equation 19.11 for zinc as an impurity in copper–zinc alloys.

19.16 **(a)** Using the data in Figure 19.8, determine the values of ρ_0 and *a* from Equation 19.10 for pure copper. Take the temperature *T* to be in degrees Celsius. **(b)** Determine the value of *A* in Equation 19.11 for nickel as an impurity in copper, using the data in Figure 19.8. **(c)** Using the results of parts a and b, estimate the electrical resistivity of copper containing 1.75 at%Ni at 100°C.

19.17 A 90 wt% Cu–10 wt% Ni alloy is known to have an electrical resistivity of 1.90×10^{-7} Ω-m at room temperature (25°C). Calculate the composition of a copper–nickel alloy that gives a room-temperature resistivity of 2.5×10^{-7} Ω-m. The room-temperature resistivity of pure copper may be determined from the data in Table 19.1; assume that copper and nickel form a solid solution.

19.18 Using information contained in Figures 19.8 and 19.9, determine the electrical conductivity of an 85 wt% Cu–15 wt% Zn alloy at -100°C (-150°F).

19.19 Is it possible to alloy copper with zinc to achieve a minimum tensile strength of 40,000 psi (275 MPa) and yet maintain an electrical conductivity of 2×10^7 $(\Omega\text{-m})^{-1}$? If not, why? If so, what concentration of zinc is required? You may want to consult Figure 7.16*a*.

19.20 Determine the electrical conductivity of a Cu–Zn alloy that has a hardness of 50 HRF. You will find Figure 7.16 helpful.

19.21 Tin bronze has a composition of 89 wt% Cu and 11 wt% Sn and consists of two phases at room temperature: an α phase, which is copper containing a very small amount of tin in solid solution, and an ϵ phase, which consists of approximately 37 wt% Sn. Compute the room temperature conductivity of this alloy given the following data:

Phase	Electrical Resistivity (Ω-m)	Density (g/cm^3)
α	1.88×10^{-8}	8.92
ϵ	5.32×10^{-7}	8.43

19.22 A cylindrical metal wire 0.08 in. (2 mm) in diameter is required to carry a current of 10 A with a minimum of 0.03 V drop per foot (305 mm) of wire. Which of the metals and alloys listed in Table 19.1 are possible candidates?

19.23 **(a)** Compute the number of free electrons and holes that exist in intrinsic germanium at room temperature, using the data in Table 19.2. **(b)** Now calculate the number of free electrons per atom for germanium and silicon (Example Problem 19.1). **(c)** Explain the difference. You will need the densities for Ge and Si, which are 5.32 and 2.33 g/cm^3, respectively.

19.24 For intrinsic semiconductors, both electron and hole concentrations depend on temperature as follows:

$$n, p \propto \exp\left(-\frac{E_g}{2kT}\right) \qquad (19.35)$$

or, taking natural logarithms,

$$\ln n, \ln p \propto -\frac{E_g}{2kT}$$

Thus a plot of the intrinsic $\ln n$ (or $\ln p$) versus $1/T$ (K)$^{-1}$ should be linear and yield a slope of $-E_g/2k$. Using this information and Figure 19.16, determine the band gap energy for silicon. Compare this value with the one given in Table 19.2.

19.25 Define the following terms as they pertain to semiconducting materials: intrinsic, extrinsic, compound, elemental. Now provide an example of each.

19.26 Is it possible for compound semiconductors to exhibit intrinsic behavior? Explain your answer.

19.27 For each of the following pairs of semiconductors, decide which will have the smaller band gap energy E_g and then cite the reason for your choice: **(a)** C (diamond) and Ge, **(b)** AlP and InSb, **(c)** GaAs and ZnSe, **(d)** ZnSe and CdTe, and **(e)** CdS and NaCl.

19.28 **(a)** In your own words, explain how donor impurities in semiconductors give rise to free electrons in numbers in excess of those generated by valence band–conduction band excitations. **(b)** Also explain how acceptor impurities give rise to holes in numbers in excess of those generated by valence band–conduction band excitations.

19.29 **(a)** Explain why no hole is generated by the electron excitation involving a donor impurity atom. **(b)** Explain why no free electron is generated by the electron excitation involving an acceptor impurity atom.

19.30 Will each of the following elements act as a donor or an acceptor when added to the indicated semiconducting material? Assume that the impurity elements are substitutional.

Impurity	Semiconductor
N	Si
B	Ge
Zn	GaAs
S	InSb
In	CdS
As	ZnTe

19.31 **(a)** At approximately what position is the Fermi energy for an intrinsic semiconductor? **(b)** At approximately what position is the Fermi energy for an n-type semiconductor? **(c)** Make a schematic plot of Fermi energy versus temperature for an n-type semiconductor up to a temperature at which it becomes intrinsic. Also note on this plot energy positions corresponding to the top of the valence band and the bottom of the conduction band.

19.32 **(a)** The room-temperature electrical conductivity of a silicon specimen is 10^3 $(\Omega\text{-m})^{-1}$. The hole concentration is known to be 1.0×10^{23} m^{-3}. Using the electron and hole mobilities for silicon in Table 19.2, compute the electron concentration. **(b)** On the basis of the result in part a, is the specimen intrinsic, n-type extrinsic, or p-type extrinsic? Why?

19.33 Using the data in Table 19.2, compute the electron and hole concentrations for intrinsic GaAs at room temperature.

19.34 Germanium to which 10^{24} m^{-3} As atoms have been added is an extrinsic semiconductor at room temperature, and virtually all the As atoms may be thought of as being ionized (i.e., one charge carrier exists for each As atom). **(a)** Is this material n-type or p-type? **(b)** Calculate the electrical conductivity of this material, assuming electron and hole mobilities of 0.1 and 0.05 m^2/V-s, respectively.

19.35 The following electrical characteristics have been determined for both intrinsic and n-type extrinsic indium phosphide (InP) at room temperature:

	σ $(\Omega\text{-m})^{-1}$	n (m^{-3})	p (m^{-3})
Intrinsic	2.5×10^{-6}	3.0×10^{13}	3.0×10^{13}
Extrinsic (n-type)	3.6×10^{-5}	4.5×10^{14}	2.0×10^{12}

Calculate electron and hole mobilities.

19.36 Compare the temperature dependence of the conductivity for metals and intrinsic semiconductors. Briefly explain the difference in behavior.

19.37 Using the data in Table 19.2, estimate the electrical conductivity of intrinsic GaAs at 275°C (548 K).

19.38 Briefly explain the presence of the factor 2 in the denominator of the second term on the right-hand side of Equation 19.19.

19.39 Using the data in Table 19.2, estimate the temperature at which the electrical conductivity of intrinsic InSb is 4×10^3 $(\Omega\text{-m})^{-1}$.

19.40 The intrinsic electrical conductivities of a semiconductor at 20 and 100°C (293 and 373 K) are 1.0 and 500 $(\Omega\text{-m})^{-1}$, respectively. Determine the approximate band gap energy for this material.

19.41 Below, the intrinsic electrical conductivity of a semiconductor at three temperatures is tabulated:

T (K)	σ $(\Omega\text{-m})^{-1}$
350	17.2
400	81.4
500	717

(a) Determine the band gap energy (in eV) for this material.
(b) Estimate the electrical conductivity at 300 K (27°C).

19.42 The slope of the extrinsic portions of the curves in Figure 19.16 is related to the position of the acceptor level in the band gap (Figure 19.14). Write an expression for the dependence of p on the position of this level.

19.43 Briefly describe electron and hole motions in a $p–n$ junction for forward and reverse biases; then explain how these lead to rectification.

19.44 How is the energy in the reaction described by Equation 19.21 dissipated?

19.45 What are the two functions that a transistor may perform in an electronic circuit?

19.46 Would you expect increasing temperature to influence the operation of $p–n$ junction rectifiers and transistors? Explain.

19.47 Cite the differences in operation and application for junction transistors and MOSFETs.

19.48 At temperatures between 540°C (813 K) and 727°C (1000 K), the activation energy and preexponential for the diffusion coefficient of Na^+ in NaCl are 173,000 J/mol and 4.0×10^{-4} m^2/s, respectively. Compute the mobility for an Na^+ ion at 600°C (873 K).

19.49 A parallel-plate capacitor using a dielectric material having an ϵ_r of 2.5 has a plate spacing of 0.04 in. (1 mm). If another material having a dielectric constant of 4.0 is used and the capacitance is to be unchanged, what must be the new spacing between the plates?

19.50 A parallel-plate capacitor with dimensions of $1\frac{1}{2}$ in. by $2\frac{1}{2}$ in. (38.1 mm by 63.5 mm) and a plate separation of 0.05 in. (1.3 mm) must have a minimum capacitance of 70 pF (7×10^{-11} F) when an ac potential of 1000 V is applied at a frequency of 1 MHz. Which of those materials listed in Table 19.4 are possible candidates? Why?

19.51 Consider a parallel-plate capacitor having an area of 5 in.^{2} (3225 mm^2) and a plate separation of 0.04 in. (1.0 mm), and with a material of dielectric constant 3.5 positioned between the plates. (a) What is the capacitance of this capacitor?

(b) Compute the electric field that must be applied for a charge of 2×10^{-8} C to be stored on each plate.

19.52 In your own words, explain the mechanism by which charge storing capacity is increased by the insertion of a dielectric material within the plates of a capacitor.

19.53 For NaCl, the ionic radii for Na^+ and Cl^- ions are 0.102 and 0.181 nm, respectively. If an externaly applied electric field produces a 5% expansion of the lattice, compute the dipole moment for each Na^+–Cl^- pair. Assume that this material is completely unpolarized in the absence of an electric field.

19.54 The polarization P of a dielectric material positioned within a parallel-plate capacitor is to be 1.0×10^{-6} C/m^2.
(a) What must be the dielectric constant if an electric field is 5×10^4 V/m is applied?
(b) What will be the dielectric displacement D?

19.55 A charge of 3.5×10^{-11} C is to be stored on each plate of a parallel-plate capacitor having an area of 0.25 in.^{2} (160 mm^2) and a plate separation of 0.14 in. (3.5 mm).
(a) What voltage is required if a material having a dielectric constant of 5.0 is positioned within the plates?
(b) What voltage would be required if a vacuum is used?
(c) What are the capacitances for parts a and b?
(d) Compute the dielectric displacement for part a.
(e) Compute the polarization for part a.

19.56 **(a)** For each of the three types of polarization, briefly describe the mechanism by which dipoles are induced and/or oriented by the action of an applied electric field. **(b)** For gaseous argon, solid LiF, liquid H_2O, and solid Si, what kind(s) of polarization is (are) possible, and why?

19.57 The dielectric constant for a soda–lime glass measured at very high frequencies (on the order of 10^{15} Hz) is approximately 2.3. What fraction of the dielectric constant at relatively low frequencies (1 MHz) is attributed to ionic polarization? Neglect any orientation polarization contributions.

19.58 Compute the magnitude of the dipole moment associated with each unit cell of $BaTiO_3$, as illustrated in Figure 19.32.

19.59 Briefly explain why the ferroelectric behavior of $BaTiO_3$ ceases above its ferroelectric Curie temperature.

19.60 Would you expect the physical dimensions of a piezoelectric material such as $BaTiO_3$ to change when it is subjected to an electric field? Why or why not?

CHAPTER 20

THERMAL PROPERTIES

This photograph shows a white-hot cube of a silica fiber insulation material, which, only seconds after having been removed from a hot furnace, can be held by its edges with the bare hands. Initially, the heat transfer from the surface is relatively rapid; however, the thermal conductivity of this material is so small that heat conduction from the interior [maximum temperature approximately 1250°C (2300°F)] is extremely slow.

This material was developed especially for the tiles that cover the space shuttles and protect and insulate them during their fiery reentry into the atmosphere. Other attractive features of this *high-temperature reusable surface insulation* (*HRSI*) include low density and a low coefficient of thermal expansion. (Photograph courtesy of Lockheed Missiles & Space Company, Inc.)

20.1 INTRODUCTION

By "thermal property" is meant the response of a material to the application of heat. As a solid absorbs energy in the form of heat, its temperature rises and its dimensions increase. The energy may be transported to cooler regions of the specimen if temperature gradients exist, and ultimately, the specimen may melt. Heat capacity, thermal expansion, and thermal conductivity are properties that are often critical in the practical utilization of solids.

20.2 HEAT CAPACITY

A solid material, when heated, experiences an increase in temperature signifying that some energy has been absorbed. **Heat capacity** is a property that is indicative of a material's ability to absorb heat from the external surroundings; it represents the amount of energy required to produce a unit temperature rise. In mathematical terms, the heat capacity C is expressed as follows:

$$C = \frac{dQ}{dT} \tag{20.1}$$

where dQ is the energy required to produce a dT temperature change. Ordinarily, heat capacity is specified per mole of material (e.g., J/mol-K, or cal/mol-K). **Specific heat** (often denoted by a lowercase c) is sometimes used; this represents the heat capacity per unit mass and has various units (J/kg-K, cal/g-K, Btu/lb$_m$-°F).

There are really two ways in which this property may be measured, according to the environmental conditions accompanying the transfer of heat. One is the heat capacity while maintaining the specimen volume constant, C_v; the other is for constant external pressure, which is denoted C_p. The magnitude of C_p is always greater than C_v; however, this difference is very slight for most solid materials at room temperature and below.

Vibrational Heat Capacity

In most solids the principal mode of thermal energy assimilation is by the increase in vibrational energy of the atoms. Again, atoms in solid materials are constantly vibrating at very high frequencies and with relatively small amplitudes. Rather than being independent of one another, the vibrations of adjacent atoms are coupled by virtue of the atomic bonding. These vibrations are coordinated in such a way that traveling lattice waves are produced, a phenomenon represented in Figure 20.1. These may be thought of as elastic waves or simply sound waves, having short wavelengths and very high frequencies, which propagate through the crystal at the velocity of sound. The vibrational thermal energy for a material consists of a series of these elastic waves, which have a range of distributions and frequencies. Only certain energy values are allowed (the energy is said to be quantized), and a single quantum of vibrational energy is called a **phonon.** (A phonon is analogous to the quantum of electromagnetic radiation, the **photon.**) On occasion, the vibrational waves themselves are termed phonons.

The thermal scattering of free electrons during electronic conduction (Section 19.7) is by these vibrational waves, and these elastic waves also participate in the transport of energy during thermal conduction (see Section 20.4).

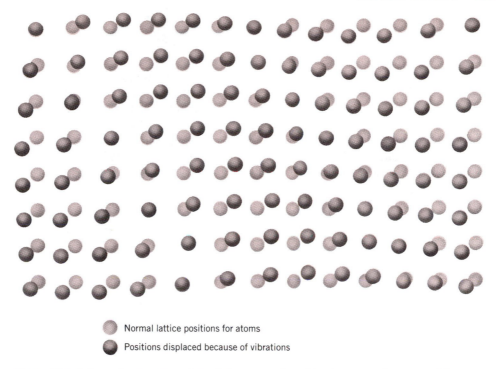

Normal lattice positions for atoms

Positions displaced because of vibrations

Figure 20.1 Schematic representation of the generation of lattice waves in a crystal by means of atomic vibrations. (Adapted from "The Thermal Properties of Materials" by J. Ziman. Copyright © 1967 by SCIENTIFIC AMERICAN, Inc. All rights reserved.)

Temperature Dependence of the Heat Capacity

The variation with temperature of the vibrational contribution to the heat capacity at constant volume for many relatively simple crystalline solids is shown in Figure 20.2. The C_v is zero at 0 K, but it rises rapidly with temperature; this corresponds to an increased ability of the lattice waves to enhance their average energy with ascending temperature. At low temperatures the relationship between C_v and the absolute

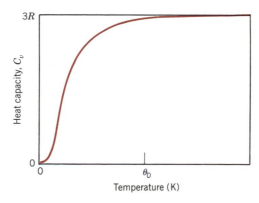

Figure 20.2 The temperature dependence of the heat capacity at constant volume; θ_D is the Debye temperature.

temperature T is

$$C_v = AT^3 \qquad (20.2)$$

where A is a temperature-independent constant. Above what is called the *Debye temperature* θ_D, C_v levels off and becomes essentially independent of temperature at a value of approximately $3R$, R being the gas constant. Thus even though the total energy of the material is increasing with temperature, the quantity of energy required to produce a one-degree temperature change is constant. The value of θ_D is below room temperature for many solid materials, and 25 J/mol-K (6 cal/mol-K) is a reasonable room-temperature approximation for C_v. Table 20.1 presents experimental specific heats for a number of materials.

TABLE 20.1 Tabulation of the Thermal Properties for a Variety of Materials

Material	c_p (J/kg-K)[a]	α_l [(°C)$^{-1}$ × 10^{-6}][b]	k (W/m-K)[c]	L [Ω-W/(K)^2 × 10^{-8}]
Metals				
Aluminum	900	23.6	247	2.24
Copper	386	16.5	398	2.27
Gold	130	13.8	315	2.52
Iron	448	11.8	80.4	2.66
Nickel	443	13.3	89.9	2.10
Silver	235	19.0	428	2.32
Tungsten	142	4.5	178	3.21
1025 Steel	486	12.5	51.9	—
316 Stainless steel	502	16.0	16.3[d]	—
Brass (70Cu–30Zn)	375	20.0	120	—
Ceramics				
Alumina (Al$_2$O$_3$)	775	8.8	30.1	—
Beryllia (BeO)	1050[d]	9.0[d]	220[e]	—
Magnesia (MgO)	940	13.5[d]	37.7[e]	—
Spinel (MgAl$_2$O$_4$)	790	7.6[d]	15.0[e]	—
Fused silica (SiO$_2$)	740	0.5[d]	2.0[e]	—
Soda–lime glass	840	9.0[d]	1.7[e]	—
Polymers				
Polyethylene	2100	60–220	0.38	—
Polypropylene	1880	80–100	0.12	—
Polystyrene	1360	50–85	0.13	—
Polytetrafluoroethylene (Teflon)	1050	135–150	0.25	—
Phenol-formaldehyde (Bakelite)	1650	68	0.15	—
Nylon 6,6	1670	80–90	0.24	—
Polyisoprene	—	220	0.14	—

[a] To convert to cal/g-K, multiply by 2.39 × 10^{-4}; to convert to Btu/lb$_m$-°F, multiply by 2.39 × 10^{-4}.
[b] To convert to (°F)$^{-1}$, multiply by 0.56.
[c] To convert to cal/s-cm-K, multiply by 2.39 × 10^{-3}; to convert to Btu/ft-h-°F, multiply by 0.578.
[d] Value measured at 100°C.
[e] Mean value taken over the temperature range 0–1000°C.

Other Heat Capacity Contributions

Other energy-absorptive mechanisms also exist that can add to the total heat capacity of a solid. In most instances, however, these are minor relative to the magnitude of the vibrational contribution. There is an electronic contribution in that electrons absorb energy by increasing their kinetic energy. However, this is possible only for free electrons—those that have been excited from filled states to empty states above the Fermi energy (Section 19.6). In metals, only electrons at states near the Fermi energy are capable of such transitions, and these represent only a very small fraction of the total number. An even smaller proportion of electrons experiences excitations in insulating and semiconducting materials. Hence, this electronic contribution is ordinarily insignificant, except at temperatures near 0 K.

Furthermore, in some materials other energy-absorptive processes occur at specific temperatures, for example, the randomization of electron spins in a ferromagnetic material as it is heated through its Curie temperature. A large spike is produced on the heat capacity-versus-temperature curve at the temperature of this transformation.

20.3 THERMAL EXPANSION

Most solid materials expand upon heating and contract when cooled. The change in length with temperature for a solid material may be expressed as follows:

$$\frac{l_f - l_0}{l_0} = \alpha_l(T_f - T_0) \tag{20.3a}$$

or

$$\frac{\Delta l}{l_0} = \alpha_l \Delta T \tag{20.3b}$$

where l_0 and l_f represent, respectively, initial and final lengths with the temperature change from T_0 to T_f. The parameter α_l is called the **linear coefficient of thermal expansion;** it is a material property that is indicative of the extent to which a material expands upon heating, and has units of reciprocal temperature $[(°C)^{-1}$ or $(°F)^{-1}]$. Of course, heating or cooling affects all the dimensions of a body, with a resultant change in volume. Volume changes with temperature may be computed from

$$\frac{\Delta V}{V_0} = \alpha_v \Delta T \tag{20.4}$$

where ΔV and V_0 are the volume change and the original volume, respectively, and α_v symbolizes the volume coefficient of thermal expansion. In many materials, the value of α_v is anisotropic; that is, it depends on the crystallographic direction along which it is measured. For materials in which the thermal expansion is isotropic, α_v is approximately $3\alpha_l$.

From an atomic perspective, thermal expansion is reflected by an increase in the average distance between the atoms. This phenomenon can best be understood by consultation of the potential energy versus interatomic spacing curve for a solid material introduced previously (Figure 2.8b), and reproduced in Figure 20.3a. The curve is in the form of a potential energy trough, and the equilibrium interatomic spacing at 0 K, r_0, corresponds to the trough minimum. Heating to successively higher

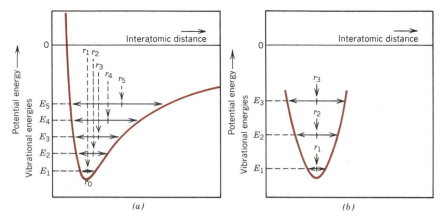

Figure 20.3 (*a*) Plot of potential energy versus interatomic distance, demonstrating the increase in interatomic separation with rising temperature. With heating, the interatomic separation increases from r_0 to r_1 to r_2, and so on. (*b*) For a symmetric potential energy-versus-interatomic distance curve, there is no increase in interatomic separation with rising temperature (i.e., $r_1 = r_2 = r_3$). (Adapted from R. M. Rose, L. A. Shepard, and J. Wulff, *The Structure and Properties of Materials,* Vol. 4, *Electronic Properties.* Copyright © 1966 by John Wiley & Sons, New York. Reprinted by permission of John Wiley & Sons, Inc.)

temperatures (T_1, T_2, T_3, etc.) raises the vibrational energy from E_1 to E_2 to E_3, and so on. The average vibrational amplitude of an atom corresponds to the trough width at each temperature, and the average interatomic distance is represented by the mean position, which increases with temperature from r_0 to r_1 to r_2, and so on.

Thermal expansion is really due to the asymmetric curvature of this potential energy trough, rather than the increased atomic vibrational amplitudes with rising temperature. If the potential energy curve were symmetric (Figure 20.3*b*), there would be no net change in interatomic separation and, consequently, no thermal expansion.

For each class of materials (metals, ceramics, and polymers), the greater the atomic bonding energy, the deeper and more narrow this potential energy trough. As a result, the increase in interatomic separation with a given rise in temperature will be lower, yielding a smaller value of α_l. Table 20.1 lists the linear coefficients of thermal expansion for several materials. With regard to temperature dependence, the magnitude of the coefficient of expansion increases with rising temperature, which increase is especially rapid at very near 0 K. The values given in Table 20.1 are taken at room temperature unless indicated otherwise.

Metals

As noted in Table 20.1, linear coefficients of thermal expansion for some of the common metals range between about 5×10^{-6} and 25×10^{-6} $(°C)^{-1}$. For some applications, a high degree of dimensional stability with temperature fluctuations is essential. This has resulted in the development of a family of iron–nickel and iron–cobalt alloys that have α_l values on the order of 1×10^{-6} $(°C)^{-1}$. One such alloy has been designed to have expansion characteristics the same as Pyrex glass; when joined to Pyrex and subjected to temperature variations, thermal stresses and possible fracture at the junction are avoided.

Ceramics

Relatively strong interatomic bonding forces are found in many ceramic materials as reflected in comparatively low coefficients of thermal expansion; values typically range between about 0.5×10^{-6} and 15×10^{-6} $(°C)^{-1}$. For noncrystalline ceramics and also those having cubic crystal structures, α_l is isotropic. Otherwise, it is anisotropic; and, in fact, some ceramic materials, upon heating, contract in some crystallographic directions while expanding in others. For inorganic glasses, the coefficient of expansion is dependent on composition. Fused silica (high-purity SiO_2 glass) has an extremely small expansion coefficient, 0.5×10^{-6} $(°C)^{-1}$. This is explained by a low atomic packing density such that interatomic expansion produces relatively small macroscopic dimensional changes. Adding impurities to the fused silica increases the coefficient of expansion.

Ceramic materials that are to be subjected to temperature changes must have coefficients of thermal expansion that are relatively low, and in addition, isotropic. Otherwise, these brittle materials may experience fracture as a consequence of nonuniform dimensional changes in what is termed **thermal shock,** as discussed later in the chapter.

Polymers

Some polymeric materials experience very large thermal expansions upon heating as indicated by coefficients that range from approximately 50×10^{-6} to 300×10^{-6} $(°C)^{-1}$. The highest α_l values are found in linear and branched polymers because the secondary intermolecular bonds are weak, and there is a minimum of cross-linking. With increased crosslinking, the magnitude of the expansion coefficient diminishes; the lowest coefficients are found in the thermosetting network polymers such as Bakelite, in which the bonding is almost entirely covalent.

20.4 THERMAL CONDUCTIVITY

Thermal conduction is the phenomenon by which heat is transported from high- to low-temperature regions of a substance. The property that characterizes the ability of a material to transfer heat is the **thermal conductivity.** It is best defined in terms of the expression

$$q = -k \frac{dT}{dx} \tag{20.5}$$

where q denotes the *heat flux,* or heat flow, per unit time per unit area (area being taken as that perpendicular to the flow direction), k is the thermal conductivity, and dT/dx is the *temperature gradient* through the conducting medium.

The units of q and k are W/m^2 (Btu/ft^2-h) and W/m-K (Btu/ft-h-°F), respectively. Equation 20.5 is valid only for steady-state heat flow, that is, for situations in which the heat flux does not change with time. Also, the minus sign in the expression indicates that the direction of heat flow is from hot to cold, or down the temperature gradient.

Equation 20.5 is similar in form to Fick's first law (Equation 5.3) for atomic diffusion. For these expressions, k is analogous to the diffusion coefficient D, and the temperature gradient parallels the concentration gradient, dC/dx.

Mechanisms of Heat Conduction

Heat is transported in solid materials by both lattice vibration waves (phonons) and free electrons. A thermal conductivity is associated with each of these mechanisms, and the total conductivity is the sum of the two contributions, or

$$k = k_l + k_e \tag{20.6}$$

where k_l and k_e represent the lattice vibration and electron thermal conductivities, respectively; usually one or the other predominates. The thermal energy associated with phonons or lattice waves is transported in the direction of their motion. The k_l contribution results from a net movement of phonons from high- to low-temperature regions of a body across which a temperature gradient exists.

Free or conducting electrons participate in electronic thermal conduction. To the free electrons in a hot region of the specimen is imparted a gain in kinetic energy. They then migrate to colder areas, where some of this kinetic energy is transferred to the atoms themselves (as vibrational energy) as a consequence of collisions with phonons or other imperfections in the crystal. The relative contribution of k_e to the total thermal conductivity increases with increasing free electron concentrations, since more electrons are available to participate in this heat transference process.

Metals

In high-purity metals, the electron mechanism of heat transport is much more efficient than the phonon contribution because electrons are not as easily scattered as phonons and have higher velocities. Furthermore, metals are extremely good conductors of heat because relatively large numbers of free electrons exist that participate in thermal conduction. The thermal conductivities of several of the common metals are given in Table 20.1; values generally range between about 20 and 400 W/m-K.

Since free electrons are responsible for both electrical and thermal conduction in pure metals, theoretical treatments suggest that the two conductivities should be related according to the *Wiedemann–Franz law:*

$$L = \frac{k}{\sigma T} \tag{20.7}$$

where σ is the electrical conductivity, T is the absolute temperature, and L is a constant. The theoretical value of L, 2.44×10^{-8} Ω-W/(K)^2, should be independent of temperature and the same for all metals if the heat energy is transported entirely by free electrons. Included in Table 20.1 are the experimental L values for these several metals; note that the agreement between these and the theoretical value is quite reasonable (well within a factor of 2).

Alloying metals with impurities results in a reduction in the thermal conductivity, for the same reason that the electrical conductivity is diminished (Section 19.8); namely, the impurity atoms, especially if in solid solution, act as scattering centers, lowering the efficiency of electron motion. A plot of thermal conductivity versus composition

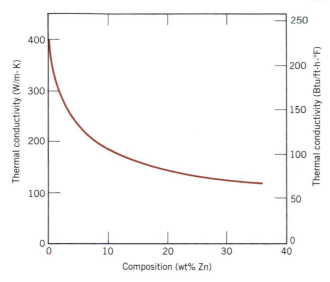

Figure 20.4 Thermal conductivity versus composition for copper–zinc alloys. (Adapted from *Metals Handbook: Properties and Selection: Nonferrous Alloys and Pure Metals,* Vol. 2, 9th edition, H. Baker, Managing Editor, American Society for Metals, 1979, p. 315.)

for copper–zinc alloys (Figure 20.4) displays this effect. Also, stainless steels, which are highly alloyed, become relatively resistive to heat transport.

Ceramics

Nonmetallic materials are thermal insulators inasmuch as they lack large numbers of free electrons. Thus the phonons are primarily responsible for thermal conduction: k_e is much smaller than k_l. Again, the phonons are not as effective as free electrons in the transport of heat energy as a result of the very efficient phonon scattering by lattice imperfections.

Thermal conductivity values for a number of ceramic materials are contained in Table 20.1; room-temperature thermal conductivities range between approximately 2 and 50 W/m-K. Glass and other amorphous ceramics have lower conductivities than crystalline ceramics, since the phonon scattering is much more effective when the atomic structure is highly disordered and irregular.

The scattering of lattice vibrations becomes more pronounced with rising temperature; hence, the thermal conductivity of most ceramic materials normally diminishes with increasing temperature, at least at relatively low temperatures (Figure 20.5). As Figure 20.5 indicates, the conductivity begins to increase at higher temperatures, which is due to radiant heat transfer: significant quantities of infrared radiant heat may be transported through a transparent ceramic material. The efficiency of this process increases with temperature.

Porosity in ceramic materials may have a dramatic influence on thermal conductivity; increasing the pore volume will, under most circumstances, result in a reduction of the thermal conductivity. In fact, many ceramics that are used for thermal

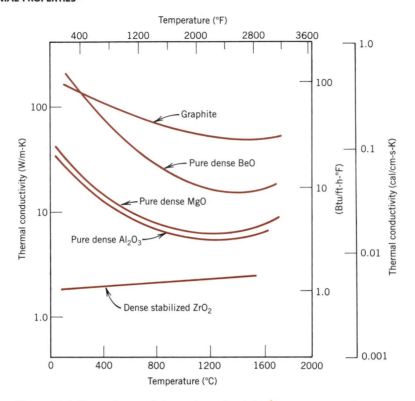

Figure 20.5 Dependence of thermal conductivity on temperature for several ceramic materials. (Adapted from W. D. Kingery, H. K. Bowen, and D. R. Uhlmann, *Introduction to Ceramics,* 2nd edition. Copyright © 1976 by John Wiley & Sons, New York. Reprinted by permission of John Wiley & Sons, Inc.)

insulation are porous. Heat transfer across pores is ordinarily slow and inefficient. Internal pores normally contain still air, which has an extremely low thermal conductivity—approximately 0.02 W/m-K. Furthermore, gaseous convection within the pores is also comparatively ineffective.

Polymers

As noted in Table 20.1, thermal conductivities for most polymers are on the order of 0.3 W/m-K. For these materials, energy transfer is accomplished by the vibration, translation, and rotation of the chain molecules. The magnitude of the thermal conductivity depends on the degree of crystallinity; a polymer with a highly crystalline and ordered structure will have a greater conductivity than the equivalent amorphous material. This is due to the more effective coordinated vibration of the molecular chains for the crystalline state.

Polymers are often utilized as thermal insulators because of their low thermal conductivities. As with ceramics, their insulative properties may be further enhanced by the introduction of small pores, which are ordinarily introduced by foaming during polymerization (Section 16.16). Foamed polystyrene (Styrofoam) is commonly used for drinking cups and insulating chests.

20.5 THERMAL STRESSES

Thermal stresses are stresses induced in a body as a result of changes in temperature. An understanding of the origins and nature of thermal stresses is important because these stresses can lead to fracture or undesirable plastic deformation.

Stresses Resulting from Restrained Thermal Expansion and Contraction

Let us first consider a homogeneous and isotropic solid rod that is heated or cooled uniformly; that is, no temperature gradients are imposed. For free expansion or contraction, the rod will be stress free. If, however, axial motion of the rod is restrained by rigid end supports, thermal stresses will be introduced. The magnitude of the stress σ resulting from a temperature change from T_0 to T_f is

$$\sigma = E\alpha_l(T_0 - T_f) = E\alpha_l \Delta T \qquad (20.8)$$

where E is the modulus of elasticity and α_l is the linear coefficient of thermal expansion. Upon heating ($T_f > T_0$), the stress is compressive ($\sigma < 0$), since rod expansion has been constrained. Of course, if the rod specimen is cooled ($T_f < T_0$), a tensile stress will be imposed ($\sigma > 0$). Also, the stress in Equation 20.8 is the same as that which would be required to elastically compress (or elongate) the rod specimen back to its original length after it had been allowed to freely expand (or contract) with the $T_0 - T_f$ temperature change.

EXAMPLE PROBLEM 20.1

A brass rod is to be used in an application requiring its ends to be held rigid. If the rod is stress free at room temperature [20°C (68°F)], what is the maximum temperature to which the rod may be heated without exceeding a compressive stress of 25,000 psi (172 MPa)? Assume a modulus of elasticity of 14.6×10^6 psi (10^5 MPa) for brass.

SOLUTION
Use Equation 20.8 to solve this problem, where the stress of 25,000 psi is taken to be negative. Also, the initial temperature T_0 is 20°C, and the magnitude of the linear coefficient of thermal expansion from Table 20.1 is 20×10^{-6} (°C)$^{-1}$. Thus solving for the final temperature T_f yields

$$T_f = T_0 - \frac{\sigma}{E\alpha_l}$$

$$= 20°C - \frac{-25{,}000 \text{ psi}}{(14.6 \times 10^6 \text{ psi})[20 \times 10^{-6} \text{ (°C)}^{-1}]}$$

$$= 20°C + 86°C = 106°C \ (223°F)$$

Stresses Resulting from Temperature Gradients

When a solid body is heated or cooled, the internal temperature distribution will depend on its size and shape, the thermal conductivity of the material, and the rate of temperature change. Thermal stresses may be established as a result of temperature gradients across a body, which are frequently caused by rapid heating or cooling, in that the outside changes temperature more rapidly than the interior; differential dimensional changes serve to restrain the free expansion or contraction of adjacent volume elements within the piece. For example, upon heating, the exterior of a specimen is hotter and, therefore, will have expanded more than the interior regions. Hence, surface stresses, being compressive, are induced and are balanced by interior tensile stresses. The interior–exterior stress conditions are reversed for rapid cooling such that the surface is put into a state of tension.

Thermal Shock of Brittle Materials

For ductile metals and polymers, alleviation of thermally induced stresses may be accomplished by plastic deformation. However, the nonductility of most ceramics enhances the possibility of brittle fracture from these stresses. Rapid cooling of a brittle body is more likely to inflict such thermal shock than heating, since the induced surface stresses are tensile. Crack formation and propagation from surface flaws are more probable when an imposed stress is tensile (Section 13.6).

The capacity of a material to withstand this kind of failure is termed its *thermal shock resistance*. For a ceramic body that is rapidly cooled, the resistance to thermal shock depends not only on the magnitude of the temperature change, but also on the mechanical and thermal properties of the material. The thermal shock resistance is best for ceramics that have high fracture strengths σ_f and high thermal conductivities, as well as a low moduli of elasticity and low coefficients of thermal expansion. The resistance of many materials to this type of failure may be approximated by a thermal shock resistance parameter *TSR*:

$$TSR \cong \frac{\sigma_f k}{E \alpha_l} \tag{20.9}$$

Thermal shock may be prevented by altering the external conditions to the degree that cooling or heating rates are reduced and temperature gradients across a body are minimized. Modification of the thermal and/or mechanical characteristics in Equation 20.9 may also enhance the thermal shock resistance of a material. Of these parameters, the coefficient of thermal expansion is probably most easily changed and controlled. For example, common soda–lime glasses, which have an α_l of approximately 9×10^{-6} $(°C)^{-1}$, are particularly susceptible to thermal shock, as anyone who has baked can probably attest. Reducing the CaO and Na_2O contents while at the same time adding B_2O_3 in sufficient quantities to form borosilicate (or Pyrex) glass will reduce the coefficient of expansion to about 3×10^{-6} $(°C)^{-1}$; this material is entirely suitable for kitchen oven heating and cooling cycles. The introduction of some relatively large pores or a ductile second phase may also improve the thermal shock characteristics of a material; both serve to impede the propagation of thermally induced cracks.

It is often necessary to remove thermal stresses in ceramic materials as a means of improving their mechanical strengths and optical characteristics. This may be accomplished by an annealing heat treatment, as discussed for glasses in Section 14.4.

SUMMARY

This chapter discussed heat absorption, thermal expansion, and thermal conduction—three important thermal phenomena. Heat capacity represents the quantity of heat required to produce a unit rise in temperature for one mole of a substance; on a per-unit mass basis, it is termed specific heat. Most of the energy assimilated by many solid materials is associated with increasing the vibrational energy of the atoms; contributions to the total heat capacity by other energy-absorptive mechanisms (i.e., increased free-electron kinetic energies) are normally insignificant.

For many crystalline solids and at temperatures within the vicinity of 0 K, the heat capacity measured at constant volume varies as the cube of the absolute temperature; in excess of the Debye temperature, C_v becomes temperature independent, assuming a value of approximately $3R$.

Solid materials expand when heated and contract when cooled. The fractional change in length is proportional to the temperature change, the constant of proportionality being the coefficient of thermal expansion. Thermal expansion is reflected by an increase in the average interatomic separation, which is a consequence of the asymmetric nature of the potential energy versus interatomic spacing curve trough. The larger the interatomic bonding energy, the lower the coefficient of thermal expansion.

The transport of thermal energy from high- to low-temperature regions of a material is termed thermal conduction. For steady-state heat transport, the flux is proportional to the temperature gradient along the direction of flow; the proportionality constant is the thermal conductivity.

For solid materials, heat is transported by free electrons and by vibrational lattice waves, or phonons. The high thermal conductivities for relatively pure metals are due to the large numbers of free electrons, and also the efficiency with which these electrons transport thermal energy. By way of contrast, ceramics and polymers are poor thermal conductors because free-electron concentrations are low and phonon conduction predominates.

Thermal stresses, which are introduced in a body as a consequence of temperature changes, may lead to fracture or undesirable plastic deformation. The two prime sources of thermal stresses are restrained thermal expansion (or contraction), and temperature gradients established during heating or cooling.

Thermal shock is the fracture of a body resulting from thermal stresses induced by rapid temperature changes. Because ceramic materials are brittle, they are especially susceptible to this type of failure. The thermal shock resistance of many materials is proportional to the fracture strength and thermal conductivity, and inversely proportional to both the modulus of elasticity and the coefficient of thermal expansion.

IMPORTANT TERMS AND CONCEPTS

Heat capacity	Phonon	Thermal shock
Linear coefficient of thermal expansion	Specific heat	Thermal stress
	Thermal conductivity	

REFERENCES

KINGERY, W. D., *Property Measurements at High Temperatures,* John Wiley & Sons, New York, 1959.

KINGERY, W. D., H. K. BOWEN, and D. R. UHLMANN, *Introduction to Ceramics,* 2nd edition, John Wiley & Sons, New York, 1976. Chapters 12 and 16.

ROSE, R. M., L. A. SHEPARD, and J. WULFF, *The Structure and Properties of Materials,* Vol. IV, *Electronic Properties,* John Wiley & Sons, New York, 1966. Chapters 3 and 8.

ZIMAN, J., "The Thermal Properties of Materials," *Scientific American,* Vol. 217, No. 3, September 1967, pp. 180–188.

QUESTIONS AND PROBLEMS

20.1 Estimate the energy required to raise the temperature of 2 kg (4.42 lb$_m$) of the following materials from 20 to 100°C (68 to 212°F): aluminum, steel, soda–lime glass, and polyethylene.

20.2 To what temperature would 10 lb$_m$ of a brass specimen at 25°C (77°F) be raised if 65 Btu of heat is supplied?

20.3 **(a)** Determine the room-temperature heat capacities at constant pressure for the following materials: copper, iron, gold, and nickel. **(b)** How do these values compare with one another? How do you explain this?

20.4 For aluminum, the heat capacity at constant volume C_v at 30 K is 0.81 J/mol-K, and the Debye temperature is 375 K. Estimate the specific heat **(a)** at 50 K and **(b)** at 425 K.

20.5 The constant A in Equation 20.2 is $12\pi^4 R/5\theta_D^3$, where R is the gas constant and θ_D is the Debye temperature (K). Estimate θ_D for copper, given that the specific heat is 0.78 J/kg-K at 10 K.

20.6 **(a)** Briefly explain why C_v rises with increasing temperature at temperatures near 0 K. **(b)** Briefly explain why C_v becomes virtually independent of temperature at temperatures far removed from 0 K.

20.7 A bimetallic strip is constructed from strips of two different metals that are bonded along their lengths. Explain how such a device may be used in a thermostat to regulate temperature.

20.8 **(a)** Explain why a brass lid ring on a glass canning jar will loosen when heated. **(b)** Suppose the ring is made of tungsten instead of brass. What will be the effect of heating the lid and jar? Why?

20.9 An aluminum wire 10 m (32.8 ft) long is cooled from 38 to −1°C (100 to 30°F). How much change in length will it experience?

20.10 A 0.1 m (3.9 in.) rod of a metal elongates 0.2 mm (0.0079 in.) on heating from 20 to 100°C (68 to 212°F). Determine the value of the linear coefficient of thermal expansion for this material.

20.11 Railroad tracks made of 1025 steel are to be laid during the time of year when the temperature averages 10°C (50°F). If a joint space of 0.180 in. (4.6 mm) is allowed between the standard 39 ft (11.9 m) long rails, what is the hottest possible temperature that can be tolerated without the introduction of thermal stresses?

20.12 Briefly explain thermal expansion using the potential energy-versus-interatomic spacing curve.

20.13 The difference between the specific heats at constant pressure and volume is described by the expression

$$c_p - c_v = \frac{\alpha_v^2 v_0 T}{\beta}$$ (20.10)

where α_v is the volume coefficient of thermal expansion, v_0 is the specific volume (i.e., volume per unit mass, or the reciprocal of density), β is the compressibility, and T is the absolute temperature. Compute the values of c_v at room temperature (293 K) for aluminum and iron using the data in Table 20.1, assuming that $\alpha_v = 3\alpha_l$ and given that the values of β for Al and Fe are 1.77×10^{-11} and 2.65×10^{-12} $(Pa)^{-1}$, respectively.

20.14 To what temperature must a cylindrical rod of tungsten 15.025 mm in diameter and a plate of 1025 steel having a circular hole 15.000 mm in diameter have to be heated for the rod to just fit into the hole? Assume that the initial temperature is 25°C.

20.15 Explain why, on a cold day, the metal door handle of an automobile feels colder to the touch than a plastic steering wheel, even though both are at the same temperature.

20.16 (a) Calculate the heat flux through a sheet of steel 10 mm (0.39 in.) thick if the temperatures at the two faces are 300 and 100°C (572 and 212°F); assume steady-state heat flow. (b) What is the heat loss per hour if the area of the sheet is 0.25 m² (2.7 ft²)? (c) What will be the heat loss per hour if soda–lime glass instead of steel is used? (d) Calculate the heat loss per hour if steel is used and the thickness is increased to 20 mm (0.79 in.).

20.17 (a) Would you expect Equation 20.7 to be valid for ceramic and polymeric materials? Why or why not? (b) Estimate the value for the Wiedemann–Franz constant L [in Ω-W/(K)²] and at room temperature (293 K) for the following nonmetals: alumina, soda–lime glass, polyethylene, and phenol-formaldehyde. Consult Tables 19.3 and 20.1.

20.18 (a) The thermal conductivity of a single-crystal specimen is slightly greater than a polycrystalline one of the same material. Why is this so? (b) The thermal conductivity of a plain carbon steel is greater than for a stainless steel. Why is this so?

20.19 Briefly explain why the thermal conductivities are higher for crystalline than noncrystalline ceramics.

20.20 Briefly explain why metals are typically better thermal conductors than ceramic materials.

20.21 (a) Briefly explain why porosity decreases the thermal conductivity of ceramic and polymeric materials, rendering them more thermally insulative. (b) Briefly explain how the degree of crystallinity affects the thermal conductivity of polymeric materials and why.

20.22 For some ceramic materials, why does the thermal conductivity first decrease and then increase with rising temperature?

20.23 For each of the following pairs of materials, decide which has the larger thermal conductivity. Justify your choices.
(a) Pure silver; sterling silver (92.5 wt% Ag–7.5 wt% Cu).
(b) Fused silica; polycrystalline silica.
(c) Linear polyethylene (M_n = 450,000 g/mol); lightly branched polyethylene (M_n = 650,000 g/mol).
(d) Atactic polypropylene (M_w = 10^6 g/mol); isotactic polypropylene (M_w = 5×10^5 g/mol).

20.24 We might think of a porous material as being a composite wherein one of the phases is a pore phase. Estimate upper and lower limits for the room-temperature thermal conductivity of an aluminum oxide material having a volume fraction of 0.25 of pores that are filled with still air.

20.25 Nonsteady-state heat flow may be described by the following partial differential equation:

$$\frac{\partial T}{\partial t} = D_T \frac{\partial^2 T}{\partial x^2}$$

where D_T is the thermal diffusivity; this expression is the thermal equivalent of Fick's second law of diffusion (Equation 5.4b). The thermal diffusivity is defined according to

$$D_T = \frac{k}{\rho c_p}$$

In this expression, k, ρ, and c_p represent the thermal conductivity, the mass density, and the specific heat at constant pressure, respectively.
(a) What are the SI units for D_T?
(b) Determine values of D_T for aluminum, steel, alumina, soda-lime glass, polystyrene, and nylon 6,6 using the data in Table 20.1. Density values are included in Appendix C.

20.26 Beginning with Equation 20.3, show that Equation 20.8 is valid.

20.27 (a) Briefly explain why thermal stresses may be introduced into a structure by rapid heating or cooling. (b) For cooling, what is the nature of the surface stresses? (c) For heating, what is the nature of the surface stresses? (d) For a ceramic material, is thermal shock more likely to occur on rapid heating or cooling? Why?

20.28 (a) If a rod of 1025 steel 0.5 m (19.7 in.) long is heated from 20° to 80°C (68° to 176°F) while its ends are maintained rigid, determine the type and magnitude of stress that develops. Assume that at 20°C the rod is stress free. (b) What will be the stress magnitude if a rod 1 m (39.4 in.) long is used? (c) If the rod in part a is cooled from 20° to -10°C (68° to 14°F), what type and magnitude of stress will result?

20.29 A copper wire is stretched with a stress of 10,000 psi (69 MPa) at 20°C (68°F). If the length is held constant, to what temperature must the wire be heated to reduce the stress to 5000 psi (34.5 MPa)?

20.30 The ends of a cylindrical rod 0.25 in. (6.4 mm) in diameter and 10 in. (254 mm) long are mounted between rigid supports. The rod is stress free at room temperature [20°C (68°F)]; and upon cooling to −60°C (−76°F), a maximum thermally induced tensile stress of 20,000 psi (138 MPa) is possible. Of which of the following metals or alloys may the rod be fabricated: aluminum, copper, brass, 1025 steel, and tungsten? Why?

20.31 If a cylindrical rod of brass 150.00 mm long and 10.000 mm in diameter is heated from 20°C to 160°C while its ends are maintained rigid, determine its change in diameter. You may want to consult Table 6.1.

20.32 The two ends of a cylindrical rod of nickel 120.00 mm long and 12.000 mm in diameter are maintained rigid. If the rod is initially at 70°C, to what temperature must it be cooled to have a 0.023-mm reduction in diameter?

20.33 **(a)** What are the units for the thermal shock resistance parameter (*TSR*)? **(b)** Rank the following ceramic materials according to their thermal shock resistance: magnesium oxide, spinel, fused silica, and soda–lime glass. For your computations take the fracture strength to be the modulus of rupture.

20.34 Equation 20.9, for the thermal shock resistance of a material, is valid for relatively low rates of heat transfer. When the rate is high, then, upon cooling of a body, the maximum temperature change allowable without thermal shock, ΔT_f, is approximately

$$\Delta T_f \cong \frac{\sigma_f}{E\alpha_l}$$

where σ_f is the fracture strength. Using the data in Tables 13.4 and 20.1, determine ΔT_f for alumina, beryllia, soda–lime glass, and fused silica.

20.35 What measures may be taken to reduce the likelihood of thermal shock of a ceramic piece?

CHAPTER 21

MAGNETIC PROPERTIES

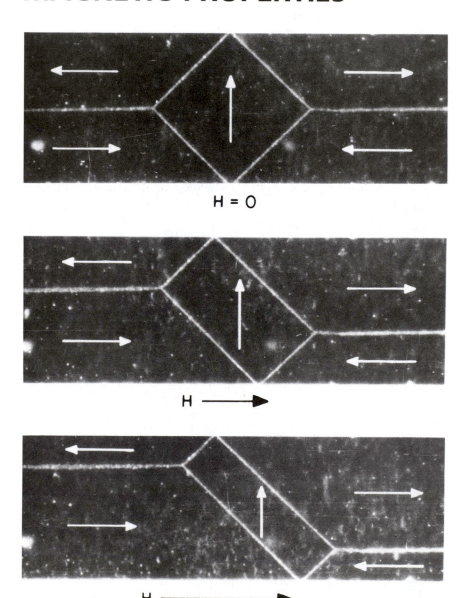

Photomicrographs of an iron single crystal, showing magnetic domains and their change in shape as a magnetic field (H) is applied. The magnetization direction of each domain is indicated by an arrow. Those domains that are favorably oriented with the applied field grow at the expense of the unfavorably oriented domains. (Photomicrographs courtesy of General Electric Research Laboratory.)

21.1 INTRODUCTION

Magnetism, the phenomenon by which materials assert an attractive or repulsive force or influence on other materials, has been known for thousands of years. However, the underlying principles and mechanisms that explain the magnetic phenomenon are complex and subtle, and their understanding has eluded scientists until relatively recent times. Many of our modern technological devices rely on magnetism and magnetic materials; these include electrical power generators and transformers, electric motors, radio, television, telephones, computers, and components of sound and video reproduction systems.

Iron, some steels, and the naturally occurring mineral lodestone are well-known examples of materials that exhibit magnetic properties. Not so familiar, however, is the fact that all substances are influenced to one degree or another by the presence of a magnetic field. This chapter provides a brief description of the origin of magnetic fields and discusses the various magnetic field vectors and magnetic parameters; the phenomena of diamagnetism, paramagnetism, ferromagnetism, and ferrimagnetism; some of the different magnetic materials; and the phenomenon of superconductivity.

21.2 BASIC CONCEPTS

Magnetic Dipoles

Magnetic forces are generated by moving electrically charged particles; they are in addition to any electrostatic forces that may prevail. Many times it is convenient to think of magnetic forces in terms of fields. Imaginary lines of force may be drawn to indicate the direction of the force at positions in the vicinity of the field source. The magnetic field distributions as indicated by lines of force are shown for a current loop and also a bar magnet in Figure 21.1.

Magnetic dipoles are found to exist in magnetic materials, which, in some respects, are analogous to electric dipoles (Section 19.17). Magnetic dipoles may be thought of as small bar magnets composed of north and south poles instead of positive and negative electric charges. In the present discussion, magnetic dipole moments are represented by arrows, as shown in Figure 21.2. Magnetic dipoles are influenced by magnetic fields in a manner similar to the way in which electric dipoles are affected by electric fields (Figure 19.27). Within a magnetic field, the force of the field itself exerts a torque that tends to orient the dipoles with the field. A familiar example is the way in which a magnetic compass needle lines up with the earth's magnetic field.

Magnetic Field Vectors

Before discussing the origin of magnetic moments in solid materials, we describe magnetic behavior in terms of several field vectors. The externally applied magnetic field, sometimes called the **magnetic field strength,** is designated by H. If the magnetic field is generated by means of a cylindrical coil (or solenoid) consisting of N closely spaced turns, having a length l, and carrying a current of magnitude I, then

$$H = \frac{NI}{l} \qquad (21.1)$$

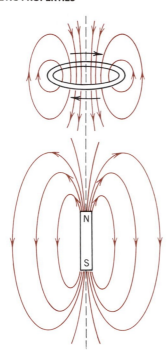

Figure 21.1 Magnetic field lines of force around a current loop and a bar magnet.

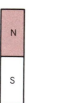

Figure 21.2 The magnetic moment as designated by an arrow.

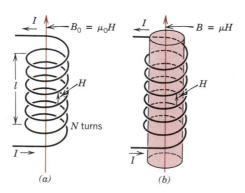

Figure 21.3 (a) The magnetic field H as generated by a cylindrical coil is dependent on the current I, the number of turns N, and the coil length l, according to Equation 21.1. The magnetic flux density B_0 in the presence of a vacuum is equal to $\mu_0 H$, where μ_0 is the permeability of a vacuum, $4\pi \times 10^{-7}$ H/m. (b) The magnetic flux density B within a solid material is equal to μH, where μ is the permeability of the solid material. (Adapted from A. G. Guy, *Essentials of Materials Science*, McGraw-Hill Book Company, New York, 1976.)

A schematic diagram of such an arrangement is shown in Figure 21.3*a*. The magnetic field that is generated by the current loop and the bar magnet in Figure 21.1 is an *H* field. The units of *H* are ampere-turns per meter, or just amperes per meter.

The **magnetic induction,** or **magnetic flux density,** denoted by *B*, represents the magnitude of the internal field strength within a substance that is subjected to an *H* field. The units for *B* are teslas [or webers per square meter (Wb/m^2)]. Both *B* and *H* are field vectors, being characterized not only by magnitude, but also by direction in space.

The magnetic field strength and flux density are related according to

$$B = \mu H \tag{21.2}$$

The parameter μ is called the **permeability,** which is a property of the specific medium through which the *H* field passes and in which *B* is measured, as illustrated in Figure 21.3*b*. The permeability has dimensions of webers per ampere-meter (Wb/A-m) or henries per meter (H/m).

In a vacuum,

$$B_0 = \mu_0 H \tag{21.3}$$

where μ_0 is the *permeability of a vacuum,* a universal constant, which has a value of $4\pi \times 10^{-7}$ (1.257×10^{-6}) H/m. The parameter B_0 represents the flux density within a vacuum as demonstrated in Figure 21.3*a*.

Several parameters may be used to describe the magnetic properties of solids. One of these is the ratio of the permeability in a material to the permeability in a vacuum, or

$$\mu_r = \frac{\mu}{\mu_0} \tag{21.4}$$

here μ_r is called the *relative permeability,* which is unitless. The permeability or relative permeability of a material is a measure of the degree to which the material can be magnetized, or the ease with which a *B* field can be induced in the presence of an external *H* field.

Another field quantity, *M*, called the **magnetization** of the solid, is defined by the expression

$$B = \mu_0 H + \mu_0 M \tag{21.5}$$

In the presence of an *H* field, the magnetic moments within a material tend to become aligned with the field and to reinforce it by virtue of their magnetic fields; the term $\mu_0 M$ in Equation 21.5 is a measure of this contribution.

The magnitude of *M* is proportional to the applied field as follows:

$$M = \chi_m H \tag{21.6}$$

and χ_m is called the **magnetic susceptibility,** which is unitless.[1] The magnetic susceptibility and the relative permeability are related as follows:

$$\chi_m = \mu_r - 1 \tag{21.7}$$

---

[1] This χ_m is taken to be the volume susceptibility in SI units, which, when multiplied by *H*, yields the magnetization per unit volume (cubic meter) of material. Other susceptibilities are also possible; see Problem 21.4.

TABLE 21.1 Magnetic Units and Conversion Factors for the SI and cgs–emu Systems

Quantity	Symbol	SI Units		cgs–emu unit	Conversion
		Derived	Primary		
Magnetic induction (flux density)	B	tesla (Wb/m^2)[a]	$kg/s\text{-}C$	gauss	$1\ Wb/m^2 = 10^4$ gauss
Magnetic field strength	H	amp-turn/m	$C/m\text{-}s$	oersted	1 amp-turn/m $= 4\pi \times 10^{-3}$ oersted
Magnetization	M (SI) I (cgs–emu)	amp-turn/m	$C/m\text{-}s$	maxwell/cm^2	1 amp-turn/m $= 10^{-3}$ maxwell/cm^2
Permeability of a vacuum	μ_0	henry/m[b]	$kg\text{-}m/C^2$	Unitless (emu)	$4\pi \times 10^{-7}$ henry/m $= 1$ emu
Relative permeability	μ_r (SI) μ' (cgs–emu)	Unitless	Unitless	Unitless	$\mu_r = \mu'$
Susceptibility	χ_m (SI) χ'_m (cgs–emu)	Unitless	Unitless	Unitless	$\chi_m = 4\pi\chi'_m$

[a] Units of the weber (Wb) are volt-seconds.
[b] Units of the henry are webers per ampere.

There is a dielectric analogue for each of the foregoing magnetic field parameters. The B and H fields are, respectively, analogous to the dielectric displacement D and the electric field $\mathscr{E}$, whereas the permeability μ parallels the permittivity ϵ (cf. Equations 21.2 and 19.30). Furthermore, the magnetization M and polarization P are correlates (Equations 21.5 and 19.31).

Magnetic units may be a source of confusion because there are really two systems in common use. The ones used thus far are SI [rationalized MKS (meter-kilogram-second)]; the others come from the $cgs-emu$ (centimeter-gram-second–electromagnetic unit) system. The units for both systems as well as the appropriate conversion factors are contained in Table 21.1.

Origins of Magnetic Moments

The macroscopic magnetic properties of materials are a consequence of *magnetic moments* associated with individual electrons. Some of these concepts are relatively complex and involve some quantum-mechanical principles beyond the scope of this discussion; consequently, simplifications have been made and some of the details omitted. Each electron in an atom has magnetic moments that originate from two sources. One is related to its orbital motion around the nucleus; being a moving charge, an electron may be considered to be a small current loop, generating a very small magnetic field, and having a magnetic moment along its axis of rotation, as schematically illustrated in Figure 21.4a.

Each electron may also be thought of as spinning around an axis; the other magnetic moment originates from this electron spin, which is directed along the spin axis as shown in Figure 21.4b. Spin magnetic moments may be only in an "up" direction or in an antiparallel "down" direction. Thus each electron in an atom may be thought of as being a small magnet having permanent orbital and spin magnetic moments.

The most fundamental magnetic moment is the **Bohr magneton** μ_B, which is of magnitude 9.27×10^{-24} A-m^2. For each electron in an atom the spin magnetic moment is $\pm \mu_B$ (plus for spin up, minus for spin down). Furthermore, the orbital magnetic moment contribution is equal to $m_l \mu_B$, m_l being the magnetic quantum number of the electron, as mentioned in Section 2.3.

In each individual atom, orbital moments of some electron pairs cancel each other; this also holds for the spin moments. For example, the spin moment of an electron with spin up will cancel that of one with spin down. The net magnetic moment, then, for an atom is just the sum of the magnetic moments of each of the constituent electrons, including both orbital and spin contributions, and taking into account

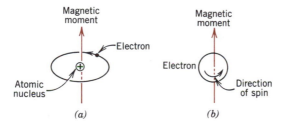

Figure 21.4 Demonstration of the magnetic moment associated with (a) an orbiting electron and (b) a spinning electron.

moment cancellation. For an atom having completely filled electron shells or subshells, when all electrons are considered, there is total cancellation of both orbital and spin moments. Thus materials composed of atoms having completely filled electron shells are not capable of being permanently magnetized. This category includes the inert gases (He, Ne, Ar, etc.) as well as some ionic materials. The types of magnetism include diamagnetism, paramagnetism, and ferromagnetism; in addition, antiferromagnetism and ferrimagnetism are considered to be subclasses of ferromagnetism. All materials exhibit at least one of these types, and the behavior depends on the response of electron and atomic magnetic dipoles to the application of an externally applied magnetic field.

21.3 DIAMAGNETISM AND PARAMAGNETISM

Diamagnetism is a very weak form of magnetism that is nonpermanent and persists only while an external field is being applied. It is induced by a change in the orbital motion of electrons due to an applied magnetic field. The magnitude of the induced magnetic moment is extremely small, and in a direction opposite to that of the applied field. Thus the relative permeability μ_r is less than unity (however, only very slightly), and the magnetic susceptibility is negative; that is, the magnitude of the B field within a diamagnetic solid is less than that in a vacuum. The volume susceptibility χ_m for diamagnetic solid materials is on the order of -10^{-5}. When placed between the poles of a strong electromagnet, diamagnetic materials are attracted toward regions where the field is weak.

Figure 21.5a illustrates schematically the atomic magnetic dipole configurations for a diamagnetic material with and without an external field; here, the arrows represent atomic dipole moments, whereas for the preceding discussion, arrows denoted only electron moments. The dependence of B on the external field H for a material that

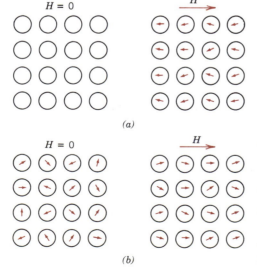

(a)

(b)

Figure 21.5 (a) The atomic dipole configuration for a diamagnetic material with and without a magnetic field. In the absence of an external field, no dipoles exist; in the presence of a field, dipoles are induced that are aligned opposite to the field direction. (b) Atomic dipole configuration with and without an external magnetic field for a paramagnetic material.

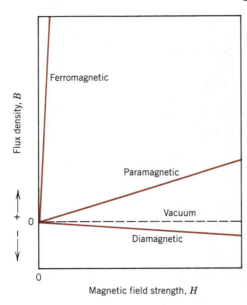

Figure 21.6 Schematic representation of the flux density B versus the magnetic field strength H for diamagnetic, paramagnetic, and ferromagnetic materials. (Adapted from A. G. Guy, *Essentials of Materials Science,* McGraw-Hill Book Company, New York, 1976.)

exhibits diamagnetic behavior is presented in Figure 21.6. Table 21.2 gives the susceptibilities of several diamagnetic materials. Diamagnetism is found in all materials; but because it is so weak, it can be observed only when other types of magnetism are totally absent. This form of magnetism is of no practical importance.

For some solid materials, each atom possesses a permanent dipole moment by virtue of incomplete cancellation of electron spin and/or orbital magnetic moments. In the absence of an external magnetic field, the orientations of these atomic magnetic moments are random, such that a piece of material possesses no net macroscopic magnetization. These atomic dipoles are free to rotate, and **paramagnetism** results when they preferentially align, by rotation, with an external field as shown in Figure 21.5b. These magnetic dipoles are acted on individually with no mutual interaction

TABLE 21.2 Room-Temperature Magnetic Susceptibilities for Diamagnetic and Paramagnetic Materials

Diamagnetics		Paramagnetics	
Material	Susceptibility χ_m (volume) (SI units)	Material	Susceptibility χ_m (volume) (SI units)
Aluminum oxide	-1.81×10^{-5}	Aluminum	2.07×10^{-5}
Copper	-0.96×10^{-5}	Chromium	3.13×10^{-4}
Gold	-3.44×10^{-5}	Chromium chloride	1.51×10^{-3}
Mercury	-2.85×10^{-5}	Manganese sulfate	3.70×10^{-3}
Silicon	-0.41×10^{-5}	Molybdenum	1.19×10^{-4}
Silver	-2.38×10^{-5}	Sodium	8.48×10^{-6}
Sodium chloride	-1.41×10^{-5}	Titanium	1.81×10^{-4}
Zinc	-1.56×10^{-5}	Zirconium	1.09×10^{-4}

between adjacent dipoles. Inasmuch as the dipoles align with the external field, they enhance it, giving rise to a relative permeability μ_r that is greater than unity, and to a relatively small but positive magnetic susceptibility. Susceptibilities for paramagnetic materials range from about 10^{-5} to 10^{-2} (Table 21.2). A schematic B-versus-H curve for a paramagnetic material is also shown in Figure 21.6.

Both diamagnetic and paramagnetic materials are considered to be nonmagnetic because they exhibit magnetization only when in the presence of an external field. Also, for both, the flux density B within them is almost the same as it would be in a vacuum.

21.4 FERROMAGNETISM

Certain metallic materials possess a permanent magnetic moment in the absence of an external field, and manifest very large and permanent magnetizations. These are the characteristics of **ferromagnetism,** and they are displayed by the transition metals iron (as BCC α-ferrite), cobalt, nickel, and some of the rare earth metals such as gadolinium (Gd). Magnetic susceptibilities as high as 10^6 are possible for ferromagnetic materials. Consequently, $H \ll M$, and from Equation 21.5 we write

$$B \cong \mu_0 M \tag{21.8}$$

Permanent magnetic moments in ferromagnetic materials result from atomic magnetic moments due to electron spin—uncancelled electron spins as a consequence of the electron structure. There is also an orbital magnetic moment contribution that is small in comparison to the spin moment. Furthermore, in a ferromagnetic material, coupling interactions cause net spin magnetic moments of adjacent atoms to align with one another, even in the absence of an external field. This is schematically illustrated in Figure 21.7. The origin of these coupling forces is not completely understood, but it is thought to arise from the electronic structure of the metal. This mutual spin alignment exists over relatively large volume regions of the crystal called **domains** (see Section 21.7).

The maximum possible magnetization, or **saturation magnetization M_s** of a ferromagnetic material represents the magnetization that results when all the magnetic dipoles in a solid piece are mutually aligned with the external field; there is also a corresponding saturation flux density B_s. The saturation magnetization is equal to the product of the net magnetic moment for each atom and the number of atoms present. For each of iron, cobalt, and nickel, the net magnetic moments per atom are 2.22, 1.72, and 0.60 Bohr magnetons, respectively.

$H = 0$

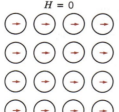

Figure 21.7 Schematic illustration of the mutual alignment of atomic dipoles for a ferromagnetic material, which will exist even in the absence of an external magnetic field.

EXAMPLE PROBLEM 21.1

Calculate (a) the saturation magnetization and (b) the saturation flux density for nickel, which has a density of 8.90 g/cm^3.

SOLUTION

(a) The saturation magnetization is just the product of the number of Bohr magnetons per atom (0.60 as given above), the magnitude of the Bohr magneton μ_B, and the number of atoms per cubic meter N, or

$$M_s = 0.60\mu_B N$$

Now, the number of atoms per cubic meter is related to the density ρ, the atomic weight A_{Ni}, and Avogadro's number N_A, as follows:

$$N = \frac{\rho N_A}{A_{Ni}}$$

$$= \frac{(8.90 \times 10^6 \text{ g/m}^3)(6.023 \times 10^{23} \text{ atoms/mol})}{58.71 \text{ g/mol}}$$

$$= 9.13 \times 10^{28} \text{ atoms/m}^3$$

Finally,

$$M_s = \left(\frac{0.60 \text{ Bohr magneton}}{\text{atom}}\right)\left(\frac{9.27 \times 10^{-24} \text{ A-m}^2}{\text{Bohr magneton}}\right)\left(\frac{9.13 \times 10^{28} \text{ atoms}}{\text{m}^3}\right)$$

$$= 5.1 \times 10^5 \text{ A/m}$$

(b) From Equation 21.8, the saturation flux density is just

$$B_s = \mu_0 M_s$$

$$= \left(\frac{4\pi \times 10^{-7} \text{ H}}{\text{m}}\right)\left(\frac{5.1 \times 10^5 \text{ A}}{\text{m}}\right)$$

$$= 0.64 \text{ tesla}$$

21.5 ANTIFERROMAGNETISM AND FERRIMAGNETISM

Antiferromagnetism

This phenomenon of magnetic moment coupling between adjacent atoms or ions occurs in materials other than those that are ferromagnetic. In one such group this coupling results in an antiparallel alignment; the alignment of the spin moments of neighboring atoms or ions in exactly opposite directions is termed **antiferromagnetism.** Manganese oxide (MnO) is one material that displays this behavior. Manganese oxide is a ceramic material that is ionic in character, having both Mn^{2+} and O^{2-} ions. No net magnetic moment is associated with the O^{2-} ions, since there is a total cancellation of both spin and orbital moments. However, the Mn^{2+} ions possess a net magnetic moment that is predominantly of spin origin. These Mn^{2+} ions are arrayed in the crystal structure such that the moments of adjacent ions are antiparallel. This arrangement is represented schematically in Figure 21.8. Obviously, the opposing

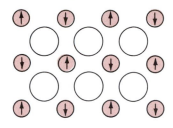

Mn²⁺

O²⁻

Figure 21.8 Schematic representation of antiparallel alignment of spin magnetic moments for antiferromagnetic manganese oxide.

magnetic moments cancel one another, and, as a consequence, the solid as a whole possesses no net magnetic moment.

Ferrimagnetism

Some ceramics also exhibit a permanent magnetization, termed **ferrimagnetism.** The macroscopic magnetic characteristics of ferromagnets and ferrimagnets are similar; the distinction lies in the source of the net magnetic moments. The principles of ferrimagnetism are illustrated with the cubic ferrites.[2] These ionic materials may be represented by the chemical formula MFe_2O_4, in which M represents any one of several metallic elements. The prototype ferrite is Fe_3O_4, the mineral magnetite, sometimes called lodestone.

The formula for Fe_3O_4 may be written as $Fe^{2+}O^{2-}–(Fe^{3+})_2(O^{2-})_3$ in which the Fe ions exist in both +2 and +3 valence states in the ratio of 1:2. A net spin magnetic moment exists for each Fe^{2+} and Fe^{3+} ion, which corresponds to 4 and 5 Bohr magnetons, respectively, for the two ion types. Furthermore, the O^{2-} ions are magnetically neutral. There are antiparallel spin-coupling interactions between the Fe ions, similar in character to antiferromagnetism. However, the net ferrimagnetic moment arises from the incomplete cancellation of spin moments.

Cubic ferrites have the inverse spinel crystal structure, which is cubic in symmetry, and similar to the spinel structure (Section 13.2). It might be thought of as having been generated by the stacking of close-packed planes of O^{2-} ions. Again, there are two types of positions that may be occupied by the iron cations, as illustrated in Figure 13.9. For one, the coordination number is 4 (tetrahedral coordination); that is, each Fe ion is surrounded by four oxygen nearest neighbors. For the other, the coordination number is 6 (octahedral coordination). With this inverse spinel structure, half the trivalent (Fe^{3+}) ions are situated in octahedral positions, the other half, in tetrahedral positions. The divalent Fe^{2+} ions are all located in octahedral positions.

---

[2] Ferrite in the magnetic sense should not be confused with the ferrite α-iron discussed in Section 9.13; in the remainder of this chapter, the term **ferrite** implies the magnetic ceramic.

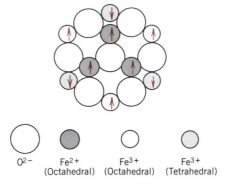

Figure 21.9 Schematic diagram showing the spin magnetic moment configuration for Fe^{2+} and Fe^{3+} ions in Fe_3O_4. (Adapted from Richard A. Flinn, and Paul K. Trojan. *Engineering Materials and Their Applications,* Third Edition. Copyright © 1986 by Houghton Mifflin Company. Used with permission.)

The critical factor is the arrangement of the spin moments of the Fe ions, as represented in Figure 21.9, and Table 21.3. The spin moments of all the Fe^{3+} ions in the octahedral positions are aligned parallel to one another; however, they are directed oppositely to the Fe^{3+} ions disposed in the tetrahedral positions, which are also aligned. This results from the antiparallel coupling of adjacent iron ions. Thus the spin moments of all Fe^{3+} ions cancel one another and make no net contribution to the magnetization of the solid. All the Fe^{2+} ions have their moments aligned in the same direction, which total moment is responsible for the net magnetization (see Table 21.3). Thus the saturation magnetization of a ferrimagnetic solid may be computed from the product of the net spin magnetic moment for each Fe^{2+} ion and the number of Fe^{2+} ions; this would correspond to the mutual alignment of all the Fe^{2+} ion magnetic moments in the Fe_3O_4 specimen.

Cubic ferrites having other compositions may be produced by adding metallic ions that substitute for some of the iron in the crystal structure. Again, from the ferrite chemical formula, $M^{2+}O^{2-}$–$(Fe^{3+})_2(O^{2-})_3$, in addition to Fe^{2+}, M^{2+} may represent divalent ions such as Ni^{2+}, Mn^{2+}, Co^{2+}, and Cu^{2+}, each of which possesses a net

TABLE 21.3 The Distribution of Spin Magnetic Moments for Fe^{2+} and Fe^{3+} Ions in a Unit Cell of Fe_3O_4[a]

Cation	Octahedral Lattice Site	Tetrahedral Lattice Site	Net Magnetic Moment
Fe^{3+}	↑ ↑ ↑ ↑ ↑ ↑ ↑ ↑	↓ ↓ ↓ ↓ ↓ ↓ ↓ ↓	Complete cancellation
Fe^{2+}	↑ ↑ ↑ ↑ ↑ ↑ ↑ ↑	—	↑ ↑ ↑ ↑ ↑ ↑ ↑ ↑

[a] Each arrow represents the magnetic moment orientation for one of the cations.

TABLE 21.4 Net Magnetic Moments for Six Cations

Cation	Net Spin Magnetic Moment (Bohr magnetons)
Fe^{3+}	5
Fe^{2+}	4
Mn^{2+}	5
Co^{2+}	3
Ni^{2+}	2
Cu^{2+}	1

spin magnetic moment different from 4; several are listed in Table 21.4. Thus by adjustment of composition, ferrite compounds having a range of magnetic properties may be produced. For example, nickel ferrite has the formula $NiFe_2O_4$. Other compounds may also be produced containing mixtures of two divalent metal ions such as $(Mn,Mg)Fe_2O_4$, in which the $Mn^{2+}:Mg^{2+}$ ratio may be varied; these are called mixed ferrites.

Ceramic materials other than the cubic ferrites are also ferrimagnetic; these include the hexagonal ferrites and garnets. Hexagonal ferrites have a crystal structure similar to the inverse spinel, with hexagonal symmetry rather than cubic. The chemical formula for these materials may be represented by $AB_{12}O_{19}$, in which A is a divalent metal such as barium, lead, or strontium, and B is a trivalent metal such as aluminum, gallium, chromium, or iron. The two most common examples of the hexagonal ferrites are $PbFe_{12}O_{19}$ and $BaFe_{12}O_{19}$.

The garnets have a very complicated crystal structure, which may be represented by the general formula $M_3Fe_5O_{12}$; here, M represents a rare earth ion such as samarium, europium, gadolinium, or yttrium. Yttrium iron garnet ($Y_3Fe_5O_{12}$), sometimes denoted YIG, is the most common material of this type.

The saturation magnetizations for ferrimagnetic materials are not as high as for ferromagnets. On the other hand, ferrites, being ceramic materials, are good electronic insulators. For some magnetic applications, such as high-frequency transformers, a low electrical conductivity is most desirable.

EXAMPLE PROBLEM 21.2

Calculate the saturation magnetization for Fe_3O_4 given that each cubic unit cell contains $8Fe^{2+}$ and $16Fe^{3+}$ ions, and that the unit cell edge length is 0.839 nm.

SOLUTION
This problem is solved in a manner similar to Example Problem 21.1, except that the computational basis is per unit cell as opposed to per atom or ion.

The saturation magnetization will be equal to the product of the number N' of Bohr magnetons per cubic meter of Fe_3O_4, and the magnetic moment per Bohr magneton μ_B,

$$M_s = N'\mu_B$$

Now, N' is just the number of Bohr magnetons per unit cell n_B divided by the unit cell volume V_C, or

$$N' = \frac{n_B}{V_C}$$

Again, the net magnetization results from the Fe^{2+} ions only. Since there are $8Fe^{2+}$ ions per unit cell and 4 Bohr magnetons per Fe^{2+} ion, n_B is 32. Furthermore, the unit cell is a cube, and $V_C = a^3$, a being the unit cell edge length. Therefore,

$$M_s = \frac{n_B\mu_B}{a^3}$$

$$= \frac{(32 \text{ Bohr magnetons/unit cell})(9.27 \times 10^{-24} \text{ A-m}^2/\text{Bohr magneton})}{(0.839 \times 10^{-9} \text{ m})^3/\text{unit cell}}$$

$$= 5.0 \times 10^5 \text{ A/m}$$

21.6 THE INFLUENCE OF TEMPERATURE ON MAGNETIC BEHAVIOR

Temperature can also influence the magnetic characteristics of materials. It may be recalled that raising the temperature of a solid results in an increase in the magnitude of the thermal vibrations of atoms. The atomic magnetic moments are free to rotate; hence, with rising temperature, the increased thermal motion of the atoms tends to randomize the directions of any moments that may be aligned.

For ferromagnetic, antiferromagnetic, and ferrimagnetic materials, the atomic thermal motions counteract the coupling forces between the adjacent atomic dipole moments, causing some dipole misalignment, regardless of whether an external field is present. This results in a decrease in the saturation magnetization for both ferro- and ferrimagnets. This saturation magnetization is a maximum at 0 K, at which temperature the thermal vibrations are a minimum. With increasing temperature, the saturation magnetization diminishes gradually and then abruptly drops to zero at what is called the **Curie temperature** T_c. The magnetization–temperature behavior for iron and Fe_3O_4 is represented in Figure 21.10. At T_c the mutual spin coupling forces are completely destroyed, such that for temperatures above T_c both ferromagnetic and ferrimagnetic materials are paramagnetic. The magnitude of the Curie temperature varies from material to material; for example, for iron, cobalt, nickel, and Fe_3O_4, the respective values are 768, 1120, 335, and 585°C.

Antiferromagnetism is also affected by temperature; this behavior vanishes at what is called the *Néel temperature*. At temperatures above this point, antiferromagnetic materials also become paramagnetic.

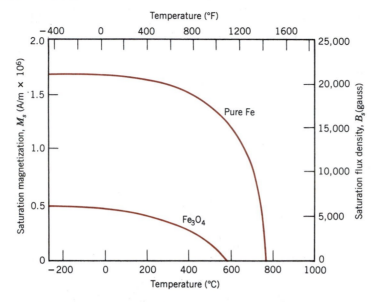

Figure 21.10 Plot of saturation magnetization as a function of temperature for iron and Fe_3O_4. [Adapted from J. Smit and H. P. J. Wijn, *Ferrites*. Copyright © 1959 by N. V. Philips Gloeilampenfabrieken, Eindhoven (Holland). Reprinted by permission.]

21.7 DOMAINS AND HYSTERESIS

Any ferromagnetic or ferrimagnetic material that is at a temperature below T_c is composed of small-volume regions in which there is a mutual alignment in the same direction of all magnetic dipole moments, as illustrated in Figure 21.11. Such a region is called a domain, and each one is magnetized to its saturation magnetization. Adjacent domains are separated by domain boundaries or walls, across which the

One domain / Another domain

Domain wall

Figure 21.11 Schematic depiction of domains in a ferromagnetic or ferrimagnetic material; arrows represent atomic magnetic dipoles. Within each domain, all dipoles are aligned, whereas the direction of alignment varies from one domain to another.

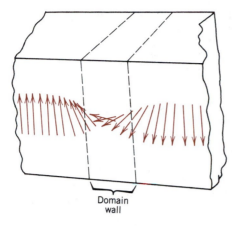

Figure 21.12 The gradual change in magnetic dipole orientation across a domain wall. (From W. D. Kingery, H. K. Bowen, and D. R. Uhlmann, *Introduction to Ceramics,* 2nd edition. Copyright © 1976 by John Wiley & Sons, New York. Reprinted by permission of John Wiley & Sons, Inc.)

direction of magnetization gradually changes (Figure 21.12). Normally, domains are microscopic in size, and for a polycrystalline specimen, each grain may consist of more than a single domain. Thus in a macroscopic piece of material, there will be a large number of domains, and all may have different magnetization orientations. The magnitude of the M field for the entire solid is the vector sum of the magnetizations of all the domains, each domain contribution being weighted by its volume fraction. For an unmagnetized specimen, the appropriately weighted vector sum of the magnetizations of all the domains is zero.

Flux density B and field intensity H are not proportional for ferromagnets and ferrimagnets. If the material is initially unmagnetized, then B varies as a function of H as shown in Figure 21.13. The curve begins at the origin, and as H is increased, the B field begins to increase slowly, then more rapidly, finally leveling off and becoming independent of H. This maximum value of B is the saturation flux density

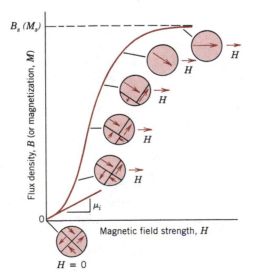

Figure 21.13 The B-versus-H behavior for a ferromagnetic or ferrimagnetic material that was initially unmagnetized. Domain configurations during several stages of magnetization are represented. Saturation flux density B_s, magnetization M_s, and initial permeability μ_i are also indicated. (Adapted from O. H. Wyatt and D. Dew-Hughes, *Metals, Ceramics and Polymers,* Cambridge University Press, 1974.)

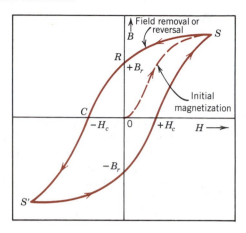

Figure 21.14 Magnetic flux density versus the magnetic field strength for a ferromagnetic material to forward and reverse saturation (points S and S'). The hysteresis loop is represented by the solid curve; the dashed curve indicates the initial magnetization. The remanence B_r and the coercive force H_c are also shown.

B_s, and the corresponding magnetization is the saturation magnetization M_s, mentioned previously. Since the permeability μ from Equation 21.2 is the slope of the B-versus-H curve, it may be noted from Figure 21.13 that the permeability changes with and is dependent on H. On occasion, the slope of the B-versus-H curve at $H = 0$ is specified as a material property, which is termed the *initial permeability* μ_i, as indicated in Figure 21.13.

As an H field is applied, the domains change shape and size by the movement of domain boundaries. Schematic domain structures are represented at several points along the B-versus-H curve in Figure 21.13. Initially, the moments of the constituent domains are randomly oriented such that there is no net B (or M) field. As the external field is applied, the domains that are oriented in directions favorable to (or nearly aligned with) the applied field grow at the expense of those that are unfavorably oriented. This process continues with increasing field strength until the macroscopic specimen becomes a single domain, which is nearly aligned with the field. Saturation is achieved when this domain, by means of rotation, becomes oriented with the H field. Alteration of the domain structure with magnetic field for an iron single crystal is shown on page 676.

From saturation, point S in Figure 21.14, as the H field is reduced by reversal of field direction, the curve does not retrace its original path. A **hysteresis** effect is produced in which the B field lags behind the applied H field, or decreases at a lower rate. At zero H field (point R on the curve), there exists a residual B field that is called the **remanence,** or remanent flux density, B_r; the material remains magnetized in the absence of an external H field.

Hysteresis behavior and permanent magnetization may be explained by the motion of domain walls. Upon reversal of the field direction from saturation (point S in Figure 21.14), the process by which the domain structure changes is reversed. First, there is a rotation of the single domain with the reversed field. Next, domains having magnetic moments aligned with the new field form and grow at the expense of the former domains. Critical to this explanation is the resistance to movement of domain walls that occurs in response to the increase of the magnetic field in the opposite direction; this accounts for the lag of B with H, or the hysteresis. When the applied field reaches zero, there is still some net volume fraction of domains oriented in the former direction, which explains the existence of the remanence B_r.

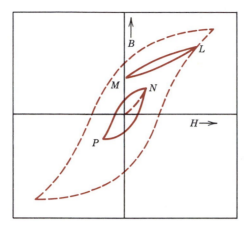

Figure 21.15 A hysteresis curve at less than saturation (curve NP) within the saturation loop for a ferromagnetic material. The B–H behavior for field reversal at other than saturation is indicated by curve LM.

To reduce the B field within the specimen to zero (point C on Figure 21.14), an H field of magnitude $-H_c$ must be applied in a direction opposite to that of the original field; H_c is called the **coercivity**, or sometimes the coercive force. Upon continuation of the applied field in this reverse direction, as indicated in the figure, saturation is ultimately achieved in the opposite sense, corresponding to point S'. A second reversal of the field to the point of the initial saturation (point S) completes the symmetrical hysteresis loop and also yields both a negative remanence $(-B_r)$ and a positive coercivity $(+H_c)$.

The B-versus-H curve in Figure 21.14 represents a hysteresis loop taken to saturation. Of course, it is not necessary to increase the H field to saturation before reversing the field direction; in Figure 21.15, loop NP is a hysteresis curve corresponding to less than saturation. Furthermore, it is possible to reverse the direction of the field at any point along the curve and generate other hysteresis loops. One such loop is indicated on the saturation curve in Figure 21.15: for loop LM, the H field is reversed to zero. One method of demagnetizing a ferromagnet or ferrimagnet is to repeatedly cycle it in an H field that alternates direction and decreases in magnitude.

21.8 SOFT MAGNETIC MATERIALS

The size and shape of the hysteresis curve for ferromagnetic and ferrimagnetic materials is of considerable practical importance. The area within a loop represents a magnetic energy loss per unit volume of material per magnetization–demagnetization cycle; this energy loss is manifested as heat that is generated within the magnetic specimen and is capable of raising its temperature.

Both ferromagnetic and ferrimagnetic materials are classified as either *soft* or *hard* on the basis of their hysteresis characteristics. Soft magnetic materials are used in devices that are subjected to alternating magnetic fields and in which energy losses must be low; one familiar example consists of transformer cores. For this reason the relative area within the hysteresis loop must be small; it is characteristically thin and narrow, as represented in Figure 21.16. Consequently, a soft magnetic material must have a high initial permeability and a low coercivity. A material possessing these

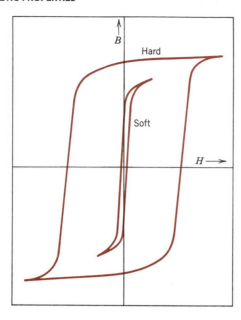

Figure 21.16 Schematic magnetization curves for soft and hard magnetic materials. (From K. M. Ralls, T. H. Courtney, and J. Wulff, *Introduction to Materials Science and Engineering.* Copyright © 1976 by John Wiley & Sons, New York. Reprinted by permission of John Wiley & Sons, Inc.)

properties may reach its saturation magnetization with a relatively low applied field (i.e., is easily magnetized and demagnetized) and still has low hysteresis energy losses.

The saturation field or magnetization is determined only by the composition of the material. For example, in cubic ferrites, substitution of a divalent metal ion such as Ni^{2+} for Fe^{2+} in $FeO-Fe_2O_3$ will change the saturation magnetization. However, susceptibility and coercivity (H_c), which also influence the shape of the hysteresis curve, are sensitive to structural variables rather than to composition. For example, a low value of coercivity corresponds to the easy movement of domain walls as the magnetic field changes magnitude and/or direction. Structural defects such as particles of a nonmagnetic phase or voids in the magnetic material tend to restrict the motion of domain walls, and thus increase the coercivity. Consequently, a soft magnetic material must be free of such structural defects.

Another property consideration for soft magnetic materials is electrical resistivity. In addition to the hysteresis energy losses described above, energy losses may result from electrical currents that are induced in a magnetic material by a magnetic field that varies in magnitude and direction with time; these are called *eddy currents*. It is most desirable to minimize these energy losses in soft magnetic materials by increasing the electrical resistivity. This is accomplished in ferromagnetic materials by forming solid solution alloys; iron–silicon and iron–nickel alloys are examples. The ceramic ferrites are commonly used for applications requiring soft magnetic materials because they are intrinsically electrical insulators. Their applicability is somewhat limited, however, inasmuch as they have relatively small susceptibilities. The properties of a half-dozen soft magnetic materials are shown in Table 21.5.

The hysteresis characteristics of soft magnetic materials may be enhanced for some applications by an appropriate heat treatment in the presence of a magnetic field. Using such a technique, a square hysteresis loop may be produced, which is desirable in some magnetic amplifier and pulse transformer applications. In addi-

TABLE 21.5 Typical Properties for Several Soft Magnetic Materials

Material	Composition (wt%)	Initial Relative Permeability μ_i	Saturation Flux Density B_s [Tesla (gauss)]	Hysteresis Loss/Cycle [J/m³ (erg/cm³)]	Resistivity ρ (Ω-m)
Commercial iron ingot	99.95Fe	150	2.14 (21,400)	270 (2700)	1.0×10^{-7}
Silicon–iron (oriented)	97Fe, 3Si	1400	2.01 (20,100)	40 (400)	4.7×10^{-7}
45 Permalloy	55Fe, 45Ni	2500	1.60 (16,000)	120 (1200)	4.5×10^{-7}
Supermalloy	79Ni, 15Fe, 5Mo, 0.5Mn	75,000	0.80 (8000)	—	6.0×10^{-7}
Ferroxcube A	48MnFe$_2$O$_4$, 52ZnFe$_2$O$_4$	1400	0.33 (3300)	~40 (~400)	2000
Ferroxcube B	36NiFe$_2$O$_4$, 64ZnFe$_2$O$_4$	650	0.36 (3600)	~35 (~350)	10^7

Source: Adapted from *Metals Handbook: Properties and Selection: Stainless Steels, Tool Materials and Special-Purpose Metals,* Vol. 3, 9th edition, D. Benjamin, Senior Editor, American Society for Metals, 1980.

tion, soft magnetic materials are used in generators, motors, dynamos, and switching circuits.

21.9 HARD MAGNETIC MATERIALS

Hard magnetic materials are utilized in permanent magnets, which must have a high resistance to demagnetization. In terms of hysteresis behavior, a hard magnetic material has a high remanence, coercivity, and saturation flux density, as well as a low initial permeability, and high hysteresis energy losses. The hysteresis characteristics for hard and soft magnetic materials are compared in Figure 21.16. Sometimes it is convenient to relate the relative hardness of a magnetic material to the product of B_r and H_c, which is roughly twice the energy required to demagnetize a unit volume of material. Thus the larger the $B_r \times H_c$ product, the harder is the material in terms of its magnetic characteristics.

Again, the hysteresis behavior is related to the ease with which the magnetic domain boundaries move; by impeding domain wall motion, the coercivity and susceptibility are enhanced, such that a large external field is required for demagnetization. Furthermore, these characteristics are interrelated to the microstructure of the material. Domain wall motion is effectively hindered when small precipitate particles are caused to form. Most permanent magnets are ferromagnetic; steels alloyed with tungsten and chromium have been used extensively in the past. These two elements, under the proper heat treating conditions, readily combine with carbon in the steel to form tungsten and chromium carbide precipitate particles, which are especially effective in obstructing domain wall motion. Some of the hard magnetic alloys that have been developed more recently contain a number of alloying elements, including iron, cobalt, nickel, aluminum, and copper. An appropriate heat treatment forms extremely small single-domain and strongly magnetic iron–cobalt particles within a nonmagnetic matrix phase.

TABLE 21.6 Typical Properties for Several Hard Magnetic Materials

Material	Composition (wt%)	Remanence B_r [Tesla (gauss)]	Coercivity H_c [amp-turn/m (oersted)]	$(BH)_{max}$ [J/m³ (gauss-oersted)]	Curie Temperature T_c [°C (°F)]	Resistivity ρ (Ω-m)
Martensitic carbon steel	98.1Fe, 0.9C, 1Mn	0.95 (9500)	4000 (50)	1600 (0.20×10^6)	—	—
Tungsten steel	92.8Fe, 6W, 0.5Cr, 0.7C	0.95 (9500)	5900 (74)	2600 (0.33×10^6)	760 (1400)	3.0×10^{-7}
Cunife	20Fe, 20Ni, 60Cu	0.54 (5400)	44,000 (550)	12,000 (1.5×10^6)	410 (770)	1.8×10^{-7}
Cunico	29Co, 21Ni, 50Cu	0.34 (3400)	54,000 (680)	6400 (0.8×10^6)	860 (1580)	2.4×10^{-7}
Sintered Alnico 8	34Fe, 7Al, 15Ni, 35Co, 4Cu, 5Ti	0.76 (7600)	123,000 (1550)	36,000 (4.5×10^6)	860 (1580)	—
Ferroxdur (oriented)	BaO-6Fe₂O₃	0.32 (3200)	240,000 (3000)	20,000 (2.5×10^6)	450 (840)	$\sim 10^4$

Source: Adapted from *Metals Handbook: Properties and Selection: Stainless Steels, Tool Materials and Special-Purpose Metals,* Vol. 3, 9th edition, D. Benjamin, Senior Editor, American Society for Metals, 1980.

Of the ferrimagnets, those most often used as hard magnets are the hexagonal ferrites. Table 21.6 presents some of the critical properties of several hard magnetic materials.

21.10 MAGNETIC STORAGE

Within the past few years magnetic materials have become increasingly important in the area of information storage. This is especially true for computers; whereas semiconductor elements serve as primary memory, magnetic disks and tapes are capable of storing larger quantities of information and at a lower cost. In addition, the recording and television industries rely heavily on magnetic tapes for the storage and reproduction of audio and video sequences.

In essence, computer bytes, sound, or visual images in the form of electrical signals are transferred to and then retained within very small segments of the magnetic storage medium. This transference to and retrieval from the tape or disk is accomplished by means of a head, which consists basically of a wire coil wound around a magnetic material core into which a gap is cut. Data are introduced (or "written") by the electrical signal within the coil, which generates a magnetic field across the gap. This field in turn magnetizes a very small area of the disk or tape within the proximity of the head. Upon removal of the field, the magnetization remains; that is, the signal has been stored.

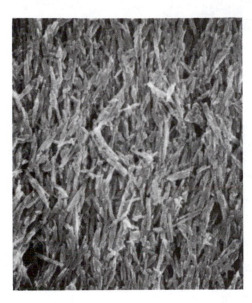

Figure 21.17 A scanning electron micrograph showing the microstructure of a magnetic storage disk. Needle-shaped particles of γ-Fe_2O_3 are oriented and embedded within an epoxy phenolic resin. 8000 $\times$. (Photograph courtesy of P. Rayner and N. L. Head, IBM Corporation.)

Furthermore, the same head may be utilized to retrieve (or "read") the stored information. A voltage is induced when there is a change in the magnetic field as the tape or disk passes by the head coil gap; this may be amplified and then converted back into its original form or character.

Most magnetic media consist of very small needlelike or acicular particles of γ-Fe_2O_3 ferrite, or doped alloys thereof; these are applied and bonded to a polymeric film (for magnetic tapes) or to a metal or polymer disk. During manufacture, these particles are aligned with their long axis in a direction that parallels the direction of motion past the head (see Figure 21.17). Each particle is a single domain that may be magnetized only with its magnetic moment lying along this axis. Two magnetic states are possible, corresponding to the saturation magnetization in one axial direction, and its opposite. These two states make possible the storage of information in digitial form, as 1's and 0's. In one system a 1 is represented by a reversal in the magnetic field direction from one small area of the storage medium to another as the numerous acicular particles of each such region pass by the head. A lack of reversal between adjacent regions is indicated by a 0.

The hysteresis loop for the magnetic storage medium should be relatively large and square. These characteristics ensure that storage will be permanent, and, in addition, magnetization reversal will result over a narrow range of applied field strengths. The saturation flux density normally ranges between 0.4 and 0.6 tesla for these materials.

21.11 SUPERCONDUCTIVITY

Superconductivity is basically an electrical phenomenon; however, its discussion has been deferred to this point because there are magnetic implications relative to the superconducting state, and, in addition, superconducting materials are used primarily in magnets capable of generating high fields.

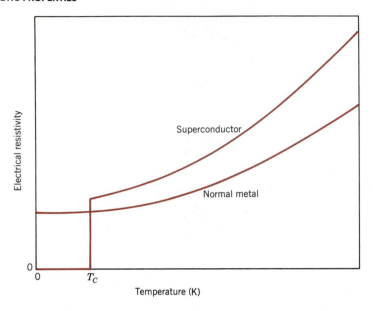

Figure 21.18 Temperature dependence of the electrical resistivity for normally conducting and superconducting materials in the vicinity of 0 K.

As most high-purity metals are cooled down to temperatures nearing 0 K, the electrical resistivity decreases gradually, approaching some small yet finite value that is characteristic of the particular metal. There are a few materials, however, for which the resistivity, at a very low temperature, abruptly plunges from a finite value to one that is virtually zero and remains there upon further cooling. Materials that display this latter behavior are called *superconductors,* and the temperature at which they attain **superconductivity** is called the critical temperature T_C.[3] The resistivity–temperature behaviors for superconductive and nonsuperconductive materials are contrasted in Figure 21.18. The critical temperature varies from superconductor to superconductor but lies between less than 1 K and approximately 20 K for metals and metal alloys. Recently, it has been demonstrated that some complex oxide ceramics have critical temperatures approaching 100 K.

At temperatures below T_C, the superconducting state will cease upon application of a sufficiently large magnetic field, termed the critical field H_C, which depends on temperature and decreases with increasing temperature. The same may be said for current density; that is, a critical applied current density J_C exists below which a material is superconductive. Figure 21.19 shows schematically the boundary in temperature–magnetic field–current density space separating normal and superconducting states. The position of this boundary will, of course, depend on the material. For temperature, magnetic field, and current density values lying between the origin

---

[3] The symbol T_c is used to represent both the Curie temperature (Section 21.6) and the superconducting critical temperature in the scientific literature. They are totally different entities and should not be confused. In this discussion they are denoted by T_c and T_C, respectively.

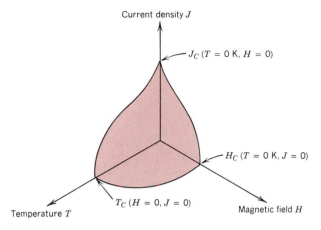

Figure 21.19 Critical temperature, current density, and magnetic field boundary separating superconducting and normal conducting states (schematic).

and this boundary, the material will be superconductive; outside the boundary, conduction is normal.

The superconductivity phenomenon has been satisfactorily explained by means of a rather involved theory. In essence, the superconductive state results from attractive interactions between pairs of conducting electrons; the motions of these paired electrons become coordinated such that scattering by thermal vibrations and impurity atoms is highly inefficient. Thus the resistivity, being proportional to the incidence of electron scattering, is zero.

On the basis of magnetic response, superconducting materials may be divided into two classifications designated as type I and type II. Type I materials, while in the superconducting state, are completely diamagnetic; that is, all of an applied magnetic field will be excluded from the body of material, a phenomenon known as the *Meissner effect*, which is illustrated in Figure 21.20. As H is increased, the material remains diamagnetic until the critical magnetic field H_C is reached. At this point conduction becomes normal, and complete magnetic flux penetration takes place.

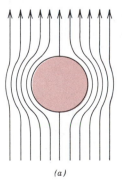

(a)

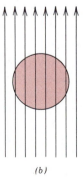

(b)

Figure 21.20 Representation of the Meissner effect. (a) While in the superconducting state, a body of material (circle) excludes a magnetic field (arrows) from its interior. (b) The magnetic field penetrates the same body of material once it becomes normally conductive.

TABLE 21.7 Critical Temperatures and Magnetic Fluxes
for Selected Superconducting Materials

Material	Critical Temperature T_C (K)	Critical Magnetic Flux Density B_C (teslas)[a]
Elements		
Aluminum	1.18	0.0105
Lead	7.19	0.0803
Mercury (α)	4.15	0.0411
Tin	3.72	0.0305
Titanium	0.40	0.0056
Tungsten	0.02	0.0001
Compounds and Alloys		
Nb–Ti alloy	10.2	12
Nb–Zr alloy	10.8	11
Nb_3Sn	18.3	22
Nb_3Al	18.9	32
Nb_3Ge	23.0	30
V_3Ga	16.5	22
$PbMo_6S_8$	14.0	45

[a] The critical magnetic flux density (μ_0H_C) for the elements was measured at 0 K. For alloys and compounds, the flux is taken as μ_0H_{C2} (in teslas), measured at 0 K.
Source: Adapted with permission from *Materials at Low Temperatures*, R. P. Reed and A. F. Clark, Editors, American Society for Metals, Metals Park, OH, 1983.

Several metallic elements including aluminum, lead, tin, and mercury belong to the type I group.

Type II superconductors are completely diamagnetic at low applied fields, and field exclusion is total. However, the transition from the superconducting state to the normal state is gradual and occurs between lower critical and upper critical fields, designated H_{C1} and H_{C2}, respectively. The magnetic flux lines begin to penetrate into the body of material at H_{C1}, and with increasing applied magnetic field, this penetration continues; at H_{C2}, field penetration is complete. For fields between H_{C1} and H_{C2}, the material exists in what is termed a mixed state—both normal and superconducting regions are present.

Type II superconductors are preferred over type I for most practical applications by virtue of their higher critical temperatures and critical magnetic fields. At present, the three most commonly utilized superconductors are niobium–zirconium (Nb–Zr) and niobium–titanium (Nb–Ti) alloys and the niobium–tin intermetallic compound Nb_3Sn. Table 21.7 lists several type I and II superconductors, their critical temperatures, and their critical magnetic flux densities.

Very recently, a family of ceramic materials which are normally electrically insulative have been found to be superconductors with inordinately high critical temperatures. Initial research has centered on yttrium barium copper oxide, $YBa_2Cu_3O_7$, which has a complex perovskite-type crystal structure (Section 13.2) and a critical temperature of about 95 K. New superconducting ceramic materials reported to have even higher critical temperatures have been and are currently being developed. The technological potential of these materials is extremely promising inasmuch as their critical temperatures are above 77 K, which permits the use of liquid nitrogen, a very inexpensive coolant, in comparison to liquid hydrogen and liquid helium. These new ceramic superconductors are not without drawbacks, chief of which is their brittle nature. This characteristic limits the ability of these materials to be fabricated into useful forms such as wires.

The phenomenon of superconductivity has many important practical implications. Superconducting magnets capable of generating high fields with low power consumption are currently being employed in scientific test and research equipment. In addition, they are also used for magnetic resonance imaging (MRI) in the medical field as a diagnostic tool. Abnormalities in body tissues and organs can be detected on the basis of the production of cross-sectional images. Chemical analysis of body tissues is also possible using magnetic resonance spectroscopy (MRS). Numerous other potential applications of superconducting materials also exist. Some of the areas being explored include (1) electrical power transmission through superconducting materials—power losses would be extremely low, and the equipment would operate at low voltage levels; (2) magnets for high-energy particle accelerators; (3) higher-speed switching and signal transmission for computers; and (4) high-speed magnetically levitated trains, wherein the levitation results from magnetic field repulsion. The chief deterrent to the widespread application of these superconducting materials is, of course, the difficulty in attaining and maintaining extremely low temperatures. Hopefully, this problem will be overcome with the development of the new generation of superconductors with reasonably high critical temperatures.

SUMMARY

The macroscopic magnetic properties of a material are a consequence of interactions between an external magnetic field and the magnetic dipole moments of the constituent atoms. Associated with each individual electron are both orbital and spin magnetic moments. The net magnetic moment for an atom is just the sum of the contributions of each of its electrons, wherein there will be spin and orbital moment cancellation for electron pairs.

Diamagnetism results from changes in electron orbital motion that are induced by an external field. The effect is extremely small and in opposition to the applied field. All materials are diamagnetic. Paramagnetic materials are those having permanent atomic dipoles, which are acted on individually and are aligned in the direction of an external field. Since the magnetizations are relatively small and persist only while an applied field is present, diamagnetic and paramagnetic materials are considered to be nonmagnetic.

Large and permanent magnetizations may be established within the ferromagnetic metals (Fe, Co, Ni). Atomic magnetic dipole moments are of spin origin, which are coupled and mutually aligned with moments of adjacent atoms.

Antiparallel coupling of adjacent cation spin moments is found for some ionic materials. Those in which there is total cancellation of spin moments are termed antiferromagnetic. With ferrimagnetism, permanent magnetization is possible because spin moment cancellation is incomplete. For cubic ferrites, the net magnetization results from the divalent ions (e.g., Fe^{2+}) that reside on octahedral lattice sites, the spin moments of which are all mutually aligned.

With rising temperature, increased thermal vibrations tend to counteract the dipole coupling forces in ferromagnetic and ferrimagnetic materials. Consequently, the saturation magnetization gradually diminishes with temperature, up to the Curie temperature, at which point it drops to near zero; above T_c these materials are paramagnetic.

Below its Curie temperature, a ferromagnetic or ferrimagnetic material is composed of domains—small-volume regions wherein all net dipole moments are mutually aligned and the magnetization is saturated. The total magnetization of the solid is just the appropriately weighted vector sum of the magnetizations of all these domains. As an external magnetic field is applied, domains having magnetization vectors oriented in the direction of the field grow at the expense of domains that have unfavorable magnetization orientations. At total saturation, the entire solid is a single domain and the magnetization is aligned with the field direction. The change in domain structure with increase or reversal of a magnetic field is accomplished by the motion of domain walls. Both hysteresis (the lag of the B field behind the applied H field) as well as permanent magnetization (or remanence) result from the resistance to movement of these domain walls.

For soft magnetic materials, domain wall movement is easy during magnetizaiton and demagnetization. Consequently, they have small hysteresis loops and low energy losses. Domain wall motion is much more difficult for the hard magnetic materials, which results in larger hysteresis loops; because greater fields are required to demagnetize these materials, the magnetization is more permanent.

Superconductivity has been observed in a number of materials, in which, upon cooling and in the vicinity of absolute zero temperature, the electrical resistivity vanishes. The superconducting state ceases to exist if temperature, magnetic field, or current density exceeds the critical value. For type I superconductors, magnetic field exclusion is complete below a critical field, and field penetration is complete once H_C is exceeded. This penetration is gradual with increasing magnetic field for type II materials. New complex oxide ceramics are being developed with relatively high critical temperatures, which allow inexpensive liquid nitrogen to be used as a coolant.

IMPORTANT TERMS AND CONCEPTS

Antiferromagnetism	Ferrite (ceramic)	Magnetization
Bohr magneton	Ferromagnetism	Paramagnetism
Coercivity	Hard magnetic material	Permeability
Curie temperature	Hysteresis	Remanence
Diamagnetism	Magnetic flux density	Saturation magnetization
Domain	Magnetic field strength	Soft magnetic material
Ferrimagnetism	Magnetic susceptibility	Superconductivity

REFERENCES

AZAROFF, L. V. and J. J. BROPHY, *Electronic Processes in Materials,* McGraw-Hill Book Company, New York, 1963, Chapter 13.

BRAILSFORD, F., *Magnetic Materials,* 3rd edition, Methuen & Company, Ltd., London (John Wiley & Sons, New York), 1960.

BROCKMAN, F. G., "Magnetic Ceramics—A Review and Status Report," *American Ceramic Society Bulletin,* Vol. 47, No. 2, February 1968, pp. 186–194.

CULLITY, B. D., *Introduction to Magnetic Materials,* Addison-Wesley Publishing Co., Reading, MA, 1972.

KEFFER, F., "The Magnetic Properties of Materials," *Scientific American,* Vol. 217, No. 3, September 1967, pp. 222–234.

ROSE, R. M., L. A. SHEPARD, and J. WULFF, *The Structure and Properties of Materials,* Vol. IV, *Electronic Properties,* John Wiley & Sons, New York, 1966, Chapters 9–11.

WERT, C. A. and R. M. THOMSON, *Physics of Solids,* 2nd edition, McGraw-Hill Book Company, New York, 1970, Chapters 20–22.

QUESTIONS AND PROBLEMS

21.1 A coil of wire 0.2 m long and having 200 turns carries a current of 10 A.
(a) What is the magnitude of the magnetic field strength H?
(b) Compute the flux density B if the coil is in a vacuum.
(c) Compute the flux density inside a bar of titanium that is positioned within the coil. The susceptibility for titanium is found in Table 21.2.
(d) Compute the magnitude of the magnetization M.

21.2 A coil of wire 0.1 m long and having 15 turns carries a current of 1.0 A.
(a) Compute the flux density if the coil is within a vacuum.
(b) A bar of an iron–silicon alloy, the B–H behavior for which is shown in Figure 21.21, is positioned within the coil. What is the flux density within this bar?
(c) Suppose that a bar of molybdenum is now situated within the coil. What current must be used to produce the same B field in the Mo as was produced in the iron–silicon alloy (part b) using 1.0 A?

21.3 Demonstrate that the relative permeability and the magnetic susceptibility are related according to Equation 21.7.

21.4 It is possible to express the magnetic susceptibility χ_m in several different units. For the discussion of this chapter, χ_m was used to designate the volume susceptibility in SI units, that is, the quantity that gives the magnetization per unit volume (m³) of material when multiplied by H. The mass susceptibility $\chi_m(\text{kg})$ yields the magnetic moment (or magnetization) per kilogram of material when multiplied by H; and, similarly, the atomic susceptibility, $\chi_m(a)$, gives the magnetization per kilogram-mole. The latter two quantities are related to χ_m through the relationships

$$\chi_m = \chi_m(\text{kg}) \times \text{mass density (kg/m}^3)$$

$$\chi_m(a) = \chi_m(\text{kg}) \times \text{atomic weight (in kg)}$$

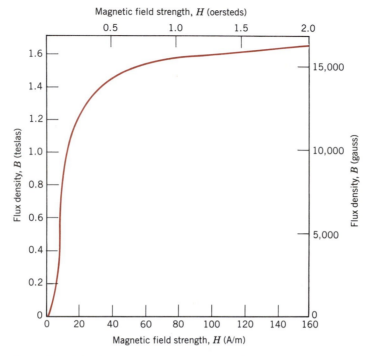

Figure 21.21 Initial magnetization B-versus-H curve for an iron–silicon alloy.

When using the cgs-emu system, comparable parameters exist, which may be designated by χ'_m, $\chi'_m(g)$, and $\chi'_m(a)$; the χ_m and χ'_m are related in accordance with Table 21.1. From Table 21.2, χ_m for silver is -2.38×10^{-5}; convert this value into the other five susceptibilities.

21.5 **(a)** Explain the two sources of magnetic moments for electrons. **(b)** Do all electrons have a net magnetic moment? Why or why not? **(c)** Do all atoms have a net magnetic moment? Why or why not?

21.6 The magnetic flux density within a bar of some material is 0.630 tesla at an H field of 5×10^5 A/m. Compute the following for this material: **(a)** the magnetic permeability, and **(b)** the magnetic susceptibility. **(c)** What type(s) of magnetism would you suggest as being displayed by this material? Why?

21.7 The magnetization within a bar of some metal alloy is 1.2×10^6 A/m at an H field of 200 A/m. Compute the following: **(a)** the magnetic susceptibility, **(b)** the permeability, and **(c)** the magnetic flux density within this material. **(d)** What type(s) of magnetism would you suggest as being displayed by this material? Why?

21.8 Compute **(a)** the saturation magnetization and **(b)** the saturation flux density for cobalt, which has a net magnetic moment per atom of 1.72 Bohr magnetons and a density of 8.90 g/cm³.

21.9 Confirm that there are 2.2 Bohr magnetons associated with each iron atom, given that the saturation magnetization is 1.70×10^6 A/m, that iron has a BCC crystal structure, and that the unit cell length is 0.2866 nm.

21.10 Assume there exists some hypothetical metal that exhibits ferromagnetic behavior and that has (1) a simple cubic crystal structure (Figure 3.22), (2) an atomic radius of 0.125 nm, and (3) a saturation flux density of 0.85 tesla. Determine the number of Bohr magnetons per atom for this material.

21.11 There is associated with each atom in paramagnetic and ferromagnetic materials a net magnetic moment. Explain why ferromagnetic materials can be permanently magnetized whereas paramagnetic ones cannot.

21.12 Cite the major similarities and differences between ferromagnetic and ferrimagnetic materials.

21.13 What is the difference between the spinel and inverse spinel crystal structures?

21.14 Consult another reference in which Hund's rule is outlined, and on its basis explain the net magnetic moments for each of the cations listed in Table 21.4.

21.15 Estimate **(a)** the saturation magnetization, and **(b)** the saturation flux density for nickel ferrite $[(NiFe_2O_4)_8]$, which has a unit cell edge length of 0.8337 nm.

21.16 The chemical formula for manganese ferrite may be written as $(MnFe_2O_4)_8$ because there are eight formula units per unit cell. If this material has a saturation magnetization of 5.6×10^5 A/m and a density of 5.00 g/cm^3, estimate the number of Bohr magnetons associated with each Mn^{2+} ion.

21.17 The formula for yttrium iron garnet $(Y_3Fe_5O_{12})$ may be written in the form $Y_3^c Fe_2^a Fe_3^d O_{12}$, in which the superscripts a, c, and d represent different sites on which the Y^{3+} and Fe^{3+} ions are located. The spin magnetic moments for the Y^{3+} and Fe^{3+} ions positioned in the a and c sites are oriented parallel to one another and antiparallel to the Fe^{3+} ions in d sites. Compute the number of Bohr magnetons associated with each Y^{3+} ion, given the following information: (1) each unit cell consists of eight formula $(Y_3Fe_5O_{12})$ units; (2) the unit cell is cubic with an edge length of 1.2376 nm; (3) the saturation magnetization for this material is 1.0×10^4 A/m; and (4) assume that there are 5 Bohr magnetons associated with each Fe^{3+} ion.

21.18 Explain why repeatedly dropping a permanent magnet on the floor will cause it to become demagnetized.

21.19 Briefly explain why the magnitude of the saturation magnetization decreases with increasing temperature for ferromagnetic materials, and why ferromagnetic behavior ceases above the Curie temperature.

21.20 Briefly describe the phenomenon of magnetic hysteresis, and why it occurs for ferromagnetic and ferrimagnetic materials.

21.21 Schematically sketch on a single plot the B-versus-H behavior for a ferromagnetic material **(a)** at 0 K, **(b)** at a temperature just below its Curie temperature, and **(c)** at a temperature just above its Curie temperature. Briefly explain why these curves have different shapes.

21.22 Schematically sketch the hysteresis behavior for a ferromagnet which is gradually demagnetized by cycling in an H field that alternates direction and decreases in magnitude.

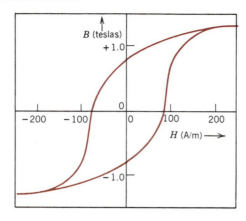

Figure 21.22 Complete magnetic hysteresis loop for steel.

21.23 Cite the differences between hard and soft magnetic materials in terms of both hysteresis behavior and typical applications.

21.24 Assume that the commercial iron (99.95 wt% Fe) in Table 21.5 just reaches the point of saturation when inserted within the coil in Problem 21.1. Compute the saturation magnetization.

21.25 Figure 21.22 shows the B-versus-H curve for a steel.
(a) What is the saturation flux density?
(b) What is the saturation magnetization?
(c) What is the remanence?
(d) What is the coercivity?
(e) On the basis of the data in Tables 21.5 and 21.6, would you classify this material as a soft or hard magnetic material? Why?

21.26 A ferromagnetic material has a remanence of 1.25 teslas and a coercivity of 50,000 A/m. Saturation is achieved at a magnetic field intensity of 100,000 A/m, at which the flux density is 1.50 teslas. Using these data, sketch the entire hysteresis curve in the range $H = -100,000$ to $+100,000$ A/m. Be sure to scale and label both coordinate axes.

21.27 The following data are for a transformer steel:

H (A/m)	B (teslas)	H (A/m)	B (teslas)
0	0	200	1.04
10	0.03	400	1.28
20	0.07	600	1.36
50	0.23	800	1.39
100	0.70	1000	1.41
150	0.92		

(a) Construct a graph of B versus H.
(b) What are the values of the initial permeability and initial relative permeability?
(c) What is the value of the maximum permeability?
(d) At about what H field does this maximum permeability occur?

(e) To what magnetic susceptibility does this maximum permeability correspond?

21.28 An iron bar magnet having a coercivity of 4000 A/m is to be demagnetized. If the bar is inserted within a cylindrical wire coil 0.15 m long and having 100 turns, what electric current is required to generate the necessary magnetic field?

21.29 A bar of an iron–silicon alloy having the B–H behavior shown in Figure 21.21 is inserted within a coil of wire 0.20 m long and having 60 turns, through which passes a current of 0.1 A.
(a) What is the B field within this bar?
(b) At this magnetic field,
    (i) What is the permeability?
    (ii) What is the relative permeability?
    (iii) What is the susceptibility?
    (iv) What is the magnetization?

21.30 It is possible, by various means (e.g., alteration of microstructure and impurity additions), to control the ease with which domain walls move as the magnetic field is changed for ferromagnetic and ferrimagnetic materials. Sketch a schematic B-versus-H hysteresis loop for a ferromagnetic material, and superimpose on this plot the loop alterations that would occur if domain boundary movement were hindered.

21.31 Briefly explain the manner in which information is stored magnetically.

21.32 For a superconducting material at a temperature T below the critical temperature T_C, the critical field $H_C(T)$, depends on temperature according to the relationship

$$H_C(T) = H_C(0)\left(1 - \frac{T^2}{T_C^2}\right) \tag{21.9}$$

where $H_C(0)$ is the critical field at 0 K.
(a) Using the data in Table 21.7, calculate the critical magnetic fields for lead at 2.5 and 5.0 K.
(b) To what temperature must mercury (α) be cooled in a magnetic field of 15,000 A/m for it to be superconductive?

21.33 Using Equation 21.9, determine which of the superconducting elements in Table 21.7 are superconducting at 2 K and in a magnetic field of 40,000 A/m.

21.34 Cite the differences between type I and type II superconductors.

21.35 Briefly describe the Meissner effect.

21.36 Cite the primary limitation of the new superconducting materials that have relatively high critical temperatures.

CHAPTER 22

OPTICAL PROPERTIES

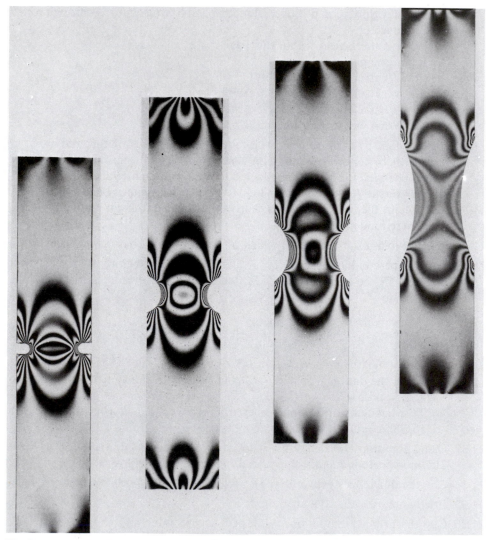

The four notched and transparent rods shown in this photograph demonstrate the phenomenon of *photoelasticity*. When elastically deformed, the optical properties (e.g., index of refraction) of a photoelastic specimen become anisotropic. Using a special optical system and polarized light, the stress distribution within the specimen may be deduced from interference fringes that are produced. These fringes within the four photoelastic specimens shown in the photograph indicate how the stress concentration and distribution change with notch geometry for an axial tensile stress. (Photograph courtesy of Measurements Group, Inc., Raleigh, North Carolina, U.S.A.)

22.1 INTRODUCTION

By "optical property" is meant a material's response to exposure to electromagnetic radiation and, in particular, to visible light. This chapter first discusses some of the basic principles and concepts relating to the nature of electromagnetic radiation and its possible interactions with solid materials. Next to be explored are the optical behaviors of metallic and nonmetallic materials in terms of their absorption, reflection, and transmission characteristics. The final sections outline luminescence, photoconductivity, and light amplification by stimulated emission of radiation (laser), and the practical utilization of these phenomena.

BASIC CONCEPTS

22.2 ELECTROMAGNETIC RADIATION

In the classical sense, electromagnetic radiation is considered to be wavelike, consisting of electric and magnetic field components that are perpendicular to each other and also to the direction of propagation (Figure 22.1). Light, heat (or radiant energy), radar, radio waves, and x-rays are all forms of electromagnetic radiation. Each is characterized primarily by a specific range of wavelengths, and also according to the technique by which it is generated. The *electromagnetic spectrum* of radiation spans the wide range from γ-rays (emitted by radioactive materials) and having wavelengths on the order of 10^{-12} m (10^{-3} nm), through x-rays, ultraviolet, visible, infrared, and finally radio waves with wavelengths as long as 10^5 m. This spectrum, on a logarithmic scale, is shown in Figure 22.2.

Visible light lies within a very narrow region of the spectrum, with wavelengths ranging between about 0.4 μm (4×10^{-7} m) and 0.7 μm. The perceived color is determined by wavelength; for example, radiation having a wavelength of approximately 0.4 μm appears violet, whereas green and red occur at about 0.5 and 0.65 μm, respectively. The spectral ranges for the several colors are included in Figure 22.2. White light is simply a mixture of all colors. The ensuing discussion is concerned primarily with this visible radiation, by definition the only radiation to which the eye is sensitive.

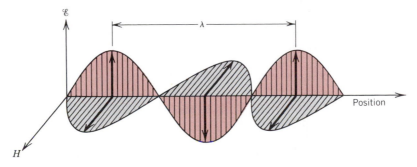

Figure 22.1 An electromagnetic wave showing electric field $\mathscr{E}$ and magnetic field H components, and the wavelength λ.

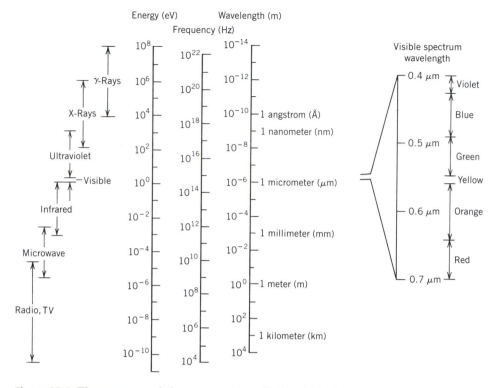

Figure 22.2 The spectrum of electromagnetic radiation, including wavelength ranges for the various colors in the visible spectrum.

All electromagnetic radiation traverses a vacuum at the same velocity, that of light, namely, 3×10^8 m/s (186,000 miles/s). This velocity, c, is related to the electrical permittivity of a vacuum ϵ_0 and the magnetic permeability of a vacuum μ_0 through

$$c = \frac{1}{\sqrt{\epsilon_0 \mu_0}} \tag{22.1}$$

Thus there is an association between the electromagnetic constant c and these electrical and magnetic constants.

Furthermore, the frequency v and the wavelength λ of the electromagnetic radiation are a function of velocity according to

$$c = \lambda v \tag{22.2}$$

Frequency is expressed in terms of hertz (Hz), and 1 Hz = 1 cycle per second. Ranges of frequency for the various forms of electromagnetic radiation are also included in the spectrum (Figure 22.2).

Sometimes it is more convenient to view electromagnetic radiation from a quantum-mechanical perspective, in that the radiation, rather than consisting of waves, is composed of groups or packets of energy, which are called **photons.** The energy E of a photon is said to be quantized, or can only have specific values, defined by the relationship

$$E = hv = \frac{hc}{\lambda} \tag{22.3}$$

where h is a universal constant called **Planck's constant,** which has a value of 6.63 × 10^{-34} J-s. Thus photon energy is proportional to the frequency of the radiation, or inversely proportional to the wavelength. Photon energies are also included in the electromagnetic spectrum (Figure 22.2).

When describing optical phenomena involving the interactions between radiation and matter, an explanation is often facilitated if light is treated in terms of photons. On other occasions, a wave treatment is more appropriate; at one time or another, both approaches are used in this discussion.

22.3 LIGHT INTERACTIONS WITH SOLIDS

When light proceeds from one medium into another (e.g., from air into a solid substance), several things happen. Some of the light radiation may be transmitted through the medium, some will be absorbed, and some will be reflected at the interface between the two media. The intensity I_0 of the beam incident to the surface of the solid medium must equal the sum of the intensities of the transmitted, absorbed, and reflected beams, denoted as I_T, I_A, and I_R, respectively, or

$$I_0 = I_T + I_A + I_R \tag{22.4}$$

Radiation intensity, expressed in watts per square meter, corresponds to the energy being transmitted per unit of time across a unit area that is perpendicular to the direction of propagation.

An alternate form of Equation 22.4 is

$$T + A + R = 1 \tag{22.5}$$

where T, A, and R represent, respectively, the transmissivity (I_T/I_0), absorptivity (I_A/I_0), and reflectivity (I_R/I_0), or the fractions of incident light that are transmitted, absorbed, and reflected by a material; their sum must equal unity, since all the incident light is either transmitted, absorbed, or reflected.

Materials that are capable of transmitting light with relatively little absorption and reflection are **transparent**—one can see through them. **Translucent** materials are those through which light is transmitted diffusely; that is, light is scattered within the interior, to the degree that objects are not clearly distinguishable when viewed through a specimen of the material. Those materials that are impervious to the transmission of visible light are termed **opaque.**

Bulk metals are opaque throughout the entire visible spectrum; that is, all light radiation is either absorbed or reflected. On the other hand, electrically insulating materials can be made to be transparent. Furthermore, some semiconducting materials are transparent whereas others are opaque.

22.4 ATOMIC AND ELECTRONIC INTERACTIONS

The optical phenomena that occur within solid materials involve interactions between the electromagnetic radiation and atoms, ions, and/or electrons. Two of the most important of these interactions are electronic polarization and electron energy transitions.

Electronic Polarization

One component of an electromagnetic wave is simply a rapidly fluctuating electric field (Figure 22.1). For the visible range of frequencies, this electric field interacts with the electron cloud surrounding each atom within its path in such a way as to induce electronic polarization, or to shift the electron cloud relative to the nucleus of the atom with each change in direction of electric field component, as demonstrated in Figure 19.29a. Two consequences of this polarization are: (1) some of the radiation energy may be absorbed, and (2) light waves are retarded in velocity as they pass through the medium. The second consequence is manifested as refraction, a phenomenon to be discussed in Section 22.5.

Electron Transitions

The absorption and emission of electromagnetic radiation may involve electron transitions from one energy state to another. For the sake of this discussion, consider an isolated atom, the electron energy diagram for which is represented in Figure 22.3. An electron may be excited from an occupied state at energy E_2 to a vacant and higher lying one, denoted E_4, by the absorption of a photon of energy. The change in energy experienced by the electron, ΔE, depends on the radiation frequency as follows:

$$\Delta E = h\nu \qquad (22.6)$$

where, again, h is Planck's constant. At this point it is important that several concepts be understood. First, since the energy states for the atom are discrete, only specific ΔE's exist between the energy levels; thus only photons of frequencies corresponding to the possible ΔE's for the atom can be absorbed by electron transitions. Furthermore, all of a photon's energy is absorbed in each excitation event.

A second important concept is that a stimulated electron cannot remain in an **excited state** indefinitely; after a short time, it falls or decays back into its **ground state**

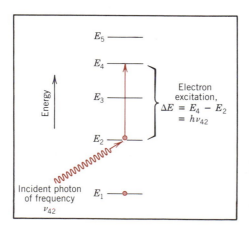

Figure 22.3 For an isolated atom, a schematic illustration of photon absorption by the excitation of an electron from one energy state to another. The energy of the photon ($h\nu_{42}$) must be exactly equal to the difference in energy between the two states ($E_4 - E_2$).

or unexcited level with a reemission of electromagnetic radiation. Several decay paths are possible, and these are discussed later. In any case, there must be a conservation of energy for absorption and emission electron transitions.

As the ensuing discussions show, the optical characteristics of solid materials that relate to absorption and emission of electromagnetic radiation are explained in terms of the electron band structure of the material (possible band structures were discussed in Section 19.5) and the principles relating to electron transitions, as outlined in the preceding two paragraphs.

OPTICAL PROPERTIES OF METALS

Consider the electron energy band schemes for metals as illustrated in Figures 19.4a and 19.4b; in both cases a high-energy band is only partially filled with electrons. Metals are opaque because the incident radiation having frequencies within the visible range excites electrons into unoccupied energy states above the Fermi energy, as demonstrated in Figure 22.4a; as a consequence, the incident radiation is absorbed, in accordance with Equation 22.6. Total absorption is within a very thin outer layer, usually less than 0.1 μm; thus only metallic films thinner than 0.1 μm are capable of transmitting visible light.

All frequencies of visible light are absorbed by metals because of the continuously available empty electron states, which permit electron transitions as in Figure 22.4a. In fact, metals are opaque to all electromagnetic radiation on the low end of the frequency spectrum, from radio waves, through infrared, the visible, and into about the middle of the ultraviolet radiation. Metals are transparent to high-frequency (x- and γ-ray) radiation.

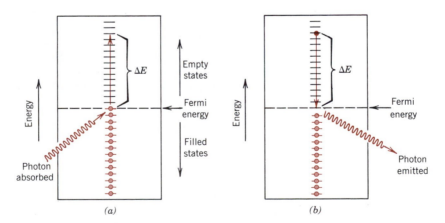

Figure 22.4 (a) Schematic representation of the mechanism of photon absorption for metallic materials in which an electron is excited into a higher energy unoccupied state. The change in energy of the electron ΔE is equal to the energy of the photon. (b) Reemission of a photon of light by the direct transition of an electron from a high to a low energy state.

Most of the absorbed radiation is reemitted from the surface in the form of visible light of the same wavelength, which appears as reflected light; an electron transition accompanying reradiation is shown in Figure 22.4b. The reflectivity for most metals is between 0.90 and 0.95; some small fraction of the energy from electron decay processes is dissipated as heat.

Since metals are opaque and highly reflective, the perceived color is determined by the wavelength distribution of the radiation that is reflected and not absorbed. A bright silvery appearance when exposed to white light indicates that the metal is highly reflective over the entire range of the visible spectrum. In other words, for the reflected beam, the composition of these reemitted photons, in terms of frequency and number, is approximately the same as for the incident beam. Aluminum and silver are two metals that exhibit this reflective behavior. Copper and gold appear red-orange and yellow, respectively, because some of the energy associated with light photons having short wavelengths is not reemitted as visible light.

OPTICAL PROPERTIES OF NONMETALS

By virtue of their electron energy band structures, nonmetallic materials may be transparent to visible light. Therefore, in addition to reflection and absorption, refraction and transmission phenomena also need to be considered.

22.5 REFRACTION

Light that is transmitted into the interior of transparent materials experiences a decrease in velocity, and, as a result, is bent at the interface; this phenomenon is termed **refraction.** The **index of refraction** n of a material is defined as the ratio of the velocity in a vacuum c to the velocity in the medium v, or

$$n = \frac{c}{v} \tag{22.7}$$

The magnitude of n (or the degree of bending) will depend on the wavelength of the light. This effect is graphically demonstrated by the familiar dispersion or separation of a beam of white light into its component colors by a glass prism. Each color is deflected by a different amount as it passes into and out of the glass, which results in the separation of the colors. Not only does the index of refraction affect the optical path of light, but also, as explained below, it influences the fraction of incident light that is reflected at the surface.

Just as Equation 22.1 defines the magnitude of c, an equivalent expression gives the velocity of light v in a medium as

$$v = \frac{1}{\sqrt{\epsilon\mu}} \tag{22.8}$$

where ϵ and μ are, respectively, the permittivity and permeability of the particular substance. From Equation 22.7, we have

$$n = \frac{c}{v} = \frac{\sqrt{\epsilon\mu}}{\sqrt{\epsilon_0\mu_0}} = \sqrt{\epsilon_r\mu_r} \tag{22.9}$$

where ϵ_r and μ_r are the dielectric constant and the relative magnetic permeability, respectively. Since most substances are only slightly magnetic, $\mu_r \cong 1$, and

$$n \cong \sqrt{\epsilon_r} \qquad (22.10)$$

Thus for transparent materials, there is a relation between the index of refraction and the dielectric constant. As already mentioned, the phenomenon of refraction is related to electronic polarization (Section 22.4) at the relatively high frequencies for visible light; thus, the electronic component of the dielectric constant may be determined from index of refraction measurements using Equation 22.10.

Since the retardation of electromagnetic radiation in a medium results from electronic polarization, the size of the constituent atoms or ions has a considerable influence on the magnitude of this effect—generally, the larger an atom or ion, the greater will be the electronic polarization, the slower the velocity, and the greater the index of refraction. The index of refraction for a typical soda–lime glass is approximately 1.5. Additions of large barium and lead ions (as BaO and PbO) to a glass will increase n significantly. For example, highly leaded glasses containing 90 wt% PbO have an index of refraction of approximately 2.1.

For crystalline ceramics that have cubic crystal structures, and for glasses, the index of refraction is independent of crystallographic direction (i.e., it is isotropic). Noncubic crystals, on the other hand, have an anisotropic n; that is, the index is greatest along the directions that have the highest density of ions. Table 22.1 gives refractive indices for several glasses, transparent ceramics, and polymers. Average values are provided for the crystalline ceramics in which n is anisotropic.

TABLE 22.1 Refractive Indices for Some Transparent Materials

Material	Average Index of Refraction
Ceramics	
Silica glass	1.458
Soda–lime glass	1.51
Pyrex glass	1.47
Dense optical flint glass	1.65
Corundum (Al_2O_3)	1.76
Periclase (MgO)	1.74
Quartz (SiO_2)	1.55
Spinel ($MgAl_2O_4$)	1.72
Polymers	
Polytetrafluoroethylene	1.35
Polyethylene	1.51
Polystyrene	1.60
Polymethyl methacrylate	1.49
Polypropylene	1.49

22.6 REFLECTION

When light radiation passes from one medium into another having a different index of refraction, some of the light is scattered at the interface between the two media even if both are transparent. The reflectivity R represents that fraction of the incident light which is reflected at the interface, or

$$R = \frac{I_R}{I_0} \tag{22.11}$$

where I_0 and I_R are the intensities of the incident and reflected beams, respectively. If the light is normal (or perpendicular) to the interface, then

$$R = \left(\frac{n_2 - n_1}{n_2 + n_1}\right)^2 \tag{22.12}$$

where n_1 and n_2 are the indices of refraction of the two media. If the incident light is not normal to the interface, R will depend on the angle of incidence. When light is transmitted from a vacuum or air into a solid s, then

$$R = \left(\frac{n_s - 1}{n_s + 1}\right)^2 \tag{22.13}$$

since the index of refraction of air is very nearly unity. Thus the higher the index of refraction of the solid, the greater is the reflectivity. For typical silicate glasses, the reflectivity is approximately 0.05. Just as the index of refraction of a solid depends on the wavelength of the incident light, so also does the reflectivity vary with wavelength. Reflection losses for lenses and other optical instruments are minimized significantly by coating the reflecting surface with very thin layers of dielectric materials such as magnesium fluoride (MgF_2).

22.7 ABSORPTION

Nonmetallic materials may be opaque or transparent to visible light; and, if transparent, they often appear colored. In principle, light radiation is absorbed in this group of materials by three basic mechanisms, which also influence the transmission characteristics of these nonmetals. One of these is electronic polarization (Section 22.4). Absorption by electronic polarization is important only at light frequencies in the vicinity of the relaxation frequency of the constituent atoms. The other two mechanisms involve electron transitions, which depend on the electron energy band structure of the material; band structures for semiconductors and insulators were discussed in Section 19.5. One of these absorption mechanisms involves the absorption as a consequence of electron excitations across the band gap; the other is related to electron transitions to impurity or defect levels that lie within the band gap.

Absorption of a photon of light may occur by the promotion or excitation of an electron from the nearly filled valence band, across the band gap, and into an empty state within the conduction band, as demonstrated in Figure 22.5a; a free electron in the conduction band and a hole in the valence band are created. Again, the energy of excitation ΔE is related to the absorbed photon frequency through Equation 22.6. These excitations with the accompanying absorption can take place only if the photon

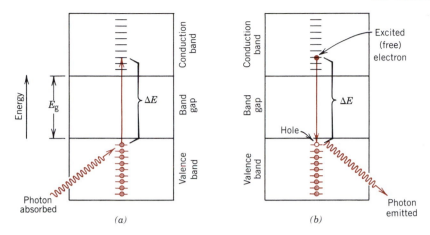

Figure 22.5 (*a*) Mechanism of photon absorption for nonmetallic materials in which an electron is excited across the band gap, leaving behind a hole in the valence band. The energy of the photon absorbed is ΔE, which is necessarily greater than the band gap energy E_g. (*b*) Emission of a photon of light by a direct electron transition across the band gap.

energy is greater than that of the band gap E_g, that is, if

$$h\nu > E_g \qquad (22.14)$$

or, in terms of wavelength,

$$\frac{hc}{\lambda} > E_g \qquad (22.15)$$

The minimum wavelength for visible light $\lambda(\text{min})$ is about 0.4 μm, and since $c = 3 \times 10^8$ m/s and $h = 4.13 \times 10^{-15}$ eV-s, the maximum band gap energy $E_g(\text{max})$ for which absorption of visible light is possible is just

$$E_g(\text{max}) = \frac{hc}{\lambda(\text{min})}$$

$$= \frac{(4.13 \times 10^{-15} \text{ eV-s})(3 \times 10^8 \text{ m/s})}{4 \times 10^{-7} \text{ m}}$$

$$= 3.1 \text{ eV} \qquad (22.16a)$$

Or, no visible light is absorbed by nonmetallic materials having band gap energies greater than about 3.1 eV; these materials, if of high purity, will appear transparent and colorless.

On the other hand, the maximum wavelength for visible light $\lambda(\text{max})$ is about 0.7 μm; computation of the minimum band gap energy $E_g(\text{min})$ for which there is absorption of visible light is according to

$$E_g(\text{min}) = \frac{hc}{\lambda(\text{max})}$$

$$= \frac{(4.13 \times 10^{-15} \text{ eV-s})(3 \times 10^8 \text{ m/s})}{7 \times 10^{-7} \text{ m}} = 1.8 \text{ eV} \qquad (22.16b)$$

This result means that all visible light is absorbed by valence band to conduction band electron transitions for those semiconducting materials that have band gap energies less than about 1.8 eV; thus these materials are opaque. Only a portion of the visible spectrum is absorbed by materials having band gap energies between 1.8 and 3.1 eV; consequently, these materials appear colored.

Every nonmetallic material becomes opaque at some wavelength, which depends on the magnitude of its E_g. For example, diamond, having a band gap of 5.6 eV, is opaque to radiation having wavelengths less than about 0.22 μm.

Absorption of light radiation can also occur in dielectric solids having wide band gaps, by other than valence band–conduction band electron transitions. If impurities or other electrically active defects are present, electron levels within the band gap may be introduced, such as the donor and acceptor levels (Section 19.11), except that they lie closer to the center of the band gap. Light radiation of specific wavelengths may be absorbed as a result of electron transitions from or to these levels within the band gap, as illustrated in Figure 22.6a.

Again, electromagnetic energy that is absorbed by electron excitations must be dissipated in some manner; several mechanisms are possible. For excitations from valence band to conduction band, this dissipation may occur via direct electron and hole recombination according to the reaction

$$\text{electron} + \text{hole} \longrightarrow \text{energy } (\Delta E) \qquad (22.17)$$

which is represented schematically in Figure 22.5b. In addition, multiple-step electron transitions may occur, which involve impurity levels lying within the band gap. One possibility, as indicated in Figure 22.6b, is the emission of two photons; one is emitted as the electron drops from a state in the conduction band to the impurity level, the other as it decays back into the valence band. Or, alternatively, one of the transitions

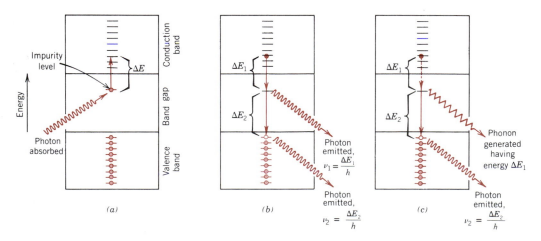

Figure 22.6 (a) The mechanism of electron excitation, from an impurity level that lies within the band gap, by absorption of a photon of light. (b) Emission of two photons involving electron decay first into an impurity state, and finally into the ground state. (c) Generation of both a phonon and a photon as an excited electron falls first into an impurity level and finally back to its ground state.

may involve the generation of a phonon (Figure 22.6c), wherein the associated energy is dissipated in the form of heat.

The intensity of the net absorbed radiation is dependent on the character of the medium as well as the path length within. The intensity of transmitted or nonabsorbed radiation I'_T continuously decreases with distance x that the light traverses:

$$I'_T = I'_0 e^{-\beta x} \qquad (22.18)$$

where I'_0 is the intensity of the nonreflected incident radiation and β, the *absorption coefficient* (in mm^{-1}), is characteristic of the particular material; furthermore, β varies with wavelength of the incident radiation. The distance parameter x is measured from the incident surface into the material. Materials that have large β values are considered to be highly absorptive.

22.8 TRANSMISSION

The phenomena of absorption, reflection, and transmission may be applied to the passage of light through a transparent solid, as shown in Figure 22.7. For an incident beam of intensity I_0 that impinges on the front surface of a specimen of thickness l and absorption coefficient β, the transmitted intensity at the back face I_T is

$$I_T = I_0(1 - R)^2 e^{-\beta l} \qquad (22.19)$$

where R is the reflectance; for this expression, it is assumed that the same medium exists outside both front and back faces. The derivation of Equation 22.19 is left as a homework problem.

Thus the fraction of incident light that is transmitted through a transparent material depends on the losses that are incurred by absorption and reflection. Again, the sum of the reflectivity R, absorptivity A, and transmissivity T, is unity according to Equation 22.5. Also, each of the variables R, A, and T depends on light wavelength.

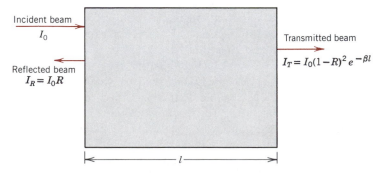

Figure 22.7 The transmission of light through a transparent medium for which there is reflection at front and back faces, as well as absorption within the medium. (Adapted from R. M. Rose, L. A. Shepard, and J. Wulff, *The Structure and Properties of Materials,* Vol. 4, *Electronic Properties.* Copyright © 1966 by John Wiley & Sons, New York. Reprinted by permission of John Wiley & Sons, Inc.)

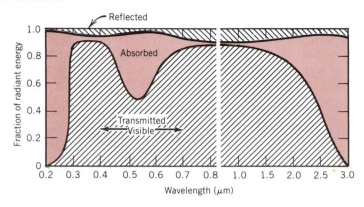

Figure 22.8 The variation with wavelength of the fractions of incident light transmitted, absorbed, and reflected through a green glass. (From W. D. Kingery, H. K. Bowen, and D. R. Uhlmann, *Introduction to Ceramics,* 2nd edition. Copyright © 1976 by John Wiley & Sons, New York. Reprinted by permission of John Wiley & Sons, Inc.)

This is demonstrated over the visible region of the spectrum for a green glass in Figure 22.8. For example, for light having a wavelength of 0.4 μm, the fractions transmitted, absorbed, and reflected are approximately 0.90, 0.05, and 0.05, respectively. However, at 0.55 μm, the respective fractions have shifted to about 0.50, 0.48, and 0.02.

22.9 COLOR

Transparent materials appear colored as a consequence of specific wavelength ranges of light that are selectively absorbed; the color discerned is a result of the combination of wavelengths that are transmitted. If absorption is uniform for all visible wavelengths, the material appears colorless; examples include high-purity inorganic glasses and high-purity and single-crystal diamonds and sapphire.

Usually, any selective absorption is by electron excitation. One such situation involves semiconducting materials that have band gaps within the range of photon energies for visible light (1.8 to 3.1 eV). Thus the fraction of the visible light having energies greater than E_g is selectively absorbed by valence band–conduction band electron transitions. Of course, some of this absorbed radiation is reemitted as the excited electrons drop back into their original lower lying energy states. It is not necessary that this reemission occur at the same frequency as that of the absorption; the frequency and associated energy may be less in cases of multiple radiative (Figure 22.6*b*) or nonradiative (Figure 22.6*c*) electron transitions. As a result, the color depends on the frequency distribution of both transmitted and reemitted light beams.

For example, cadmium sulfide (CdS) has a band gap of about 2.4 eV; hence, it absorbs photons having energies greater than about 2.4 eV, which correspond to the blue and violet portions of the visible spectrum; some of this energy is reradiated as light having other wavelengths. Nonabsorbed visible light consists of photons having

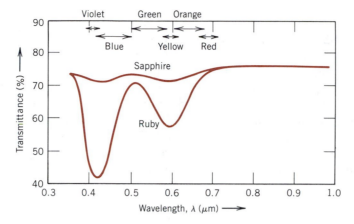

Figure 22.9 Transmission of light radiation as a function of wavelength for sapphire (single-crystal aluminum oxide) and ruby (aluminum oxide containing some chromium oxide). The sapphire appears colorless, while the ruby has a red tint due to selective absorption over specific wavelength ranges. (Adapted from "The Optical Properties of Materials," by A. Javan. Copyright © by SCIENTIFIC AMERICAN, Inc. All rights reserved.)

energies between about 1.8 and 2.4 eV. Cadmium sulfide takes on a yellow-orange color because of the composition of the transmitted beam.

With insulator ceramics, specific impurities also introduce electron levels within the forbidden band gap, as discussed above. Photons having energies less than that of the band gap may be absorbed as a consequence of electron excitations involving impurity atoms or ions as demonstrated in Figure 22.6a; of course, some reemission will probably occur. Again, the color of the material is a function of the distribution of wavelengths that are found in the transmitted beam.

For example, high-purity and single-crystal aluminum oxide or sapphire is colorless. Ruby, which has a brilliant red color, is simply sapphire to which has been added 0.5 to 2% of chromium oxide (Cr_2O_3). The Cr^{3+} ion substitutes for the Al^{3+} ion in the Al_2O_3 crystal structure and, furthermore, introduces impurity levels within the wide energy band gap of the sapphire. Specific wavelengths are absorbed preferentially as a consequence of electron transitions to or from these impurity levels. The transmittance as a function of wavelength for sapphire and ruby is presented in Figure 22.9. For the sapphire, transmittance is relatively constant with wavelength over the visible spectrum, which accounts for the colorlessness of this material. However, strong absorption peaks (or minima) occur for the ruby, one in the blue-violet region (at about 0.4 μm), and the other for yellow-green light (at about 0.6 μm). That nonabsorbed or transmitted light mixed with reemitted light imparts to ruby its deep-red color.

Inorganic glasses are colored by incorporating transition or rare earth ions while the glass is still in the molten state. Representative color–ion pairs include Cu^{2+}, blue-green; Co^{2+}, blue-violet; Cr^{3+}, green; Mn^{2+}, yellow; and Mn^{3+}, purple.

22.10 OPACITY AND TRANSLUCENCY IN INSULATORS

The extent of translucency and opacity for inherently transparent dielectric materials depends to a great degree on their internal reflectance and transmittance characteristics. Many dielectric materials that are intrinsically transparent may be made translucent or even opaque because of interior reflection and refraction. A transmitted light beam is deflected in direction and appears diffuse as a result of multiple scattering events. Opacity results when the scattering is so extensive that virtually none of the incident beam is transmitted, undeflected, to the back surface.

This internal scattering may result from several different sources. Polycrystalline specimens in which the index of refraction is anisotropic normally appear translucent. Both reflection and refraction occur at grain boundaries, which causes a diversion in the incident beam. This results from a slight difference in index of refraction n between adjacent grains that do not have the same crystallographic orientation.

Scattering of light also occurs in two-phase materials in which one phase is finely dispersed within the other. Again, the beam dispersion occurs across phase boundaries when there is a difference in the refractive index for the two phases; the greater this difference, the more efficient is the scattering.

As a consequence of fabrication or processing, many ceramic pieces contain some residual porosity in the form of finely dispersed pores. These pores also effectively scatter light radiation.

Figure 22.10 demonstrates the difference in optical transmission characteristics of single-crystal, fully dense polycrystalline, and porous ($\sim 5\%$ porosity) aluminum oxide specimens. Whereas the single crystal is totally transparent, polycrystalline and porous materials are, respectively, translucent and opaque.

For intrinsic polymers (without additives and impurities), the degree of translucency is influenced primarily by the extent of crystallinity. Some scattering of visible

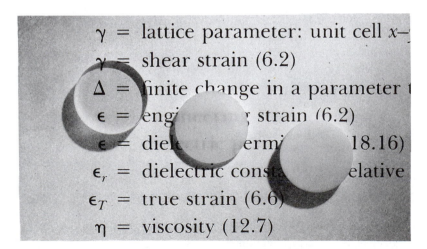

Figure 22.10 Photograph showing the light transmittance of three aluminum oxide specimens. From left to right: single-crystal material (sapphire), which is transparent; a polycrystalline and fully dense (nonporous) material, which is translucent; and a polycrystalline material that contains approximately 5% porosity, which is opaque. (Specimen preparation, P. A. Lessing; photography by J. Telford.)

light occurs at the boundaries between crystalline and amorphous regions, again, as a result of different indices of refraction. For highly crystalline specimens, this degree of scattering is extensive, which leads to translucency, and, in some instances, even opacity. Highly amorphous polymers are completely transparent.

APPLICATIONS OF OPTICAL PHENOMENA

22.11 LUMINESCENCE

Some materials are capable of absorbing energy and then reemitting visible light in a phenomenon called **luminescence.** Photons of emitted light are generated from electron transitions in the solid. Energy is absorbed when an electron is promoted to an excited energy state; visible light is emitted when it falls back to a lower energy state if 1.8 eV $< hv < 3.1$ eV. The absorbed energy may be supplied as higher-energy electromagnetic radiation (causing valence band–conduction band transitions) such as ultraviolet light, or other sources such as high energy electrons, or by heat, mechanical, or chemical energy. Furthermore, luminescence is classified according to the magnitude of the delay time between absorption and reemission events. If reemission occurs for times much less than one second, the phenomenon is termed **fluorescence;** for longer times, it is called **phosphorescence.** A number of materials can be made to fluoresce or phosphoresce; these include some sulfides, oxides, tungstates, and a few organic materials. Ordinarily, pure materials do not display these phenomena, and to induce them, impurities in controlled concentrations must be added.

Luminescence has a number of commercial applications. Fluorescent lamps consist of a glass housing, coated on the inside with specially prepared tungstates or silicates. Ultraviolet light is generated within the tube from a mercury glow discharge, which causes the coating to fluoresce and emit white light. The picture viewed on a television screen is the product of luminescence. The inside of the screen is coated with a material that fluoresces as an electron beam inside the picture tube very rapidly traverses the screen. Detection of x-rays and γ-rays is also possible; certain phosphors emit visible light or glow when introduced into a beam of the radiation which is otherwise invisible.

Some p–n rectifying junctions, as described in Section 19.13, may also be used to generate visible light in a process termed **electroluminescence.** When a forward-biased potential is applied across the device, electrons and holes will annihilate one another within the recombination region according to Equation 22.17. Under some circumstances the energy produced will appear as visible light. Such diodes that luminesce visible light are the familiar *light-emitting diodes* (*LEDs*), which are used for digital displays. The characteristic color of an LED depends on the particular semiconducting material that is used.

22.12 PHOTOCONDUCTIVITY

The conductivity of semiconducting materials depends on the number of free electrons in the conduction band and also the number of holes in the valence band, according to Equation 19.13. Thermal energy associated with lattice vibrations can promote electron excitations in which free electrons and/or holes are created, as described in

Section 19.6. Additional charge carriers may be generated as a consequence of photon-induced electron transitions in which light is absorbed; the attendant increase in conductivity is called **photoconductivity.** Thus when a specimen of a photoconductive material is illuminated, the conductivity increases.

This phenomenon is utilized in photographic light meters. A photoinduced current is measured, and its magnitude is a direct function of the intensity of the incident light radiation, or the rate at which the photons of light strike the photoconductive material. Of course, visible light radiation must induce electronic transitions in the photoconductive material; cadmium sulfide is commonly utilized in light meters.

Sunlight may be directly converted into electrical energy in solar cells, which also employ semiconductors. The operation of these devices is, in a sense, the reverse of that for the light-emitting diode. A *p–n* junction is used in which photoexcited electrons and holes are drawn away from the junction, in opposite directions, and become part of an external current.

22.13 LASERS

All the radiative electron transitions heretofore discussed are spontaneous; that is, an electron falls from a high energy state to a lower one without any external provocation. These transition events occur independently of one another and at random times, producing radiation that is incoherent; that is, the light waves are out of phase with one another. With lasers, however, coherent light is generated by electron transitions initiated by an external stimulus; in fact, **"laser"** is just the acronym for light amplification by stimulated emission of radiation.

Although there are several different varieties of laser, the principles of operation are explained using the solid state ruby laser. Ruby is simply a single crystal of Al_2O_3 (sapphire) to which has been added on the order of 0.05% Cr^{3+} ions. As previously explained (Section 22.9), these ions impart to ruby its characteristic red color; more important, they provide electron states that are essential for the laser to function. The ruby laser is in the form of a rod, the ends of which are flat, parallel, and highly polished. Both ends are silvered such that one is totally reflecting and the other partially transmitting.

The ruby is illuminated with light from a xenon flash lamp (Figure 22.11). Before this exposure, virtually all the Cr^{3+} ions are in their ground states; that is, electrons

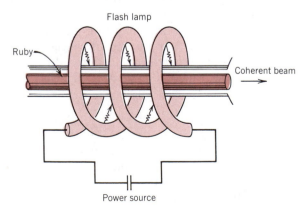

Figure 22.11 Schematic diagram of the ruby laser and xenon flash lamp. (From R. M. Rose, L. A. Shepard, and J. Wulff, *The Structure and Properties of Materials,* Vol. 4, *Electronic Properties.* Copyright © 1966 by John Wiley & Sons, New York. Reprinted by permission of John Wiley & Sons, Inc.)

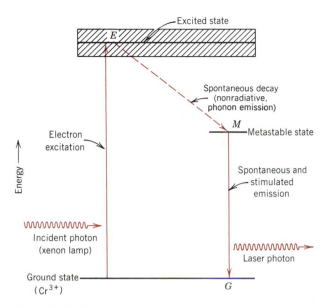

Figure 22.12 Schematic energy diagram for the ruby laser, showing electron excitation and decay paths.

fill the lowest energy levels, as represented schematically in Figure 22.12. However, photons of wavelength 0.56 μm from the xenon lamp excite electrons from the Cr^{3+} ions into higher energy states. These electrons can decay back into their ground state by two different paths. Some fall back directly; associated photon emissions are not part of the laser beam. Other electrons decay into a metastable intermediate state (path EM, Figure 22.12), where they may reside for up to 3 ms before spontaneous emission (path MG). In terms of electronic processes, 3 ms is a relatively long time, which means that a large number of these metastable states may become occupied. This situation is indicated in Figure 22.13b.

The initial spontaneous photon emission by a few of these electrons is the stimulus that triggers an avalanche of emissions from the remaining electrons in the metastable state (Figure 22.13c). Of the photons directed parallel to the long axis of the ruby rod, some are transmitted through the partially silvered end; others, incident to the totally silvered end, are reflected. Photons that are not emitted in this axial direction are lost. The light beam repeatedly travels back and forth along the rod length, and its intensity increases as more emissions are stimulated. Ultimately, a high intensity, coherent, and highly collimated laser light beam of short duration is transmitted through the partially silvered end of the rod (Figure 22.13e). This monochromatic red beam has a wavelength of 0.6943 μm.

A variety of substances may be used for lasers, including some gases and glasses, as well as semiconductor junction diodes. Laser applications are diverse. Since laser beams may be focused to produce localized heating, they are used in some surgical procedures and for cutting and machining metals. Lasers are also used as light sources for optical communication systems. Furthermore, because the beam is highly coherent, they may be utilized for making very precise distance measurements.

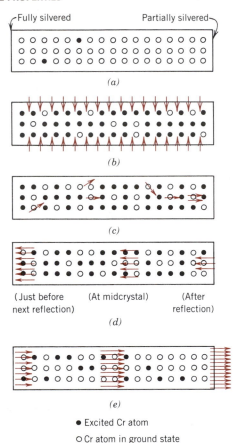

(a)

(b)

(c)

(Just before   (At midcrystal)   (After
next reflection)   reflection)

(d)

(e)

● Excited Cr atom

○ Cr atom in ground state

Figure 22.13 Schematic representations of the stimulated emission and light amplification for a ruby laser. (*a*) The chromium ions before excitation. (*b*) Electrons in some chromium atoms are excited into higher energy states by the xenon light flash. (*c*) Emission from metastable electron states is initiated or stimulated by photons that are spontaneously emitted. (*d*) Upon reflection from the silvered ends, the photons continue to stimulate emissions as they traverse the rod length. (*e*) The coherent and intense beam is finally emitted through the partially silvered end. (From R. M. Rose, L. A. Shepard, and J. Wulff, *The Structure and Properties of Materials,* Vol. 4, *Electronic Properties.* Copyright © 1966 by John Wiley & Sons, New York. Reprinted by permission of John Wiley & Sons, Inc.)

SUMMARY

The optical behavior of a solid material is a function of its interactions with electromagnetic radiation having wavelengths within the visible region of the spectrum. Possible interactive phenomena include refraction, reflection, absorption, and transmission of incident light.

Metals appear opaque as a result of the absorption and then reemission of light radiation within a thin outer surface layer. Absorption occurs via the excitation of electrons from occupied energy states to unoccupied ones above the Fermi energy level. Reemission takes place by decay electron transitions in the reverse direction. The perceived color of a metal is determined by the spectral composition of the reflected light.

Nonmetallic materials are either intrinsically transparent or opaque. Opacity results in relatively narrow band gap materials as a result of absorption whereby a photon's energy is sufficient to promote valence band–conduction band electron transitions. Transparent nonmetals have band gaps greater than about 3 eV.

Light radiation experiences refraction in transparent materials; that is, its velocity is retarded and the light beam is bent at the interface. Index of refraction is the ratio of the velocity of light in a vacuum to that in the particular medium. The phenomenon of refraction is a consequence of electronic polarization of the atoms or ions, which is induced by the electric field component of the light wave.

When light passes from one transparent medium to another having a different index of refraction, some of it is reflected at the interface. The degree of the reflectance depends on the indices of refraction of both media, as well as the angle of incidence.

Some light absorption occurs in even transparent materials as a consequence of electronic polarization and electron transitions to impurity electron states that lie within the band gap. These materials appear colored because of the selective absorption of light wavelength ranges.

Normally transparent materials may be made translucent or even opaque if the incident light beam experiences interior reflection and/or refraction. Translucency and opacity as a result of internal scattering may occur (1) in polycrystalline materials that have an anisotropic index of refraction, (2) in two-phase materials, (3) in materials containing small pores, and (4) in highly crystalline polymers.

This chapter concluded with a discussion of three other important optical phenomena: luminescence, photoconductivity, and light amplification by stimulated emission of radiation (lasers). With luminescence, energy is absorbed as a consequence of electron excitations and then reemitted as visible light. The electrical conductivity of some semiconductors may be enhanced by photoinduced electron transitions, whereby additional free electrons and holes are generated. Coherent and high-intensity light beams are produced in lasers by stimulated electron transitions.

IMPORTANT TERMS AND CONCEPTS

Absorption	**Laser**	**Planck's constant**
Color	**Luminescence**	**Reflection**
Electroluminescence	**Opaque**	**Refraction**
Excited state	**Phosphorescence**	**Translucent**
Fluorescence	**Photoconductivity**	**Transmission**
Ground state	**Photon**	**Transparent**
Index of refraction		

REFERENCES

AZAROFF, L. V. and J. J. BROPHY, *Electronic Processes in Materials,* McGraw-Hill Book Company, New York, 1963, Chapter 14.

JAVAN, A., "The Optical Properties of Materials," *Scientific American,* Vol. 217, No. 3, September 1967, pp. 238–248.

KINGERY, W. D., H. K. BOWEN, and D. R. UHLMANN, *Introduction to Ceramics,* 2nd edition, John Wiley & Sons, New York, 1976, Chapter 13.

RALLS, K. M., T. H. COURTNEY, and J. WULFF, *Introduction to Materials Science and Engineering,* John Wiley & Sons, New York, 1976, Chapter 27.

QUESTIONS AND PROBLEMS

22.1 Briefly discuss the similarities and differences between photons and phonons.

22.2 Electromagnetic radiation may be treated from the classical or the quantum-mechanical perspective. Briefly compare these two viewpoints.

22.3 Visible light having a wavelength of 6×10^{-7} m appears orange. Compute the frequency and energy of a photon of this light.

22.4 Distinguish between materials that are opaque, translucent, and transparent in terms of their appearance and light transmittance.

22.5 (a) Briefly describe the phenomenon of electronic polarization by electromagnetic radiation. (b) What are two consequences of electronic polarization in transparent materials?

22.6 (a) In ionic materials, how does the size of the component ions affect the extent of electronic polarization? (b) Which of the following oxide materials when added to fused silica (SiO_2) will increase its index of refraction: BaO, CaO, Na_2O, K_2O? Why? You may find Table 13.2 helpful.

22.7 (a) Briefly explain why metals are opaque to electromagnetic radiation having photon energies within the visible region of the spectrum. (b) Why are metals transparent to high-frequency x-ray and γ-ray radiation?

22.8 Can a material have an index of refraction less than unity? Why or why not?

22.9 Compute the velocity of light in diamond, which has a dielectric constant ϵ_r of 5.5 and a magnetic susceptibility χ_m of -2.17×10^{-5}.

22.10 The indices of refraction of fused silica and a soda–lime glass within the visible spectrum are 1.458 and 1.51, respectively. For each of these materials determine the fraction of the relative dielectric constant at 60 Hz that is due to electronic polarization, using the data of Table 19.4. Neglect any orientation polarization effects.

22.11 Using the data in Table 22.1, estimate the dielectric constants for silica glass (fused silica), soda–lime glass, polytetrafluoroethylene, polyethylene, and polystyrene, and compare these values with those cited in Table 19.4. Briefly explain any discrepancies.

22.12 Briefly describe the phenomenon of dispersion in a transparent medium.

22.13 It is desired that the reflectivity of light at normal incidence to the surface of a transparent medium be less than 4.5%. Which of the following materials in Table 22.1 are likely candidates: soda–lime glass, dense optical flint glass, corundum, quartz, polymethyl methacrylate, and polyethylene? Justify your selections.

22.14 Briefly explain how reflection losses of transparent materials are minimized by thin surface coatings.

22.15 Briefly describe the three absorption mechanisms in nonmetallic materials.

22.16 Will the elemental semiconductors silicon and germanium be transparent to visible light? Why or why not? Their band gap energies are given in Table 19.2.

22.17 Zinc telluride has a band gap of 2.26 eV. Over what range of wavelengths of visible light is it transparent?

22.18 Briefly explain why the magnitude of the absorption coefficient (β in Equation 22.18) depends on the radiation wavelength.

22.19 The fraction of nonreflected radiation that is transmitted through a 10-mm thickness of a transparent material is 0.90. If the thickness is increased to 20 mm, what fraction of light will be transmitted?

22.20 Derive Equation 22.19, starting from other expressions given in the chapter.

22.21 The transmissivity T of a transparent material 20 mm thick to normally incident light is 0.85. If the index of refraction of this material is 1.6, compute the thickness of material that will yield a transmissivity of 0.75. All reflection losses should be considered.

22.22 Briefly explain what determines the characteristic color of **(a)** a metal and **(b)** a transparent nonmetal.

22.23 Briefly explain why some transparent materials appear colored while others are colorless.

22.24 The index of refraction of quartz is anisotropic. Suppose that visible light is passing from one grain to another of different crystallographic orientation and at normal incidence to the grain boundary. Calculate the reflectivity at the boundary if the indices of refraction for the two grains are 1.544 and 1.553.

22.25 Briefly explain why amorphous polymers are transparent, while predominantly crystalline polymers appear opaque or, at best, translucent.

22.26 Some semicrystalline polymers such as polyethylene and polypropylene have a milky, semitranslucent appearance. However, after having been plastically deformed in tension, they become completely transparent. Briefly explain this change in optical characteristics.

22.27 **(a)** In your own words describe briefly the phenomenon of luminescence. **(b)** What is the distinction between fluorescence and phosphorescence?

22.28 Gallium arsenide (GaAs) and gallium phosphide (GaP) are compound semiconductors that have room-temperature band gap energies of 1.42 and 2.25 eV, respectively, and form solid solutions in all proportions. Furthermore, the band gap of the alloy increases approximately linearly with GaP additions (in mol%). Alloys of these two materials are used for light-emitting diodes wherein light is generated by conduction band-to-valence band electron transitions. Determine the composition of a GaAs–GaP alloy that will emit red light having a wavelength of 0.68 μm.

22.29 **(a)** In your own words, briefly describe the phenomenon of photoconductivity. **(b)** Would the semiconductor zinc sulfide, which has a band gap of 3.6 eV, be photoconductive when exposed to visible light radiation? Why or why not?

22.30 Briefly explain the operation of a photographic lightmeter.

22.31 In your own words, describe how a ruby laser operates.

22.32 Compute the difference in energy between metastable and ground electron states for the ruby laser.

THE INTERNATIONAL SYSTEM OF UNITS (SI)

Units in the *International System of Units* fall into two classifications: base and derived. Base units are fundamental and not reducible. Table A.1 lists the base units of interest in the discipline of materials science and engineering.

Derived units are expressed in terms of the base units, using mathematical signs for multiplication and division. For example, the SI units for density are kilogram per cubic meter (kg/m^3). For some derived units, special names and symbols exist; for example, N is used to denote the newton, the unit of force, which is equivalent to 1 kg-m/s^2. Table A.2 contains a number of the important derived units.

It is sometimes necessary, or convenient, to form names and symbols that are decimal multiples or submultiples of SI units. Only one prefix is used when a multiple of an SI unit is formed, which should be in the numerator. These prefixes and their approved symbols are given in Table A.3. Symbols for all units used in this book, SI or otherwise, are contained inside the front cover.

TABLE A.1 The SI Base Units

Quantity	Name	Symbol
Length	meter, metre	m
Mass	kilogram	kg
Time	second	s
Electric current	ampere	A
Thermodynamic temperature	kelvin	K
Amount of substance	mole	mol

TABLE A.2 Some of the SI Derived Units

Quantity	Name	Formula	Special Symbol[a]
Area	square meter	m^2	—
Volume	cubic meter	m^3	—
Velocity	meter per second	m/s	—
Density	kilogram per cubic meter	kg/m^3	—
Concentration	moles per cubic meter	mol/m^3	—
Force	newton	$kg\text{-}m/s^2$	N
Energy	joule	$kg\text{-}m^2/s^2$, N-m	J
Stress	pascal	$kg/m\text{-}s^2$, N/m^2	Pa
Strain	—	m/m	—
Power, radiant flux	watt	$kg\text{-}m^2/s^3$, J/s	W
Viscosity	pascal-second	$kg/m\text{-}s$	Pa-s
Frequency (of a periodic phenomenon)	hertz	s^{-1}	Hz
Electric charge	coulomb	A-s	C
Electric potential	volt	$kg\text{-}m^2/s^2\text{-}C$	V
Capacitance	farad	$s^2\text{-}C/kg\text{-}m^2$	F
Electric resistance	ohm	$kg\text{-}m^2/s\text{-}C^2$	Ω
Magnetic flux	weber	$kg\text{-}m^2/s\text{-}C$	Wb
Magnetic flux density	tesla	$kg/s\text{-}C$, Wb/m^2	(T)

[a] Special symbols in parentheses are approved in the SI but not used in this text; here, the name is used.

TABLE A.3 SI Multiple and Submultiple Prefixes

Factor by Which Multiplied	Prefix	Symbol
10^9	giga	G
10^6	mega	M
10^3	kilo	k
10^{-2}	centi[a]	c
10^{-3}	milli	m
10^{-6}	micro	μ
10^{-9}	nano	n
10^{-12}	pico	p

[a] Avoided when possible.

ELECTRONIC CONFIGURATIONS FOR THE ELEMENTS

Number of Electrons

Atomic Number	Element	Symbol	1s	2s	2p	3s	3p	3d	4s	4p	4d	4f	5s	5p	5d	5f	6s	6p	6d	6f	7s
1	Hydrogen	H	1																		
2	Helium	He	2																		
3	Lithium	Li	2	1																	
4	Beryllium	Be	2	2																	
5	Boron	B	2	2	1																
6	Carbon	C	2	2	2																
7	Nitrogen	N	2	2	3																
8	Oxygen	O	2	2	4																
9	Fluorine	F	2	2	5																
10	Neon	Ne	2	2	6																
11	Sodium	Na	2	2	6	1															
12	Magnesium	Mg	2	2	6	2															
13	Aluminum	Al	2	2	6	2	1														
14	Silicon	Si	2	2	6	2	2														
15	Phosphorus	P	2	2	6	2	3														
16	Sulfur	S	2	2	6	2	4														
17	Chlorine	Cl	2	2	6	2	5														
18	Argon	Ar	2	2	6	2	6														
19	Potassium	K	2	2	6	2	6	—	1												
20	Calcium	Ca	2	2	6	2	6	—	2												
21	Scandium	Sc	2	2	6	2	6	1	2												
22	Titanium	Ti	2	2	6	2	6	2	2												
23	Vanadium	V	2	2	6	2	6	3	2												
24	Chromium	Cr	2	2	6	2	6	5	1												
25	Manganese	Mn	2	2	6	2	6	5	2												
26	Iron	Fe	2	2	6	2	6	6	2												
27	Cobalt	Co	2	2	6	2	6	7	2												
28	Nickel	Ni	2	2	6	2	6	8	2												
29	Copper	Cu	2	2	6	2	6	10	1												
30	Zinc	Zn	2	2	6	2	6	10	2												
31	Gallium	Ga	2	2	6	2	6	10	2	1											
32	Germanium	Ge	2	2	6	2	6	10	2	2											
33	Arsenic	As	2	2	6	2	6	10	2	3											

Electron configuration table (subshell occupancies)

No.	Symbol	Name	1s	2s	2p	3s	3p	3d	4s	4p	4d	4f	5s	5p	5d	6s
34	Se	Selenium	2	2	6	2	6	10	2	4						
35	Br	Bromine	2	2	6	2	6	10	2	5						
36	Kr	Krypton	2	2	6	2	6	10	2	6						
37	Rb	Rubidium	2	2	6	2	6	10	2	6			1			
38	Sr	Strontium	2	2	6	2	6	10	2	6			2			
39	Y	Yttrium	2	2	6	2	6	10	2	6	1		2			
40	Zr	Zirconium	2	2	6	2	6	10	2	6	2		2			
41	Nb	Niobium	2	2	6	2	6	10	2	6	4		1			
42	Mo	Molybdenum	2	2	6	2	6	10	2	6	5		1			
43	Tc	Technetium	2	2	6	2	6	10	2	6	5		2			
44	Ru	Ruthenium	2	2	6	2	6	10	2	6	7		1			
45	Rh	Rhodium	2	2	6	2	6	10	2	6	8		1			
46	Pd	Palladium	2	2	6	2	6	10	2	6	10		0			
47	Ag	Silver	2	2	6	2	6	10	2	6	10		1			
48	Cd	Cadmium	2	2	6	2	6	10	2	6	10		2			
49	In	Indium	2	2	6	2	6	10	2	6	10		2	1		
50	Sn	Tin	2	2	6	2	6	10	2	6	10		2	2		
51	Sb	Antimony	2	2	6	2	6	10	2	6	10		2	3		
52	Te	Tellurium	2	2	6	2	6	10	2	6	10		2	4		
53	I	Iodine	2	2	6	2	6	10	2	6	10		2	5		
54	Xe	Xenon	2	2	6	2	6	10	2	6	10		2	6		
55	Cs	Cesium	2	2	6	2	6	10	2	6	10		2	6		1
56	Ba	Barium	2	2	6	2	6	10	2	6	10		2	6		2
57	La	Lanthanum	2	2	6	2	6	10	2	6	10		2	6	1	2
58	Ce	Cerium	2	2	6	2	6	10	2	6	10	2	2	6		2
59	Pr	Praesodymium	2	2	6	2	6	10	2	6	10	3	2	6		2
60	Nd	Neodymium	2	2	6	2	6	10	2	6	10	4	2	6		2
61	Pm	Promethium	2	2	6	2	6	10	2	6	10	5	2	6		2
62	Sm	Samarium	2	2	6	2	6	10	2	6	10	6	2	6		2
63	Eu	Europium	2	2	6	2	6	10	2	6	10	7	2	6		2
64	Gd	Gadolinium	2	2	6	2	6	10	2	6	10	7	2	6	1	2
65	Tb	Terbium	2	2	6	2	6	10	2	6	10	9	2	6		2
66	Dy	Dysprosium	2	2	6	2	6	10	2	6	10	10	2	6		2
67	Ho	Holmium	2	2	6	2	6	10	2	6	10	11	2	6		2
68	Er	Erbium	2	2	6	2	6	10	2	6	10	12	2	6		2
69	Tm	Thulium	2	2	6	2	6	10	2	6	10	13	2	6		2

Number of Electrons

Atomic Number	Element	Symbol	1s	2s	2p	3s	3p	3d	4s	4p	4d	4f	5s	5p	5d	5f	6s	6p	6d	6f	7s
70	Ytterbium	Yb	2	2	6	2	6	10	2	6	10	14	2	6	—	—	2				
71	Lutetium	Lu	2	2	6	2	6	10	2	6	10	14	2	6	1	—	2				
72	Hafnium	Hf	2	2	6	2	6	10	2	6	10	14	2	6	2	—	2				
73	Tantalum	Ta	2	2	6	2	6	10	2	6	10	14	2	6	3	—	2				
74	Tungsten	W	2	2	6	2	6	10	2	6	10	14	2	6	4	—	2				
75	Rhenium	Re	2	2	6	2	6	10	2	6	10	14	2	6	5	—	2				
76	Osmium	Os	2	2	6	2	6	10	2	6	10	14	2	6	6	—	2				
77	Iridium	Ir	2	2	6	2	6	10	2	6	10	14	2	6	7	—	2				
78	Platinum	Pt	2	2	6	2	6	10	2	6	10	14	2	6	9	—	1				
79	Gold	Au	2	2	6	2	6	10	2	6	10	14	2	6	10	—	1				
80	Mercury	Hg	2	2	6	2	6	10	2	6	10	14	2	6	10	—	2				
81	Thallium	Tl	2	2	6	2	6	10	2	6	10	14	2	6	10	—	2	1			
82	Lead	Pb	2	2	6	2	6	10	2	6	10	14	2	6	10	—	2	2			
83	Bismuth	Bi	2	2	6	2	6	10	2	6	10	14	2	6	10	—	2	3			
84	Polonium	Po	2	2	6	2	6	10	2	6	10	14	2	6	10	—	2	4			
85	Astatine	At	2	2	6	2	6	10	2	6	10	14	2	6	10	—	2	5			
86	Radon	Rn	2	2	6	2	6	10	2	6	10	14	2	6	10	—	2	6			
87	Francium	Fr	2	2	6	2	6	10	2	6	10	14	2	6	10	—	2	6	—	—	1
88	Radium	Ra	2	2	6	2	6	10	2	6	10	14	2	6	10	—	2	6	—	—	2
89	Actinium	Ac	2	2	6	2	6	10	2	6	10	14	2	6	10	—	2	6	1	—	2
90	Thorium	Th	2	2	6	2	6	10	2	6	10	14	2	6	10	—	2	6	2	—	2
91	Protactinium	Pa	2	2	6	2	6	10	2	6	10	14	2	6	10	2	2	6	1	—	2
92	Uranium	U	2	2	6	2	6	10	2	6	10	14	2	6	10	3	2	6	1	—	2
93	Neptunium	Np	2	2	6	2	6	10	2	6	10	14	2	6	10	4	2	6	1	—	2
94	Plutonium	Pu	2	2	6	2	6	10	2	6	10	14	2	6	10	6	2	6	—	—	2
95	Americium	Am	2	2	6	2	6	10	2	6	10	14	2	6	10	7	2	6	—	—	2
96	Curium	Cm	2	2	6	2	6	10	2	6	10	14	2	6	10	7	2	6	1	—	2
97	Berkelium	Bk	2	2	6	2	6	10	2	6	10	14	2	6	10	9	2	6	—	—	2
98	Californium	Cf	2	2	6	2	6	10	2	6	10	14	2	6	10	10	2	6	—	—	2
99	Einsteinium	Es	2	2	6	2	6	10	2	6	10	14	2	6	10	11	2	6	—	—	2
100	Fermium	Fm	2	2	6	2	6	10	2	6	10	14	2	6	10	12	2	6	—	—	2
101	Mendelevium	Md	2	2	6	2	6	10	2	6	10	14	2	6	10	13	2	6	—	—	2
102	Nobelium	No	2	2	6	2	6	10	2	6	10	14	2	6	10	14	2	6	—	—	2
103	Lawrencium	Lw	2	2	6	2	6	10	2	6	10	14	2	6	10	14	2	6	1	—	2

PROPERTIES OF SELECTED ENGINEERING MATERIALS

This appendix is meant to be a convenient compilation of frequently used properties of a number of common engineering materials. Data are tabulated according to material type for ferrous and nonferrous alloys, ceramic materials, thermoplastic and thermosetting polymers, and elastomeric materials.

TABLE C.1 Room-Temperature Properties of Selected Ferrous Alloys in an Annealed State

Material	Density (g/cm³)	Modulus of Elasticity [psi × 10⁶ (GPa)]	Yield Strength [ksi (MPa)]	Tensile Strength [ksi (MPa)]	Ductility (%EL in 2 in.)	Poisson's Ratio	Electrical Conductivity [(Ω-m)⁻¹ × 10⁶]	Thermal Conductivity (W/m-K)	Coefficient of Thermal Expansion [(°C)⁻¹ × 10⁻⁶]	Melting Temperature or Range (°C)
Iron	7.87	30 (207)	19 (130)	38 (260)	45	0.29	10	80	11.8	1538
Gray cast iron	7.15	Variable	—	18 (125)	—	Variable	~1	46	10.8	a
Nodular cast iron	7.12	24 (165)	40 (275)	60 (415)	18	0.28	~1.5	33	11.8	a
Malleable cast iron	7.20–7.45	25 (172)	32 (220)	50 (345)	10	0.26	0.25–0.35	51	11.9	a
Low-carbon steel (1020)	7.86	30 (207)	43 (295)	57 (395)	37	0.30	5.9	52	11.7	1495–1520
Medium-carbon steel (1040)	7.85	30 (207)	51 (350)	75 (520)	30	0.30	5.8	52	11.3	1495–1505
High-carbon steel (1080)	7.84	30 (207)	55 (380)	89 (615)	25	0.30	5.6	48	11.0	1385–1475
Stainless steels										
Ferritic, type 446	7.50	29 (200)	50 (345)	80 (552)	20	0.30	1.5	21	10.4	1425–1510
Austenitic, type 316	8.00	28 (193)	30 (207)	80 (552)	60	0.30	1.4	16	16.0	1370–1400
Martensitic, type 410	7.80	29 (200)	40 (275)	70 (483)	30	0.30	1.8	25	9.9	1480–1530

a Dependent on composition.

TABLE C.2 Room-Temperature Properties of Nonferrous Alloys in an Annealed State

Material	Density (g/cm^3)	Modulus of Elasticity [psi × 10^6 (GPa)]	Yield Strength [ksi (MPa)]	Tensile Strength [ksi (MPa)]	Ductility (%EL in 2 in.)	Poisson's Ratio	Electrical Conductivity [(Ω-m)$^{-1}$ × 10^6]	Thermal Conductivity (W/m-K)	Coefficient of Thermal Expansion [(°C)$^{-1}$ × 10^{-6}]	Melting Temperature or Range (°C)
Aluminum (>99.5)	2.71	10 (69)	2.5 (17)	8 (55)	25	0.33	36	231	23.6	646–657
Aluminum alloy 2014	2.80	10.5 (72)	14 (97)	27 (186)	18	0.33	29	192	22.5	507–638
Copper (99.95)	8.94	16 (110)	10 (69)	32 (220)	45	0.35	58	398	16.5	1085
Brass (70Cu–30Zn)	8.53	16 (110)	11 (75)	44 (303)	68	0.35	16	120	20.0	915–955
Bronze (92Cu–8Sn)	8.80	16 (110)	22 (152)	55 (380)	70	0.35	7.5	62	18.2	880–1020
Magnesium (>99)	1.74	6.5 (45)	6 (41)	24 (165)	14	0.29	17.5	122	27.0	650
Molybdenum (>99)	10.22	47 (324)	82 (565)	95 (655)	35	—	19.2	142	4.9	2610
Nickel (>99)	8.90	30 (207)	20 (138)	70 (483)	40	0.31	11.8	80	13.3	1454
Silver (>99)	10.49	11 (76)	8 (55)	18 (125)	48	0.37	56	418	19.0	961
Titanium (>99)	4.51	15.5 (107)	35 (240)	48 (330)	30	0.34	2.0	17	9.0	1670

TABLE C.3 Properties of Selected Ceramic Materials. Data are for Fully Dense Materials and at Room Temperature Unless Noted Otherwise

Material	Density (g/cm³)	Modulus of Elasticity [psi × 10⁶ (GPa)]	Poisson's Ratio	Approximate Hardness (Knoop)	Modulus of Rupture [ksi (MPa)]	Electrical Resistivity (Ω-m)	Thermal Conductivity (W/m-K)	Coefficient of Thermal Expansion [(°C)⁻¹ × 10⁶]	Melting Temperature or Range [°C (°F)]
Alumina (Al_2O_3)	3.97	57 (393)	0.27	2100	40–80 (275–550)	$>10^{12}$	30	8.8[a]	2050 (3720)
Magnesia (MgO)	3.58	30 (207)	0.36	370	15[b] (105)	$>10^{12}$	48	13.5[a]	2850 (5160)
Spinel ($MgAl_2O_4$)	3.55	36 (284)	—	1600	12–32[b] (83–220)	—	15.0[a]	7.6[a]	2135 (3875)
Zirconia[c] (ZrO_2)	5.56	22 (152)	0.32	1200	20–35[b] (138–240)	—	2.0	10.0[a]	2500–2600 (4530–4710)
Fused silica (SiO_2)	2.2	11 (75)	0.16	500	16 (110)	$>10^{18}$	1.3	0.5[a]	—
Soda–lime glass	2.5	10 (69)	0.23	550	10 (69)	$>10^{10}$	1.7	9.0[d]	—
Borosilicate glass	2.23	9 (62)	0.20	—	10 (69)	$\sim10^{13}$	1.4	3.3[d]	—
Silicon carbide (SiC)	3.22	60 (414)	0.19	2500	65–75[b] (450–520)	$\sim10^{-2}$	90	4.7	2300–2500[e] (4170–4530)
Silicon nitride (Si_3N_4)	3.44	44 (304)	0.24	2200	60–80[b] (414–580)	$>10^{12}$	16–33[a]	3.6[a]	~1900[e] (3450)
Titanium carbide (TiC)	4.92	67 (462)	—	2600	40–65[b] (275–450)	$\sim10^{-6}$	17.2	7.4	3160 (5720)

[a] Mean value taken over the temperature range 0–1000°C.
[b] Sintered and containing approximately 5% porosity.
[c] Stabilized with CaO.
[d] Mean value taken over the temperature range 0–300°C.
[e] Sublimes.

TABLE C.4 Properties of Selected Thermoplastic and Thermosetting Polymers

Repeat Unit	State	Density (g/cm³)	Tensile Modulus [ksi (GPa)]	Tensile Strength at Break [ksi (MPa)]	Elongation at Break (%)	T_g (°C)	T_m (°C)	Thermal Conductivity (W/m-K)	Electrical Resistivity (Ω-m)	Coefficient of Thermal Expansion [(°C)$^{-1}$ × 10^{-6}]
Thermoplastics										
Polyethylene $[-CH_2-CH_2-]_n$	High density, 70–80% crystalline	0.952–0.965	155–158 (1.07–1.09)	3.2–4.5 (22–31)	10–1200	−90	130–137	0.48	10^{13}–10^{17}	60–110
	Low density, 40–50% crystalline	0.917–0.932	25–41 (0.17–0.28)	1.2–4.5 (8.3–31.0)	100–650	−110	98–115	0.33	10^{13}–10^{17}	100–220
Polytetrafluoroethylene $[-CF_2-CF_2-]_n$	50–70% crystalline	2.14–2.20	58–80 (0.40–0.55)	2.0–5.0 (14–34)	200–400	−90	327	0.25	$>10^{16}$	70–120
Polyvinyl chloride	Highly amorphous	1.30–1.58	350–600 (2.4–4.1)	6.0–7.5 (41–52)	40–80	75–105	212	0.18	—	50–100
Polypropylene	50–60% crystalline	0.90–0.91	165–225 (1.14–1.55)	4.5–6.0 (31–41)	100–600	−20	168–175	0.12	$>10^{15}$	80–100
Polystyrene	Amorphous	1.04–1.05	330–475 (2.28–3.28)	5.2–7.5 (36–52)	1.2–2.5	74–105	—	0.13	$>10^{14}$	50–83
Polymethyl methacrylate	Amorphous	1.17–1.20	325–470 (2.24–3.24)	7–11 (48–76)	2–10	85–105	—	0.21	$>10^{12}$	50–90

(Continued)

TABLE C.4 (*continued*)

Repeat Unit	State	Density (g/cm³)	Tensile Modulus [ksi (GPa)]	Tensile Strength at Break [ksi (MPa)]	Elongation at Break (%)	T_g (°C)	T_m (°C)	Thermal Conductivity (W/m-K)	Electrical Resistivity (Ω-m)	Coefficient of Thermal Expansion [$(°C)^{-1} \times 10^{-6}$]
Nylon 6,6 Poly(hexamethylene adipamide) $\left[-N-(CH_2)_6-N-C-(CH_2)_4-C- \right]_n$ H H O O	30–40% crystalline	1.13–1.15	230–550 (1.58–3.79)	11–13.7 (76–94)	15–300	57	255–265	0.24	10^{12}–10^{13}	80
Polyethylene terephthalate (PET, a polyester) [structure]	0–30% crystalline	1.29–1.40	400–600 (2.76–4.14)	7.0–10.5 (48–72)	30–300	73–80	245–265	0.14	10^{12}	65
Polycarbonate (Polybisphenol-A carbonate) [structure]	Amorphous	1.20	345 (2.38)	9.5 (65.5)	110	150	—	0.20	10^{15}	68
Thermosets										
Epoxy	Complex network, amorphous	1.11–1.40	350 (2.41)	4.0–13.0 (28–90)	3–6	—	—	0.19	$\sim 10^{14}$	45–65
Phenolic	Complex network, amorphous	1.24–1.32	400–700 (2.76–4.83)	5–9 (34–62)	1.5–2.0	—	—	0.15	10^{9}–10^{10}	68
Polyester	Complex network, amorphous	1.04–1.46	300–640 (2.07–4.41)	6–13 (41–90)	<2	—	—	0.19	10^{11}	55–100

Source: Adapted from *Modern Plastics Encyclopedia* 1988. Copyright 1987, McGraw-Hill, Inc. Reprinted with permission.

TABLE C.5 Properties of Selected Elastomers

Name/Repeat Unit	Specific Gravity	Tensile Strength [ksi (MPa)]	Maximum Elongation (%)	Modulus, at 100% Elongation [psi (MPa)]	Minimum Service Temperature [°C (°F)]	Maximum Service Temperature [°C (°F)]	Abrasion Resistance	Tear Resistance	Oxidation Resistance
Natural polyisoprene (natural rubber, NR)	0.92–1.037	3.5–4.6 (24–32)	500–760	480–850 (3.3–5.9)	−60 (−75)	120 (250)	Excellent	Excellent	Good
Styrene-butadiene (SBR, GRS)	0.94	1.8–3.0 (12–21)	450–500	300–1500 (2.1–10.3)	−60 (−75)	120 (250)	Excellent	Fair	Good
Acrylonitrile-butadiene (Nitrile, Buna A, NBR)	0.98	1.0–3.5 (7–24)	400–600	490 (3.4)	−50 (−60)	150 (300)	Excellent	Good	Fair-Good
Chloroprene (Neoprene, CR)	1.23–1.25	0.5–3.5 (3.5–24)	100–800	100–3000 (0.7–20)	−50 (−60)	105 (225)	Excellent	Good	Very good
Polybutadiene (BR)	0.91	2.0–2.5 (14–17)	450	300–1500 (2.1–10.3)	−100 (−150)	90 (200)	Excellent	Good	Good
Polyurethane	1.02–1.25	0.8–8.0 (5.5–55)	250–800	25–5000 (0.17–34.5)	−55 (−65)	120 (250)	Excellent	Outstanding	Excellent
Polydimethylsiloxane (Silicone)	1.1–1.6	1.5 (10)	100–800	—	−115 (−175)	315 (600)	Poor	Fair	Excellent

Source: Adapted from *Materials Engineering*, a Penton publication.

GLOSSARY

Abrasive. A hard and wear-resistant material (commonly a ceramic) that is used to wear, grind, or cut away other material.

Acceptor level. For a semiconductor or insulator, an energy level lying within yet near the bottom of the energy band gap, which may accept electrons from the valence band, leaving behind holes. The level is normally introduced by an impurity atom.

Activation energy (Q). The energy required to initiate a reaction, such as diffusion.

Activation polarization. The condition wherein the rate of an electrochemical reaction is controlled by the one slowest step in a sequence of steps that occur in series.

Addition (or chain reaction) polymerization. The process by which bifunctional monomer units are attached one at a time, in chainlike fashion, to form a linear polymer macromolecule.

Adhesive. A substance that bonds together the surfaces of two other materials (termed adherends).

Age hardening. See **Precipitation hardening.**

Allotropy. The possibility of existence of two or more different crystal structures for a substance (generally an elemental solid).

Alloy. A metallic substance that is composed of two or more elements.

Alloy steel. A ferrous (or iron-based) alloy that contains appreciable concentrations of alloying elements (other than C and residual amounts of Mn, Si, S, and P). These alloying elements are usually added to improve mechanical and corrosion resistance properties.

Alternating copolymer. A copolymer in which two different mer units alternate positions along the molecular chain.

Amorphous. Having a noncrystalline structure.

Anelastic deformation. Time-dependent elastic (nonpermanent) deformation.

Anion. A negatively charged, nonmetallic ion.

Anisotropic. Exhibiting different values of a property in different crystallographic directions.

Annealing. A generic term used to denote a heat treatment wherein the microstructure and, consequently, the properties of a material are altered. "Annealing" frequently refers to a heat treatment whereby a previously cold-worked metal is softened by allowing it to recrystallize.

Annealing point (glass). That temperature at which residual stresses in a glass are eliminated within about 15 min; this corresponds to a glass viscosity of about 10^{12} Pa-s (10^{13} P).

Anode. The electrode in an electrochemical cell or galvanic couple that experiences oxidation, or gives up electrons.

Antiferromagnetism. A phenomenon observed in some materials (e.g., MnO); complete magnetic moment cancellation occurs as a result of antiparallel coupling of adjacent atoms or ions. The macroscopic solid possesses no net magnetic moment.

Artificial aging. For precipitation hardening, aging above room temperature.

Atactic. A type of polymer chain configuration wherein side groups are randomly positioned on one side of the chain or the other.

Athermal transformation. A reaction that is not thermally activated, and usually diffusionless, as with the martensitic transformation. Normally, the tranformation takes place with great speed (i.e., is independent of time), and the extent of reaction depends on temperature.

Atom percent (at%). Concentration specification on the basis of the number of moles (or atoms) of a particular element relative to the total number of moles (or atoms) of all elements within an alloy.

Atomic mass unit (amu). A measure of atomic mass; one twelfth of the mass of an atom of C^{12}.

Atomic number (Z). For a chemical element, the number of protons within the atomic nucleus.

Atomic packing factor (APF). The fraction of the volume of a unit cell that is occupied by "hard sphere" atoms or ions.

Atomic weight (A). The weighted average of the atomic masses of an atom's naturally occurring isotopes. It may be expressed in terms of atomic mass units (on an atomic basis), or the mass per mole of atoms.

Austenite. Face-centered cubic iron; also iron and steel alloys that have the FCC crystal structure.

Austenitizing. Forming austenite by heating a ferrous alloy above its upper critical temperature—to within the austenite phase region from the phase diagram.

Bainite. An austenitic transformation product found in some steels and cast irons. It forms at temperatures between those at which pearlite and martensite transformations occur. The microstructure consists of α-ferrite and a fine dispersion of cementite.

Bifunctional. Designating monomer units that have two active bonding positions.

Block copolymer. A linear copolymer in which identical mer units are clustered in blocks along the molecular chain.

Body-centered cubic (BCC). A common crystal structure found in some elemental metals. Within the cubic unit cell, atoms are located at corner and cell center positions.

Bohr atomic model. An early atomic model, in which electrons are assumed to revolve around the nucleus in discrete orbitals.

Bohr magneton (μ_B). The most fundamental magnetic moment, of magnitude 9.27×10^{-24} A-m^2.

Boltzmann's constant (k). A thermal energy constant having the value of 1.38×10^{-23} J/atom-K $(8.62 \times 10^{-5}$ eV/atom-K). See also **Gas constant.**

Bonding energy. The energy required to separate two atoms that are chemically bonded to each other. It may be expressed on a per-atom basis, or per mole of atoms.

Bragg's law. A relationship (Equation 3.9) which stipulates the condition for diffraction by a set of crystallographic planes.

Branched polymer. A polymer having a molecular structure of secondary chains that extend from the primary main chains.

Brass. A copper-rich copper–zinc alloy.

Brazing. A metal joining technique that uses a molten filler metal alloy having a melting temperature greater than about 425°C (800°F).

Brittle fracture. Fracture that occurs by rapid crack propagation and without appreciable macroscopic deformation.

Bronze. A copper-rich copper–tin alloy; aluminum, silicon, and nickel bronzes are also possible.

Burgers vector (b). A vector that denotes the magnitude and direction of lattice distortion associated with a dislocation.

Calcination. A high-temperature reaction whereby one solid material dissociates to form a gas and another solid. It is one step in the production of cement.

Capacitance (C). The charge-storing ability of a capacitor, defined as the magnitude of charge stored on either plate divided by the applied voltage.

Carburizing. The process by which the surface carbon concentration of a ferrous alloy is increased by diffusion from the surrounding environment.

Case hardening. Hardening of the outer surface (or "case") of a steel component by a carburizing or nitriding process; used to improve wear and fatigue resistance.

Cast iron. Generically, a ferrous alloy, the carbon content of which is greater than the maximum solubility in austenite at the eutectic temperature. Most commercial cast irons contain between 3.0 and 4.5 wt% C, and between 1 and 3 wt% Si.

Cathode. The electrode in an electrochemical cell or galvanic couple at which a reduction reaction occurs; thus the electrode that receives electrons from an external circuit.

Cathodic protection. A means of corrosion prevention whereby electrons are supplied to the structure to be protected from an external source such as another more reactive metal or a dc power supply.

Cation. A positively charged metallic ion.

Cement. A substance (often a ceramic) that by chemical reaction binds particulate aggregates into a cohesive structure. With hydraulic cements the chemical reaction is one of hydration, involving water.

Cementite. Iron carbide (Fe_3C).

Ceramic. A compound of metallic and nonmetallic elements, for which the interatomic bonding is predominantly ionic.

Cermet. A composite material consisting of a combination of ceramic and metallic materials. The most common cermets are the cemented carbides, composed of an extremely hard ceramic (e.g., WC, TiC), bonded together by a ductile metal such as cobalt or nickel.

Chain-folded model. For crystalline polymers, a model that describes the structure of platelet crystallites. Molecular alignment is accomplished by chain folding that occurs at the crystallite faces.

Charpy test. One of two tests (see also **Izod test**) that may be used to measure the impact energy or notch toughness of a standard notched specimen. An impact blow is imparted to the specimen by means of a weighted pendulum.

Cis. For polymers, a prefix denoting a type of molecular structure. For some unsaturated carbon chain atoms within a mer unit, a side atom or group may be situated on one side of the chain or directly opposite at a 180° rotation position. In a cis structure, two such side groups within the same mer reside on the same chain side (e.g., *cis*-isoprene).

Coarse pearlite. Pearlite for which the alternating ferrite and cementite layers are relatively thick.

Coercivity (or coercive field, H_c). The applied magnetic field necessary to reduce to zero the magnetic flux density of a magnetized ferromagnetic or ferrimagnetic material.

Cold working. The plastic deformation of a metal at a temperature below that at which it recrystallizes.

Colorant. An additive that imparts a specific color to a polymer.

Component. A chemical constituent (element or compound) of an alloy, which may be used to specify its composition.

Composition. The concentrations of all components or elements that are found in an alloy.

Concentration (C_i). The relative content of a particular element or constituent (i) within an alloy, usually expressed in weight percent or atom percent.

Concentration gradient (dC/dx). The slope of the concentration profile at a specific position.

Concentration polarization. The condition wherein the rate of an electrochemical reaction is limited by the rate of diffusion in the solution.

Concentration profile. The curve that results when the concentration of a chemical species is plotted versus position in a material.

Concrete. A composite material consisting of aggregate particles bound together in a solid body by a cement.

Condensation (or step reaction) polymerization. The formation of polymer macromolecules by an intermolecular reaction involving at least two monomer species, usually with the production of a by-product of low molecular weight, such as water.

Conduction band. The lowest lying electron energy band that is not completely filled with electrons. Conduction electrons are those that have been excited to states within this band.

Congruent transformation. A transformation of one phase to another of the same composition.

Continuous cooling transformation (CCT) diagram. A plot of temperature versus the logarithm of time for a steel alloy of definite composition. Used to indicate when transformations occur as the initially austenitized material is continuously cooled at a specified rate; in addition, the final microstructure and mechanical characteristics may be predicted.

Coordination number. The number of atomic or ionic nearest neighbors.

Copolymer. A polymer that consists of two or more dissimilar mer units in combination along its molecular chains.

Corrosion. Deteriorative loss of a metal as a result of dissolution environmental reactions.

Corrosion fatigue. A type of failure that results from the simultaneous action of a cyclic stress and chemical attack.

Corrosion penetration rate (CPR). Thickness loss of material per unit of time as a result of corrosion; usually expressed in terms of mils per year or millimeters per year.

Coulombic force. A force between charged particles such as ions; the force is attractive when the particles are of opposite charge.

Covalent bond. A primary interatomic bond that is formed by the sharing of electrons between neighboring atoms.

Creep. The time-dependent permanent deformation that occurs under stress; for most materials it is important only at elevated temperatures.

Crevice corrosion. A form of corrosion that occurs within narrow crevices and under deposits of dirt or corrosion products (i.e., in regions of localized depletion of oxygen in the solution).

Critical resolved shear stress (τ_{crss}). That shear stress, resolved within a slip plane and direction, which is required to initiate slip.

Crosslinked polymer. A polymer in which adjacent linear molecular chains are joined at various positions by covalent bonds.

Crystalline. The state of a solid material characterized by a periodic and repeating three-dimensional array of atoms, ions, or molecules.

Crystallinity. For polymers, the state wherein a periodic and repeating atomic arrangement is achieved by molecular chain alignment.

Crystallite. A region within a crystalline polymer in which all the molecular chains are ordered and aligned.

Crystal structure. For crystalline materials, the manner in which atoms or ions are arrayed in space. It is defined in terms of the unit cell geometry and the atom positions within the unit cell.

Crystal system. A scheme by which crystal structures are classified according to unit cell geometry. This geometry is specified in terms of the relationships between edge lengths and interaxial angles. There are seven different crystal systems.

Curie temperature (T_c). That temperature above which a ferromagnetic or ferrimagnetic material becomes paramagnetic.

Defect structure. Relating to the kinds and concentrations of vacancies and interstitials in a ceramic compound.

Degradation. A term used to denote the deteriorative processes that occur with polymeric materials. These processes include swelling, dissolution, and chain scission.

Degree of polymerization. The average number of mer units per polymer chain molecule.

Devitrification. The process in which a glass (noncrystalline or vitreous solid) transforms to a crystalline solid.

Diamagnetism. A weak form of induced or nonpermanent magnetism for which the magnetic susceptibility is negative.

Dielectric. Any material that is electrically insulating.

Dielectric constant (ϵ_r). The ratio of the permittivity of a medium to that of a vacuum. Often called the relative dielectric constant or relative permittivity.

Dielectric displacement (D). The magnitude of charge per unit area of capacitor plate.

Dielectric (breakdown) strength. The magnitude of an electric field necessary to cause significant current passage through a dielectric material.

Diffraction (x-ray). Constructive interference of x-ray beams that are scattered by atoms of a crystal.

Diffusion. Mass transport by atomic motion.

Diffusion coefficient (D). The constant of proportionality between the diffusion flux and the concentration gradient in Fick's first law. Its magnitude is indicative of the rate of atomic diffusion.

Diffusion flux (J). The quantity of mass diffusing through and perpendicular to a unit cross-sectional area of material per unit time.

Dipole (electric). A pair of equal yet opposite electrical charges that are separated by a small distance.

Dislocation. A linear crystalline defect around which there is atomic misalignment. Plastic deformation corresponds to the motion of dislocations in response to an applied shear stress. Edge, screw, and mixed dislocations are possible.

Dislocation density. The total dislocation length per unit volume of material; alternately, the number of dislocations that intersect a unit area of a random surface section.

Dislocation line. The line that extends along the end of the extra half-plane of atoms for an edge dislocation, and along the center of the spiral of a screw dislocation.

Dispersed phase. For composites and some two-phase alloys, the discontinuous phase that is surrounded by the matrix phase.

Dispersion strengthening. A means of strengthening materials wherein very small particles (usually less than 0.1 μm) of a hard yet inert phase are uniformly dispersed within a load-bearing matrix phase.

Domain. A volume region of a ferromagnetic or ferrimagnetic material in which all atomic or ionic magnetic moments are aligned in the same direction.

Donor level. For a semiconductor or insulator, an energy level lying within yet near the top of the energy band gap, and from which electrons may be excited into the conduction band. It is normally introduced by an impurity atom.

Doping. The intentional alloying of semiconducting materials with controlled concentrations of donor or acceptor impurities.

Drawing (metals). A forming technique used to fabricate metal wire and tubing. Deformation is accomplished by pulling the material through a die by means of a tensile force applied on the exit side.

Drawing (polymers). A deformation technique wherein polymer fibers are strengthened by elongation.

Driving force. The impetus behind a reaction, such as diffusion, grain growth, or a phase transformation. Usually attendant to the reaction is a reduction in some type of energy (e.g., free energy).

Ductile fracture. A mode of fracture that is attended by extensive gross plastic deformation.

Ductile iron. A cast iron that is alloyed with silicon and a small concentration of magnesium and/or cerium and in which the free graphite exists in nodular form. Sometimes called nodular iron.

Ductile-to-brittle transition. The transition from ductile to brittle behavior with a decrease in temperature exhibited by BCC alloys; the temperature range over which the transition occurs is determined by Charpy and Izod impact tests.

Ductility. A measure of a material's ability to undergo appreciable plastic deformation before fracture; it may be expressed as percent elongation (%EL) or percent area reduction (%AR) from a tensile test.

Edge dislocation. A linear crystalline defect associated with the lattice distortion produced in the vicinity of the end of an extra half-plane of atoms within a crystal. The Burgers vector is perpendicular to the dislocation line.

Elastic deformation. Deformation that is nonpermanent, that is, totally recovered upon release of an applied stress.

Elastomer. A polymeric material that may experience large and reversible elastic deformations.

Electrical conductivity (σ). The proportionality constant between current density and applied electric field; also a measure of the ease with which a material is capable of conducting an electric current.

Electric field ($\mathscr{E}$). The gradient of voltage.

Electroluminescence. The emission of visible light by a p–n junction across which a forward-biased voltage is applied.

Electrolyte. A solution through which an electric current may be carried by the motion of ions.

Electromotive force (emf) series. A ranking of metallic elements according to their standard electrochemical cell potentials.

Electron configuration. For an atom, the manner in which possible electron states are filled with electrons.

Electron energy band. A series of electron energy states that are very closely spaced with respect to energy.

Electron state (level). One of a set of discrete, quantized energies that are allowed for electrons. In the atomic case each state is specified by four quantum numbers.

Electronegative. For an atom, having a tendency to accept valence electrons. Also, a term used to describe nonmetallic elements.

Electroneutrality. The state of having exactly the same numbers of positive and negative electrical charges (ionic and electronic), that is, of being electrically neutral.

Electropositive. For an atom, having a tendency to release valence electrons. Also, a term used to describe metallic elements.

Electron volt (eV). A convenient unit of energy for atomic and subatomic systems. It is equivalent to the energy acquired by an electron when it falls through an electric potential of 1 volt.

Endurance limit. See **Fatigue limit.**

Energy band gap (E_g). For semiconductors and insulators, the energies that lie between the valence and conduction bands; for intrinsic materials, electrons are forbidden to have energies within this range.

Equilibrium (phase). The state of a system wherein the phase characteristics remain constant over indefinite time periods. At equilibrium the free energy is a minimum.

Erosion-corrosion. A form of corrosion that arises from the combined action of chemical attack and mechanical wear.

Eutectic phase. One of the two phases found in the eutectic structure.

Eutectic reaction. A reaction wherein, upon cooling, a liquid phase transforms isothermally and reversibly into two intimately mixed solid phases.

Eutectic structure. A two-phase microstructure resulting from the solidification of a liquid having the eutectic composition; the phases exist as lamellae which alternate with one another.

Eutectoid reaction. A reaction wherein, upon cooling, one solid phase transforms isothermally and reversibly into two new solid phases that are intimately mixed.

Excited state. An electron energy state, not normally occupied, to which an electron may be promoted (from a lower energy state) by the absorption of some type of energy (e.g., heat, radiative).

Extrinsic semiconductor. A semiconducting material for which the electrical behavior is determined by impurities.

Extrusion. A forming technique whereby material is forced, by compression, through a die orifice.

Face-centered cubic (FCC). A crystal structure found in some of the common elemental metals. Within the cubic unit cell, atoms are located at all corner and face-centered positions.

Fatigue. Failure, at relatively low stress levels, of structures that are subjected to fluctuating and cyclic stresses.

Fatigue limit. For fatigue, the maximum stress amplitude level below which a material can endure an essentially infinite number of stress cycles and not fail.

Fatigue strength. The maximum stress level that a material can sustain, without failing, for some specified number of cycles.

Fermi energy (E_f). For a metal, the energy corresponding to the highest filled electron state in the valence band at 0 K.

Ferrimagnetism. Permanent and large magnetizations found in some ceramic materials. It results from antiparallel spin coupling and incomplete magnetic moment cancellation.

Ferrite (ceramic). Ceramic oxide materials composed of both divalent and trivalent cations (e.g., Fe^{2+} and Fe^{3+}), some of which are ferrimagnetic.

Ferrite (iron). Body-centered cubic iron; also iron and steel alloys that have the BCC crystal structure.

Ferroelectric. A dielectric material that may exhibit polarization in the absence of an electric field.

Ferromagnetism. Permanent and large magnetizations found in some metals (e.g., Fe, Ni, and Co), which result from the parallel alignment of neighboring magnetic moments.

Fiber. Any polymer, metal, or ceramic that has been drawn into a long and thin filament.

Fiber reinforcement. Strengthening or reinforcement of a relatively weak material by embedding a strong fiber phase within the weak matrix material.

Fick's first law. The diffusion flux is proportional to the concentration gradient. This relationship is employed for steady-state diffusion situations.

Fick's second law. The time rate of change of concentration is proportional to the second derivative of concentration. This relationship is employed in nonsteady-state diffusion situations.

Filler. An inert foreign substance added to a polymer to improve or modify its properties.

Fine pearlite. Pearlite for which the alternating ferrite and cementite layers are relatively thin.

Firing. A high temperature heat treatment that increases the density and strength of a ceramic piece.

Flame retardant. A polymer additive that increases flammability resistance.

Fluorescence. Luminescence that occurs for times much less than a second after an electron excitation event.

Foam. A polymer that has been made porous (or spongelike) by the incorporation of gas bubbles.

Forging. Mechanical forming of a metal by heating and hammering.

Forward bias. The conducting bias for a p–n junction rectifier such that electron flow is to the n side of the junction.

Fracture mechanics. A technique of fracture analysis used to determine the stress level at which preexisting cracks of known size will propagate, leading to fracture.

Fracture toughness (K_c). Critical value of the stress intensity factor for which crack extension occurs.

Free electron. An electron that has been excited into an energy state above the Fermi energy (or into the conduction band for semiconductors and insulators) and may participate in the electrical conduction process.

Free energy. A thermodynamic quantity that is a function of both the internal energy and entropy (or randomness) of a system. At equilibrium, the free energy is at a minimum.

Frenkel defect. In an ionic solid, a cation–vacancy and cation–interstitial pair.

Full annealing. For ferrous alloys, austenitizing, followed by cooling slowly to room temperature.

Galvanic corrosion. The preferential corrosion of the more chemically active of two metals that are electrically coupled and exposed to an electrolyte.

Galvanic series. A ranking of metals and alloys as to their relative electrochemical reactivity in seawater.

Gas constant (R). Boltzmann's constant per mole of atoms. R = 8.31 J/mol-K (1.987 cal/mol-K).

Gibbs phase rule. For a system at equilibrium, an equation (Equation 9.13) that expresses the relationship between the number of phases present and the number of externally controllable variables.

Glass–ceramic. A fine-grained crystalline ceramic material that was formed as a glass and subsequently devitrified (or crystallized).

Glass transition temperature (T_g). That temperature at which, upon cooling, a noncrystalline ceramic or polymer transforms from a supercooled liquid to a rigid glass.

Graft copolymer. A copolymer wherein homopolymer side branches of one mer type are grafted to homopolymer main chains of a different mer.

Grain. An individual crystal in a polycrystalline metal or ceramic.

Grain boundary. The interface separating two adjoining grains having different crystallographic orientations.

Grain growth. The increase in average grain size of a polycrystalline material; for most materials, an elevated-temperature heat treatment is necessary.

Grain size. The average grain diameter as determined from a random cross section.

Gray cast iron. A cast iron alloyed with silicon in which the graphite exists in the form of flakes. A fractured surface appears gray.

Green ceramic body. A ceramic piece, formed as a particulate aggregate, that has been dried but not fired.

Ground state. A normally filled electron energy state from which electron excitation may occur.

Hard magnetic material. A ferrimagnetic or ferromagnetic material that has large coercive field and remanence values, normally used in permanent magnet applications.

Hardenability. A measure of the depth to which a specific ferrous alloy may be hardened by the formation of martensite upon quenching from a temperature above the upper critical temperature.

Hardness. The measure of a material's resistance to deformation by surface indentation or by abrasion.

Heat capacity (C_p, C_v). The quantity of heat required to produce a unit temperature rise per mole of material.

Hexagonal close-packed (HCP). A crystal structure found for some metals. The HCP unit cell is of hexagonal geometry and is generated by the stacking of close-packed planes of atoms.

High polymer. A solid polymeric material having a molecular weight greater than about 10,000 g/mol.

High-strength, low-alloy (HSLA) steels. Relatively strong, low-carbon steels, with less than about 10 wt% total of alloying elements.

Hole (electron). For semiconductors and insulators, a vacant electron state in the valence band that behaves as a positive charge carrier in an electric field.

Homopolymer. A polymer having

a chain structure in which all mer units are of the same type.

Hot working. Any metal forming operation that is performed above a metal's recrystallization temperature.

Hybrid composite. A composite that is fiber reinforced by two or more types of fibers (e.g., glass and carbon).

Hydrogen bond. A strong secondary interatomic bond which exists between a bound hydrogen atom (its unscreened proton) and the electrons of adjacent atoms.

Hydroplastic forming. The molding or shaping of clay-based ceramics that have been made plastic and pliable by adding water.

Hypereutectoid alloy. For an alloy system displaying a eutectoid, an alloy for which the concentration of solute is greater than the eutectoid composition.

Hypoeutectoid alloy. For an alloy system displaying a eutectoid, an alloy for which the concentration of solute is less than the eutectoid composition.

Hysteresis (magnetic). The irreversible magnetic flux density-versus-magnetic field strength (B-versus-H) behavior found for ferromagnetic and ferrimagnetic materials; a closed B–H loop is formed upon field reversal.

Impact energy (notch toughness). A measure of the energy absorbed during the fracture of a specimen of standard dimensions and geometry when subjected to very rapid (impact) loading. Charpy and Izod impact tests are used to measure this parameter, which is important in assessing the ductile-to-brittle transition behavior of a material.

Index of refraction (n). The ratio of the velocity of light in a vacuum to the velocity in some medium.

Inhibitor. A chemical substance that, when added in relatively low concentrations, retards a chemical reaction.

Insulator (electrical). A nonmetallic material that has a filled valence band at 0 K and a relatively wide energy band gap. Consequently, the room-temperature electrical conductivity is very low, less than about 10^{-10} $(\Omega\text{-m})^{-1}$.

Integrated circuit. Thousands of electronic circuit elements (transistors, diodes, resistors, capacitors, etc.) incorporated on a very small silicon chip.

Interdiffusion. Diffusion of atoms of one metal into another metal.

Intergranular corrosion. Preferential corrosion along grain boundary regions of polycrystalline materials.

Intergranular fracture. Fracture of polycrystalline materials by crack propagation along grain boundaries.

Intermediate solid solution. A solid solution or phase having a composition range that does not extend to either of the pure components of the system.

Intermetallic compound. A compound of two metals that has a distinct chemical formula. On a phase diagram it appears as an intermediate phase that exists over a very narrow range of compositions.

Interstitial diffusion. A diffusion mechanism whereby atomic motion is from interstitial site to interstitial site.

Interstitial solid solution. A solid solution wherein relatively small solute atoms occupy interstitial positions between the solvent or host atoms.

Intrinsic semiconductor. A semiconductor material for which the electrical behavior is characteristic of the pure material; that is, electrical conductivity depends only on temperature and the band gap energy.

Invariant point. A point on a binary phase diagram at which three phases are in equilibrium.

Ionic bond. A coulombic interatomic bond that exists between two adjacent and oppositely charged ions.

Isomerism. The phenomenon whereby two or more polymer molecules or mer units have the same composition but different structural arrangements and properties.

Isomorphous. Having the same structure. In the phase diagram sense, isomorphicity means having the same crystal structure or complete solid solubility for all compositions (see Figure 9.2a).

Isotactic. A type of polymer chain configuration wherein all side groups are positioned on the same side of the chain molecule.

Isothermal. At a constant temperature.

Isothermal transformation (T–T–T) diagram. A plot of temperature versus the logarithm of time for a steel alloy of definite composition. Used to determine when transformations begin and end for an isothermal (constant-temperature) heat treatment of a previously austenitized alloy.

Isotopes. Atoms of the same element that have different atomic masses.

Isotropic. Having identical values of a property in all crystallographic directions.

Izod test. One of two tests (see also **Charpy test**) that may be used to measure the impact energy of a standard notched specimen. An impact blow is imparted to the specimen by a weighted pendulum.

Jominy end-quench test. A standardized laboratory test that is used to assess the hardenability of ferrous alloys.

Junction transistor. A semiconducting device composed of appropriately biased n–p–n or p–n–p junctions, used to amplify an electrical signal.

Kinetics. The study of reaction rates and the factors that affect them.

Laminar composite. A series of two-dimensional sheets, each having a preferred high-strength direction, fastened one on top of the other at different orientations; strength in the plane of the laminate is highly isotropic.

Large-particle composite. A type of particle-reinforced composite wherein particle–matrix interactions cannot be treated on an atomic level; the particles reinforce the matrix phase.

Laser. Acronym for light amplification by stimulated emission of radiation—a source of light that is coherent.

Lattice. The regular geometrical arrangement of points in crystal space.

Lattice parameters. The combination of unit cell edge lengths and interaxial angles that defines the unit cell geometry.

Lattice strains. Slight displacements of atoms relative to their normal lattice positions, normally imposed by crystalline defects such as dislocations, and interstitial and impurity atoms.

Lever rule. Mathematical expression, such as Equation 9.1b or Equation 9.2b, whereby the relative phase amounts in a two-phase alloy at equilibrium may be computed.

Linear coefficient of thermal expansion. See **Thermal expansion coefficient, linear.**

Linear polymer. A polymer in which each molecule consists of bifunctional mer units joined end to end in a single chain.

Liquidus line. On a binary phase diagram, that line or boundary separating liquid and liquid + solid phase regions. For an alloy, the liquidus temperature is that temperature at which a solid phase first forms under conditions of equilibrium cooling.

Longitudinal direction. The lengthwise dimension. For a rod or fiber, in the direction of the long axis.

Lower critical temperature. For a steel alloy, the temperature below which, under equilibrium conditions, all austenite has transformed to ferrite and cementite phases.

Luminescence. The emission of visible light as a result of electron decay from an excited state.

Macromolecule. A huge molecule made up of thousands of atoms.

Magnetic field strength (H). The intensity of an externally applied magnetic field.

Magnetic flux density (B). The magnetic field produced in a substance by an external magnetic field.

Magnetic induction (B). See **Magnetic flux density.**

Magnetic susceptibility (χ_m). The proportionality constant between the magnetization M and the magnetic field strength H.

Magnetization (M). The total magnetic moment per unit volume of material. Also, a measure of the contribution to the magnetic flux by some material within an H field.

Malleable cast iron. White cast iron that has been heat treated to convert the cementite into graphite clusters; a relatively ductile cast iron.

Martensite. A metastable iron phase supersaturated in carbon that is the product of a diffusionless (athermal) transformation from austenite.

Matrix phase. The phase in a composite or two-phase alloy microstructure that is continuous or completely surrounds the other (or dispersed) phase.

Matthiessen's rule. The total electrical resistivity of a metal is equal to the sum of temperature-, impurity-, and cold work-dependent contributions.

Melting point (glass). The temperature at which the viscosity of a glass material is 10 Pa-s (100 P).

Mer. The group of atoms that constitutes a polymer chain repeat unit.

Metal. The electropositive elements and alloys based on these elements. The electron band structure of metals is characterized by a partially filled valence band.

Metallic bond. A primary interatomic bond involving the nondirectional sharing of nonlocalized valence electrons ("sea of electrons") which are mutually shared by all the atoms in the metallic solid.

Metastable. Nonequilibrium state that may persist for a very long time.

Microscopy. The investigation of microstructural elements using some type of microscope.

Microstructure. The structural features of an alloy (e.g., grain and phase structure) that are subject to observation under a microscope.

Miller indices. A set of three integers (four for hexagonal) that designate crystallographic planes, as

determined from reciprocals of fractional axial intercepts.

Mixed dislocation. A dislocation that has both edge and screw components.

Mobility (electron, μ_e, and hole, μ_h). The proportionality constant between the carrier drift velocity and applied electric field; also, a measure of the ease of charge carrier motion.

Modulus of elasticity (E). The ratio of stress to strain when deformation is totally elastic; also a measure of the stiffness of a material.

Modulus of rupture (σ_{mr}). Stress at rupture from a bend test.

Molarity (M). Concentration in a liquid solution, in terms of the number of moles of a solute dissolved in 10^6 mm^3 (10^3 cm^3) of solution.

Molding (plastics). Shaping a plastic material by forcing it, under pressure and at an elevated temperature, into a mold cavity.

Mole. The quantity of a substance corresponding to 6.023×10^{23} atoms or molecules.

Molecular chemistry (polymer). With regard only to composition, not the structure of a mer.

Molecular structure (polymer). With regard to atomic arrangements within and interconnections between polymer molecules.

Molecular weight. The sum of the atomic weights of all the atoms in a molecule.

Molecule. A group of atoms that are bound together by primary interatomic bonds.

Monomer. A molecule consisting of a single mer.

MOSFET. Metal−oxide−silicon field effect transistor, an integrated circuit element.

n-Type semiconductor. A semiconductor for which the predominant charge carriers responsible for electrical conduction are electrons. Normally, donor impurity atoms give rise to the excess electrons.

Natural aging. For precipitation hardening, aging at room temperature.

Network polymer. A polymer composed of trifunctional mer units that form three-dimensional molecules.

Nodular iron. See **Ductile iron.**

Noncrystalline. The solid state wherein there is no long-range atomic order. Sometimes the terms *amorphous, glassy,* and *vitreous* are used synonymously.

Normalizing. For ferrous alloys, austenitizing above the upper critical temperature, then cooling in air. The objective of this heat treatment is to enhance toughness by refining the grain size.

Nucleation. The initial stage in a phase transformation. It is evidenced by the formation of small particles (nuclei) of the new phase, which are capable of growing.

Octahedral position. The void space among close-packed, hard sphere atoms or ions for which there are six nearest neighbors. An octahedron (double pyramid) is circumscribed by lines constructed from centers of adjacent spheres.

Opaque. Being impervious to the transmission of light as a result of absorption, reflection, and/or scattering of incident light.

Overaging. During precipitation hardening, aging beyond the point at which strength and hardness are at their maxima.

Oxidation. The removal of one or more electrons from an atom, ion, or molecule.

p-**Type semiconductor.** A semiconductor for which the predominant charge carriers responsible for electrical conduction are holes. Normally, acceptor impurity atoms give rise to the excess holes.

Paramagnetism. A relatively weak form of magnetism that results from the independent alignment of atomic dipoles (magnetic) with an applied magnetic field.

Particle-reinforced composite. A composite for which the dispersed phase is equiaxed.

Passivity. The loss of chemical reactivity, under particular environmental conditions, by some active metals and alloys.

Pauli exclusion principle. The postulate that for an individual atom, at most two electrons, which necessarily have opposite spins, can occupy the same state.

Pearlite. A two-phase microstructure found in some steels and cast irons; it results from the transformation of austenite of eutectoid composition and consists of alternating layers (or lamellae) of α-ferrite and cementite.

Periodic table. The arrangement of the chemical elements with increasing atomic number according to the periodic variation in electron structure. Nonmetallic elements are positioned at the far right-hand side of the table.

Peritectic reaction. A reaction wherein, upon cooling, a solid and a liquid phase transform isothermally and reversibly to a solid phase having a different composition.

Permeability (magnetic, μ). The proportionality constant between B and H fields. The value of the permeability of a vacuum (μ_0) is 1.257×10^{-6} H/m.

Permittivity (ϵ). The proportionality constant between the dielec-

tric displacement D and the electric field $\mathscr{E}$. The value of the permittivity ϵ_0 for a vacuum is 8.85×10^{-12} F/m.

Phase. A homogeneous portion of a system that has uniform physical and chemical characteristics.

Phase diagram. A graphical representation of the relationships between environmental constraints (e.g., temperature and sometimes pressure), composition, and regions of phase stability, ordinarily under conditions of equilibrium.

Phase transformation. A change in the number and/or character of the phases that constitute the microstructure of an alloy.

Phonon. A single quantum of vibrational or elastic energy.

Phosphorescence. Luminescence that occurs at times greater than on the order of a second after an electron excitation event.

Photoconductivity. Electrical conductivity that results from photon-induced electron excitations in which light is absorbed.

Photomicrograph. The photograph made with a microscope, which records a microstructural image.

Photon. A quantum unit of electromagnetic energy.

Piezoelectric. A dielectric material in which polarization is induced by the application of external forces.

Pilling–Bedworth ratio (P–B ratio). The ratio of metal oxide volume to metal volume; used to predict whether or not a scale that forms will protect a metal from further oxidation.

Pitting. A form of very localized corrosion wherein small pits or holes form, usually in a vertical direction.

Plain carbon steel. A ferrous alloy in which carbon is the prime alloying element.

Planck's constant (h). A universal constant that has a value of 6.63×10^{-34} J-s. The energy of a photon of electromagnetic radiation is the product of h and the radiation frequency.

Plane strain. The condition, important in fracture mechanical analyses, wherein, for tensile loading, there is zero strain in a direction perpendicular to both the stress axis and the direction of crack propagation; this condition is found in thick plates, and the zero-strain direction is perpendicular to the plate surface.

Plane strain fracture toughness (K_{Ic}). The critical value of the stress intensity factor (i.e., at which crack propagation occurs) for the condition of plane strain.

Plastic. A solid material the primary ingredient of which is an organic polymer of high molecular weight; it may also contain additives such as fillers, plasticizers, flame retardants, and the like.

Plastic deformation. Deformation that is permanent or nonrecoverable after release of the applied load. It is accompanied by permanent atomic displacements.

Plasticizer. A low molecular weight polymer additive that enhances flexibility and workability and reduces stiffness and brittleness.

Point defect. A crystalline defect associated with one or, at most, several atomic sites.

Poisson's ratio (ν). For elastic deformation, the negative ratio of lateral and axial strains that result from an applied axial stress.

Polar molecule. A molecule in which there exists a permanent electric dipole moment by virtue of the asymmetrical distribution of positively and negatively charged regions.

Polarization (*P*). The total electric dipole moment per unit volume of dielectric material. Also, a measure of the contribution to the total dielectric displacement by a dielectric material.

Polarization (corrosion). The displacement of an electrode potential from its equilibrium value as a result of current flow.

Polarization (electronic). For an atom, the displacement of the center of the negatively charged electron cloud relative to the positive nucleus, which is induced by an electric field.

Polarization (ionic). Polarization as a result of the displacement of anions and cations in opposite directions.

Polarization (orientation). Polarization resulting from the alignment (by rotation) of permanent electric dipole moments with an applied electric field.

Polycrystalline. Referring to crystalline materials that are composed of more than one crystal or grain.

Polymer. A solid, nonmetallic (normally organic) compound of high molecular weight the structure of which is composed of small repeat (or mer) units.

Polymorphism. The ability of a solid material to exist in more than one form or crystal structure.

Powder metallurgy (P/M). The fabrication of metal pieces having intricate and precise shapes by the compaction of metal powders, followed by a densification heat treatment.

Precipitation hardening. Hardening and strengthening of a metal alloy by extremely small and uniformly dispersed particles that precipitate from a supersaturated solid solution; sometimes also called *age hardening*.

Precipitation heat treatment. A heat treatment used to precipitate a new phase from a supersaturated solid solution. For precipitation hardening, it is termed *artificial aging*.

Prepreg. Continuous fiber reinforcement preimpregnated with a polymer resin which is then partially cured.

Prestressed concrete. Concrete into which compressive stresses have been introduced using steel wires or rods.

Primary phase. A phase that exists in addition to the eutectic structure.

Principle of combined action. The supposition, often valid, that new properties, better property combinations, and/or a higher level of properties can be fashioned by the judicious combination of two or more distinct materials.

Process annealing. Annealing of previously cold-worked products (commonly steel alloys in sheet or wire form) below the lower critical (eutectoid) temperature.

Proeutectoid cementite. Primary cementite that exists in addition to pearlite for hypereutectoid steels.

Proeutectoid ferrite. Primary ferrite that exists in addition to pearlite for hypoeutectoid steels.

Property. A material trait expressed in terms of the measured response to a specific imposed stimulus.

Proportional limit. The point on a stress–strain curve at which the straight line proportionality between stress and strain ceases.

Quantum mechanics. A branch of physics that deals with atomic and subatomic systems; it allows only discrete values of energy that are separated from one another. By contrast, for classical mechanics, permissible energy values are continuous.

Quantum numbers. A set of four numbers, the values of which are used to label possible electron states. Three of the quantum numbers are integers, which also specify the size, shape, and spatial orientation of an electron's probability density; the fourth number designates spin orientation.

Random copolymer. A polymer in which two different mer units are randomly distributed along the molecular chain.

Recovery. The relief of some of the internal strain energy of a previously cold-worked metal, usually by heat treatment.

Recrystallization. The formation of a new set of strain-free grains within a previously cold-worked material; normally an annealing heat treatment is necessary.

Recrystallization temperature. For a particular alloy, the minimum temperature at which complete recrystallization will occur within approximately one hour.

Rectifying junction. A semiconductor *p–n* junction that is conductive for a current flow in one direction and highly resistive for the opposite direction.

Reduction. The addition of one or more electrons to an atom, ion, or molecule.

Refraction. Bending of a light beam upon passing from one medium into another; the velocity of light differs in the two media.

Refractory. A metal or ceramic that may be exposed to extremely high temperatures without deteriorating rapidly or without melting.

Reinforced concrete. Concrete that is reinforced (or strengthened in tension) by the incorporation of steel rods, wires, or mesh.

Relative magnetic permeability (μ_r). The ratio of the magnetic permeability of some medium to that of a vacuum.

Relaxation frequency. The reciprocal of the minimum reorientation time for an electric dipole within an alternating electric field.

Relaxation modulus $[E_r(t)]$. For viscoelastic polymers, the time-dependent modulus of elasticity. It is determined from stress relaxation measurements as the ratio of stress (taken at some time after the load application—normally 10 s) to strain.

Remanence (remanent induction, B_r). For a ferromagnetic or ferrimagnetic material, the magnitude of residual flux density that remains when a magnetic field is removed.

Residual stress. A stress that persists in a material that is free of external forces or temperature gradients.

Resilience. The capacity of a material to absorb energy when it is elastically deformed.

Resistivity (ρ). The reciprocal of electrical conductivity, and a measure of a material's resistance to the passage of electric current.

Resolved shear stress. An applied tensile or compressive stress resolved into a shear component along a specific plane and direction within that plane.

Reverse bias. The insulating bias for a p–n junction rectifier; electrons flow into the p side of the junction.

Rolling. A metal-forming operation that reduces the thickness of sheet stock; also, elongated shapes may be fashioned using grooved circular rolls.

Rule of mixtures. The properties of a multiphase alloy or composite material are a weighted average (usually on the basis of volume) of the properties of the individual constituents.

Rupture. Failure that is accompanied by significant plastic deformation; often associated with creep failure.

Sacrificial anode. An active metal or alloy that preferentially corrodes and protects another metal or alloy to which it is electrically coupled.

Safe stress (σ_w). A stress used for design purposes; for ductile metals, it is the yield strength divided by a factor of safety.

Sandwich panel. A type of structural composite consisting of two stiff and strong outer faces that are separated by a lightweight core material.

Saturated. A term describing a carbon atom that participates in only single covalent bonds with four other atoms.

Saturation magnetization, flux density (M_s, B_s). The maximum magnetization (or flux density) for a ferromagnetic or ferrimagnetic material.

Scanning electron microscope (SEM). A microscope that produces an image by using an electron beam that scans the surface of a specimen; an image is produced by reflected electron beams. Examination of surface and/or microstructural features at high magnifications is possible.

Schottky defect. In an ionic solid, a defect consisting of a cation–vacancy and anion–vacancy pair.

Scission. A polymer degradation process whereby molecular chain bonds are ruptured by chemical reactions or by exposure to radiation or heat.

Screw dislocation. A linear crystalline defect associated with the lattice distortion created when normally parallel planes are joined together to form a helical ramp. The Burgers vector is parallel to the dislocation line.

Selective leaching. A form of corrosion wherein one element or constituent of an alloy is preferentially dissolved.

Self-diffusion. Atomic migration in pure metals.

Self-interstitial. A host atom or ion that is positioned on an interstitial lattice site.

Semiconductor. A nonmetallic material that has a filled valence band at 0 K and a relatively narrow energy band gap. The room temperature electrical conductivity ranges between about 10^{-6} and $10^4 \, (\Omega\text{-m})^{-1}$.

Shear. A force applied so as to cause or tend to cause two adjacent parts of the same body to slide relative to each other, in a direction parallel to their plane of contact.

Shear strain (γ). The tangent of the shear angle that results from an applied shear load.

Shear stress (τ). The instantaneous applied shear load divided by the original cross-sectional area across which it is applied.

Single crystal. A crystalline solid for which the periodic and repeated atomic pattern extends throughout its entirety without interruption.

Sintering. Particle coalescence of a powdered aggregate by diffusion that is accomplished by firing at an elevated temperature.

Slip. Plastic deformation as the result of dislocation motion; also, the shear displacement of two adjacent planes of atoms.

Slip casting. A forming technique used for some ceramic materials. A slip, or suspension of solid particles in water, is poured into a porous mold. A solid layer forms on the inside wall as water is absorbed by the mold, leaving a shell (or ultimately a solid piece) having the shape of the mold.

Slip system. The combination of a crystallographic plane and, within that plane, a crystallographic direction along which slip (i.e., dislocation motion) occurs.

Softening point (glass). The maximum temperature at which a glass piece may be handled without permanent deformation; this corresponds to a viscosity of approximately 4×10^6 Pa-s (4×10^7 P).

Soft magnetic material. A ferromagnetic or ferrimagnetic material having a small B versus H hysteresis loop, which may be magnetized and demagnetized with relative ease.

Soldering. A technique for joining metals using a filler metal alloy that has a melting temperature less than about 425°C (800°F). Lead–tin alloys are common solders.

Solid solution. A homogeneous crystalline phase that contains two or more chemical species. Both substitutional and interstitial solid solutions are possible.

Solid-solution hardening. Hardening and strengthening of metals that result from alloying in which a solid solution is formed. The presence of impurity atoms restricts dislocation mobility.

Solidus. On a phase diagram, the locus of points at which solidification is complete upon equilibrium cooling, or at which melting begins upon equilibrium heating.

Solubility limit. The maximum concentration of solute that may be added without forming a new phase.

Solute. One component or element of a solution present in a minor concentration. It is dissolved in the solvent.

Solution heat treatment. The process used to form a solid solution by dissolving precipitate particles. Often, the solid solution is supersaturated and metastable at ambient conditions as a result of rapid cooling from an elevated temperature.

Solvent. The component of a solution present in the greatest amount. It is the component that dissolves a solute.

Solvus. The locus of points on a phase diagram representing the limit of solid solubility as a function of temperature.

Specific heat (c_p, c_v). The heat capacity per unit mass of material.

Specific modulus (specific stiffness). The ratio of elastic modulus to specific gravity for a material.

Specific strength. The ratio of tensile strength to specific gravity for a material.

Spheroidite. Microstructure found in steel alloys, consisting of sphere-like cementite particles within an α-ferrite matrix. It is produced by an appropriate elevated-temperature heat treatment of pearlite, bainite, or martensite, and is relatively soft.

Spheroidizing. For steels, a heat treatment carried out at a temperature just below the eutectoid in which the spheroidite microstructure is produced.

Spherulite. An aggregate of ribbonlike polymer crystallites radiating from a common center, which crystallites are separated by amorphous regions.

Spinning. The process by which fibers are formed. A multitude of fibers are spun as molten material is forced through many small orifices.

Stabilizer. A polymer additive that counteracts deteriorative processes.

Standard half-cell. An electrochemical cell consisting of a pure metal immersed in a $1M$ aqueous solution of its ions, which is electrically coupled to the standard hydrogen electrode.

Steady-state diffusion. The diffusion condition for which there is no net accumulation or depletion of diffusing species. The diffusion flux is independent of time.

Stereoisomerism. Polymer isomerism in which side groups within mer units are bonded along the molecular chain in the same order, but in different spatial arrangements.

Stoichiometry. For ionic compounds, the state of having exactly the ratio of cations to anions specified by the chemical formula.

Strain, engineering (ϵ). The change in gauge length of a specimen (in the direction of an applied stress) divided by its original gauge length.

Strain hardening. The increase in hardness and strength of a ductile metal as it is plastically deformed below its recrystallization temperature.

Strain point (glass). The maximum temperature at which glass fractures without plastic deformation; this corresponds to a viscosity of about 3×10^{13} Pa-s (3×10^{14} P).

Stress concentration. The concentration or amplification of an applied stress at the tip of a notch or small crack.

Stress corrosion (cracking). A form of failure that results from the combined action of a tensile stress and a corrosion environment; it occurs at lower stress levels than are required when the corrosion environment is absent.

Stress, engineering (σ). The instantaneous load applied to a specimen divided by its cross-sectional area before any deformation.

Stress intensity factor (K). A factor used in fracture mechanics to specify the stress intensity at the tip of a crack.

Stress raiser. A small flaw (internal or surface) or a structural discontinuity at which an applied tensile stress will be amplified and from which cracks may propagate.

Stress relief. A heat treatment for the removal of residual stresses.

Structural clay products. Ceramic products made principally of clay and used in applications where structural integrity is important (e.g., bricks, tiles, pipes).

Structure. The arrangement of the internal components of matter: electron structure (on a subatomic level), crystal structure (on an atomic level), and microstructure (on a microscopic level).

Structural composite. A composite the properties of which depend on the geometrical design of the structural elements. Laminar composites and sandwich panels are two subclasses.

Substitutional solid solution. A solid solution wherein the solute atoms replace or substitute for the host atoms.

Superconductivity. A phenomenon observed in some materials: the disappearance of the electrical resistivity at temperatures approaching 0 K.

Supercooling. Cooling to below a phase transition temperature without the occurrence of the transformation.

Superheating. Heating to above a phase transition temperature without the occurrence of the transformation.

Syndiotactic. A type of polymer chain configuration in which side groups regularly alternate positions on opposite sides of the chain.

System. Two meanings are possible: (1) a specific body of material that is being considered, and (2) a series of possible alloys consisting of the same components.

Temper designation. A letter–digit code used to designate the mechanical and/or thermal treatment to which a metal alloy has been subjected.

Tempered martensite. The microstructural product resulting from a tempering heat treatment of a martensitic steel. The microstructure consists of extremely small and uniformly dispersed cementite particles embedded within a continuous α-ferrite matrix. Toughness and ductility are enhanced significantly by tempering.

Tempering (glass). See **Thermal tempering.**

Tensile strength (TS). The maximum engineering stress, in tension, that may be sustained without fracture. Often termed *ultimate (tensile) strength*.

Terminal solid solution. A solid solution that exists over a composition range extending to either composition extremity of a binary phase diagram.

Tetrahedral position. The void space among close-packed, hard sphere atoms or ions for which there are four nearest neighbors.

Thermal conductivity (k). For steady-state heat flow, the proportionality constant between the heat flux and the temperature gradient. Also, a parameter characterizing the ability of a material to conduct heat.

Thermal expansion coefficient, linear (α_l). The fractional change in length divided by the change in temperature.

Thermal fatigue. A type of fatigue failure wherein the cyclic stresses are introduced by fluctuating thermal stresses.

Thermal shock. The fracture of a brittle material as a result of stresses that are introduced by a rapid temperature change.

Thermal stress. A residual stress introduced within a body resulting from a change in temperature.

Thermal tempering. Increasing the strength of a glass piece by the introduction of residual compressive stresses within the outer surface using an appropriate heat treatment.

Thermally activated. A reaction that depends on atomic thermal fluctuations; the atoms having energies greater than an activation energy will spontaneously react or transform. The rate of this type of transformation depends on temperature according to Equation 10.3.

Thermoplastic (polymer). A polymeric material that softens when heated and hardens upon cooling. While in the softened state, articles may be formed by molding or extrusion.

Thermosetting (polymer). A polymeric material that, once having cured (or hardened) by a chemical reaction, will not soften or melt when subsequently heated.

Tie line. A horizontal line constructed across a two-phase region of a binary phase diagram; its intersections with the phase boundaries on either end represent the equilibrium compositions of the respective phases at the temperature in question.

Time-temperature-transformation (*T–T–T*) diagram. See **Isothermal transformation diagram.**

Toughness. A measure of the amount of energy absorbed by a material as it fractures. Toughness is indicated by the total area under the material's tensile stress–strain curve.

Trans. For polymers, a prefix denoting a type of molecular structure. To some unsaturated carbon chain atoms within a mer unit, a single side atom or group may be situated on one side of the chain, or directly opposite at a 180° rotation position. In a trans structure, two such side groups within the same mer reside on opposite chain sides (e.g., *trans*-isoprene).

Transformation rate. The reciprocal of the time necessary for a reaction to proceed halfway to its completion.

Transgranular fracture. Fracture of polycrystalline materials by crack propagation through the grains.

Translucent. Having the property of transmitting light only diffusely; objects viewed through a translucent medium are not clearly distinguishable.

Transmission electron microscope (TEM). A microscope that produces an image by using electron beams that are transmitted (pass through) the specimen. Examination of internal features at high magnifications is possible.

Transparent. Having the property of transmitting light with relatively little absorption, reflection, and scattering, such that objects viewed through a transparent medium can be distinguished readily.

Transverse direction. A direction that crosses (usually perpendicularly) the longitudinal or lengthwise direction.

Trifunctional monomer. Designating monomer units that have three active bonding positions.

True strain (ϵ_T). The natural logarithm of the ratio of instantaneous gauge length to original gauge length of a specimen being deformed by a uniaxial force.

True stress (σ_T). The instantaneous applied load divided by the instantaneous cross-sectional area of a specimen.

Ultimate (tensile) strength. See **Tensile strength.**

Unit cell. The basic structural unit of a crystal structure. It is generally defined in terms of atom (or ion) positions within a parallelepiped volume.

Unsaturated. A term describing carbon atoms that participate in double or triple covalent bonds and, therefore, do not bond to a maximum of four other atoms.

Upper critical temperature. For a steel alloy, the minimum temperature above which, under equilibrium conditions, only austenite is present.

Vacancy. A normally occupied lattice site from which an atom or ion is missing.

Vacancy diffusion. The diffusion mechanism wherein net atomic migration is from lattice site to an adjacent vacancy.

Valence band. For solid materials, the electron energy band that contains the valence electrons.

Valence electrons. The electrons in the outermost occupied electron shell, which participate in interatomic bonding.

van der Waals bond. A secondary interatomic bond between adjacent molecular dipoles, which may be permanent or induced.

Viscoelasticity. A type of deformation exhibiting the mechanical characteristics of viscous flow and elastic deformation.

Viscosity (η). The ratio of the magnitude of an applied shear stress to the velocity gradient that it produces; that is, a measure of a noncrystalline material's resistance to permanent deformation.

Vitrification. During firing of a ceramic body, the formation of a liquid phase that upon cooling becomes a glass-bonding matrix.

Vulcanization. Nonreversible chemical reaction involving sulfur or other suitable agent wherein crosslinks are formed between molecular chains in rubber materials. The rubber's modulus of elasticity and strength are enhanced.

Wave-mechanical model. Atomic model in which electrons are treated as being wavelike.

Weight percent (wt%). Concentration specification on the basis of weight (or mass) of a particular element relative to the total alloy weight (or mass).

Weld decay. Intergranular corrosion that occurs in some welded stainless steels at regions adjacent to the weld.

Welding. A technique for joining metals in which actual melting of the pieces to be joined occurs in the vicinity of the bond. A filler metal may be used to facilitate the process.

Whisker. A very thin, single crystal of high perfection which has an extremely large length-to-diameter ratio. Whiskers are used as the reinforcing phase in some composites.

White cast iron. A low-silicon and very brittle cast iron, in which the carbon is in combined form as cementite; a fractured surface appears white.

Whiteware. A clay-based ceramic product that becomes white after high temperature firing; white-wares include porcelain, china, and plumbing sanitary ware.

Working point (glass). The temperature at which a glass is easily deformed, which corresponds to a viscosity of 10^3 Pa-s (10^4 P).

Wrought alloy. A metal alloy that is relatively ductile and amenable to hot working or cold working during fabrication.

Yielding. The onset of plastic deformation.

Yield strength (σ_y). The stress re-quired to produce a very slight yet specified amount of plastic strain; a strain offset of 0.002 is commonly used.

Young's modulus. See **Modulus of elasticity.**

ANSWERS TO SELECTED PROBLEMS

CHAPTER 2

2.3 (a) 1.66×10^{-24} g/amu;
(b) 2.73×10^{26} atoms/lb-mol

2.13

$$r_0 = \left(\frac{A}{nB}\right)^{1/(1-n)}$$

$$E_0 = -\frac{A}{\left(\dfrac{A}{nB}\right)^{1/(1-n)}} + \frac{B}{\left(\dfrac{A}{nB}\right)^{n/(1-n)}}$$

2.14 (c) $r_0 = 0.236$ nm; $E_0 = -5.32$ eV

2.18 73.4% for MgO; 14.8% for CdS

CHAPTER 3

3.3 $V_C = 6.62 \times 10^{-29}$ m³

3.8 (b) APF = 0.52

3.10 $R = 0.136$ nm

3.13 (a) $V_C = 1.06 \times 10^{-28}$ m³;
(b) $a = 0.296$ nm, $c = 0.468$ nm

3.16 Metal B: simple cubic

3.20 (a) $n = 4$; (b) $\rho = 7.31$ g/cm³

3.23 $V_C = 6.64 \times 10^{-2}$ nm³

3.28 (a) Direction 1: [012]; (b) plane 1: (020)

3.30 Direction A: [$\bar{1}$10]; direction C: [0$\bar{1}$2]

3.31 Direction B: [$\bar{4}$0$\bar{3}$]; direction D: [$\bar{1}$1$\bar{1}$]

3.32 (b) [$\bar{1}$00], [010], [0$\bar{1}$0]

3.34 Plane A: (11$\bar{1}$) or ($\bar{1}\bar{1}$1)

3.35 Plane B: (122)

3.36 Plane B: (02$\bar{1}$)

3.38 (a) (1$\bar{2}$11)

3.42 (a) (100) and (0$\bar{1}$0)

3.43 (c) [$\bar{1}$10]

3.44 [100]: LD = 0.71

3.45 [111]: LD = 1.0

3.46 (100): PD = 0.79

3.47 (110): PD = 0.83

3.53 $2\theta = 45.88°$

3.54 $d_{111} = 0.1655$ nm

3.56 (a) $d_{211} = 0.1348$ nm; (b) $R = 0.1429$ nm

3.58 $d_{200} = 0.2488$ nm; $d_{311} = 0.1495$ nm;
$a = 0.496$ nm

CHAPTER 4

4.1 $N/N_v = 2.4 \times 10^{-5}$

4.3 $Q_v = 1.10$ eV/atom

4.5 $r = 0.41R$

4.7 $C'_{Cu} = 41.9$ at%; $C'_{Zn} = 58.1$ at%

4.9 $C'_{Ag} = 87.7$ at%; $C'_{Cu} = 12.3$ at%

4.11 6.02×10^{28} m^{-3}

4.13 $N_{Ni} = 9.15 \times 10^{20}$ atoms/cm³

4.17 (a) FCC: $\mathbf{b} = a/2$ [110]; (b) Cu: $|\mathbf{b}| = 0.2556$ nm

4.22 $d \approx 0.07$ mm

4.24 (a) $N = 8$

CHAPTER 5

5.6 $dM/dt = 2.6 \times 10^{-3}$ kg/h

5.8 $D = 3.9 \times 10^{-11}$ m^2/s

5.11 $t = 13.9$ h

5.15 $t = 40$ h

5.18 $T = 1148$ K (875°C)

5.21 (a) $Q_d = 252$ kJ/mol (60.4 kcal/mol);
    $D_0 = 2.2 \times 10^{-5}$ m^2/s;
    (b) $D = 5.3 \times 10^{-15}$ m^2/s

5.24 $T = 900$ K (627°C)

5.29 $x = 15.1$ mm

5.32 $C_s = 28.5$ wt% Ni

CHAPTER 6

6.2 $l_0 = 10$ in. (254 mm)

6.5 (a) $F = 10,000$ lb$_f$ (44,850 N);
    (b) $\Delta l = 0.01$ in. (0.25 mm)

6.7 $\Delta l = 0.070$ mm (0.003 in.)

6.10

$$\left(\frac{dF}{dr}\right)_{r_0} = -\frac{2A}{\left(\dfrac{A}{nB}\right)^{3/(1-n)}} + \frac{nB[n+1]}{\left(\dfrac{A}{nB}\right)^{(n+2)/(1-n)}}$$

6.12 (a) $\Delta l = 0.02$ in. (0.50 mm);
    (b) $\Delta d = 6.2 \times 10^{-4}$ in. (1.6×10^{-2} mm);
    decrease

6.13 $F = 4190$ lb$_f$ (18,000 N)

6.14 $v = 0.367$

6.16 $E = 10^5$ MPa (14.7×10^6 psi)

6.19 (a) $\Delta l = 0.0056$ in. (0.14 mm);
    (b) $\epsilon_x = 1.96 \times 10^{-4}$ in. (4.9×10^{-3} mm)

6.22 Steel

6.25 (a) Both elastic and plastic; (b) $\Delta l = 0.14$ in. (3.6 mm)

6.27 (b) $E = 8.7 \times 10^6$ psi (6.0×10^4 MPa);
    (c) $\sigma_y = 41,500$ psi (286 MPa);
    (d) $TS = 53,500$ psi (370 MPa);
    (e) %EL = 16.4%;
    (f) $U_r = 96.6$ in.-lb$_f$/in.^{3} (6.6×10^5 J/m^3);
    (g) $\sigma_w = 20,750$ psi (143 MPa)

6.30 Figure 6.11: $U_r = 45.3$ in.-lb$_f$/in.^{3}
    (3.3×10^5 J/m^3)

6.32 $\sigma_y = 134,000$ psi (925 MPa)

6.37 $\epsilon_T = 0.237$

6.39 $\sigma_T = 63,700$ psi (440 MPa)

6.41 Toughness = 5.29×10^5 in.-lb$_f$/in.^{3}
    (3.65×10^9 J/m^3)

6.43 $n = 0.13$

6.45 (a) ϵ (elastic) $\cong 0.0025$; ϵ (plastic) $\cong 0.0020$;
    (b) $l_i = 24.04$ in. (611.1 mm)

6.47 (a) 125 HB (70 HRB)

6.52 Figure 6.11: $\sigma_w = 18,000$ psi (125 MPa)

CHAPTER 7

7.11 $\cos \lambda \cos \phi = 0.408$

7.13 (b) $\tau_{crss} = 130$ psi (0.90 MPa)

7.14 $\tau_{crss} = 60.6$ psi (0.42 MPa)

7.21 $d = 0.014$ mm

7.22 (a) $C \cong 30$ wt% Zn–70 wt% Cu

7.25 $r_d = 8.80$ mm

7.27 $r_0 = 0.424$ in. (10.8 mm)

7.29 Not possible

7.33 $\tau_{crss} = 910$ psi (6.3 MPa)

7.39 Cold work to between 26.5 and 28%CW [to $d'_0 \cong 0.35$ in. (8.9 mm)], anneal, then cold work to give a final diameter of 0.30 in. (7.6 mm)

7.42 (a) $t = 3500$ min

7.43 (b) $d = 0.085$ mm

CHAPTER 8

8.3 $\sigma_m = 354,000$ psi (2404 MPa)

8.6 $\sigma_c = 32.6$ MPa

8.8 (a) $\sigma_x = \sigma_y = 316$ MPa (45,800 psi);
    (d) $\sigma_x = 84.6$ MPa (12,300 psi),
    $\sigma_y = 176.3$ MPa (25,600 psi)

8.10 (a) $\sigma_m = 36,000$ psi (250 MPa)

8.11 (a) $\sigma_m = 12,500$ psi (86.3 MPa)

8.13 2024-T351 Al: $B \geq 30.7$ mm (1.23 in.); 4340 steel (tempered at 260°C): $B \geq 2.3$ mm (0.093 in.)

8.15 Fracture will occur

8.17 $a = 0.72$ in. (18.2 mm)

8.19 Is subject to detection since $a \geq 3$ mm

8.23 (b) -99°C; (c) -107°C

8.26 (a) $\sigma_{max} = 40,000$ psi (280 MPa),
    $\sigma_{min} = -20,000$ psi (-140 MPa);
    (b) $R = -0.50$; (c) $\sigma_r = 60,000$ psi (420 MPa)

8.28 $N_f = 1.0 \times 10^7$ cycles

8.30 (b) $S = 100$ MPa; (c) $N_f \cong 5.7 \times 10^5$ cycles

8.31 (a) $\tau = 74$ MPa; (c) $\tau = 103$ MPa

8.33 (a) $N_f = 30$ min; (c) $N_f = 28$ h

8.40 $\sigma_{max} = 350$ MPa

8.42 $a_c = 0.20$ in.

8.46 $\Delta\epsilon/\Delta t = 7 \times 10^{-3}$ min^{-1}

8.47 $\Delta l = 5.9$ in. (5.4 mm)

8.49 $t_r = 10^4$ h

8.51 427°C: $n = 5.9$

8.52 (a) $Q_c = 196$ kJ/mol

8.54 $\dot{\epsilon}_s = 4.7 \times 10^{-2}$ h^{-1}

8.56 $T = 1197$ K (924°C)

8.57 $\sigma \cong 42,000$ psi (280 MPa)

CHAPTER 9

9.5 (a) $\alpha + \beta$; $C_\alpha = 4$ wt% Sn–96 wt% Pb, $C_\beta \cong 100$ wt% Sn;
(c) $\beta + L$; $C_\beta = 92$ wt% Ag–8 wt% Cu, $C_L = 78$ wt% Ag–22 wt% Cu;
(e) α; $C_\alpha = 8.2$ wt% Sn–91.8 wt% Pb;
(g) $L + Mg_2Pb$; $C_L = 93$ wt% Pb–7 wt% Mg, $C_{Mg_2Pb} = 81$ wt% Pb–19 wt% Mg

9.7 (a) $W_\alpha = 0.89$, $W_\beta = 0.11$; (c) $W_\beta = 0.53$, $W_L = 0.47$; (e) $W_\alpha = 1.0$; (g) $W_L = 0.96$, $W_{Mg_2Pb} = 0.04$

9.9 (a) $T \cong 300$°C (570°F)

9.10 (a) $m_s = 2846$ g; (b) $C_L = 64$ wt% sugar; (c) $m_s = 1068$ g

9.14 Not possible

9.17 (a) $T = 1315$°C (2400°F);
(b) $C_\alpha = 63$ wt% Ni–37 wt% Cu;
(c) $T = 1270$°C (2320°F);
(d) $C_L = 36$ wt% Ni–64 wt% Cu

9.20 (a) $T \cong 550$°C (1020°F); (b) $C_\alpha = 26$ wt% Pb; $C_L = 54$ wt% Pb

9.21 $C_\alpha = 90$ wt% A–10 wt% B; $C_\beta = 20$ wt% A–80 wt% B

9.23 Possible at $T \cong 800$°C

9.25 $V_\alpha = 0.27$; $V_\beta = 0.73$

9.28 Not possible because different C_0 required for each situation

9.31 $C_0 = 25.2$ wt% Ag–74.8 wt% Cu

9.34

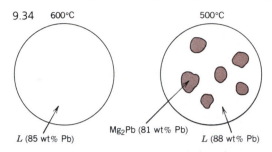

600°C   500°C

Mg₂Pb (81 wt% Pb)

L (85 wt% Pb) L (88 wt% Pb)

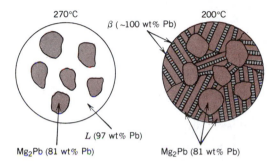

270°C   200°C

β (~100 wt% Pb)

L (97 wt% Pb)

Mg₂Pb (81 wt% Pb)   Mg₂Pb (81 wt% Pb)

9.38 Eutectics: (1) 10 wt% Au, 217°C, $L \rightarrow \alpha + \beta$;
(2) 80 wt% Au, 280°C, $L \rightarrow \delta + \zeta$
Congruent melting point: 62.5 wt% Au, 418°C, $L \rightarrow \delta$
Peritectics: (1) 30 wt% Au, 252°C, $L + \gamma \rightarrow \beta$; (2) 45 wt% Au, 309°C, $L + \delta \rightarrow \gamma$; (3) 92 wt% Au, 490°C, $L + \eta \rightarrow \zeta$
Eutectoids: none

9.41 For point A, $F = 2$

9.47 $C_0 = 0.42$ wt% C

9.50 (a) α-ferrite; (b) 2.26 kg ferrite, 0.24 kg Fe₃C; (c) 0.40 kg proeutectoid ferrite, 2.1 kg pearlite

9.52 $C_0 = 0.55$ wt% C

9.54 $C_0 = 1.42$ wt% C

9.57 Not possible

9.60 Two answers are possible: $C_0 = 0.92$ wt% C and 0.75 wt% C

9.63 HB (alloy) = 128

9.66 (a) T (eutectoid) = 650°C (1200°F); (b) ferrite; (c) $W_{\alpha'} = 0.68$, $W_p = 0.32$

CHAPTER 10

10.2 $t = 305$ s

10.4 $r = 8.76 \times 10^{-3}$ min^{-1}

10.6 $y = 0.65$

10.7   (c) $t \cong 250$ days

10.10 (b) 265 HB (27 HRC)

10.14 (a) 50% coarse pearlite, 50% martensite; (d) 100% martensite; (e) 40% bainite, 60% martensite; (g) 100% fine pearlite

10.16 (a) proeutectoid cementite and martensite; (c) bainite; (e) cementite, medium pearlite, bainite, and martensite; (g) bainite and martensite

10.19 (a) martensite

10.24 (a) martensite; (c) martensite, proeutectoid ferrite, and bainite

10.34 (b) 180 HB (87 HRB); (g) 265 HB (27 HRC)

10.36 (c) $TS = 132{,}500$ psi (915 MPa)

10.37 Yes; coarse pearlite

10.40 (b) Rapidly cool to about 630°C (1170°F), hold for about 25 s, then cool to room temperature

10.41 (a) $t \cong 650$ s

10.42 Temper at between 400 and 450°C (750 and 840°F) for 1 h

CHAPTER 11

11.3   (a) 890–920°C (1640–1690°F)

11.4   (b) 790–815°C (1450–1500°F)

11.12 4340, 4140, and 8640 alloys satisfy the criteria

11.16 Maximum diameter = 3 in. (75 mm)

11.18 Maximum diameter = 3.65 in. (93 mm)

11.21 For about 0.4 h at 204°C, or between 9 and 25 h at 149°C

CHAPTER 12

12.12 $V_{Gr} = 11.1$ vol%

CHAPTER 13

13.6   (a) Cesium chloride; (c) sodium chloride

13.8   APF = 0.79

13.10 (a) FCC; (b) tetrahedral; (c) one half

13.12 (a) tetrahedral; (b) one half

13.15 (a) $a = 0.421$ nm; (b) $a = 0.424$ nm

13.17 ρ (calculated) = 4.11 g/cm^3; ρ (measured) = 4.10 g/cm^3

13.19 (a) $\rho = 4.20$ g/cm^3

13.21 Sodium chloride and zinc blende

13.23 APF = 0.84

13.25 APF = 0.68

13.29 (a) $C = 45.9$ wt% Al_2O_3–54.1 wt% SiO_2

13.33 (a) O^{2-} vacancy; one O^{2-} vacancy for every two Li^+ added; (b) Ca^{2+} vacancy; one Ca^{2+} vacancy for every two Cl^- added.

13.36 (a) 7.7% of Mg^{2+} vacancies

13.38 $\rho_t = 0.46$ nm

13.41 $R = 0.38$ in. (9.6 mm)

13.43 $F = 3870$ lb$_f$ (17,200 N)

13.46 $E_0 = 33.1 \times 10^6$ psi (23.2×10^4 MPa)

CHAPTER 14

14.7   (b) $Q_{vis} = 365$ kJ/mol (87.2 kcal/mol)

14.24 (a) $T = 2170$°C (3940°F)

14.26 (a) $W_L = 0.73$

14.27 (a) 1828°C (3320°F); between ~73 and 100 wt% Al_2O_3

CHAPTER 15

15.4 $n = 23{,}700$

15.6   (a) $\overline{M}_n = 33{,}040$ g/mol; (c) $n_n = 785$

15.9   (a) $C_{Cl} = 20.3$ wt%

15.11 $L = 2682$ nm; $r = 22.5$ nm

15.16 8530 of both styrene and butadiene mers

15.18 PVC

15.21 $f(\text{isoprene}) = 0.88$; $f(\text{isobutylene}) = 0.12$

15.25 $\rho = 0.998$ g/cm^3

15.27 (a) $\rho_a = 0.870$ g/cm^3, $\rho_c = 0.998$ g/cm^3; (b) % crystallinity = 65.7%

CHAPTER 16

16.20 $E_r(10) = 520$ psi (3.7 MPa)

16.30 (a) $m(\text{ethylene glycol}) = 8.50$ kg; (b) $m(\text{polyester}) = 26.03$ kg

16.38 20%

16.40 One sulfur crosslink per isoprene mer

CHAPTER 17

17.4 $k_{max} = 41.0$ W/m-K (0.098 cal/s-cm-K); $k_{min} = 34.1$ W/m-K (0.082 cal/s-cm-K)

17.9 $\tau_c = 87.5$ MPa

17.11 Carbon and aramid

17.13 Not possible

17.15 Not possible

17.17 $E_f = 10^7$ psi $(7.1 \times 10^4$ MPa);
$E_m \cong 4 \times 10^5$ psi $(2.76 \times 10^3$ MPa)

17.20 (a) $F_f/F_m = 23.3$; (b) $F_f = 9588$ lb$_f$ (42,570 N),
$F_m = 412$ lb$_f$ (1830 N); (c) $\sigma_f = 64,000$ psi
(440 MPa), $\sigma_m = 1180$ psi (8.1 MPa);
(d) $\epsilon = 3.36 \times 10^{-3}$

17.22 $(TS)_c = 948$ MPa (137,500 psi)

17.24 $(TS)_c = 764$ MPa (109,740 psi)

17.32 (b) $E_c = 18.6 \times 10^6$ psi $(12.75 \times 10^4$ MPa)

CHAPTER 18

18.5 (a) $\Delta V = 0.031$ V;
(b) $Fe^{2+} + Cd \rightarrow Fe + Cd^{2+}$

18.7 $[Pb^{2+}] = 2.5 \times 10^{-2} M$

18.12 $t = 2.5$ yr

18.15 CPR = 36.4 mpy

18.19 (a) $r = 4.56 \times 10^{-12}$ mol/cm²-s;
(b) $V_C = -0.0166$ V

18.34 Mg: P–B ratio = 0.81; nonprotective

18.36 (a) Parabolic kinetics; (b) $W = 3.70$ mg/cm²

CHAPTER 19

19.2 $d = 1.3$ mm

19.5 (a) $R = 4.7 \times 10^{-3} \Omega$; (b) $I = 10.6$ A;
(c) $J = 1.5 \times 10^6$ A/m²;
(d) $\mathscr{E} = 2.5 \times 10^{-2}$ V/m

19.12 $\sigma = 0.096$ $(\Omega\text{-m})^{-1}$

19.13 (a) $n = 1.25 \times 10^{29}$ m^{-3};
(b) 1.48 free electrons/atom

19.16 (a) $\rho_0 = 1.6 \times 10^{-8}$ Ω-m,
$a = 6.5 \times 10^{-11}$ Ω-m/°C;
(b) $A = 1.2 \times 10^{-6}$ Ω-m;
(c) $\rho = 4.24 \times 10^{-8}$ Ω-m

19.18 $\rho = 3.9 \times 10^{-8}$ Ω-m

19.21 $\sigma = 5.7 \times 10^6$ $(\Omega\text{-m})^{-1}$

19.23 (b) for Si, 2.7×10^{-13}; for Ge, 5.6×10^{-10}

19.32 (a) $n = 8.9 \times 10^{21}$ m^{-3}; (b) p-type extrinsic

19.35 $\mu_e = 0.50$ m²/V-s; $\mu_h = 0.02$ m²/V-s

19.40 $E_g = 1.46$ eV

19.41 (b) $\sigma = 2.16$ $(\Omega\text{-m})^{-1}$

19.49 $l = 1.6$ mm (0.064 in.)

19.53 $p_i = 2.26 \times 10^{-30}$ C-m

19.55 (a) $V = 17.3$ V; (b) $V = 86.5$ V;
(e) $P = 1.75 \times 10^{-7}$ C/m²

19.57 Fraction of ϵ_r due to $P_i = 0.67$

CHAPTER 20

20.2 $T_f = 65.1$°C (149.2°F)

20.4 (a) $C_v = 139$ J/kg-K; (b) $C_v = 925$ J/kg-K

20.9 $\Delta l = -9.2$ mm $(-0.36$ in.)

20.11 $T_f = 41$°C (106°F)

20.14 $T_f = 233.5$°C

20.16 (b) $dQ/dt = 9.3 \times 10^8$ J/h

20.24 k(upper) = 22.6 W/m-K

20.28 (a) $\sigma = 22,500$ psi (155 MPa); compression

20.29 $T_f = 39$°C (102°F)

20.31 $\Delta d = 0.0378$ mm

20.34 Al$_2$O$_3$: $\Delta T_f = 83$°C; soda–lime glass:
$\Delta T_f = 111$°C

CHAPTER 21

21.1 (a) $H = 10,000$ A-turns/m;
(b) $B_0 = 1.257 \times 10^{-2}$ tesla;
(c) $B = 1.257 \times 10^{-2}$ tesla;
(d) $M = 1.81$ A/m

21.6 (a) $\mu = 1.26 \times 10^{-6}$ H/m; (b) $\chi_m = 2.39 \times 10^{-3}$

21.8 (a) $M_s = 1.45 \times 10^6$ A/m

21.16 4.6 Bohr magnetons/Mn^{2+} ion

21.24 $M_s = 1.7 \times 10^6$ A/m

21.27 (b) $\mu_i \cong 3 \times 10^{-3}$ H/m, $\mu_{ri} \cong 2400$;
(c) μ(max) $\cong 1.1 \times 10^{-2}$ H/m

21.29 (b) (i) $\mu \cong 1.1 \times 10^{-2}$ H/m; (iii) $\chi_m \cong 7954$

21.32 (a) 2.5 K: 5.62×10^4 A/m; (b) 3.05 K

CHAPTER 22

22.9 $v = 1.28 \times 10^8$ m/s

22.10 Silica: 0.53; soda–lime glass: 0.33

22.11 Fused silica: $\epsilon_r = 2.12$; polyethylene: $\epsilon_r = 2.28$

22.19 $I_T'/I_0' = 0.81$

22.21 $l = 67.3$ mm

22.32 $\Delta E = 1.78$ eV

INDEX

Page numbers in *italics* refer to the glossary.

UNIT CONVERSION FACTORS

Length

$$1 \text{ m} = 10^{10} \text{ Å} \qquad\qquad 1 \text{ Å} = 10^{-10} \text{ m}$$
$$1 \text{ m} = 10^{9} \text{ nm} \qquad\qquad 1 \text{ nm} = 10^{-9} \text{ m}$$
$$1 \text{ m} = 10^{6} \text{ } \mu\text{m} \qquad\qquad 1 \text{ } \mu\text{m} = 10^{-6} \text{ m}$$
$$1 \text{ m} = 10^{3} \text{ mm} \qquad\qquad 1 \text{ mm} = 10^{-3} \text{ m}$$
$$1 \text{ m} = 10^{2} \text{ cm} \qquad\qquad 1 \text{ cm} = 10^{-2} \text{ m}$$
$$1 \text{ mm} = 0.0394 \text{ in.} \qquad 1 \text{ in.} = 25.4 \text{ mm}$$
$$1 \text{ cm} = 0.394 \text{ in.} \qquad 1 \text{ in.} = 2.54 \text{ cm}$$
$$1 \text{ m} = 3.28 \text{ ft} \qquad\qquad 1 \text{ ft} = 0.3048 \text{ m}$$

Area

$$1 \text{ m}^2 = 10^4 \text{ cm}^2 \qquad\qquad 1 \text{ cm}^2 = 10^{-4} \text{ m}^2$$
$$1 \text{ mm}^2 = 10^{-2} \text{ cm}^2 \qquad 1 \text{ cm}^2 = 10^2 \text{ mm}^2$$
$$1 \text{ m}^2 = 10.76 \text{ ft}^2 \qquad\qquad 1 \text{ ft}^2 = 0.093 \text{ m}^2$$
$$1 \text{ cm}^2 = 0.1550 \text{ in.}^2 \qquad 1 \text{ in.}^2 = 6.452 \text{ cm}^2$$

Volume

$$1 \text{ m}^3 = 10^6 \text{ cm}^3 \qquad\qquad 1 \text{ cm}^3 = 10^{-6} \text{ m}^3$$
$$1 \text{ mm}^3 = 10^{-3} \text{ cm}^3 \qquad 1 \text{ cm}^3 = 10^3 \text{ mm}^3$$
$$1 \text{ m}^3 = 35.32 \text{ ft}^3 \qquad\qquad 1 \text{ ft}^3 = 0.0283 \text{ m}^3$$
$$1 \text{ cm}^3 = 0.0610 \text{ in.}^3 \qquad 1 \text{ in.}^3 = 16.39 \text{ cm}^3$$

Mass

$$1 \text{ Mg} = 10^3 \text{ kg} \qquad\qquad 1 \text{ kg} = 10^{-3} \text{ Mg}$$
$$1 \text{ kg} = 10^3 \text{ g} \qquad\qquad\quad 1 \text{ g} = 10^{-3} \text{ kg}$$
$$1 \text{ kg} = 2.205 \text{ lb}_m \qquad\quad 1 \text{ lb}_m = 0.4536 \text{ kg}$$
$$1 \text{ g} = 2.205 \times 10^{-3} \text{ lb}_m \qquad 1 \text{ lb}_m = 453.6 \text{ g}$$

Density

$$1 \text{ kg/m}^3 = 10^{-3} \text{ g/cm}^3 \qquad 1 \text{ g/cm}^3 = 10^3 \text{ kg/m}^3$$
$$1 \text{ Mg/m}^3 = 1 \text{ g/cm}^3 \qquad\quad 1 \text{ g/cm}^3 = 1 \text{ Mg/m}^3$$
$$1 \text{ kg/m}^3 = 0.0624 \text{ lb}_m/\text{ft}^3 \qquad 1 \text{ lb}_m/\text{ft}^3 = 16.02 \text{ kg/m}^3$$
$$1 \text{ g/cm}^3 = 62.4 \text{ lb}_m/\text{ft}^3 \qquad 1 \text{ lb}_m/\text{ft}^3 = 1.602 \times 10^{-2} \text{ g/cm}^3$$
$$1 \text{ g/cm}^3 = 0.0361 \text{ lb}_m/\text{in.}^3 \qquad 1 \text{ lb}_m/\text{in.}^3 = 27.7 \text{ g/cm}^3$$

Force

$$1 \text{ N} = 10^5 \text{ dynes} \qquad 1 \text{ dyne} = 10^{-5} \text{ N}$$
$$1 \text{ N} = 0.2248 \text{ lb}_f \qquad 1 \text{ lb}_f = 4.448 \text{ N}$$

Stress

$$1 \text{ MPa} = 145 \text{ psi} \qquad\qquad 1 \text{ psi} = 6.90 \times 10^{-3} \text{ MPa}$$
$$1 \text{ MPa} = 0.102 \text{ kg/mm}^2 \qquad 1 \text{ kg/mm}^2 = 9.806 \text{ MPa}$$
$$1 \text{ Pa} = 10 \text{ dynes/cm}^2 \qquad 1 \text{ dyne/cm}^2 = 0.10 \text{ Pa}$$
$$1 \text{ kg/mm}^2 = 1422 \text{ psi} \qquad 1 \text{ psi} = 7.03 \times 10^{-4} \text{ kg/mm}^2$$

Fracture Toughness

$$1 \text{ psi} \sqrt{\text{in.}} = 1.099 \times 10^{-3} \text{ MPa}\sqrt{\text{m}} \qquad 1 \text{ MPa} \sqrt{\text{m}} = 910 \text{ psi} \sqrt{\text{in.}}$$

Energy

$$1 \text{ J} = 10^7 \text{ ergs} \qquad\qquad 1 \text{ erg} = 10^{-7} \text{ J}$$
$$1 \text{ J} = 6.24 \times 10^{18} \text{ eV} \qquad 1 \text{ eV} = 1.602 \times 10^{-19} \text{ J}$$
$$1 \text{ J} = 0.239 \text{ cal} \qquad\qquad 1 \text{ cal} = 4.184 \text{ J}$$
$$1 \text{ J} = 9.48 \times 10^{-4} \text{ Btu} \qquad 1 \text{ Btu} = 1054 \text{ J}$$
$$1 \text{ J} = 0.738 \text{ ft-lb}_f \qquad\quad 1 \text{ ft-lb}_f = 1.356 \text{ J}$$
$$1 \text{ eV} = 3.83 \times 10^{-20} \text{ cal} \qquad 1 \text{ cal} = 2.61 \times 10^{19} \text{ eV}$$
$$1 \text{ cal} = 3.97 \times 10^{-3} \text{ Btu} \qquad 1 \text{ Btu} = 252.0 \text{ cal}$$